Architecture

Drafting and Design

Seventh Edition

Donald E. Hepler
President, Technical Writing and Design Service Inc.
Leesburg, Florida

Paul Ross Wallach
Architecture Instructor, Cañada College and
Technical Writer and Consultant
Burlingame, California

Dana J. Hepler
President, Hepler Associates, Architects and Land Planners
Massapequa, New York

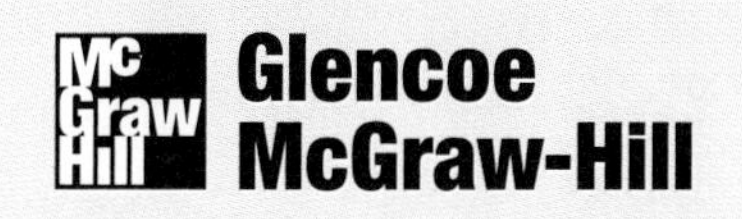

New York, New York Columbus, Ohio Woodland Hills, California Peoria, Illinois

Glencoe/McGraw-Hill

A Division of The ***McGraw-Hill*** *Companies*

Send all inquiries to:
Glencoe/McGraw-Hill
3008 W. Willow Knolls Drive
Peoria, IL 61614

ISBN: 0-02-637067-0 (Student Workbook)

ISBN 0-02-637069-7 (Student Workbook)
ISBN 0-02-637068-9 (Teacher's Resource Binder)

Printed in the United States of America

4 5 6 7 8 9 10 071/046 06 05 04 03 02 01 00

Acknowledgments

The publisher and authors gratefully acknowledge the cooperation and assistance received from many individuals and companies during the development of *Architecture: Drafting and Design.* Special thanks to the hundreds of drafting teachers who responded to our survey and helped us develop already excellent educational materials into an even more comprehensive, up-to-date, and effective architectural drafting and design program.

Contributors

Jody James
Oviedo, Florida

Bettye King
LeRoy, Illinois

John Kingston
Vice President, Key Bank
Burlington, Vermont

James J. Kirkwood
Department of Industry and Technology
Ball State University
Muncie, Indiana

Susan L. Meisel
Encino, California

Reviewers

Michael Bobbitt
Technology Education Teacher
Hominy High School
Hominy, Oklahoma

Douglas W. Cotton
Technology Education Teacher
Lyon County Junior/Senior High School
Eddyville, Kentucky

J. David Miller
Drafting Teacher
Harlandale High School
San Antonio, Texas

Holly Leigh Pipkins
Technology Teacher
Foley Middle School
Foley, Alabama

Design Photos

Cover:
Alan Goldstein Photography

Table of Contents:
John B. Scholz, Architect, page 4
Foto—Graphics, Inc., page 5
David Karram, page 6 (top)
Edna Carvin, Remax, page 6 (bottom)
HL Stud Corp., page 7
Arnold & Brown, page 8

Part 1: IMHOTEP, Boca Raton, Florida — Design by Alfred Karram/©Tom Knibbs, pages 12-13
Zigy Kaluzny/Tony Stone Images, page 13

Part 2: Comstock Inc., pages 44-45
©Stephen Simpson/FPG International Corp., page 45

Part 3: Alan Goldstein Photography, pages 114-115
David Rigg/Tony Stone Images, page 115

Part 4: Rhoda Sidney/PhotoEdit, pages 228-229
Michael Newman/PhotoEdit, page 229

Part 5: ©Telegraph Colour Library/FPG International Corp., pages 378-379
David Karram, page 379

Part 6: Jeremy Walker/Tony Stone Images, pages 422-423
Comstock Inc./Bob Pizaro, page 423

Part 7: Kevin Morris/Tony Stone Images, pages 526-527
Gary Conner/Index Stock Photography Inc., page 527

Part 8: Signature Properties, Winter Park, FL, pages 614-615
Comstock Inc., page 615

Part 9: Ken Fisher/Tony Stone Images, pages 686-687
©Ron Chapple/FPG International Corp., page 687

Part 10: Jon Riley/Tony Stone Images, pages 730-731
Michael Newman/PhotoEdit, page 731

Table of Contents

PREFACE . . . 9

ABOUT THE AUTHORS . . . 11

PART 1—Introduction to Architecture . . . 13

CHAPTER 1, ARCHITECTURAL HISTORY AND STYLES . . .14
■ The Development of Architectural Forms ■ How Architectural Styles Develop
■ Influences on Early American Architecture ■ Early American Styles
■ Later American Styles ■ New Challenges in Architecture

CHAPTER 2, FUNDAMENTALS OF DESIGN . . . 30
■ Architecture and Design ■ Elements of Design ■ Principles of Design
■ *Careers in Architecture: Architect*

PART 2—Architectural Drafting Fundamentals . . . 45

CHAPTER 3, DRAFTING SCALES AND INSTRUMENTS . . . 46
■ Scales ■ Guides for Straight Lines ■ Instruments for Curved Lines
■ Drafting and Lettering Tools ■ Papers and Drawing Surfaces
■ Correction Equipment ■ Timesaving Aids and Devices for Drafting

CHAPTER 4, ARCHITECTURAL DRAFTING CONVENTIONS . . . 71
■ Architectural Drawings ■ Architectural Conventions
■ Architectural Drawing Techniques

CHAPTER 5, INTRODUCTION TO COMPUTER-AIDED DRAFTING AND DESIGN . . . 86
■ Advantages of CAD ■ CAD Limitations ■ Components of a CAD System
■ CAD Hardware ■ CAD Software ■ Types of CAD Drawings ■ Using a CAD System
■ Architectural Applications ■ *Careers in Architecture: Architectural Drafter*

PART 3—Basic Area Design . . . 115

CHAPTER 6, ENVIRONMENTAL DESIGN FACTORS . . . 116
■ Orientation ■ Ergonomic Planning ■ Ecology

CHAPTER 7, INDOOR LIVING AREAS . . . 136
■ Living Area Plans ■ Living Room ■ Dining Rooms ■ Family Rooms ■ Recreation Rooms ■ Special-Purpose Rooms ■ *Careers in Architecture: Interior Designer*

CHAPTER 8, OUTDOOR LIVING AREAS . . . 160
■ Porches ■ Patios ■ Lanais ■ Swimming Pools

CHAPTER 9, TRAFFIC AREAS AND PATTERNS . . . 176
■ Traffic Patterns ■ Halls ■ Stairs ■ Entrances

CHAPTER 10, KITCHENS . . . 188
■ Kitchen Design Considerations ■ Kitchen Planning Guidelines

CHAPTER 11, GENERAL SERVICE AREAS . . . 199
■ Utility Rooms ■ Garages and Carports ■ Driveways ■ Workshops ■ Storage Areas

CHAPTER 12, SLEEPING AREAS . . . 213
■ Bedrooms ■ Baths

PART 4—Basic Architectural Drawings . . . 229

CHAPTER 13, DESIGNING FLOOR PLANS . . . 230
■ Floor Plan Development ■ The Design Process ■ Functional Space Planning ■ Developing Plans to Accommodate Special Needs ■ *Careers in Architecture: CAD Design Specialist*

CHAPTER 14, DRAWING FLOOR PLANS . . . 260
■ Types of Floor Plans ■ Floor-Plan Symbols ■ Steps in Drawing Floor Plans ■ Multiple-Level Floor Plans ■ Reversed Plans ■ Reflected Ceiling Plans ■ Floor-Plan Dimensioning ■ Guidelines for Dimensioning ■ CAD Dimensioning ■ Metric Dimensioning ■ Modular Dimensions

CHAPTER 15, DESIGNING ELEVATIONS . . . 284
■ Relationship with the Floor Plan ■ Elements of Design and Elevations ■ Design Sequence for Creating Elevations

CHAPTER 16, DRAWING ELEVATIONS . . . 299
■ Elevation Projection ■ Drawing Elevations from a Floor Plan ■ Elevation Symbols ■ Interior Elevations ■ Elevation Dimensioning ■ Landscape on Elevation Drawings

CHAPTER 17, SECTIONAL, DETAIL, AND CABINETRY DRAWINGS . . . 322
■ Sectional Drawings ■ Full Sections ■ Detail Sections ■ Cabinetry and Built-in Component Drawings

CHAPTER 18, SITE DEVELOPMENT PLANS . . . 345
■ Site Analysis ■ Zoning Ordinances ■ Survey Plans ■ Plot Plans
■ Landscape Plans ■ Landscape Rendering ■ Site Details and Schedules
■ *Careers in Architecture: Landscape Architect*

PART 5—Presentation Methods . . . 379

CHAPTER 19, PICTORIAL DRAWINGS . . . 380
■ Types of Pictorial Projection ■ Perspective Drawings
■ Projection of Exterior Perspective Drawings
■ Projection of Interior Perspective Drawings

CHAPTER 20, ARCHITECTURAL RENDERINGS . . . 396
■ Choosing Media for Rendering ■ Showing the Effects of Light ■ Texture ■ Entourage ■ Landscape
■ Steps in Preparing a Rendering
■ *Careers in Architecture: Architectural Illustrator*

CHAPTER 21, ARCHITECTURAL MODELS . . . 410
■ Design Study Models
■ Presentation Models
■ Steps in Constructing a Model ■ Computer Modeling

PART 6—Foundations and Construction Systems . . . 423

CHAPTER 22, PRINCIPLES OF CONSTRUCTION . . . 424
■ Structural Design ■ Modular Construction

CHAPTER 23, FOUNDATIONS AND FIREPLACE STRUCTURES . . . 440
■ Foundation Materials and Components ■ Types of Foundations
■ Fireplaces ■ Foundation Drawings

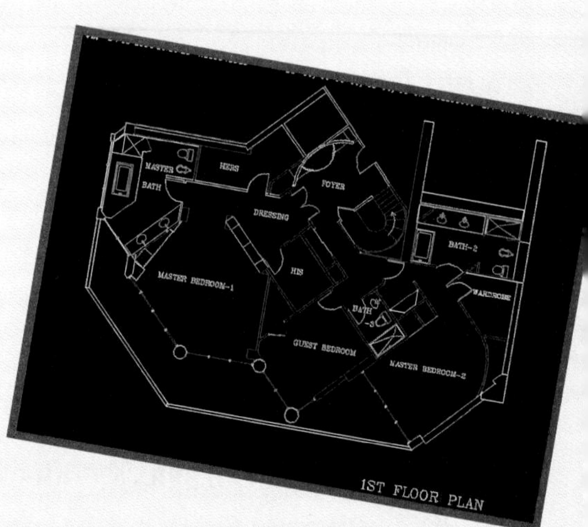

CHAPTER 24, WOOD-FRAME SYSTEMS . . . 470
■ Skeleton-Frame Construction ■ Post-and-Beam Construction

CHAPTER 25, MASONRY AND CONCRETE SYSTEMS . . . 490
■ Masonry Construction Systems ■ Concrete Construction Systems

CHAPTER 26, STEEL AND REINFORCED-CONCRETE SYSTEMS . . . 506
■ Steel Building Construction ■ Steel Structural Members
■ Steel Fasteners and Intersections ■ Structural Steel Drawing Conventions ■ *Careers in Architecture: Structural Engineer*

CHAPTER 27, DISASTER PREVENTION DESIGN . . . 519
■ Preventing Wind Damage ■ Preventing Earthquake Damage
■ Preventing Gas Leakage ■ Fire Prevention and Control
■ Water and Air Purification ■ Other Safety Considerations

PART 7—Framing Systems . . . 527

CHAPTER 28, FLOOR FRAMING DRAWINGS . . . 528
■ Types of Platform Floor Systems ■ Floor Framing Members
■ Floor Framing Plans

CHAPTER 29, WALL FRAMING DRAWINGS . . . 551
■ Exterior Walls ■ Interior Walls ■ Stud Layouts

CHAPTER 30, ROOF FRAMING DRAWINGS . . . 581
■ Roof Function ■ Roof Framing Members ■ Roof Framing Components ■ Roof Framing Drawings ■ Roof Pitch ■ Roof Framing Methods for Wood ■ Roof Framing Types ■ Steel and Concrete Roof Framing Methods ■ Roof-covering Materials ■ Roof Appendages

PART 8—Electrical and Mechanical Design and Drawings . . . 615

CHAPTER 31, ELECTRICAL DESIGN AND DRAWINGS . . . 616
■ Electrical Principles ■ Lighting Design ■ Developing and Drawing Electrical Plans
■ Electrical Working Drawings ■ Electronic Systems ■ *Careers in Architecture: Lighting Designer*

CHAPTER 32, COMFORT-CONTROL SYSTEMS (HVAC) . . . 646
■ HVAC Plans and Conventions ■ Principles of Heat Transfer ■ Heating Systems
■ Cooling Systems ■ Heat Pumps ■ Ventilation ■ Air Filtration ■ HVAC Control Devices
■ Humidity Control ■ Passive Solar Systems ■ Active Solar Systems

CHAPTER 33, PLUMBING DRAWINGS . . . 671
■ Plumbing Conventions and Symbols ■ Plumbing Systems ■ Plumbing Drawings

PART 9—Checking Plans and Using Support Services . . . 687

CHAPTER 34, DRAWING COORDINATION AND CHECKING . . . 688
■ Checking Drawing ■ Corrections and Changes

CHAPTER 35, SCHEDULES AND SPECIFICATIONS . . . 700
■ Schedules ■ Material Lists ■ Specifications

CHAPTER 36, BUILDING COSTS AND FINANCIAL PLANNING . . . 714
■ Building Costs ■ Financial Planning ■ *Careers in Architecture: Estimator*

CHAPTER 37, CODES AND LEGAL DOCUMENTS . . . 724
■ Building Codes ■ Legal Documents

PART 10—Careers in Architecture and Related Fields . . . 731

CHAPTER 38, ARCHITECTURE-RELATED CAREERS . . . 732
■ Careers in Architectural Design ■ Careers in Construction-Related Engineering Design
■ Careers in Construction ■ *Careers in Architecture: New Opportunities*

CHAPTER 39, PREPARING FOR A CAREER IN ARCHITECTURE . . . 739
■ Opportunities for Education and Training ■ Educational Requirements

APPENDIX A, BASIC MATHEMATICAL CALCULATIONS . . . 744

APPENDIX B, ARCHITECTURAL ABBREVIATIONS . . . 767

APPENDIX C, ARCHITECTURAL SYNONYMS . . . 771

GLOSSARY . . . 773

INDEX . . . 787

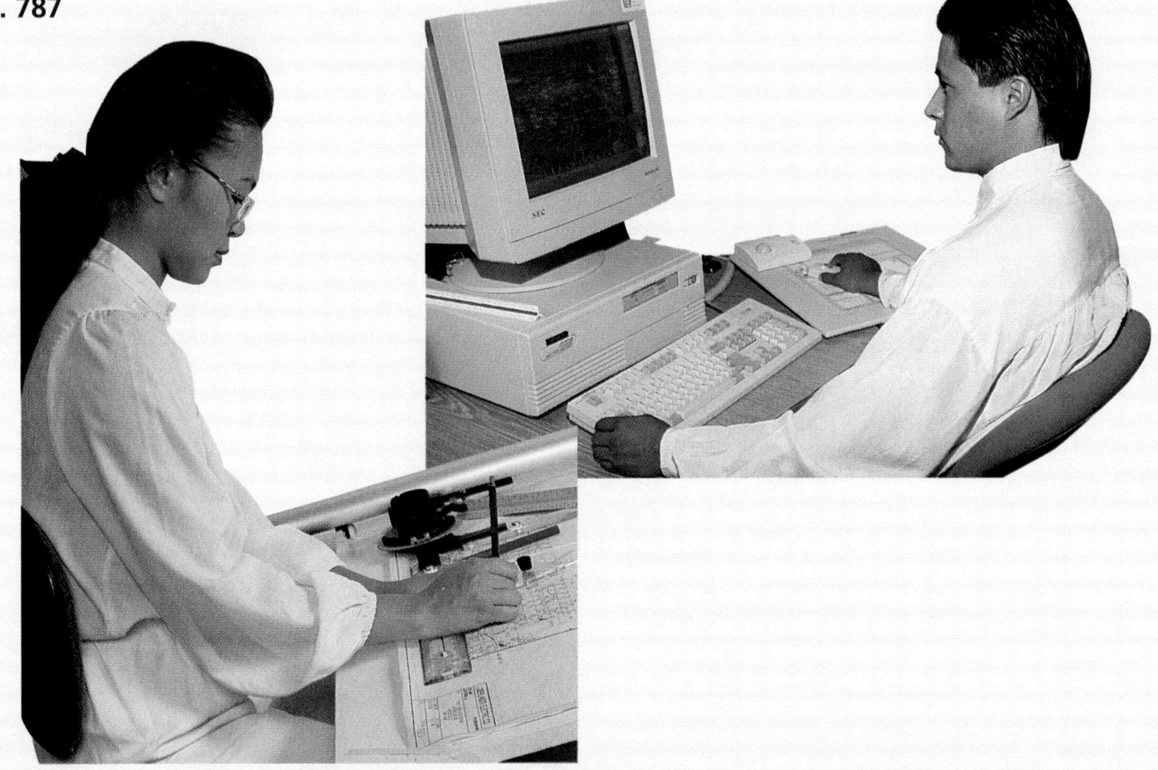

Preface

Architecture: Drafting and Design is a comprehensive textbook designed for use in a first course in architectural or construction drafting. Its purpose is to help students learn the fundamental skills and concepts necessary for architectural planning, designing, and drawing.

Organization and Content

Material in this seventh edition has been organized into ten major parts consisting of thirty-nine chapters.

Parts

- **Part One,** "Introduction to Architecture," provides background information on the history and development of major architectural styles, with excellent examples shown of both past and present designs. It also covers the basic principles and elements of architectural design.
- **Part Two,** "Architectural Drafting Fundamentals," provides basic information on the use of scales, drafting instruments, and CAD systems and explains the various architectural drafting conventions used in creating working drawings. Information in this part is needed to apply the information covered in subsequent specific drafting and design chapters. CAD information has been rewritten, updated, and expanded for this edition.
- **Part Three,** "Basic Area Design," covers the environmental and functional design factors needed to plan specific areas of a structure. This includes the design considerations necessary for effective solar orientation, efficient energy use, and ergonomic and ecological planning. Major considerations include the function, location, decor, size, and shape of the various areas.
- **Part Four,** "Basic Architectural Drawings," presents the design process and drafting methods used to combine areas into composite, functional, and effective architectural plans. Procedures for designing and drawing floor plans, elevations, sectional, detail, cabinetry, and site development drawings are explained. Guidelines for designing structures for persons with physical impairments are also included in this part. The information on site development and design factors and procedures has been completely revised and updated. New information on cabinetry drawings has been added.
- **Part Five,** "Presentation Methods," shows the different methods used to present architectural designs to non-technical personnel such as marketing staffs, financial supporters, and prospective buyers. Step-by-step instructions for preparing pictorial drawings, renderings, and three-dimensional models are provided.
- **Part Six,** "Foundations and Construction Systems," begins with an overview of the basic scientific and modular principles upon which construction systems are based. Each major construction system is then explained as students are introduced to the specialized drawings needed to complete detailed descriptions of the structural design. Types of drawings included are those used to describe foundations and fireplaces and wood-frame, masonry, concrete, steel, and reinforced-concrete systems. This part includes a new chapter which covers disaster prevention design features needed to reduce structural failure due to earthquakes, tornadoes, floods, and hurricanes.
- **Part Seven,** "Framing Systems," explains and shows in detail how to design and draw the framing systems for the major construction components of a building: floors, walls, and roofs.

- **Part Eight,** "Electrical and Mechanical Design and Drawings," includes the principles and procedures for preparing working drawings to describe the electrical, comfort-control (HVAC), and plumbing systems of a structure. Passive and active solar heating and cooling systems are also explained.
- **Part Nine,** "Checking Plans and Using Support Services," describes how architectural plans are checked and combined into sets and how drawings are interrelated to other drawings, details, and documents such as schedules, specifications, cost estimates, financial plans, codes, and contracts. A complete set of working drawings is presented.
- **Part Ten,** "Careers in Architecture and Related Fields," describes the major career opportunities in architecture and construction, including information on preparing for a career in these fields. Career information has been rewritten and expanded for this edition.
- **The Appendixes** provide reference materials which apply to the preparation of architectural drawings. Information includes mathematical calculations and architectural abbreviations and synonyms. A glossary of architectural terms follows the appendixes.

Chapters

Each chapter is introduced with listings of the major objectives and important terms defined and explained in the chapter. It ends with a set of exercises.

Since communication in the field of architectural drafting and design depends largely on understanding the vocabulary of architecture, new terms, abbreviations, and symbols are defined or explained where they first appear. This learning is reinforced throughout the remainder of the text.

The exercises that appear at the end of each chapter are organized to provide the maximum amount of reinforcement of the concepts covered. Exercises are flexible, ranging from the very simplest, which can be completed in a few minutes, to the more complex, which require considerable time and detailed application of the principles of architectural drafting and design. Exercises that require original design work are marked with a house symbol. Completion of these exercises by the student will result in the creation of a complete set of related architectural plans and documents. Exercises that are specifically designed or most appropriate for CAD use are also marked with a symbol.

Architecture: Drafting and Design is very well illustrated. All the illustrations have been specifically selected and/or prepared to reinforce and amplify the principles and procedures described in the text.

Special Features

Math and science concepts are integral to the study of architecture. Math formulas and calculations are presented as applied throughout the text. The sections in which they appear are marked with the symbol shown at the right.

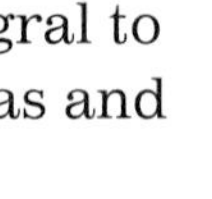

Math Connection

Science concepts are explained as appropriate. Those sections are marked with this symbol:

Architecture offers many exciting and challenging career opportunities for students to consider. To help them see architecture "in action," career profiles of professionals in architecture and construction are presented in related chapters throughout the book in a feature called "Careers in Architecture."

Ancillary Materials

Ancillary materials related to this edition include:

- Student Workbook—A collection of drawing exercises which provide students with opportunities to apply concepts learned in the text and improve drafting skills. Exercises include CAD tutorials for different programs.
- Teacher's Resource Binder—A comprehensive collection of valuable teaching aids and resource materials.
- Full-size set of architectural drawings, packaged with the Teacher's Resource Binder.
- Disk of CAD Exercises—A collection of drawing exercises to help students expand their computer skills as they apply concepts taught in the text. The disk is packaged with the Binder.

About the Authors

Donald E. Hepler completed his undergraduate work at California State College, California, Pennsylvania, and his graduate work at the University of Pittsburgh, Pennsylvania. He has been an architectural designer and drafter for several architectural firms, has served as an officer with the United States Army Corps of Engineers, and has taught architecture, design, and drafting at both the secondary and college levels. He is the former publisher of McGraw-Hill's technical education program and is currently devoting full time to technical authorship.

Paul Ross Wallach received his undergraduate education at the University of California at Santa Barbara and did his graduate work at California State University, Los Angeles. He has acquired extensive experience in the drafting, designing, and construction phases of architecture and has taught architecture and engineering drawing for many years in Europe and California at the secondary and post-secondary levels. He currently teaches architecture at Cañada College in Redwood City, California, and also does technical writing and consulting.

Dana J. Hepler received his bachelor's degree from Ohio State University. He is an ASLA licensed landscape architect and a member of the Construction Specifications Institute. He has been associated with several of the largest architectural firms in the world as designer, director of planning, and construction manager. He has received several national and international awards for his designs. Presently he is President of Hepler Associates, Architects and Land Planners, Massapequa, New York, and Hackensack, New Jersey.

PART 1

Introduction to Architecture

CHAPTER

1 Architectural History and Styles
2 Fundamentals of Design

Architectural History and Styles

Objectives

In this chapter you will learn to:

- recognize historical architectural styles and identify several distinct characteristics of each style.
- relate how the development of materials and construction methods influenced architectural styles.

Terms

arch
bearing walls
buttress
dome
Early American style
English style
French style
Gothic arch
Italian style
keystone
Mediterranean style
Mid-Atlantic style
New England Colonial
post-and-lintel construction
ranch style
skeleton frame construction
vault

The study of architectural history is more vast and complex than can be covered in one textbook—much less in one chapter. However, an overview of architectural forms is an excellent way to begin. Architecture is dynamic. As societies change and develop, so does architecture. This chapter gives a background for evaluating and studying the broad range of architectural styles and forms.

■ **Architecture is a blending of technology and art.** *Westlight/W. Cody*

The Development of Architectural Forms

When humans were nomadic, shelter consisted of natural caves or portable tents made of animal skins. As people began to settle in fixed locations, there was a greater need to draw or plan the construction of dwellings. They began to construct permanent tents and adobe huts, and to modify caves or shelters with existing natural materials.

The addition of more permanent dwellings near fertile areas gave rise to villages. Village life created a need for still more planning, such as for public areas. The art and science of architecture began with the planning and construction of the first dwellings and public areas. The field of architectural drafting began when people first drew the outline of a shelter or a village in the sand or dirt. They planned how to build structures with existing materials.

As centuries passed and civilizations developed, human needs expanded. Lifestyles and cultures began to develop and change. More complete, accurate, and detailed drawings became necessary, and basic principles of architecture began to be developed. Most early architecture used bearing walls for support. **Bearing walls** are solid walls that provide support for each other and for the roof. One of the first major problems in architectural drafting and design was how to provide door and window openings in these supporting walls without sacrificing the needed support.

Post and Lintel

One solution to the problem was simply to place a horizontal beam, called a *lintel,* across two vertical posts. This early type of construction became known as **post-and-lintel construction.** This method was used by the Egyptians and later by the ancient Greeks. See Fig. 1-1.

Since most ancient people used stone as the primary building material, architectural designs were limited. Because of the great weight of the stone, stone post-

Fig. 1-1 ■ The Parthenon (Athens, Greece c. 447 B.C.) is a classic example of post-and-lintel construction. *North Wind Picture Archives*

and-lintel construction could not support wide openings. Therefore, many posts (or columns) were placed close together to provide the needed support. Refer again to Fig. 1-1. The Greeks and, later, the Romans expanded their architectural designs by creating several styles of these columns. See Fig. 1-2.

Oriental architects also made effective use of the post and lintel. They were able to construct buildings with greater space between the posts under the lintel because they used lighter materials, such as wood. With these lighter materials, they developed a style of architecture that was very open and graceful.

Fig. 1-2 ■ Column orders of architecture.

The Arch

The Romans, who used stone, began a new trend in the design of wall openings when they developed the **arch.** Arch construction overcame several limitations of the post and lintel. Arches were easier to erect because they were constructed from many smaller, lighter blocks of stone. Each stone is supported by leaning on the keystone in the center. The **keystone** is a wedge-shaped stone that locks the other stones in place. See Fig. 1-3. This construction has the advantage of spanning greater areas, instead of being limited by the size of the one stone used for the lintel. The Romans combined arches and columns extensively in their architecture.

Fig. 1-3 ■ A keystone supports both sides of an arch.

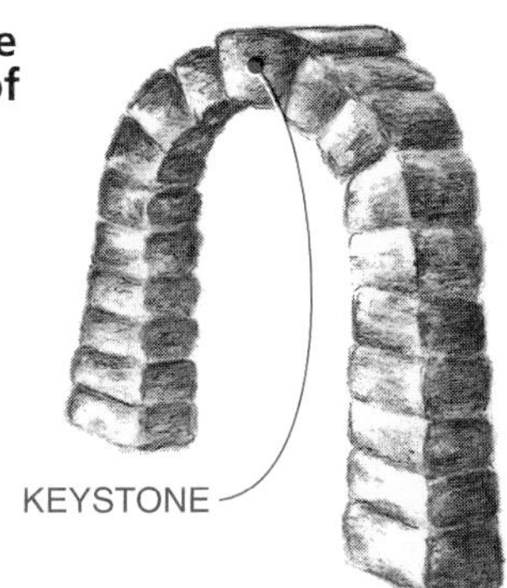

The Vault

The success of the arch led to the development of the vault. A **vault** can be viewed as a series of arches that forms a continuous arched covering. The term may also refer to an arched underground passageway or a space, such as a room, that is covered by arches. See Fig. 1-4.

The Dome

A **dome** is a further refinement of the arch. A dome is made of many arches arranged so that their bases form a circle and the tops meet in the center. The Romans viewed the dome as a symbol of power. Throughout the world, domes have often been used in religious and governmental structures. See Fig. 1-5.

The Gothic Arch

A variation of the arch was a defining characteristic of the Gothic style of architecture that spread throughout Europe during the Middle Ages. The pointed arch was called the **Gothic arch** and became a very popular feature in cathedrals. The emphasis on vertical lines created a sense

Fig. 1-4 ■ Types of vaults.

A. A barrel vault is like a series of connected arches.

B. The intersection of two barrel vaults creates a cross vault.

Fig. 1-5 ■ Examples of domed structures.

A. The Pantheon is a Roman temple completed about 27 B.C. *North Wind Picture Archives*

B. The U. S. Capitol Building in Washington, D.C. is a contemporary example of a domed structure. *David R. Frazier Photolibrary, Inc./Mark Burnett*

of height and aspiration. However, the pointed arch posed the same problem as did other arches—namely, of spreading at the bottom because of the weight above.

To add support to an arch or wall, a protruding structure called a **buttress,** or pilaster, was added at the base, as shown in Fig. 1-6. As the style evolved, buttresses were connected to higher areas of the walls and came to be called flying buttresses. A flying buttress helps support the sides of a wall without adding additional weight. Thinner walls and more windows could be used. Large structures such as the high-arched cathedral in Fig. 1-7 incorporate flying buttresses in their design.

Fig. 1-6 ■ Buttresses (pilasters) support bearing walls.

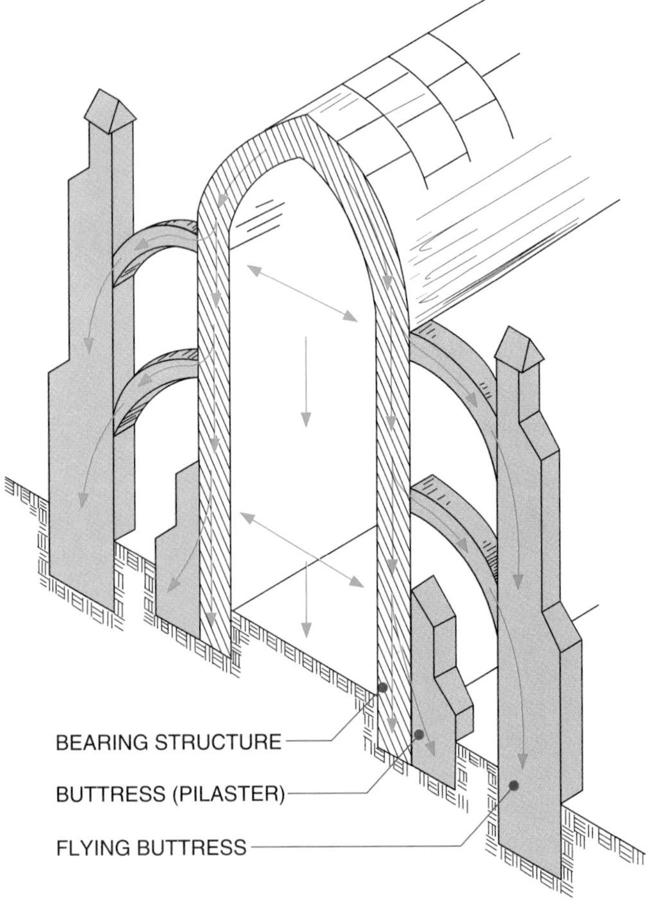

Fig. 1-7 ■ Notre Dame Cathedral in Paris, France, is famous for its pointed arches, high, stained-glass windows, and the flying buttresses that help support the graceful, thin walls. *David R. Frazier Photolibrary, Inc.*

How Architectural Styles Develop

The Gothic (pointed) arch is one of many features that were developed over time. The development of one architectural solution and a resulting style in one culture often causes changes in the architecture of another culture. Transitions occur from one time period to another as well as from one part of the country and world to another.

Few structures, past or present, are pure examples of one specific style. In fact, to identify an architectural style as one that originated in only one country or in only one time period is difficult, if not inaccurate. Nonetheless, architectural styles *are* categorized and labeled by their most common and outstanding features. The label "Early American style," for example, is something of a misnomer, since all styles that found their way to America during colonial times can be labeled Early American. The overlapping of characteristics of architectural design is also typical among European styles. Nonetheless, labels are applied and used as a frame of reference with which to study and compare architectural styles.

Influences on Early American Architecture

To understand the development of American architecture, an overview of the following European styles provides an important background. The English, French, Spanish, and Italians have provided the most significant influences on Early American architecture.

English Architecture

English style architecture includes several variations of some common architectural features. For example, some features that many English structures share are: high-pitched roofs, massive chimneys, half-timber siding, small windows, and exterior stone walls. Within this frame of reference, variations in architectural style range from the very simple to the very lavish. Wood may replace the stone on the exterior walls. The most commonly adapted English styles include the Elizabethan and the Tudor. See Fig. 1-8.

French Architecture

There are several distinct types of **French style** architecture. *French Provincial* houses contain steeply pitched hip

Fig. 1-8 ■ Examples of English architecture.

A. The Elizabethan style is distinguished by its half-timber construction. *Michelle Walsh Photographer*

B. The Tudor style has multiple gables. *Scholz Design*

roofs, long projecting windows, corner quoins, curved lintels, and towers. See Fig. 1-9. *French Chateau* houses are symmetrical with Mansard roofs. The Mansard roof was originally designed by the French architect François Mansard.

Spanish and Italian Architecture

Spanish and Italian architecture share several similarities—arches, low-pitched roofs of ceramic tile, and stucco exterior walls, as shown in Fig. 1-10. A distinguishing feature of a Spanish home is an open courtyard patio. Two-story Spanish homes contain open balconies often with grillwork trim.

Although **Italian style** architecture is very similar to Spanish architecture, a few features are particular to Italian styles. Columns and arches are generally part of an entrance, as in Fig. 1-11, and windows or balconies open onto a loggia. A loggia is an open passage covered by a roof.

Fig. 1-9 ■ Example of French Provincial architecture. *Country Club of the South*

Fig. 1-10 ■ Example of contemporary Spanish architecture. *Tucson Realty*

Classical moldings around first floor windows also help to distinguish the Italian style from the Spanish. Despite these distinguishing features, both of these styles are generally classified as **Mediterranean style** or Southern European style architecture.

Early American Styles

The early colonists came to this country from many different cultures and were familiar with many different styles of architecture. **Early American style** architecture refers to all styles that developed in various regions of the colonies.

Fig. 1-11 ■ A contemporary adaptation of Italian architecture. *John Henry, Architect*

The European styles that primarily dominated Early American residential architecture were brought from England and France. French styles, however, were brought to this country in the eighteenth century, much later than English styles. The English styles had greater impact on colonial residential architecture.

New England Colonial

The colonists who settled the New England coastal areas brought the strong influence of English architectural styles with them. Because they lacked time and equipment and depended on the locally available building materials, the colonists had to greatly simplify the English styles. This adapted style came to be known as **New England Colonial** architecture.

One of the most popular of the New England Colonial styles was called the Cape Cod. See Fig. 1-12. This one-and-one-half-story gabled-roof house has a central front entrance, a large central chimney, exterior walls of clapboard or bevel siding, and may include dormers. The floor plan is generally symmetrical. Cold New England winters also influenced the development of varied design features for added warmth, such as window shutters, small window areas, and enclosed breezeways.

Mid-Atlantic Colonial

The availability of brick, a seasonal climate, and the influence of the architecture of Thomas Jefferson led to the development of the **Mid-Atlantic style** of architecture. In colonial days, Mid-Atlantic style buildings, located from Virginia to New Jersey, were formal, massive, and ornate. This style is also known as *classical revival* because it was influenced by early Greek and Roman architecture.

Fig. 1-12 ■ The Cape Cod style is one example of the New England Colonial styles. Dormers, the structures with windows that project from the sloping roof, were not part of the original Cape Cod designs. These were added later and have become a very popular feature. *Home Planners, Inc.*

Fig. 1-13 ■ Example of the Mid-Atlantic Colonial style. *Scholz Design*

It also included adaptations of many urban English designs. See Fig. 1-13. The *Georgian* style is a simplified version of this design which includes many elements of the New England Colonial style.

Dutch Colonial

Many colonists from the Netherlands and Germany settled in New York and Pennsylvania. A gambrel roof was a typical part of their colonial style of architecture. This style was originally described as "Deutsch." However, the German term was given the English form by the colonists who called the masonry farmhouse style "Dutch" Colonial. See Fig. 1-14.

Fig. 1-14 ■ The gambrel roof is a distinguishing characteristic of Dutch Colonial architecture. *H. Armstrong Roberts, Inc./G. Hampfler*

Southern Colonial

When the early settlers migrated south, warmer climates and outdoor living activities led them to develop the Southern Colonial style of architecture. The house became the center of plantation living. Southern Colonial homes were usually larger than most English houses. A second story was added, often with a veranda or porch. See Fig. 1-15.

Fig. 1-15 ■ In Southern Colonial style homes, two-story columns are used to support the front-roof overhang. *John Henry, Architect*

Later American Styles

After the colonial period, other architectural styles continued to evolve. Styles were influenced by climate, availability of land, and industrial developments—as well as by other architectural styles.

Victorian

The new machinery developed during the Industrial Revolution in the late 18th century led to the addition of intricate house decorations (gingerbread). Ornate finials, lintels, parapets, and balconies were added to existing designs, as shown in Fig. 1-16.

Ranch Style

As settlers moved west of the colonies, they adapted architectural styles to meet their needs. The availability of land eliminated the need for second floors. Since the needed space was spread horizontally rather than vertically, a rambling plan called **ranch style** resulted.

The Spanish and Mexican influence in the Southwest led to the popularization of the *southwestern ranch,* with a U-shaped plan and a patio in the center. One-story Spanish/Mexican homes were the forerunners of the present ranch-style homes that were developed in the Southwest. See Fig. 1-17.

Influences on Contemporary Styles

Advances in architecture throughout history have depended on using available building materials. In American colonial times, builders had only wood, stone, and some ceramic materials, such as glass, with which to work. Early American architecture reflects the reliance on these

Fig. 1-16 ■ Victorian styles are very ornate. *Home Planners, Inc.*

Fig. 1-17 ■ A southwestern ranch style home. *Home Planners, Inc.*

materials, and they continue to be used in contemporary buildings.

With today's improved technological developments, lighter and safer buildings can be designed in forms, sizes, and shapes never before possible. See Fig. 1-18. With so many choices, designers can create many new combinations of styles and materials. These must be carefully combined using the basic principles and elements of design which are presented in Chapter 2.

Historical styles continue to influence contemporary architecture. Advancements in technology, however, have freed designers from many design restrictions of the past. Present day designers must often decide how many contemporary features can or should be incorporated into the design of a particular architectural style. See Fig. 1-19.

NEW MATERIALS Many "new" materials are actually old materials manufactured in new ways or in different forms. For example, wood is one of the oldest materials used in construction. Yet the manufacture of new structural wood with new synthetic materials has enabled shapes and sizes that have revolutionized the use of wood in architecture.

The development and use of plastics and new glass products have also changed the nature of design. Glass was once available only in small, single panes. Today, large double and triple glass sheets and structural forms are used in many ways.

Developments in the manufacture of another old material, concrete, have also contributed to changes in design. Factory-made concrete shapes are now used not only for concrete floors but also for structural designs of roofs and walls.

Fig. 1-18 ■ Example of a contemporary style home.
Edna Carvin, Remax, Danville, CA

Fig. 1-19 ■ Various contemporary features have been incorporated into the design of this French Provincial style home. *John Henry, Architect*

Fig. 1-20 ■ Comparison of heights of the world's largest structures.

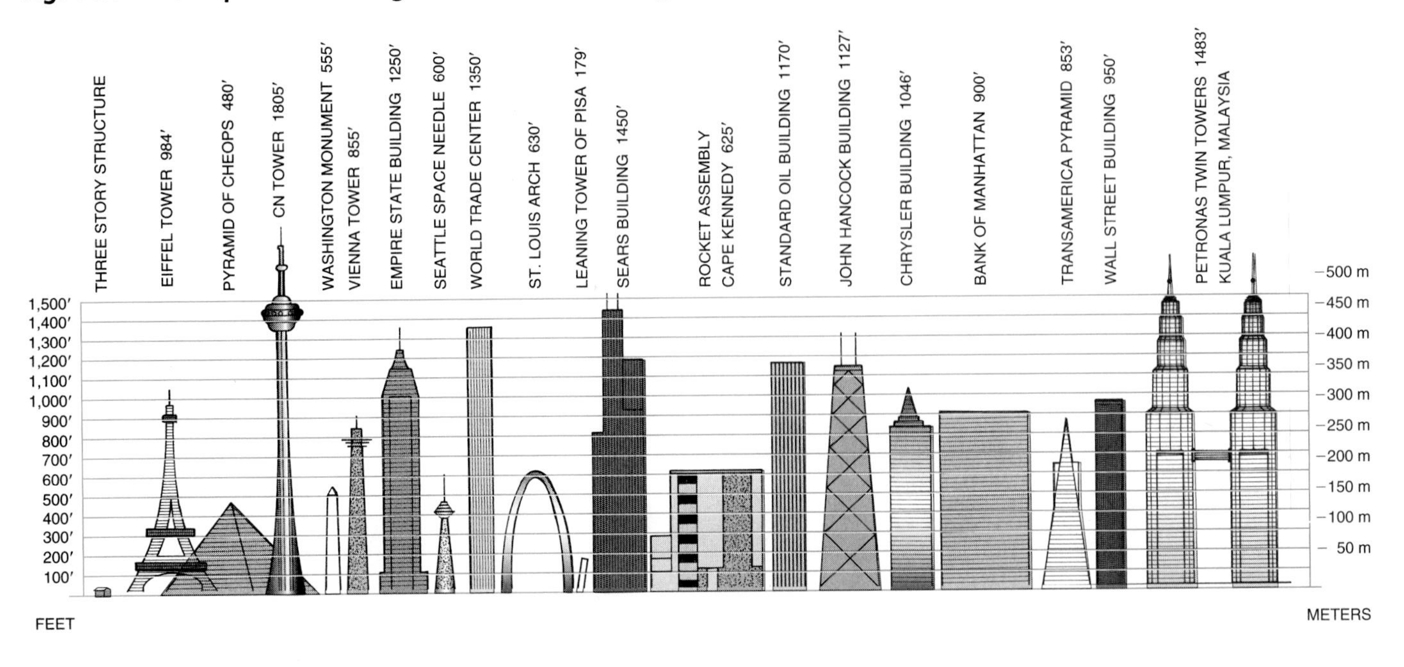

A manufactured material that has had one of the most profound impacts on architectural styles is steel. Small structures can now be built on locations and in shapes that were previously impossible without the structural stability of steel. Without the use of steel, construction of most large high-rise buildings would be impossible. See Fig. 1-20. Aluminum is used for framing and covering lightweight structures.

NEW CONSTRUCTION METHODS The development of new materials also affects the development of new construction methods. One of the first methods developed to employ modern materials is the **skeleton frame construction** method. This type of construction has an open frame to which wall coverings are attached. The frame provides the support, and the covering provides shelter. See Fig. 1-21.

Today, large pre-manufactured components can be used. Also, contemporary structures tend to have fewer structural restrictions. Lines may be simpler, bolder, and less cluttered. Other contemporary buildings can be constructed with more diversified structural shapes. No longer are architectural shapes simply squares or cubes. Shapes such as triangles, octagons, pentagons, circles, and spheres are now used extensively.

New Challenges in Architecture

Today's architecture stems from a rich historical background. Architectural styles continually change and develop. The future of architecture will certainly continue to be influenced by the development of new materials and new construction methods, as well as by the way people live in society.

Fig. 1-21 ■ The skeleton frame construction method makes good use of new materials.

A. Skeleton frame of a house.

B. The completed house.

With technological advancements, designers have greater freedom of choice to create diverse and exciting architecture. Greater freedom, however, and more choices also mean greater challenges. One of the primary challenges of architectural design is to blend art and technology. The role of the architect is to create a relationship between art and technology that enables all types of buildings to be both technically appropriate and aesthetically acceptable.

CHAPTER 1 Exercises

1. Explain what is meant by the term "architectural styles."
2. Describe post-and-lintel construction. Give examples of it in three different cultures. Compare it to the development of the arch.
3. Find examples of several different styles of architecture in your area. Photograph or sketch the structures and list their distinguishing characteristics.
4. Which European countries and styles had the greatest impact on Early American architecture? List two or three characteristics of each architectural style.
5. Describe a contemporary structure—imagined or real. Tell why you consider it to be a contemporary style.

6. Considering the styles in this chapter, what style of home do you prefer for your own home? Explain a few advantages and disadvantages of building this style.
7. Do research to identify and prepare a report on an architectural style not mentioned in this text. Share what you learn with your class.
8. Identify houses in your town or city that are examples of various architectural styles.

CHAPTER

2

Fundamentals of Design

Objectives

In this section you will learn to:

- relate design concepts to architecture.
- identify six elements of design.
- apply design principles to a work of architecture.

Terms

aesthetic value
elements of design
functionalism
organic design
principles of design

Opinions about art and design are often very personal. Individuals may respond quite differently to one single work of art or architecture. One person may favor dynamic art, while another may prefer art that is restful. Professional designers too may have strong preferences, but they can explain why one design works well and another one doesn't. They approach design with a knowledge of the "ingredients" and principles of design.

■ Architects apply the basic principles of design to create structures that are pleasing in appearance as well as functional. *Mark Romine*

Architecture and Design

Louis Sullivan, an important American architect in the late 1800s and early 1900s, wrote, "Our architecture reflects us as truly as a mirror." Architecture reflects the people, society, and culture of a given time. For example, modern architecture reflects our freedom and our technological advances.

Form Follows Function

"Form follows function" is a design concept conceived by Louis Sullivan but largely identified with Frank Lloyd Wright, probably the most famous of all 20th century architects. This concept has now been accepted by most designers.

"Form follows function" means that any architectural form (shape, object) should have an intended practical purpose and should perform a function. This concept distinguishes architecture from other art forms, such as sculpture. A piece of sculpture's primary purpose is its **aesthetic value.** Its value is in the *appreciation* of its form, its beauty, or its uniqueness. **Functionalism** is the quality of being useful, of serving a purpose other than adding beauty or aesthetic value. Functionalism in architecture led to the development of the organic concept. In the organic approach all materials, functions, forms, and surroundings are completely coordinated and in harmony with nature. See Fig. 2-1.

Interior Design

Interior construction and furnishings should be closely related to—if not determined by—the exterior architecture. Both inside and outside designs must be coordinated to obtain a desired consistent design. See Fig. 2-2.

Fig. 2-1 ■ This Frank Lloyd Wright house, Fallingwater, was designed and built more than 60 years ago. Using materials and structural shapes to relate structures to the environment still characterizes contemporary architecture.

While coordinating exterior and interior design activities, the architect creates a relationship between art and technology. The goal is to create buildings that are designed well—both technically and aesthetically.

Designers need to consider that styles and individual tastes change. An effective, creative designer recognizes the difference between *trends* (general developments) and *fads* (temporary popular fashions).

Creativity in Architectural Design

Creativity in architecture involves the ability to imagine forms before they exist. Creative imagination often involves arranging familiar objects and patterns in new ways. Creativity and imagination are both needed to bring many isolated and unrelated factors together into arrangements of cohesive unity and beauty. Every part of a building should be designed in relation to its function. Architects and

Fig. 2-2A&B ■ In what ways is this contemporary interior consistent with its exterior design?
John Henry, Architect

interior designers apply the elements and principles of design to a building's function to make it aesthetically pleasing as well.

Elements of Design

Like a mixture composed of many ingredients, a design is composed of many elements. The basic elements of design include: *line, form, space, color, light* (value), *texture*, and *materials*.

Line

Lines enclose space and provide the outline or contour of forms. Straight lines are either horizontal, vertical, or diagonal. Refer back to Fig. 2-2B. Curved lines have an infinite number of variations. The element of *line* can produce a sense of movement or produce a greater sense of length or height.

Horizontal lines emphasize width as the eye moves horizontally. These lines suggest relaxation and calm. Vertical lines create the impression of height because they lead the eye upward. These lines create a feeling of strength and alertness. Diagonal lines create a feeling of restlessness or transition. Vertical and horizontal lines tend to dominate architectural designs, giving a sense of stability. Curved lines indicate soft, graceful, and flowing movements. See Fig. 2-3. As in any art form, it is often the combination of straight and curved lines in patterns that create the most pleasing design.

Fig. 2-3 ■ The curved lines of the cabinetry combine with the straight lines of the screen and modular units to form a pleasing design. *Gold Coast Inc.*

Form

Lines joined together can produce a *form* and create the *shape* of an area. More than two straight lines joined together can produce triangles, rectangles, squares, and other geometric shapes. Closed curved lines can form circles, ovals, and ellipses, as well as free-form closed curves. The relationships of these forms or shapes is an important factor in design.

Circles and ovals convey a feeling of completeness. Squares and rectangles produce a feeling of mathematical precision. See Fig. 2-4. Whether the form of an object is closed, open, solid, or hollow, the form of the structure should always be determined by its function.

Space

Space surrounds form and is contained within it. A design can create a feeling of space as in Fig. 2-5. Architectural design includes the art of defining space and space relationships. Space is as important a consideration as the actual objects and materials.

Color

Choices of *color* have a strong influence on the final appearance of any design. In architecture, color can strengthen or diminish interest. It can also distinguish one part from another. Color may be an integral part of an architectural material such as natural wood. Manufactured products, such as synthetic wall panels, may have color added to create a desired effect. To create effective designs, designers need to understand the nature and relationships of colors.

THE COLOR SPECTRUM Colors in the spectrum are divided into primary, secondary, and tertiary colors. *Primary* colors are red, yellow, and blue. These cannot be made from any other color. The primary colors and combinations of colors are illustrated on the color wheel in Fig. 2-6.

Fig. 2-4 ■ Rectangles suggest precision. *Audio Tec Designs*

Fig. 2-5 ■ In design, space is an important element.
Marvin Windows & Doors

A *secondary* color can be made from equal mixtures of two primaries. Green is a combination of yellow and blue. Violet is a combination of blue and red. Orange is a combination of red and yellow.

A *tertiary* color is the combination of a primary color and a neighboring secondary color. Refer again to Fig. 2-6. The tertiary colors are red-orange, yellow orange, yellow-green, blue-green, blue-violet, and red-violet.

A *neutral* does not show color in the ordinary sense of the word. The neutrals are white, gray, and black. The three primary colors, if mixed in equal strengths, will produce black. When colors cancel each other out in this manner, they are neutralized.

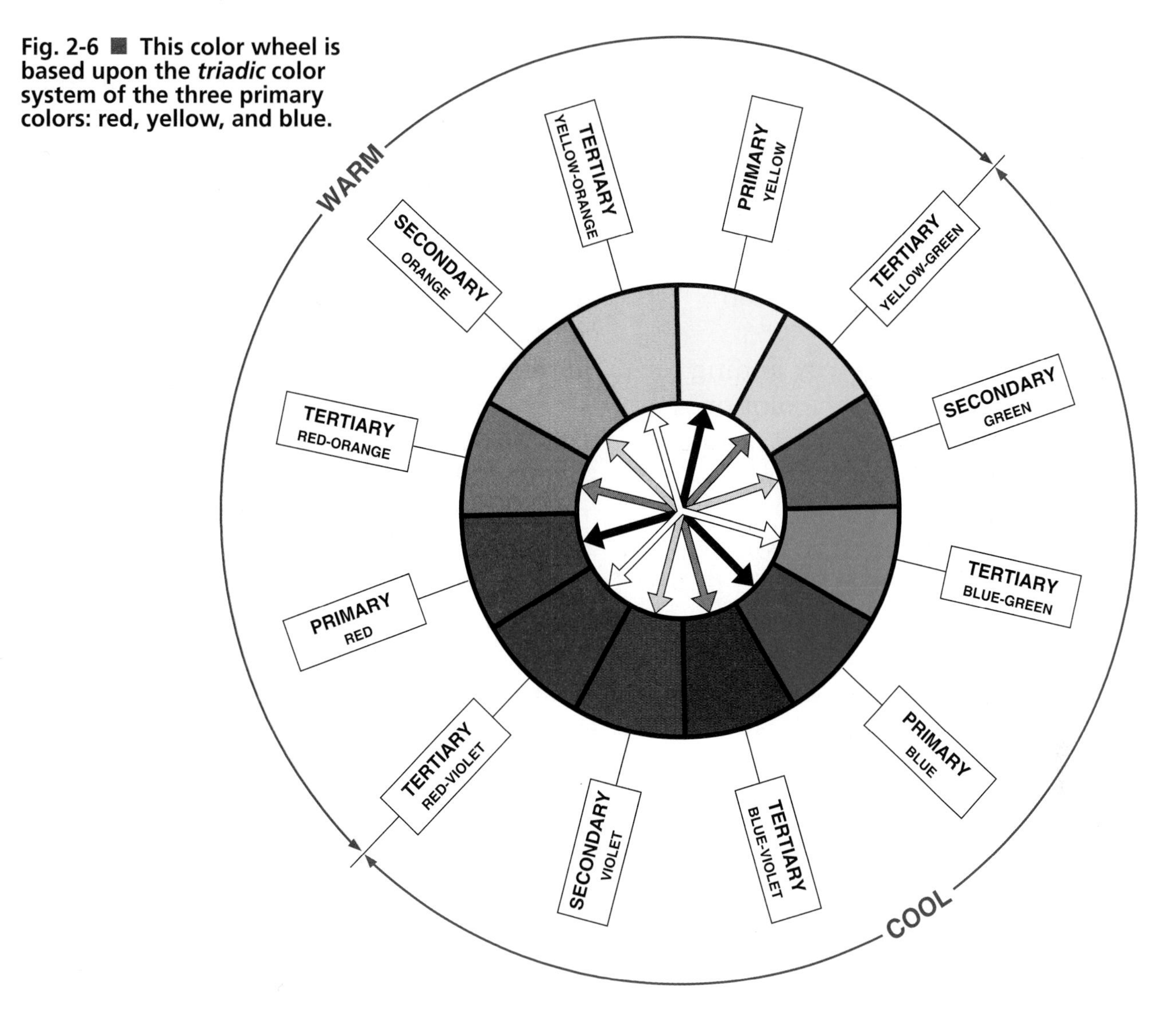

Fig. 2-6 ■ This color wheel is based upon the *triadic* color system of the three primary colors: red, yellow, and blue.

COLOR QUALITY For greater accuracy in describing a color's exact appearance, colorists distinguish three qualities: hue, value, and intensity.

The *hue* of a color is its basic consistent identity. A color hue may be identified as being yellow, yellow-green, blue, blue-green, and so forth. Even when a color is made lighter or darker, the hue remains the same.

The *value* of a color refers to the lightness or darkness of a hue. See Fig. 2-7. A great many degrees of value can be obtained. Varying the value of colors can dramatically change the mood of a room. A *tint* is lighter (or higher) in value than

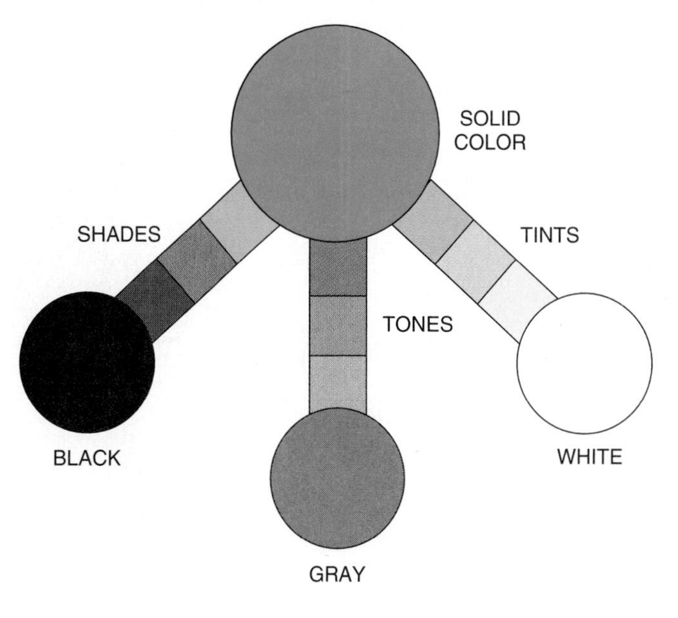

Fig. 2-7 ■ Color is used with black and white to form tints, tones, and shades.

the normal value of a color. It is produced by adding white to a color. A lighter tint of a hue will make a room look larger in area.

A *shade* is darker (or lower) in value than the normal value of the color. A shade is produced by adding black to the normal color. A dark shade will often make a room look smaller.

A *tone* is usually produced by adding gray to the normal color. Each color on a color wheel can have a value that is equivalent to another color if they have the same amount of gray in them.

The intensity (strength) of a color is its degree of purity (or brightness), that is, its freedom from neutralizing factors. This quality is also referred to in color terminology as *chroma*. A color entirely free of neutral elements is called a saturated color. The intensity of a color can be changed, without changing the color's value, by mixing that color with a gray of the same value.

Color *harmonies* are groups of colors which relate to each other in a predictable manner. The basic color harmonies are *complementary, monochromatic,* and *triadic*. Refer again to Fig 2-6. Complementary colors are opposite each other on the color wheel. Monochromatic colors are side by side. Triadic colors are three colors which are an equal distance apart on the color wheel.

USES OF COLOR The use of color has a very strong effect on the atmosphere of a building in several ways. The perceived level of formality, temperature, and mood are all influenced by the color design. Colors such as red, yellow, and orange create a feeling of warmth, informality, cheer, and exuberance. Colors such as blue and green create a feeling of quiet, formality, restfulness, and coolness. See Fig. 2-8.

Color is also used to change the apparent visual dimensions of a building. It is used to make rooms appear higher or

Fig. 2-8 ■ Imagine how the effect of this fireplace would change if the colors were cool instead of warm. *John Henry, Architect*

Fig. 2-9 ■ Bold colors advance and pale colors recede.

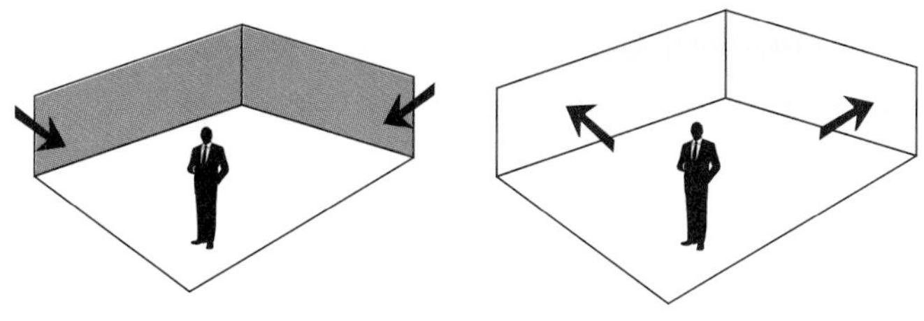

longer, lower or shorter. Warm bold colors, such as red, create the illusion of advancement. Cool and pale colors (including pastels) tend to recede. See Fig. 2-9.

Light and Shadow

Light reflects from the surfaces of forms. Shadows appear in areas that light cannot reach. Light and shadow both give a sense of depth to any structure. The effective designer plans the relationship of light and dark areas accordingly. See Fig. 2-10.

To achieve a dramatic effect, the designer must consider which surfaces reflect (instead of absorb) light and which surfaces refract (bend) light as it passes through materials. The designer must also remember that with continued exposure to light, visual sensitivity decreases. People become adapted to degrees of darkness or lightness after extended exposure. Thus, a designer should plan for a variety of levels of light in a room or building.

Texture and Materials

Materials are the raw substances with which designers create. Materials possess their own unique properties, such as color, form, dimension, degree of hardness, and texture. *Texture* is a significant factor in the selection of appropriate materials. *Texture* refers to the surface finish of an object—its roughness,

Fig. 2-10 ■ Notice the different light and shadow patterns in this computer simulation. *Intergraph Corp.*

A. Full sunlight.

B. Moonlight or dusk.

C. Artificial light on a dark night.

smoothness, coarseness, or fineness. Surfaces of materials such as concrete, stone, and brick are rough and dull and suggest strength and informality. Smoother surfaces, such as those of glass, aluminum, and plastics, create a feeling of luxury and formality. The designer must be careful not to include too many textures of a similar nature, such as stone and brick. When positioned close together, they tend to compete with each other. Textures, such as wood and stone, are more pleasing when contrasted with other surfaces. See Fig. 2-11.

Rough surfaces reduce the apparent height of a ceiling or distance of a wall and may appear darker. Smooth surfaces increase the apparent height of a ceiling or wall and reflect more light, thus making colors appear brighter.

Fig. 2-11 ■ Different textures combined to create a pleasing effect. *Scholz Homes Inc.*

Principles of Design

The basic **principles of design** are the guidelines for how to combine the elements of design. For buildings to be aesthetically pleasing, as well as functional, the basic principles of design should be applied. These are: *balance, rhythm, repetition, emphasis, subordination, proportion, unity, variety, opposition,* and *transition*.

Balance

Equilibrium (feeling of stability) in design is known as balance. Buildings are *informally balanced* if they are asymmetrical and *formally balanced* if they are symmetrical. See Figs. 2-12 and 2-13. Whether a design is formal or informal, balance requires a harmonious relationship in the distribution of space, form, line, color, light, and materials.

Fig. 2-12 ■ The design shown in this rendering is informally balanced (asymmetrical). *Architect: Randall E. Stoff; Builder: Bryason Homes, Inc.; Rendering: Hoffman Illustrations.*

Fig. 2-13 ■ This formally balanced design is so symmetrical that one side could be a mirror image of the other. *John Henry, Architect*

Rhythm and Repetition

When lines, planes, or surface treatments are repeated in a regular sequence, the order or arrangement creates a sense of rhythm. *Rhythm* creates motion and carries the viewer's eyes to various parts of the space. This may be accomplished by the repetition of lines, colors, and patterns. *Repetition* is designed into the house shown in Fig. 2-14, creating harmony by repeating the shapes.

Emphasis and Subordination

The principle of *emphasis*, or giving something importance, means drawing a viewer's attention to an area or subject. In architectural design, some emphasis or *focal point* (center of attention) should be

Fig. 2-14 ■ Which architectural features create a sense of repetition and rhythm? *John Henry, Architect*

designed into each exterior and interior space. Directing attention to a point of emphasis, the focal point, can be accomplished by arrangement of features, contrast of colors, line direction, variations in light, space relationships, and change in materials or texture. See Fig. 2-15.

Subordination occurs when emphasis is achieved through design. Other features become subordinate. They have less emphasis or importance.

Proportion

Proportion means the relationship of one part to another or ratio. The early Greeks found that the proportions of a rectangle in the ratio of 2 to 3, 3 to 5, and 5 to 8 were more pleasing than other ratios. For example, a room or a rug with dimensions of 9′ × 15′ or 10′ × 16′ will have the proportions 3 to 5 and 5 to 8.

The proportion (ratio) of interior space, furniture, and accessories should be harmonious. Large bulky components in

Fig. 2-15 ■ The design of this media-oriented room makes the screen the focal point. *Audio Tec Design*

small rooms should be avoided, just as small components in large rooms should not be used. Areas can appear completely different depending upon how the proportional division of space within the area is allocated. See Figs. 2-16 and 2-17.

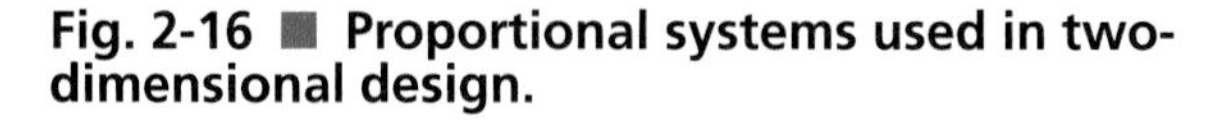

Fig. 2-16 ■ Proportional systems used in two-dimensional design.

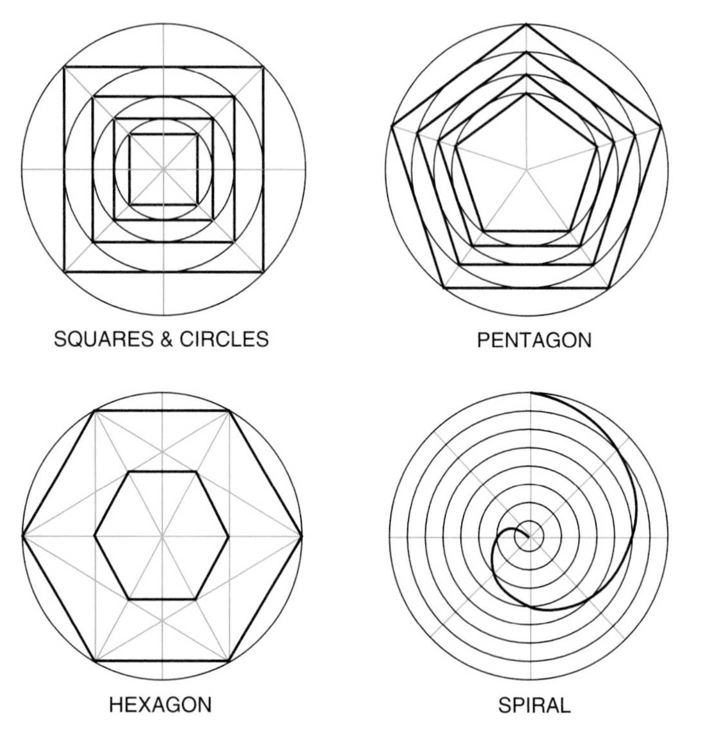

Fig. 2-17 ■ The division of space can make the same size rectangle appear quite different.

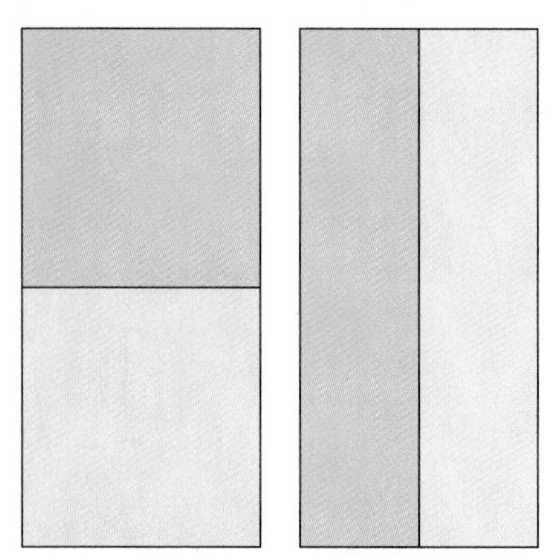

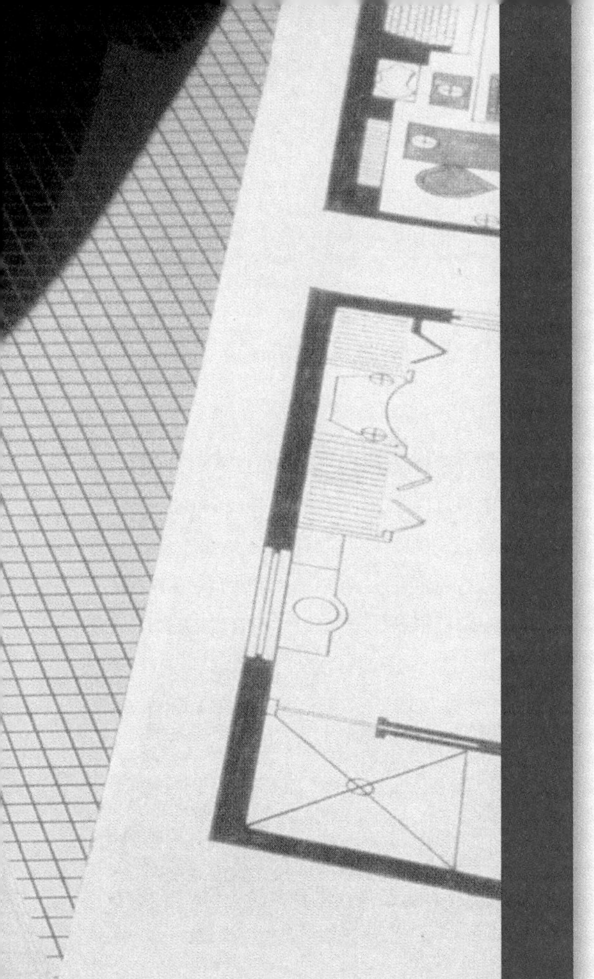

Architectual Drafting Fundamentals

CHAPTER

3 Drafting Scales and Instruments

4 Architectural Drafting Conventions

5 Introduction to Computer-Aided Drafting and Design

CHAPTER

3

Drafting Scales and Instruments

Objectives

In this chapter you will learn to:

- measure and prepare drawings with different scales.
- draw with drafting instruments.
- select and use appropriate types of paper and other drafting supplies.
- use timesaving devices.

Terms

architect's scale
civil engineer's scale
compass
dividers
drafting brush
drafting machine
erasing shield
flexible curve
French curve
metric scale
overlay
parallel slide
protractor
stamps
technical pens
templates
triangles
T square
underlays
vellum

To make certain a structure is built and completed the way it is designed, a set of dimensioned drawings is needed. Architectural designers and drafters use an assortment of instruments and supplies to create the drawings, including a large, adjustable drawing table or board and an overhead adjustable lamp. They also need scales and a variety of pens, pencils, papers, and special devices. Drafting tools help drafters and designers produce architectural drawings that are accurate and readable.

■ **The creation of clear and accurate architectural drawings requires skillful use of drafting instruments.** *Uniphoto Picture Agency*

Scales

In architectural drawing, the term "scale" is used in three ways. A scale may be a proportion (or ratio), a drafting instrument, or the equal interval of marks along the edges of a drafting instrument or machine.

When making a scaled (proportional) drawing, one measurement is used to represent another. Scaled drawings allow objects of all sizes to be proportionally reduced or enlarged to show the correct relationship of all parts. Different ratios are required depending on the size of the object.

Without the use of *reduced* scales no object larger than a sheet of paper could be accurately drawn. For example, if the earth were drawn to the same scale used on a typical architectural drawing, 1/4″ = 1′-0″ (or 1″ = 4′-0″, a ratio of 1 to 48), the drawing sheet would need to be 165 miles wide! See Fig. 3-1.

Conversely, small objects, even those which may be invisible or barely visible to the human eye, can be proportionally enlarged to show details. In architectural drawing, a small item such as a hinging mechanism may need to be drawn at an *enlarged* scale.

In the preparation of scaled drawings, instruments called scales are used. The three main types of scales are architect's scale, civil engineer's scale, and metric scale. All have scale markings along the edges.

Architect's Scale

The ability to use **architect's scales** accurately is required not only in preparing drawings but also in checking existing architectural plans and details. The architect's scale is also needed in a variety of related architectural jobs such as bidding, estimating, and model building.

Whether to reduce a structure's size so that it can be drawn to fit on paper or to enlarge a small detail for clarity and accurate dimensions, a drafter needs to use the appropriate scale divisions.

Architect's scales are either open-divided or fully divided. In fully divided scales, each main unit on the scale is fully subdivided into smaller units along the full length scale. On open-divided scales, only the main units of the scale

Fig. 3-1 ■ Scaled drawings are needed to show the size and shape of large objects. Note the different scales necessary to fit these large objects into a one-inch space.

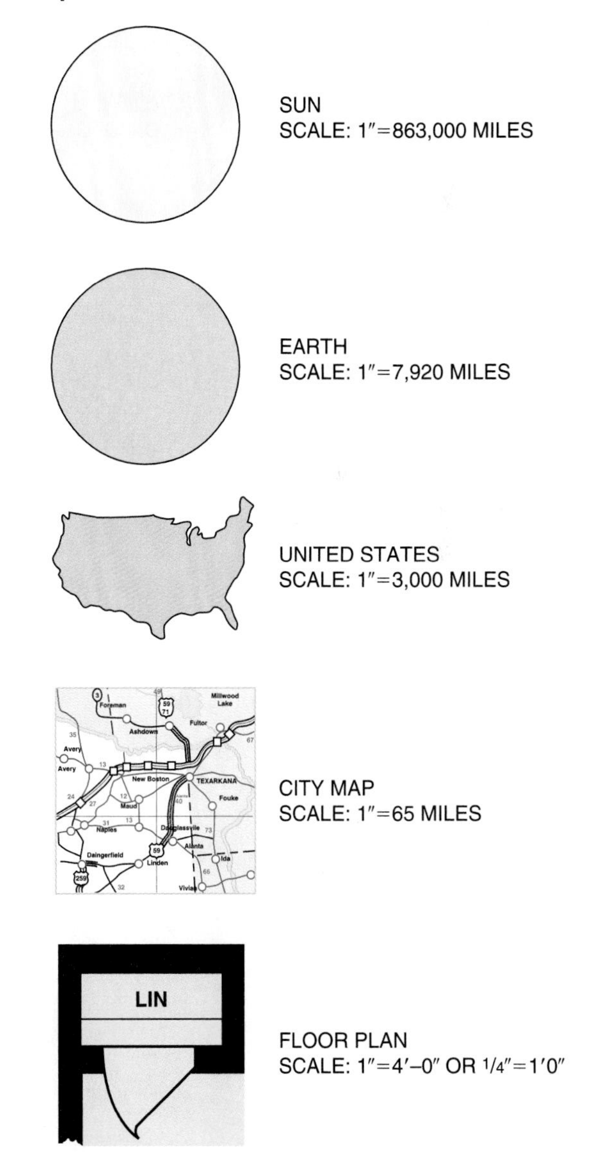

are graduated (marked off) all along the scale. There is a fully subdivided unit at the start of each scale. See Fig. 3-2.

The main function of an architect's scale is to enable the architect, designer, or drafter to plan accurately and make drawings in proportion to the actual size of the structure. For example, when a drawing is prepared to a reduced scale of ¼″= 1′-0″, a line that is drawn ¼″ long is thought of by the drafter as 1′-0″, not as ¼″. See Fig. 3-3.

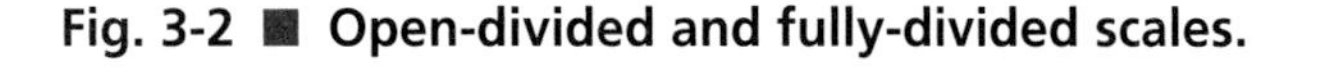

Fig. 3-2 ■ Open-divided and fully-divided scales.

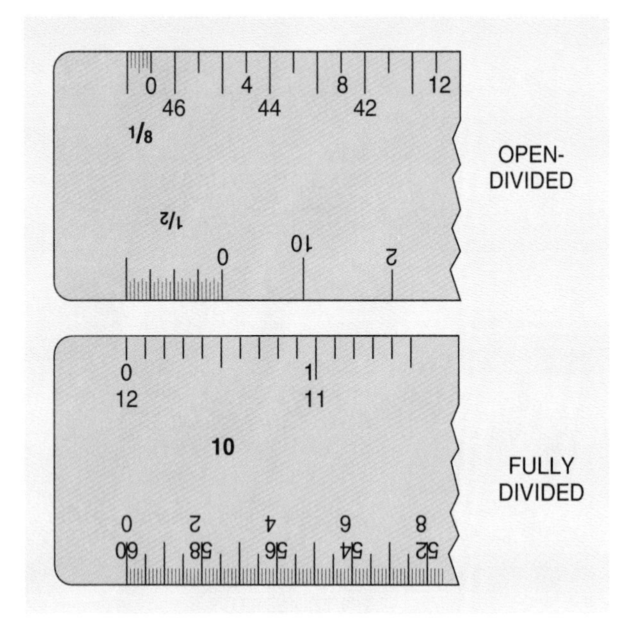

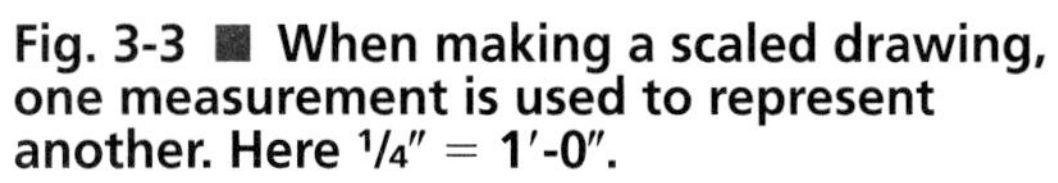

Fig. 3-3 ■ When making a scaled drawing, one measurement is used to represent another. Here ¼″ = 1′-0″.

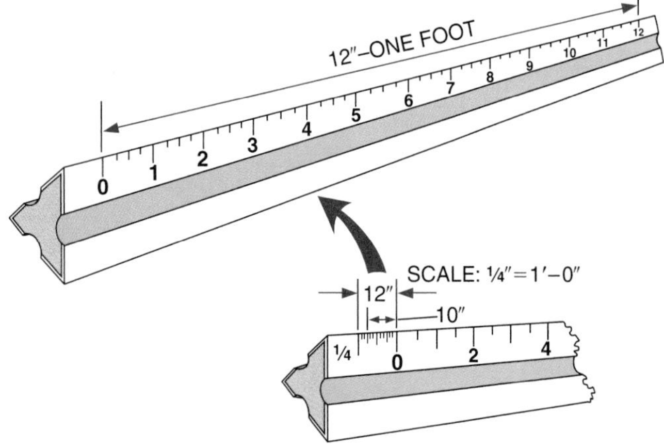

USING A SCALE Architect's scales may be either bevel or triangular style. See Fig. 3-4. Note that the triangular style has three sides and six edges. It accommodates eleven different scales (marked units of measure). One edge is a full-size scale of 12″, divided into 16 parts per inch. The other five edges contain open-divided scales paired to include: 3⁄16 with 3⁄32, ¼ with ⅛, ¾ with ⅜, 1 with ½, and 3 with 1½. Locating two scales on each edge maximizes the use of space. One scale reads from left to right. The opposite scale, which is twice as large, reads from right to left. For example, the ¼″ scale and the ⅛″ scale are placed on the same edge but are read from opposite directions. Be sure you are reading the scale numbers in the correct direction when using an open-divided scale. Otherwise, your measurement could be wrong. See Fig. 3-5.

The architect's scale can be used to make the divisions of the scale equal 1″ or

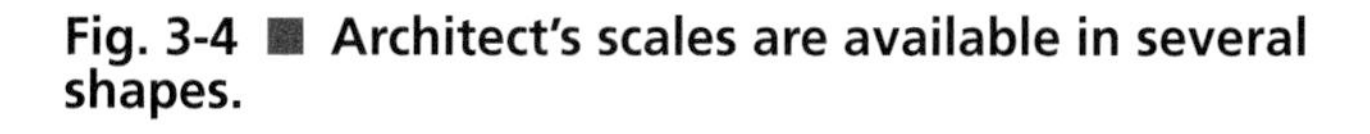

Fig. 3-4 ■ Architect's scales are available in several shapes.

SINGLE BEVEL
ONE SCALE EDGE

ALTERNATIVE DOUBLE BEVEL
TWO SCALE EDGES

DOUBLE BEVEL
TWO SCALE EDGES

QUADRUPLE BEVEL
FOUR SCALE EDGES

TRIANGULAR
SIX SCALE EDGES

Fig. 3-5 ■ Scales which read from left to right are half as large as scales which read from right to left.

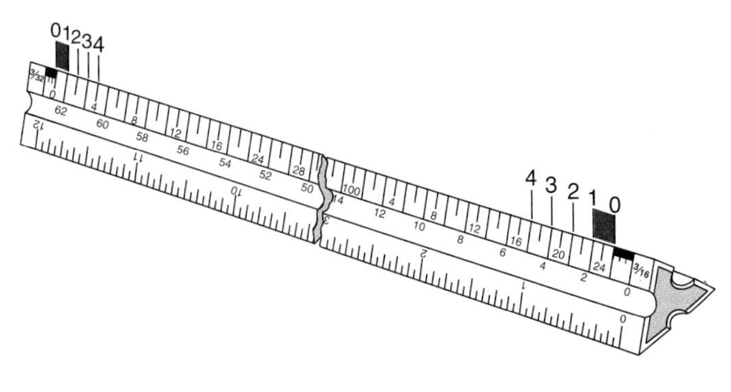

1′-0″. For example, ½″ can equal 1″ or 1′-0″ or any unit of measurement such as yards or miles. See Fig. 3-6.

Since buildings are large, most major architectural drawings use a scale that relates the parts of an inch to 1′-0″. Architectural details, such as cabinet construction and joints, often use the parts of an inch to represent 1″. On open-divided scales, the divided section at the end of the scale is not a part of the numerical scale. This divided section is an additional length to show smaller subdivisions of the larger unit. When measuring with the scale, start at zero, not at the end of the fully divided section. First measure the number of larger units (for example, feet) and then measure the additional smaller units (inches) in the subdivided area. Look at Fig. 3-7. The distance of 4′-11″ is established by measuring from the division line 4 to 0 for feet. Then, measure on the subdivided area 11″ past 0. On this scale, each line in the subdivided part equals 1″. On smaller scales, these lines may equal only 2″. On larger scales, they may equal a fractional part of an inch. Figure 3-8 shows a further use of the architect's scale.

To further understand the architect's scale, compare one specific distance shown on different scales. See Figs. 3-9 and 3-10.

Fig. 3-6 ■ On a half-inch scale, 1/2″ may equal 1″ or 1′-0″.

A. If 1/2″ = 1″, then 1/4″ = 1/2″.

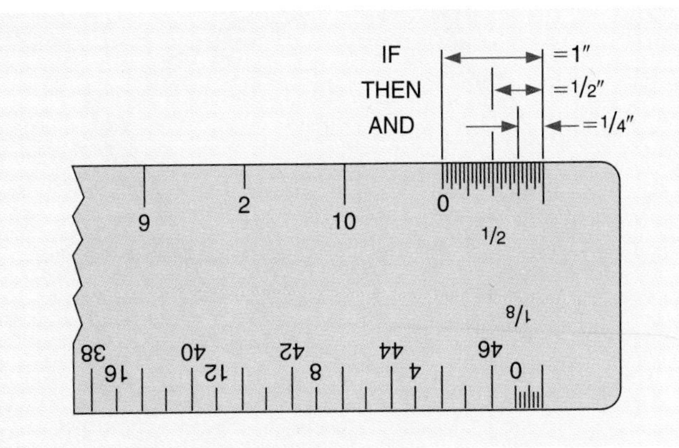

B. If 1/2″ = 1′-0″, then 1/4″ = 6″.

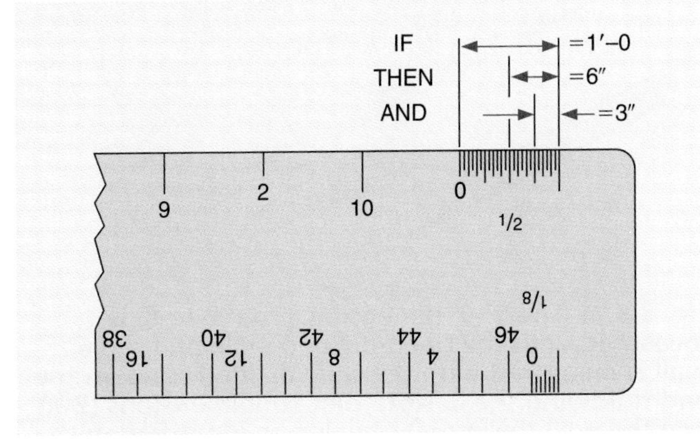

Fig. 3-7 ■ The subdivisions in the fully divided area of an open-divided scale are used for inches and fractions of an inch.

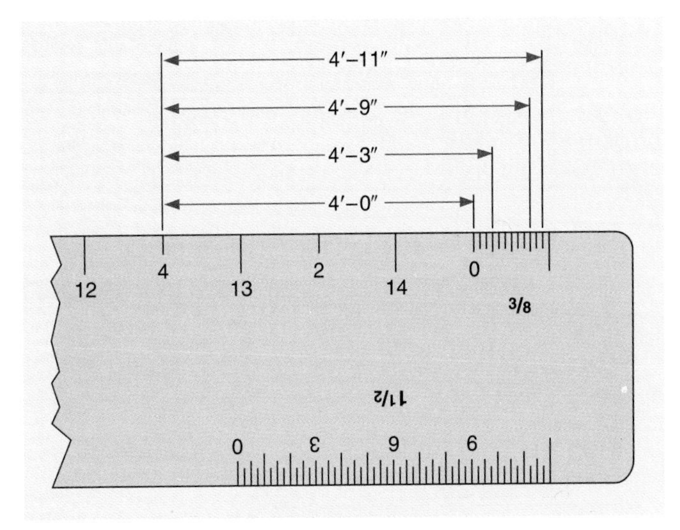

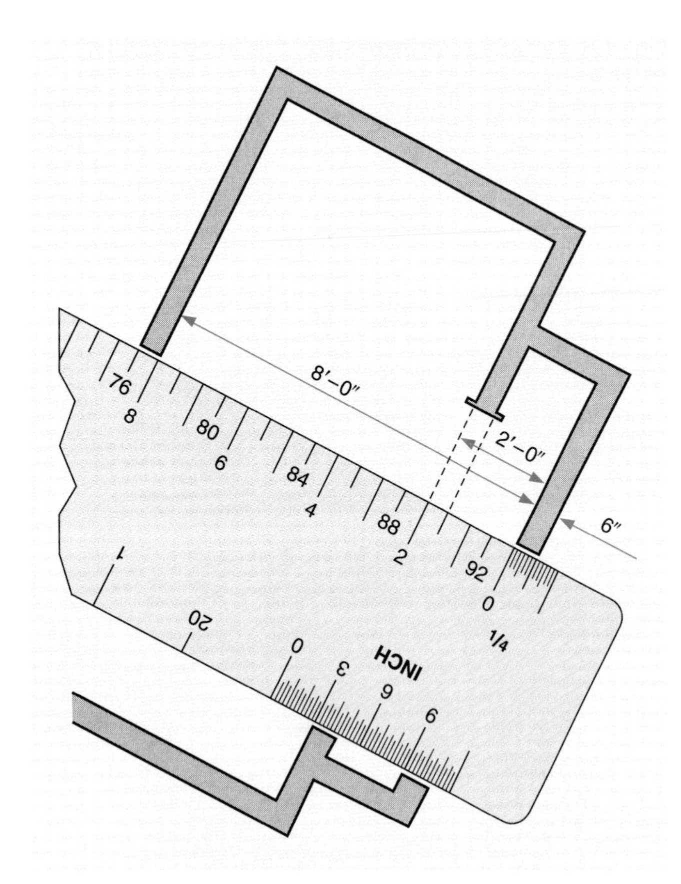

Fig. 3-8 ■ The 8′-0″ dimension (from 0 to 8) is measured with the open-divided scale and the 6″ wall is measured with the fully divided scale on the end following the 0. Note that subdimensions (2′-0″ and the wall thickness) can also be identified with the overall dimensions.

The architect's scale is only as accurate as its user. In using the scale, always lay out the overall dimensions of the drawing first. If the width and length are correct, only minor errors in subdimensions may occur. Moreover, if overall dimensions are correct, it's easier to check subdimensions. If one is inaccurate, another will be also.

Remember, an architect's scale is a measuring device, not a drawing instrument. Never use a scale as a straightedge for drawing. The fine increment lines on a scale will be worn down or removed if misused, making accurate measuring difficult.

SELECTING A SCALE (PROPORTION) The selection of the proper scale is sometimes difficult. If the structure to be drawn is extremely large, a small scale must be used. Small structures can be drawn to a larger scale, since they will not take up as much space on the drawing sheet. Most plans that show major parts of residences (floor plans, elevations, and foundation plans) are drawn to ¼″ = 1′-0″ scale. Construction details pertaining to these drawings are often drawn to ½″, ¾″, or even 1″ = 1′-0″.

Fig. 3-9 ■ The drawn distance varies depending upon the scale used. Here 5′-6″ is shown using different scales.

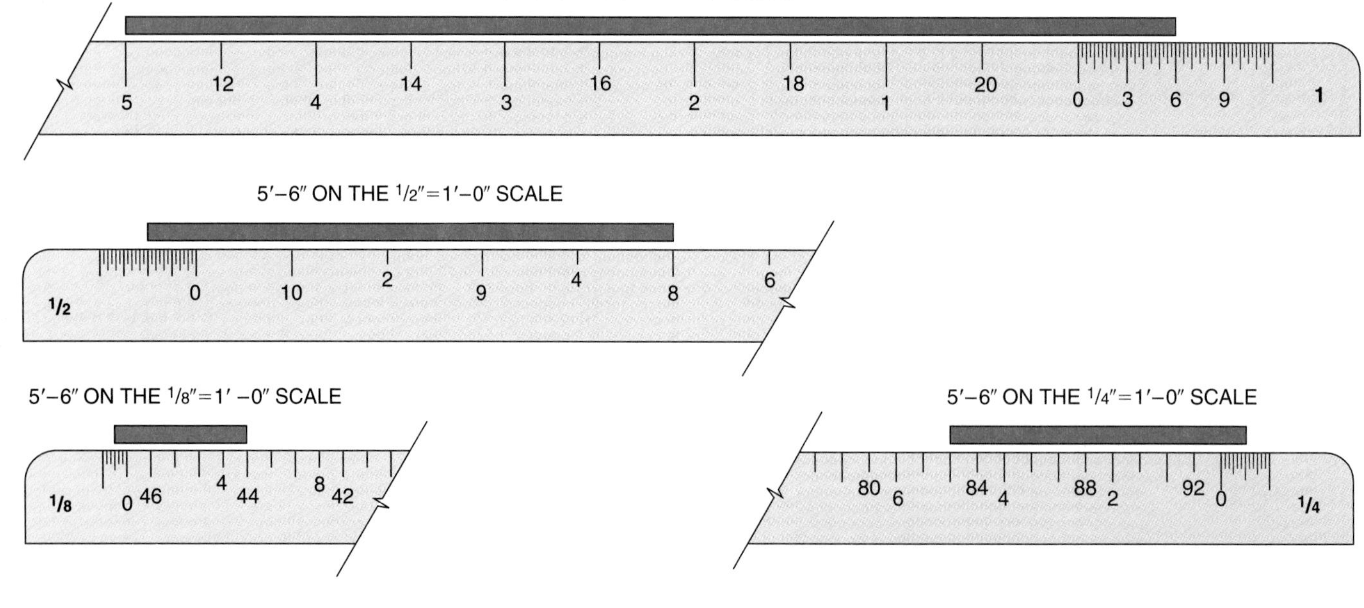

Remember that as the scale changes, not only does the length of each line increase or decrease, but also the width of each wall increases or decreases. See Fig. 3-11. A wall drawn to the scale of 1/16″ = 1′-0″ is small and little detail can be shown. The ½″ = 1′-0″ wall would probably cover too large an area on the drawing if the building were very large. Therefore, the ¼″ and ⅛″ scales are used most often for drawing floor plans and elevations.

The **full scale** (1) is used to draw objects full size 1″ = 1″ or 1′ = 1′ or to a scale of 1″ = 1′-0″. An open-divided full scale is 12″, and each inch may be divided into ½″, ¼″, ⅛″, and 1/16″ units. The end of the full scale is divided into 48 units. The fully divided full scale is divided into 16 units throughout.

The full architect's scale is also useful in dividing any area into an equal number of parts by following these steps:

1 Scale the distance required. See Fig. 3-12.
2 Place the zero point of the architect's scale on one side line.
3 Count off the correct number of spaces, using any convenient unit, such as 1″ or ½″.
4 Place the last unit mark on the line opposite the zero-point line. In Fig. 3-12 the 8″ mark is used since 1″ divisions are the most convenient.

Fig. 3-10 ■ Reduction resulting from the use of scales

Scale						
1″ = 1′-0″	will **decrease**	drawing	size	by	12	times
1½″ = 1′-0″	"	"	"	"	8	"
3″ = 1′-0″	"	"	"	"	4	"
½″ = 1′-0″	"	"	"	"	24	"
¼″ = 1′-0″	"	"	"	"	48	"
⅛″ = 1′-0″	"	"	"	"	96	"

Fig. 3-11 ■ A typical corner wall drawn at 1/16″, 1/8″, 1/4″, and 1/2″ = 1′-0″ scale.

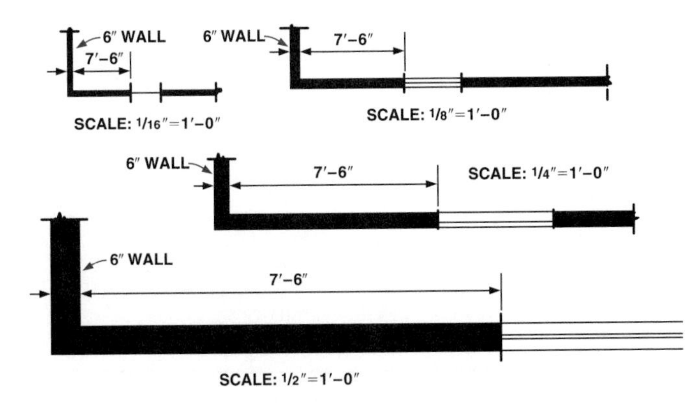

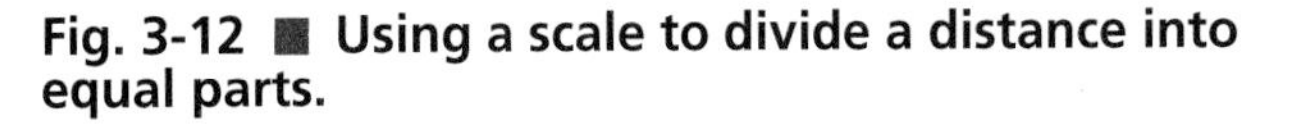
Fig. 3-12 ■ Using a scale to divide a distance into equal parts.

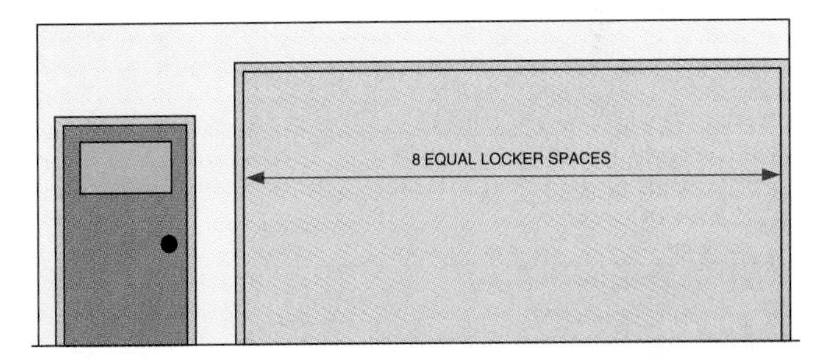

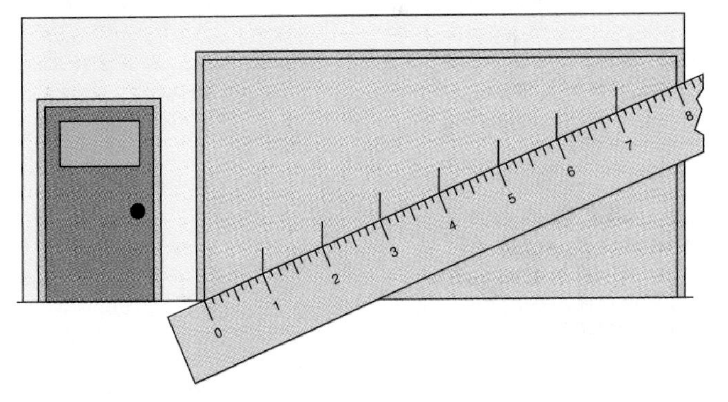

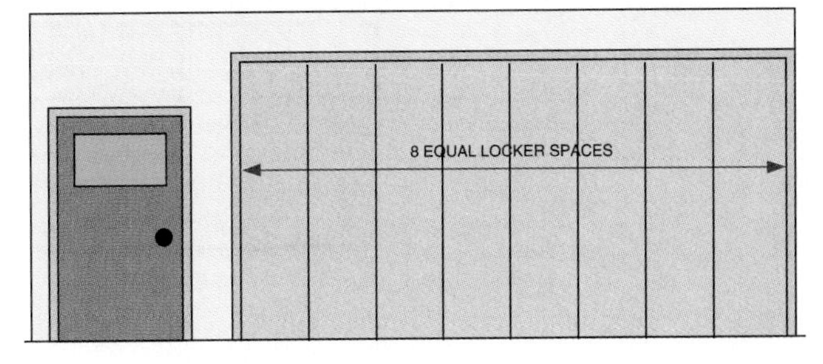

5 Mark each division and draw the dividing lines.
6 Measure the horizontal distance between the lines to find the actual spacing.
7 Multiply by the number of spaces to check for the overall dimension.

Civil Engineer's Scale

Although the architect's scale is used for most architectural drawings, the **civil engineer's scale** is often used for plans that show the size and features of the land surrounding a building (plot plans, site plans, landscape plans). A civil engineer's scale divides the inch into decimal parts. These parts are 10, 20, 30, 40, 50, and 60 parts per inch. See Fig. 3-13. Each one of these units can represent any distance, such as an inch, a foot, a yard, or a mile, depending on the final drawing size.

The civil engineer's scale can also be used to draw floor plans. The scale 1/4″ = 1′-0″ (a 1:48 ratio) is the same ratio as 1″ = 4′-0″ (also a 1:48 ratio). A civil engineer's scale does not use feet and inches. A civil engineer's scale includes feet and decimal parts of a foot. Thus 2′-6″ reads 2.5′ and 7′-3″ reads 7.25′ on an engineer's scale. See Fig. 3-14.

Fig. 3-13 ■ Civil engineer's scales (decimals).

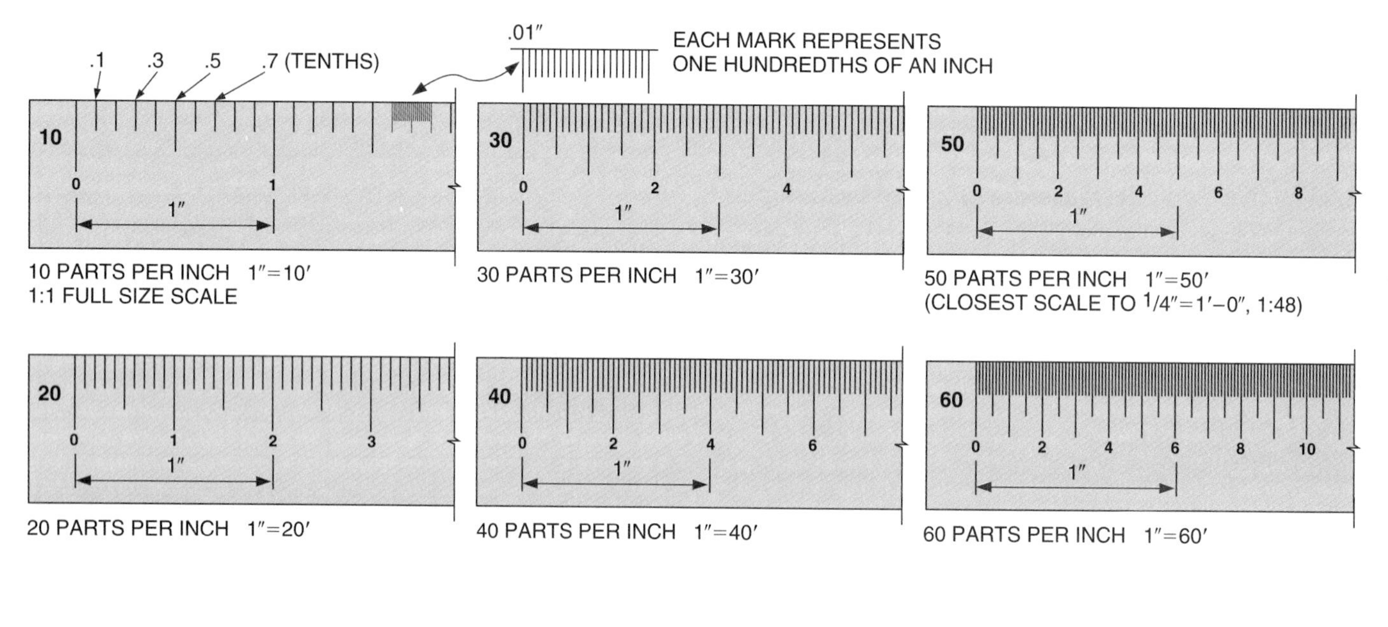

Fig. 3-14 ■ A civil engineer's scale of 1″ = 4′-0″ is the same as the architect's scale 1/4″ = 1′-0″.

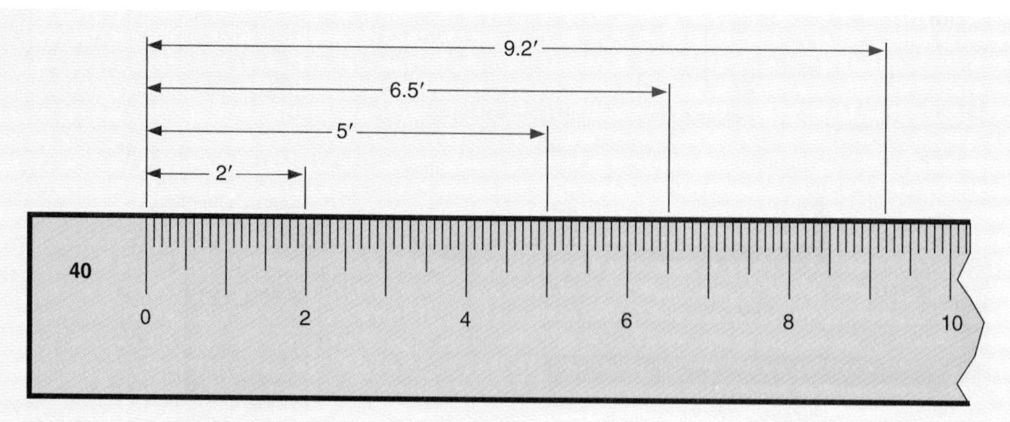

A civil engineer's scale of 1″ = 10′ is normally used. If a land site is very large, a scale of 1″ = 20′ or 1″ = 30′ may be needed to allow the plan to fit the sheet size. See Fig. 3-15.

SELECTING A SCALE (INSTRUMENT) Different scales are designed for a broad range of applications. Before a drawing is started, determine the actual size of the area to be covered. Then select an appropriate scale—whether an architect's or civil engineer's scale—that will provide the greatest detail and yet fit appropriately on the drawing sheet. See Fig. 3-16.

Fig. 3-15 ■ Distances represented by 2″ on different civil engineer's scales.

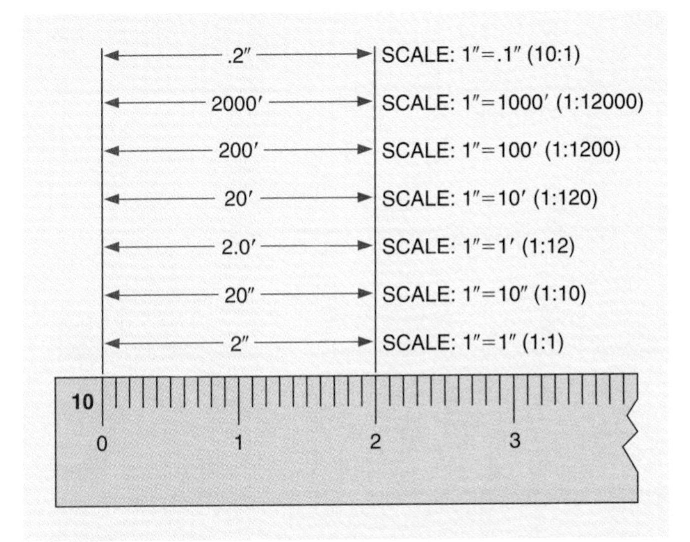

Metric Scales

Math Connection

Metric scales such as those shown in Fig. 3-17 are used in the same manner as the architect's scale to prepare reduced-size drawings. Metric scales, however, use ratios in increments of 10 rather than the fractional ratios of 12 used in architect's scales. The metric system of measure is a decimal system. Units are related by tens.

Most measurements used on architectural drawings are linear distances. The basic unit of measure in the metric system for distance is the *meter* (m). Prefixes are used to change the base (meter) to larger or smaller amounts by *units of 10*. Prefixes that represent subdivisions of less than

Fig. 3-16 ■ Range of scales on architectural drawings.

Drawing type	U.S. Customary Architect's scales (feet/inches)	ISO Metric scales (millimeters)	Civil Engineer's scales (feet/decimal)
Site plans	1⁄8″ = 1′–0″ or 1″ = 10′	1:100 or 1:150	1″ = 10′ or 1″ = 20′
Floor plans	1⁄4″ = 1′–0″ or 1⁄8″ = 1′–0″	1:50 or 1:00	1″ = 40′ or 1″ = 50′
Foundation plans	1⁄4″ = 1′–0″	1:50	1″ = 40′ or 1″ = 50′
Exterior elevations	1⁄4″ = 1′–0″ or 1⁄8″ = 1′–0″	1:50 or 1:00	1″ = 40′ or 1″ = 50′
Interior elevations	1⁄2″ = 1′–0″	1:20	1″ = 20′ or 1″ = 10′
Construction details	1 1⁄2″ = 1′–0″ thru 3⁄4″ = 1′–0″	1:5 thru 1:10	1″ = 10′ or 1″ = 5′
Cabinet details	1⁄2″ = 1′–0″	1:20	1″ = 1′–0″

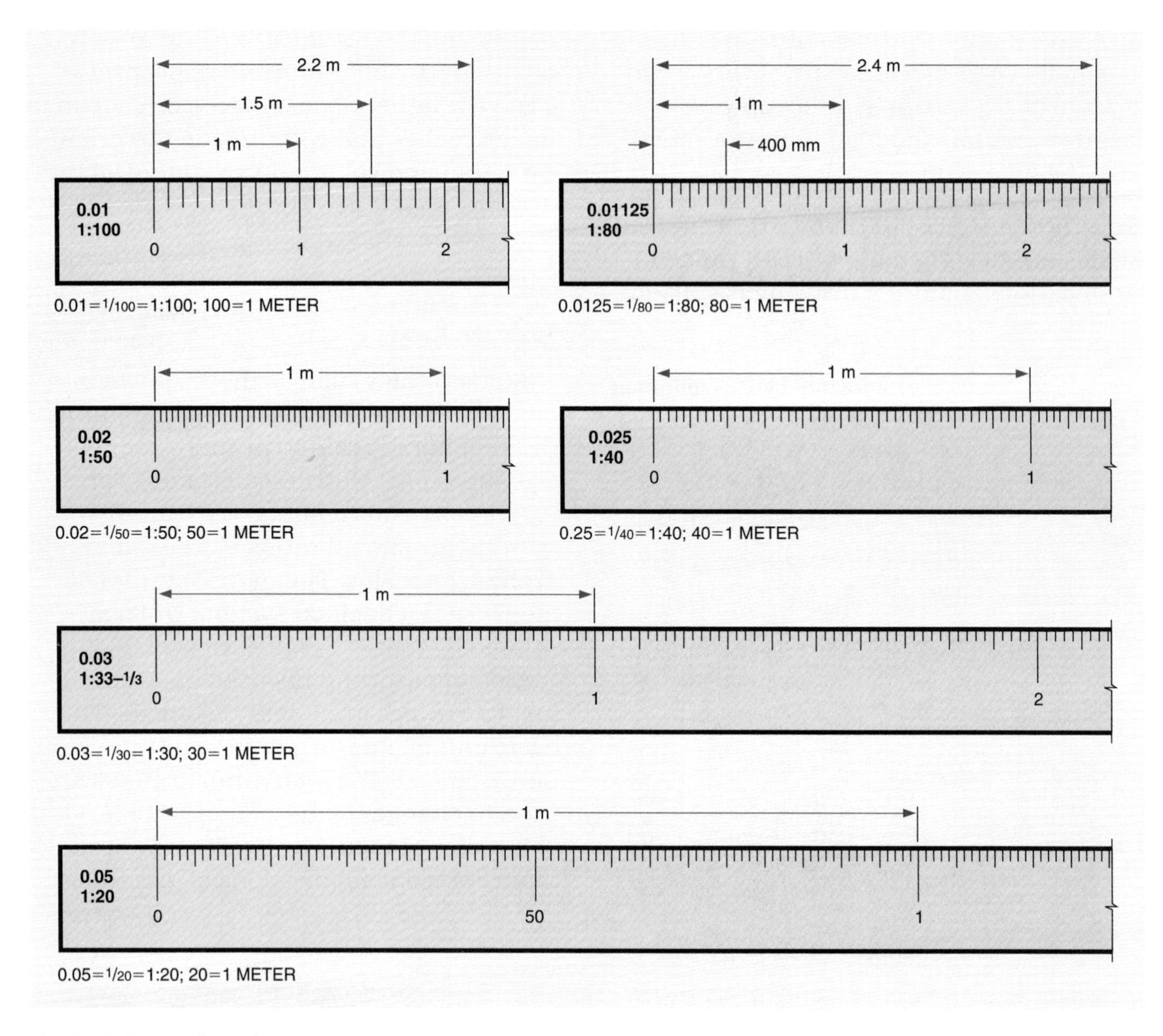

Fig. 3-17 ■ Metric scales.

one meter are deci-, centi-, and milli-. A decimeter equals one-tenth (0.1) of a meter. A centimeter equals one one-hundredth (0.01) of a meter. A millimeter equals one one-thousandth (0.001) of a meter. See Fig. 3-18. The most commonly used subdivisions of a meter are the centimeter and the millimeter. The most commonly used multiple of the meter is the *kilo*meter. A kilometer equals 1000 meters.

The numbers on a meter scale mark every tenth line to represent centimeters. Each single line represents millimeters. Refer back to Fig. 3-17. Note that there are 10 millimeters within each centimeter.

Just as with any other scale the ratio chosen depends on the size of the drawing compared with the full size of the object. Figure 3-19 shows some common metric ratios and the various types of architectural drawings for which they are used.

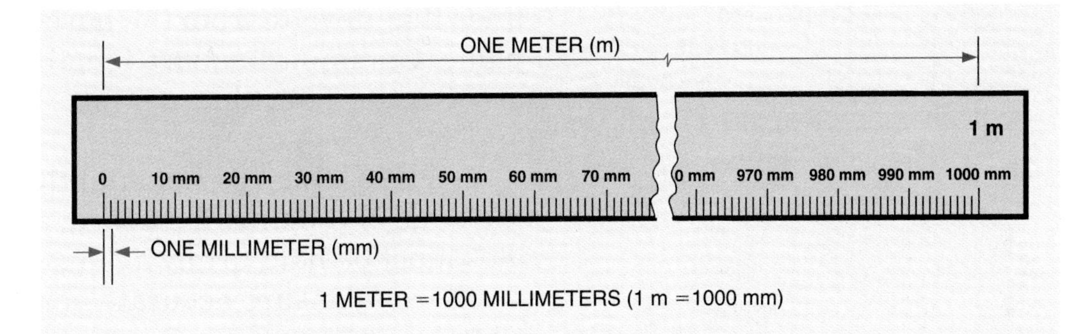

Fig. 3-18 ■ **A millimeter is one one-thousandth of a meter.**

Fig. 3-19 **Use of metric ratios.**

Use	Ratio	Comparison to 1 meter
CITY MAP	1:2500 1:1250	(0.4 mm equals 1 m) (0.8 mm equals 1 m)
PLAT PLANS	1:500 1:200	(2 mm equals 1 m) (5 mm equals 1 m)
PLOT PLANS	1:100 1:80	(10 mm equals 1 m) (12.5 mm equals 1 m)
FLOOR PLANS	1:75 1:50 1:40	(13.3 mm equals 1 m) (20 mm equals 1 m) (25 mm equals 1 m)
DETAILS	1:20 1:10 1:5	(50 mm equals 1 m) (100 mm equals 1 m) (200 mm equals 1 m)

Fig. 3-20 ■ **Examples of meter dimensions carried to three decimal places.**

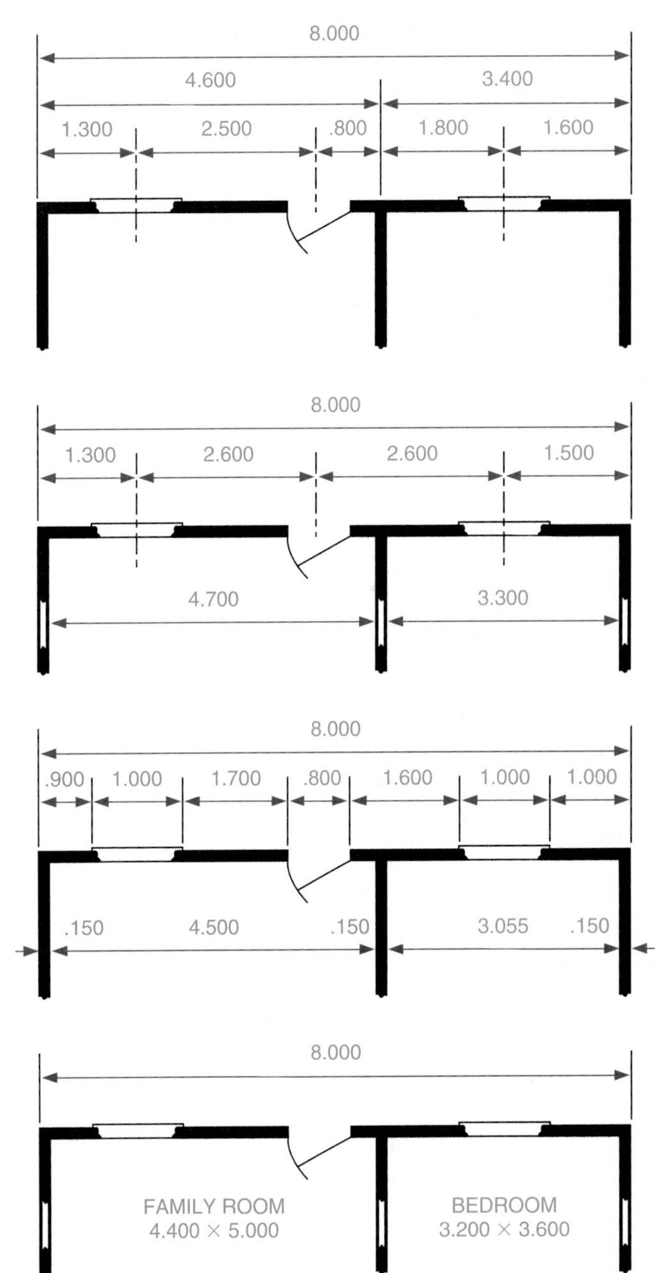

In architectural drawing, ISO standards recommend that only millimeters be used. However, in practice, linear metric sizes used on basic architectural drawings such as floor plans and elevations may be expressed in meters and decimal parts of a meter. Dimensions on these plans are usually carried to three decimal places. See Fig. 3-20. Small-detail drawings normally use millimeters. This eliminates the use of decimal points. See Fig. 3-21. Note that the dimension 4.5 m on Fig. 3-20 could also be read as 4500 mm.

Some drawings prepared with fractional dimensions need to be converted to metric dimensions. To convert inch

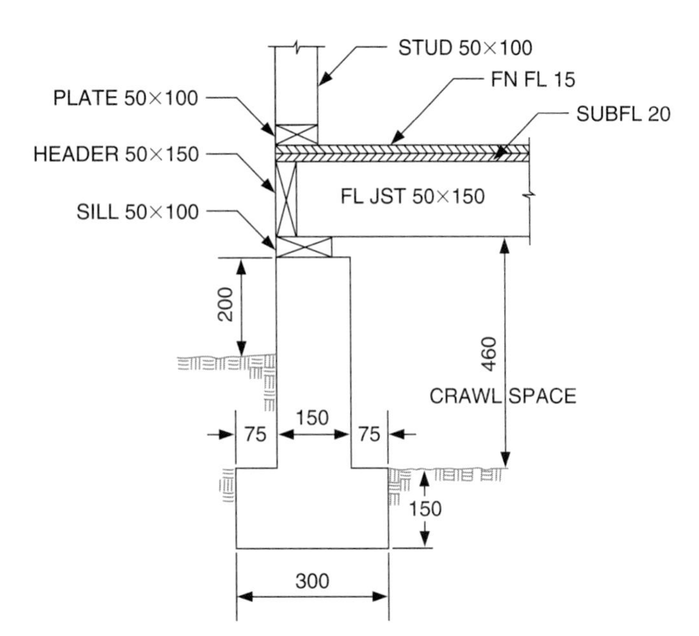

Fig. 3-21 ■ **The millimeter is used for detail dimensioning.**

dimensions to millimeter dimensions, use the following formula:

Formula: in. × 25.4 = mm
(inches × 25.4 = millimeters)
Example: 6′-6″ = 78″
78″ × 25.4 = 1981.2 mm

Prepare all drawings in a set using either metric ratios or the customary fractional system. *Do not mix metric and customary units.* If approximate conversion from one system to the other is necessary, refer to the appendix. When very accurate conversion from customary to metric units is necessary, consult a handbook or use Metric Practice Guide, ASME.

Guides for Straight Lines

T Square

The **T square** serves several purposes. It is used primarily as a guide for drawing horizontal lines. It also serves as a base for a triangle that is used to draw vertical and inclined lines. The T square is particularly useful for drawing extremely long lines that deviate from the horizontal plane. Common T square lengths for architectural drafting are 18″, 24″, 36″, and 42″.

T squares must be held tightly against the edge of the drawing board, and triangles must be held firmly against the T square to ensure accurate horizontal and vertical lines. See Fig. 3-22A. Since only one end of the T square is held against the drawing board, some sag may occur when long T squares are not held securely. Figure 3-22B shows the difference between right-handed and left-handed use of a T square. Horizontal lines

Fig. 3-22 ■ Drawing horizontal and vertical lines with a T square and a triangle.

A. Right-handed method.

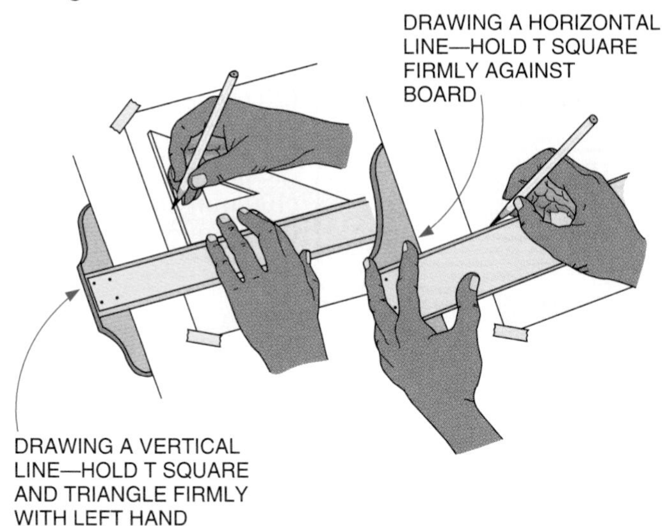

B. Left-handed method.

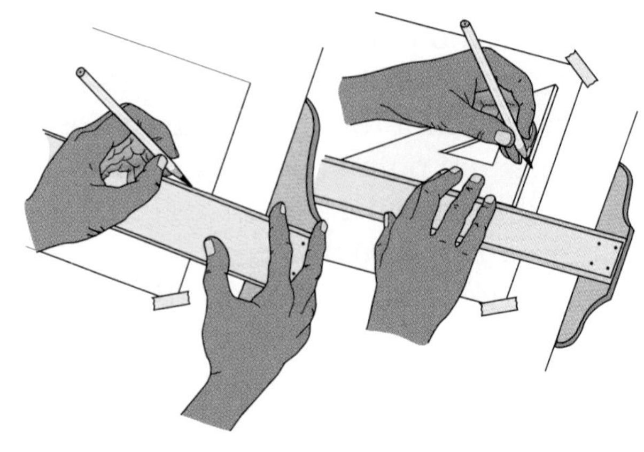

are drawn from left to right by right-handed people and right to left by left-handed people. Vertical lines are made by pulling the pencil or pen upward.

Parallel Slide

The **parallel slide** (or parallel rule) performs the same function as the T square. It is also used as a guide for drawing horizontal lines and as a base for aligning triangles.

Extremely long lines are common in many architectural drawings such as floor plans and elevations. Since most of these lines should be drawn continuously, the parallel slide is used extensively by architectural drafters.

A parallel slide is anchored at both sides of a drawing board. See Fig. 3-23. This attachment eliminates the possibility of sag at one end, which is a common objection to the use of the T square. Another advantage of using the parallel slide is that the drawing board can be tilted to a very steep angle without causing the slide to fall to the bottom of the board. If the parallel slide is adjusted correctly, it will stay in the exact position in which it is placed.

Fig. 3-23 ■ Using a parallel slide.

Triangles

Triangles are used to draw vertical and diagonal or inclined lines with either a T square or other horizontal guide. A variety of possible combinations produce numerous angles, as shown in Figs. 3-24 and 3-25.

The 8″, 45° triangle and the 10″, 30°–60° triangle are preferred for architectural work. Refer back to Fig. 3-22 to see the correct method of using the T square and triangle to draw vertical lines.

Fig. 3-24 ■ Using triangles to draw angles.

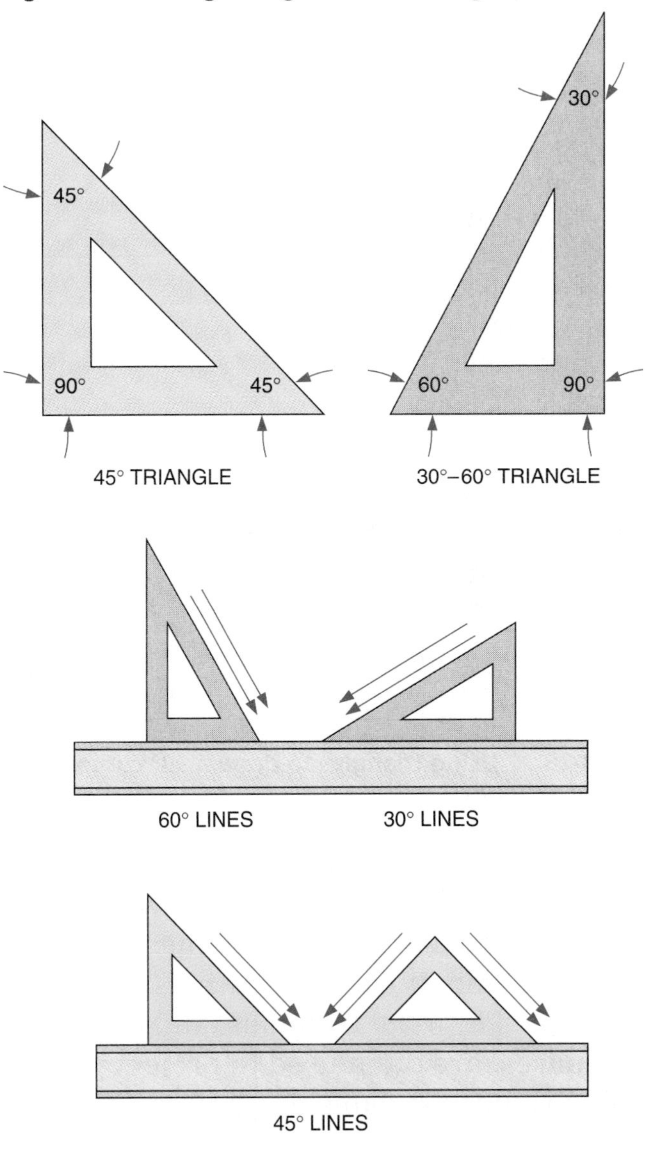

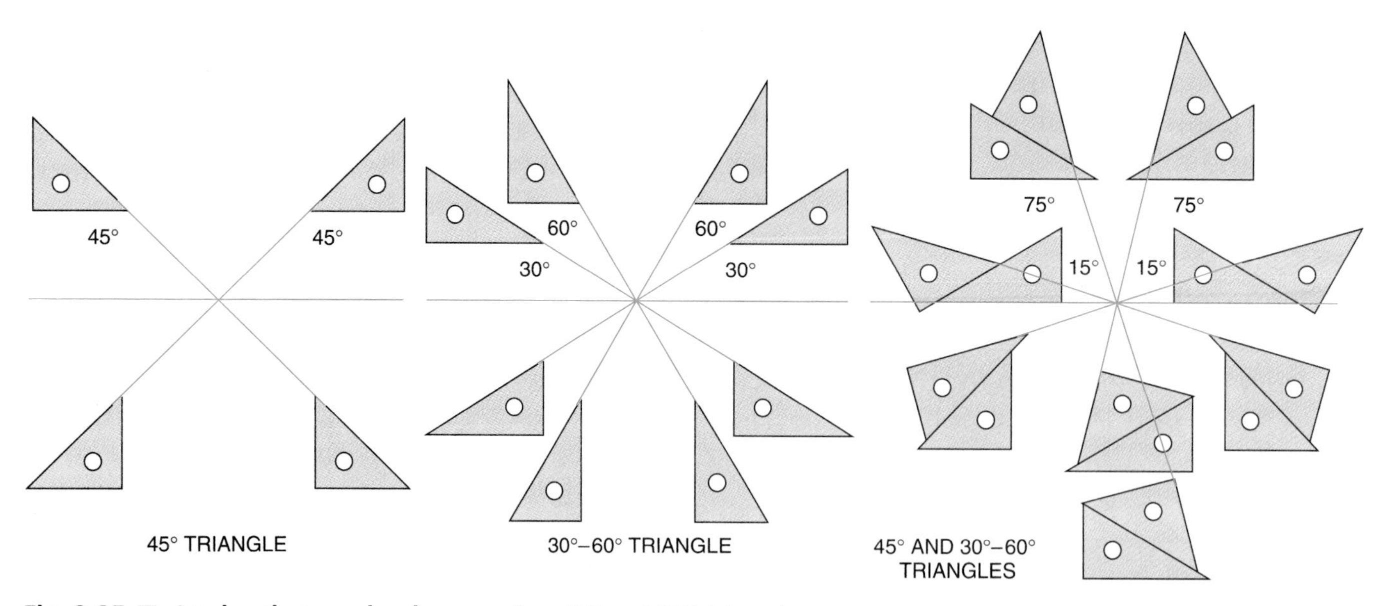

Fig. 3-25 ■ **Angles that can be drawn using 45° and 30° triangles.**

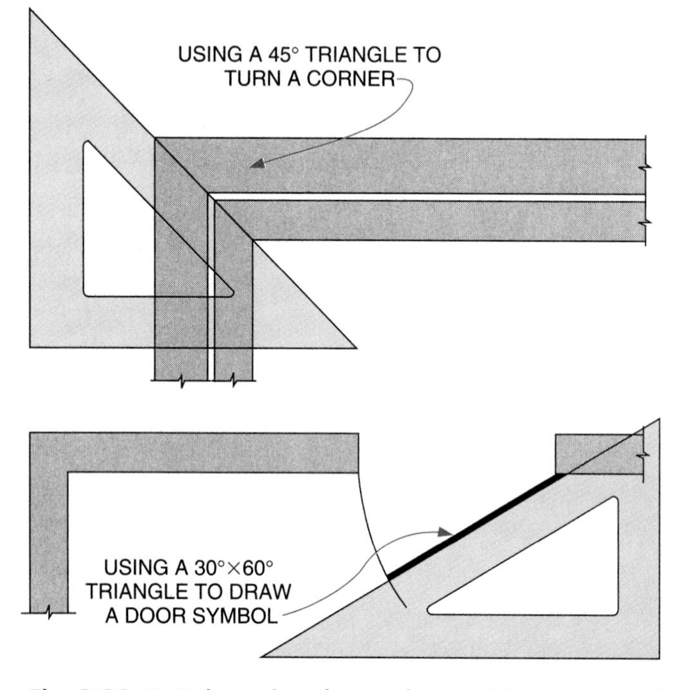

Fig. 3-26 ■ **Using triangles to draw a 90° corner and a door symbol.**

Fig. 3-27 ■ **An adjustable triangle is used to draw lines at any angle.**

The 45° triangle is used to draw (45°) lines that turn angles of buildings. Also, various symbols can be drawn using triangles. See Fig. 3-26. Triangles (and inverted T squares) are often used to project perspective lines to vanishing points. (You will learn about perspective drawing in Chapter 19.)

An adjustable triangle is shown in Fig. 3-27. These triangles are used to draw angles that cannot be laid out by combining the 45° and 30°–60° triangles.

Drafting Machine

A **drafting machine** is a mechanical tool that can serve as an architect's scale, triangle, protractor, T square, or parallel slide all in one. A drafting machine consists of a "head" to which two graduated scales are attached perpendicular to each other. See Fig. 3-28. The scales (arms) of the drafting machine are usually made of aluminum or plastic. The horizontal scale performs the function of a T square or parallel slide in drawing horizontal lines. The vertical scale performs the function of a triangle in drawing vertical lines.

The head of the drafting machine can be rotated so that either scale can be used to draw lines at any angle. When the indexing thumbpiece is depressed and then released, the protractor head of the drafting machine will lock into position every 15°. See Fig. 3-29. If the indexing thumbpiece remains depressed, the protractor head can be aligned to any degree. The protractor brake wing nut is used to lock the head in position.

A degree can be subdivided into 60 minutes. If accuracy in minutes is desired, the vernier scale is used to set the head at the desired angle. The vernier clamp locks the head in the exact position when the desired setting is achieved.

The drafting machine is not often used for large floor plans and elevations which require long horizontal and vertical lines. It is most often used for smaller architectural detail drawings.

Large drawings can be made using track drafting machines like the one in Fig. 3-30. These machines are smoother and faster than elbow-type machines. However, the operation of the head is identical. The protractor head of a track machine is mounted on a vertical track which is attached to a horizontal track.

Fig. 3-28 ■ **Using a drafting machine.** *Teledyne Post.*

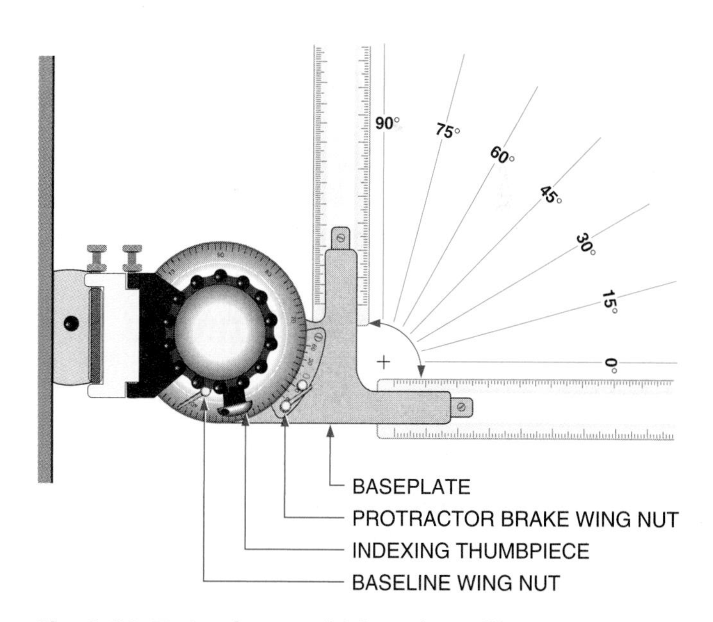

Fig. 3-29 ■ **Angles at which scales will automatically lock into place.** *Vemco.*

Fig. 3-30 ■ Track drafting machine used to make large drawings.

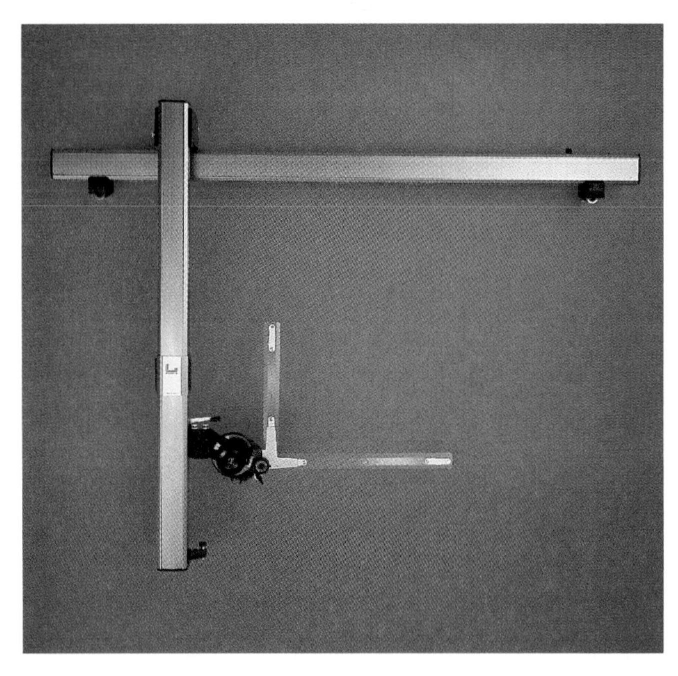

Fig. 3-31 ■ A protractor is used to draw an azimuth.

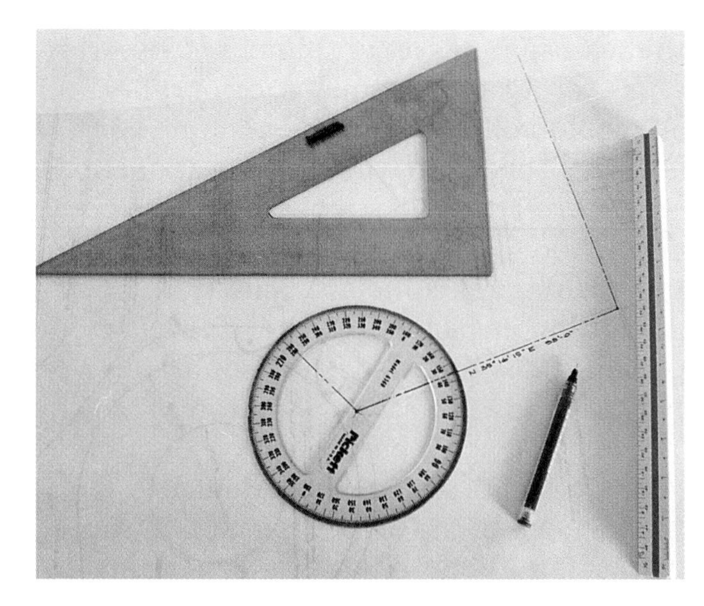

Protractor

A **protractor** provides a graduated scale applicable to all 360° of a circle. To draw an angle by compass degree, align the center of the protractor with the degree located on the circumference of the protractor. Then draw a line between the center and the degree mark. Figure 3-31 shows how a compass degree line (azimuth) is projected.

Instruments for Curved Lines

Compass

A **compass** is used in architectural work to draw circles, arcs, radii, and parts of many symbols. Small circles are drawn with a *bow compass*. To use the bow, set it to the desired radius, and hold the stem between the thumb and forefinger. Rotate the compass with a forward clockwise motion and forward inclination as shown in Fig. 3-32.

Large circles on architectural drawings, such as those used to show the radius of driveways, walks, patios, and stage outlines, are drawn with a large *beam compass*. See Fig. 3-33. Figure 3-34 shows the use of the compass in drawing door symbols. Very small circles on architectural drawings are drawn with either a *drop-bow compass* or a *circle template*. (Templates are described later in this chapter.) Figure 3-35 shows the correct method of sharpening and setting compass leads.

Fig. 3-32 ■ Method of using a compass.

Fig. 3-33 ■ Large circles are drawn with a beam compass.

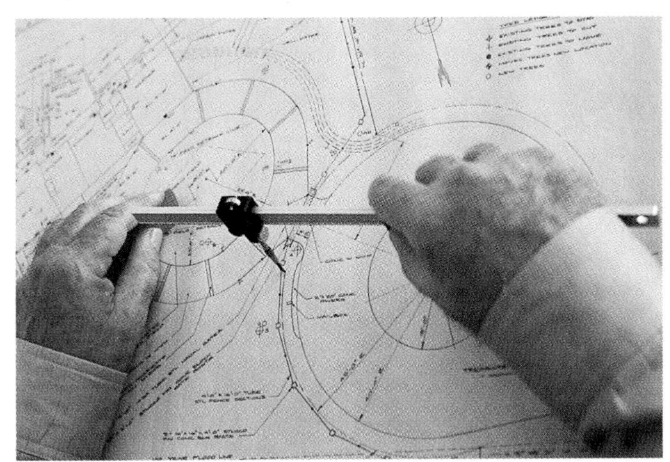

Dividers

Dividing an area into an equal number of parts is a common task performed by architectural drafters. In addition to the architect's scale, **dividers** are used for this purpose. See Fig. 3-36.

To divide an area equally by the trial-and-error method, first adjust the dividers until they appear to represent the desired division of the area. Then place one point at the end of the area and "step off" the distance with the dividers. If the divisions turn out to be too short, increase the opening on the dividers. If the divisions are too long, decrease the setting. Repeat the process until the line is equally divided. See Fig. 3-37.

Dividers are also used frequently to transfer dimensions and to enlarge or reduce the size of a drawing. See Fig. 3-38.

Irregular Curve Instruments

Many architectural drawings contain irregular lines that must be repeated. A **flexible curve** is shown in Fig. 3-39. These curves are used to repeat irregular curves that have no true radius or series of radii and cannot be drawn with a compass. Curved lines that are not part of an arc can also be drawn with a **French (irregular) curve**. See Fig. 3-40.

Fig. 3-34 ■ Method of drawing a door symbol with a compass.

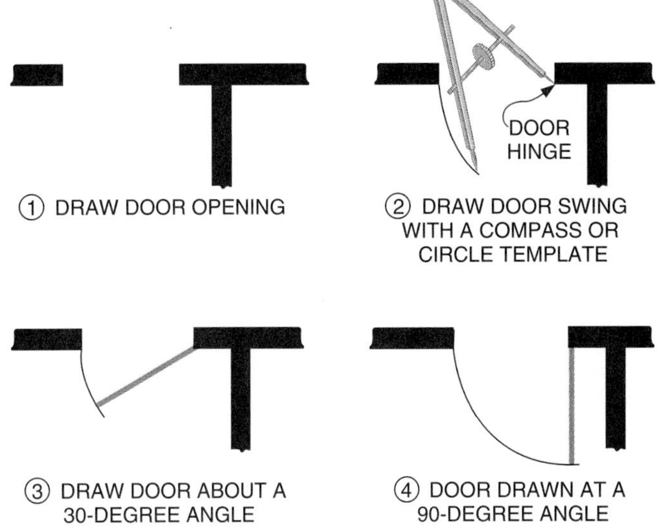

Fig. 3-35 ■ Compass lead point and the position of lead in a compass.

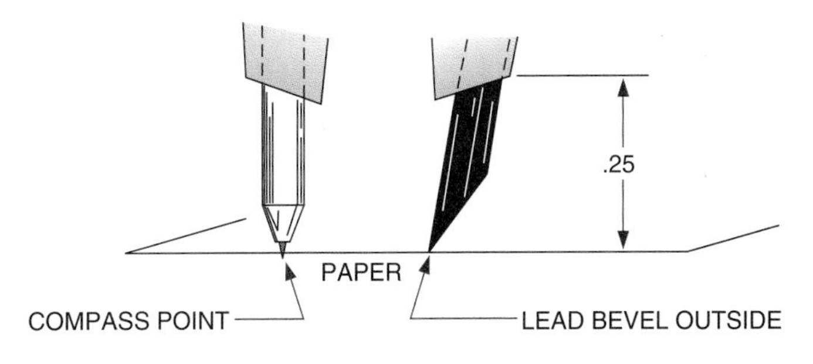

Fig. 3-36 ■ Types of dividers. *K & E.*

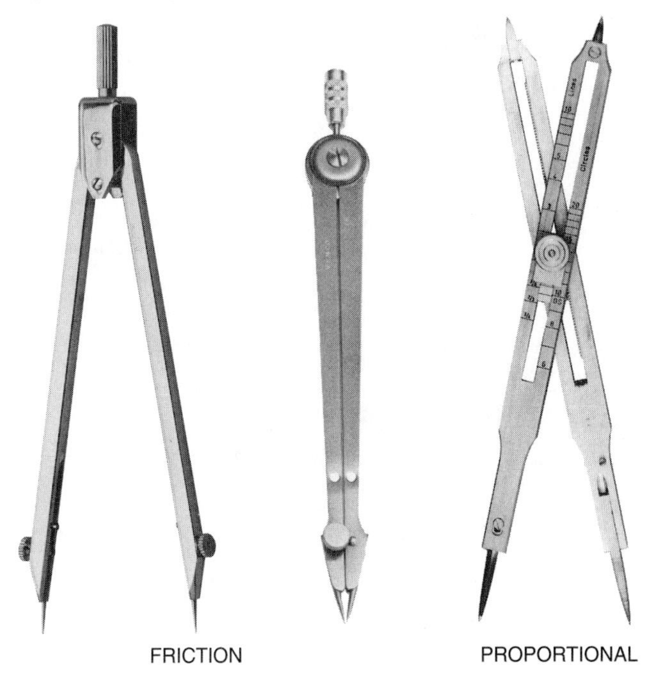

Fig. 3-37 ■ Using dividers to estimate the division of a line.

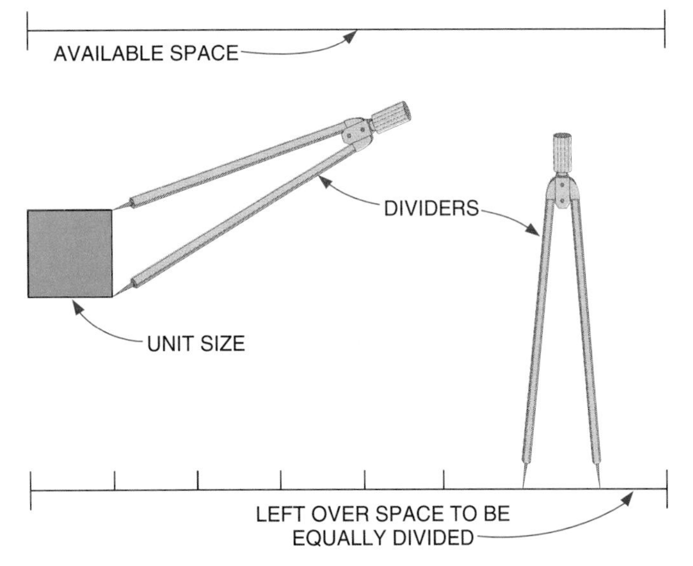

Fig. 3-39 ■ Using a flexible rule.

Fig. 3-38 ■ To double the size of a floor plan, set dividers to each distance on the plan and then step the distance twice on the new plan.

Fig. 3-40 ■ How to use an irregular curve.

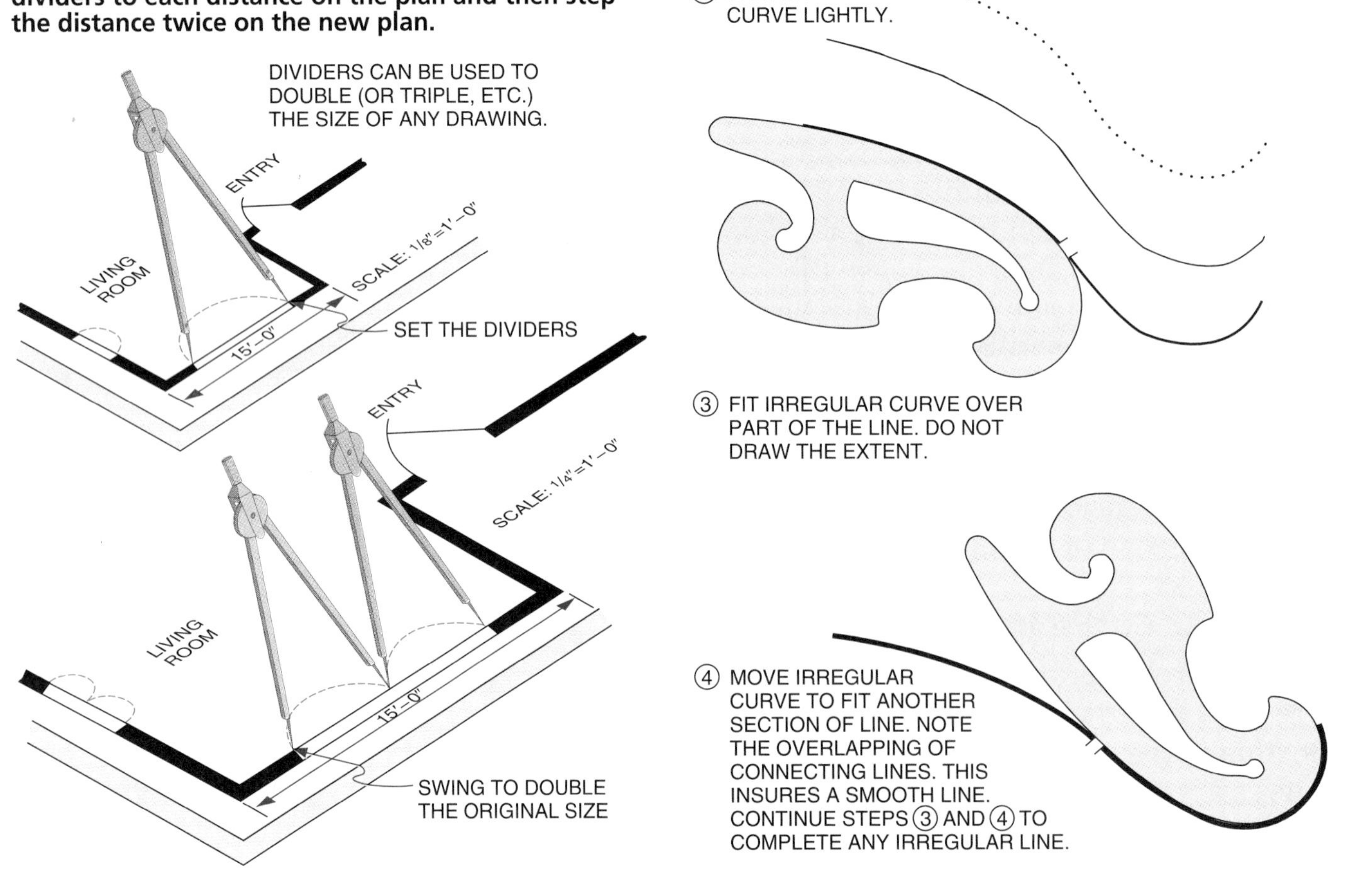

Drafting and Lettering Tools

Drafting pencils and pens are used with other drafting instruments to produce accurate, readable, and consistent architectural lines and symbols.

Drafting Pencils

Pencils used for drafting are either wood encased or mechanical. The width and density of the line produced depends on the degree of hardness and the point of the pencil's lead. See Figs. 3-41 and 3-42.

Fig. 3-41 ■ Types of drafting pencils.

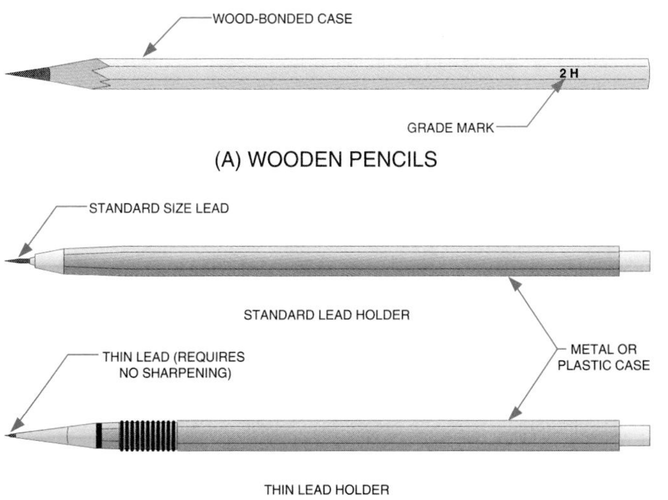

Although referred to as "lead," the core of a drawing pencil is composed mainly of graphite. Hard pencils (3H, 4H), are often used to begin architectural layout work. Medium pencils (2H, H, F), are used for most of the lines in a completed drawing. Soft pencils (HB, B, 2B), are used for lettering and thick cutting-plane lines, as well as for shading in pictorial drawings.

Drafting pencils can be sharpened to several types of points depending on the type of line desired. See Fig. 3-43. Regardless of the type of pencil point used, care must be taken to produce an even point. When uneven points are used, such as a chisel point, uneven lines will result. See Fig. 3-44. Pencil lines used on architectural drawings vary in width, but should not vary in density. Thin lines should be just as black and dense as thick lines.

Technical Pens

Ink pens used for drafting are called **technical pens.** Their points range in thickness from 0.1 mm to 0.9 mm. See Fig. 3-45. One reason for using pens rather than pencils is to create very dense and consistent lines. Working with pens, however, tends to slow down drawing speed, and ink lines are difficult to erase.

Fig. 3-42 ■ Degrees of hardness of pencil leads and matching line weights.

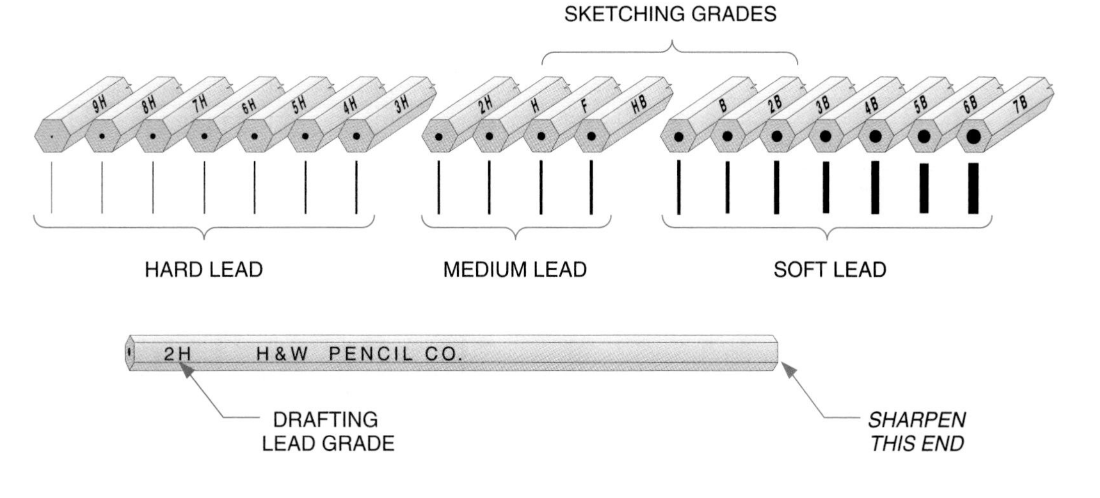

Fig. 3-43 ■ Types of pencil point shapes.

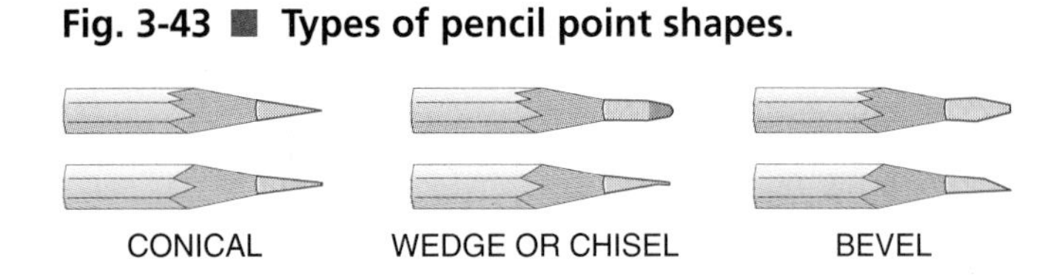

Fig. 3-44 ■ Pencil points and line quality.

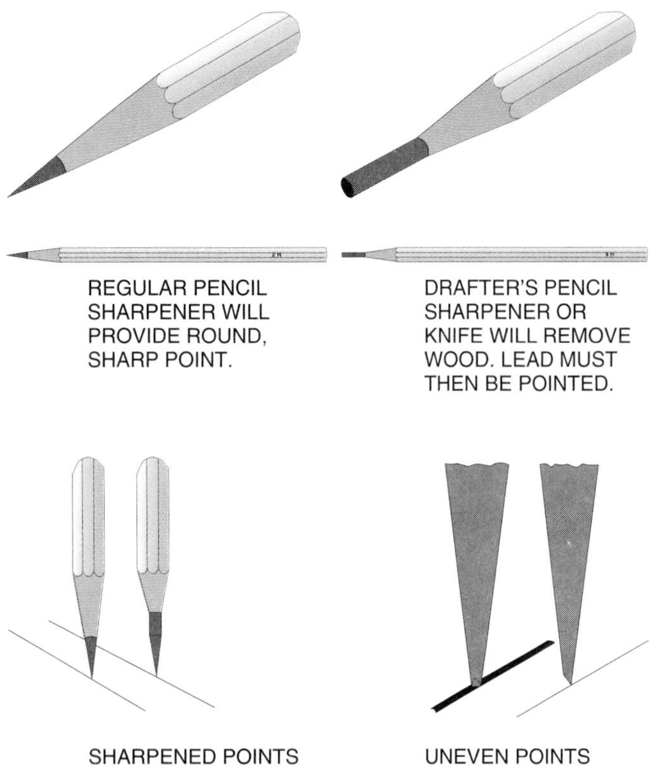

Fig. 3-45 ■ Sizes of technical pen points and widths of lines in millimeters.

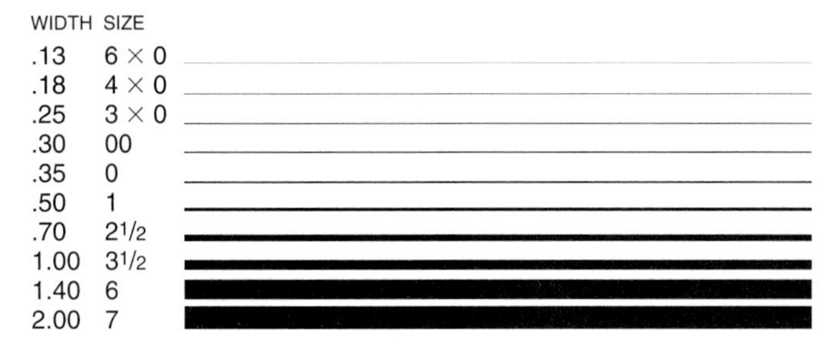

Papers and Drawing Surfaces

Most architectural drawings are prepared on **vellum** or a good-quality tracing paper. Preliminary design work and progressive sketches are usually done on extremely thin tracing paper ("bum wad," "flimsy," "trash"). These preliminary drawings are eventually discarded.

A wide variety of drafting papers is available. Some vellum papers have nonreproducible grid lines that do not reproduce when the original drawing is duplicated. See Fig. 3-46. Grid papers are also printed with nonreproducible angles and lines for perspective drawings (pictorial drawings). See Fig. 3-47.

The size of the drawing surface is determined at the beginning of a project. The drawing format selected should be larger than the largest drawing in the set. Figure 3-48 shows the standard sizes of paper or vellum used for architectural drawings.

The type of paper, as well as the writing instrument, greatly affects the line quality. Different pencil grades of hardness or

Fig. 3-46 ■ Design drawn directly on nonreproducible grid paper and then printed.

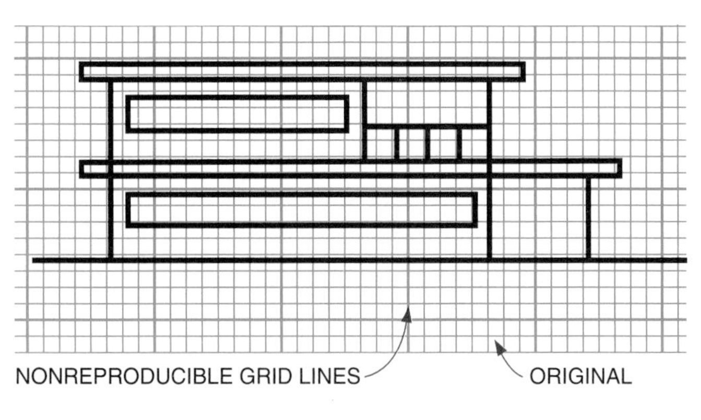

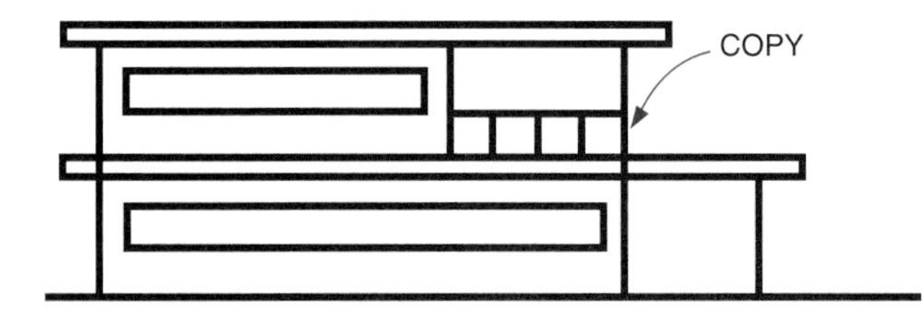

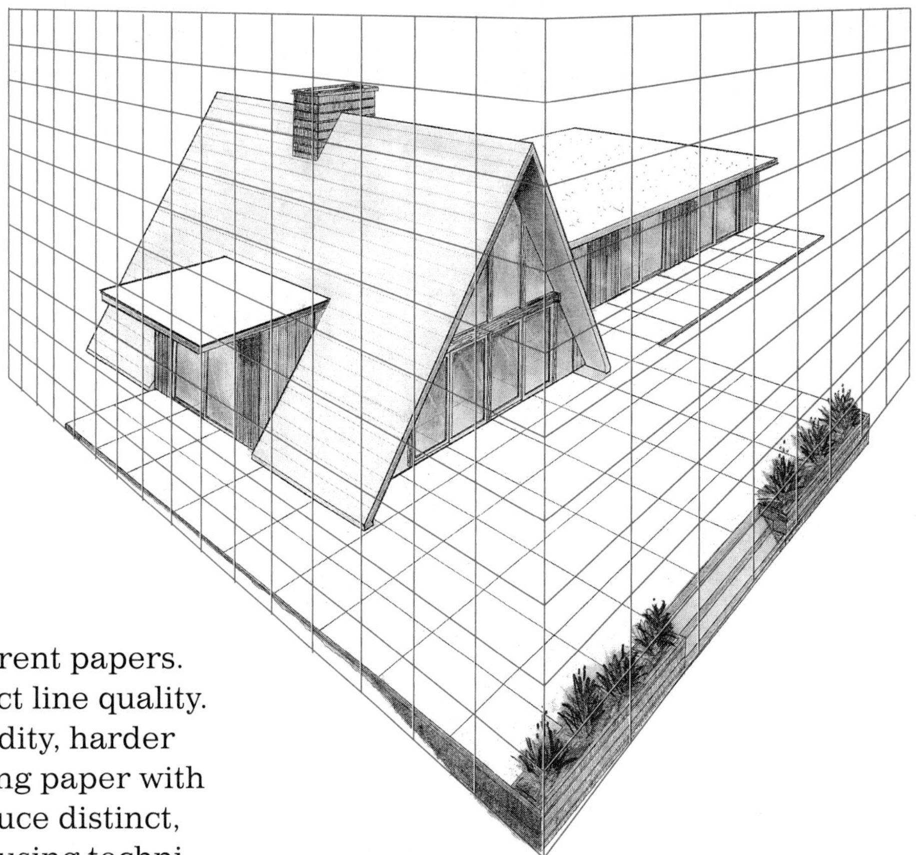

Fig. 3-47 ■ Use of preprinted perspective grid.

softness are needed for different papers. Weather conditions also affect line quality. During periods of high humidity, harder pencils must be used. Drawing paper with a hard surface helps to produce distinct, clean lines, especially when using technical pens. Soft surfaces absorb too much ink and result in feathered lines. See Fig. 3-49.

Fig. 3-48 ■ Standard paper sizes for drawings.

Customary inches	Metric mm
8″ × 10″	
8″ × 11″	
*8.5″ × 11″ (A size)	210 × 297 mm (A4)
*9″ × 12″	
11″ × 14″	297 × 420 mm (A3)
*11″ × 17″ (B size)	
*12″ × 18″	
14″ × 17″	
15″ × 20″	
*17″ × 22″ (C size)	420 × 594 mm (A2)
*18″ × 24″	
19″ × 24″	
21″ × 27″	
*22″ × 34″ (D size)	594 × 841 mm (A1)
*24″ × 36″	
*34″ × 44″ (E size)	841 × 1189 mm (A0)
*36″ × 48″	

* Most commonly used.

Fig. 3-49 ■ The types of paper and pen used affect line quality.

SHARP INK LINES

FEATHERED INK LINES

NONSTICKING INK LINES

Correction Equipment

Mistakes and corrections are part of every drawing process. Designers employ a variety of erasers and ways of keeping drawings clean. *Basic erasers* are used for general purposes. *Gum erasers* are used for light lines. *Electric erasers* are very fast and do not damage the surface of the drawing paper. A very light touch is used to eradicate lines. *Kneaded erasers* pick up loose graphite by dabbing.

To keep drawings clean, *dry cleaner bags* are used to remove smudges. *Powder* sprinkled on the drawing reduces smudging and keeps instruments clean. It also enables drafting instruments to move freely.

Erasing shields are thin pieces of metal or plastic with a variety of small, different-shaped openings. See Fig. 3-50. The appropriate opening is positioned over a line to be erased. The shield covers the surrounding area. Lines can be erased without disturbing nearby lines that are to remain on the drawing.

A **drafting brush** is used periodically to remove eraser and graphite particles and to keep them from being redistributed on the drawing. Do not blow or use your hand to remove debris.

Fig. 3-50 ■ When using an erasing shield, identify the area to be erased, position the shield over the area, and erase through the opening in the shield. Other lines will not be disturbed.

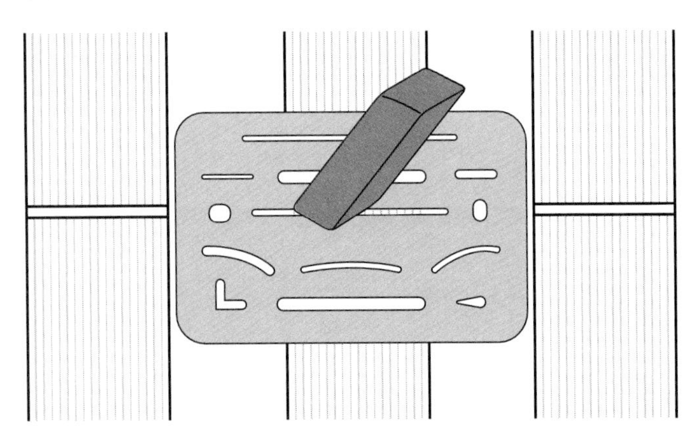

Timesaving Aids and Devices for Drafting

Construction often begins immediately upon completion of the working drawings. Under these conditions, speed in the preparation of drawings is of utmost importance. To work quickly, many timesaving devices are employed by architectural drafters. These devices eliminate unnecessary time on the drawing board without sacrificing the quality of the drawing.

Architectural Templates

Templates are usually made of sheet plastic, but may be made of paper, cardboard, or metal. Openings in the template are shaped to represent various symbols and fixtures. A symbol or fixture is traced on the drawing by following the outline with a pencil or pen as shown in Fig. 3-51. This procedure eliminates the repetitious task of measuring and laying out the symbol each time it is to be used on the drawing. Remember the template scale must always be the same as the scale of the drawing.

A wide assortment of templates is available. See Fig. 3-52. Many are used to draw only one type of symbol. Some are designed specifically for furniture, doors,

Fig. 3-51 ■ Using a door-symbol template.

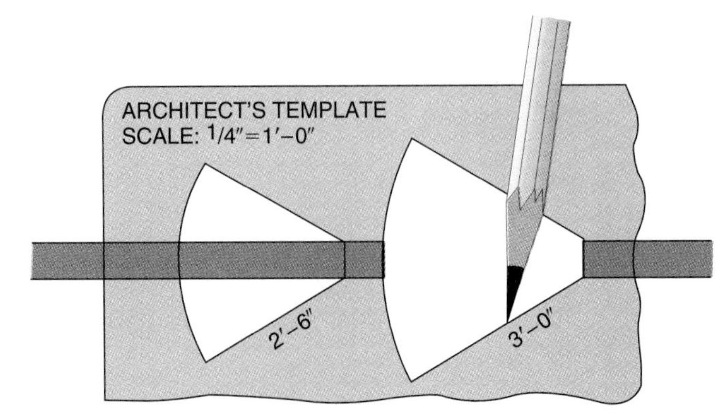

Fig. 3-52 ■ Using a typical architectural template.

windows, or landscape features. Others provide electrical or plumbing symbols. Some serve as lettering, circle, or ellipse guides.

When a symbol needs to be aligned with the lines of the drawing, it is necessary to use a T square, drafting machine, or parallel slide as a guide. Alignment is made by resting one true edge of the template against the blade of the guide and then tracing the symbol. This procedure is also necessary when symbols must be repeated in an aligned pattern.

Overlays

An **overlay** is any sheet that is placed over an original drawing. The information placed on an overlay becomes a visual part of the original drawing. Some overlays remain separate sheets and some are permanently affixed to the base drawing.

SHEET OVERLAYS Most separate *sheet overlays* are made by drawing on transparent or translucent material such as acetate or drafting film or translucent material such as vellum. Overlays are used in the design process to add to or change features of the original drawing, without marking the original drawing.

Overlays are also used to add features to a drawing that would normally complicate the original drawing. Lines that are hidden and many other details can be made clear by drawing this information on an overlay. See Fig. 3-53. Some sheet overlays are also used to aid in the presentation of an architectural design concept.

PRESSURE-SENSITIVE OVERLAYS Pressure-sensitive overlays adhere directly to the surface of the drawing. Details that are often repeated on other drawings or projects are frequently reproduced on pressure-sensitive "appliqué paper."

The transparent appliqué is then attached to any drawing without repeated redrawing.

In addition to fixture and symbol overlays, continuous material symbols are often used. Section-lining overlays are self-adhering and can be cut to any desired size or shape with a sharp stylus. See Fig. 3-54.

Temporary changes can be added to a drawing by writing or drawing the note, symbol, or change on a translucent matte-surface overlay. If the drawing is

Fig. 3-53 ■ The colored portion of this drawing was added by placing it on an overlay.

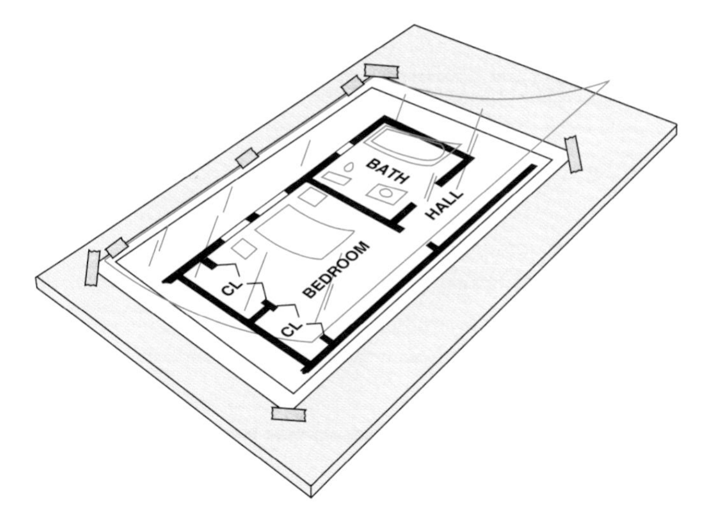

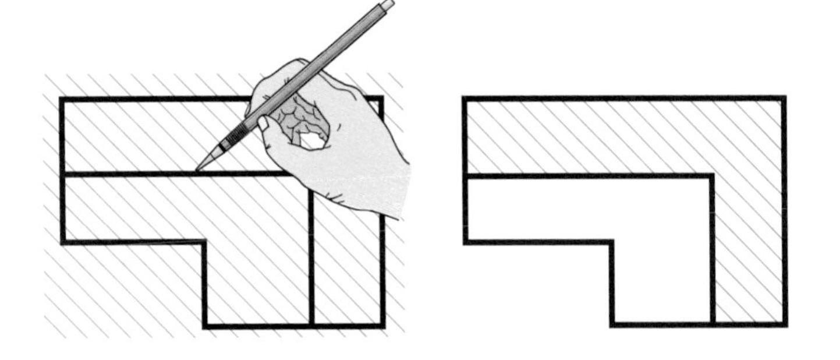

Fig. 3-54 ■ Use of section-lining overlay.

changed, the overlay can be removed and a new symbol added. Otherwise the symbol remains part of the drawing.

On a drawing, a title block is an area in which information is provided about the drawing and the project. Title-block appliqués are often used to ensure the correct spacing of lettering. However, drawing paper preprinted with title blocks is preferred.

Tapes

Many types of *printed pressure-sensitive tapes* can be substituted for drawn lines and symbols on architectural drawings. These are used to produce lines and symbols that otherwise would be difficult and time-consuming to construct. See Figs. 3-55 and 3-56. A special roll-on applicator

Fig. 3-55 ■ Typical architectural line symbol tapes. *Chart-Pak Inc.*

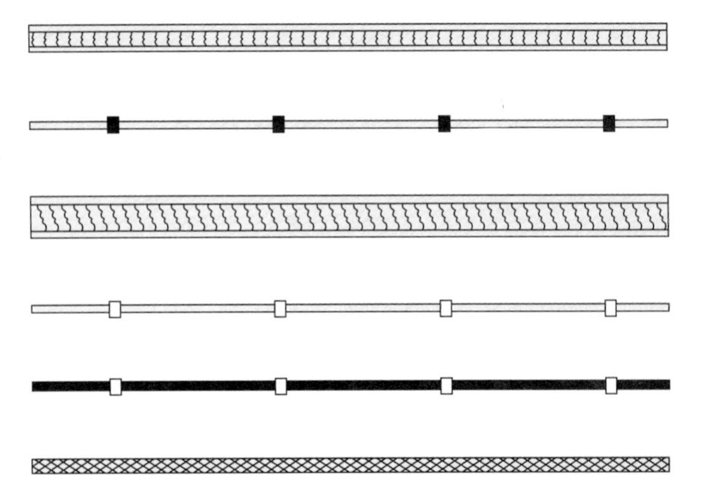

Fig. 3-56 ■ Pressure-sensitive tape used to mark irregular lines on an overlay for a map. *Chart-Pak Inc.*

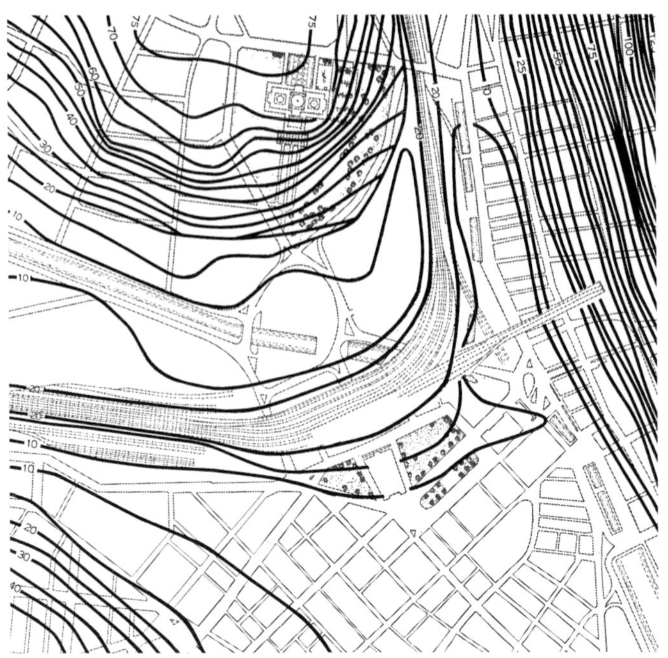

aids in using line tape, as shown in Fig. 3-57.

Drafting tape is used to attach drawings to a drawing board. In addition, strips of drafting tape can also help ensure the equal length of lines when ruling many close lines. Strips of tape are placed on the drawing to mask the areas not being lined. The lines are then drawn on the paper and extended onto the tape. When the tape is removed, the ends of the lines are even and sharp. See Fig. 3-58. This procedure eliminates the careful starting and stopping of the pencil stroke with each line.

Stamps

For architectural features that are often repeated, **stamps** are effective timesavers. Stamps can be used with any color ink. They can also be used in faint colors to provide an outline that can then be rendered with pencil or ink. Stamps are used most often for symbols that do not require

Fig. 3-57 ■ Applying pressure-sensitive tape.

Fig. 3-59 ■ Typical symbols placed on stamps.

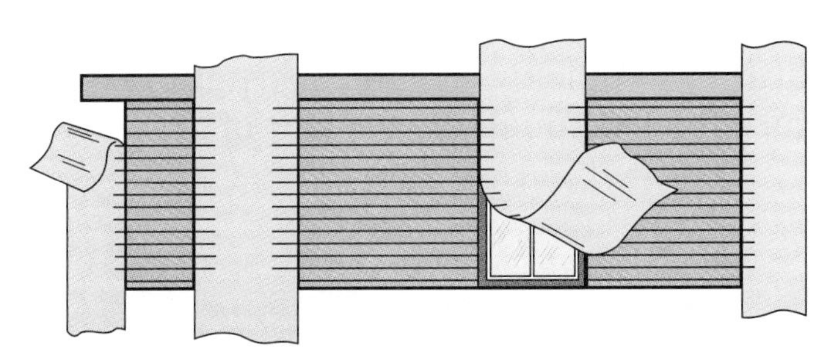

Fig. 3-58 ■ Use of drafting tape to mask areas.

Fig. 3-60 ■ A wall thickness may be layed out using a scale, an underlay, or dividers.

precise positioning on the drawing, such as landscape features, people, and cars. Stamps may also be used for furniture outlines and labels. See Fig. 3-59. Images on stamps are also available as underlays or on pressure-sensitive material.

Underlays

Underlays are drawings or parts of drawings that are placed under the original drawing and traced onto the original. Architects use them as master drawings many times. To be effective, a master drawing must be prepared to the correct scale and aligned carefully each time it is used.

An underlay is first positioned under the drawing and aligned. Then it is traced on the drawing. The underlay can then be removed or moved to a new location to trace the symbol or feature again if necessary.

Underlays do not necessarily eliminate the need for instruments or scales in original design work. See Fig. 3-60. The use of the underlay in this case is only possible after the original wall dimensions have been established by the use of the scale.

SYMBOL UNDERLAYS Many symbols and features of buildings are drawn more than once. Many drafters prepare a series of underlays of the features repeated most often on their drawings. Underlays are commonly prepared for doors, windows, fireplaces, trees, walls, and stairs.

GUIDELINE UNDERLAYS Guidelines for lettering are frequently prepared on underlays. When placed under the drawing, the drafter can trace the lines instead of measuring each one. If the underlay remains under the drawing while it is being lettered, there is no need to draw guidelines on the drawing. Other lines, such as cross-hatching and brick-symbol lines, are also prepared on underlays.

Squared Paper

Graph or squared paper often serves as underlay guidelines for architectural drawings. These grid sheets are printed in gradations of 4, 8, 16, and 32 squares per inch. Squared paper is also available in decimal-divided increments of 10, 20, and 30 or more squares per inch. This paper is used for the layout of survey and plot plans. Metric graph paper is usually ruled in units of 2 or 5 mm. Grid sheet underlays are used under the drawing or tracing paper and are removed after the drawing is finished.

Burnishing Plates

Burnishing plates are embossed sheets with raised areas that represent an outline of a symbol or texture. The plates are placed under a drawing. Then a soft pencil is rubbed over the raised portions of the plate onto the surface of the drawing. The use of burnishing plates allows the drafter to quickly create consistent texture lines throughout a series of drawings.

CHAPTER 3 Exercises

1. Draw the following four lines using a scale of ¼″ = 1′0″:
   a. 5′-0″
   b. 7′-6″
   c. 9′-10″
   d. 11′-3″.
2. Measure the distances you drew in Exercise 1, using the ⅛″ = 1″ scale.
3. Using a 2D CAD system, learn the commands for producing different scaled drawings.
4. Measure a book, desk, car, and room, using a metric scale. Record your results. Compare your measurements with customary measurements.
5. Convert the following dimensions to millimeters: 5′-6″, 6′-8″, 10′-4″, 11′-7″, 15′-3″.

6. List the scales (proportional measures) you will use in drawing plans of a residence you are designing. Tell why in relation to the size of the paper you are using.
7. Using a T square and triangle, parallel slide and triangle, or drafting machine, draw the walls shown in Fig. 3-11. Use a scale of ¼″ = 1′-0″.
8. Practice drawing lines using all pencil grades on vellum and paper. Compare the results.
9. Select the drawing sheet size you will use to draw the residence you are designing. Explain how you determined the selected size.

Architectural Drafting Conventions

CHAPTER 4

Builders follow a set of working drawings in order to make a designer's idea a reality. In order to clearly communicate information about the project to be built, standards and conventions in the preparation of these drawings have been established. Drafters and designers must not only know and apply these conventions, but they must also develop good lettering skills and drawing techniques to make the plans readable and understandable.

■ **When preparing working drawings, drafters and designers use drafting conventions to help them clearly communicate design ideas.** *Tony Stone Images/Rosemary Weller*

Objectives

In this chapter you will learn:

- to differentiate between the types and purposes of architectural drawings.
- to produce the line conventions used on architectural drawings.
- to develop good lettering techniques.
- to sketch lines, patterns, and a floor plan.

Terms

coding system
construction documents
detail drawings
elevations
general purpose drawings
line conventions
models
plans
renderings
sections
title block
working drawings

Architectural Drawings

The design of a structure is interpreted through the use of several types of architectural drawings, including floor plans, elevations, details, and pictorial drawings. Drawings may vary from simple to complex. The number of drawings needed to construct a building depends on the complexity of the structure and on the degree to which the designer needs or wants to control the methods and details of construction. A *minimum* set of plans provides the builder with great latitude in selection of materials and processes. A *maximum* set of plans will assure, to the greatest degree possible, agreement between the wishes of the designer and the final constructed building. See Fig. 4-1.

Even though a building may be relatively simple, as many drawings as necessary should be prepared. Any detailed working drawing that is omitted forces the builder into the role of the designer. For some buildings and some builders, this may be acceptable. For others, this is highly unacceptable. The more plans, details, and specifications developed accurately for a structure, the closer the finished building will be to what was conceived by the designer.

Fig. 4-1 Types of drawings in sets of plans.

Drawings	Size of set of plans		
	Min.	Aver.	Max.
FLOOR PLANS	x	x	x
FRONT ELEVATION	x	x	x
REAR ELEVATION		x	x
RIGHT ELEVATION	x	x	x
LEFT ELEVATION		x	x
AUXILIARY ELEVATIONS			x
INTERIOR ELEVATIONS		x	x
EXTERIOR PICTORIAL RENDERINGS		x	x
INTERIOR RENDERINGS			x
PLOT PLAN		x	x
LANDSCAPE PLAN			x
SURVEY PLAN	x	x	x
FULL SECTION	x	x	x
DETAIL SECTIONS		x	x
FLOOR-FRAMING PLANS			x
EXTERIOR-WALL FRAMING PLANS			x
INTERIOR-WALL FRAMING PLANS			x
STUD LAYOUTS			x
ROOF-FRAMING PLAN			x
ELECTRICAL PLAN		x	x
AIR-CONDITIONING PLAN			x
PLUMBING DIAGRAM			x
SCHEDULES			x
SPECIFICATIONS			x
COST ANALYSIS			x
SCALE MODEL			x

Types of Drawings

Architectural drawings are often called "the plans." However, specific architectural drawings show certain views of a structure and only some of those details are called *plans*. The following list includes brief descriptions of the various types of architectural drawings.

- **Plans** (or plan view) are views from the top down, a "bird's-eye" view. An example is a floor plan. See Fig. 4-2.
- **Elevations** are flat two-dimensional views. Depth is indicated by a change in line weight. Heavier line weights are used for major building lines and areas close to the viewing plane. Lighter line weights are used for component details and areas farther from the viewing plane. See Fig. 4-3.
- **Sections** show a view of one "slice" of a planned structure. It's as if an imaginary line were cut vertically at a particular place, showing the parts of the structure and components along the plane of that cut. See Fig. 4-4.
- **Detail drawings** are prepared at a larger scale than other types of

Fig. 4-2 ■ Floor plan.

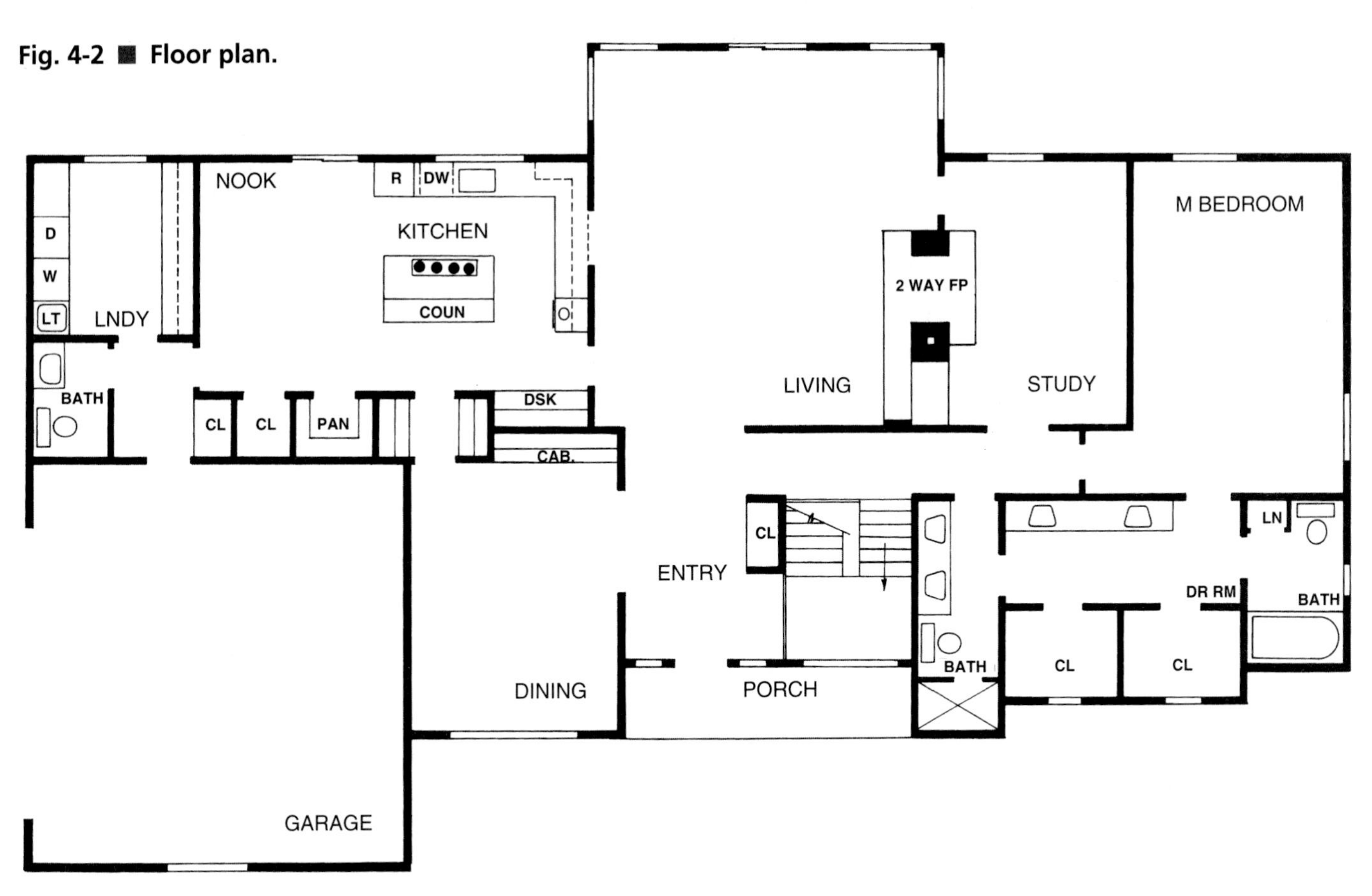

Fig. 4-3 ■ Elevation drawings.

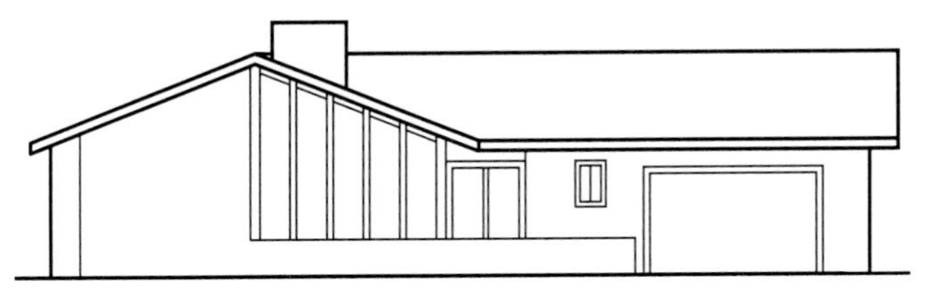

A. FRONT ELEVATION

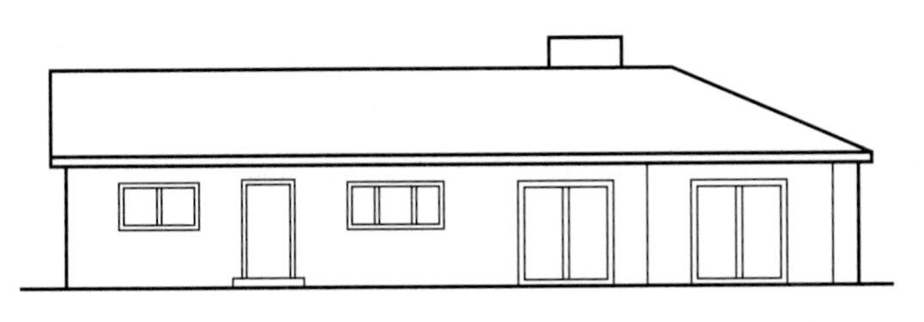

B. REAR ELEVATION

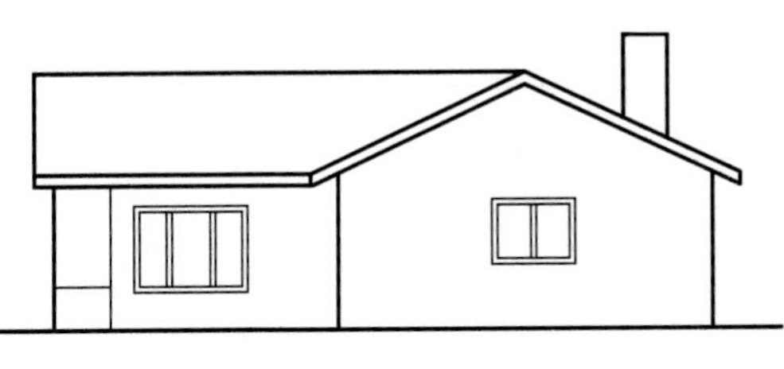

C. RIGHT-SIDE ELEVATION

Fig. 4-4 ■ Section drawing.

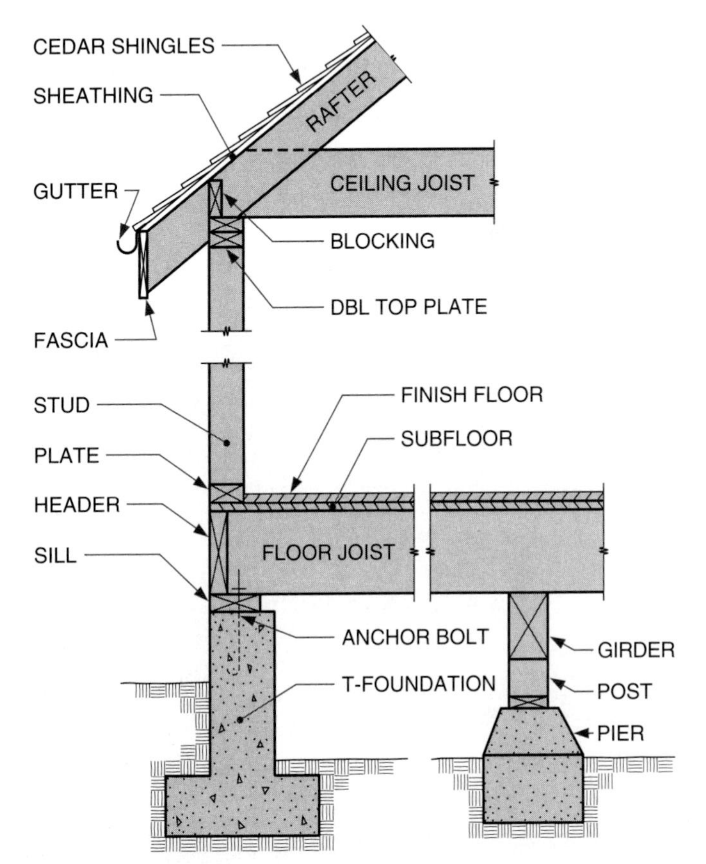

drawings. They are drawn to reveal precise information about construction methods and materials. Details may be prepared in plan view, in pictorial form, or as sections. See Fig. 4-5.

- **Renderings** are usually 1-, 2-, or 3-point perspective drawings. Often called *pictorials,* these show how the finished product is expected to look. Renderings are made of both interior and exterior portions of a building. See Fig. 4-6.

Fig. 4-5 ■ Detail drawing.

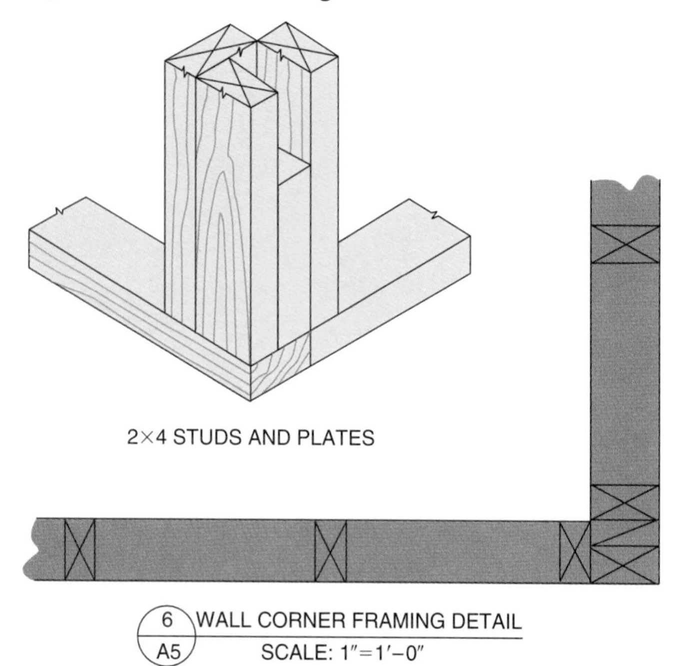

Fig. 4-6 ■ Exterior rendering. *Tangerine Bay Club*

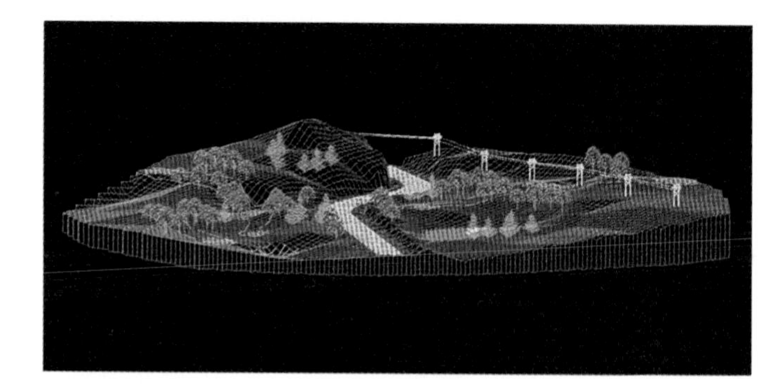

Fig. 4-7 ■ 3-D computer model.

- **Models** are 3-D proportional replicas of a structure. Although actual miniature models are often made, many models are now 3-D computer images. See Fig. 4-7.

Drawings and Documents

Information about the architectural design and its construction is provided basically in three ways: general purpose drawings, working drawings, and construction documents.

Architectural drawings used for sales promotion or preliminary planning purposes are known as **general purpose drawings.** Drawings of this type usually consist of only approximate room sizes and dimensions on a single line floor plan. See Fig. 4-8. Pictorial drawings of exterior and/or front elevation views are also used for these purposes.

Drawings used during the building process are known as **working drawings.** Working drawings should contain all the information needed to completely construct a building: the dimensions, materials, and drawings of the building's shape. Complete floor plans and elevations are required. A full set of working drawings also includes specialized drawings, such as framing, electrical, plumbing, and landscape plans.

Even with a set of drawings, the building information is still incomplete. So

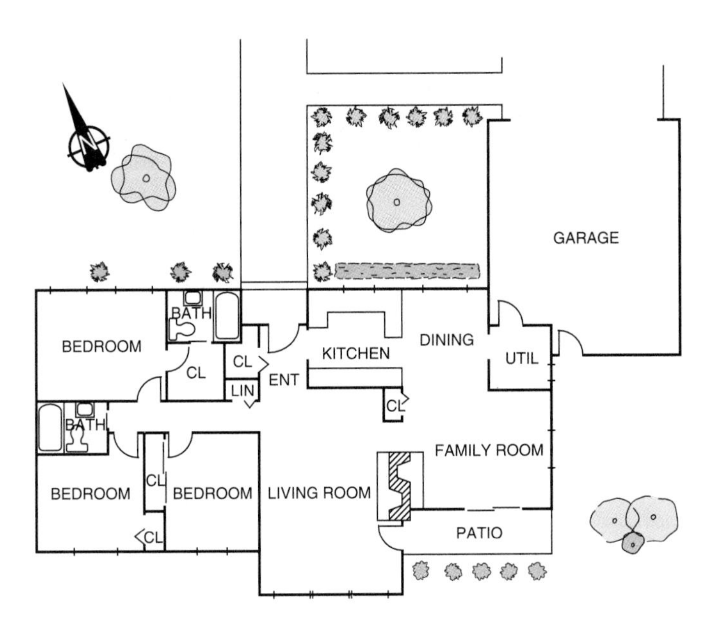

Fig. 4-8 ■ Preliminary floor plan sketched as a single line floor plan.

much information is necessary for building a structure that not all of it can be put on a set of working drawings. For this reason **construction documents** are prepared that contain hundreds of facts and figures, plus legal and financial information related to the building process. Documents such as building specifications and schedules eliminate guesswork and specify exactly which processes, materials, and building components are to be used.

Reading Architectural Drawings

A small number of working drawings and documents may be sufficient and easy to use for one residence. However, for a very large project, several sets of complicated plans may be needed that require many different views, as well as details and documents containing very specific information.

CODING SYSTEM To make a large number of drawings manageable and easy to use, a **coding system** is often necessary. The coding system identifies every specific drawing and detail. It is also a method of keeping similar drawings together and organized in a working drawing set.

Most architects follow the American Institute of Architects' (AIA) coding system. In the AIA's coding system, drawings are identified by letters and numbers for ease of referencing. See Figs. 4-9 and 4-10.

The group number always remains the same, no matter how many drawings are within it. More drawings may be added within groups without interrupting the alphanumerical order in the set.

Fig. 4-9 ■ Letters used to identify sets of drawings.

A	Architectural
C	Civil
D	Interior design (color schemes, furniture, furnishings)
E	Electrical
F	Fire protection (sprinkler, standpipes, CO_2, and so forth)
G	Graphics
K	Dietary (food service)
L	Landscape
M	Mechanical (heating, ventilating, air-conditioning)
P	Plumbing
S	Structural
T	Transportation/conveying systems

Fig. 4-10 ■ In this code, "A" indicates that the drawing belongs to a set of *architectural* working drawings. The number "2" identifies the group to which the drawing belongs within the set, and the "1" after the period shows that the drawing is the first one in the group.

SYSTEM CODE:
A 2. 1
Drawing Number
Group Number
Discipline Prefix

Fig. 4-11 ■ Title blocks provide a means of quick identification of drawings in a set.

A. Information in a typical title block.

- PROJECT TITLE AND NUMBER
- DRAWING SHEET TITLE
- NAME AND ADDRESS OF CLIENT
- NAME AND ADDRESS OF ARCHITECT OR FIRM
- NAME AND ADDRESS OF CONTRACTOR (if known)
- INITIALS OF DESIGNER
- INITIALS OF DRAFTER
- INITIALS OF CHECKER
- REVISION BLOCK INCLUDING
  - Title
  - Number
  - Preparer
- PROFESSIONAL SEAL SPACE
- SCALE
- DATE
- SHEET NUMBER – Using AIA code & showing number of sheets in the set.

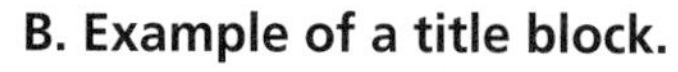

B. Example of a title block.

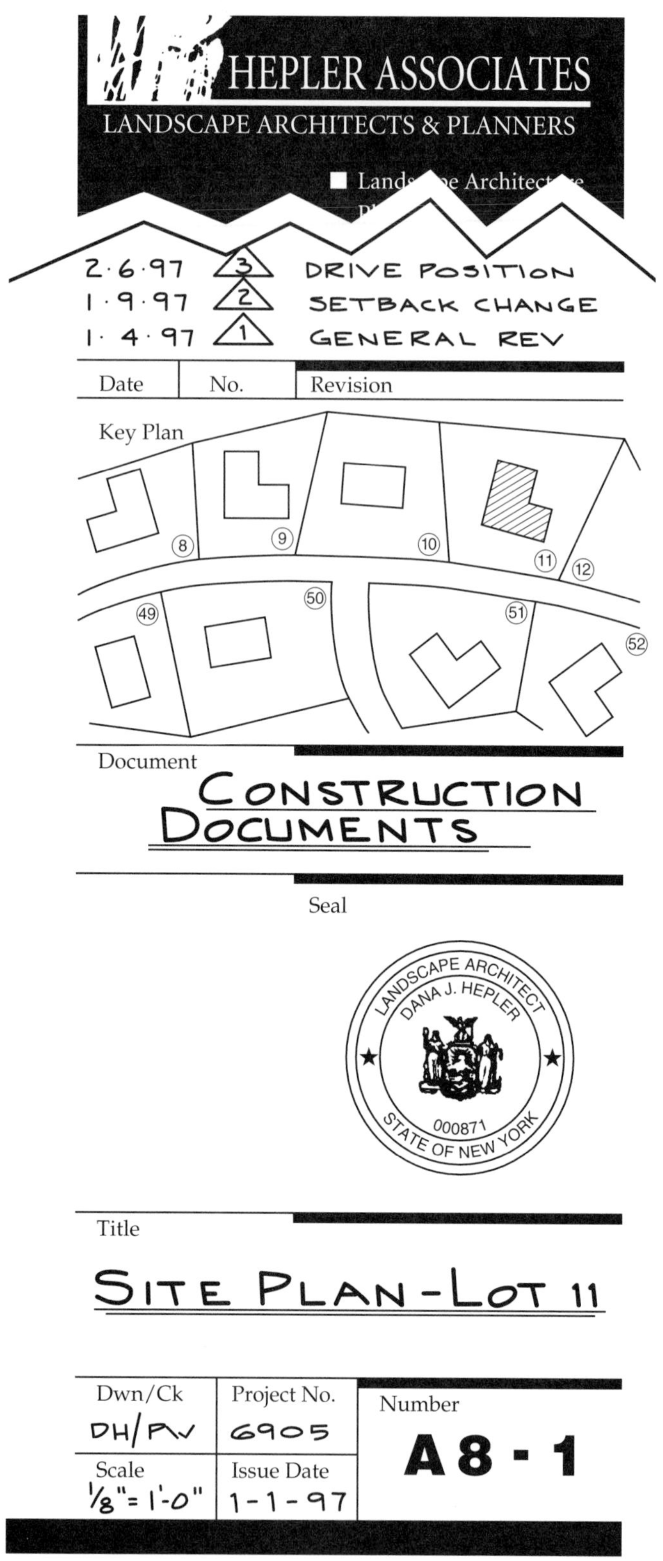

TITLE BLOCKS Similar to other kinds of written information, architectural drawings are identified by titles. In any drawing system **title blocks** identify drawings in a consistent and convenient format. See Fig. 4-11. Title blocks are usually located on the bottom and/or right side of each drawing sheet.

Because the number of revisions varies, revision entries are made in sequence from bottom to top. Revisions are shown with a number inside a triangle on each drawing change.

CROSS-REFERENCING Drawing all views, sections, or details of features on one floor plan or one elevation is usually impossible. For example, floor plans do not show height details and dimensions. Elevation drawings do not show all horizontal dimensions. It is therefore often necessary to provide cross-references in order to guide the reader from one drawing to another. Numbered symbols are generally used for this purpose. See Fig. 4-12.

Fig. 4-12 ■ An example of cross-referencing symbols that work two ways between a floor plan and related elevation drawing.

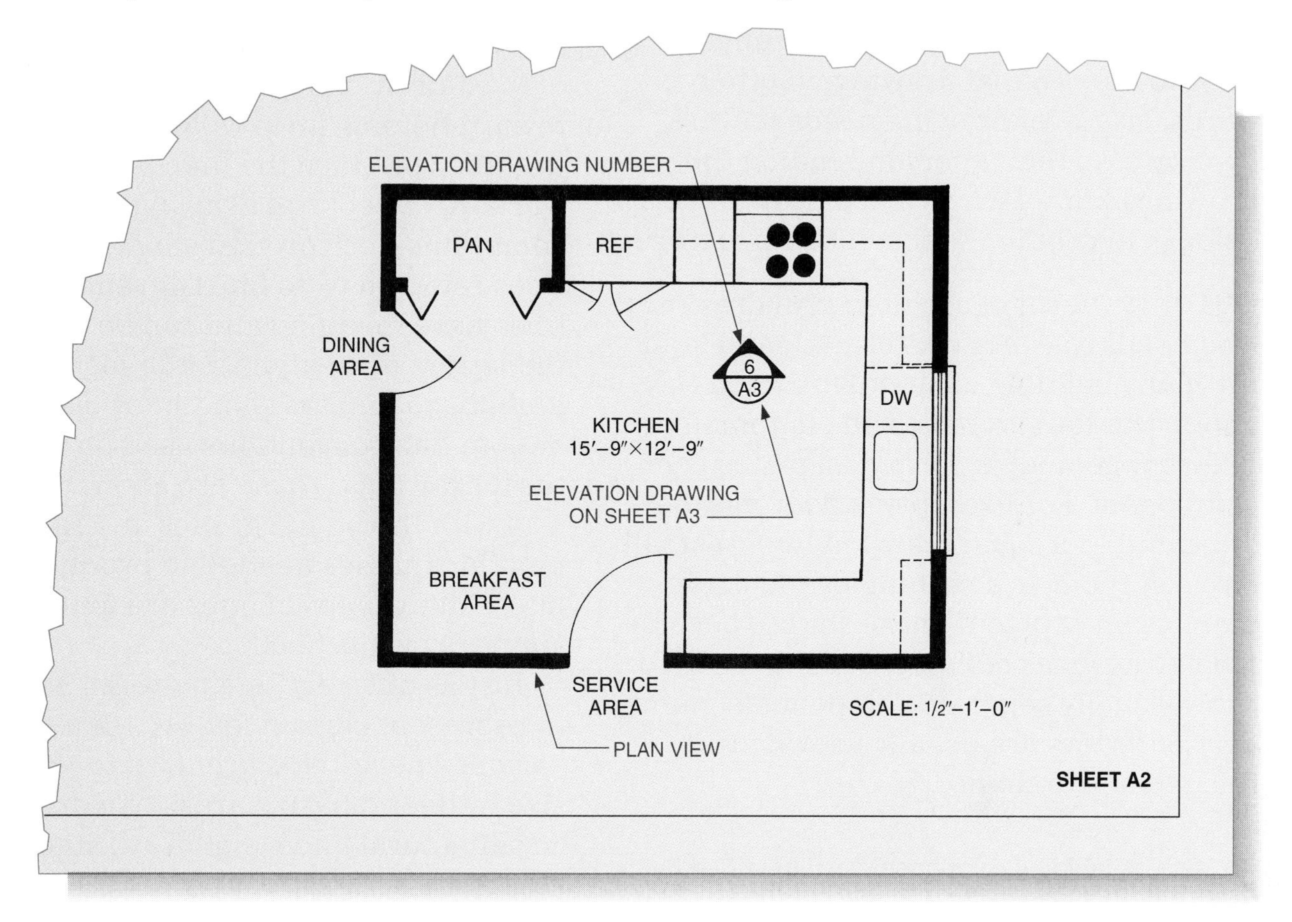

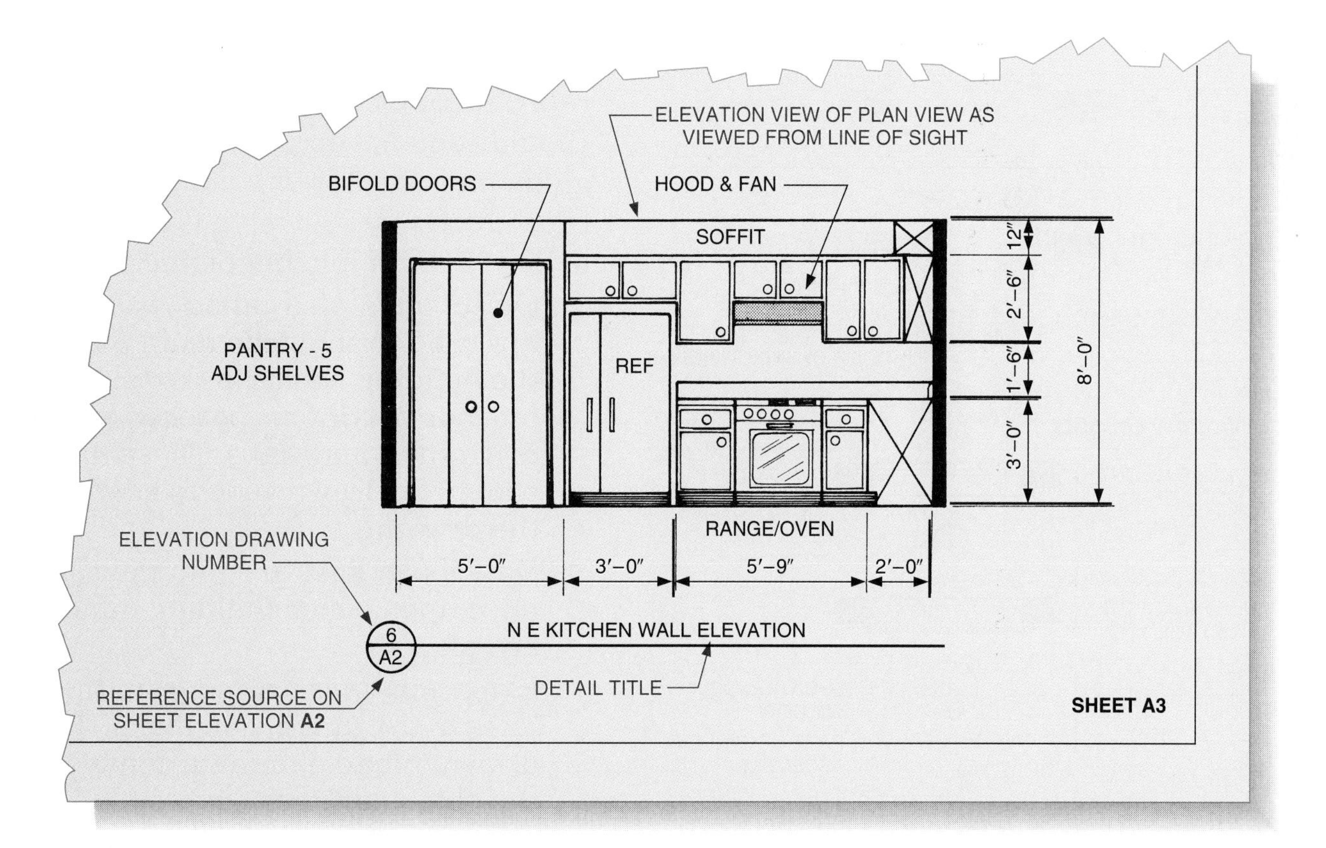

Pages with sections and detail drawings that have been referenced *from* other plans also need to be cross-referenced *back* to the original drawing. In other words, cross-referencing needs to work two ways, so that a person reading the drawings knows where each drawing belongs in relation to the entire structure.

CALLOUTS A set of architectural drawings also needs to contain information to identify many building components, such as doors, windows, rooms, and equipment. A different geometric form designates each component. For example, a door may be indicated as a square, a window as a small circle, a room as a rectangle, and equipment as an octagon. These shapes become labels known as *callouts*. See Fig. 4-13. To show visually separate components, numbers or letters are usually shown inside the geometric shape.

Fig. 4-13 ■ Building components are identified by geometric symbols called callouts.

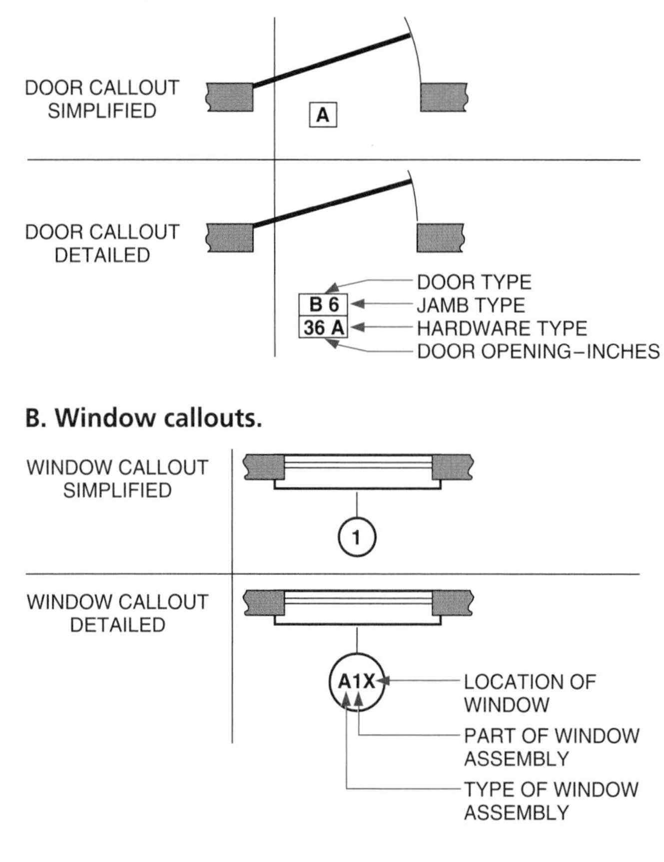

Architectural Conventions

Architectural Line Conventions

Most drafting equipment is aimed toward helping an architect, drafter, or designer produce the highest quality line work. Architectural drawings are mainly communicated through a language of lines referred to as **line conventions**. The lines have meaning and can be read like the letters of an alphabet. In fact, the term *alphabet of lines* is sometimes used to denote line conventions used on architectural drawings. These are shown in Fig. 4-14 with the pencil grades and technical pen thicknesses needed to produce these lines. Many types of lines are found on a single drawing. See Fig. 4-15.

Just as different line patterns are used to represent certain features of a drawing, various *line weights* are used to emphasize or deemphasize areas of a drawing. Architectural line weights are standardized to provide for consistent interpretation of architectural drawings. All architectural line weights must be very dark (opaque) so they will make clear reproduction copies. The only lines that should remain very light are layout and guide lines so they will *not* be seen on the reproductions. Following is a list of types of lines in the alphabet of lines.

1. *Object lines*, also called *visible lines*, are used to show the main outline of the building, including exterior walls, interior partitions, porches, patios, driveways, and walls. These lines should be drawn wide to stand out on the drawing.
2. *Dimension lines* are thin unbroken lines upon which building dimensions are placed.
3. *Extension lines* extend from the object lines to the dimension lines. They are drawn thin to eliminate confusion with the object outlines.

Fig. 4-14 ■ Architectural line conventions.

Name of Lines	Line Symbols	Line Width	Pencil	Pen Sizes	
1. Object lines		Thick	H,F	2	0.50 mm
2. Hidden lines		Medium	2H,H	0	0.35 mm
3. Center lines		Thin	2H,3H,4H	0	0.35 mm
4. Long break lines		Thin	2H,3H,4H	0	0.35 mm
5. Short break lines		Thick	H,F	2	0.50 mm
6. Phantom lines		Thin	2H,3H,4H	0	0.35 mm
7. Stitch lines		Thin	2H,3H,4H	0	0.35 mm
8. Border lines		Very thick	F,HB	3	0.80 mm
9. Extension lines		Thin	2H,3H,4H	00	0.25 mm
10. Dimension lines					
11. Leader lines		Thin	2H,3H,4H	00	0.25 mm
12. Cutting plane lines		Very thick	F,HB	3	0.80 mm
13. Section lines		Thin	2H,3H,4H	00	0.25 mm
14. Layout lines		Very thin light	4H		
15. Guidelines					
16. Lettering	ARCHITECTURAL	Thick	H,F	1	0.40 mm

Fig. 4-15 ■ Architectural line conventions as used on an architectural floor plan.

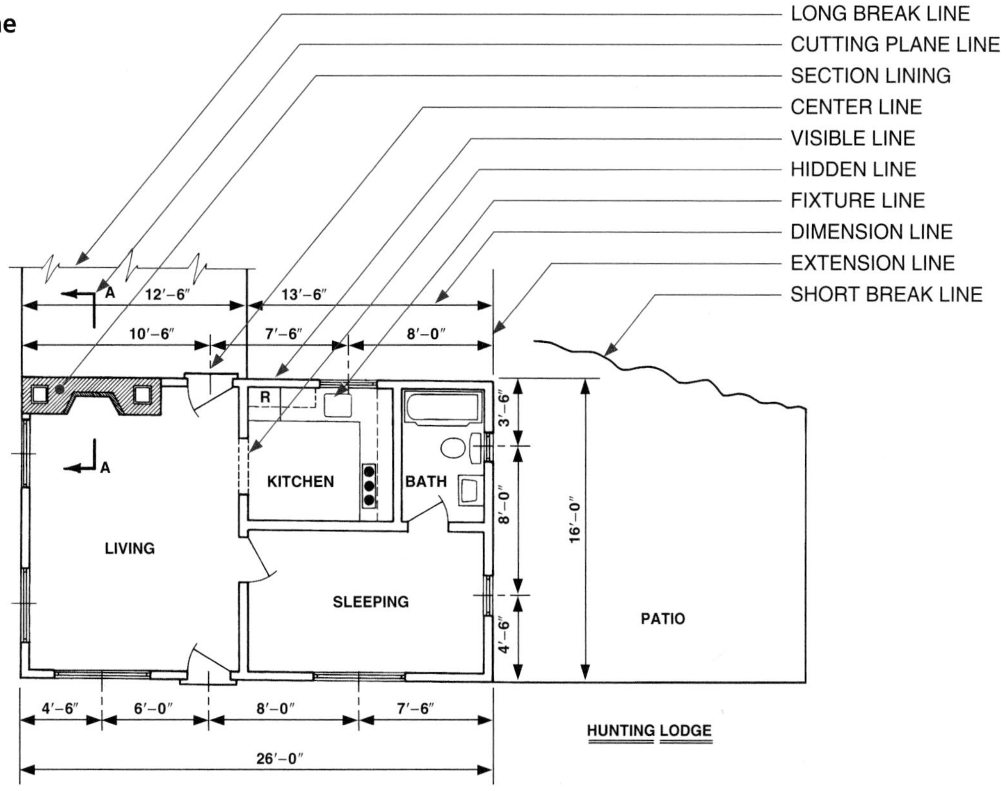

4. *Hidden lines* are used to show areas that are not visible on the surface, but that exist behind the plane of projection. Hidden lines are also used in floor plans to show objects above the floor section, such as wall cabinets, arches, and beams. Hidden lines are drawn thin.
5. *Centerlines* denote the centers of symmetrical objects such as exterior doors and windows. These lines are usually necessary for dimensioning purposes. Centerlines are drawn thin.
6. *Cutting-plane lines* are very wide lines used to denote an area to be sectioned. In this case, the only part of the cutting-plane line drawn is the extreme ends of the line. This is because the cutting-plane line would interfere with other lines on the drawing.
7. *Break lines* are used when an area cannot or should not be drawn entirely. A ruled line with freehand breaks is used for long, straight breaks. The long break line is thin. A wavy, uneven freehand line is used for smaller, irregular breaks. The short break line is wide.
8. *Phantom lines* are used to indicate alternate positions of moving parts, adjacent positions of related parts, and repeated detail. The phantom line is thin.
9. *Fixture lines* outline the shape of kitchen, laundry, and bathroom fixtures, or built-in furniture. These lines are thin to eliminate confusion with object lines.
10. *Leaders* are used to connect a note or dimension to a symbol or to part of the building. They are drawn thin and sometimes are curved to eliminate confusion with other lines.
11. *Section lines* are used to indicate the cut surface in sectional drawings. A different symbol pattern is used for each building material. The section lining patterns are drawn thin.
12. *Border lines* are the heaviest lines used on a drawing and are often preprinted with the title block. Border lines define the active area of a drawing sheet.
13. *Guidelines* are drawn to provide a horizontal guide for lettering to keep letters and numbers aligned. These are very light lines so they do not reproduce on a finished blueprint.
14. *Construction lines* are very light preliminary layout lines which do not become part of the finished drawing when reproduced. These lines are the lightest on any drawing.

Architectural Lettering

STYLES Without lettering, a plan does not communicate a complete description of the materials, type, size, and location of the various components. All labels, notes, dimensions, and descriptions must be legible on architectural drawings if they are to be an effective means of graphic communication.

Architectural designs are often personalized. Likewise, lettering styles may reflect the individuality of various architects and drafters. Architectural drafters often develop their own style of lettering to work quickly, yet maintain accurate and attractive drawings. Nevertheless, personalized styles are all based on the American National Standard Alphabet shown in Fig. 4-16.

Fig. 4-16 ■ The American National Standard Alphabet is recommended for architectural drawings.

A. Straight.

ABCDEFGHIJKLMNOPQRSTUVWXYZ 1234567890

B. Inclined.

ABCDEFGHIJKLMNOPQRSTUVWXYZ
1234567890

No personalized style should be used that is difficult to read or easily misinterpreted. Errors of this type can be very costly, especially if numbers used for dimensioning are misread.

HOW TO DEVELOP GOOD LETTERING SKILLS

Practice is necessary to develop the skills needed to letter effectively. Although architectural lettering styles may be very different, all professional drafters follow certain basic techniques for lettering. If you follow these recommendations, you will develop accuracy, consistency, and speed in lettering your drawings.

1. Always use guidelines in lettering. See Fig. 4-17.
2. Choose one style of lettering, and practice the formation of the letters of that style until you master it. See Fig. 4-18.
3. Make letters bold and distinctive. Avoid a delicate, fine touch.
4. Make each line quickly from the beginning to the end of the stroke. Do not try to develop speed at first. Make each stroke quickly, but take your time between letters and between strokes until you have mastered each letter. See Fig. 4-19. Then gradually increase your speed. You will soon be able to letter almost as fast as you can write script.
5. Practice with larger letters (about ¼″, or 6 mm), and gradually reduce the size until you can letter effectively at ⅛″ (3 mm).
6. Aim for uniform and even spacing of areas between letters by practicing words and writing sentences, not alphabets. See Fig. 4-20.
7. Practice lettering whenever possible—as you take notes, address envelopes, or write your name.
8. Use only the CAPITAL alphabet. Lowercase letters are rarely used in architectural work.
9. If your lettering has a tendency to slant in one direction or the other, practice making a series of vertical lines, as shown in Fig. 4-21.

Fig. 4-17 ■ Notice the difference in lettering when guidelines are used.

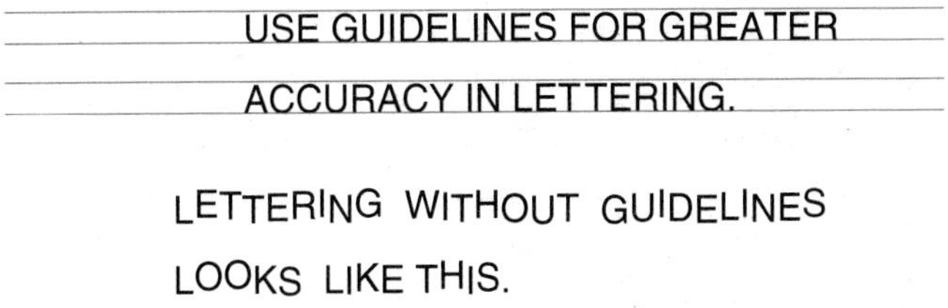

USE GUIDELINES FOR GREATER
ACCURACY IN LETTERING.

LETTERING WITHOUT GUIDELINES
LOOKS LIKE THIS.

Fig. 4-18 ■ Avoid inconsistent lettering.

ABCDEFGHIJKLMNOPQRSTUVWXYZ 1234567890

Fig. 4-19 ■ Practice will help you increase your lettering speed.

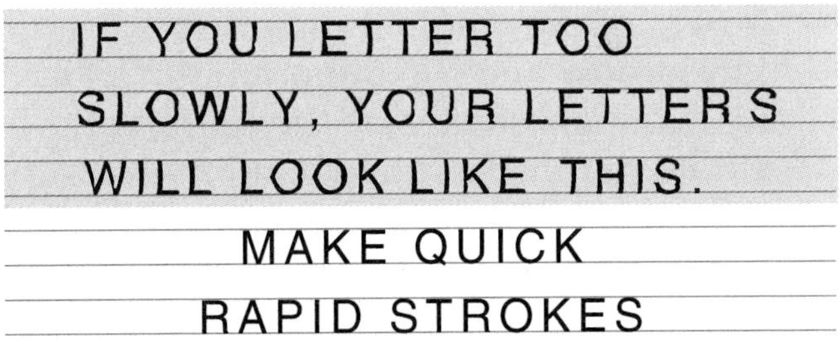

IF YOU LETTER TOO
SLOWLY, YOUR LETTERS
WILL LOOK LIKE THIS.

MAKE QUICK
RAPID STROKES

Fig. 4-20 ■ Maintain consistent spacing between letters.

UNIFORM SPACING

UNEVEN SPACING

Fig. 4-21 ■ Practice vertical and horizontal lines.

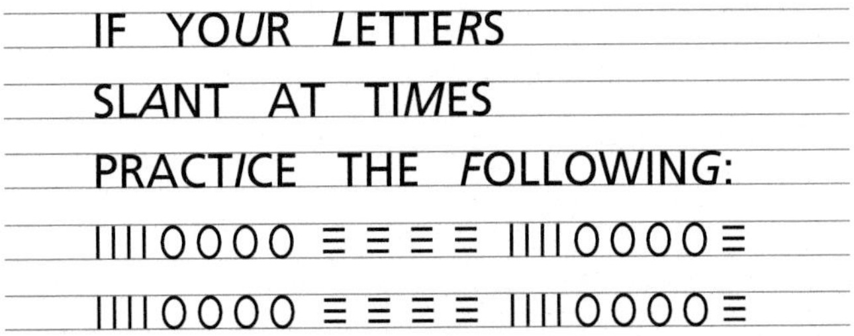

IF YOUR LETTERS
SLANT AT TIMES
PRACTICE THE FOLLOWING:
IIIII OOOO ≡ ≡ ≡ ≡ IIIII OOOO ≡
IIIII OOOO ≡ ≡ ≡ ≡ IIIII OOOO ≡

10. If slant lettering *is* desired, practice slanting the horizontal strokes at approximately 68°. The problem with most slant lettering is that it is difficult to maintain the same degree of slant continually. The tendency is for more slant to creep into the style.
11. Letter the drawing last to avoid smudges and overlapping with other areas of the drawing. This procedure will enable you to space out your lettering and to avoid lettering through important drawing details.
12. Use a soft pencil, preferably an HB or F. A soft lead pencil will glide and is more easily controlled than a hard lead pencil.
13. Numerals used in architectural drawing should be adapted to the same style as the letters. Fractions also should be made consistent with the style. Fractions are 1⅔ times the height of the whole number. The numerator and the denominator of a fraction are each ⅔ of the height of the whole number as shown in Fig. 4-22A. Notice also that in the expanded style, the fraction is slashed to conserve vertical space. The fraction takes the same amount of space as the whole number. See Fig. 4-22B.
14. The size of the lettering should be related to the importance of the labeling. See Fig. 4-23.
15. Specialized lettering templates can also be used.

Fig. 4-22 ■ Lettering fractions.

A. Proportions for fractions.

4′–3 1/2″ 6′–6 3/4″ 8′–9 1/4″

B. Alternate fraction style used to conserve space.

5 units 3 units $4'\text{–}6\frac{1}{2}''$ $6'\text{–}3\frac{1}{8}''$ $9'\text{–}9\frac{3}{4}''$

Fig. 4-23 ■ Lettering height is related to the importance of the label.

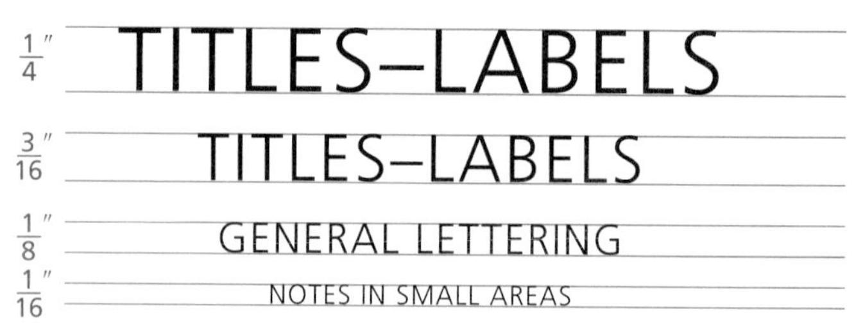

TYPESET LETTERING Various kinds of typeset lettering are available. Pressure-sensitive letters are applied one at a time. These letters are used primarily for major labels on architectural drawings.

Words, sentences, and dimensions can be set in type on opaque paper or transparent tape. See Fig. 4-24. The type is then attached to the drawing with rubber cement or wax-backed paper. In architectural work, transparent tape labels can be positioned without covering the lines that pass close to or through the label. Although more consistent in size and style than hand lettering, typeset letters take more time to apply.

In computer-aided drafting (CAD), lettering is produced using the keyboard and the CAD program's "text" features.

Fig. 4-24 ■ Applying a transparent tape label.

Menu items include the option of selecting type font (style), slope angle, line weight (pen size), width, height, and spacing of characters.

Architectural Drawing Techniques

Drawing (as a verb) is an overall term for the creation of all types of graphic forms. *Drafting* is drawing with the use of mechanical devices.

Using the Appropriate Pencil

Floor plans and elevation drawings are prepared primarily for the builder. These drawings must be accurately scaled and dimensioned. The accuracy, effectiveness, and appearance of a finished drawing depend largely on the selection of the correct pencil and the point of that pencil.

The degrees of hardness of drawing pencils range from 9H, extremely hard, to 7B, extremely soft. Pencils in the hard range are used for layout work. Basic architectural drawings are usually drawn with pencils in a medium range. If the pencil is too soft, it will produce a line that smudges. If the pencil is too hard, a groove will be left in the paper, and the line will be very difficult to erase. Since the density of the line also depends on the hardness of the lead, it is necessary to select the correct degree of hardness for pencil tracings. Different degrees of hardness react differently, depending on the type of drawing medium (paper or other material) used.

Sharpen the drawing pencil by exposing approximately ½″ of lead. Then form the lead into a sharp conical point by slowly rotating the pencil as you rub it over a sanding pad. A mechanical lead pointer or smooth single-cut file may also be used. Mechanical pencils with leads of 0.3, 0.5, 0.7 and 0.9 mm are very thin and do not need to be sharpened because the lead width matches the correct line width.

Since working drawings are prepared for builders' use, they must be drawn with precision. Appropriate line weights and consistency of lines assure a greater degree of accuracy. To provide the control needed for precise drawings, practice the technique shown in Fig. 4-25. Use a straightedge (such as a T square, triangle, drafting machine, or parallel slide) to guide the pencil.

Care must be taken in drawing corner intersections. Overlapping corner lines may intersect another material part or dimension and create confusion. When corner lines do not meet, no corner exists for interpretation or measurement. See Fig. 4-26 for the preferred method.

Fig. 4-25 ■ Technique for using a pencil with a straightedge.

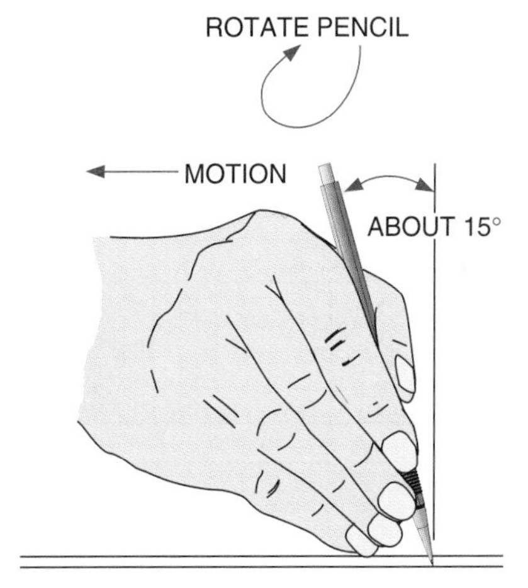

Fig. 4-26 ■ Be precise when drawing line intersections.

PREFERRED
A

ACCEPTABLE
B

NOT ACCEPTABLE
C

Rendering and Sketching Techniques

In addition to the precise technical line work on floor plans and elevations, other line techniques are used for rendering and sketching. Some drawings are *rendered* to provide the prospective customer with a better idea of the final appearance of the building. These drawings show no dimensions but may include items such as plantings, floor surfaces, shade and shadows, and material textures.

Some of the line techniques used for renderings are simply variations—in the distance between lines, the width of lines, or the blending of lines. Dots, gray tones, or solid black areas are other ways of showing materials, texture, contrast between areas, or light and shadow patterns. See Fig. 4-27. Just varying the interval between dots or lines drawn with pencil or pen can indicate texture, light, and pattern. Screens, lines, and patterns are also available as appliqués.

When less precise line identification is desired, *wash drawing* (watercolor) is another effective technique. Some drawings are rendered with soft pencils or with a combination of wash techniques over a line drawing done in ink or pencil.

Fig. 4-27 ■ Types of line techniques in architectural drawings.

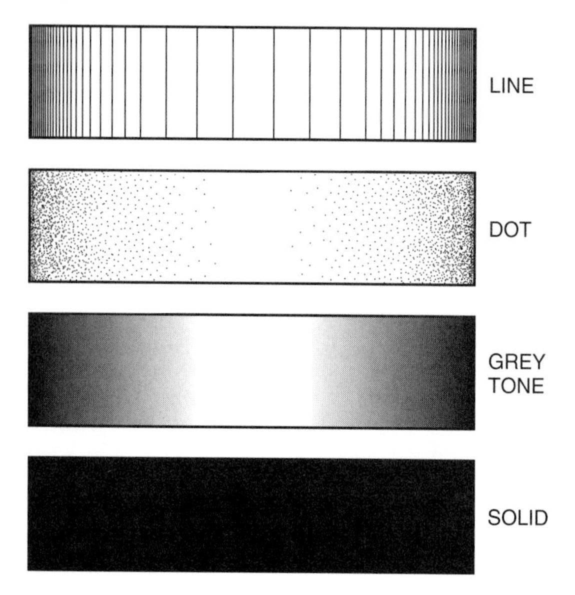

Fig. 4-28 ■ Sketching techniques.

A. Sketching horizontal lines.

Sketching is a means of communicating that is used constantly by designers. In fact, most designers begin with a sketch. Sketches, or rough drafts drawn freehand, are used to record dimensions and the placement of existing objects and features prior to beginning a final drawing. Many times alternatives to a design problem are shown with sketches. Sketches also help record ideas on the job site and help the designer remember unique features about a structure or site. Then the actual design activity can continue in a different location.

When sketching, use a soft pencil. Hold the pencil comfortably. Draw with the pencil; do not push it. Position the paper so your hand can move freely. Sketch in short, rapid strokes. Long, continuous lines tend to arc when drawn freehand. See Figs. 4-28A and 4-28B. Sketching on graph paper helps increase speed and accuracy. See Fig. 4-28C.

(Fig. 4-28 ■ *Continued*)

B. Sketching diagonal lines.

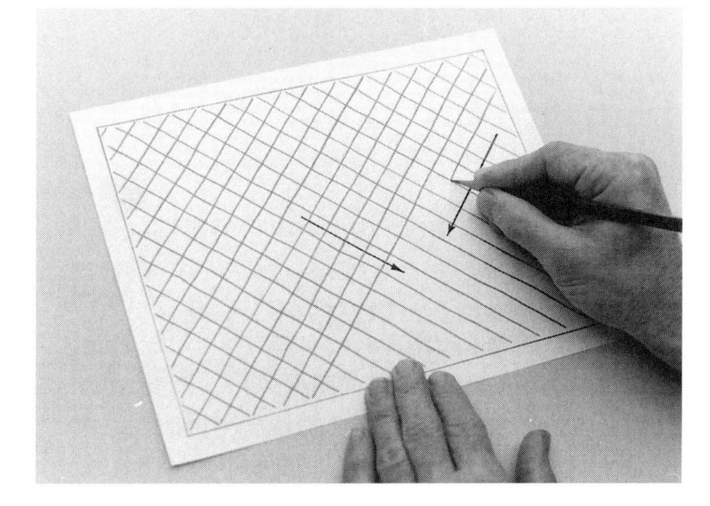

C. Sketching on graph paper.

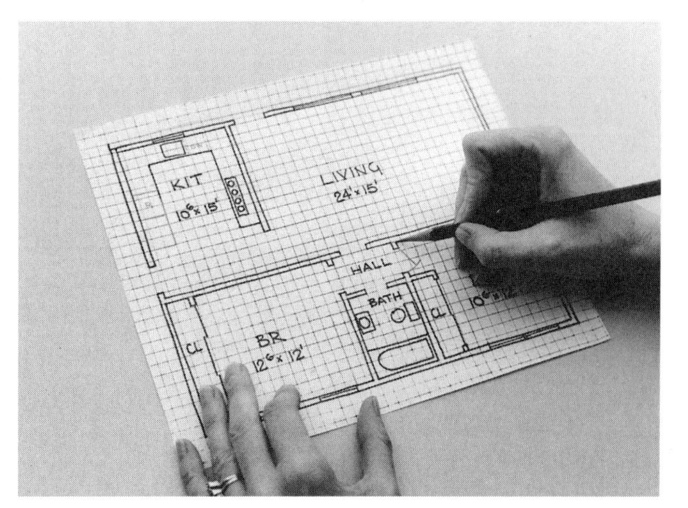

CHAPTER 4 Exercises

1. Describe six types of architectural drawings in terms of the type of information that is communicated in each type. List the ones you would use for your own set of drawings.
2. How are drawings used during the planning and construction of a building?
3. Explain the purpose of a coding system and cross-referencing.
4. Using a CAD system, draw a reference symbol and a callout. Label them.
5. Find and obtain a sample title block that is used by a local design or architectural office. Design your own version of a title block.

6. Select a lettering style from a CAD text menu and create all the data you would use in a title block.

7. Practice drawing each of the lines shown in Fig. 4-14 using a T square, triangle, or line tasks on a CAD system.
8. Use three different grades of pencil to draw five of the lines in Fig. 4-14. Compare your results.
9. Copy the rules for lettering using any lettering style you choose.

10. Select a lettering style to be used on your own set of plans. Complete three practice sheets. Critique and improve each.
11. Draw the line and shade forms shown in Fig. 4-27.

CHAPTER 5

Introduction to Computer-Aided Drafting and Design

Objectives

In this chapter you will learn:

- to use a computer to prepare architectural drawings.
- the different kinds of hardware and their functions.
- to evaluate CAD software programs.

Terms

Cartesian coordinate system
CD-ROM
central processing unit (CPU)
commands
computer-aided design and drafting (CADD)
computer-aided drafting (CAD)
digitizing tablet
graphic accelerator
hardware
modem
RAM
rendering
ROM
software
solid model
surface model
symbol library
wireframe drawing

Many architectural and engineering companies today depend on computer-aided drafting (CAD) systems to prepare the highly accurate drawings they need. A CAD system is a combination of computer software (programming) and hardware (machinery) that allows a designer or drafter to create drawings and store them electronically in a computer. Every phase of the design and engineering process, from preliminary concept drawings to the final working drawings, can be completed using a CAD system.

■ **A CAD system enables the designer to present the same area in a number of different ways.** *KETIV Technologies Inc., Portland, OR*

Advantages of CAD

A **computer-aided drafting (CAD)** system is a sophisticated drafting tool that is based on the speed and accuracy of a computer. See Fig. 5-1. A CAD system can be compared to a conventional drafting machine in the same way a slide rule can be compared to an electronic calculator. Both tools do the job well, but the CAD system, like the calculator, can perform many of the same tasks faster and with less effort.

One of the advantages of a CAD system is its capacity to perform repetitious tasks, such as drawing lines and inserting symbols that are used over and over again. By automating such tasks, the computer enables designers and drafters to increase their productivity and frees them to spend more time concentrating on the creative design process. (Note that when a computer system is used for both design and drafting purposes, it is more properly called a **computer-aided design and drafting (CADD)** system. However, in general use, many companies refer to both the design and the drafting simply as CAD.)

CAD provides several other advantages compared to manual drafting. For example, because of the speed, accuracy, and consistency of CAD systems, architectural firms often produce an increased number of detail drawings. This helps to eliminate many construction site errors.

CAD systems can also produce consistent lettering for a drawing much more quickly than the lettering could be produced manually. The designer or drafter "types" the letters on a keyboard to be printed on drawings and related documents.

Often the greatest advantage of using CAD, however, is not the preparation of original drawings, which requires time and skill, but the speed and ease with which the drawings can be revised. CAD drawings are stored electronically as drawing files. They can be retrieved ("called up") at any time for quick reference, for revisions and design changes, or for duplication.

Fig. 5-1 ■ Computer-aided drafting system.
Compaq Computer Corp.

CAD revisions are very clean. The CAD software allows the drafter to erase parts of the drawing, add to the drawing, or manipulate individual geometry as needed. The revised drawing shows no sign of erasures. The next time the drawing is printed, it looks like (and is) an original drawing.

CAD Limitations

In spite of these advantages, remember that just as typewriters do not create stories, CAD systems do not create drawings. People create drawings using tools—whether with CAD systems or with manual drafting instruments.

Although CAD systems can perform many tasks, it is important to remember that CAD systems cannot think or make decisions. CAD systems cannot create and design. They perform only those tasks that humans instruct them to do. Therefore, they cannot automatically eliminate all human errors. The phrase "garbage in, garbage out" is a catchy reminder that the computer is only as accurate as the information it is given.

Components of a CAD System

CAD systems, like all computer systems, consist of hardware and software components. The physical devices used in a computer system are known as the **hardware.** The programs the operator uses to do work (such as creating a CAD drawing or a word processing document) are known as **software.**

CAD Hardware

The hardware for a basic computer system includes the **central processing unit (CPU),** input devices such as a keyboard and a mouse, storage devices, and output devices such as monitors and printers. Other equipment is usually added to this basic system to create CAD workstations. See Fig. 5-2. The diagram in Fig. 5-3 shows the basic relationships among these components.

The CPU

The CPU is the "brain" of all computer systems. It contains a microprocessor chip, such as a Pentium chip, which determines the speed and power of the computer. See Fig. 5-4. Most CAD software requires a fast, powerful computer to perform properly.

The Monitor

A monitor is a hardware device resembling a television screen. It allows the operator to see the results of commands given to the computer. Refer again to Fig. 5-2. CAD systems should be equipped with the largest monitors possible to avoid eyestrain while working with complex drawings. A 17-inch monitor is the smallest monitor recommended for long-term CAD work. Many CAD workstations have 20- or 21-inch monitors.

Memory and Storage

Computers contain two types of memory: **ROM** (read-only memory) and **RAM** (random-access memory). ROM contains the fixed data that the computer uses while it is operating. It contains instructions that keep the "operating system," which coordinates instructions between the software and hardware, operating smoothly.

RAM is the computer's temporary memory. The amount of RAM determines the amount of software data the CPU can process at one time. To operate properly, most CAD software requires a minimum of 8 megabytes (MB), but 12 MB may be needed for large drawings such as architectural floor plans. Larger software programs require 24 or more megabytes of RAM.

Memory in a computer should not be confused with storage capacity. The computer *memory* enables the computer to function. To store files, computers need *storage* devices. The most common storage devices are floppy disk drives and hard disk drives. These are standard equipment on almost all computer models.

Hard disk drives are generally the main storage device on a computer. They store the main operating system, the software,

Fig. 5-2 ■ Typical hardware components for a CAD system.

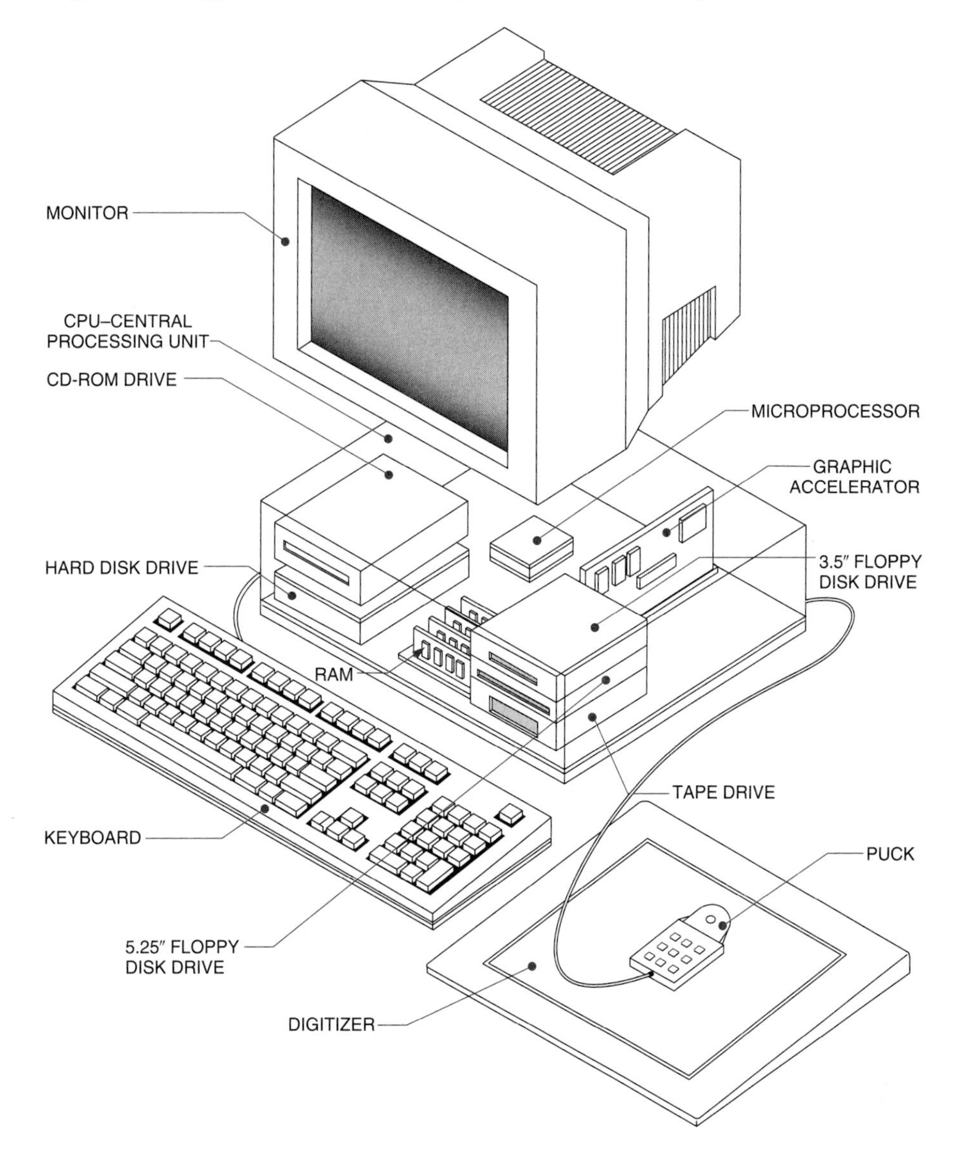

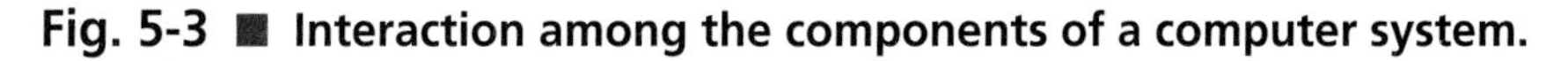

Fig. 5-3 ■ Interaction among the components of a computer system.

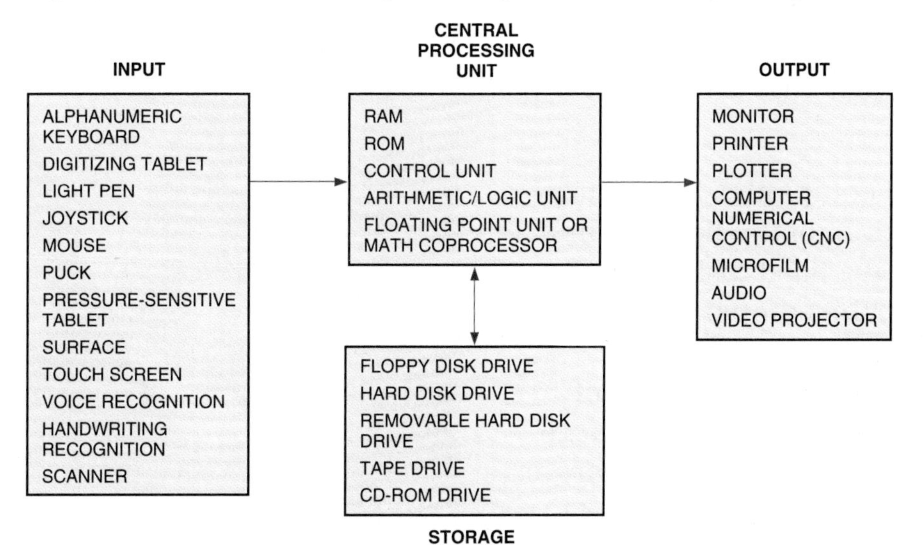

Fig. 5-4 ■ The Pentium is an example of a powerful microprocessor chip.

and any electronic files, such as drawing files, that the operator creates and saves. CAD programs generally take up a large amount of hard disk space, and architectural CAD files can take up a large percentage also. Therefore, most CAD systems include at least a 500 MB hard disk drive. Today, hard disk capacities of up to 1.6 gigabytes (1600 MB) are common on CAD systems.

Floppy disks (diskettes) that can be inserted and removed from a floppy disk drive are used primarily to store data. Floppy disks often serve as a backup copy (an extra copy) of files that are copied from a hard disk.

BACKUPS Careful companies keep a complete, current backup of the files on their hard disks. Then, if a hard disk becomes corrupted or fails, the company can retrieve its important files from the backup. Floppy disks are an inefficient means for completely backing up a large hard disk drive, however. Although larger capacities are under development, 1.44 MB is currently the maximum that can be stored on a single floppy disk without compressing the data. To back up a 500 MB hard drive, you would need 348 floppy disks!

To solve the need for comprehensive backups, many companies now use *tape drives.* The magnetic tapes used in these tape drives can hold large amounts of data—from 200 MB to more than 4 gigabytes (GB). Other companies are beginning to use recordable CD-ROMs as an alternative for backup storage. Recordable CD-ROM is an optical (laser) technology that can keep data safe indefinitely, whereas magnetic tape drives are usually guaranteed for only up to 5 years.

Still other companies use *removable hard disk drives* such as Syquest or Bernoulli drives. These drives have characteristics of both hard and floppy drives. Cartridges inserted in these drives have a capacity of from 44 to 250 MB. Unlike stationary hard drives, however, the cartridges can be removed from the hard drive, so that any number of cartridges can be used. See Fig. 5-5.

Fig. 5-5 ■ Companies now have many options for backing up important computer files. *Arnold & Brown*

Input Devices

Data or information (*input*) can be entered into the computer using various input devices. The alphanumeric *keyboard* is a standard input device that resembles a typewriter. Many computer systems also include a "pointing device" called a *mouse* that can be used to move the arrow (called the *cursor*) around the screen.

CAD systems are usually outfitted with another input device called a **digitizing tablet,** often called simply a *digitizer.* See Fig. 5-6. The CAD operator can use the digitizer to enter CAD commands as well as to digitize, or create an electronic form of, a drawing that has been created manually. Digitizers require the use of a *puck,* which usually looks similar to a mouse but is much more advanced. The puck contains crosshairs to allow the CAD operator to pick points very accurately. It may also contain up to 16 buttons that can be pressed to perform various functions.

Other input devices for computers are less widely used with CAD systems. *Light pens,* for example, can be used directly on the monitor's glass screen to activate commands. However, because of the lack of accuracy, light pens are used only for general design functions.

A newer development in computer technology provides another means of entering data—by *voice.* Voice recognition devices enable a computer operator to enter commands by speaking into a microphone connected to the computer. Voice recognition is still a young technology. It has not yet been developed to the point that it recognizes "natural speech." However, it is becoming widely used in special environments such as the "clean" rooms used to manufacture certain products such as microchips. Voice recognition also provides an alternative for physically challenged people who cannot easily manipulate a mouse, puck, or keyboard.

Output Devices

Paper copies (hard copies) of CAD drawings require the use of a plotter or printer. *Plotters* produce high-quality drawings using ink pens, felt tips, ballpoints, or pencils of various colors. See Fig. 5-7. One advantage of plotters is that

Fig. 5-6 ■ Digitizing tablet.

Fig. 5-7 ■ Flatbed plotter.

they are available in all standard drafting sizes, including E size (48″ × 36″). However, they are relatively slow.

Laser printers offer a fast method of producing high-quality graphics and typeset reproductions. Their small format, however, is a drawback for most architectural work that requires large sheets of paper. See Fig. 5-8. Most laser printers can print an area no larger than 11″ × 14″ inches.

Ink-jet printers are available in all sizes and offer good graphic quality at low cost. See Fig. 5-9. These printers spray tiny jets of black and/or colored ink to produce a good-quality drawing. Because of their versatility and low price, ink-jet printers are becoming standard equipment in architectural offices and other businesses that produce CAD drawings.

Fig. 5-9 ■ Ink-jet printer.

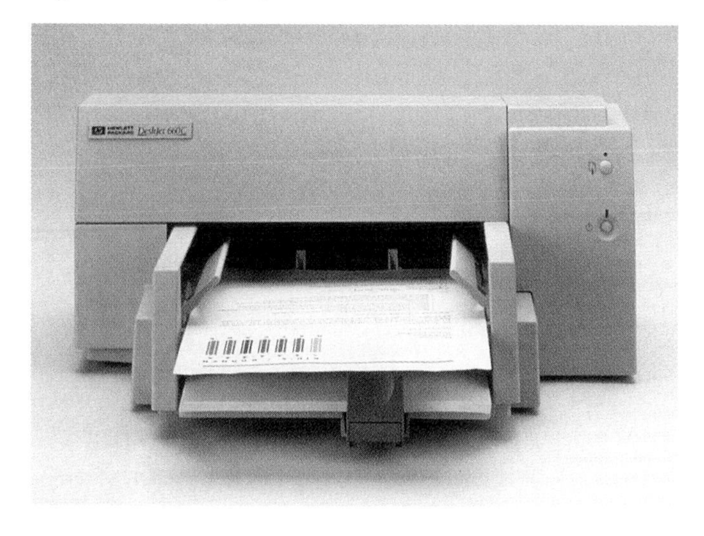

Modems

A **modem** (*mo*dulator/*dem*odulator) is a telecommunications device that allows computer operators to send and receive information over standard telephone lines. See Fig. 5-10. In this sense, modems are both input devices and output devices. Modems are often installed in CAD systems, particularly in architectural offices that may have more than one location. They allow large CAD drawings to be sent directly from one branch to another or from the designer to a major client for approval. This reduces the amount of time involved in creating a hard copy and transporting it to its destination.

Fig. 5-10 ■ A modem allows computers in different locations to communicate electronically over the telephone lines.

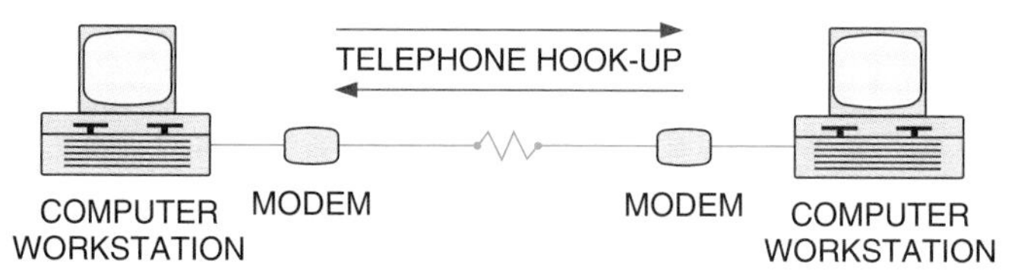

Fig. 5-8 ■ Laser printer.

Multimedia Computers

Multimedia is a term used to indicate the combining of more than one medium. Computers can now incorporate text, audio, graphics, animation, and full-motion video. For example, CAD drawings can be combined with other media to pro-

Fig. 5-11 ■ Typical components for a multimedia computer system.

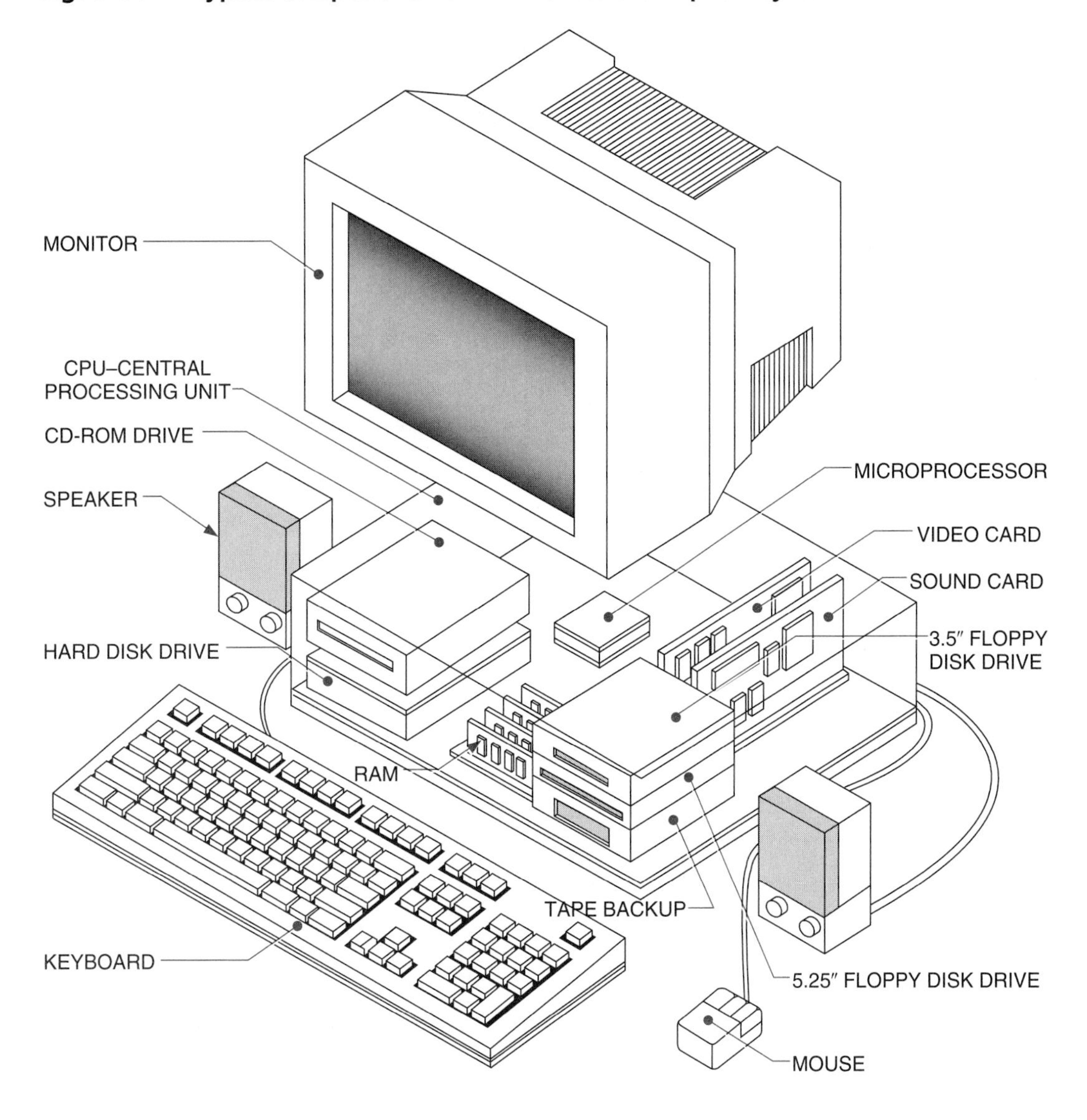

duce a complete product. Figure 5-11 shows the components typically included in a multimedia computer.

Multimedia computers are well-suited to architectural sales presentations. Drafters can prepare colorful presentations that include "mood music," actual photographs of buildings they have designed in the past, as well as floor plans and rendered images of how rooms in the buildings might look. At conventions, for example, the entire presentation can be programmed to run over and over on the computer.

The components necessary for multimedia computers differ somewhat from those for CAD systems. For example, digitizers are rarely needed on a multimedia computer. Instead, the computer requires special hardware that controls the various media.

CD-ROM All multimedia computers contain CD-ROM (compact disc read-only memory) drives. **CD-ROM** discs are similar to the compact discs used for music, except that CD-ROMs can contain text and graphics as well as audio.

A single CD-ROM can contain more than 600 MB of data. This makes it a good choice for multimedia applications, which typically require a large amount of storage space.

GRAPHIC ACCELERATORS Graphics—including digital photos and illustrations in multimedia systems as well as CAD drawings—typically take a long time to process and display on a monitor. To speed up the processing, some companies install graphic accelerators in their computers. A **graphic accelerator** is a circuit board ("card") that can be added to a computer to increase the amount of RAM and processing power dedicated to displaying images on the monitor.

SOUND CARDS Most computers have an elementary speaker system. However, to enhance the audio of a multimedia presentation, a sound card is usually added to the computer. The sound card contains electronics that allow the computer to be connected to external speakers. This produces sound of much better quality than the internal computer speaker can produce.

CAD Software

Computer software is the programming that directs the computer hardware to perform specific tasks. CAD software directs the computer to perform drawing tasks as instructed by a drafter. See Fig. 5-12. It allows the drafter to create very accurate drawings by placing geometry such as lines and circles in precise locations. It also provides editing tools that the drafter can use to change the size, shape, and position of the geometry.

Fig. 5-12 ■ Examples of CAD software. *Arnold & Brown*

Hundreds of CAD (and CADD) software programs are currently available. These range from very simple CAD programs to extremely complex architectural engineering and construction (AEC) programs. In general, the simple, low-cost programs require more input for each task than expensive and sophisticated systems. For example, in some simple programs, the drafter must position each wall line precisely. In more sophisticated programs, wall corners are attached automatically, and extra lines are eliminated. Room sizes can be stretched or shrunk to the correct dimension and scale. More advanced programs may automatically align appliances on a wall, develop material lists from drawing input, or signal the user if a feature will not fit where it was placed.

Selecting software programs that contain expensive and unneeded features is wasteful. However, selecting programs that do not include required drawing features makes the completion of working drawings difficult and more time-consuming. It is therefore important to understand, as much as possible, exactly what features and levels of sophistication are required before you select a specific software program.

Types of CAD Drawings

The type of CAD drawing a drafter creates depends on the purpose of the drawing. The two major types of drawings are two-dimensional (2D) drawings and three-dimensional (3D) drawings.

Two-Dimensional Drawings

Traditional architectural working drawings are usually two-dimensional. Only two dimensions (width and length, width and height, or length and height) are shown on one drawing. Two-dimensional drawings created on CAD systems include floor plans, elevation drawings, detail drawings, and plot plans or site drawings.

CAD floor plans can be fully dimensioned working drawings or layered design layouts. A good quality CAD plan should appear the same as a good quality conventionally drawn plan. CAD plans can be quickly layered to show levels or components in different colors. See Fig. 5-13.

CAD elevation drawings can be created from a floor plan by inputting elevation heights at key intersections on a floor plan. See Fig. 5-14. CAD elevation drawings can be converted to pictorial drawings by changing the viewing angle, as shown in Fig. 5-15. Consistent surface material symbols, shading, and shadows can also be added to elevation drawings. See Fig. 5-16.

Detail drawings and site drawings, as covered in Chapters 17 and 18, are prepared using the same CAD procedures used in preparing floor plans and elevations.

Fig. 5-13 ■ Computer-generated site plan of Research Hatchery Aquarium created with DATACAD 6.0. *David Karram*

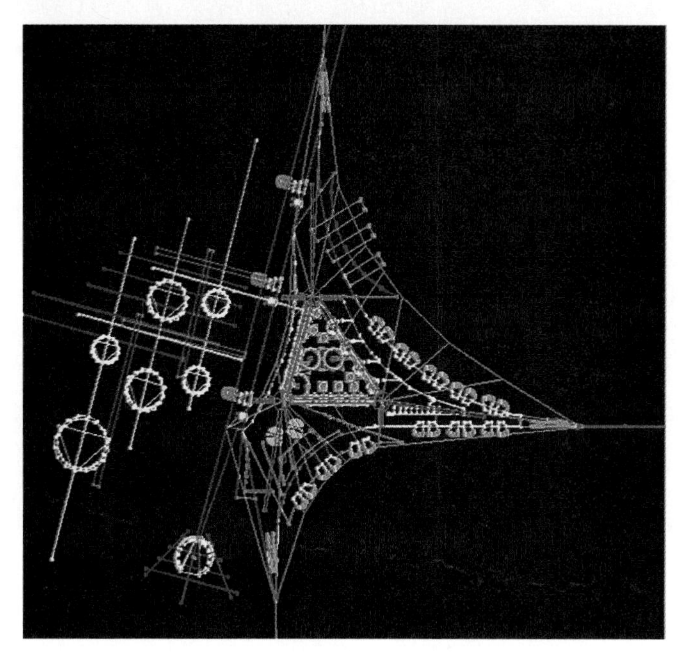

CAD software allows the drafter to create 2D drawings by locating important points such as the endpoints of lines and arcs and the centers of circles. To place geometry accurately, the drafter specifies the Cartesian coordinates of the points.

The **Cartesian coordinate system** is an imaginary grid that employs two axes (for two-dimensional drawings) called the *X axis* and the *Y axis*. These two axes are at right angles to each other. The point at which they meet is called the origin. See Fig. 5-17A. Note also that each axis has a positive (+) side and a negative (−) side.

In Cartesian geometry, the axes divide a drawing into four imaginary quadrants. To locate a point in any of these quadrants, you need only specify two numbers. The two numbers are called a *coordinate pair.* For example, the coordinate pair (2,1) are both positive numbers. They identify a point in quadrant 1 two units to the right of the origin on the X axis and one unit above the origin on the Y axis. See Fig. 5-17B. Note that the X coordinate is

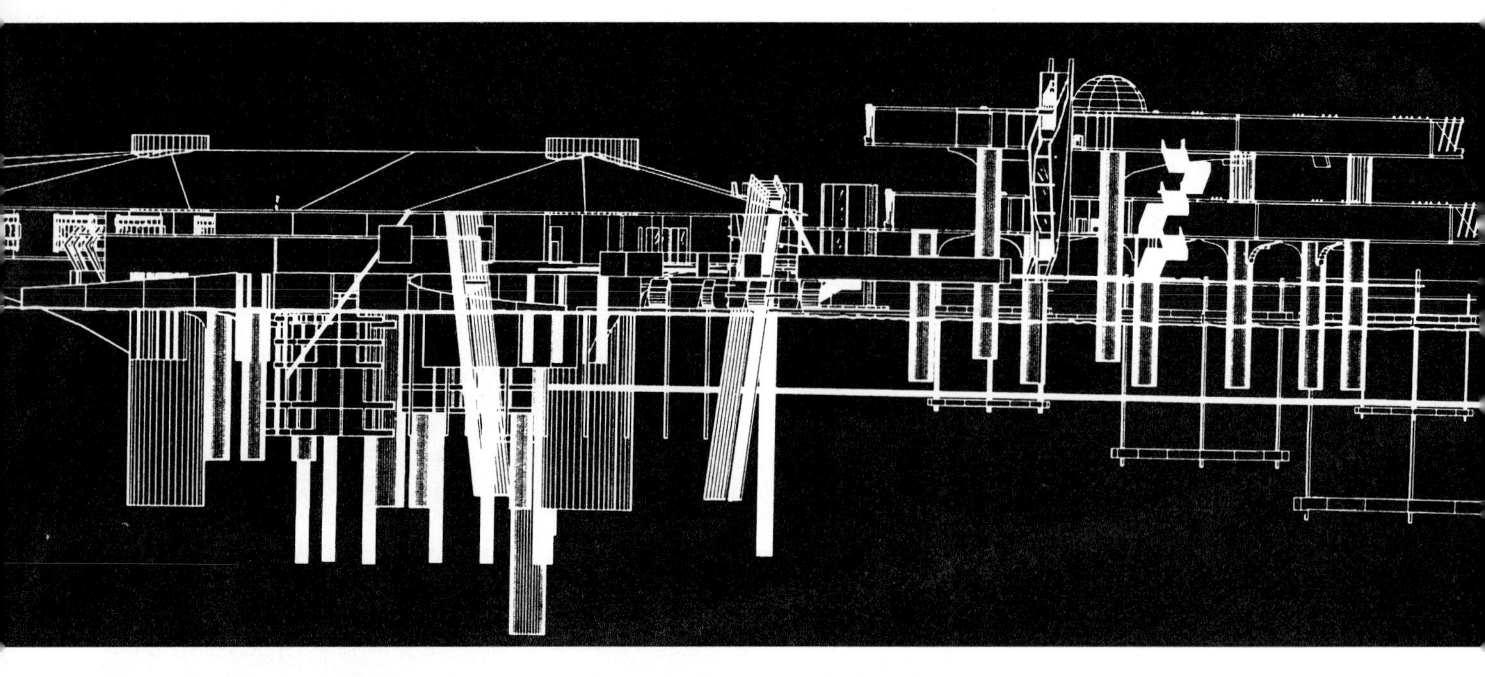

Fig. 5-14 ■ **Elevation drawing created from the floor plan in Fig. 5-13.**
David Karram

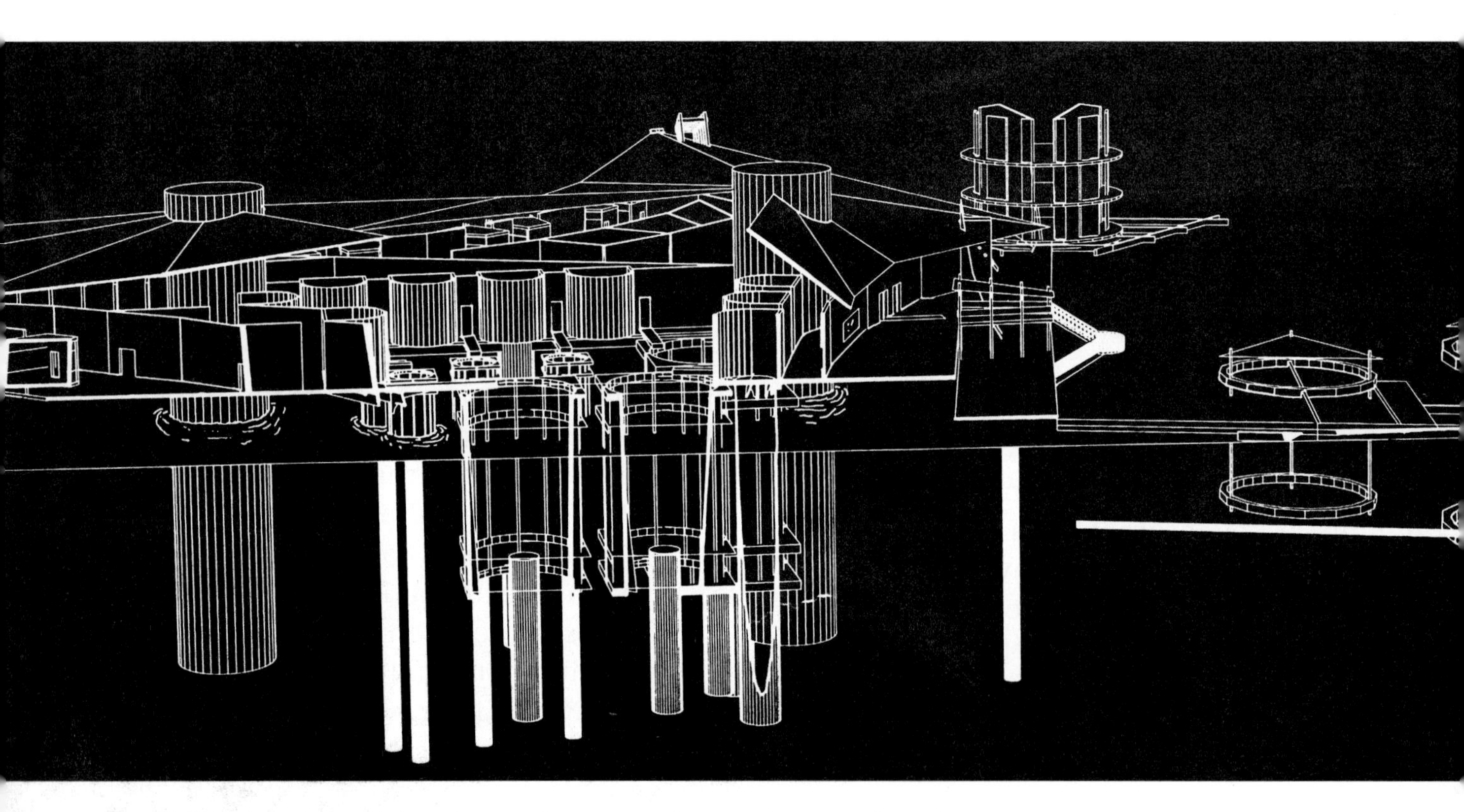

Fig. 5-15 ■ **A pictorial drawing can be created by rotating the elevation drawing in Fig. 5-14.** *David Karram*

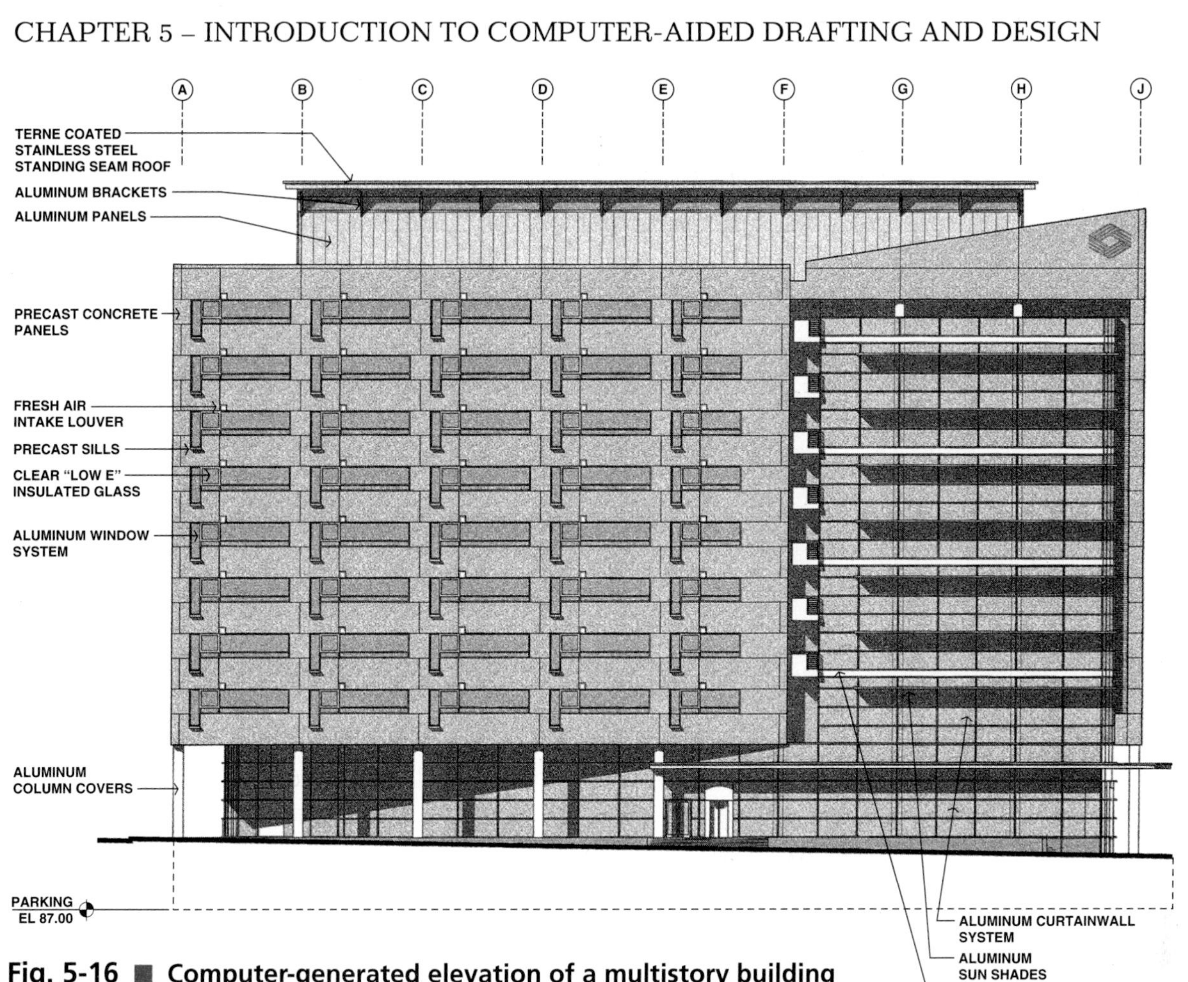

Fig. 5-16 ■ Computer-generated elevation of a multistory building with siding materials, shading, and shadows added. *Wilmington Trust, the Hillier Group*

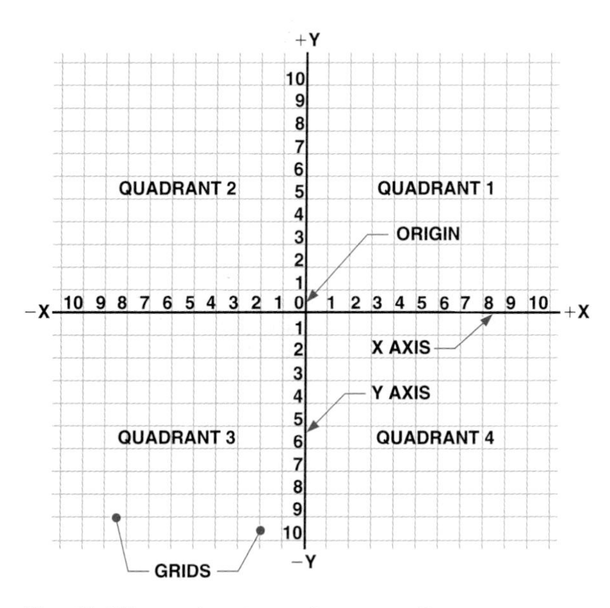

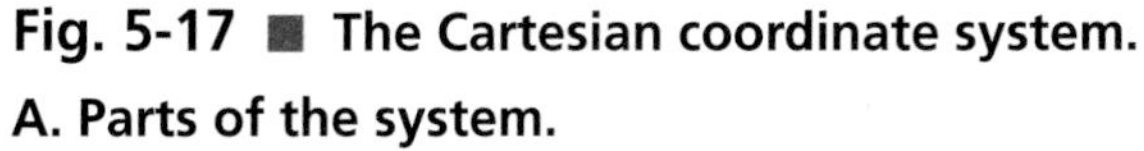

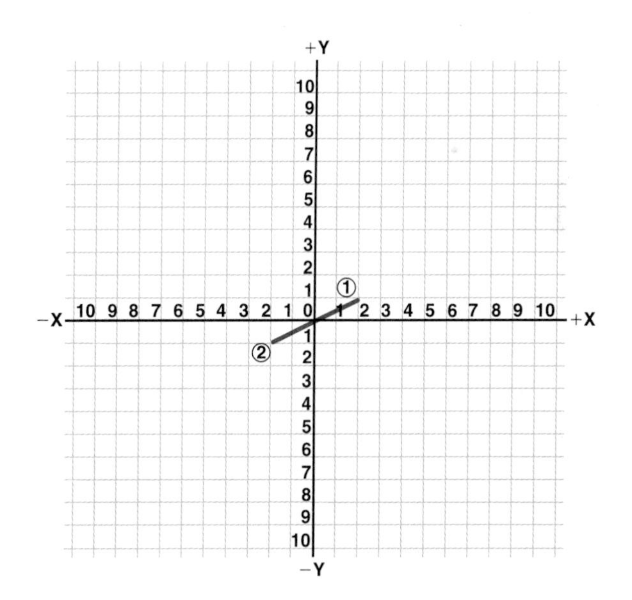

Fig. 5-17 ■ The Cartesian coordinate system. A. Parts of the system.

B. The circled number 1 shows the point located by the coordinate pair (2, 1). The first number (2) is the X coordinate and the second number (1) is the Y coordinate. The circled number 2 shows the point located by the coordinate pair (−2, −1). It is located in quadrant 3 because both numbers in the pair are negative.

Fig. 5-18 ■ Using Cartesian coordinates to plot a rectangle.

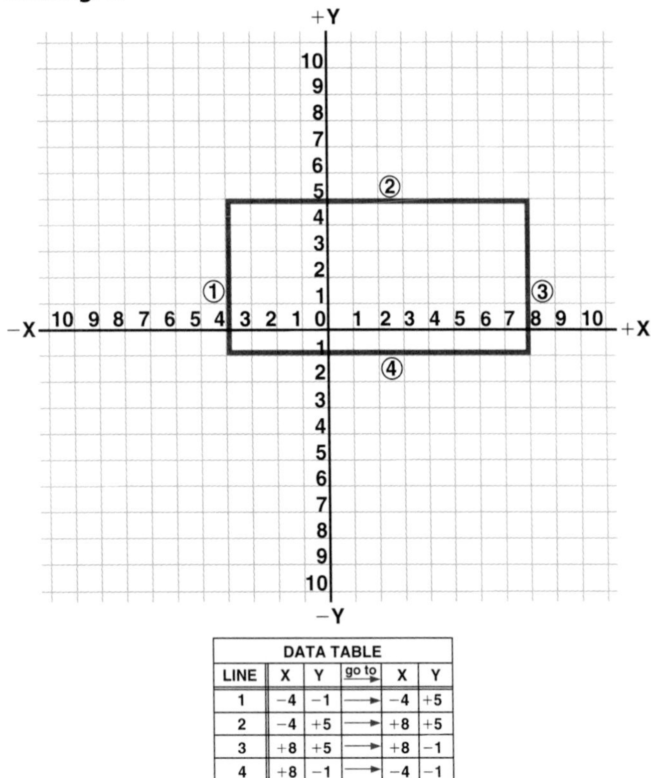

DATA TABLE					
LINE	X	Y	go to	X	Y
1	−4	−1	→	−4	+5
2	−4	+5	→	+8	+5
3	+8	+5	→	+8	−1
4	+8	−1	→	−4	−1

always the first number in a coordinate pair, and the Y coordinate is always the second number. Figures 5-18 and 5-19 show how various figures can be created using the Cartesian coordinate system.

Three-Dimensional Drawings

Most current architectural software programs can create three-dimensional drawings directly from two-dimensional drawings. This is done by adding a third axis (Z) to the Cartesian coordinate system, as shown in Fig. 5-20. The Z coordinate allows the addition of depth to the drawing.

Three different types of three-dimensional drawings can be produced on a CAD system. These include wire-frame drawings, surface models, and solid models.

Fig. 5-19 ■ Using Cartesian coordinates to plot a curve.

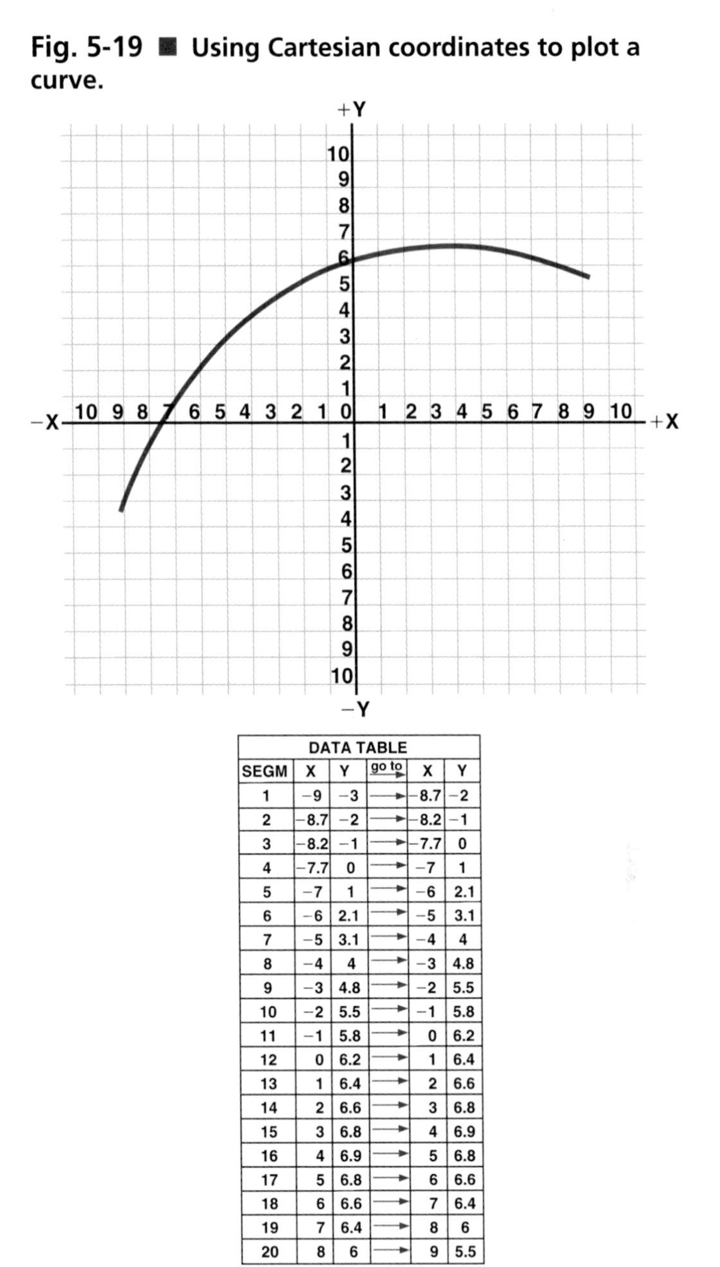

DATA TABLE					
SEGM	X	Y	go to	X	Y
1	−9	−3	→	−8.7	−2
2	−8.7	−2	→	−8.2	−1
3	−8.2	−1	→	−7.7	0
4	−7.7	0	→	−7	1
5	−7	1	→	−6	2.1
6	−6	2.1	→	−5	3.1
7	−5	3.1	→	−4	4
8	−4	4	→	−3	4.8
9	−3	4.8	→	−2	5.5
10	−2	5.5	→	−1	5.8
11	−1	5.8	→	0	6.2
12	0	6.2	→	1	6.4
13	1	6.4	→	2	6.6
14	2	6.6	→	3	6.8
15	3	6.8	→	4	6.9
16	4	6.9	→	5	6.8
17	5	6.8	→	6	6.6
18	6	6.6	→	7	6.4
19	7	6.4	→	8	6
20	8	6	→	9	5.5

Fig. 5-20 ■ X, Y, and Z coordinates are needed to produce three-dimensional drawings. The Z axis is perpendicular to both the X axis and the Y axis.

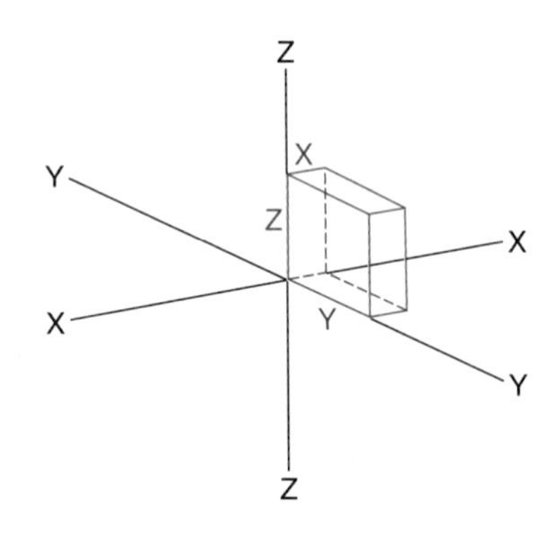

WIREFRAME DRAWINGS **Wireframe drawings** are basically see-through stick drawings in which some or all hidden areas are exposed. They show an object's width, length, and height (X, Y, and Z) dimensions. These can be hard to understand unless some or all of the hidden lines are "removed" (temporarily not displayed). See Fig. 5-21A and Fig. 5-22A.

SURFACE MODELS **Surface models** are drawings that consist of solid plane surfaces instead of the connected lines used in wireframe drawings, as shown in Fig. 5-22B. Surface models are more advanced than wireframe drawings and are often used in 3D architectural work.

Figures 5-21B and 5-22B show related surface models of the wireframe drawings shown in Figs. 5-21A and 5-22A.

Fig. 5-21 ■ Wireframe drawing and surface model of a residence. *David Karram*

A. Wireframe.

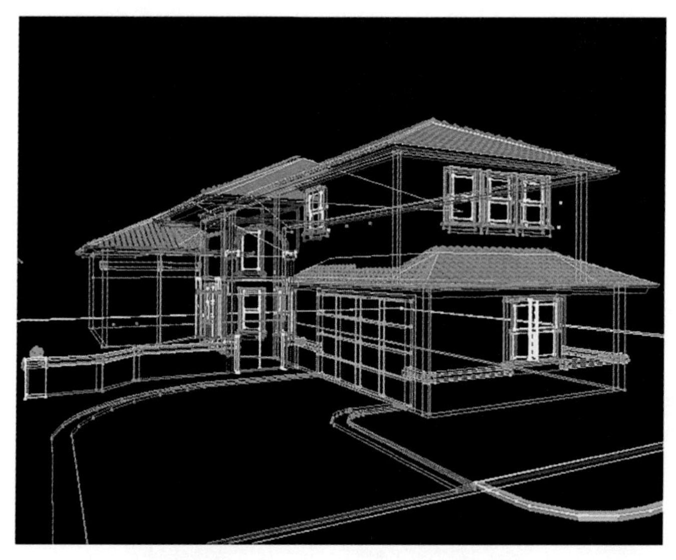

B. Surface model.

Fig. 5-22 ■ Wireframe and surface model of part of the Research Hatchery Complex in Fig. 5-13. *David Karram*

A. Wireframe.

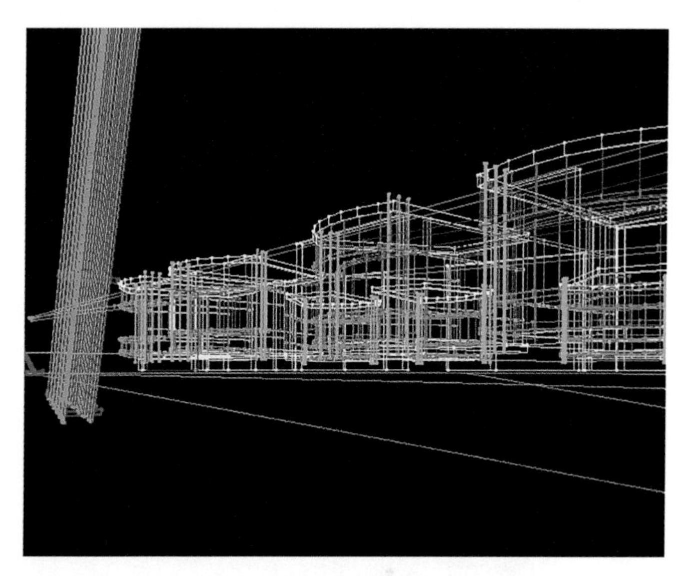

B. Surface model.

SOLID MODELS The most advanced form of 3D drawing is the **solid model.** Although a solid model may look similar to a surface model or a wireframe with hidden lines removed, it is actually quite different. A solid model has mass properties, which other forms of 3D drawings do not have. For example, you can assign a material, such as copper, aluminum, or steel, to a solid model. Then the CAD software can measure the mass, density, and other properties of the model. Therefore, solid models are often used by engineers and designers to help predict the strength and capabilities of a proposed product. Architects often use solid models to perform structural analyses on buildings and other structures.

In addition to their mass properties, solid models can be rendered so that they look almost like photographs. In CAD, **rendering** is a process of adding shading and lights to a drawing so that it looks more realistic. Figure 5-23 shows a rendered architectural solid model.

Figure 5-24 is a three-dimensional rendering of the Research Hatchery Aquarium developed from the floor plan and elevation shown in Fig. 5-13 and the wireframe and surface models shown in Fig. 5-22. This realistic still-image simulation was created through the use of CAD lighting, camera settings, and color assignments and is part of a total automation sequence. Figure 5-25 shows an enlarged solid-model rendering of a part of this complex.

Solid models can also be used to create interior views of an object. Most CAD software contains one or more commands

Fig. 5-23 ■ **A rendered architectural model like this one can help customers see how the finished building will look.** *Robert McNeel & Associates*

Fig. 5-24 ■ **A three-dimensional still image of the research complex.** *David Karram*

that "slice" the model to expose the internal structures. However, the solid model must be carefully constructed so that the inside as well as the outside features are accurate.

VIRTUAL REALITY SYSTEMS Virtual reality is a natural extension of three-dimensional drawing technology. A *virtual reality* system creates a computer-generated "world" or "reality" in which the user seems to be immersed in the computer images. See Fig. 5-26. This technology can be applied to allow architects to take clients through buildings that have not yet been built. The architect creates a 3D model of the building using a CAD system and then enhances the drawing using specialty software to "paint" the walls and add details.

Fig. 5-25 ■ **"Zoomed" (enlarged) area of the rendering in Fig. 5-24, created using the same CAD simulation methods.** *David Karram*

Fig. 5-26 ■ Virtual reality allows people to explore buildings before they have been built.

When they put on the equipment of the virtual reality system (typically a helmet and a data glove), the clients seem to be in the building and can move about freely to inspect walls, ceiling height, the location of doors and windows, and so on. Some architectural applications allow the architect to make real-time changes to the building. For example, if a client decides a window is too high, the architect can "reach out" within the virtual environment and move it down. The change is reflected in the CAD drawing file as well as in the virtual environment.

Using sophisticated rendering programs, architects can even show clients potential lighting and color schemes in the virtual environment. How will this wallpaper look with that carpet color? How will this dark paneling look in the late afternoon sun? Is one spotlight sufficient for this showcase? How should it be aimed? All of these questions—and many more—can be answered directly using virtual reality.

The use of virtual reality in architectural drafting is currently not as widespread as it could be for two reasons. First, the technology is still fairly new and is, therefore, extremely expensive. Also, computers today, as fast as they are, are still not fast enough to present excellent detailed images in real time. The two solutions to this are to move very slowly or to use less detail in the virtual environment. For general purposes, the rooms in virtual buildings look more like cartoons than real rooms. These less-detailed rooms generate faster, so they are within the capacity of today's computers. As computers become faster and virtual technology becomes more mature, more and more architectural companies may begin using this powerful tool.

Using a CAD System

To use a CAD system effectively, a CAD operator must have a thorough understanding of the principles and practices of architectural drafting and design. The background knowledge required is the same as that for creating drawings manually. Only the means of producing the drawings differ.

CAD systems require the drafter to enter **commands** that tell the software what function to perform. The commands used in CAD software include drawing commands, editing commands, and general utility commands. To enter the commands, the drafter can use the keyboard or a digitizing tablet. See Fig. 5-27. Most CAD software also includes on-screen menus that can be used to enter the commands.

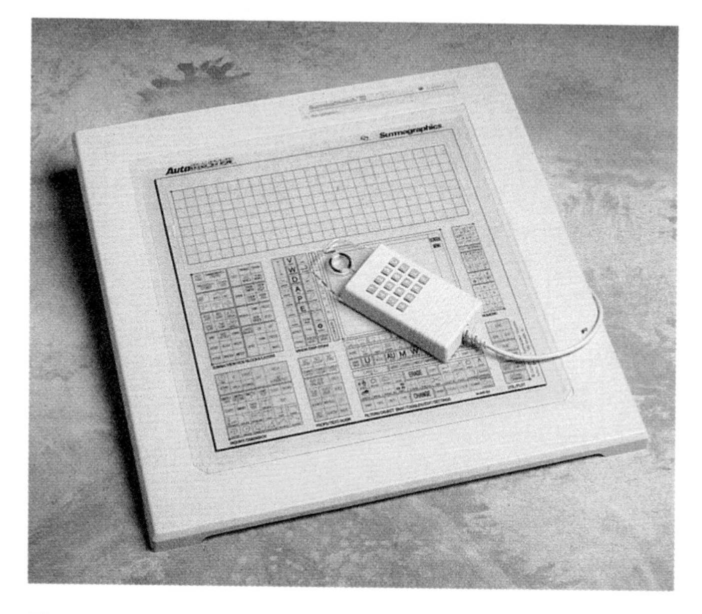

Fig. 5-27 ■ Special templates (graphic overlays) such as this one are available for architectural applications. The template fits over the digitizing tablet and allows the user to enter special symbols and commands.

Drawing Commands

The drawing commands in a CAD system are the basic commands that allow the drafter to create geometry. The individual lines, circles, etc., are commonly called *entities* or *objects*. Following is a list of a few of the more common drawing commands:

- The *Line* command is probably the most-used CAD command. Using this command, the drafter can draw lines by connecting points with a cursor (controlled by a mouse or puck) or by entering the coordinates at the keyboard.
- The *Circle* command creates circles of any size using various methods. For example, drafters can specify a circle using the center and a radius or diameter, or by specifying two or three points on the edge of the circle. See Fig. 5-28.
- The *Arc* command creates arcs of any radius and any length. To create an arc, drafters can specify the center and angle or radius, or they can specify the start, middle, and end points of the arc. Most CAD software provides several other options in addition to these basic ones. See Fig. 5-29.
- The *Text* command creates text in a manner similar to a word processor. In CAD programs, the text handling features are usually not as well developed

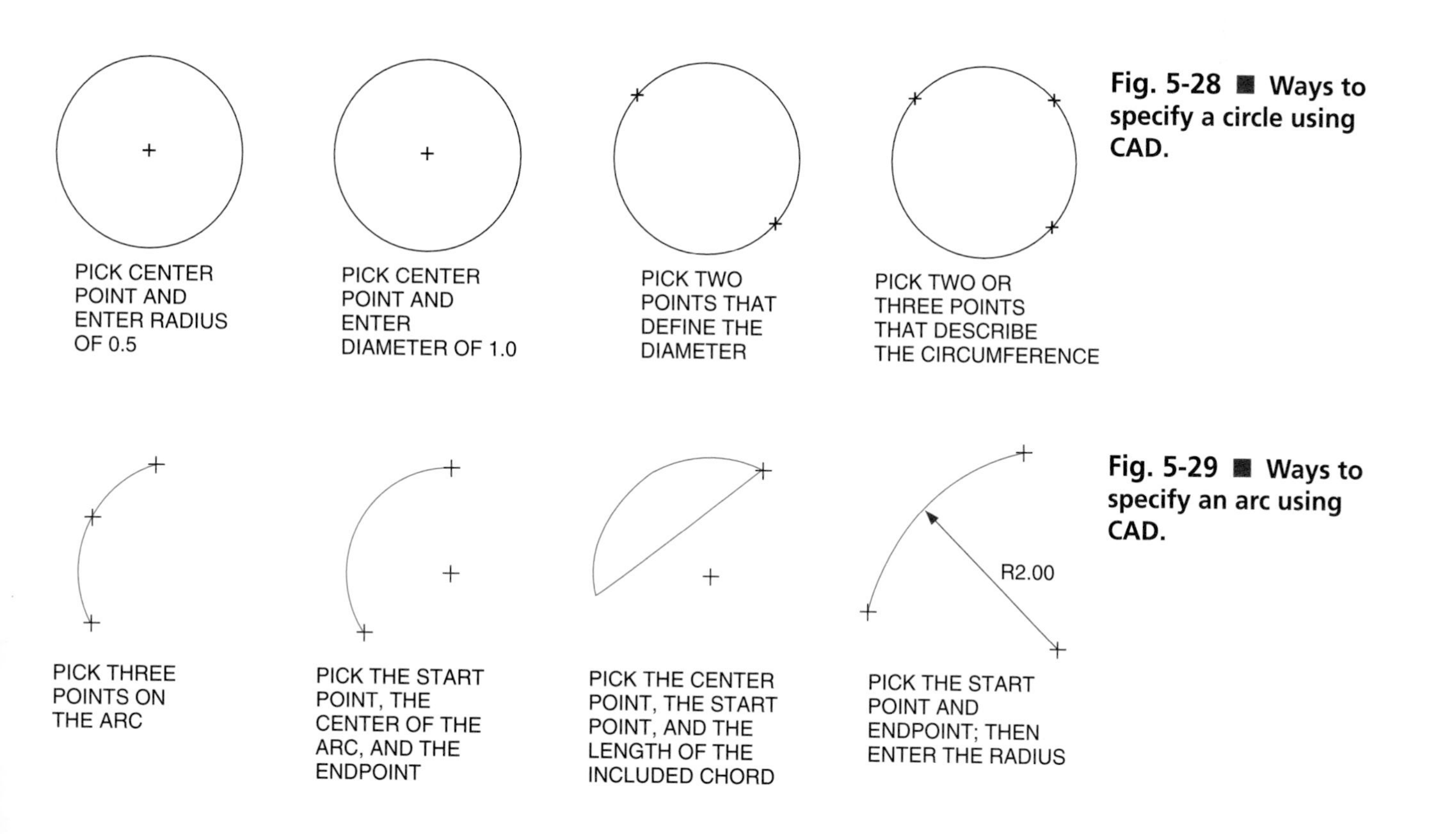

Fig. 5-28 ■ Ways to specify a circle using CAD.

Fig. 5-29 ■ Ways to specify an arc using CAD.

as in standard word processing programs. However, the Text command is an important tool for adding labels and other text to a drawing.

In addition to the commands described above, most CAD software provides a set of dimensioning commands that automatically calculate the dimensions of objects in a drawing. Also, commands are available to save parts of the drawing as "blocks" and then insert them again and again. This can save a great amount of time when a drawing contains many of the same part, such as the doors and windows on an architectural floor plan.

Editing Commands

CAD software provides many commands that allow the drafter to change and manipulate a basic drawing. In addition to being critical for drawing revisions, many of these commands are used by experienced drafters to help create the original drawing. Following are a few examples:

- The *Delete* or *Erase* command erases individual entities from a drawing.
- The *Move* command allows the drafter to select one or more entities and move them to another location in the drawing.
- The *Copy* command is similar to the Move command, except that it produces a copy of the selected entities and places them elsewhere in the drawing. The original entities remain as they were originally placed.
- The *Rotate* command allows one or more selected entities to be rotated around a point defined by the drafter. See Fig. 5-30.
- The *Mirror* command produces an exact mirror image of selected entities. See Fig. 5-31.

Fig. 5-30 ■ To rotate this simple window symbol to fit on a different wall of a floor plan, specify the base point of the rotation and the angle of rotation.

ROTATION ANGLE: 90 DEGREES
BASE POINT OF ROTATION
THE FINISHED ROTATION

Fig. 5-31 ■ The Mirror command can save the drafter a great amount of time. In this case, the drafter drew one half of the roof truss, then mirrored it to create the other half.

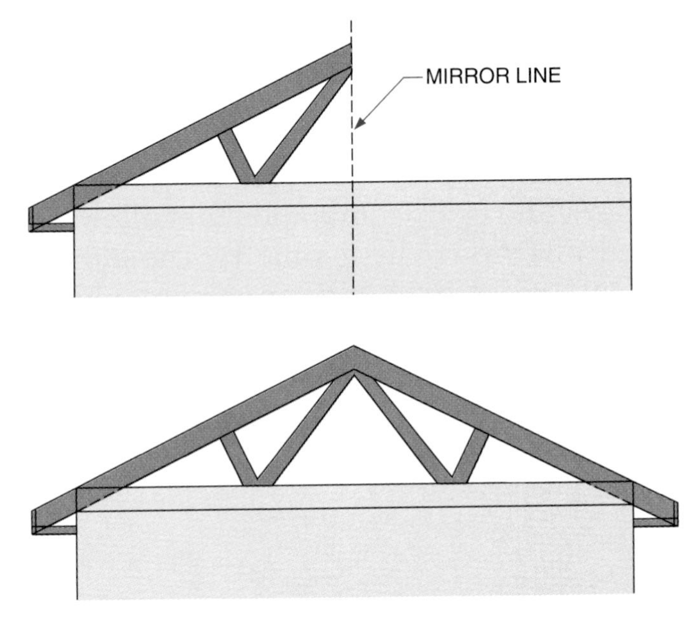

- The *Offset* command produces an exact copy of a line at a specified distance from a given line. See Fig. 5-32.

Utility Commands

Some of the commands available in a CAD program do not manipulate the drawing directly. Instead, they assist the

Fig. 5-32 ■ The Offset command creates a line, arc, or circle that is exactly the same distance at all points from a specified line, arc, or circle. In this case, the drafter used Offset to create the ledge thickness for a custom swimming pool.

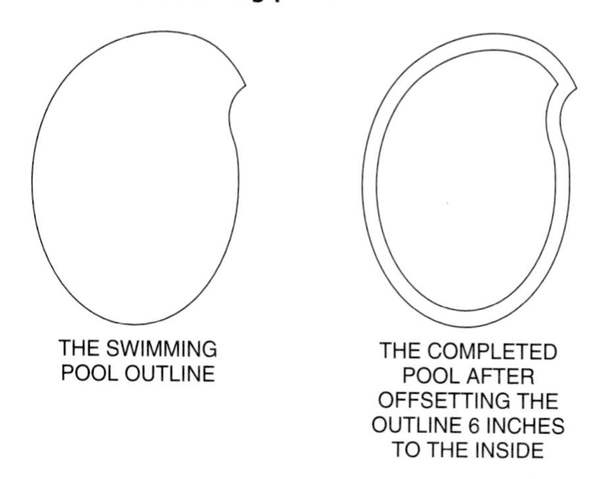

drafter in various ways. Examples of these utility commands are listed below.

- The *Zoom* command magnifies or reduces the size of a drawing on the screen. See Fig. 5-33. It does not affect the actual dimensions of the drawing, however. The drafter can zoom in to view details that are too small to see when the entire drawing shows on the screen, or zoom out to see more of the drawing at one time.
- The *Pan* command moves the drawing horizontally and vertically on the screen without changing the current zoom percentage. See Fig. 5-34.
- The *Print* or *Plot* command produces a hard copy of the drawing, assuming that the computer is connected to a printer or plotter and that all the equipment is turned on.
- The *ID* and *Measure* commands, and others similar to them, allow the drafter to identify specific coordinates and distances on the drawing.

Keep in mind also that different CAD programs may have slightly different names for some of the commands. However, in most cases, the names are self-

Fig. 5-33 ■ The Zoom command allows you to see the entire drawing or to "zoom in" to see detail on a smaller portion of the drawing, such as a window.

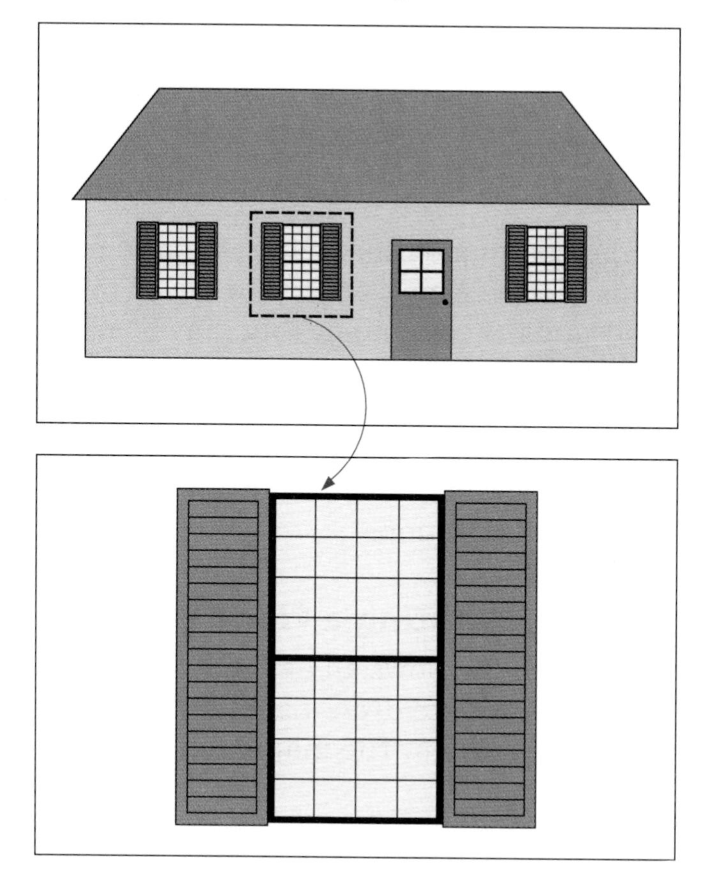

Fig. 5-34 ■ Specifying the first and second points of displacement as shown for the Pan command moves the view on the monitor screen from the window to the upper part of the door, at the same level of magnification.

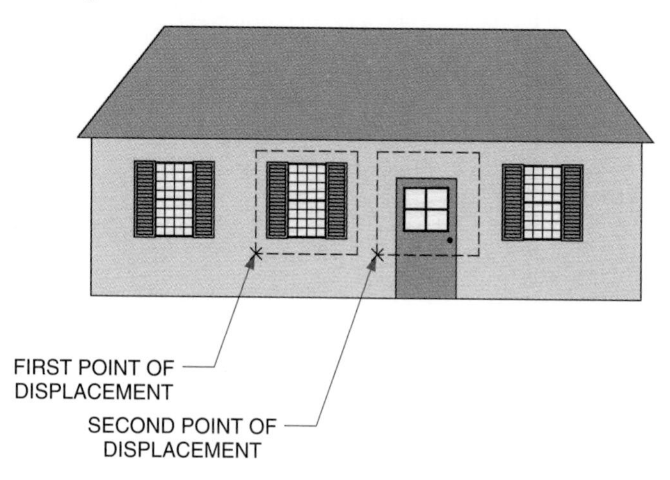

explanatory. Also, the examples of drawing, editing, and utility commands presented in this chapter are just a few of the many commands available in CAD software.

Architectural Applications

The commands summarized in the previous section represent only a few of the thousands of commands that may be used in combination to complete a full set of architectural drawings. The following sequences provide examples of how the commands can be used to create floor plans and elevations.

Steps in Creating a Floor Plan

Figure 5-35 shows an example of a sequence used to draw a simple floor plan on a CAD system. The major steps follow this order:

1. Draw the outer walls (perimeter) of the floor plan. In most cases, walls are represented by double lines to define the thickness of the walls.
2. Draw the inner walls to divide the floor plan into individual rooms.
3. Insert the windows into the floor plan at the correct locations. Most architectural firms use **symbol libraries** that contain various ready-to-use window symbols. These symbols can be inserted easily into any drawing.
4. Insert the doors. These, too, are available in symbol libraries.
5. Insert the kitchen and bath fixtures such as the stove, sink, cabinets, bathtub, and so on. Again, you may use a commercial symbol library to insert these items.
6. Add notes and labels to the drawing as necessary.
7. Dimension the drawing using the CAD program's automatic dimensioning commands.

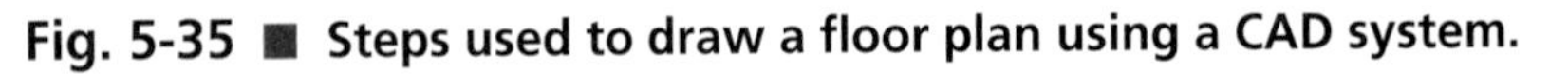

Fig. 5-35 ■ Steps used to draw a floor plan using a CAD system.

1.

2.

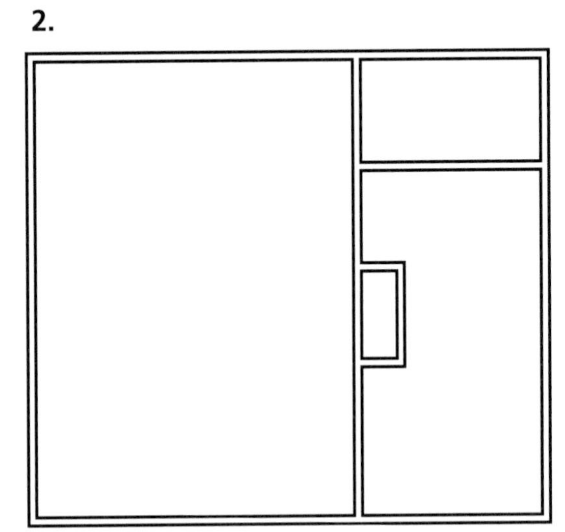

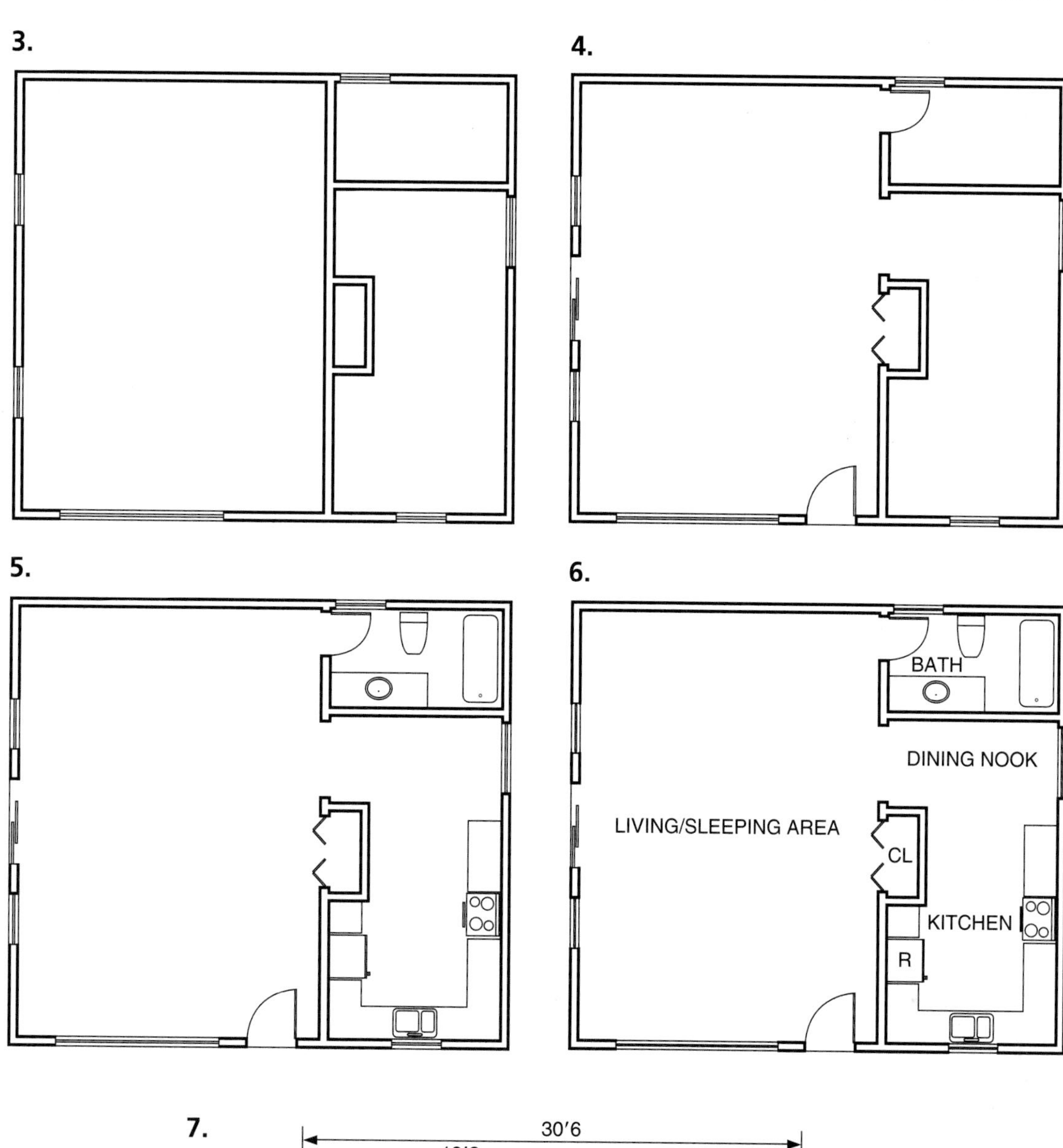
3.
4.
5.
6.
BATH
DINING NOOK
LIVING/SLEEPING AREA
CL
KITCHEN
R

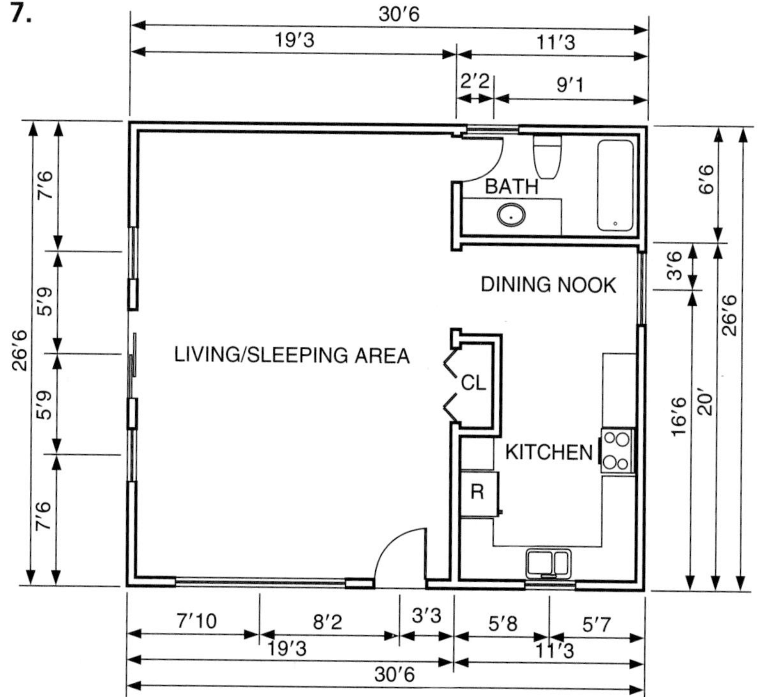
7.
30′6
19′3
11′3
2′2
9′1
BATH
DINING NOOK
LIVING/SLEEPING AREA
CL
KITCHEN
R
7′6
5′9
5′9
7′6
26′6
6′6
3′6
16′6
20′
26′6
7′10
8′2
3′3
5′8
5′7
19′3
11′3
30′6

Steps in Creating an Elevation

Figures 5-36A through 5-36R show an example of a sequence used to draw a simple elevation using a CAD system. This example uses only a few of the many graphic tasks available in most CAD software packages.

Fig. 5-36 ■ Steps used to draw an elevation using a CAD system.

A. Make a sketch of how the elevation should look.

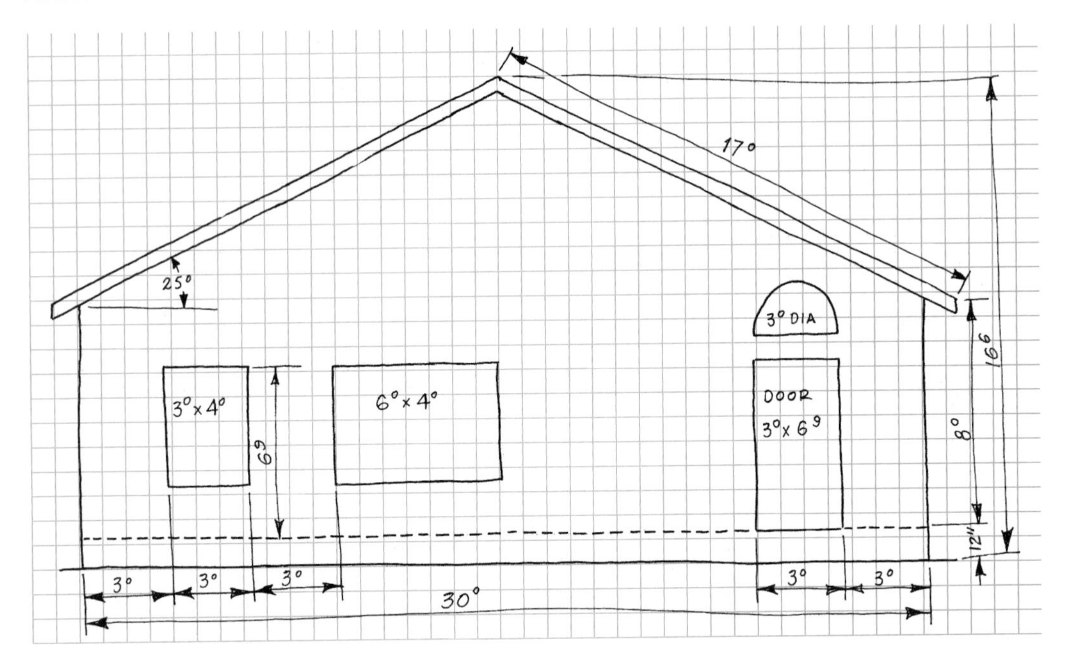

B. Turn on the 1/2″ grids and pick the starting point for the elevation. Each grid will represent 2′.

C. Key in or pick the 30′ length for the elevation. Use the Line command for all line work.

D. Key in or pick the ceiling's height (8′-6″ from the ground line).

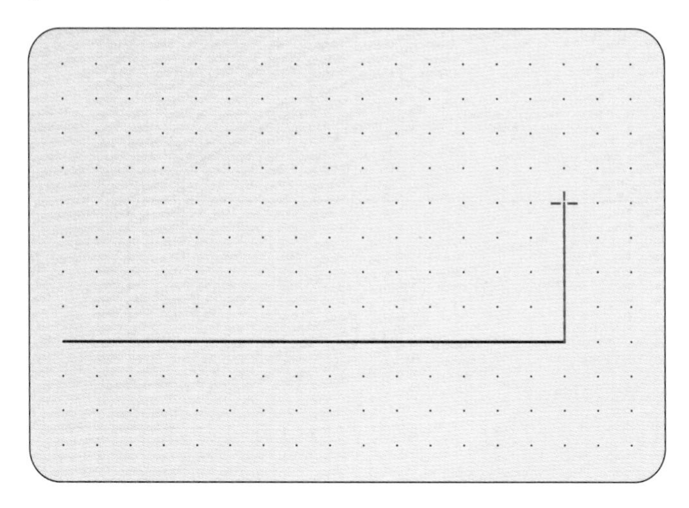

E. Key in or pick the 30′ length for the ceiling.

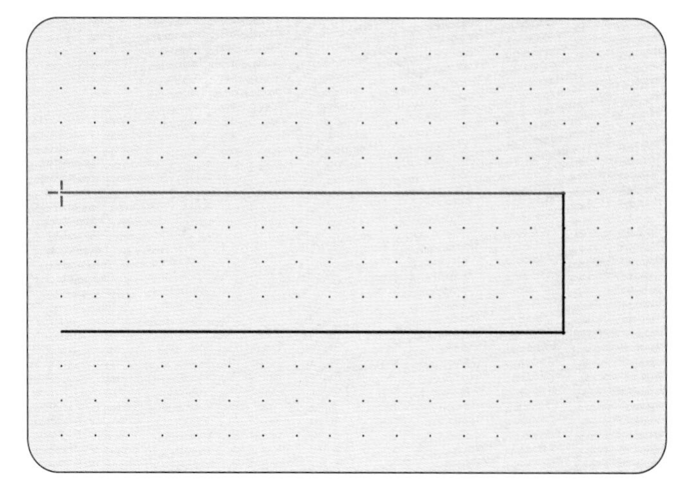

F. Close the outline of the elevation wall.

G. Key in or pick the top of the roof's ridge board.

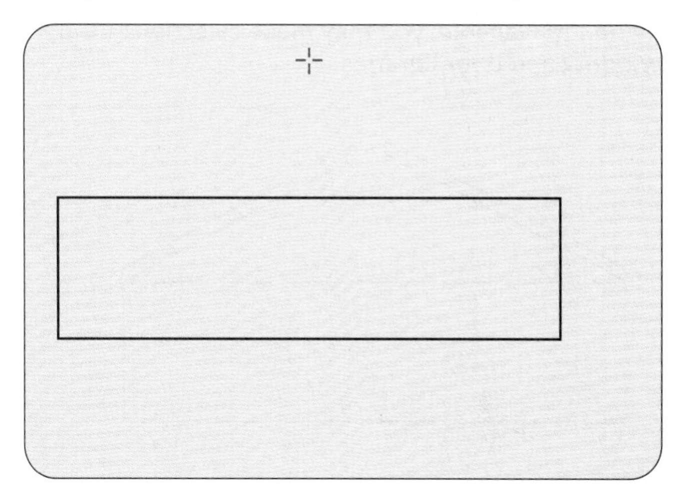

H. Key in or pick the required amount of overhang and connect that point with the roof's peak.

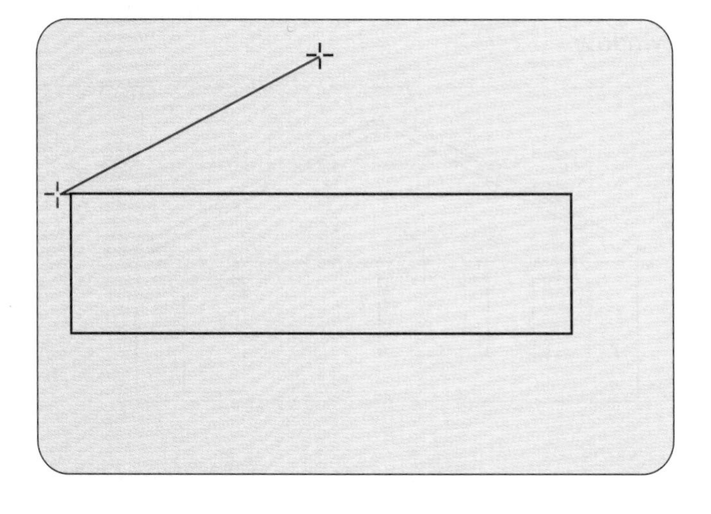

I. If entering the roof's angular pitch, use the Polar Coordinates command and enter the angle and length. Note that the angles are read counterclock-wise.

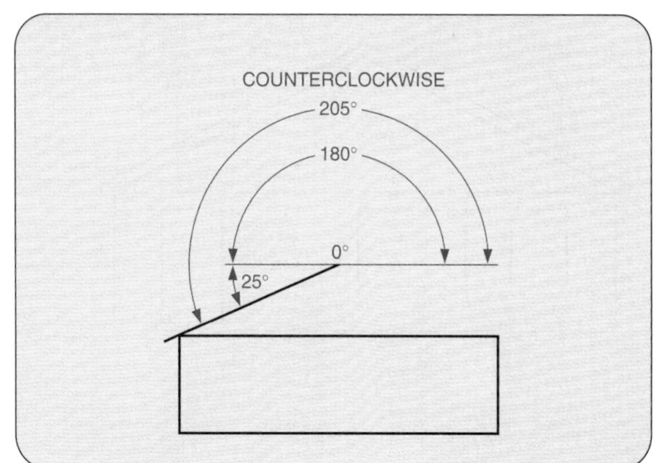

J. Repeat step H or I for the opposite side of the roof.

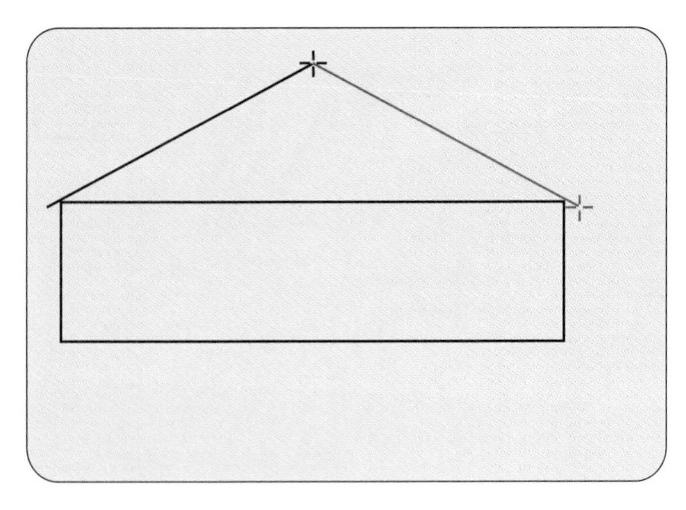

K. With the Line command, key in or pick the depth of the end (barge) rafter.

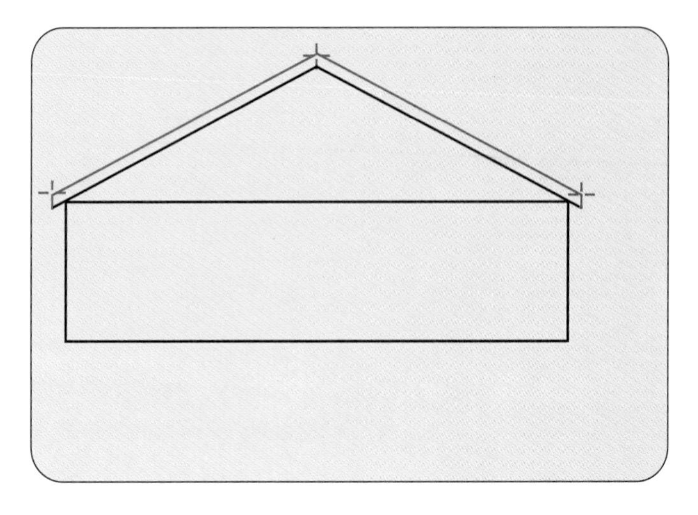

L. With the Rectangle command, key in the size for each window and door and place them in the correct position. (It may be necessary to use the Move command.) If available, you may make selections from window and door libraries.

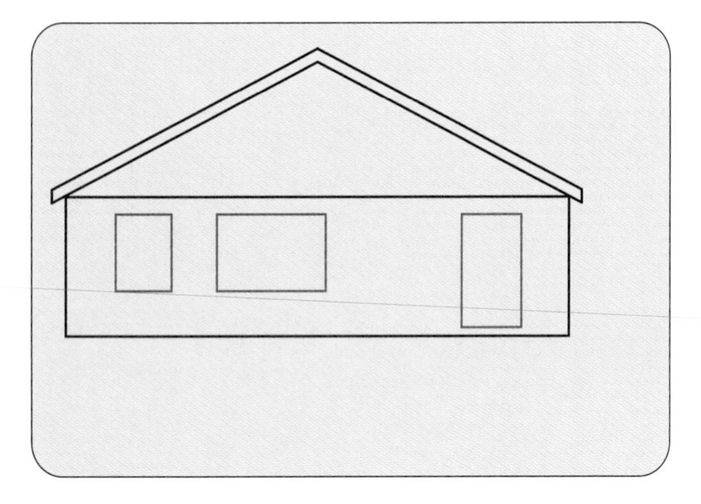

M. When the line types are loaded, select the hidden line and draw the floor line 6″ above the ground line for a slab foundation.

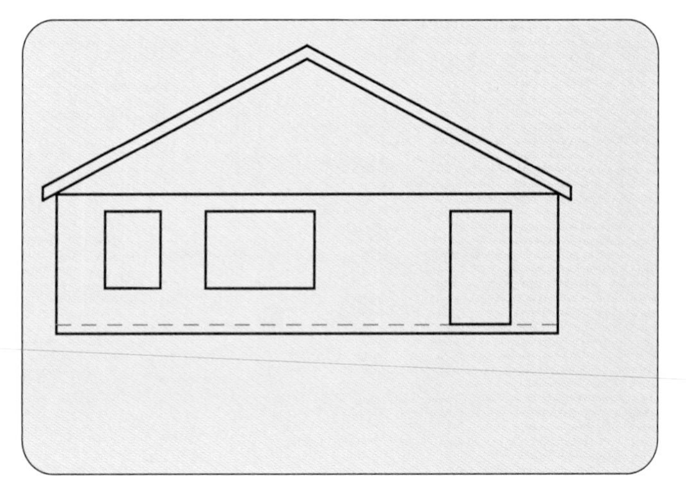

N. With the Arc command, pick the end points and the radius for the fan window over the front door.

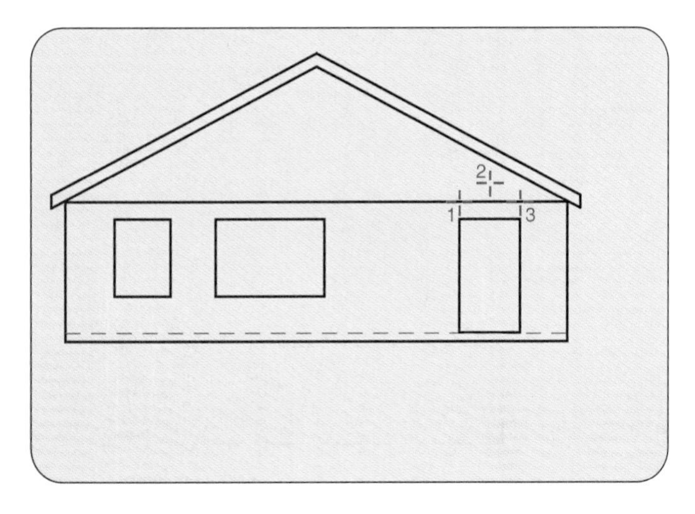

O. With the Line command, complete the fan window.

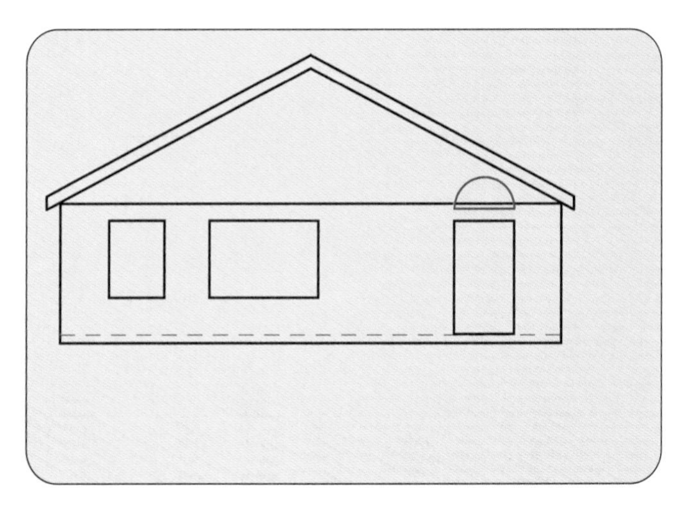

P. Add the necessary text with the Text command.

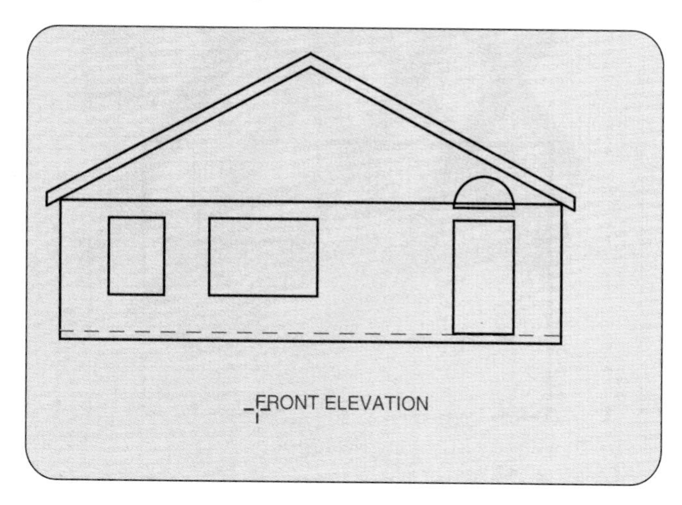

Q. With the Erase command, delete the ceiling line.

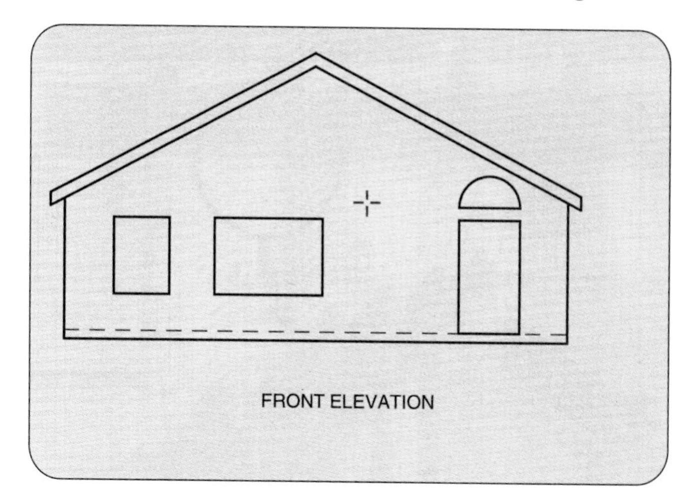

R. Add details with the Line, Circle, and Text commands. (Trim, porch landing, chimney, attic vent, scale, etc.)

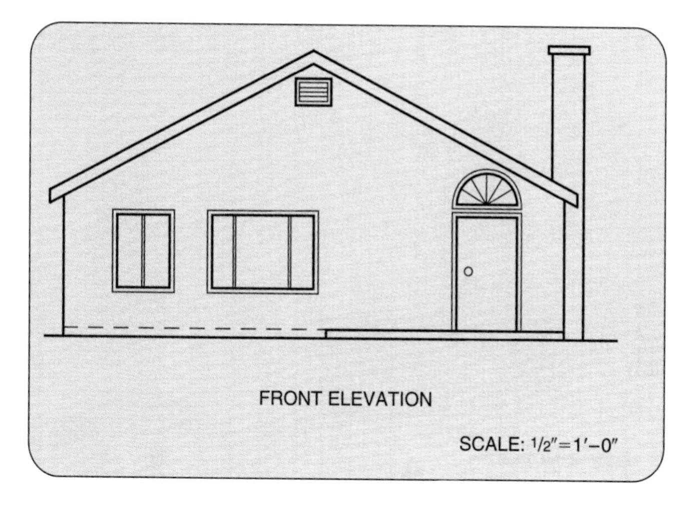

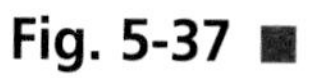

Fig. 5-37 ■

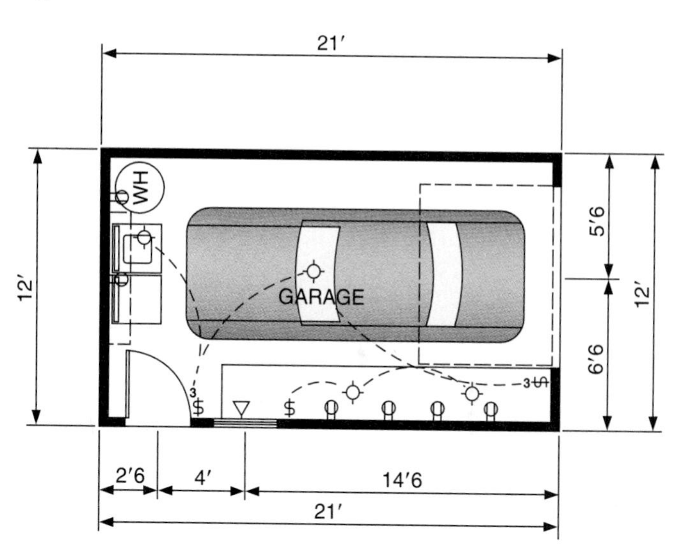

CHAPTER 5 Exercises

1. List the hardware components in a typical CAD system.
2. List the advantages of using a CAD system instead of manual drafting.
3. Explain why the basic principles and practices in architectural drafting must be learned in addition to CAD operations.
4. With a CAD system, redesign the garage shown in Fig. 5-37.
5. Use a CAD system to recreate the details shown in Fig. 5-38, page 112.
6. With a CAD system, redesign and draw the floor plan shown in Fig. 5-39.
7. Develop the CAD drawing shown in Fig. 5-40 using a digitizing tablet for input.
8. Digitize the points in Fig. 5-41.

Fig. 5-38 ■ Exercise 5, page 111.

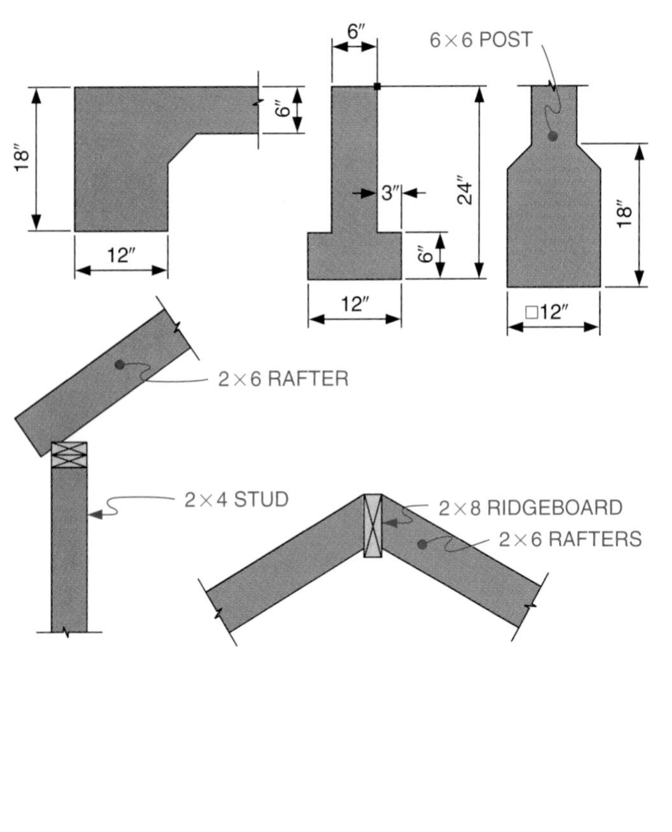

Fig. 5-39 ■ Exercise 6.

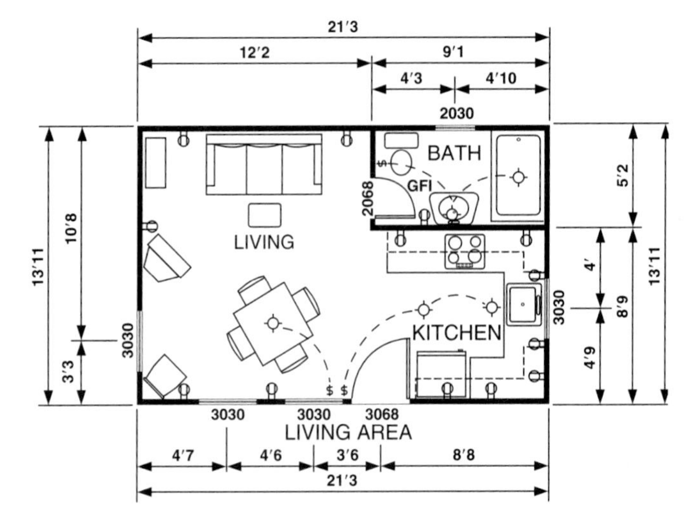

Fig. 5-40 ■ Exercise 7.

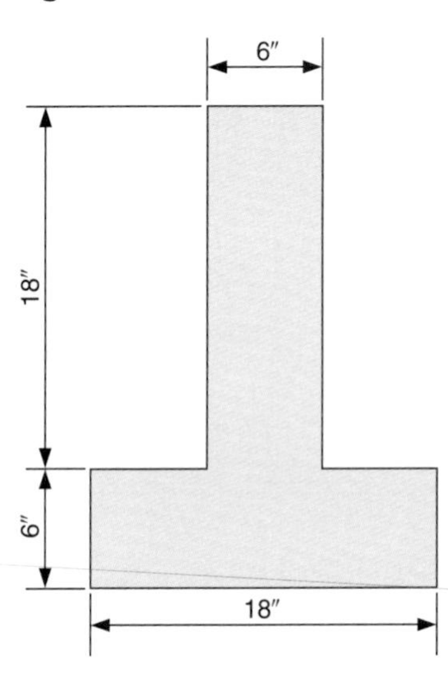

Fig. 5-41 ■ Exercise 8.

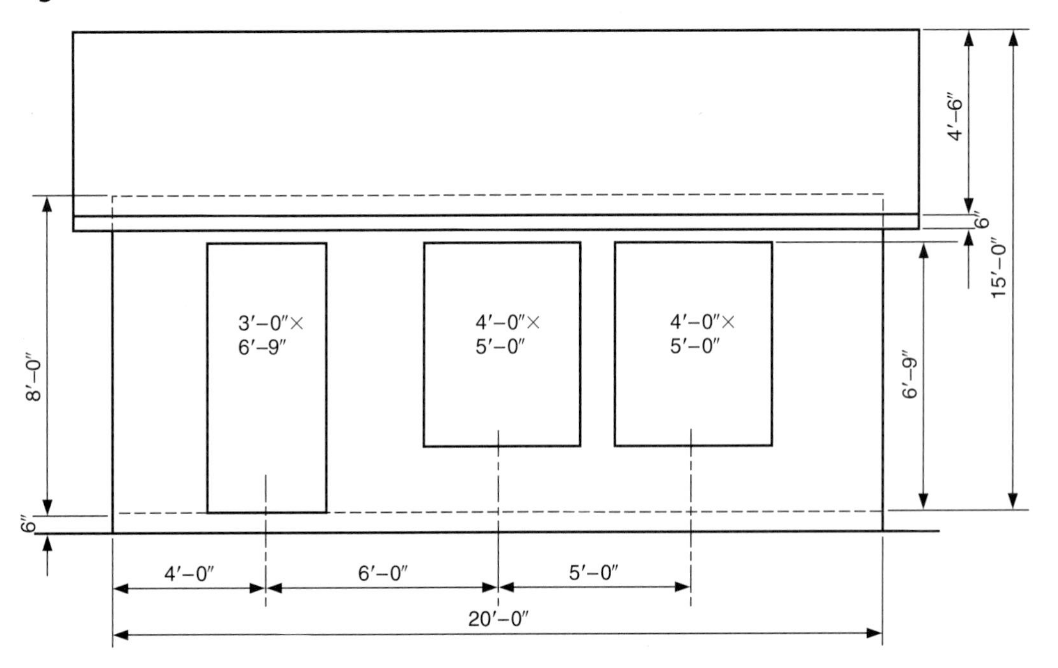

Architectural Drafter

Architectural drafters prepare all types of architectural drawings and documents. They must be able to follow instructions carefully and work as a member of a team. An architectural drafter may become an architect by getting a college degree in architecture, by gaining architectural drafting and design experience, by obtaining letters of recommendation, and by passing a state examination.

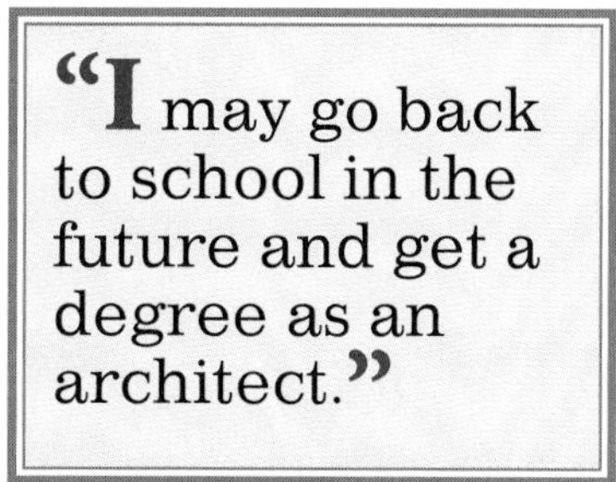

"I may go back to school in the future and get a degree as an architect."

Cassidy Loesch

Before her 21st birthday, Cassidy Loesch had realized her long-time dream of working as a designer. Architects typically undergo a five-year undergraduate program. Cassidy has a two-year Associate Degree in Applied Arts in Architectural Drafting from the Northwest Technical Institute in Eden Prairie, Minnesota. "I chose to go this way because I wanted to start getting work experience right away," says Cassidy. "With an Associate Degree, I can draw practically everything that architects draw, particularly residential drawings that don't require an engineer's seal or an architect's seal."

Cassidy enjoys her job because she is able to use her talents and skills to help other people. Cassidy works for LifeStyle Home Design Services, sister company to Homestyles Publishing and Marketing which publishes home plans. "Our company was formed from a need to modify the plans for specific customer needs," Cassidy explains. "Not only do I do the drafting both on the board and on CAD, but I also work with customers on the phone. I bid and estimate jobs, work out changes with the customers, and help them with design concepts." She also assists less-experienced drafters as they learn to work with software and try to cope with residential design and drafting challenges.

Cassidy is working as an intern towards certification as a designer through a professional association called the American Institute of Building Design (AIBD). After completing internship, she must take a test to become certified. The test is an eighteen-hour test, spread over two days. It involves not only answering questions, but drafting as well.

In following her dream, Cassidy set a goal and worked to achieve it. She has found that working in architectural drafting is not only satisfying to her, but may also help prepare her for other career opportunities in architecture.

Photo by Mark Englund

PART 3

Basic Area Design

CHAPTER

6 Environmental Design Factors
7 Indoor Living Areas
8 Outdoor Living Areas
9 Traffic Areas and Patterns
10 Kitchens
11 General Service Areas
12 Sleeping Areas

CHAPTER

6

Environmental Design Factors

Objectives

In this section you will learn:

- to orient a house on a lot to take best advantage of solar energy and features of the lot.
- to design structures ergonomically.
- ways to prevent pollution (ecology).

Terms

active solar design
earth-sheltered homes
ecological planning
ergonomics
orientation
passive solar systems

A wide range of factors must be considered to develop a fully functional architectural design—from a building's geographical area to the dimensions of an average adult. The designer must carefully study such environmental factors as the climate, the land, and energy sources. A building's design is also influenced by the needs of persons who will occupy the building. Considering these factors, the designer should also attempt to protect and improve the environment.

■ **Buildings can be designed and oriented to take advantage of heat and light from the sun.** *Mark Romine*

Orientation

A building must be positioned to maximize desirable features and minimize the negative aspects of the environment. This is accomplished through effective **orientation**. A building's orientation is its relationship to its environment.

Energy Orientation and Sources

Local resources and climatic conditions have always affected uses of energy—heating, cooling, and lighting. For example, early Native Americans built adobe houses under overhanging cliffs. The cliffs provided protection from the hot sun during summer and the cool wind during winter. See Fig. 6-1. The adobe material in these houses absorbed heat during the day and released it at night to warm the area. People soon found they could use other natural resources for heat. By burning renewable fuel such as wood, twisted grass, or blubber, people became less dependent on architectural designs and materials for protection from harsh environmental conditions. Later, when inexpensive fossil fuels (e.g., coal) appeared to be an endless energy source, people began to rely almost exclusively on those. Designers controlled inside environments artificially.

Fig. 6-1 ■ Early use of natural heating and cooling resources.

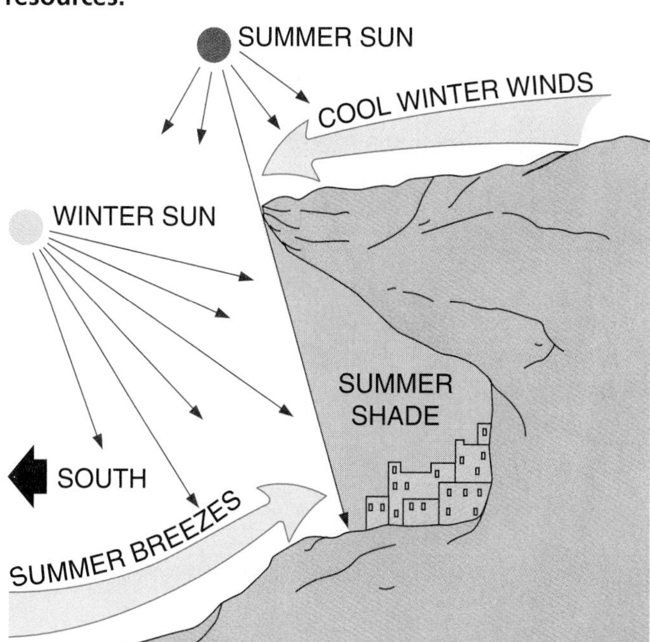

Today we know that the supply of fossil fuels is finite. They could soon be used up. There is a need to apply energy-efficient principles in building design to control indoor environments.

Energy must be obtained from sources and methods besides the burning of fossil fuels. Solar, wind, hydroelectric, and nuclear energy are available for use today. The combustion of natural materials such as reclaimed waste, wood, and other organic materials can also be used. Except for energy from the sun, all of these sources require special equipment to effectively heat and cool buildings. Mechanical or electrical devices can be added to control the sun's energy. Such systems are called **active solar design** systems. However, carefully designed buildings can use the power of the sun without mechanical devices and still provide much environmental control. These systems are called **passive solar systems.** They use only the design features and orientation of a building to gain and control the sun's energy. (Detailed discussions of both types of solar systems are presented in Chapter 32.)

Solar Orientation

The first step in designing the orientation of a building is to consider its relationship to the sun. Because the earth's axis is tilted, the angle of sunlight changes from summer to winter as the earth revolves around the sun. See Fig. 6-2. In the northern hemisphere, the south and west sides of a structure are warmer than

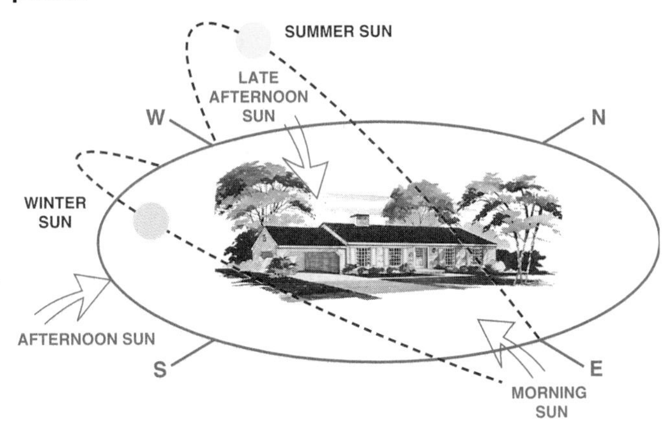

Fig. 6-2 ■ Summer sun paths compared to winter paths.

the east and north sides. The south side of a building is the warmest side, because of the nearly constant exposure to the sun during its periods of intense radiation. Ideally, a building should be oriented and windows positioned to absorb this southern-exposure heat in winter and to repel the excess heat in summer. See Fig. 6-3.

A structure needs to be located and oriented to ensure that the areas requiring the most solar exposure will be correctly positioned in reference to the sun. Keep

Fig. 6-3 ■ Effective solar planning for summer and winter.

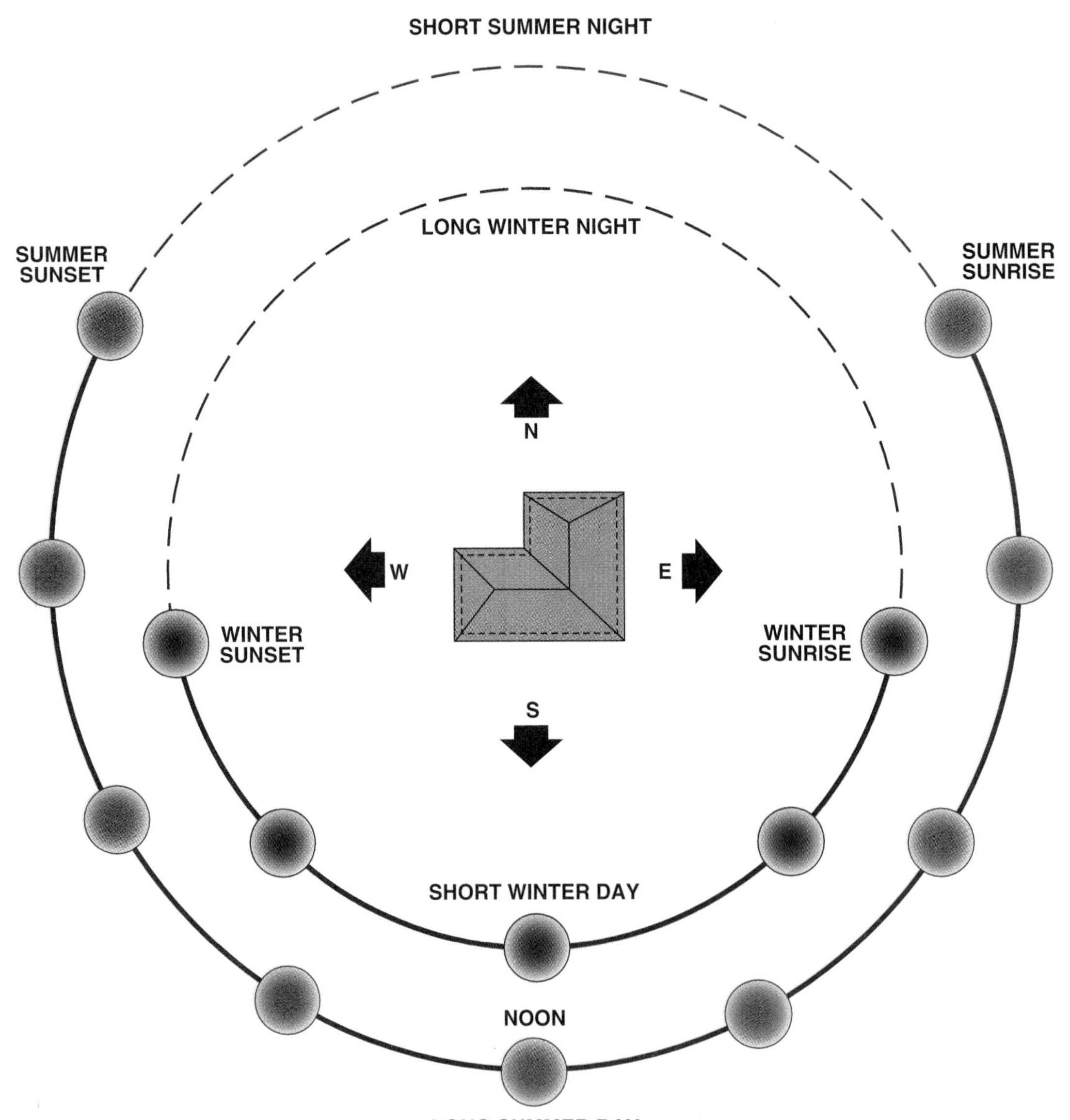

in mind that effective solar orientation should not only provide the greatest heating or cooling effect, but should also be planned to provide the greatest amount of natural sunlight where needed. For example, designers should consider the placement of windows and skylights as a source of both heat and light. See Fig. 6-4.

Walls, floors, and furniture can absorb and store heat from the sun. This heat is naturally released when the temperature is cooler or at night. Open interiors and high ceilings encourage ventilation and cooler temperatures. Low ceilings and closed floor plans tend to increase temperatures.

Fig. 6-4 ■ Skylights can be used effectively in a passive solar system. *Potlach*

ROOM AND OUTDOOR AREA LOCATIONS

Rooms should be located to absorb the heat of the sun or to be baffled (shielded) from the heat of the sun. Consider not only the function of the room, but the seasons and the time of day the room will be used most. The location of each room should also make maximum use of the light from the sun. See Fig. 6-5.

Generally, sunshine should be available in the kitchen and dining areas during the early morning and should reach the living room by afternoon. To accomplish this, kitchen and dining areas should be placed on the south or east side of the house. Living room areas are placed on the south or west side to receive the late-day rays of the sun, when the room is most likely to be used. The north side is the most appropriate side for placing sleeping areas. It provides the greatest darkness in the morning and evening and is also the coolest side. Northern light is also consistent and diffused and has little glare.

The same principles apply to planning outdoor living areas. Those areas that require sun in the morning should be located on the east side. Those requiring the sun in the evening should be placed on the west side. For both indoor and outdoor areas, remember that in the northern hemisphere, the north sides of buildings receive no direct sunlight.

OVERHANG AND BAFFLE PROTECTION Roof overhangs and baffles (shields) should be designed to allow the maximum amount of sunlight and heat to penetrate the inside of a building in winter. Conversely, the maximum amount of sun and heat should be shielded from entering the interior during a summer midday.

Since the angle of the sun differs in summer and in winter, roof overhangs can be designed with a length, height, and angle that will shade windows in summer and allow the sun to enter during the winter. See Fig. 6-6. The edge of the overhang also needs to be related to the height of the window. See Fig. 6-7. Figure 6-8 shows several effective overhang baffling systems. Baffling should be accomplished without blocking out natural light.

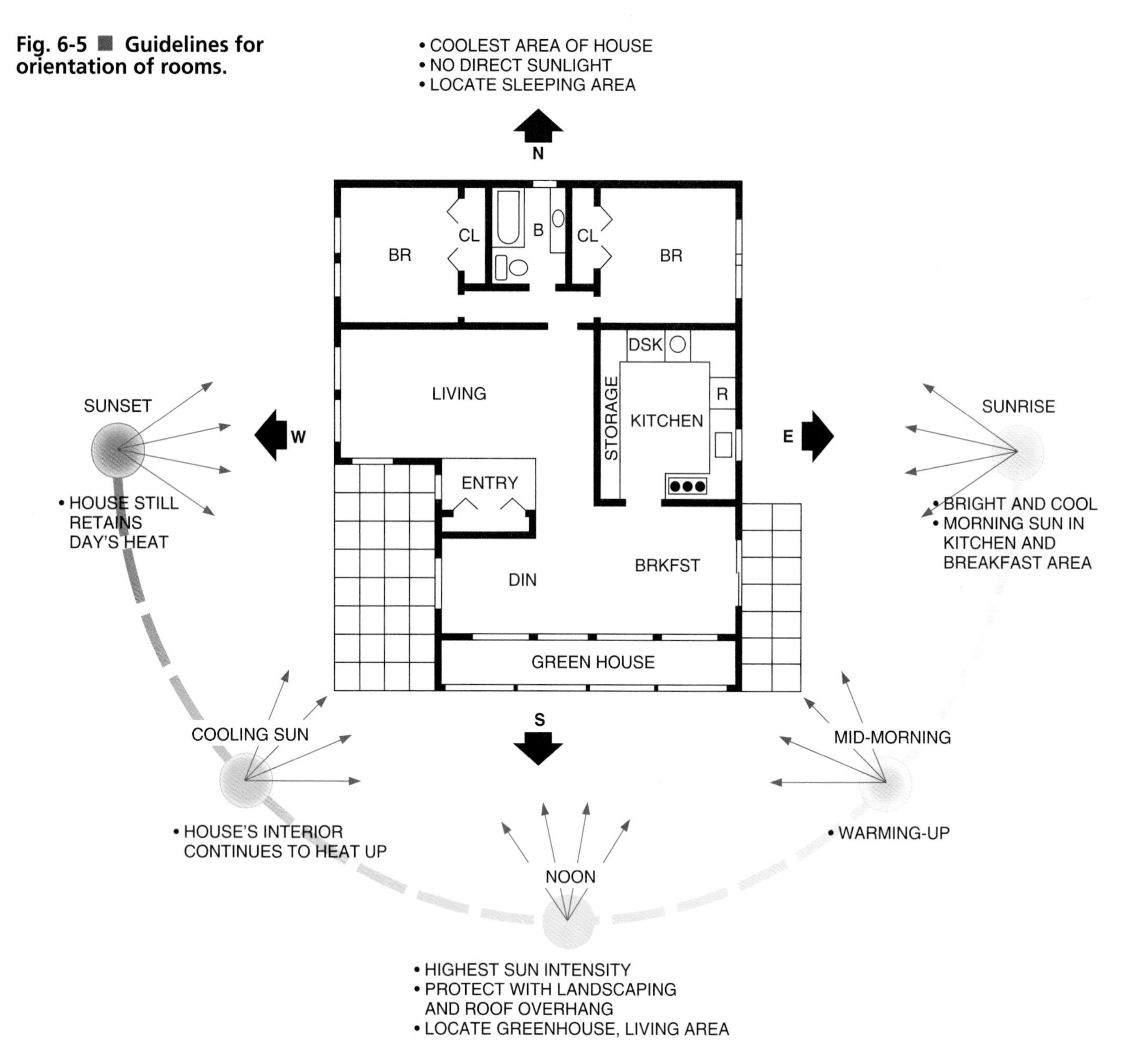

Fig. 6-5 ■ Guidelines for orientation of rooms.

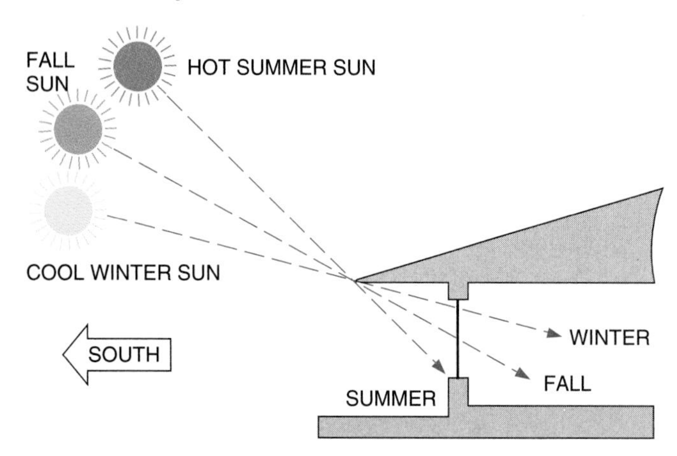

Fig. 6-6 ■ Use of overhang designed to control seasonal sun exposure.

These considerations apply primarily to the *south* side. The morning sun (east) and the late afternoon sun (west) will not be as affected by the sun, and, as you know, the north side receives no direct sunlight.

Land and a Structure

A particular plan may be compatible with one site and yet appear totally out of place in another location. The success of a design depends on how well the structure is integrated with its surroundings.

Fig. 6-7 ■ Calculating roof overhang based on latitude and window height.

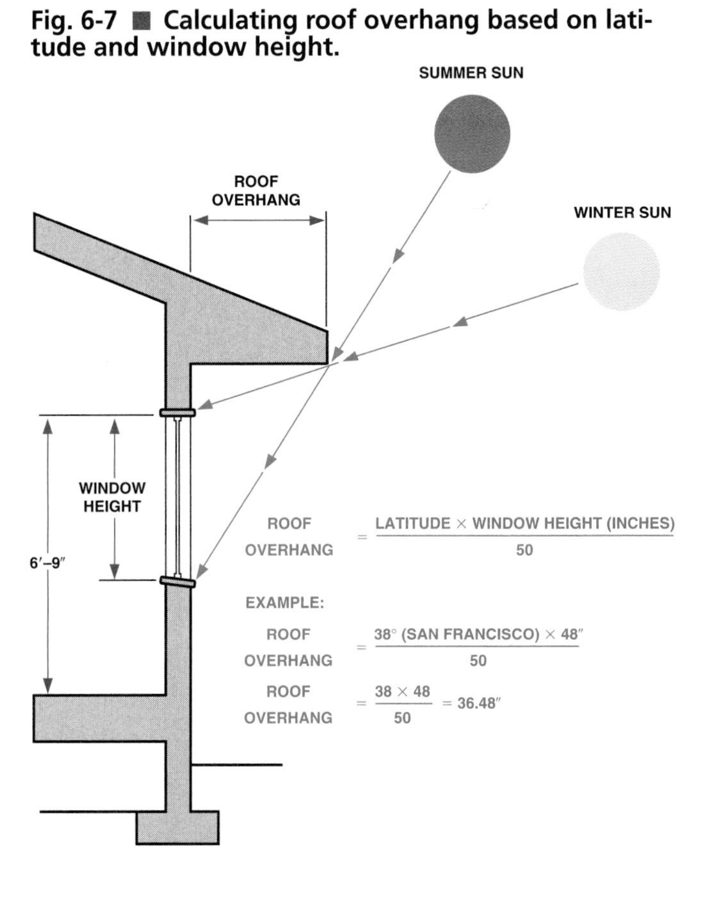

CHARACTERISTICS OF A SITE Every structure should be designed as an integral part of the site, regardless of the shape or size of the terrain (land surface). Buildings should not appear as appendages ("add-ons") to the land but as a functional part of the landscape. For the indoor and outdoor areas to effectively function as parts of the same plan, the building and the site must be designed together.

Before orienting and designing a structure, consider the specific physical characteristics of the site, such as hills, valleys, fences, other buildings, and trees. Physical features such as these may affect wind patterns and the amount and direction of available sunlight in different seasons.

Large bodies of water may also affect air temperature and air movements. Also, surrounding pavement areas and buildings can raise or lower temperatures, because concrete and asphalt collect and store the sun's heat.

Fig. 6-8 ■ Types of overhead sun baffling.

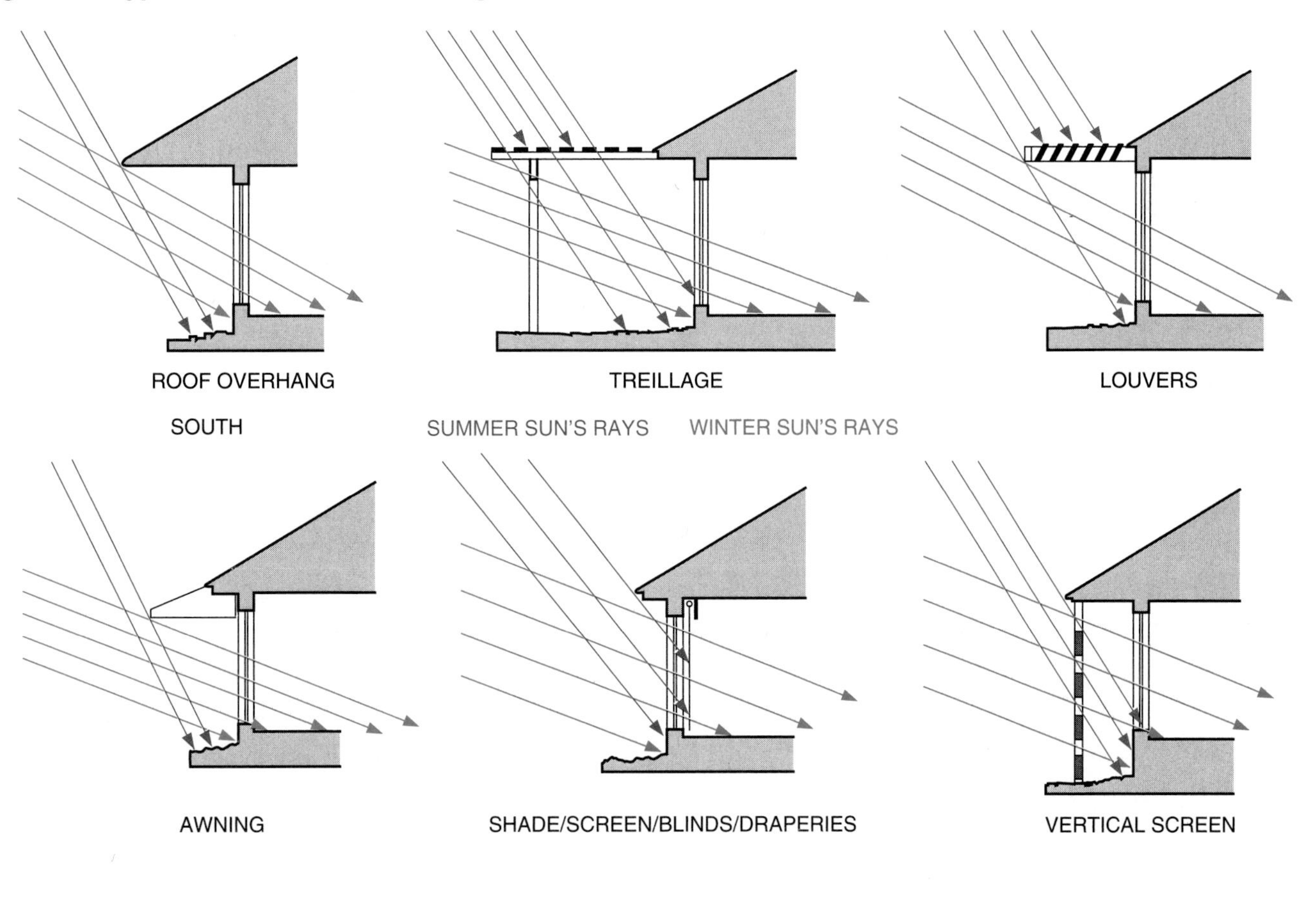

Fig. 6-9 ■ This house is oriented to take advantage of spectacular sunset views.

Consider the view options in orienting the house. Orientation of specific areas toward the best view, or away from an objectionable view, usually means careful planning of the position of the various areas of the building. See Fig. 6-9.

LOT AREAS Building sites are sold and registered as *lots*. The size and shape of a lot affects the flexibility of choice in locating structures. For planning purposes, lots are divided into three areas according to their function: public, private, and service areas. See Fig. 6-10. The placement of all structures on a lot determines the relative size and relationship of the three areas. Remember, the features of the lot should be an integral part of the total design. The site design is as important as the basic floor plan design of a structure.

EARTH-SHELTERED HOMES Characteristics of the land (site) and soil conditions are of utmost importance in the design of **earth-sheltered homes**. These types of homes are designed to be partially covered with earth. See Fig. 6-11.

The thought of living partly underground may at first seem oppressive.

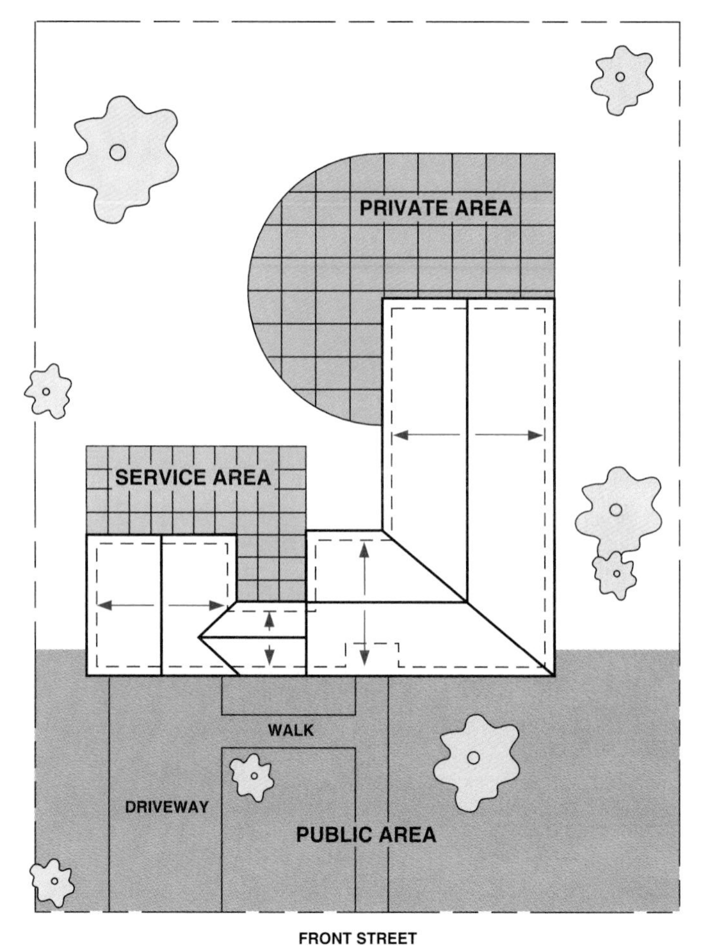

Fig. 6-10 ■ Areas of a lot.

However, with careful planning and proper orientation, adequate light can be achieved.

There are several advantages to earth-sheltered homes. Regardless of how high or low the outside temperature is, the soil just a short distance below the surface remains at a comfortable and constant temperature. Heating and cooling units can be smaller and are used less often than those in conventional structures. The underground location also avoids the problems of wind resistance and winter storm winds.

Construction costs for earth-sheltered homes can be less than those for conventional types of construction, if experienced builders are employed. The major concern in building earth-sheltered

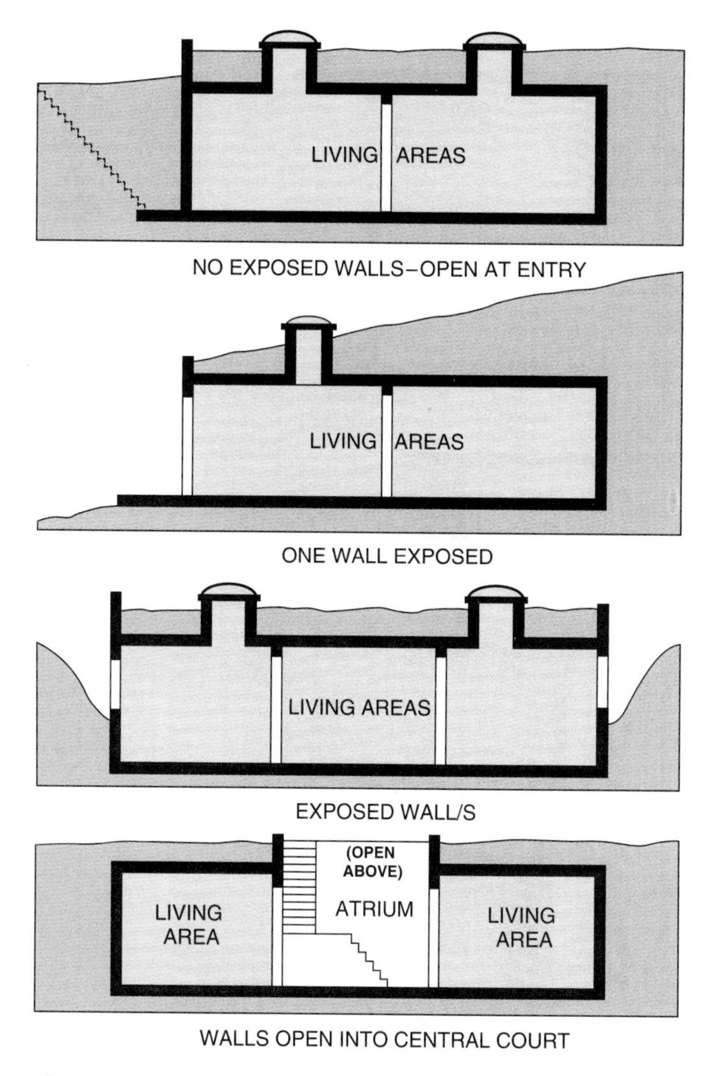

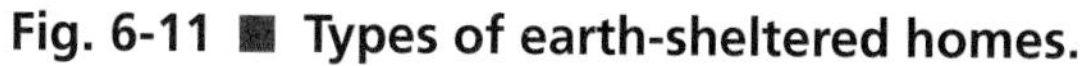
Fig. 6-11 ■ Types of earth-sheltered homes.

homes is waterproofing, but this can be accomplished with waterproof paint, sealants, membrane blankets, and proper drainage. The structure of an earth-sheltered home also needs to be heavier than that of conventional buildings, to support the heavy soil loads.

Earth-sheltered structures generally require little maintenance. Groundwater accounts for most maintenance problems. However, with careful study of drainage patterns during the planning phase and effective waterproofing during construction, such problems can be avoided.

Soil is an important consideration for designers of earth-sheltered buildings.

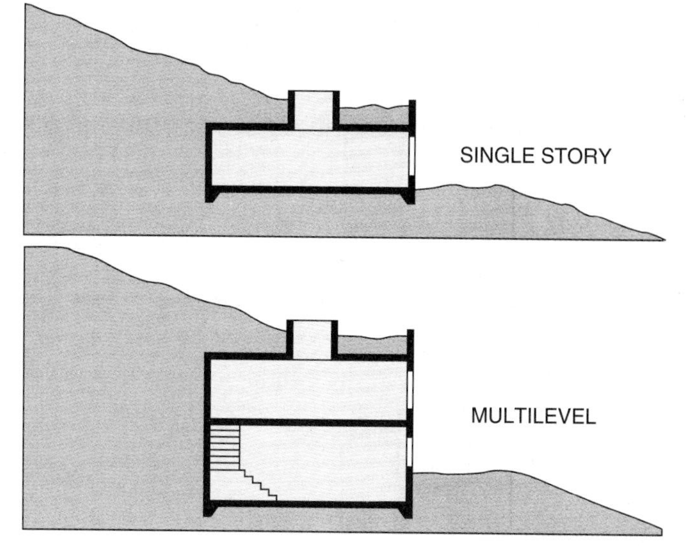

Fig. 6-12 ■ Either single-story or multilevel earth-sheltered homes can be built into a sloping site.

Soils used to cover roofs and walls must be carefully selected to avoid frost heave, excessive swelling, and runoff tendencies. Soil that supports vegetation (organic soil) must be used for surface areas.

The best site location for an earth-sheltered home is on a gentle downward slope. See Fig. 6-12. The exposed walls should make maximum use of glass to capture as much light and heat as possible. This requires a southern exposure. Windowed areas should also be oriented away from prevailing winds but should face the best possible view.

Rooms requiring the most light, such as the living room, should be located in windowed areas. Seldom-used rooms or rooms not requiring windows can be located in ground-locked areas as shown in Fig. 6-13. Skylights can be used in otherwise dark areas if they are effectively sealed and drained.

Although sloping lots are best, earth-sheltered designs can be placed on relatively level lots. Mounds of earth called *berms* can be created to provide protection.

Fig. 6-13 ■ Floor plan and section of an earth-sheltered home. *Home Planners, Inc.*

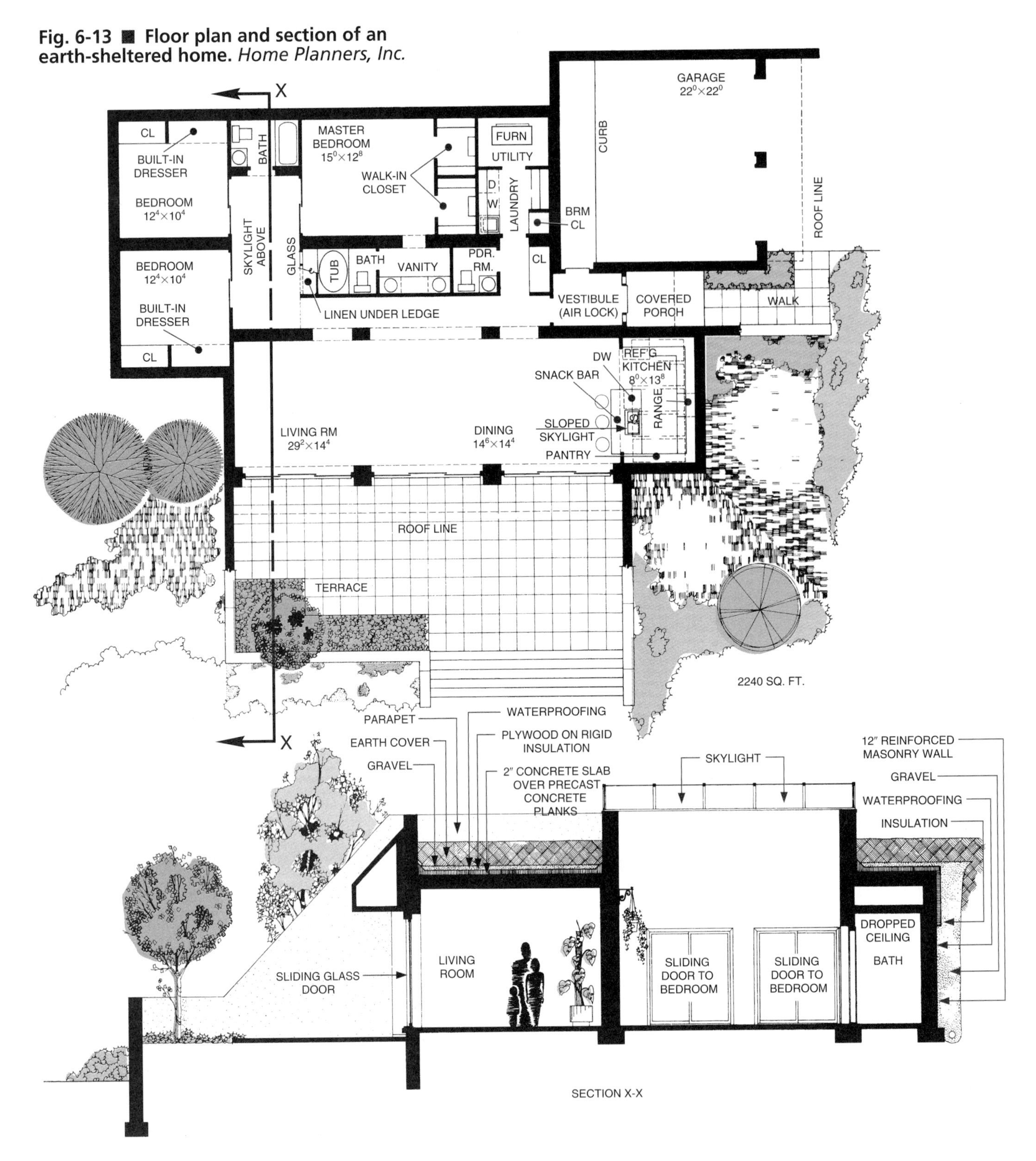

VEGETATION Vegetation of all types greatly aids in heat, light, wind, humidity, and noise control. Trees, shrubs, and ground-cover foliage, when effectively used, can also baffle undesirable views and enhance attractive scenes.

Deciduous trees maximize summer cooling and winter heating. They provide shade in the summer and then lose their leaves in winter which allows the sun's warmth to penetrate the building. See Fig. 6-14. Dense, coniferous (evergreen) trees

Fig. 6-14 ■ Deciduous trees help maximize summer cooling and winter solar heating.

and shrubs are most effective for blocking or redirecting north or northwest storm winds to help protect a building during all seasons.

Vegetation is an important design consideration. However, it should not be used as a substitute for appropriate orientation design to control a building's environment.

Wind Control

One of the functions of effective orientation is wind control. While the sun can provide natural energy to a structure, wind can easily diminish the sun's effect. Outdoor living can also be seriously curtailed by excessive wind.

Existing indoor heat can be lost very rapidly when cold air is forced into buildings through minute crevices, usually around doors and windows. Heat also escapes by *windchill* loss through walls, windows, roofs, and foundations. The windchill effect is the loss of internal stored heat. Building orientation, vegetation, and the features of the land can be used to control or minimize the effects of prevailing winds.

WIND PATTERNS Air movements at a site should be studied during different parts of the day and during different seasons. Once wind patterns are known, buildings should be oriented to take full advantage of (or offer full protection from) the cooling effect of wind directions.

Some wind patterns are relatively common. Desirable summer breezes usually flow from one direction and winter winds from the opposite direction. Also, cool air will always move to replace rising warm air. For this reason, air above a body of water usually moves toward land during the day and from land toward water at night. See Fig. 6-15. The wind pattern on southern sloping sites is generally a movement up the slope during the day and down the slope at night. See Fig. 6-16. Strong prevailing winds can change these movements, however.

PROTECTIVE MEASURES Gentle breezes are usually desirable, but harsh winds are not. Protection from wind can be provided by locating buildings in sheltered valleys or opposite the windward side of hills. Existing wooded areas can have a baffling effect on wind. See Fig. 6-17. If no wooded areas exist, or if vegetation is young or not

Fig. 6-15 ■ Effect of large bodies of water on air movement.

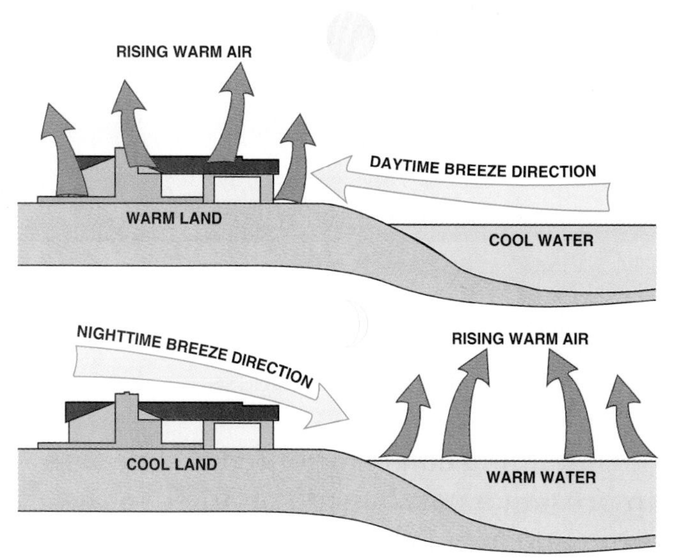

Fig. 6-16 ■ Movement of air on a southern sloping site.

A. Uphill during the day.

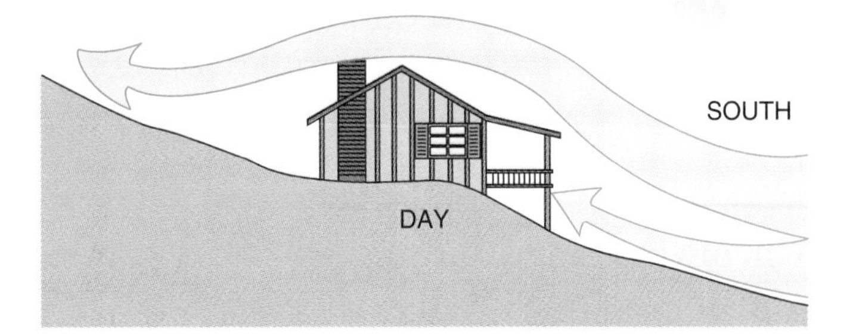

B. Downhill during the night.

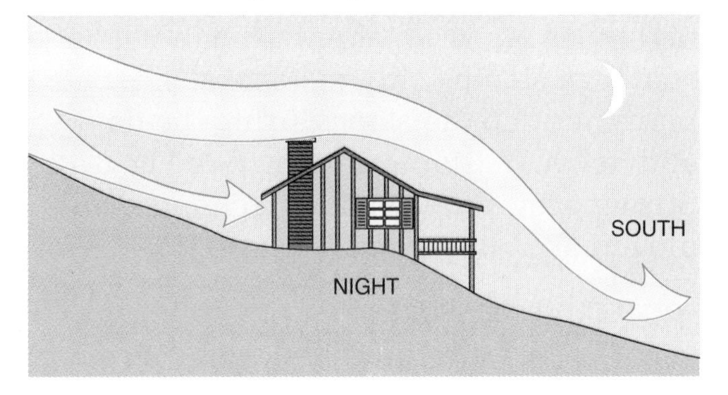

Fig. 6-17 ■ Wooded areas provide an effective baffle from wind.

Fig. 6-18 ■ Use of walls and fences to deflect wind around a home.

A. Detached.

WIND WIND

B. Attached.

WIND CALM WIND CALM WIND

available, construction baffles such as fences or walls may be necessary. See Fig. 6-18.

A building can be oriented at an angle to present a narrow side or angle to the wind and avoid direct, right-angle wind impact. See Fig. 6-19. Low roof angles can also help deflect wind over a structure.

WIND EFFECTS In planning urban buildings or isolated clusters of buildings, care must be taken to avoid the creation of turbulent wind eddies. This effect is caused by high-velocity winds striking the upper floors of high-rise buildings and being forced downward and back against lower buildings, creating turbulence on the surface.

Fig. 6-19 ■ Use of building position to reduce wind effect.

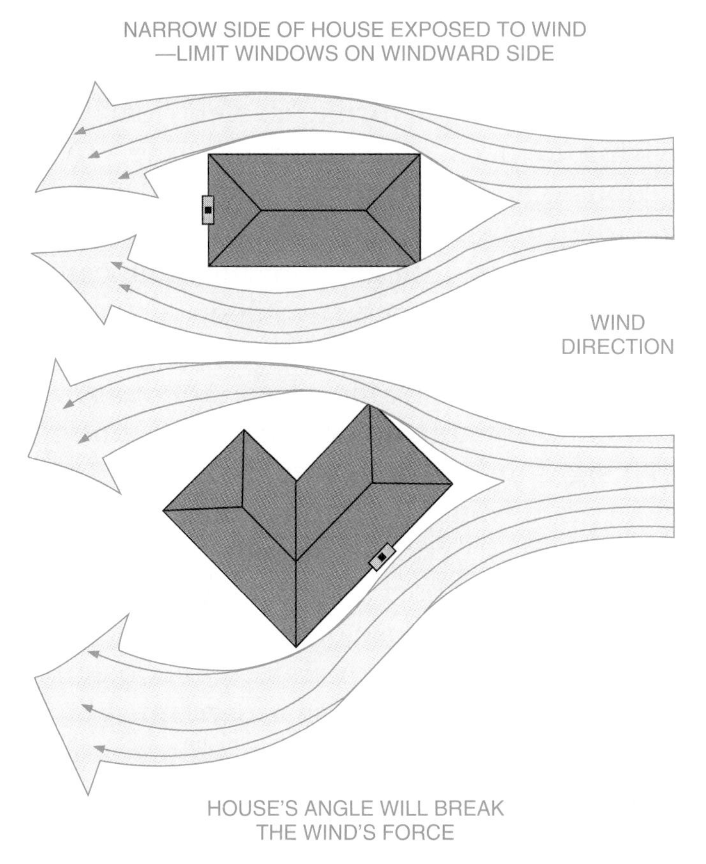

Fig. 6-20 ■ Structures can affect wind currents.

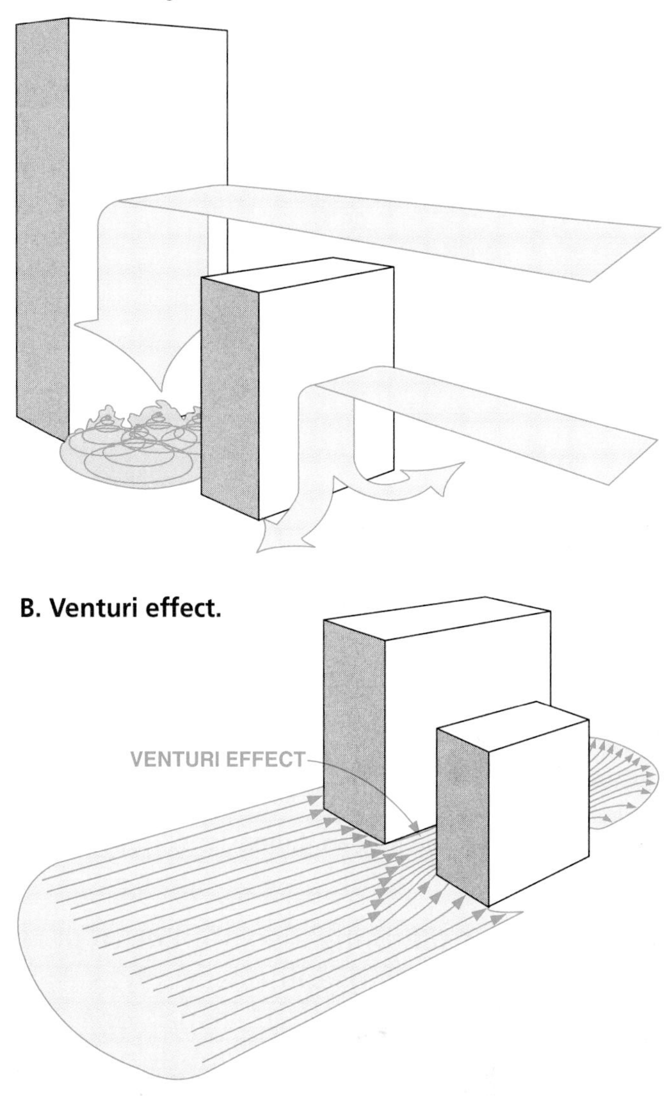

A. Wind eddy effect.

B. Venturi effect.

See Fig. 6-20A. This effect could also be caused on a smaller scale by winds trapped in courtyards and patios.

In urban situations, the *venturi effect* also complicates wind control. The venturi (wind-tunnel) effect is created as large amounts of moving air are forced into narrow openings. The reduced area through which the wind must pass creates a partial vacuum, and the air picks up speed as it is pulled through the opening. See Fig. 6-20B. The venturi effect can be partially controlled by avoiding the alignment of streets or buildings with the direction of prevailing winds.

Ergonomic Planning

Buildings are for people. Therefore, buildings must be ergonomically (biotechnically) planned. **Ergonomics** is a science that deals with designing and arranging things that people use. In architectural design, this means the design must match the size, shape, reach, and mobility of all residents. See Fig. 6-21.

Human Dimensions

Human dimensions are especially critical in planning the size and position of cabinets, shelves, and work counters. Traffic areas, door openings, and windows are all based upon human dimensions. When buildings (such as schools) are designed primarily for children, obviously the scale must be adjusted.

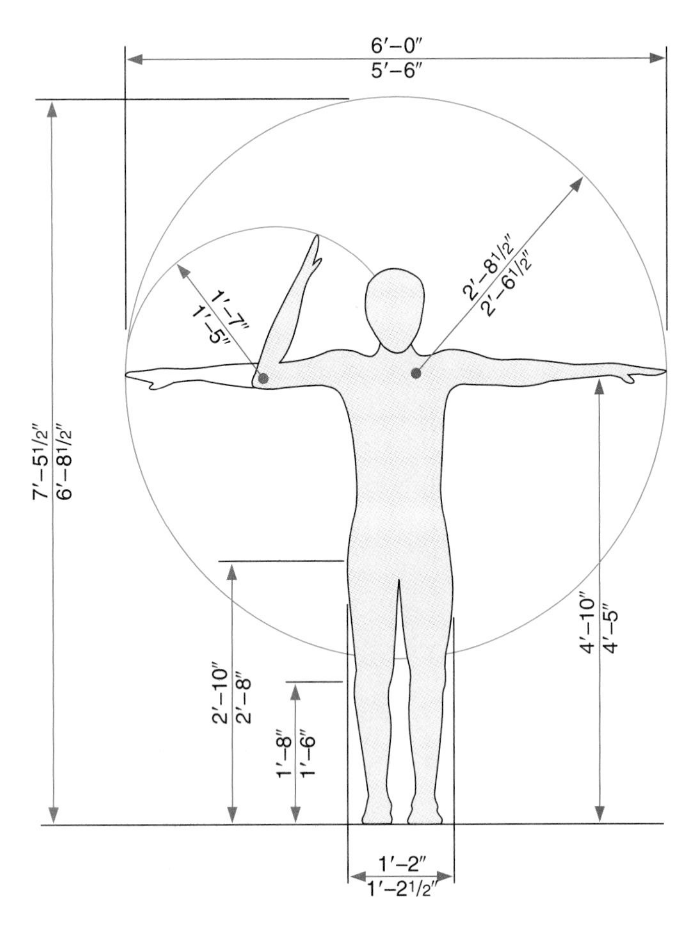

Fig. 6-21 ■ Average human dimensions.

Ergonomic planning also applies to persons with physical disabilities, such as those with hearing, speech, visual, or mobility impairment. For example, door openings need to be wider and countertops lower for convenient use by persons in wheelchairs. (Detailed design requirements for persons with disabilities are provided in Chapter 13 of this text.)

Safety Factors

Safety factors in design cover a vast array of concerns for any person who occupies a building, whether it be a home or a public or commercial structure. The designer must ensure that no design feature creates a health or safety risk, such as hidden steps or low headroom clearances. Air for an environmentally safe building should be electronically filtered to provide for the elimination of harmful pollutants. Safety precautions include specifying appropriate mechanical equipment, such as gas furnaces, electrical wiring, devices, and machinery. Hazardous materials (such as asbestos) and accident-causing materials, such as extra smooth floors, thin glass, or unstable ceiling coverings, must be avoided.

Design also involves the safe arrangement of outdoor traffic areas. Adequate vehicular turning angles, fire lanes, and exit signs are a few examples. Of course, safety in design also implies that a building will be structurally sound and will adequately support all anticipated weight.

The increasing use of technology is creating another concern. Before the design process is begun, an assessment must be made to determine what special technology needs are anticipated. Automated buildings (called "smart buildings") now contain such built-in electronic features and accommodations as TV cables, high- and low-voltage circuits, and computer equipment. Magnetic or radio-wave interferences may need to be blocked to enable sensitive electronic equipment to operate effectively. Distance between persons and electromagnetic forces (EMF) should be considered for safety.

Ecology

In the past one hundred years, the number of people on the earth has more than tripled. This population increase, combined with the shift from an agrarian (farming) society to an industrial one, has led to the creation of environmental problems previously unknown. The ever-increasing material needs of our technological economy have created enormous pollution problems that must be solved if humanity is to survive. Designers must plan in ways that eliminate or reduce pollutants.

A prime requirement in the creation of every design is to preserve our supply of clean air, pure water, and fertile land. The contemporary architect or designer must be sure that structures do not interfere with natural ecological balances. Good **ecological planning** means protecting or improving the environment without sacrificing the qualities of good design. It requires knowledge of the problems of pollution and possible solutions.

Land Pollution

Land is polluted by the discharge of solid and liquid wastes on land surfaces. Pollutants originate from industrial, agricultural, and residential waste. When they exist in excessive quantities, pollutants create health hazards, contribute to soil erosion, cause unpleasant odors, and overwork sewage-treatment plants. Another form of land pollution is the removal of topsoil, vegetation, or trees from large tracts of land. These pollution problems *can* be avoided. See Fig. 6-22.

Architecturally related ways of reducing land pollution include recycling waste material, compacting (condensing) waste material to reduce volume, providing for sanitary landfills, and providing for minimum removal of vegetation, especially trees. See Fig. 6-23. Figure 6-24 lists several land pollution problems and solutions.

Sanitary landfill is a practice that involves placing solid waste material in

Fig. 6-22 ■ Good planning can help avoid land pollution problems. This site plan is designed to protect and preserve the environment. *Norwood East Hill Tract—Environetics*

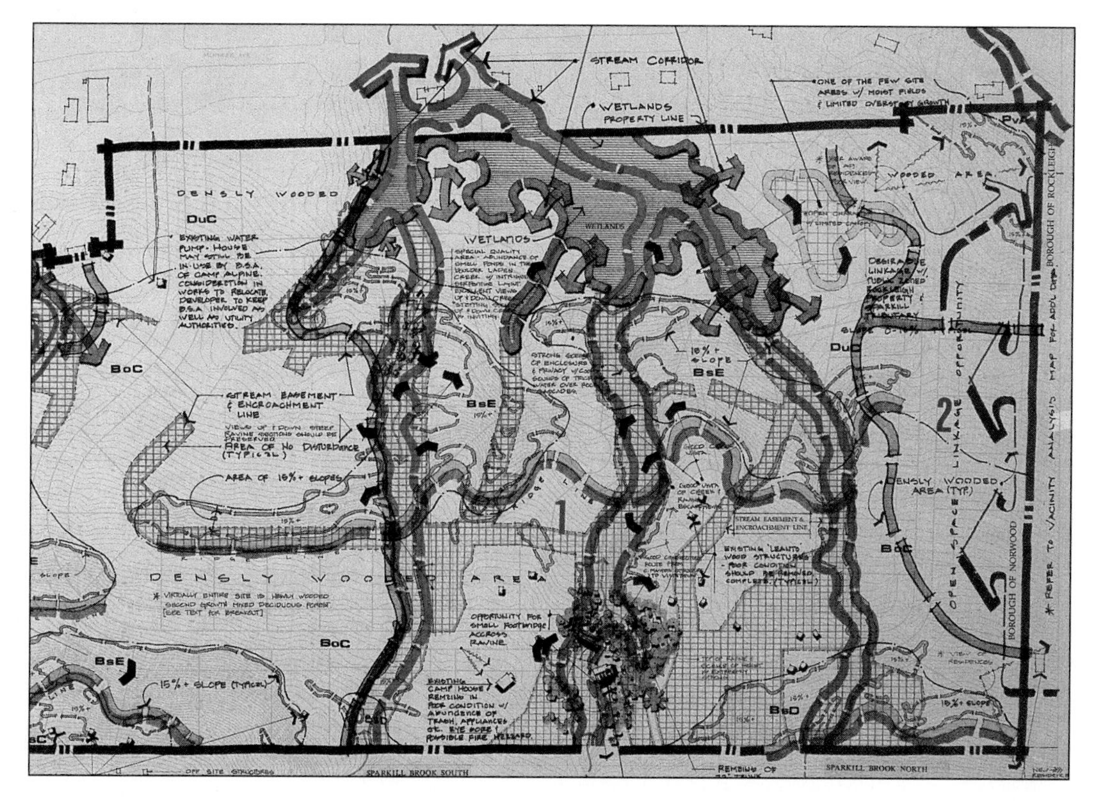

Fig. 6-23 ■ Residence integrated into a site, with natural vegetation preserved.

low-lying areas, compacting it into layers (about 10′ thick), and covering it with clean soil. Thorough landfill projects also include the planting of ground cover—trees and shrubs—on the filled area. Such practice restores the land to its original condition (or better), both ecologically and aesthetically.

Air Pollution

The discharge of industrial wastes is a cause of air pollution. Some pollutants are nontoxic and are not suspended in the air for long periods of time. They present an annoyance but usually do not endanger life. Other materials, such as sulfur oxides and organic gases composed of hydrocarbons, oxides of nitrogen, and carbon monoxide are extremely dangerous to both animal and plant life. These pollutants are emitted primarily as by-products of manufacturing plants, heating devices, and the exhaust systems of vehicles.

To avoid ecological problems caused by vehicles, designers can reduce pollution by planning heavy traffic patterns away

Fig. 6-24 Land pollution problems and solutions.

Causes	Effects	Cures or Prevention
Dumping industrial wastes	Poisons introduced into human and animal life cycles	Recycle materials
Dumping sewage	Soil erosion	Stop overpackaging of products
Using agricultural insecticides and pesticides	Loss of topsoil	Repair products—stop throwaway
Dumping residential waste	Overtaxed sewer treatment plants	Use biodegradable products
Landscape removal	Additional costs to taxpayers	Use natural, untreated building materials
Overgrazing	Loss of natural environments	Improve landscape planning
Overdevelopment		Eliminate manufacture of disposable items
Dumping nuclear wastes		Curtail use of insecticides and pesticides
Industrial use of toxic materials		Reserve land for open spaces
Destruction of natural settings—rivers, mountains, valleys		Preserve oxygen-producing trees and vegetation
		Tree-planting programs
		Engineered landfills
		Stop dumping toxic waste

from densely populated areas. Designers must also plan or specify electronic air filters or other waste removal systems to eliminate particles before they become airborne.

Designs must include features that promote energy conservation. Conserving energy not only diminishes pollution by decreasing fuel consumption that pollutes the air, but it also helps reduce the operating costs of buildings. See Fig. 6-25.

Water Pollution

Water pollution is caused mainly by the dumping of sewage, industrial chemicals, and agricultural wastes into oceans, lakes, rivers, and streams. These wastes include pathogens (such as disease-causing bacteria), unstable organic solids, mineral compounds, plant nutrients, and insecticides. Water pollution results in the destruction of marine life and presents very serious potential health hazards to animal and human life.

Fig. 6-25 Air pollution problems and solutions.

Causes	Effects	Cures or Prevention
Exhausts from manufacturing industries	Excessive energy consumption	Develop and use nonpolluting energy sources
Exhausts from vehicles	Crowded traffic	Reduce vehicular traffic
Exhausts from gas and oil heating	Serious health hazards	Create efficient vehicular traffic patterns
Removal of oxygen-producing plants	Offensive odors	Use fuel-efficient vehicles
Overuse of toxic materials in home, industry, and construction	Smog	Lower heating and raise cooling levels for buildings
Overuse of aerosol sprays	Reduction of outdoor visual ceiling	Add landscape plantings
Smoking tobacco		Curtail airborne wastes
Toxic fumes from building products		Improve mass transit
		Use natural building materials
		Eliminate burning of wastes
		Curtail tree-cutting
		Curtail use of toxic chemicals with deadly vapors
		Eliminate aerosol sprays
		Eliminate gas-engine gardening equipment
		Curtail toxic chemical spraying
		Develop solar and wind energy sources
		Use biologically sound pest controls
		Use solar heating
		Use sealed combustion chambers for heaters
		Control indoor humidity

Effective architectural planning can help reduce water pollution through the design of sewage-treatment systems with each new construction. Proper planning, so that water sources are not overused for fresh water or waste disposal, also helps reduce water pollution. Provisions must be made for the removal of industrial wastes, pesticides, and insecticides without sending them into waterways. Plans must also eliminate excessive runoff of topsoil into rivers and streams. See Fig. 6-26.

Visual Pollution

Many air, water, and land pollutants, such as unsanitary garbage dumps and smog-producing agents, are not only unhealthy but also visually undesirable. Other sources of visual pollution, such as junkyards, exposed utility lines, public litter, barren land, and large billboards, may not create health or safety hazards. They are aesthetically objectionable, however, and must be avoided in the architectural design process.

The best safeguard against visual pollution consists of following closely the basic principles of design, not only for structures but for the entire landscape. Also, laws related to building and the environment must be enacted and enforced. See Fig. 6-27.

Noise Pollution

Sound requires a path. Sound energy travels in all directions from a source. It weakens through distance or by the interruption of its path with a physical screen.

Fig. 6-26 Water pollution problems and solutions.

Causes	Effects	Cures or Prevention
Excessive sewage	Destruction of marine life	Eliminate dumping into bodies of water
Excessive industrial wastes	Serious health hazards to humans and animals	Curtail use of insecticides and pesticides
Excess use of agricultural insecticides and pesticides	Destruction of marine plants	Build more sewage treatment facilities
Watershed removal	Poor quality of water	Plan for less population congestion
Human waste products	Crop destruction	Water quality control
High population density		Stop all dumping near aquifer water supplies
Overuse of toxic chemicals in home, industry, and construction		Curtail use of toxic chemicals in all products
Dumping wastes into storm drains		Eliminate dumping of waste materials into sewer systems and storm drains
		Eliminate dumping of waste materials into flood-control areas
		Reuse gray-water (dishwater, etc.)

Fig. 6-27 Visual pollution problems and solutions.

Causes	Effects	Cures or Prevention
Poor municipal planning	Barren land	Long-range city planning
Poor building designs	Offensive-looking structures	Organic building designs
Exposed utility wires	Elimination of desirable views	Underground telephone, electrical, TV cable wires
Landscape removal	Visual confusion	Landscape preservation and planting programs
Site destruction	Lower property values	Limitation of public advertising spaces
Poor waste landfill design	Public safety hazards	Adequate trash receptacles
Excessive advertising signs	Psychological disorders	Community committee for structure approval
Littering		Graffiti-resistant surface covers
Crowded structures on building site		
Overbuilt areas		
Graffiti		

Sound levels are measured in *decibels*. Exposure to excessive decibel levels, over 80, or to even moderate levels, over 60, for long periods creates stress and can result in neurosis, irritability, and hearing loss. Excessive noise can also create hazardous environments by eliminating people's ability to identify and discriminate between sounds, especially those that warn of danger. See Fig. 6-28.

There are many architectural ways to reduce noise to acceptable levels. Among them are designing effective floor plans, landscaping to provide ample noise-buffer space, and using acoustical (soundproof) wall panels. Insulation also helps reduce excessive noise from the outside.

Rooms that require a quiet area, such as bedrooms, should be located on the side away from any major source of noise. Fabrics such as carpets, drapes, and upholstery help reduce noise by absorbing sound within a room. See Fig. 6-29.

Fig. 6-28 ■ Decibel levels of common sounds.

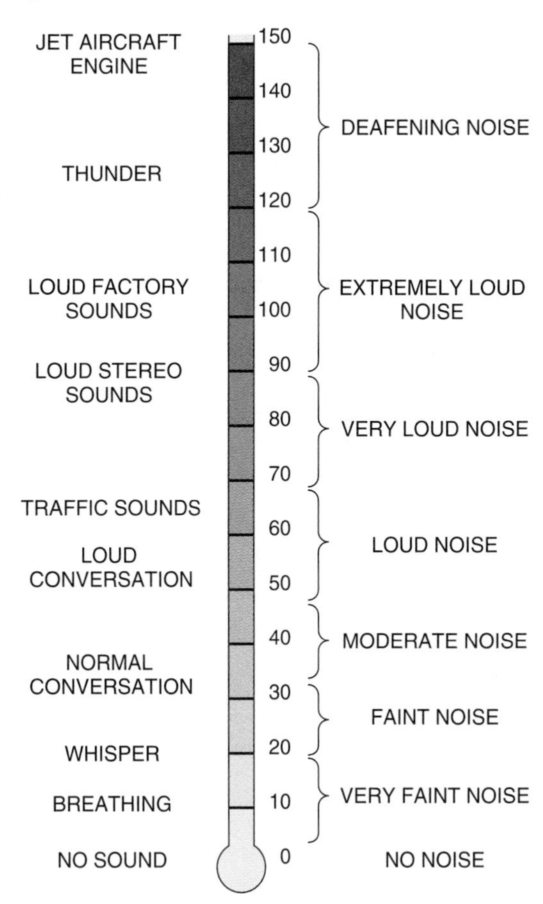

Fig. 6-29 Noise pollution problems and solutions.

Causes	Effects	Cures or Prevention
Factories	Human stress	Regulation of loud noises
Vehicles	Sleep impairment	Building orientation
Aircraft	Relaxation impairment	Vegetation barriers
Power tools	Psychological disorders	Walls
Radios, TVs	Hearing impairment	Zoning restrictions
Gas-engine gardening tools		Separation of industrial and residential areas
Barking dogs		Elimination of gas-engine garden tools
		Full sound insulation in structures
		Less traffic/Slower traffic
		Good-neighbor policy

Electronic Hazards

In addition to the danger of electrical shock from faulty electrical products or connections, continuous exposure to excessive amounts of electromagnetic force (EMF) is considered unsafe. Close proximity to high-power voltage lines, faulty microwave appliances, and concentrations of television cables and wiring may pose a threat to occupants' health.

Fig. 6-30 ■ The natural vegetation on the site helps provide a pleasant living environment for both humans and the wildlife of the area. *Lindal Cedar Homes, Inc.*

Homes should be located safe distances from voltage lines. Designing safe distances or shields between persons and radiation fields is the best architectural solution to these potential problems.

Preventive Measures

Many pollution problems can be prevented through effective architectural design. For example, effective landscape planning which preserves existing vegetation can help maintain acceptable oxygen-nitrogen cycles, provide wildlife habitats, and preserve and enhance the natural beauty of the site. See Fig. 6-30. Effective landscaping can also provide summer shade, help retain groundwater, and reduce noise. Architectural design professionals can also help solve or prevent many pollution problems by becoming involved with local planning boards to help initiate realistic, environmentally related ordinances.

CHAPTER 6 Exercises

1. Explain why certain side(s) of a house receive the most light and heat. Compare winter and summer changes.
2. Draw a rectangle to represent a floor plan of a house. Label each side N, E, S, W. Then indicate where you'd place each room.
3. Using a CAD system, draw the outline of an earth-sheltered home.
4. Sketch a 75′ × 110′ property. Label one side as a hill with a 10° slope upward. Sketch a house on the property in the most desirable location.
5. Tell how you would use solar planning, wind control, and vegetation for a residence of your own design.
6. Find magazine and newspaper articles or advertisements about products for buildings that are designed to control pollution.
7. Identify buildings that emit pollutants into the air, water, or land. List ways of correcting these conditions.

8. List the ecological factors to be considered in planning a residence of your own design.
9. Using a CAD system, make a bar graph showing the decibel levels of deafening and very loud noises, pleasant sounds, and quiet.

CHAPTER

7

Indoor Living Areas

Objectives

In this chapter you will learn:

- to identify the functions of indoor living areas.
- to design the location, decor, size, and shape of indoor living areas.
- how a room's orientation, walls, floors, windows, ceilings, lighting, and furniture can contribute to room function and appearance.
- to design indoor living areas and work them into a convenient floor plan.

Terms

atrium
closed plan
decor
den
family room
living area
open plan
partition
recreation room
serving walls
studio
study
templates (furniture)

Your first impression of a home is probably the image you retain of the *living area*. In fact, this is the only indoor area of the home that most guests observe. The **living area** is where the family entertains, relaxes, dines, listens to music, watches television, enjoys hobbies, and participates in other recreational activities.

■ **A well-designed, tastefully decorated living area provides a pleasant atmosphere for residents and guests alike.** *Mark Romine*

Living Area Plans

In most two-story dwellings, the living area is normally located on the first floor and adjacent to the foyer or entrance. However, in split-level homes or one-story homes with basements, part of the living area may be located on the lower level.

The total living area is divided into rooms or smaller areas that serve specific purposes. These subdivisions may include the living room, dining room, family room, recreation room, and special-purpose rooms such as a den, office, or studio. In some homes, particularly smaller ones, rooms may serve two or more functions. For example, the living room and dining room are often combined. In other homes, the entire living area may be one room.

The subdivisions of a living area are called rooms even though they are not always separated by a **partition** or a wall. The subdivision areas perform the function of a room, whether there is a complete separation, a partial separation, or no separation.

When partitions do not totally divide the rooms of an area, the arrangement is called an **open plan.** When rooms are completely separated by partitions and doors, the plan is known as a **closed plan.**

Open Plan

In an open plan living area, the living room, dining room, and entrance may be part of one open area. Instead of walls, separation of the living room from other rooms is accomplished in different ways. See Fig. 7-1.

Placement of area rugs or furniture will not completely separate areas of a room, but will create a functional and visual separation. If occasional privacy is desired in an open plan, folding doors can be installed to close off part of the area.

Fig. 7-1 ■ A two-sided fireplace with open access on both sides can be used to separate the living room from the dining room. *James Eismont Photo*

An open-plan effect can be designed with glass walls to separate functions. See Fig. 7-2. The glass walls separate the living room from the adjacent **atrium** (indoor garden) and yet maintain the open view and allow light to come into the room.

Closed Plan

In a closed plan, rooms are completely separated from the other rooms by means of walls. See Fig. 7-3. Access is through doors, arches, or relatively small openings in partitions. Closed plans are found most frequently in traditional or period-type homes.

Fig. 7-2 ■ These glass walls not only create a feeling of openness, they protect the living room from the humidity created in the atrium. *Scholz Homes Inc.*

Fig. 7-3 ■ A closed-plan living room. *John Berenson Interior Design Inc.*

Combined Plans

Some large contemporary living area plans include both closed and open rooms. In combined plans there is a closed, formal living room and an open-plan living area which functions as an all-purpose *great room*. See Fig. 7-4.

Living Rooms

The living room is the center of the living area in most homes. Thus its functions, location, decor, size, and shape are extremely important and affect the design and appearance of the entire residence.

Function

A designer begins a living area design by first determining the functions of rooms required in terms of the residents' needs. A living room is the key room in the design because it can serve many purposes. For example, it can be an entertainment center, a library, a social room, and perhaps a dining center. Its particular functions depend on the living habits of the residents. A living room needs to be designed in relation to its functions and activities.

Sometimes a living room can be designed by the process of elimination. For example, if a separate recreation room is planned, then planning for a TV in the living room might not be necessary. If a separate den or study is provided for reading and for storing large numbers of books, then this function might also be eliminated from the living room. On the other hand, if many living area activities are to be combined in one room, then a great room may be designed. Regardless of the exact, specific activities anticipated, the living room should be planned as an integral part of the home for family and guests.

Fig. 7-4 ■ The living area shown in this plan is completely open except for the media room. Rooms are separated by levels and by the large fireplace. (This is the floor plan of the house shown in the Part 1 opening photo.) *IMHOTEP, Boca Raton, FL, design by Alfred Karram*

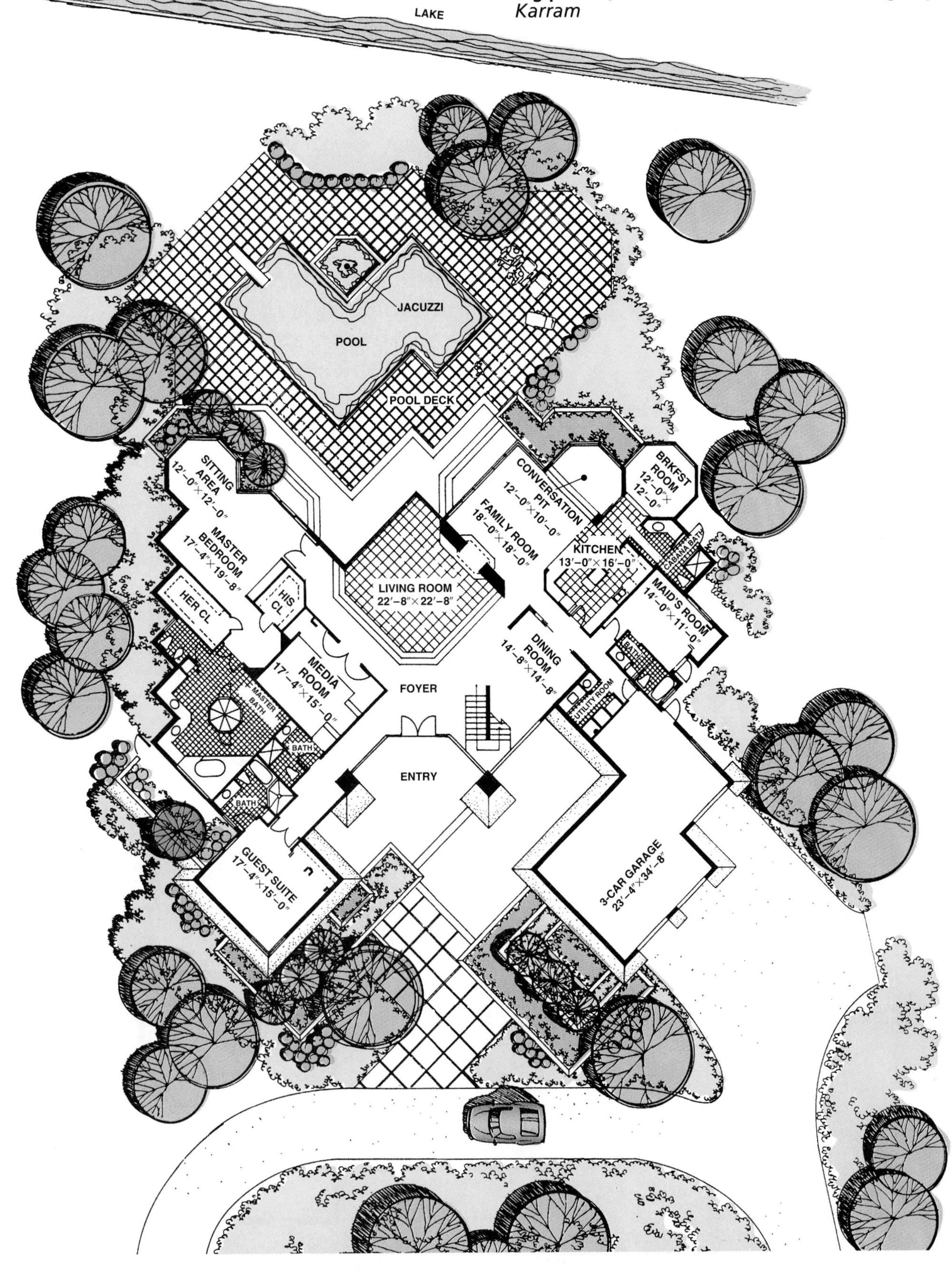

Location

Ideally, the living room should be centrally located and adjacent to the outside entrance. See Fig. 7-5. In smaller residences, the entrance may open directly into the living room. Whenever possible, however, this arrangement should be avoided. Since the living room and dining room function together, the living room is usually adjacent to the dining room, as well as to the outside entrance. The living room should not be the only "traffic lane" to the sleeping and service areas of the house. Guests could be disturbed.

Orientation

Careful consideration should be given to the placement of the living room in relation to its surroundings, including other rooms. The living room should be oriented to take full advantage of the position of the sun and the most attractive view. Since the living room is used primarily in the afternoon and evening, it should be located to receive the afternoon sun. See Fig. 7-6.

Decor

The general **decor** (pattern of decoration) of the living room depends primarily on the tastes, habits, and personalities of the residents. If their tastes are contemporary, then the wall, ceiling, window, and floor treatments should be consistent with the clean, smooth lines often found in contemporary architecture and contemporary furniture. See Fig. 7-7. If the residents prefer colonial or another style of archi-

Fig. 7-5 ■ **Floor plan with a centrally located living room.** *Scholz Homes Inc.*

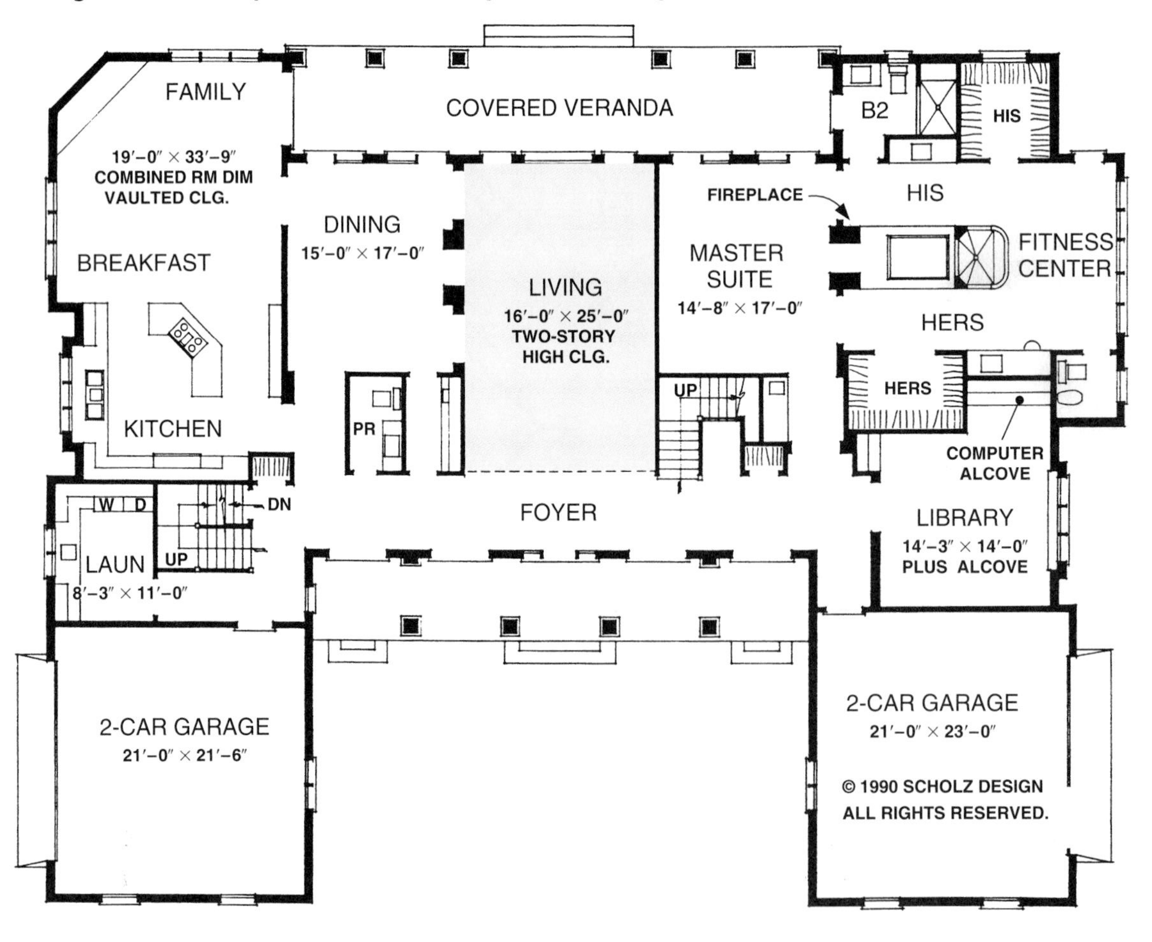

Fig. 7-6 ■ The orientation of this living room is designed to take advantage of the view during the day and the sunset during the evening. *Eagle Windows*

tecture, then this style should be the theme of the decor.

Appropriate color choices, effective lighting techniques, and the tasteful selection of materials for walls, floor covering, and ceiling can make a room appear inviting. See Fig. 7-8. The selection and placement of well-designed furniture can contribute to the appearance of comfort in a living room. The use of mirrors, floor-to-ceiling drapes, and arrangements of furniture can create a spacious effect in a relatively small room. Decorating a room, like selecting clothing, should minimize faults and emphasize good points.

WALLS The appearance of walls depends on more than wall coverings. The design and placement of doors, windows, fireplaces, chimneys, or built-in furniture along the walls of the living room will influence the entire room's appearance and should be designed as integral parts of the room. A designer considers these features in conjunction with the kind of wall-covering materials selected. Many materials are available, including plaster, gypsum, wallboard, wood paneling, brick, stone, tile, plastics, paper, and glass.

Fig. 7-7 ■ The simple, geometric lines of this fireplace are consistent with the contemporary decor of the surrounding area. *Majestic*

WINDOWS Just as the placement of a window in a living-room wall should become an integral part of a wall, the view from the window should become part of the living room decor. See Fig. 7-9. When planning windows, also consider the various seasonal changes in landscape features.

Fig. 7-8 ■ The well-planned decor of this living room gives the room an elegant, inviting atmosphere. Note particularly the effective use of lighting. *Presidential Place interior design by Alfred Karram*

Although windows themselves can be decorative items, the primary function of a window is twofold: to admit light and to provide a pleasant view of the landscape. Under some conditions, however, only the admission of light is desirable. If the view from the window is unpleasant or is restricted by other buildings, translucent glass can be used to allow the natural light to enter without showing the unwanted view.

FIREPLACES The primary function of a fireplace is to provide heat, but it can also be used as a room partition or as a major decorative feature. See Fig. 7-10. A fireplace and the chimney masonry can cover an entire wall and can become the focal point of the living area. A massive freestanding fireplace, however, also can function as a partition between the living room and the dining room. Refer back to

Fig. 7-9 ■ These windows not only correlate with the shape of the wall, but also enhance the room with the interesting landscape view. *Eagle Windows*

Fig. 7-10 ■ The unusual placement of this fireplace in a window wall adds interest to the decor of this living room. *Heatilator*

Fig. 7-1. Like all elements of a room's decor, a fireplace should correlate with the architectural style of the room.

FLOORS The living-room floor should reinforce and blend with the color scheme, textures, and overall style of the living room. Exposed hardwood flooring, room-size carpeting, wall-to-wall carpeting, throw rugs, and sometimes polished flagstone are appropriate for living-room use.

CEILINGS Most conventional ceilings are flat surfaces covered with plaster or gypsum board. However, new building materials, such as laminated beams and arches, and new construction methods have resulted in greater varieties and improvements in ceiling design. Higher ceilings allow for better air circulation and hot air exhaustion. They also create a feeling of spaciousness that a low ceiling over the same amount of floor space does not. Refer back to Fig. 7-9.

LIGHTING Appropriate lighting is essential to a room's atmosphere and comfort. Living room lighting generally comprises three types of lighting arrangements: general lighting, local lighting, and decorative lighting.

General lighting refers to illuminating the entire room and is often accomplished through the use of ceiling fixtures, wall spotlights, and cove lighting. Refer back to Fig. 7-8. *Local lighting* is light for a specific purpose, such as reading, drawing, or sewing. Local lighting may be table lamps, wall lamps, pole lamps, or floor lamps. *Decorative lighting* is used to improve the appearance of a room, create a mood, or to enhance a particularly attractive feature in the room.

FURNITURE Furniture for the living room should reflect the motif (theme) and architectural style of the home. Whether freestanding or built-in, furniture should maintain lines consistent with the entire wall treatment and blend functionally into the total decor of the room.

A living room designed primarily for conversation is often called a *formal* living room (also known as a *parlor*). This type of living room is usually closed and small and would, of course, require furniture different from an *informal* living room designed for television viewing, dining, and other activities. Houses that have formal living rooms usually also have family rooms with more informal furniture.

Size and Shape

Ideally, when designing a living room, the type and amount of living room furniture should be determined *before* the size of the living room is established. One of the most difficult aspects of planning the size and shape of a living room, or any other room, is to provide sufficient wall space for the effective placement of furniture. Continuous wall space is needed for the placement of many kinds of furniture, especially musical equipment, bookcases, chairs, and sofas. The placement of fireplaces, doors, or openings to other rooms should be planned to conserve as much wall space as possible for furniture placement.

Figure 7-11 shows typical ranges of size for living room furniture. In addition to standard sizes, furniture may be custom

Fig. 7-11 ■ Typical sizes of living room furniture.

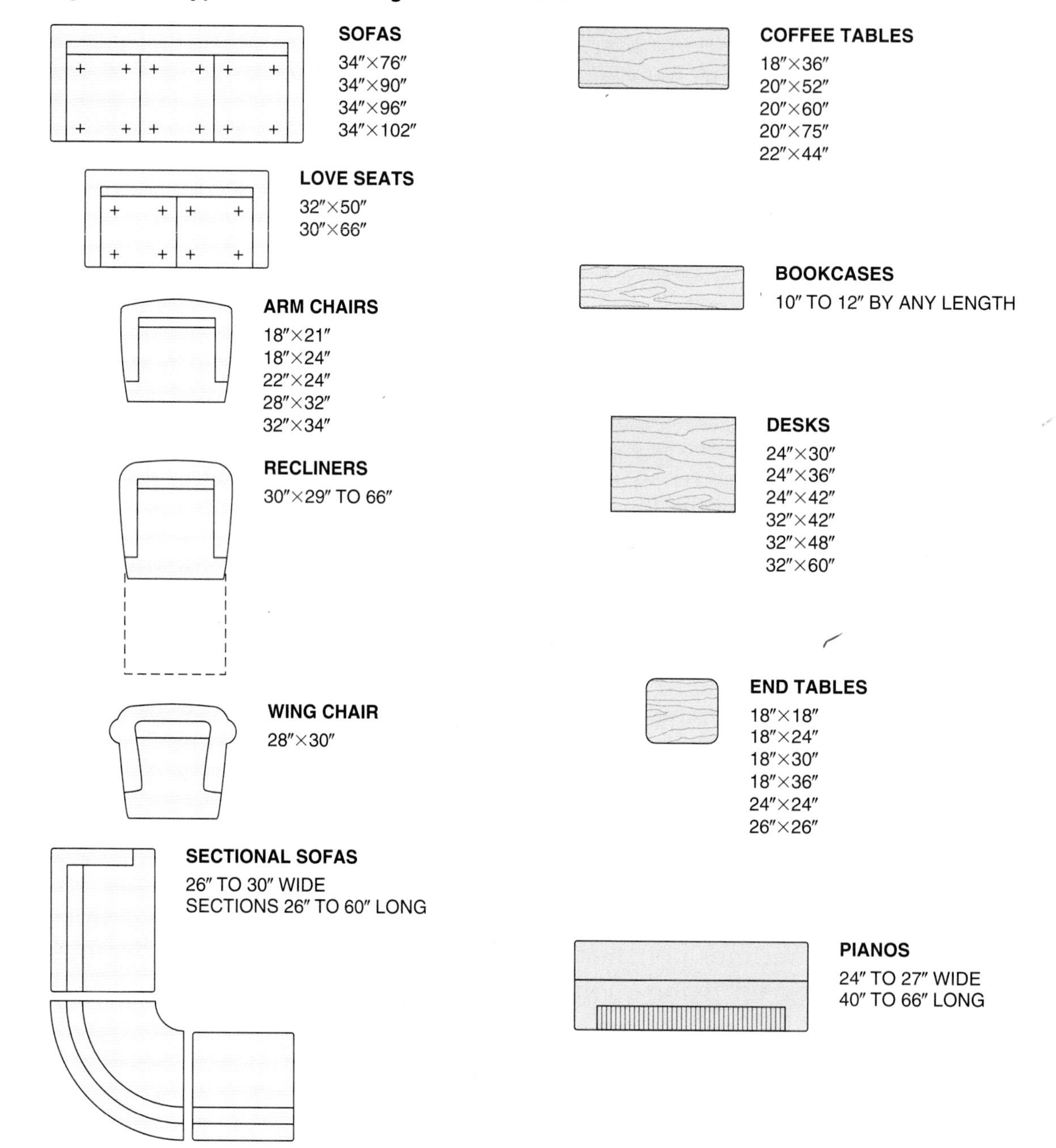

Fig. 7-12 ■ This furniture was designed to match the contour of the exterior window wall. *Eagle Windows*

designed (specially made) for specific locations and shapes. See Fig. 7-12.

To assure that the room will accommodate the necessary furniture, **templates,** as shown in Fig. 7-13, and a room plan are made to the same scale. The templates represent the width and length of each piece of furniture. The room plan represents the size and shape of the room or area. By seeing the amount of space occupied by the furniture, a designer can then more effectively evaluate the design of the entire room. Rectangular rooms are generally easier to plan and to place furniture in than are square rooms.

Living rooms vary greatly in size. A room 12′ × 18′ would be considered a small or minimum-sized living room. A living room of average size is approximately 16′ × 20′, and a large living room would be 20′ × 26′ or more.

Dining Rooms

The design of dining facilities for a residence depends greatly on the dining habits of the family. A separate dining room may be large and formal. An informal dining area may consist of a dining alcove in a living area or even a breakfast nook in the kitchen. Large homes may contain several dining facilities in different areas.

Function and Location

The function of a dining area is to provide a place for the family and guests to eat breakfast, lunch, or dinner, whether in casual or formal situations. When possible, a separate dining area capable of seating from 6 to 12 persons for dinner should be provided in addition to breakfast and lunch facilities.

Dining facilities may be located in many different areas, depending on the residents' needs and preferences. Regardless of the exact position of the dining area, it should be adjacent to the kitchen. The ideal dining location is one that requires few steps from the kitchen to the dining table.

In a closed plan, a separate dining room is usually located between the living room and kitchen, as shown in Fig 7-14. In an open plan, several different dining locations are possible. Open-area dining facilities can be in the kitchen or the living rooms. See Fig. 7-15. However, the preparation of food and other kitchen activities should not be in direct view from the dining area.

To baffle (separate or hide) an area, many design options are possible. For example, folding or sliding doors can separate and hide the kitchen from the

Fig. 7-13 ■ Furniture templates are helpful when planning the size and shape of a room.

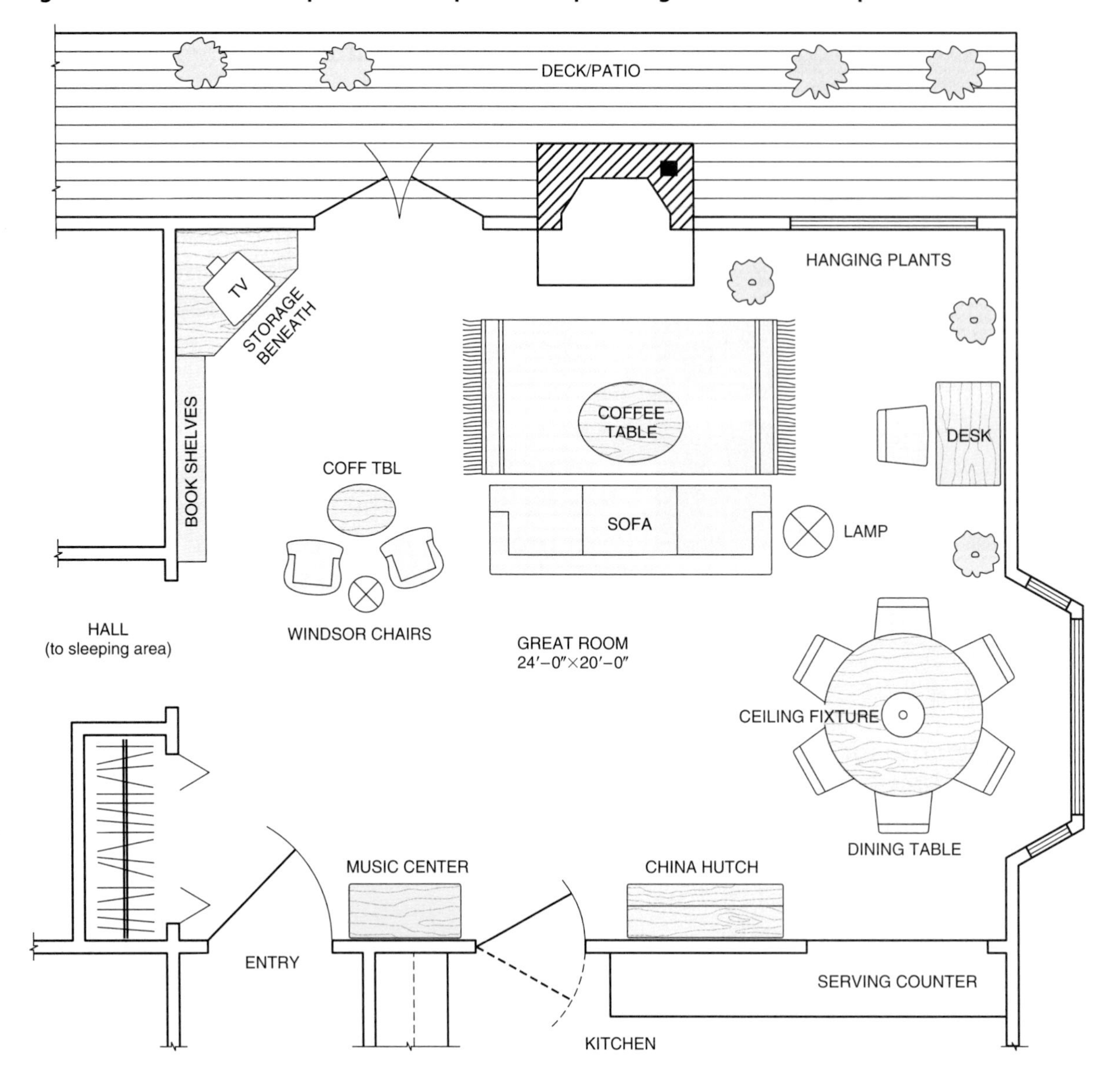

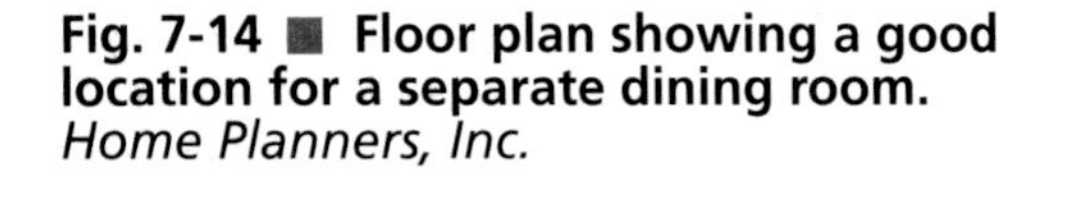
Fig. 7-14 ■ Floor plan showing a good location for a separate dining room. *Home Planners, Inc.*

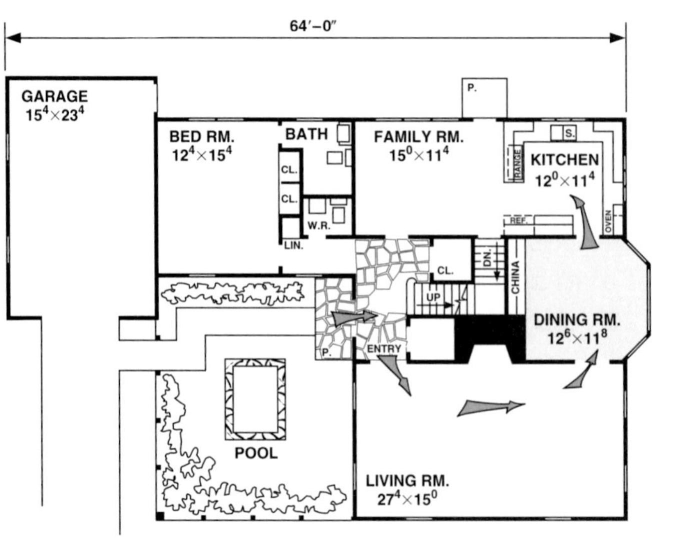

Fig. 7-15 ■ Placement of dining areas in open plans. A. With the kitchen. *Marvin Windows*

B. With the living room. *Interior Design by Daphne Weiss Inc—Maxwell MacKenzie Photographer*

Fig. 7-16 ■ Folding doors used to close off the dining area.

dining area. See Fig. 7-16. A two-sided fireplace offers another option for separating the dining room from other parts of the living area.

When dining facilities are not located in the living room, the dining area should be located adjacent to the living room. Family and guests normally enter the dining room from the living room and use both rooms jointly.

A partial separation of the dining room and the living room can be accomplished by different floor levels or by dividing the rooms with common half walls. **Serving walls** also provide a functional separation between the dining room and kitchen. These walls have openings with countertops for passing items through. Aesthetic structures can partially enclose the dining area of an open plan with creatively designed partial partitions, as shown in Fig. 7-17.

Some families enjoy dining outdoors on a porch or a patio. If so, the porch or patio should be near the kitchen and directly accessible to it. Locating the patio or dining porch directly outside the dining room

Fig. 7-17 ■ Partially enclosed dining area. *Edna Carvin, Remax, Danville, CA*

Fig. 7-18 ■ Glass walls help create a decorative effect. *Scholz Homes Inc.*

or kitchen provides maximum use of the facilities. This location minimizes the distance from the kitchen and possible inconveniences of outside dining facilities.

Decor

The decor of the dining room should blend with the remainder of the house. The floor, walls, and ceiling treatment of the dining area should work well with the decor of the living area. If a dining porch or a dining patio is used, its decor should also be considered part of the dining-room decor.

To create a partially closed dining area, the decor of any kind of partial divider wall should be considered as well as its purpose. A divider may be a planter wall; glass wall; half wall of brick, stone, or wood panels; fireplace; or grillwork. See Fig. 7-18.

Controlled lighting is another means to greatly enhance the decor of the dining room. General illumination that can be subdued or intensified can provide the appropriate atmosphere for any occasion. This type of lighting is controlled by a *rheostat*, commonly known as a dimmer switch. In addition to general illumination, local lighting should be provided directly over the dining table. See Fig. 7-19.

Size and Shape

The size and shape of the dining area are determined by the size of the family, the size and amount of furniture, and the clearances and traffic areas needed between pieces of furniture. The dining area should be planned for the largest group that will dine in it. There is little advantage in having a dining-room table that expands if the room is not large enough to accommodate the expansion. One advantage of the open plan is that the dining facilities can be expanded into the living area.

Fig. 7-19 ■ Lighting in a formal dining area. *Presidential Place interior design by Alfred Karram*

Dining-room furniture may include an expandable table, side chairs, armchairs, buffet, server or serving cart, china closet, and serving bar. In most situations, a rectangular dining room will accommodate the furniture better than a square dining room.

Regardless of the furniture arrangement, a *minimum* space of 2′ (610 mm) should be allowed between a chair and the wall or other furniture when the chair is pulled out. This allowance will permit serving traffic behind chairs, and will allow persons to approach or leave the table without difficulty.

Another space consideration is the distance *between* people when seated at the table. A distance of 2′-3″ or 27″ (686 mm), between persons should be allowed. This spacing is accomplished by allowing 27″ (686 mm) from the centerline of one chair to the centerline of another. See Fig. 7-20. Figure 7-21 illustrates the average space required for adults when seated for dining. The shapes and sizes of typical dining room furniture are shown in Fig. 7-22.

Fig. 7-20 ■ Space allowances in a dining room.

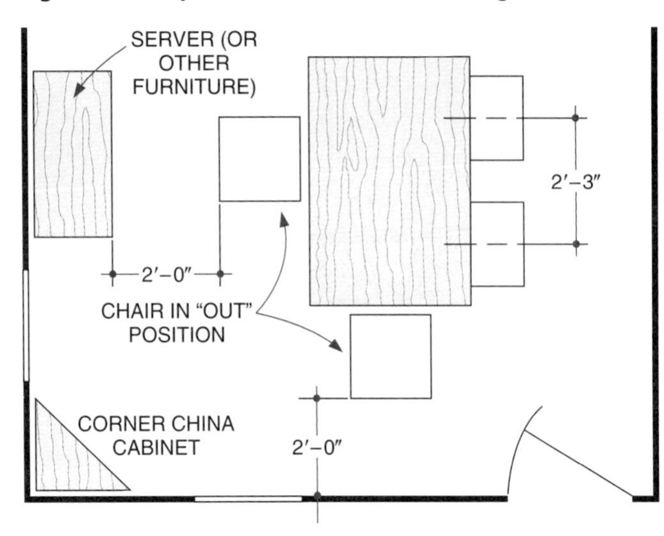

Fig. 7-21 ■ Space requirements for adults.

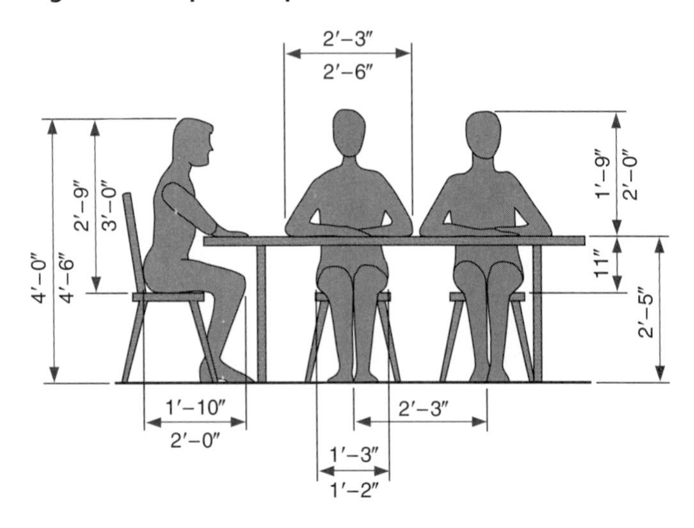

A dining room that would accommodate the minimum amount of furniture—a table, four chairs, and a buffet—would be approximately 10′ × 12′ (3048 mm × 3658 mm). A minimum-sized dining room that would accommodate a dining table, six or eight chairs, a buffet, a china closet, and a server would be approximately 12′ × 15′ (3658 mm × 4572 mm). A large dining room would be 14′ × 18′ (4267 mm × 5486 mm) or larger. See Fig. 7-23.

Fig. 7-22 ■ Typical sizes of dining room furniture.

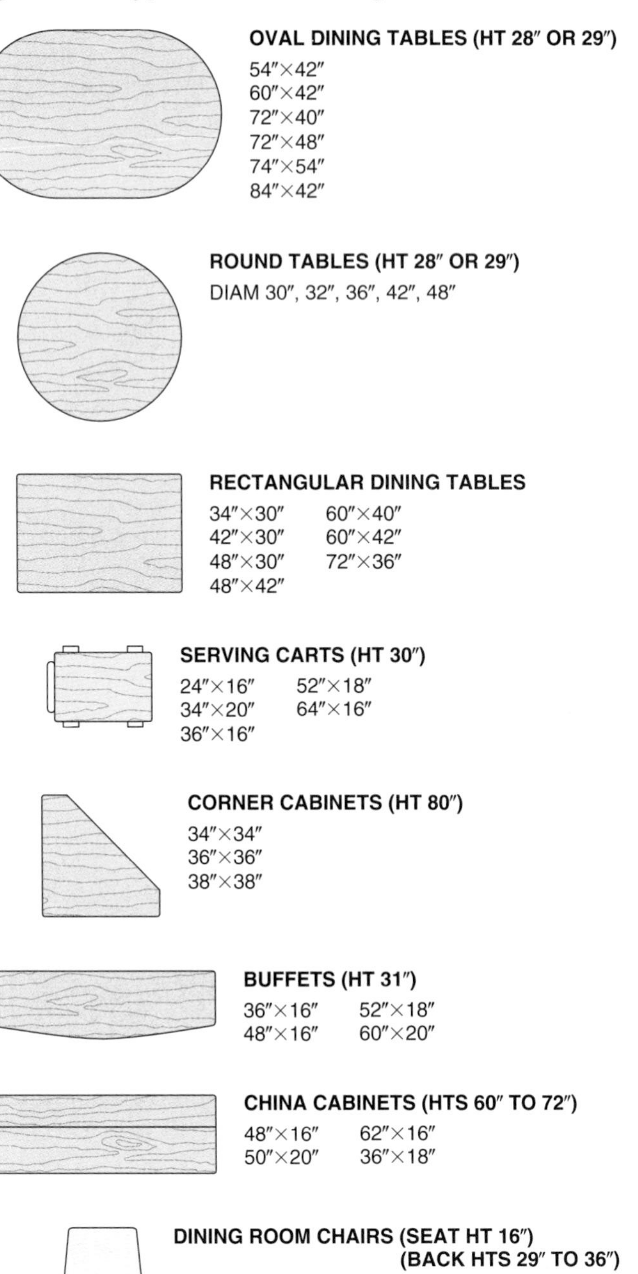

Family Rooms

Because of the trend toward more informal living, the majority of homes today are designed to include a family room.

Fig. 7-23 ■ Dining room furniture templates were used to determine the dining room size and shape.

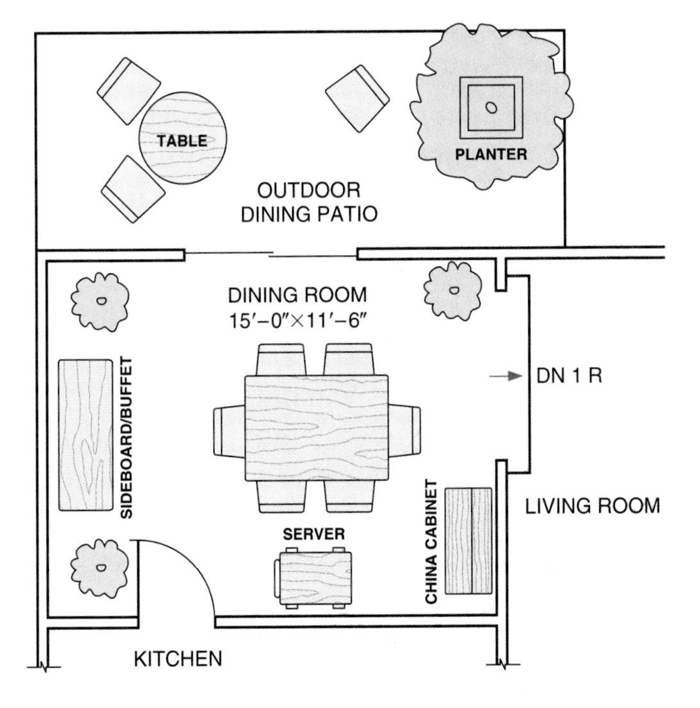

Fig. 7-24 ■ This family room designed for young children will be redesigned as the children mature and family interests change. *Lisanti Inc.*

Function

The purpose of the **family room,** as the name implies, is to provide facilities for family-centered activities. In extremely large residences, special-purpose rooms may be provided for specific types of activities, such as playing music, sewing, or painting. Typically however, additional rooms cannot fit into the plan, and facilities and equipment must be provided for a wide variety of activities. For this reason, the family room is also known as the *activities* or *multiactivities* room.

Basically sedentary activities take place in the family room. These may include playing or listening to music, pursuing hobbies such as painting or sewing, watching TV, or playing board, computer, or video games.

Family rooms should be designed to be flexible. As the family's needs and interests change, the family room will need to change also. See Fig. 7-24.

Location

Activities in the family room often result in the accumulation of hobby materials and clutter. Thus, the family room should be easily accessible, but not visible, from the rest of the living area.

Commonly, the family room is located adjacent to the kitchen. This location revives and expands the idea of the old country kitchen as the room in which most family activities were centered. See Fig. 7-25.

When the family room is located adjacent to the living room or dining room, it becomes an extension of those rooms for social affairs. In this location, the family room is often separated from the other rooms by folding doors, screens, or sliding doors. See Fig. 7-26.

Another popular location for the family room is between the service area and the living area. This location is especially appropriate if a home workshop is assigned to the family room.

Fig. 7-25 ■ Family room combined with a kitchen. *Whirlpool Corp.*

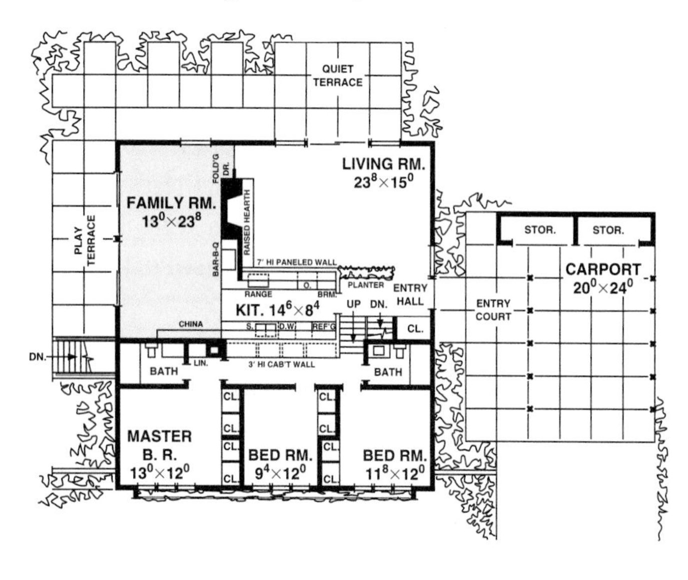

Fig. 7-26 ■ This family room is located next to the kitchen and yet is accessible from the living room when the folding door is open. *Home Planners Inc.*

Decor

Decoration of the family room should provide a vibrant atmosphere. Ease of maintenance should be one of the chief considerations. Furniture materials such as plastic, leather, and wood are easy to care for and provide great flexibility in color and style. Family-room furniture should be informal and suited to all members of the family.

The floor should be resilient—able to keep its original shape or condition despite hard use. Linoleum or tile made of asphalt, rubber, or vinyl will best resist the abuse normally given a family-room floor. If rugs are used, they should be the kind that will stand up under rough treatment. They should also be washable.

For walls, soft, easily damaged materials such as wallpaper and gypsum board should be avoided. Materials such as tile and paneling are most functional. Chalkboards, bulletin boards, cupboards, and toy-storage cabinets should be used when appropriate. Work areas that fold into the wall when not in use conserve space. Such areas perform a dual function if the cover wall can also be used as a chalkboard or a bulletin board.

Since a variety of hobby and game materials will be used in the family room, sufficient storage space must be provided. This includes the use of built-in facilities such as cabinets, closets, and drawer storage.

Acoustical ceilings are recommended. These help keep the noise of the various activities from spreading to other parts of the house. This feature is especially important if the family room is located on a lower level.

Size and Shape

The size and shape of the family room depend upon the equipment needed for the planned activities. The room may vary from a minimum-sized room of approximately 150 sq. ft. to a very spacious family room of 300 sq. ft. or more. Most family rooms require a size that ranges between these two extremes. See Fig. 7-27.

Fig. 7-27 ■ A family room designed for a variety of activities.

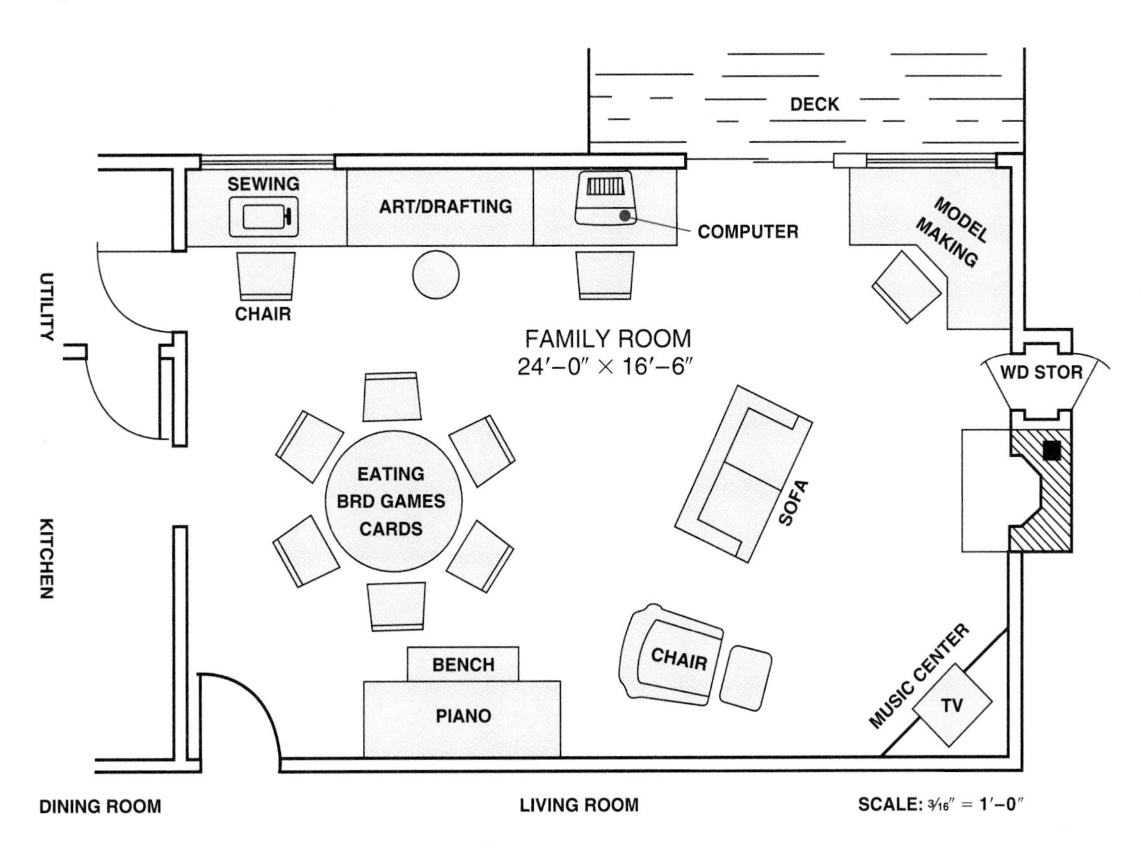

Recreation Rooms

The **recreation room** is also called a game room or playroom. As the name implies, it is a room designed specifically for active play, exercise, and recreation.

Function

The function of the recreation room often overlaps that of the family room. Overlapping occurs when a multipurpose room is designed to provide for recreational activities such as table tennis and billiards, but also includes facilities for quieter activities such as knitting, model-building, and other hobbies.

The design of the recreation room depends on the number and arrangement of the facilities needed for the various pursuits. Activities for which many recreation rooms are designed include billiards, chess, checkers, table tennis, darts, watching television, eating, and dancing. See Fig. 7-28.

Location

The recreation room should be located away from the quiet areas of the house. Most often, it is located in the basement or on the ground level.

A basement location uses space that might otherwise be wasted. Also, basement recreation rooms can often provide more space for activities that require large areas and equipment, such as table tennis, billiards, and shuffleboard. A good ground level location would allow activities to be expanded onto a patio or terrace.

Fig. 7-28 ■ Billiards oriented recreation room. *Armstrong World Industries*

Decor

Designers take more liberties in decorating the recreation room than with any other room. They do so primarily because of the active, informal atmosphere that characterizes the recreation room. This atmosphere lends itself readily to unconventional furniture, fixtures, and color schemes.

Bright, warm colors can reflect a party mood. Furnishings and accessories can be used to accent a variety of central themes. Regardless of the theme, recreation room furniture should be comfortable and easy to maintain. The same decorating guidelines that apply to the family room also apply to recreation room walls, floors, and ceilings.

Size and Shape

The size and shape of a recreation room may depend on whether the room occupies basement space or an area on the main level. If basement space is used, the only restrictions on the size are the other facilities there, such as the laundry, utility, or workshop areas. See Fig. 7-29.

Special-Purpose Rooms

Specialized activities can be planned for areas in a living room, great room, family room, or even bedroom. However, separate rooms can be designed for such activities.

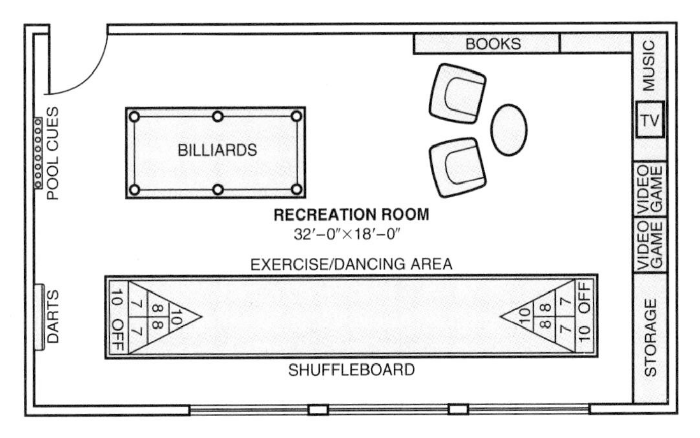

Fig. 7-29 ■ A recreation room designed for several recreational activities.

Fig. 7-30 ■ A study with built-in storage and bookshelves. *Haas Cabinet Co.*

Function

Special-purpose rooms are designed to facilitate specific activities of residents. For example, because of recent technological advances, a major trend in today's society is for persons to work at home. Many people now require an office or computer room in their homes. Other special-purpose rooms include a den, study, music room, exercise room, home theater, and studio.

Activities can often be combined in a special-purpose room. For example, some residents may want a home office and computer room to be in the same room. Others may choose to have a studio that could occasionally be used as a guest bedroom.

A **den** or **study** may function basically as a reading room, writing room, hobby room, or professional office. See Fig. 7-30. For the teacher, writer, or lawyer, it may be basically a reading and writing room. For the engineer, architect, drafter, or artist, the den or study may function primarily as a **studio.** The den or study often doubles as a guest room.

In designing special-purpose rooms, a complete list of the sizes of all special furniture, equipment, and storage space needed is necessary. Then the room can be designed accordingly.

Location

The location of a special-purpose room depends on how it will be used. If activities are quiet, such as reading or writing, the room may be located in the sleeping area of the house. If activities involve more action, it may function better in the living area. For example, if the room is used as a professional office by a person whose clients call at home, then it should be located in an accessible area near an entrance. See Fig. 7-31. If it is only for private use, then it can be located in any preferred area. Often a basement or attic space can be used that would otherwise be wasted.

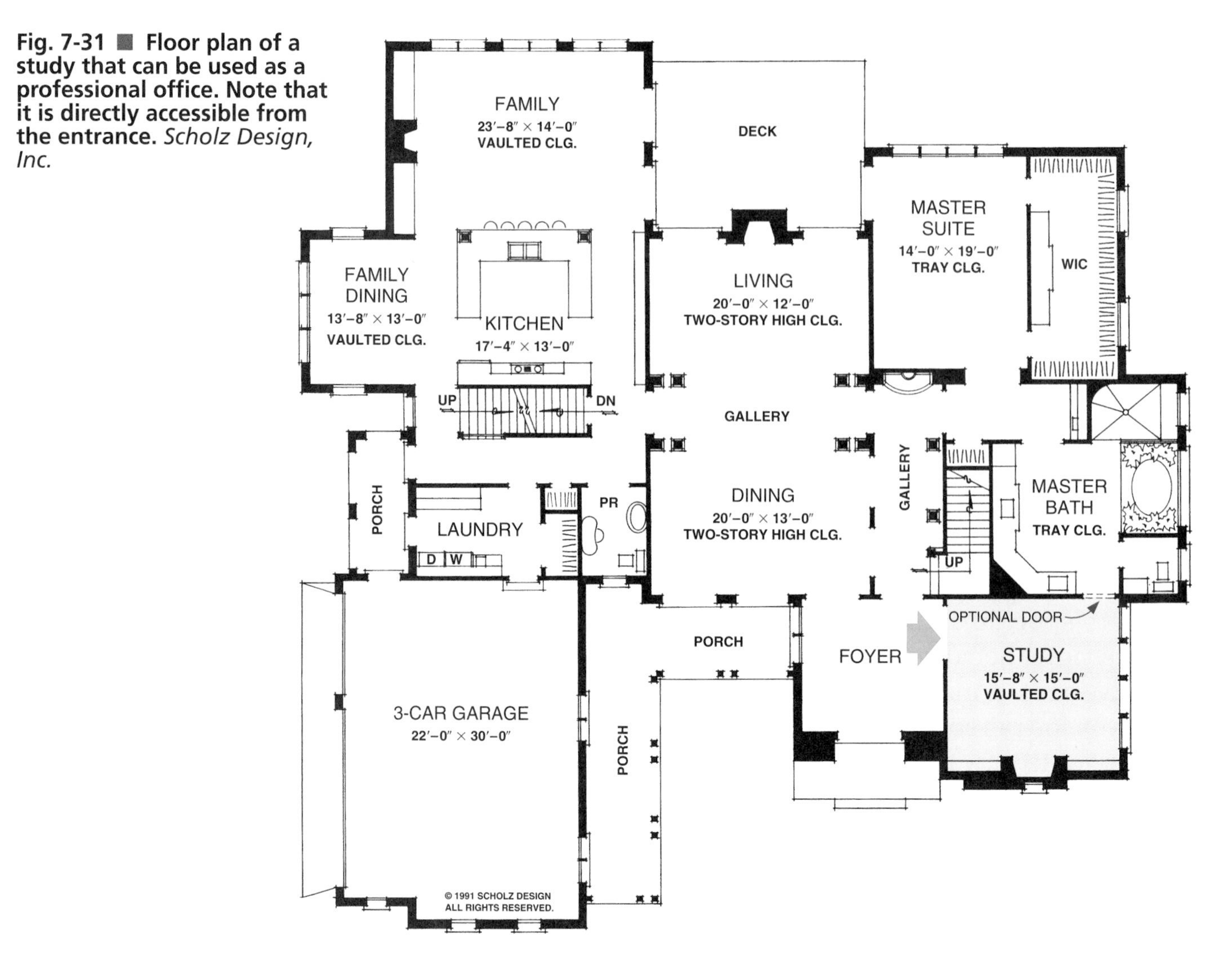

Fig. 7-31 ■ Floor plan of a study that can be used as a professional office. Note that it is directly accessible from the entrance. *Scholz Design, Inc.*

Decor and Lighting

The decor of special-purpose rooms should reflect the main activity and should allow for well-diffused general lighting and glareproof local lighting. Windows positioned above eye level admit the maximum amount of light without exposing distracting eye-level images from the outside. Such positioning also allows more wall space for furniture and storage. Figure 7-32 shows the furniture and equipment sizes to be considered in planning a home office or study.

Size and Shape

The size and shape of special-purpose rooms depend on their planned functions. For example, will one or two persons use the room privately, or should it provide a meeting place for business clients? Rooms can range in size from just enough space for a desk and chair in a small corner to a large amount of space with a diversity of furnishings.

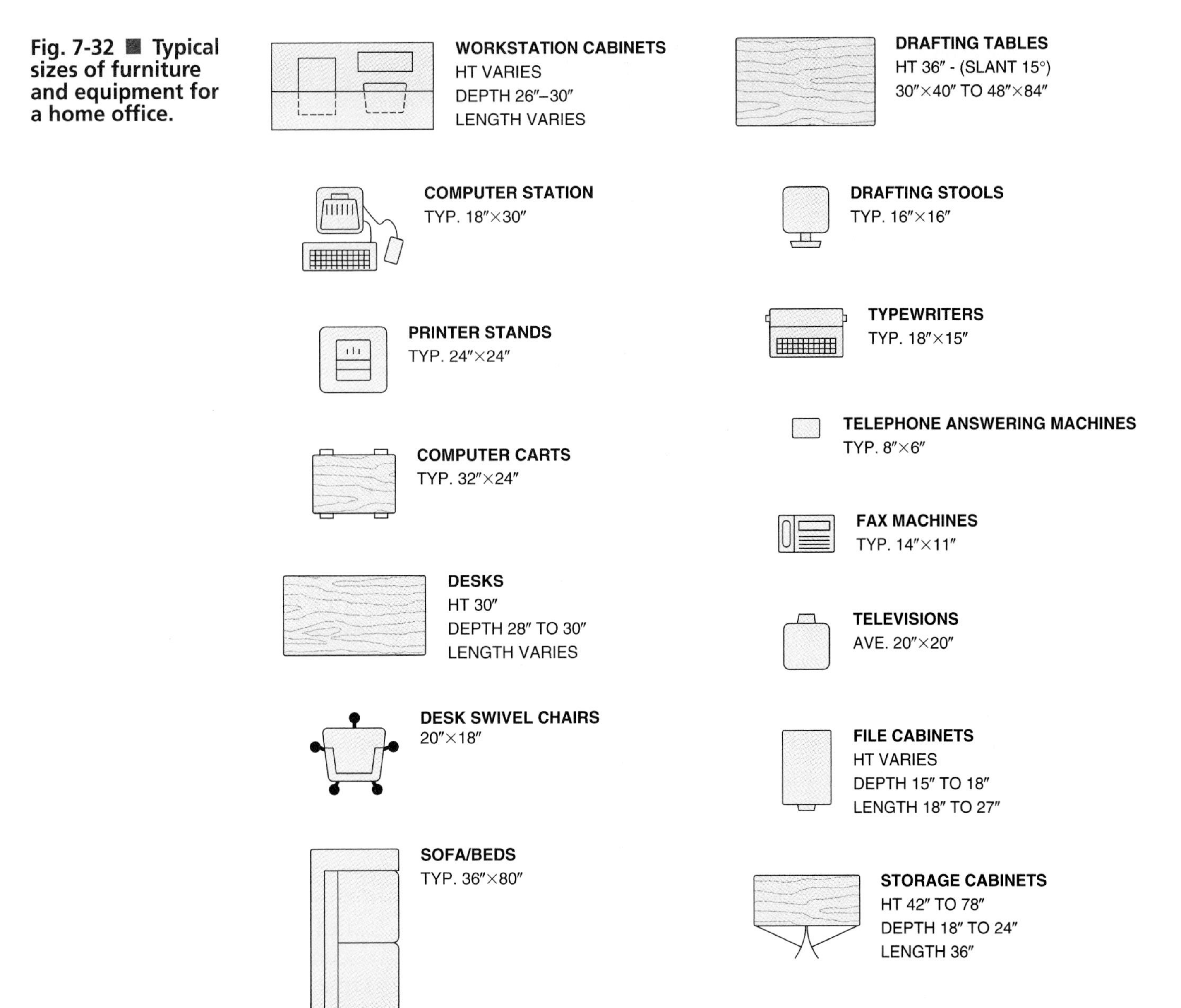

Fig. 7-32 ■ Typical sizes of furniture and equipment for a home office.

CHAPTER 7 Exercises

1. List the functions you want in a living room for yourself or an imaginary client.
2. Draw a simple sketch of an open-plan living room for the functions you listed in Exercise 1. Indicate the location of doors, windows, a fireplace, foyer, entrance, and dining room and label them accordingly.
3. Sketch a closed-plan living room for the same functions as in Exercise 1. Show the position of adjacent rooms. Explain the reasons for your choices.

4. Using LINE commands on a CAD system, draw the living room outline you prefer from those that you designed in Exercises 2 and 3. Indicate doors and windows.
5. Begin a picture file of your own for different rooms. Cut out pictures showing the decor you like of living rooms—including furniture, lights, etc.—from catalogs, newspapers, or magazines. Then list the kinds of furniture you would include in your design, including color and materials.
6. Sketch the dining area of your own home. Then make another sketch changing the design (open or closed plan) without changing any *outside* walls. (Make any needed templates to help create a different arrangement.)
7. Sketch a dining room to scale showing the position of all furniture you would like to include in the dining room of a house of your own design.
8. Collect pictures of dining areas and furniture and accessories.

9. Using a CAD system, draw a floor plan of a dining room shown in this chapter.
10. Calculate the minimum-sized dining room you would need to seat six people at a 60″ × 42″ table.
11. Refer back to Fig. 7-27. This plan is drawn to 3/16″ = 1′-0″ scale. Tell what the dimensions of the room are. Evaluate the size and the arrangement in relation to its functions.
12. Sketch a family room, recreation room, special-purpose room, or office you would like to include in a home of your own design. List the activities and furniture needed. Show the location of all furniture and facilities (scale 1/2″ = 1′-0″).
13. Design a family room primarily for children's activities. Include the furniture needed. Describe the colors and materials you would select for this room.
14. Redesign the room in Exercise 13 to accommodate teen activities. Describe the changes you made and explain why.
15. Using a CAD system, draw an outline of a family room adjacent to a kitchen.

16. Using a CAD system, draw a floor plan of the recreation room shown in Fig. 7-29.
17. Use the typical dimensions of the office equipment in Fig. 7-32 to sketch a plan to accommodate the following facilities: desk, chair, bookcases, drafting table, and lounge.

18. Collect pictures of rooms and furnishings that you would like to use in a house of your own design.

Interior Designer

Interior designers have special expertise in interior planning. They are often involved in layout and plan the finishes of interior ceilings, floors, and walls including window treatments. They are also responsible for selecting and placing furniture. The result of the work of the interior designer is probably the most visible part of the interiors of buildings.

> "A client might come to me and say, 'I'm building a 50-story skyscraper. What's it going to cost me completely furnished?'"

Allison Lanier-Jones

Camp Twin Lakes in Rutledge, Georgia, is a camp for children who are physically and mentally challenged. One of the people who helped design it was Allison Lanier-Jones, an interior designer. She believes the design of the interior space was critical to the project because the children's basic needs are so important.

Allison works on a variety of architectural projects and coordinates the six-person interior design group for her firm. "Half of our interior design group were educated as architects and the other half were educated directly as interior designers," she notes. Allison spends her day on "anything from drawing on the computer, to meeting with clients, to selecting finishes for architectural projects, to space planning, to furniture selection, and actual design work." She also spends time working with potential clients, writing project proposals and planning budgets.

An interior designer is an important part of an architectural design team. "When we are working on large projects, it's truly a team effort. We present our individual ideas to the team for consideration. Rarely does an idea survive without being modified. A lot of what we do is problem solving. A total design is like a complicated jigsaw puzzle, and we put the pieces together."

Allison wanted a career that would incorporate art, math, and science, her favorite subjects in school. She believes she's found it. "There's art in the creative aspect in architecture, math in the design of the structure, and physics related not only to the structure, but to electrical and mechanical engineering."

Allison enjoys seeing her designs become reality. "I like to visualize interiors and to create designs and make them happen." She also finds satisfaction in providing a service for people, just as she did for the children in Camp Twin Lakes.

Lord, Aeck & Sargent, Inc./Ted Mishima

CHAPTER

8

Outdoor Living Areas

Objectives

In this chapter you will learn:

- **to design and sketch a porch, patio, and lanai.**
- **to design and sketch a swimming pool.**
- **to calculate the area and volume of swimming pools.**

Terms

balcony
deck
lanai
marquee
patio
porch
stoop
swim-out
veranda

A home's living areas may be extended to the outdoors. Porches, patios, and other features provide space for dining, entertaining, playing, exercising, or relaxing. When planning an outdoor living area, consider the area's function, location, decor, size, and shape. A well-designed area will look and function like a natural extension of the house.

■ **Outdoor living areas should be planned carefully so that they coordinate with the home's overall design.** © *1987 Tina Freeman*

Porches

A **porch** is a covered platform leading into an entrance of a building. Porches are commonly enclosed by glass, screen, or posts and railings. Balconies and decks are actually elevated porches. (A **deck**, however, usually refers to an open, elevated platform.) Similar to a porch, a **stoop** is a projection from a building. However, a stoop does not provide sufficient space for activities. It provides only shelter and an access to or landing surface for the entrance of the building.

Sometimes porches are confused with patios. Although a patio may also be adjacent to a house and seemingly attached to it, a patio is directly on the ground, even if it has a finished surface. The main difference between a porch and a patio is that a porch is attached *structurally* to a house.

Fig. 8-1 ■ A Southern Colonial home with a two-level veranda. *Dixie Pacific*

Function and Types

The classic front porch and back porch that characterized most homes built in this country during the 1920s and 1930s were built merely as places in which to sit. Little effort was made to use the porch for any other activities. However, *Southern Colonial* homes were designed with **verandas**, large porches, extending around several sides of the home. See Fig. 8-1. Outdoor life often centered on the veranda.

Today, porches serve a variety of purposes. They may be used for dining and entertaining, as well as for just sitting and relaxing. See Fig. 8-2. Sometimes a porch expands or extends the entrance to a house or patio, providing additional protection. In a contemporary home, a porch may even be designed for specific activities. For example, it may serve as a game area.

Fig. 8-2 ■ A deck designed for relaxation. *Georgia Pacific*

A **balcony** is a porch suspended from an upper level of a structure. There is usually no access from the outside. Balconies often extend a living area. See Fig. 8-3. Others serve as a private extension of a bedroom. The balcony protects the lower level from the sun and precipitation.

Fig. 8-3 ■ Multiple balconies extending from each living area. *Eagles Nest—Ponte Vedra Beach, FL; Robert C. Boward, Architect; E. Joyce Reesh—Arvida*

Fig. 8-4 ■ In apartment design, balconies give residents a wider outdoor view. *David Garris*

Hillside lots lend themselves to vertical plans and allow maximum flexibility for such outdoor living areas. See Fig. 8-4. Spanish- and Italian-style architecture is typically characterized by numerous balconies that integrate indoor and outdoor living areas. New developments in building materials have increased the recent popularity of balconies in many styles of architecture.

Location

The location of the porch depends upon its purpose or function. The family's preferences for use of the porch should also be considered when designing its orientation. For example, if daytime use is anticipated for the porch and direct sunlight is desirable, then a southern exposure should be planned. If little sun is wanted during the day, a northern exposure is preferable. If morning sun is desirable, an eastern exposure is best, and for the afternoon sun and sunset, a western exposure.

A continuous porch is often designed to function with the living area and/or with the sleeping area. The porch shown in Fig. 8-5 continues on three sides of the house to include all three areas: living, sleeping, and service areas.

Fig. 8-5 ■ This continuous porch (balcony) has a continuous patio below it. *Lindal Cedar Homes*

Decor

The porch should be designed as an integral and functional part of the total structure. A blending of roof styles and

Fig. 8-6 ■ Multiple level decks with overhang sun baffles. *L'Ermitage at Grey Oaks Country Club—Naples, FL*

major lines of the porch roof and house roof is especially important.

A porch can be made consistent with the rest of the house by extending the lines of the roof to provide sufficient roof overhang, or projection, over the porch area. See Fig. 8-6. A similar consistency should characterize the vertical columns or support members of the porch and the railings.

Porch railings can provide adequate ventilation and also offer semiprivacy and safety. Various materials and styles can be used, depending on the degree of privacy or sun and wind protection needed. Railings on elevated porches, such as balconies, should be higher than 3′ (914 mm) for general safety, as well as to discourage the use of the top rail as a place to sit. By code, most *balusters* (vertical posts) must be spaced closely enough (usually 4″) to prevent a child's head from going through.

Porch furniture should withstand any kind of weather. The covering material should be waterproof, stain-resistant, and washable. Nonetheless, protection from wind and rain should be planned.

Size and Shape

Porches range in size from the very large veranda to rather modest-sized stoops. A porch approximately 6′ × 8′ (1829 mm × 2438 mm) is considered minimum-sized. An 8′ × 12′ (2438 mm × 3658 mm) porch is about average. Porches larger than 12′ × 18′ (3658 mm × 5486 mm) are considered large. The shape of the porch depends greatly upon how the porch can be integrated into the overall design of the house.

Patios

The word **patio** is Spanish for courtyard, an open space enclosed wholly or partly by buildings. Courtyard living was an important part of Spanish culture. Therefore courtyard design was an important component of early Spanish architecture.

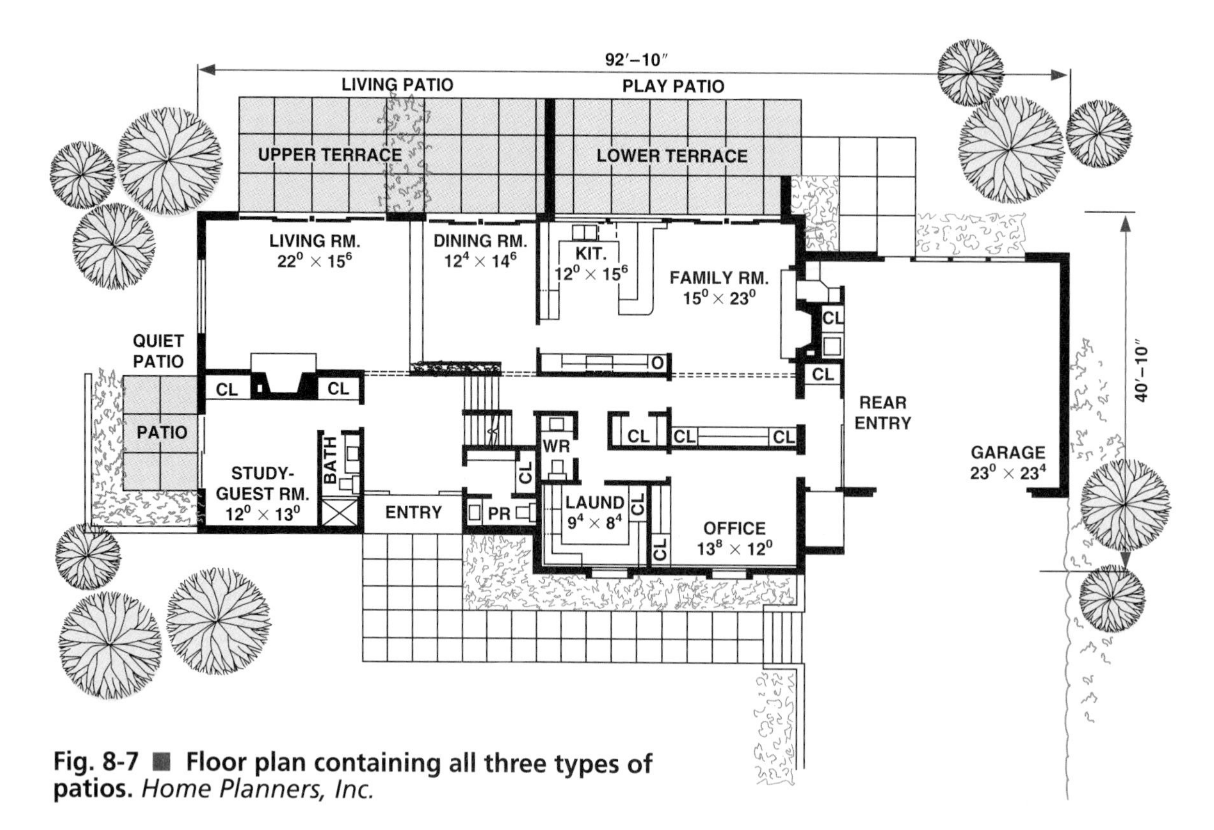

Fig. 8-7 ■ Floor plan containing all three types of patios. *Home Planners, Inc.*

Function and Types

The patio may perform all the functions outdoors that the living room, dining room, recreation room, kitchen, and family room perform indoors. The patio may be referred to by other names, such as loggia, breezeway, and terrace.

Patios are divided into three main types according to function: living patios, play patios, and quiet patios. See Fig. 8-7. Regardless of the type of patio, it should be secluded from the street and from neighboring residences if possible.

Location

The type of patio affects its location in relation to other rooms in the home. For instance, living patios should be located close to the living room or dining room. When dining is anticipated on the patio, access should be provided from the kitchen or dining room.

A children's play patio, or play terrace, for physical activities is not necessarily associated with the living area. Sometimes a play patio is located next to the service area so that it can double as a service terrace. Notice how the location of the play terrace in Fig. 8-7 is related to the service area and the family room in the living area to enable adult visual observation.

A quiet patio can become an extension of the bedroom for relaxation or sleeping. A quiet terrace should be secluded from the normal traffic of the home.

Often the design of the house will allow these separately functioning patios to be combined into one large, continuous patio. See Fig. 8-8. Similar to a continuous porch, this type of patio may be accessible from the playroom, living room, bedrooms, and/or kitchen. Other designs divide large patios and porches into different areas by different levels.

Fig. 8-8 ■ Different patio types can be combined into a continuous patio. *John B. Scholz, Architect*

Patios can be placed at the end of a building, between corners, or along the exterior form of the structure. They may also be placed in the center, such as in the center of a U-shaped house. A courtyard patio offers complete privacy from all sides.

A patio can be located completely apart from the house. A shady, wooded area, a beautiful view, or a unique feature of terrain may determine an ideal location for a patio. The designer should take full advantage of the most pleasing view and should restrict the view of undesirable sights. If the patio is located a short distance from the house, it should still be easily accessible to the house with paths or walkways.

When the patio is placed on the north side of the house, the house itself can be used to shade the patio. If sunlight is desired, the patio should be located on the south side of the house. By locating patios on different sides of a building, sun exposure and sun protection can be available during some part of the day. Protection from the sun may also be controlled with planned landscaping and fences.

Decor

The materials used in the decor of the patio should be consistent with the lines and materials used in the home. Patios should not appear to be designed as an afterthought. They should appear and function as an integral part of the total design.

PATIO SURFACE The patio surface, or deck, should be constructed from materials that are permanent and maintenance-free. Flagstone, redwood, concrete, and brick are among the best materials for use on patio decks. Wood creates a warm appearance. On some wood decks, slats may be spaced to provide drainage.

Brick-surface patios are very popular because bricks can be placed in a variety of arrangements. The area between the bricks may be filled with concrete, gravel, sand, or grass. A concrete patio is effective when a smooth, unbroken surface is desired. Concrete works well for patios where bouncing-ball games are played or where a poolside cover (roof) is desired. However, where patio surfaces also function as pool decks, a non-heat absorbing material is preferable.

PATIO COVER The manner in which a patio is covered, or not covered, is closely related not only to the decor but also to the sunlight. Patios need not be covered if the house naturally shades the patio. Since a patio is designed to provide outdoor living, too much cover can defeat the purpose of the patio.

Coverings can be graded, or tilted, to allow light to enter the patio when the sun is high and block the sun's rays when the sun is lower. Straight or slanted louvers can be placed to admit the high sun and block the low sun or *vice versa*. See Fig. 8-9.

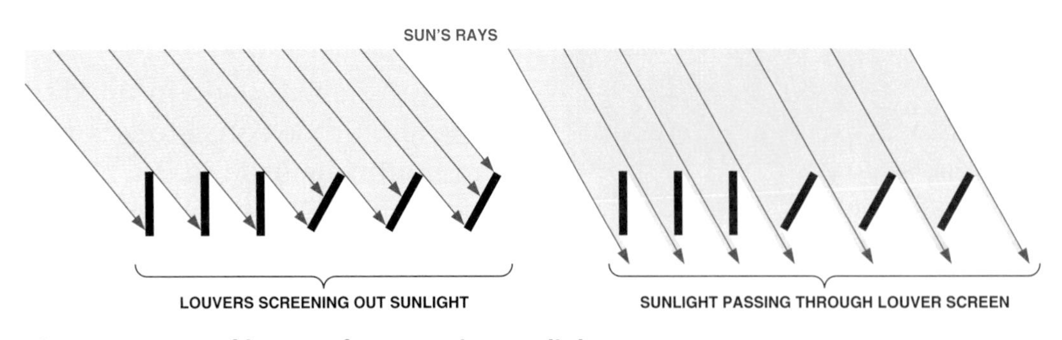

Fig. 8-9 ■ Use of louvers for screening sunlight.

Plastic, fiberglass, and other translucent materials used to cover patios admit sunlight and yet provide protection from the direct rays of the sun and from rain. When such translucent coverings are used, it is often desirable to cover only part of the patio. This arrangement provides sun for part of the patio and shade for other parts and also allows rising heat to escape. Balconies can also be used effectively to provide shade and control light on a patio.

PATIO WALLS AND BAFFLES Patios are designed for outdoor living, but outdoor living does not mean living in public. Some privacy is usually desirable. Natural landforms can sometime provide privacy. See Fig. 8-10. Solid walls can often be used effectively to baffle, or shield, the patio from a street view, from wind, and from the long shadows created by low rays of the sun. Baffling devices include solid or slatted fences, concrete blocks, post-and-rail, brick or stone walls, and hedges or other landscaping. Figure 8-11 shows a patio with maximum protection both from the sides and overhead.

A solid baffle wall is often undesirable because it restricts the view, eliminates the circulation of air, and makes the patio appear smaller. When possible, natural vegetation or a sloped baffle is preferred.

In mild climates, a patio may be enclosed with solid walls to make the patio function as another room. In such an enclosed patio, some opening should nonetheless be provided to allow light and air to enter. Grillwork openings are an effective and aesthetically pleasing solution to this problem. Where wind and blowing sand or dust may be a problem, glass windscreens may be used to protect the patio.

Fig. 8-10 ■ Natural landforms can sometimes provide privacy. *Paul Peart*

Fig. 8-11 ■ **Well-protected patio. Note the combined use of a roof overhang, overhead slats, and vegetation to provide privacy.** *Julius Shulman*

LIGHTING The patio should be designed so that it can be used both day and night. This means using general and local electrical lighting as well as natural lighting. If the walls between the house and patio are designed correctly, light from inside the house can be utilized on the patio at night. See Fig. 8-12. The combination of wall and lighting design can extend the number of hours the patio can be used.

Size and Shape

As with other rooms, the function influences the size of a patio. Patios vary greatly in size. An oriental garden terrace, for example, can be small because it often has no furniture and is designed primarily to provide a baffle and a beautiful

Fig. 8-12 ■ **Patio designed for night use.** *Western Wood Products Assoc.*

view. A patio used for recreation needs to be large enough for equipment and furnishings, such as picnic tables and benches, lounge chairs, serving carts, game apparatus, and barbecue pits. See Fig. 8-13. Adequate space for the storage of games, apparatus, and fixtures also needs to be considered.

Patios tend to vary more in length than in width. Some patios may extend along the entire length of the house. A patio 12′ × 12′ (3658 mm x 3658 mm) is considered a minimum-sized patio. Patios with dimensions of 20′ × 30′ (6096 mm × 9144 mm) or more are considered large.

Lanais

Lanai is the Hawaiian word for porch, but it also refers to a covered exterior passageway. Large lanais often double as patios.

Function

Lanais actually function as exterior hallways. They provide shelter for the exterior passageways of a building.

Lanais that are parallel to exterior walls are usually created by extending the roof overhang to cover a traffic area where people walk. See Fig. 8-14. A typical

Fig. 8-13 ■ Typical sizes of patio furniture.

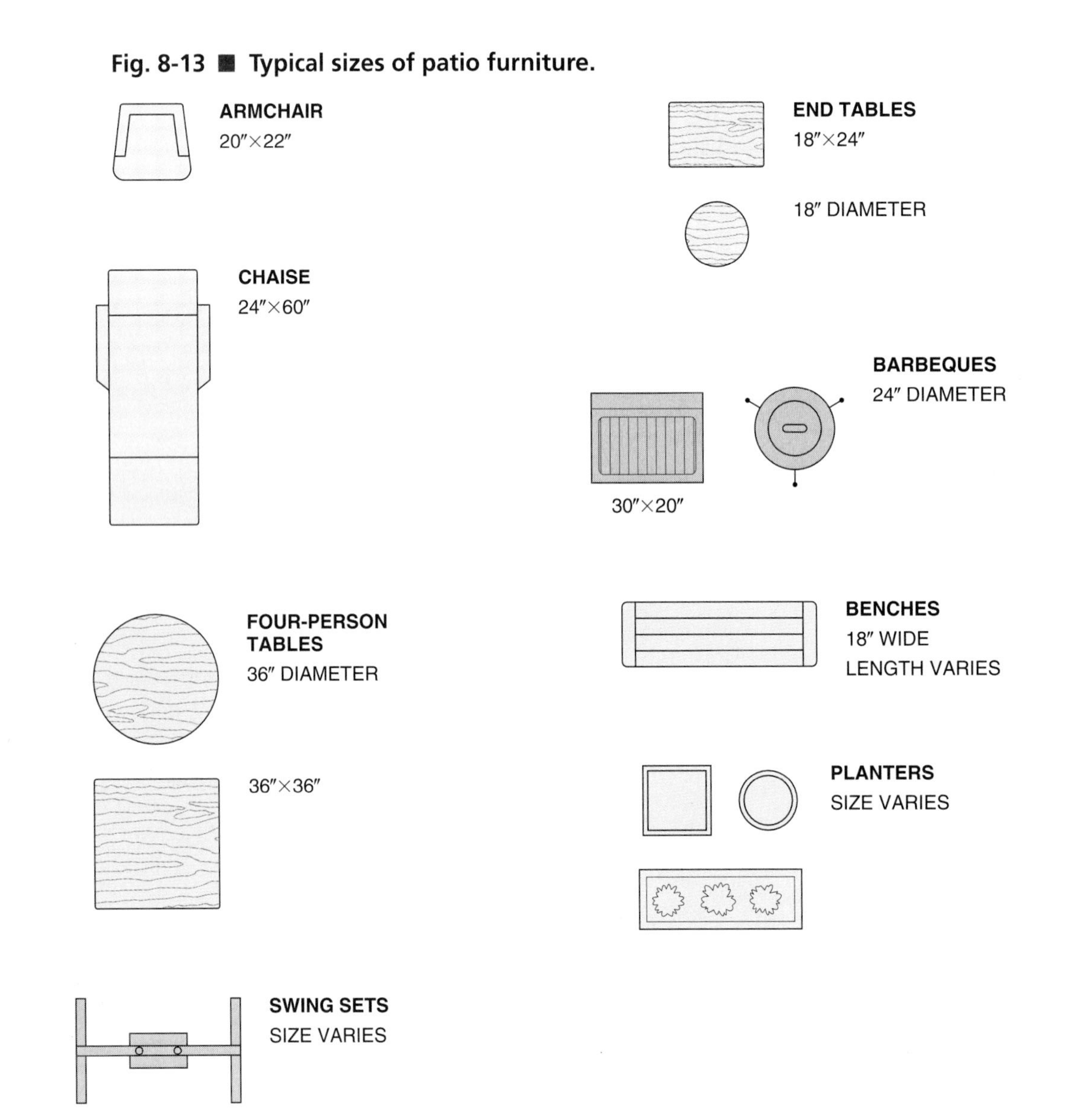

Fig. 8-14 ■ A large roof overhang supported by columns provides shelter for the lanai walkway and adjacent living area windows.

lanai eliminates the need for more costly interior halls. Lanais are used extensively in warm climates.

Location

In residence planning, a lanai can be used most effectively to connect opposite areas of a home. Lanais are commonly located between the garage and the kitchen, the patio and the kitchen or living area, and the living area and service area. U-shaped buildings are especially suitable for lanais because it is natural to connect the ends of the U.

When lanais are carefully located, they can also function as sheltered access from inside areas to outside facilities such as

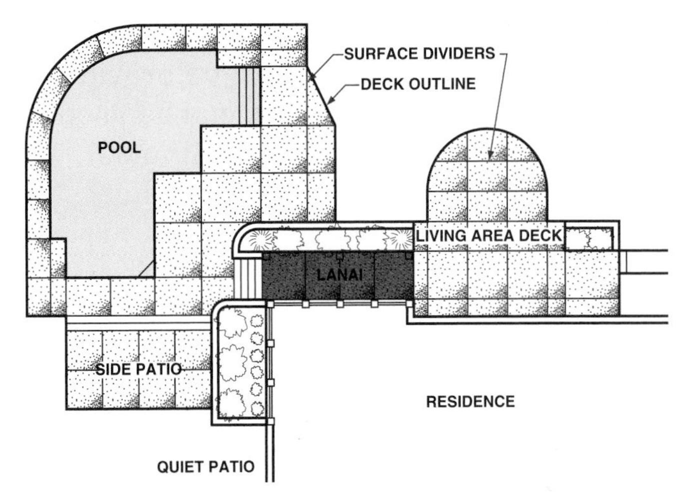

Fig. 8-15 ■ Lanai connecting living area to pool deck and patio.

patios, pools, outdoor cooking areas, or courtyards. See Fig. 8-15. A covered or partially covered patio is also considered a lanai when it doubles as a major access from one area of a structure to another.

A lanai can also be semi-enclosed and provide not only traffic access but privacy, as well as sun and wind shielding. When a lanai connects the building with the street it is called a **marquee**.

Decor

The lanai should be a consistent, integral part of the structure's design. The lanai cover may be an extension of the roof overhang or supported by columns. Refer back to Fig. 8-14. Columns also provide a visual boundary without blocking the view. If glass is placed between the columns, the lanai becomes an interior hallway rather than an exterior one. This feature is sometimes the only difference between a lanai and an interior hall.

It is often desirable to design and locate the lanai to provide access from one end of an extremely long building to the other end. The lines of this kind of lanai strengthen and reinforce the basic horizontal and vertical lines of the building.

If a lanai is to be utilized extensively at night, effective lighting must be provided. Light from within the house can be used when drapes are open, but additional lighting fixtures are used for the times when drapes are closed.

Size and Shape

Lanais may extend the full length of a building and may be designed for maximum traffic loads. They may be as small as the area under a roof overhang. However, a lanai at least 4′ (1219 mm) wide is desirable. The length and type of cover is limited only by the location of areas to be covered.

Swimming Pools

Swimming pools are becoming more popular as an integral part of residential design. Designed for exercise and relaxation, pools can also enhance the design of a house.

Pools add a great deal of expense to a home, especially in cooler climates where pools are used only part of the year. In warmer climates, pools are more common since year-round use is possible, but pools are still a luxury item.

Function

The ideal pool should provide for all functions: exercise, relaxation, and enhancement of the site decor. The pool in Fig. 8-16 belongs to a large, luxurious home and has separate whirlpool tubs, fountains, and a cabaña.

Although pools are primarily used during daytime hours, a lighted pool, such as the very dramatic one in Fig. 8-17, expands the living area by making the pool area inviting at night.

Location and Orientation

Several factors affect the location of residential pools: the relationship with the house, sun exposure, and privacy. The

Fig. 8-16 ■ Pool complex with facilities for all functions. *Jackson-Watson Homes*

Fig. 8-17 ■ Lights direct attention to the pool and beyond to the lake view. *Greg Wilson Photo*

Fig. 8-18 ■ Trees near a pool can provide both shade and privacy. Note the beautiful tile around the top of the pool and the wood deck that surrounds it.

pool should be located as close to the living area as possible, allowing pool deck and patio space between the house and the pool. Refer back to Fig. 8-8.

Most building codes require controlled access to the pool. This means you must enter the pool area from the house or through a fence gate.

The orientation of a pool should be considered in relation to the sun. A pool should be positioned to allow the option of full sun exposure or shade through an adjacent shade-escape area. Shade for the pool deck or connecting patio may come from the north side of a house. On wooded sites, the orientation can be designed so trees can supply the needed shade. See Fig. 8-18. However, the trees should not block the sun from the pool during most of the day.

Enclosure is a major consideration in the design of pools for privacy and safety. Figure 8-16 shows how a pool area can be attractively separated with buildings and landscape design. The pool in Fig. 8-18 is built into an area surrounded by the natural vegetation of the dense woods.

Screened walls and overhead enclosures have the advantage of blocking bugs and debris, but they also reduce the amount of direct solar heat on the water. Enclosures should be planned during the floor plan and elevation design phases (Chapters 13 and 15) to ensure consistency with the lines of the house.

Pool Construction

Math Connection

MATERIALS The frames of pools are constructed with concrete, wood, or steel. Pool surfaces that you see beneath the water may be covered with any of a variety of materials: vinyl sheets, marble composite, pebble aggregate, or paint.

Pool decks are constructed using non-heat-absorbing concrete mixtures, acrylic composites, or wood slats on elevated surfaces. Again, see Fig. 8-18. The use of spray-on deck surfaces which are heat-resistant, slip-resistant, waterproof, and "mildew proof" are the most popular. Flagstone and pure concrete are not recommended for pool deck surfaces because these materials retain heat and become slippery when wet.

POOL SHAPES With the development of dry-mix concrete, pool walls of almost any shape, including free-form, can be created. See Fig. 8-19. More than one shape can fit a pool site, depending on the contour of the site. Patio shapes around or near the pool also need to be considered.

In addition to the surface shape of a pool, the depth must be considered. It must first be determined if the pool is to be all shallow (3′ to 4′ deep), all deep (6′ to 10′), or a combination. Combinations of 3′ to 4′ on one end dropping to depths of 6′ to 10′ on the opposite end are most popular. If a diving board is to be included, then the deep end must include a *diving well* at least 8′ deep extending out 10′ horizontally from the end of the board as shown in Fig. 8-20.

CALCULATING POOL SIZES Residential pools range in size from about 200 sq. ft. to 800 sq. ft. or more. For example, a small 12′ × 18′ pool is 216 sq. ft. and a large 20′ × 40′ pool is 800 sq. ft. To calculate the area of a rectangular pool, use the following formula:

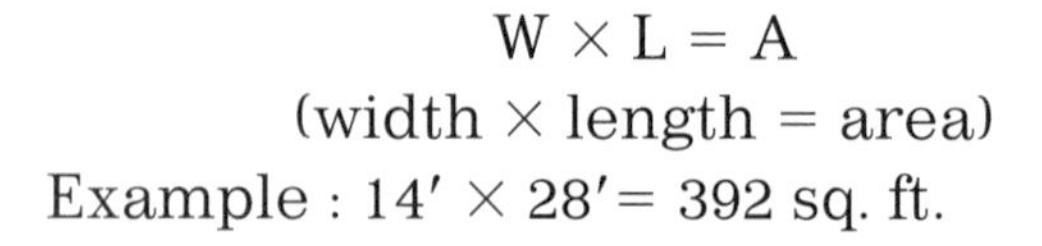

$$W \times L = A$$

(width × length = area)

Example : 14′ × 28′= 392 sq. ft.

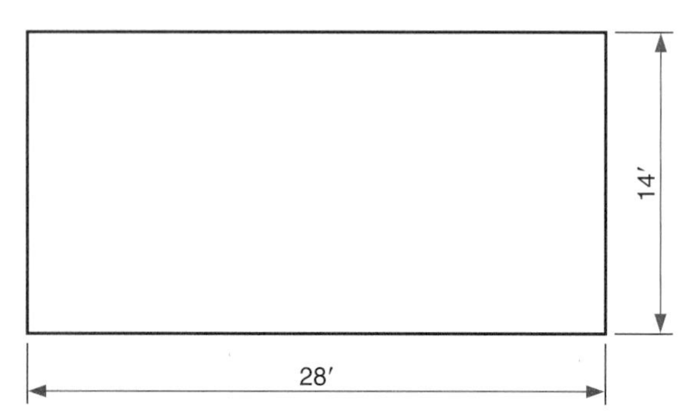

Fig. 8-19 ■ **Pools can be designed in practically any shape to fit the site and complement the building's design.** *John Henry, Architect*

Fig. 8-20 ■ **Pool dimensions necessary when a diving board is included in the plan.**

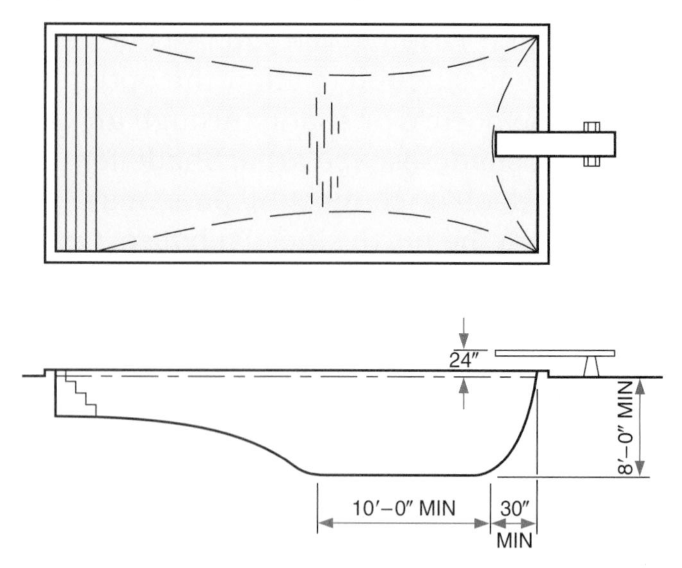

To calculate the area of a round pool, use the formula for determining the area of a circle:

$$\pi r^2 = A$$

(pi × radius squared = area)

Example : 3.14 × (15′ × 15′) = 706.5 sq. ft.

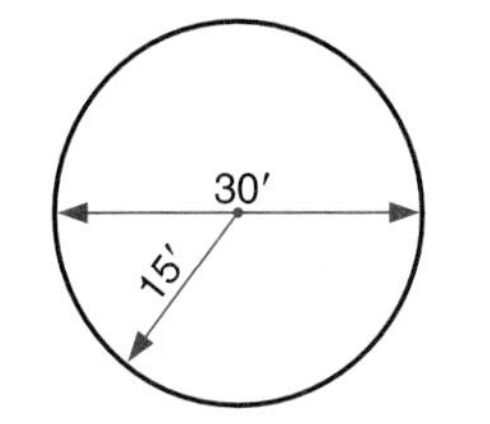

To calculate the area of a pool, such as the one in Fig. 8-21, that has a combination of circular and rectangular areas, divide the entire pool area into smaller round and rectangular segments. First calculate the area of each part, round to even square feet, and add them together. For example, to find the area of the pool shown in Fig. 8-21:

(Rectangle) Area A	16 × 30	=	480
(Rectangle) Area B	6 × 10	=	60
(Half-circle) Area C	3.14 × 64 × $\frac{1}{2}$	=	100
TOTAL SQ. FT.			640

Fig. 8-21 ■ Measurements needed to calculate the square foot area of this pool which has a combination of shapes.

For estimating cost and water capacity, cubic area is used to define the size of a pool. For example, an 18′ × 38′ pool that is 6′ deep contains 18′ × 38′ × 6′, or 4104 cu. ft., of space. The same size pool 8′ deep contains 5472 cu. ft. To find the volume of a container, use the following formula:

$$V = W \times L \times D$$

volume (cu. ft.) = width × length × depth

Example : 14′ × 28′ × 8′ = 3136 cu. ft.

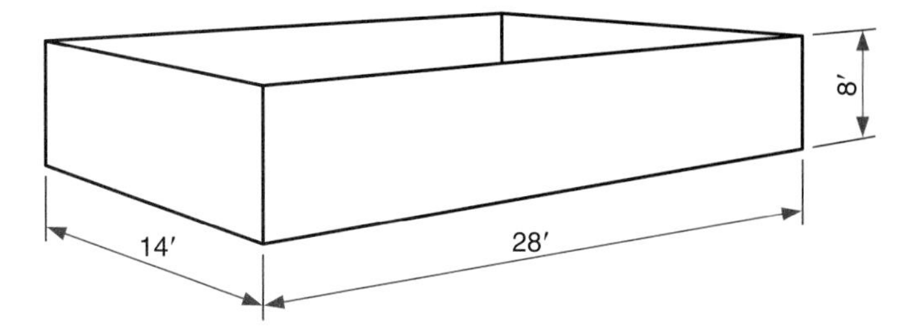

Since pools usually slope from shallow to deep, most cubic foot calculations involve dividing the pool into segments that have the same depth, as shown in Fig. 8-22.

To determine the volume of a pool with a combination of cylindrical and cubic shapes, divide the pool area into separate cylindrical and cubic segment areas according to identifiable shapes. Then

Fig. 8-22 ■ Measurements needed to calculate the volume of this pool.

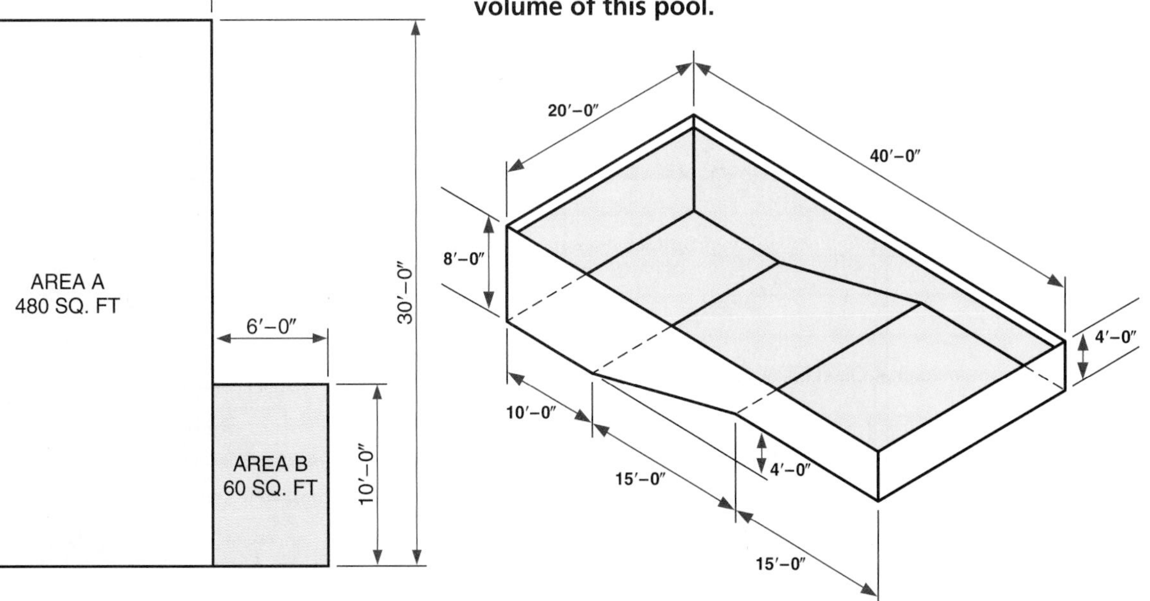

calculate the volume of each and add them together. For example, to find the volume of the pool shown in Fig. 8-21, use the appropriate formula for each segment area. Remember, you can multiply in any sequence. Assume the bottom of each area is flat. The depth of area A is 8′, B is 3′, and C is 6′.

(Rectangle) Area A	$8' \times 16' \times 30' =$	3840
(Rectangle) Area B	$3' \times 6' \times 10' =$	180
(Half-circle) Area C	$6' \times 3.14 \times 32' =$	603
TOTAL CU. FT.		4623

Safety Devices

Over 600 people drown in residential pools each year. Visibility from house to pool is important, but certain pool design features can help reduce this number dramatically. These include: minimum 4′ fencing, self-latching gates, latched house doors and windows, strong pool covers, alarms, ladders, steps and/or swim-outs. A **swim-out** is an elevated platform below the water level which allows the swimmer to get out of the pool without using a ladder.

Other safety devices and equipment beyond the basic design features include: clip-on child alarms, rope and float line, filter basket cover, posted emergency information, outside telephone, and portable infant fences.

Pool Equipment

Pools are simply cavities in the ground filled with water. To make a pool function properly, water must be circulated, filtered, purified, and sometimes heated. All of these functions require operating equipment. See Fig. 8-23.

Pool water is circulated through a series of filters and purifiers to keep the water sufficiently pure and clean for swimming. A water pump pulls water from the pool through a series of pipes connected to a skimmer device and drain. The pump moves the water through the filter, purifier, and sometimes a heater. After the water passes through these devices the pure, clean, and heated water returns to the pool through pipes. These pipes are connected to *outlets* in the pool walls *under* the waterline. The number of outlets spaced throughout the pool determines the amount and balance of water circulation. Small pools may need only

Fig. 8-23 ■ Pool equipment for water circulation.

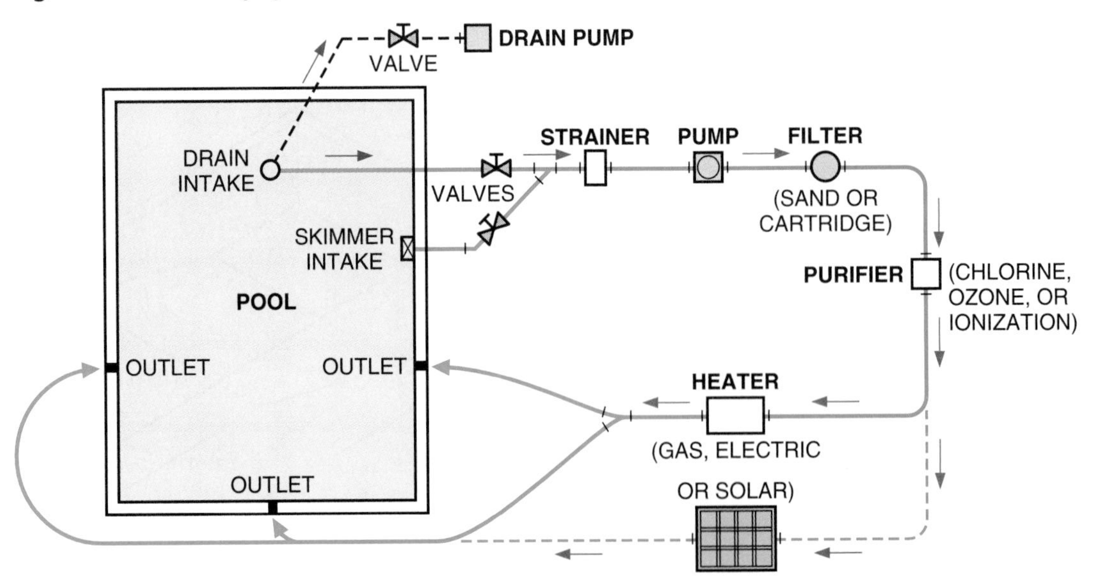

one outlet, while larger pools may need four or more.

Timing devices are recommended to control the amount of time the pump operates each day. Normally the pump is set to operate during the daylight hours because the pool equipment produces some noise. For this reason, it is better to locate the equipment away from lounging areas.

Additional (luxury) features to consider when designing a pool include a diving board, whirlpool spas, screened enclosures, and decorative fountains. Diving boards require additional foundation thickness.

Spas can be designed into the total pool layout. See Fig. 8-24. They can be included in the same pool circulation system. However, if extremely hot water is anticipated (hot tub), a by-pass system needs to be used to avoid overheating the pool water.

Fig. 8-24 ■ A hot tub spa can be located near a pool. *Emerald Pools Distribution, Inc.*

CHAPTER 8 Exercises

1. From catalogs, newspapers, and magazines, cut out pictures of porch furniture that you particularly like (you would choose for your own porch).

2. Plan a porch and/or patio for a house of your own design. Sketch the basic outline and the facilities.

3. Using a CAD system, draw the outline of the porch shown in Fig. 8-6.

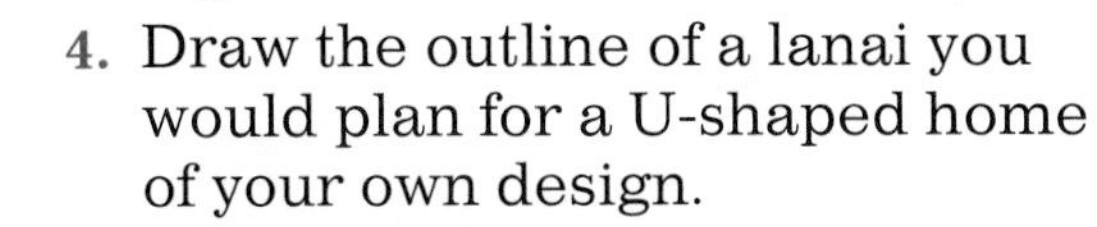

4. Draw the outline of a lanai you would plan for a U-shaped home of your own design.

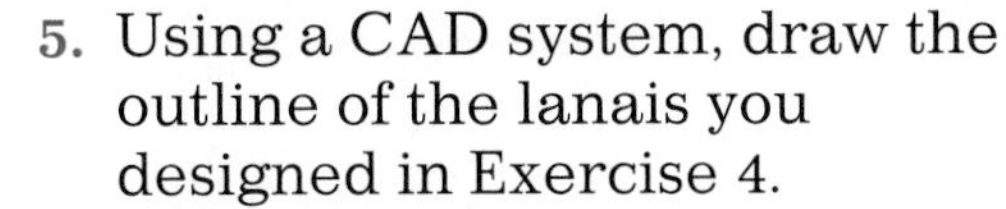

5. Using a CAD system, draw the outline of the lanais you designed in Exercise 4.
6. Sketch a floor plan of your own home. Add a lanai to connect two of the areas, such as the sleeping and living areas.
7. Explain the purposes of a lanai and describe two different plans where lanais would function well.
8. Name the required operating equipment needed for residential pools. What equipment is optional?
9. List the factors to consider in designing a pool.
10. Design a pool deck and patio area for a home you are designing. Locate the position of all operating equipment.

11. Using a CAD system, draw the outline of a free-form pool shape with deck and patio areas.

CHAPTER

9

Traffic Areas and Patterns

Objectives

In this chapter you will learn:

- to determine the effectiveness of a traffic pattern in a house.
- to plan hallways that function efficiently.
- guidelines for designing stairs.
- to calculate the correct space needed for stairways and stairwells.
- the kinds and functions of entrances.
- guidelines for entrance design.
- to design a foyer and entry.

Terms

apron
dividers
foyer
landings
riser
stairs
traffic areas
tread

The **traffic areas** of any building provides passage from one room or area to another and within a room or area. Planning the traffic areas of a residence is not extremely complex because relatively few people are involved. Nevertheless, efficient allocation of space is important. The main traffic areas of a residence include the halls, entrances or foyers, stairs, and areas of rooms that are part of a traffic pattern.

■ A well-designed, attractive stairway. Note the use of both artificial and natural lighting.
© E. "Manny" Abraben Architectural Photography

Traffic Patterns

Traffic patterns of a residence should be carefully considered when designing room layout. A minimum amount of space should be devoted to traffic areas. Extremely long halls and corridors should be avoided. These are difficult to light and provide no living space. Traffic patterns that require passage through one room to get to another should also be avoided, especially in the sleeping area.

The traffic pattern shown in the plan in Fig. 9-1A is efficient and functional. It contains a minimum amount of wasted hall space without creating a boxed-in appearance. It also provides access between areas and from the entrance without passing through other areas. Compare this pattern with that of the poorly designed plan in Fig. 9-1B.

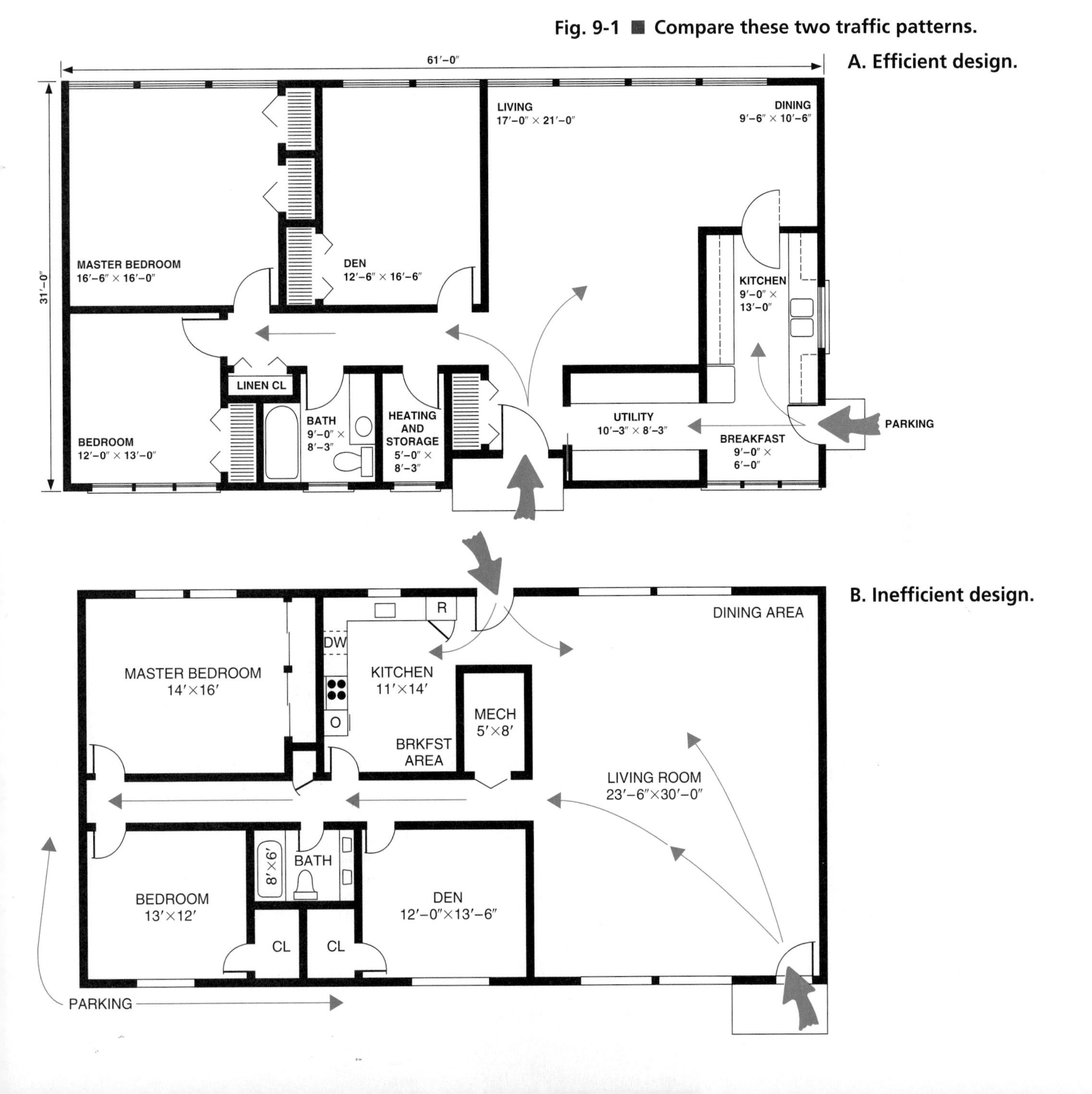

Fig. 9-1 ■ Compare these two traffic patterns.

A. Efficient design.

B. Inefficient design.

One method of determining the effectiveness of the traffic pattern of a house is to imagine yourself moving through the house by placing your pencil on the floor plan and tracing your route through the house as you perform a whole day's activities. Do the same for other members of the household. You will be able to see graphically where the heaviest traffic occurs and whether the traffic areas have been planned effectively.

Halls

Halls are the highways and streets inside the home. They provide a controlled path that connects the various areas of the house. Halls should be planned to eliminate or minimize the passage of traffic through rooms. Long, dark, tunnel-like halls should be avoided. Halls should be well lighted, light in color and texture, and planned with the decor of the whole house in mind. See Fig. 9-2.

Minimum hall widths are determined by building codes. Halls must also be wide enough for furniture movement and perhaps for wheelchair access.

One method of channeling hall traffic without the use of solid walls is with the use of **dividers.** Planters, half walls, louvered walls, and even furniture can be used as dividers. Figure 9-2 shows the use of furniture components in dividing the living area from the hall. This arrangement enables both the hall and the living area to share ventilation, light, and heat. Another method of designing halls and corridors as an integral part of an area is with the use of movable partitions. The plans in Fig. 9-3 illustrate some of the basic principles of efficient hall design.

Fig. 9-2 ■ Although this hall is long, the space is broken up by bookshelves, a gallery, and by light variations.

Fig. 9-3 ■ Design techniques to reduce hall length.

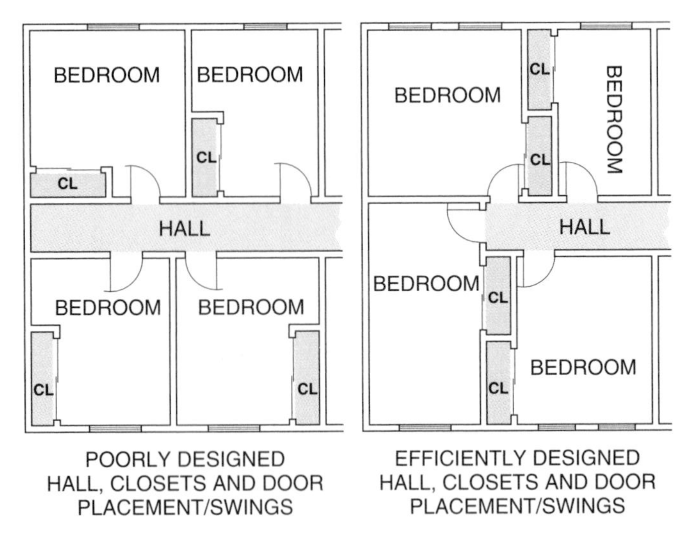

Stairs

Stairs are inclined hallways that provide access from one level to another. Stairs may lead without a change of direction, or they may turn 90° or 180° by means of **landings.** There are several types of stairs. See Fig. 9-4.

Materials and Lighting

With the use of newer, stronger building materials and new techniques, stairs can now be supported by many different devices. Stairs no longer need to be enclosed in areas that restrict light and ventilation. See Fig. 9-5.

Fig. 9-4 ■ Basic types of stairs.

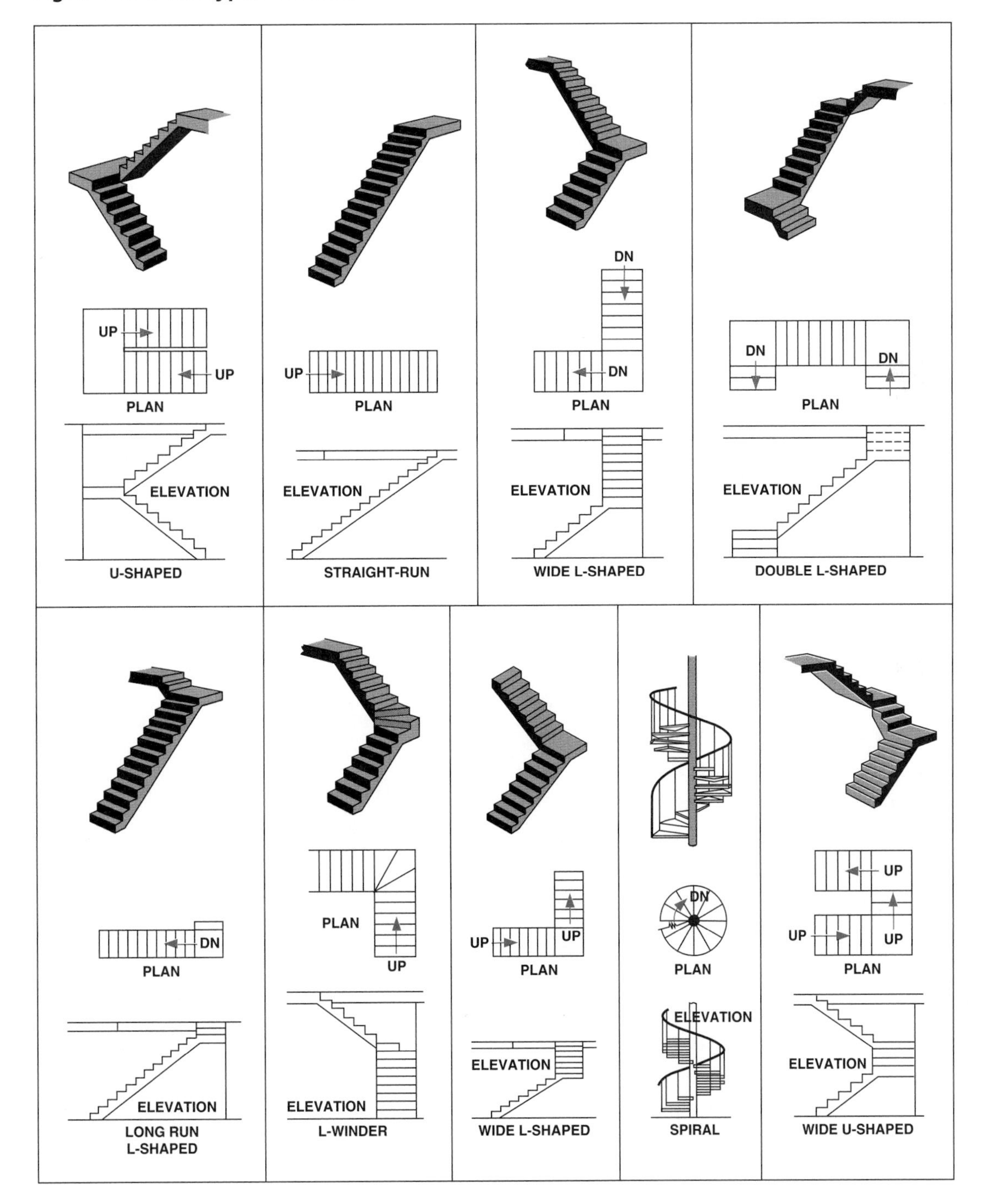

Fig. 9-5 ■ Spiral stair system in an open area. *Scholz Homes Inc.*

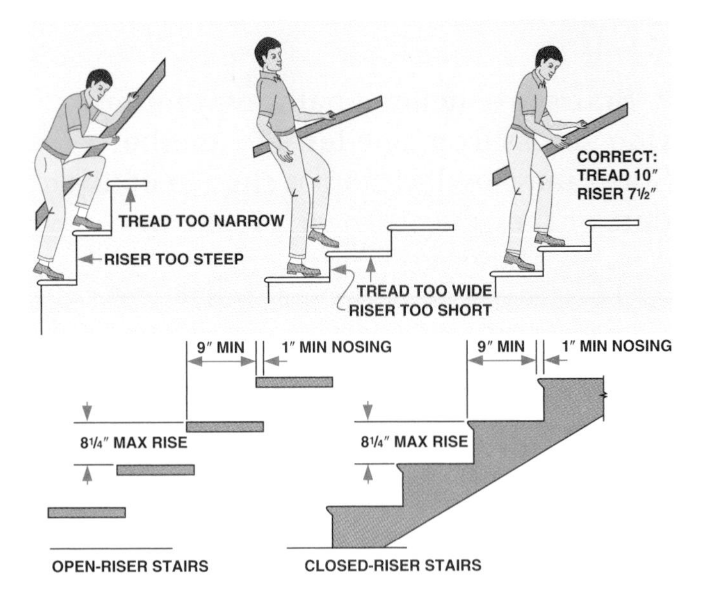

Fig. 9-6 ■ Correct tread and riser dimensions are important to stair design.

Stairwells (areas for stairs) should be lighted at all times when in use. Natural light is the most energy-efficient. Thus windows should be utilized to provide natural light for stairs wherever possible. Three-way switches should be provided at the top and bottom of the stairwell to control the stair lighting. (For details, see Chapter 31, Electrical Drawings.)

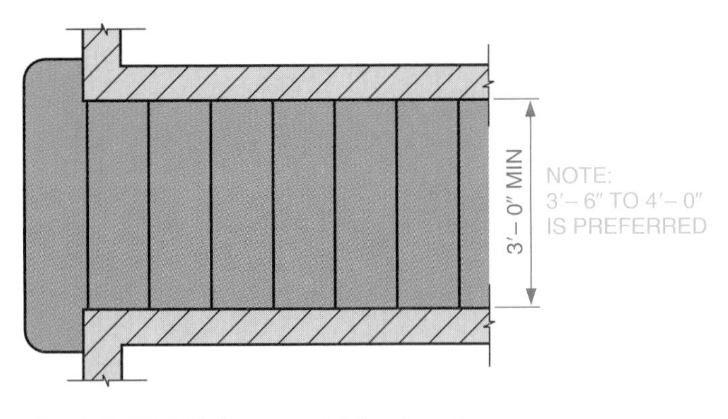

Fig. 9-7 ■ Minimum width of stairs.

Size and Shape

There are many variables to consider in designing stairs. The **tread** is the horizontal part of the stair, the "step," or the part upon which you walk. Treads must be made of or covered with non-slip surfaces. The average depth, or distance from front to back, of the tread is 10″ (254 mm). The **riser** is the vertical part of the stair. The average riser height is 7¼″ (184 mm). See Fig. 9-6.

The overall width of the stairs is the distance between the stair railings. A minimum of 3′ (914 mm) should be allowed for the total stair width. However, a width of 3′-6″ (1067 mm) or even 4′ (1219 mm) is preferred to accommodate the movement of furniture. See Fig. 9-7.

Headroom is the vertical distance between the top of each tread and the top of the stairwell ceiling. A minimum headroom distance of 6′-6″ (1981 mm) should be allowed. However, distances of 7′ (2134 mm) are more desirable.

The tread width, the riser width, the width of the stairwell opening, and the headroom all help to determine the total length of the stairwell. Landing dimensions are generally determined by the size

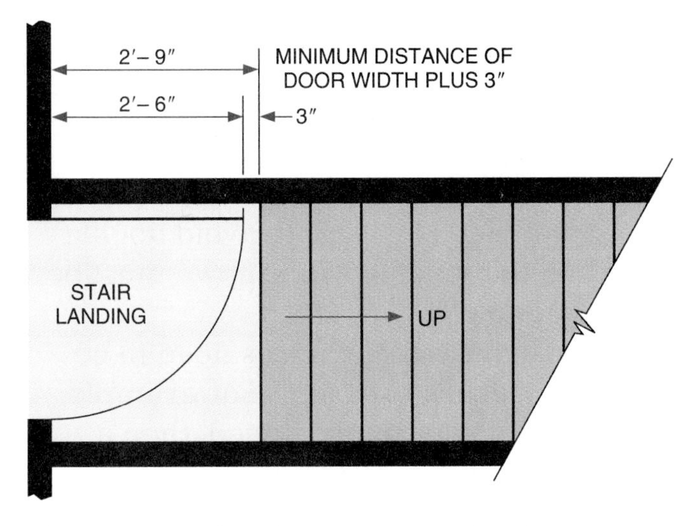

Fig. 9-8 ■ Landing dimensions to allow for door swing.

of the stairs and the space for the stairwell. More clearance must be allowed where a door opens onto a landing. See Fig. 9-8. A landing should be planned for stair systems that have more than 16 risers. It should be located at the center between levels to eliminate long runs.

Entrances

Entrances are divided into several different types: the site entrance, the main building entrance, the service entrance, and special-purpose entrances. House entrances usually have an outside waiting area (porch, marquee, lanai), a separation (door), and an inside waiting area (**foyer,** entrance hall). See Fig. 9-9.

Function and Types

Entrances provide for and control the flow of traffic into and out of a building. Different types of entrances have different functions depending on the design of the structure.

SITE ENTRANCE To design a site entrance, attention must first be given to the space from the street or road to the house. A site entrance includes the driveway, walkway, and adjacent parking or turnaround space for vehicles. See Fig. 9-10.

Fig. 9-9 ■ Floor plan showing the types of entrances. *Home Planners Inc.*

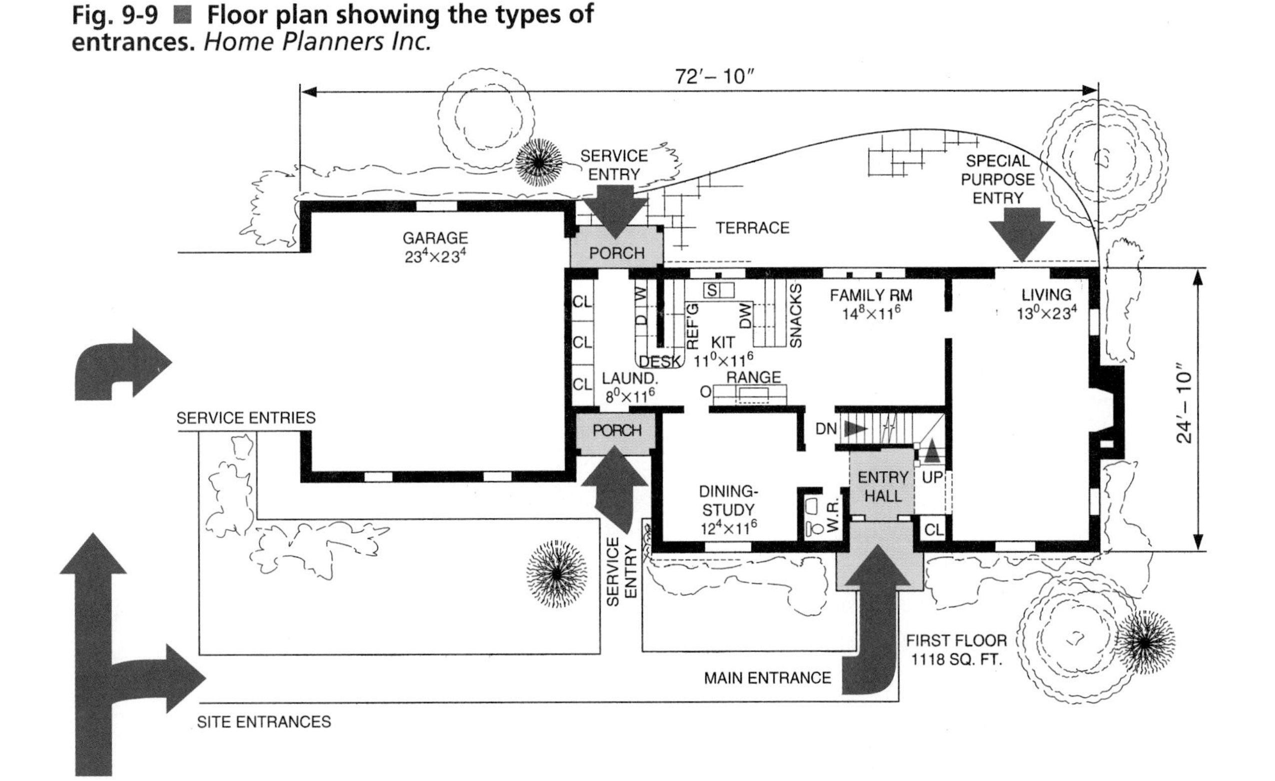

Fig. 9-10 ■ A driveway as a residential site entrance.

Driveways connect the street or road to a walkway and to a garage or carport and should be easily identified from the street. Different shapes of driveways are designed to lead directly to the garage and to blend well with the site.

Driveways may be straight or curved, depending upon the landscaping and/or land contours. See Fig. 9-11. Circular or semicircular driveways allow a car to return to the street without driving in reverse or turning around.

Fig. 9-11 ■ Plan for a circular drive.

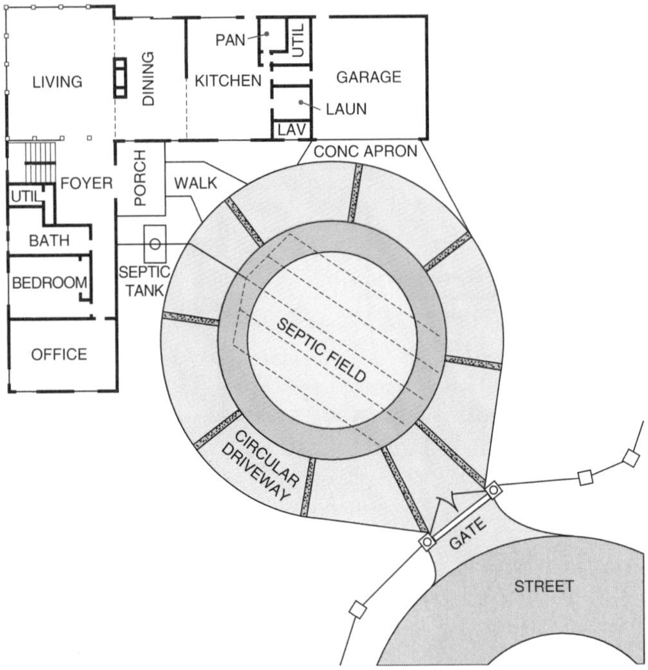

A turning and parking **apron** (area leading to garage) provides a means to exit a driveway without backing up onto the street. See Fig. 9-12. To avoid double backing, the turning radii shown must be strictly followed.

Some driveway entrances need to be gated for security reasons. For example, if a pool is not separately fenced, then a perimeter fence along the property borders is required to have a drive entrance gate. See Fig. 9-13.

Walkways leading to a front entrance may either connect the house entrance directly with the street or sidewalk, or lead to the driveway, or both. See Fig. 9-14.

MAIN HOUSE ENTRANCE The main entrance provides access to the house. It is the entrance through which guests are welcomed and from which all major traffic patterns radiate. The main entrance should be readily identifiable. It should provide shelter for anyone awaiting entrance.

Some provision should be made in the main entrance wall to see callers from the

Fig. 9-12 ■ Minimum dimensions for turning and parking.

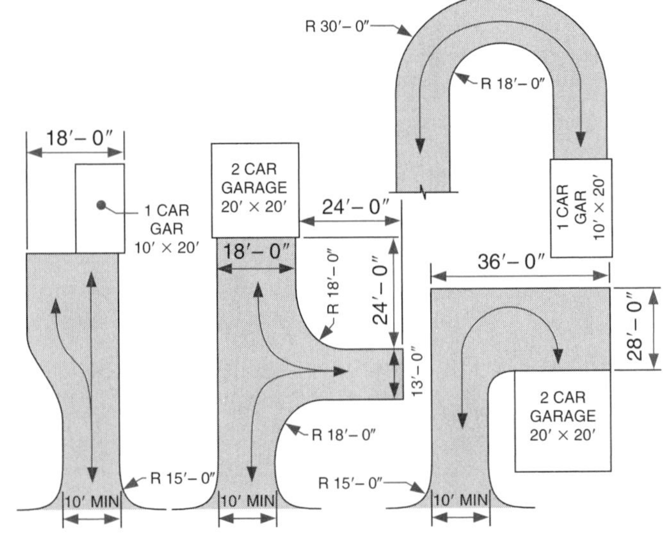

Fig. 9-13 ■ A gated fence leading to a circular drive. *Glenn Wright Construction—J. Brian Acker, Photographer*

Fig. 9-14 ■ Walkway connecting an entrance with the driveway. (The plan for this walkway and drive are shown in Fig. 9-11.)

inside before admitting them. Side panels, lights (panes) in the door or windows which face the side of the entrance are ways to view someone outside. See Fig. 9-14.

The main entrance should be planned to create a desirable first impression. A direct view of other areas of the house from the foyer should be baffled but not sealed off. The exterior parking areas should also be baffled from view.

The entrance foyer should include a closet for the storage of outdoor clothing. This foyer closet should be large enough for both family and guests to use it.

SERVICE ENTRANCE The service entrance is to be used for any entry or exit that would be inappropriate and inconvenient at the main entrance. A person should be able to pass through the service entrance and enter parts of the service area, such as the garage, laundry, or workshop. Supplies can also be delivered to the service areas without going through other parts of the house.

SPECIAL-PURPOSE ENTRANCES Special-purpose entrances and exits do not provide for outside traffic. Instead they are intended for movement from the inside living area of the house to the outside living areas. A sliding door from the living area to the patio is a special-purpose entrance. It is not an entrance through which street or sidewalk traffic would have access. See Fig. 9-15.

Location

The main entrance should be centrally located to provide easy access to each area. It should be conveniently accessible from driveways, sidewalks, or street.

The service entrance should be located close to the driveway and garage, and near the kitchen or food-storage areas.

Special-purpose entrances and exits are often located between the bedroom and the quiet patio, between the living room and the living patio, and between the dining room or kitchen and the dining patio.

Decor

To create a desirable first impression, a main entrance should be easily identifiable and yet be an integral part of the architectural style.

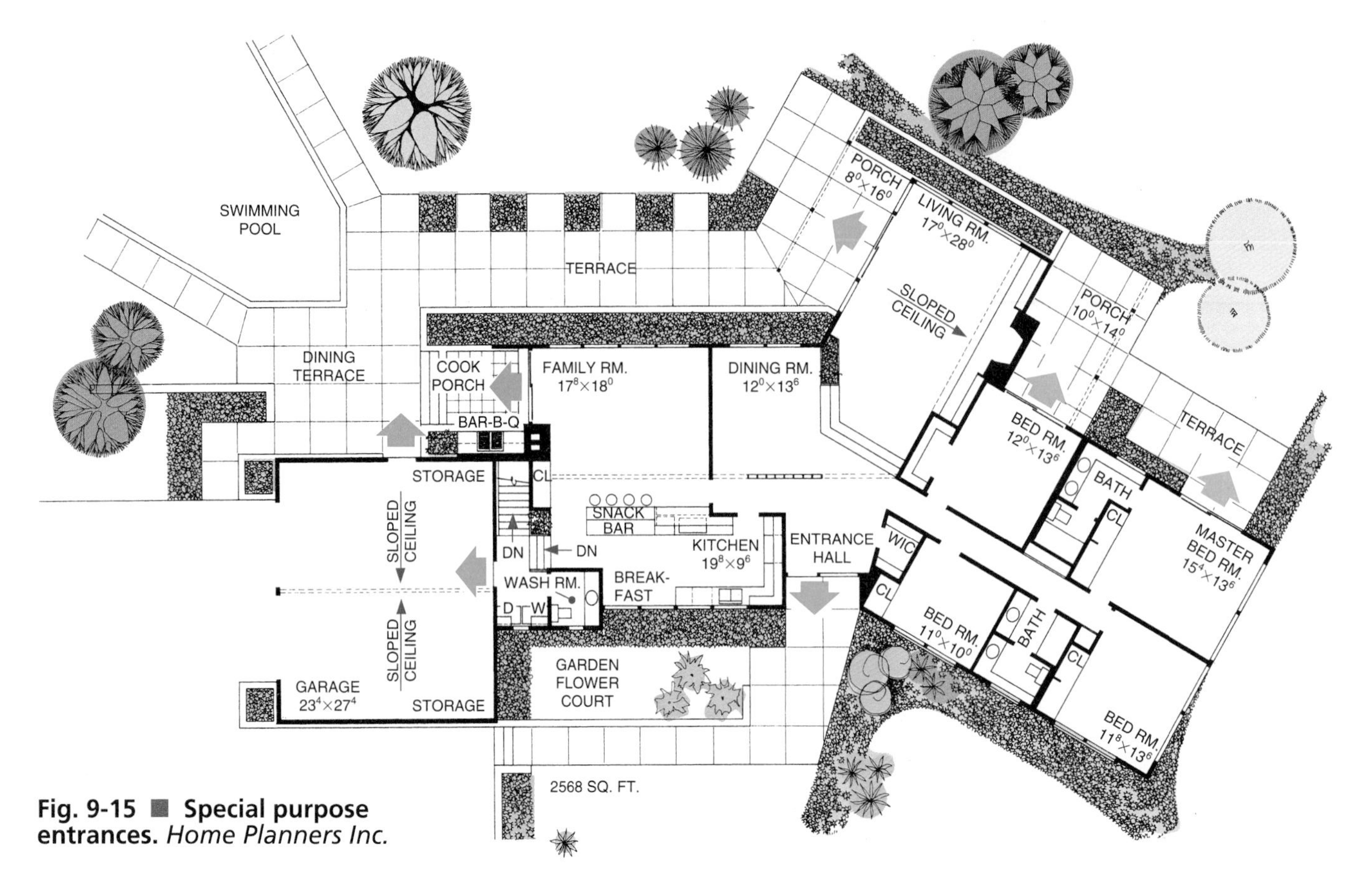

Fig. 9-15 ■ Special purpose entrances. *Home Planners Inc.*

The total design of the entrance should be consistent with the overall design of the house. That means the design of the door, the side panel, and the deck and cover should be directly related to the lines of the house. See Fig. 9-16.

OPEN AND CLOSED PLANNING Open planning is desirable for entrances. This means the view from the main entrance to the living area should be baffled without creating a boxed-in appearance. The foyer should not appear to be a dead end. The extensive use of glass, effective lighting, and carefully placed baffle walls can create an open and inviting impression. See Fig. 9-17.

Open planning between the entrance foyer and the living areas can also be accomplished by the use of louvered walls or planter walls. These provide a relief or change in the line of sight but not a complete separation. Lowering or elevating the foyer or entrance can also produce the desired effect of separation without enclosing the area. Foyers are not bounded in open planning. In formal or closed plans, the foyer is either partially or fully closed off. See Fig. 9-18.

SURFACE MATERIALS The outside portion of the entrance should be weather-resistant wood, stone, brick, or concrete. The foyer deck should be easily maintained and be resistant to mud, water, and dirt brought in from the outside. Asphalt, vinyl or rubber tile, stone, flagstone, marble, and terrazzo are most frequently used for the foyer deck. The use of a different material in the foyer area helps to define the area when no other separation exists.

Paneling, masonry, murals, and glass are used extensively for entrance foyer walls. The walls of the exterior portion of

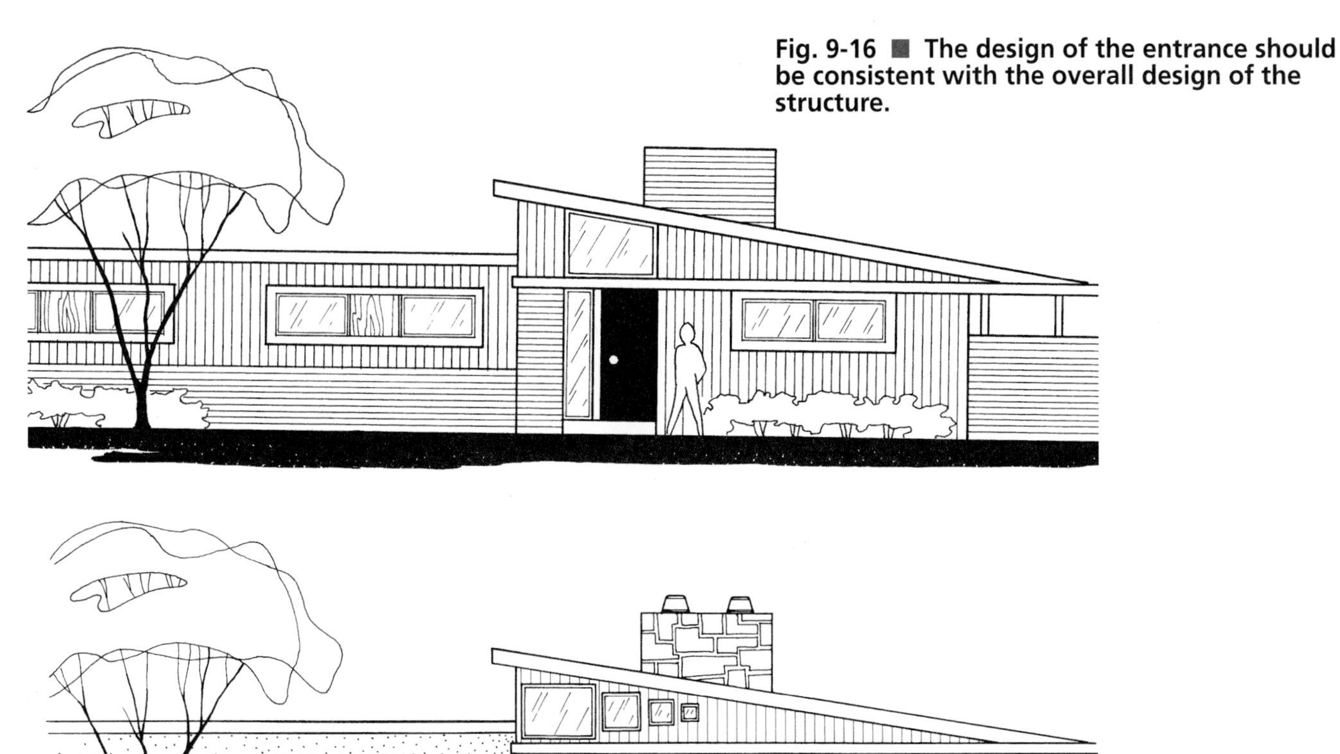

Fig. 9-16 ■ The design of the entrance should be consistent with the overall design of the structure.

Fig. 9-17 ■ A large, well-designed open plan entrance foyer. Note the double doors, two-level ceiling, and extensive use of open space. *John Henry, Architect*

Fig. 9-18 ■ Closed plan foyer that provides access to all areas.

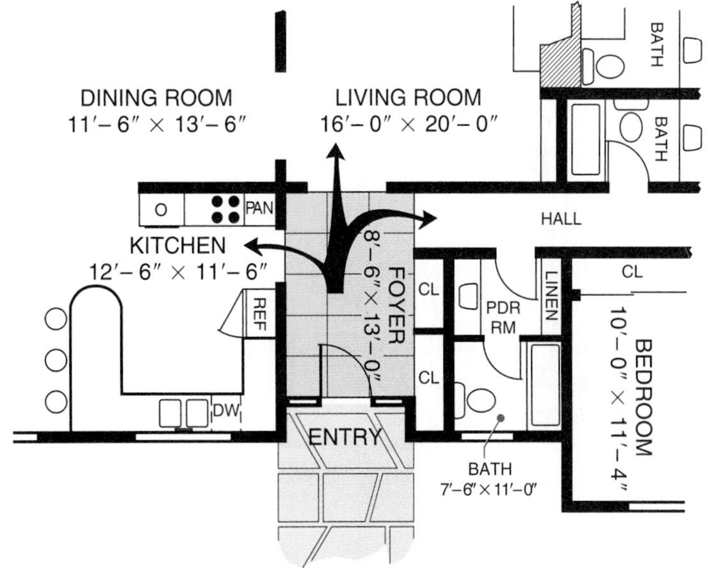

the entrance should be consistent with the other materials used on the exterior of the house.

LIGHTING An entrance must be designed to function day and night. Natural lighting should be planned for lighting entrance areas during daylight hours. General lighting, spot lighting, and all-night lighting are effective after dark.

Lighting can be used to accent distinguishing features or to illuminate the pattern of a wall. This type of lighting actually provides more light by reflection and helps to identify and accentuate the entrance at night. See Fig. 9-19.

Size and Shape

The size and shape of the areas inside and outside the entrance depend on the budget and the type of plan. The outside covered portion of the entrance should be large enough to shelter several people and at the same time provide the amount of space needed to open a storm door. Outside shelter areas are the same range in size and shape as porches and patios. (Refer back to Chapter 8.)

The inside of the entrance foyer should be sufficiently large to allow several people to enter at the same time, remove their coats, and put their things in the closet. A 6′ × 6′ (1829 mm × 1829 mm) foyer is considered minimum for this function. A foyer 8′ × 10′ (2438 mm × 3048 mm) is average, but a more desirable size is 8′ × 15′ (2438 mm × 4572 mm).

A foyer arrangement must allow for the swing of the entrance door or doors. If the foyer is too shallow, passage will be blocked when the door is open, and only one person can enter at a time. Figure

Fig. 9-19 ■ Well-planned lighting illuminates and accentuates an entrance at night. *John Henry, Architect*

9-18 is a typical foyer plan showing a single door swing. Use this checklist for the design of a main entrance:

- Adequate space to handle traffic flow.
- Access to all three areas of a home.
- A guest closet.
- Bathroom access for guests.
- Consistent decor.
- Outside weather protection.
- Effective lighting day and night.
- Avoid traffic through the living room center.

CHAPTER 9 Exercises

1. Sketch the floor plan of a home of your design. Plan the most efficient traffic pattern by tracing the route of your daily routine.
2. Using a CAD system, draw the plan view of one of the stair systems shown in this chapter.
3. Draw or sketch a plan view of a stair system in your home or school.
4. Name the types of stairs that turn 90°, 180°, 360°, and 0°.
5. List the types of entrances and tell the function of each type.

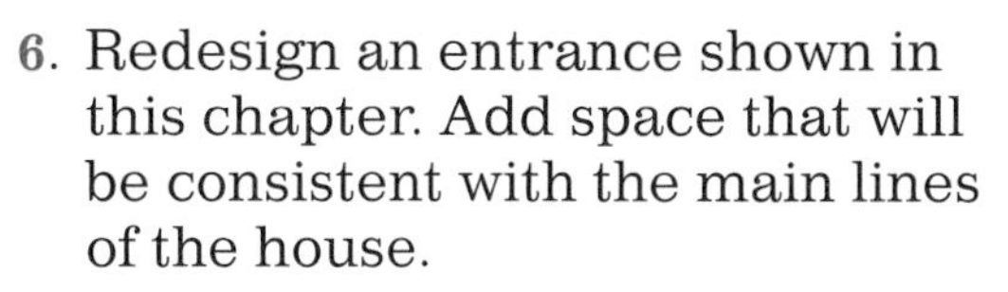

6. Redesign an entrance shown in this chapter. Add space that will be consistent with the main lines of the house.
7. Redesign and enlarge the foyer for the living area shown in Fig. 9-18. Label the materials you select for the outside deck, overhang, access walk, foyer floor, and walls.

8. Draw a foyer to the plan of a house you are designing.
9. Using a CAD system, draw a plan of the foyer you redesigned in Exercise 7.

CHAPTER 10

Kitchens

Objectives

In this chapter you will learn to:

- apply guidelines to efficient kitchen design.
- determine the best shape, size, and location for the kitchen.
- plan and draw a work triangle for a kitchen.
- design an aesthetically consistent decor for a kitchen.
- sketch small and large kitchens of the basic kitchen shapes.

Terms

corridor kitchen
family kitchen
island kitchen
L-shaped kitchen
one-wall kitchen
peninsula kitchen
U-shaped kitchen
work triangle

A well-planned kitchen is one that functions efficiently, and yet is attractive and easy to maintain. To design an efficient kitchen, the designer must consider the room's function, location, decor, size, and shape just as with other rooms. However, because a kitchen requires so much equipment, the design of a kitchen entails additional considerations and decisions.

■ **A well-designed kitchen is both efficient and pleasing in appearance.**
Mark Romine

Kitchen Design Considerations

Understanding the functions of a kitchen is the first step in planning a kitchen's design.

Functions

Food preparation is, of course, the primary function of the kitchen. However, the kitchen may also be used as a dining area. The proper placement of appliances is important in a well-planned kitchen. Locating appliances in an efficient pattern eliminates wasted motion. An efficient kitchen has three basic areas or centers: the storage center, the cooking center, and the cleanup center. A fourth area, mixing, is combined into one or more of the others, usually storage. See Fig. 10-1.

STORAGE AND MIXING CENTER The refrigerator is the major appliance in the storage and mixing center. The refrigerator may be freestanding, built-in, or even suspended from a wall. Cabinets for the storage of utensils and food ingredients, as well as a countertop work area, are also included at this center.

Fig. 10-1 ■ Basic kitchen areas. *Whirlpool Corp.*

COOKING CENTER The major appliances in the cooking center are the range and oven. The range and oven may be combined into one appliance or be separated into two appliances, with the burners installed in the countertop (cooktop) as one appliance and an oven built into a cabinet. The cooking center should have countertop work space, as well as storage space for minor appliances and cooking utensils. An adequate supply of electrical or gas outlets for using appliances is necessary.

CLEANUP CENTER At the cleanup center, the sink is the major appliance. Sinks are available in one-, two-, or three-bowl models with a variety of cabinet arrangements, countertops, and drainboard areas. The cleanup center may also include a waste-disposal unit, an automatic dishwasher, a waste compactor, and cabinets for storing cleaning supplies.

THE WORK TRIANGLE If you draw a line connecting the three centers of the kitchen, a triangle is formed. See Fig. 10-2. This is called the **work triangle**. The perimeter of an efficient kitchen work triangle should be no more than 22′ (6706 mm). Although the size of the work triangle is an indication of kitchen efficiency, the triangle is primarily useful as a starting point in kitchen design. The triangle should not be rigidly maintained at the expense of flexibility and creativity.

The arrangements of the three areas of the work triangle may vary greatly. However, efficient arrangements can be designed in each of the seven basic types of kitchens described here.

Fig. 10-2 ■ The work triangle is the area between the basic centers. *Jenn-Air*

Fig. 10-3 ■ A U-shaped kitchen with cooking and eating areas on one side. *Frigidaire Corp.*

Types of Kitchens

U-SHAPED KITCHEN The **U-shaped kitchen** is very efficient and popular. The sink is located at the bottom of the U, and the range and the refrigerator are at the opposite ends. In this arrangement, traffic passing through the kitchen is completely separated from the work triangle. The open space in the U between the sides should be 4′ (1219 mm) or 5′ (1524 mm). This arrangement produces a very efficient small kitchen. See Fig. 10-3. Figure 10-4 shows various U-shaped-kitchen designs and the planned work triangles.

When designing U-shaped kitchens, special attention must be given to door hinges and drawer positions. Design cabinet doors and drawers to open without interfering with each other, especially at cabinet corners.

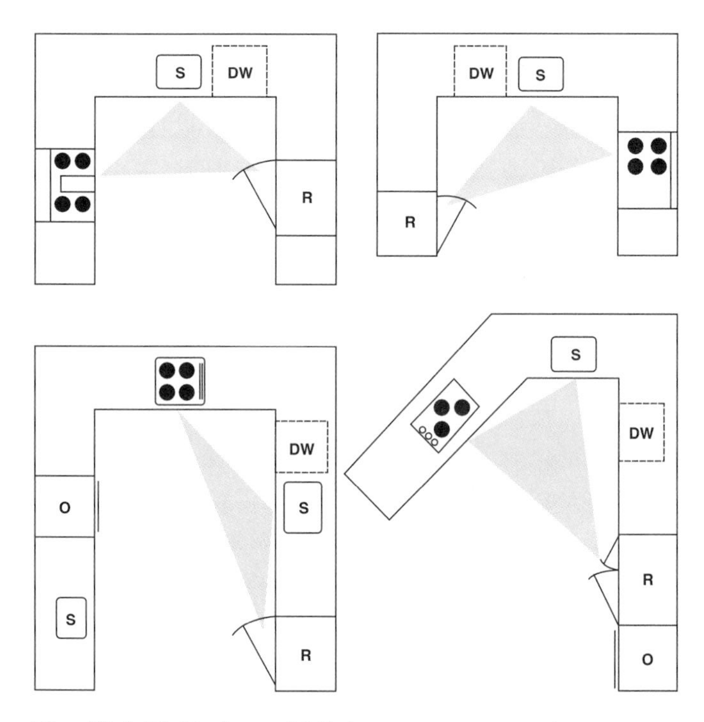

Fig. 10-4 ■ U-shaped kitchen arrangements.

PENINSULA KITCHEN The **peninsula kitchen** is similar to the U-shaped kitchen, but one end of the U is not adjacent to a wall. It projects into the room like a piece of land (peninsula) into a body of water. This peninsula is often used for the cooking center. However, it may serve several other functions as well. The peninsula is often used for an eating area as well as for food preparation. See Fig. 10-5. It may join the kitchen to the dining room or family room. Figure 10-6 shows various arrangements of peninsula kitchens and the resulting work triangles.

Fig. 10-5 ■ Different levels are used on this peninsula to separate the food preparation area from the eating area. *DuPont Co.*

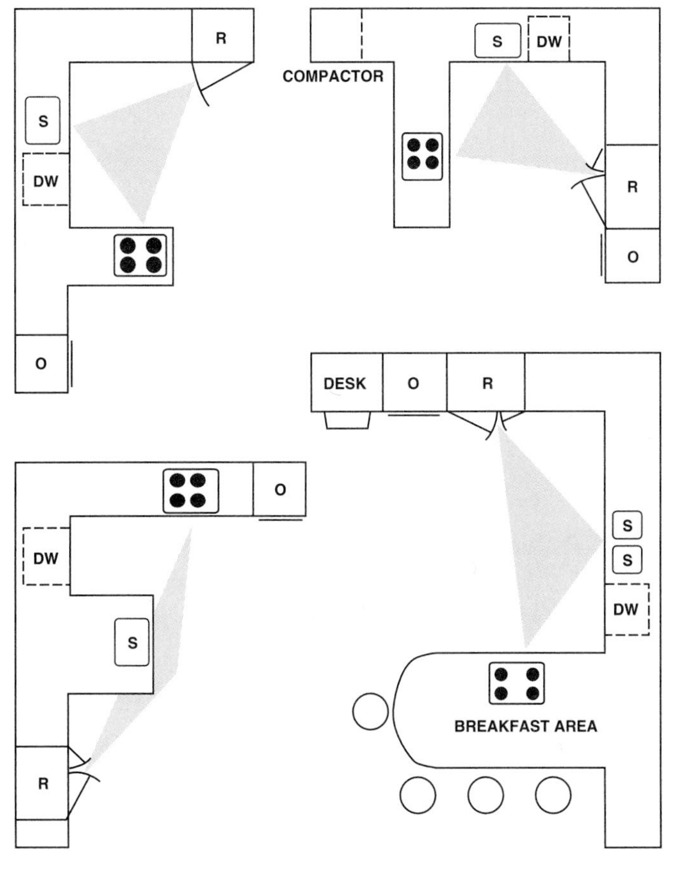

Fig. 10-6 ■ Peninsula kitchen arrangements.

Most peninsula kitchens contain large countertops for work space. Peninsulas may contain only lower or base cabinets, but some may include upper cabinets suspended from ceilings.

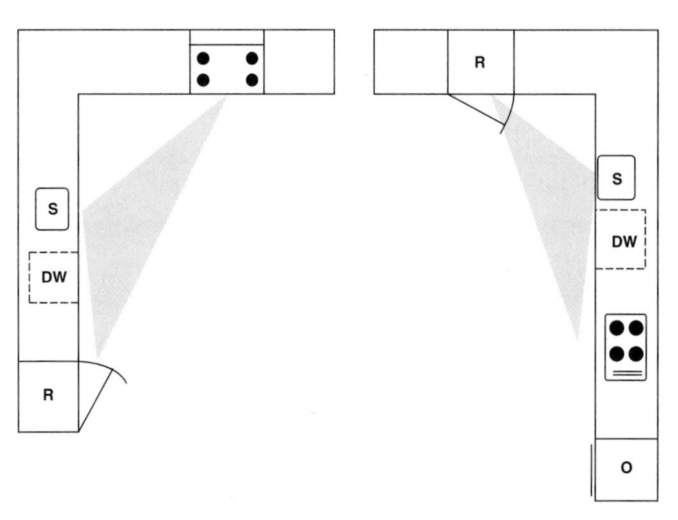

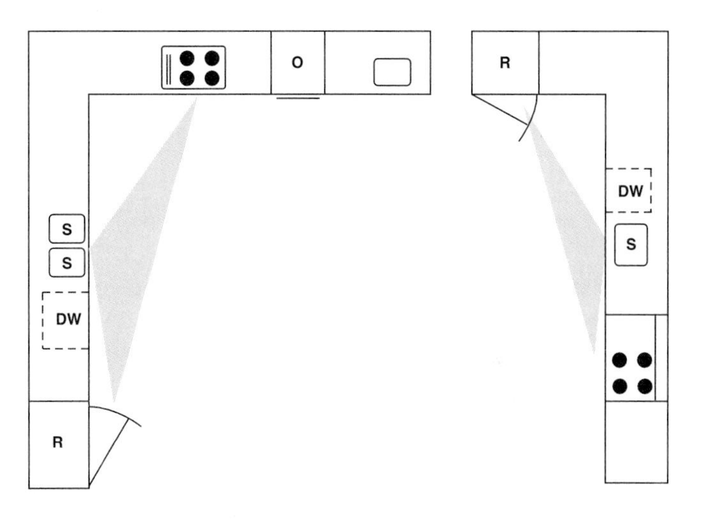

Fig. 10-7 ■ L-shaped kitchen arrangements.

L-SHAPED KITCHEN The **L-shaped kitchen** has continuous counters, appliances, and equipment located on two adjoining, perpendicular walls. Two work centers are usually located on one wall and the third center is on the other wall. See Fig. 10-7. The work triangle is not in the traffic pattern. If the walls of an L-shaped kitchen are too long, the compact efficiency of the kitchen is destroyed.

An L-shaped kitchen requires less space than the U-shaped kitchen. The remaining open space often created by an L-shaped arrangement can serve as an eating area, without taking space from the work areas. If the center area is used for eating, a minimum of 36″ (914 mm) must be allowed as an aisle between cabinets and chairs.

CORRIDOR KITCHEN Two-wall **corridor kitchens** are very efficient arrangements for long, narrow rooms. See Fig. 10-8. They are very popular for small apartments, but are used extensively anywhere space is limited. A corridor kitchen produces a very efficient work triangle, as long as traffic does not need to pass through that work triangle. The corridor space between cabinets (not walls) should be no smaller than 4′ (1219 mm), preferably 6′ (1829 mm). One of the best work arrangements locates the refrigerator and sink on one wall and the range on the opposite wall.

ONE-WALL KITCHEN A **one-wall kitchen** is an excellent plan for small apartments, cabins, or houses in which little space is available. The work centers are located along one line rather than in a triangular shape, but this design still produces an efficient arrangement. See Fig. 10-9.

Fig. 10-8 ■ Corridor kitchen arrangements.

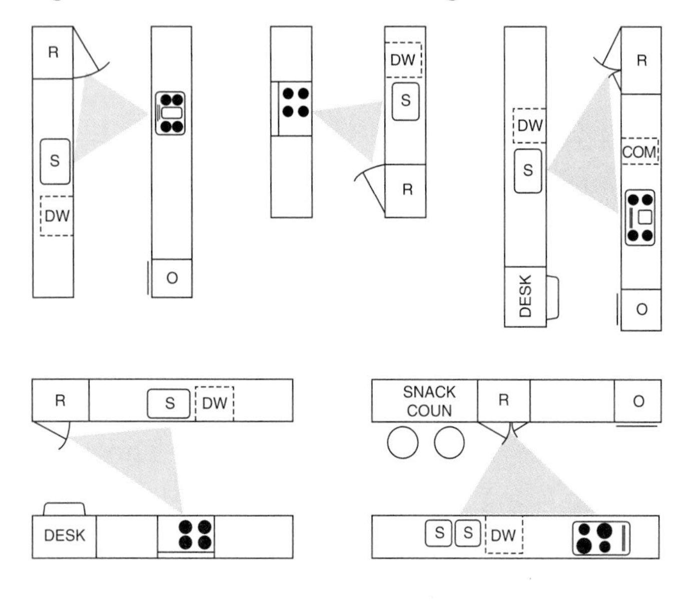

Fig. 10-9 ■ One-wall kitchen arrangements.

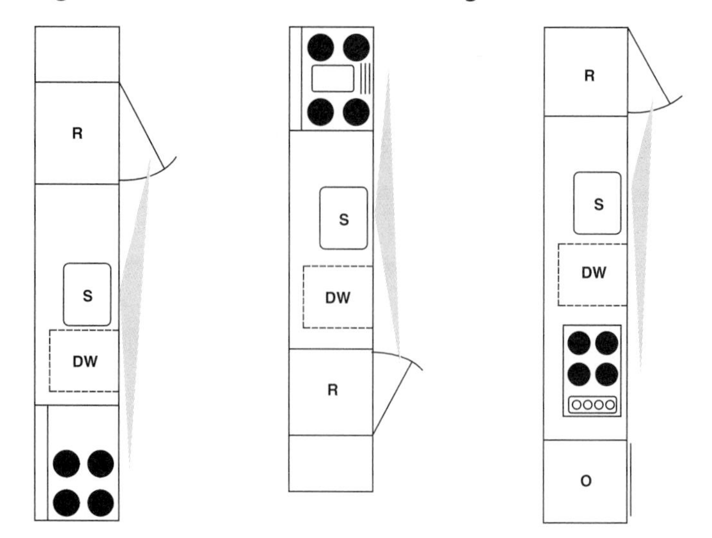

When planning a one-wall kitchen, the designer must be careful to avoid creating walls that are too long. Adequate storage facilities need to be well planned also, since space is often limited in a one-wall kitchen.

ISLAND KITCHEN The **island kitchen**, another geographically-named arrangement, has a separate, freestanding structure in the kitchen that is usually located in the central part of the room. An island in the kitchen is accessible on all sides. It usually has a rangetop or sink, or both. See Fig. 10-10. Other facilities are sometimes located in the island, such as a mixing center, work table, serving counter, extra sink, and/or snack center. See Fig. 10-11. Figure 10-12 shows examples of other island facilities. The island design is especially convenient when two or more persons work in the kitchen at the same time.

When an island contains a range or grill, allow at least 16″ (406 mm) on the sides for utensil space. Also consider the use of a downdraft exhaust system which pulls vapors down and out rather than up to eliminate the need for overhead

Fig. 10-10 ■ An island kitchen with a two-level island. *Kraftmaid*

Fig. 10-11 ■ This island provides both a snack center and a sink. *Norcraft Companies, Inc.*

Fig. 10-12 ■ Island kitchen arrangements.

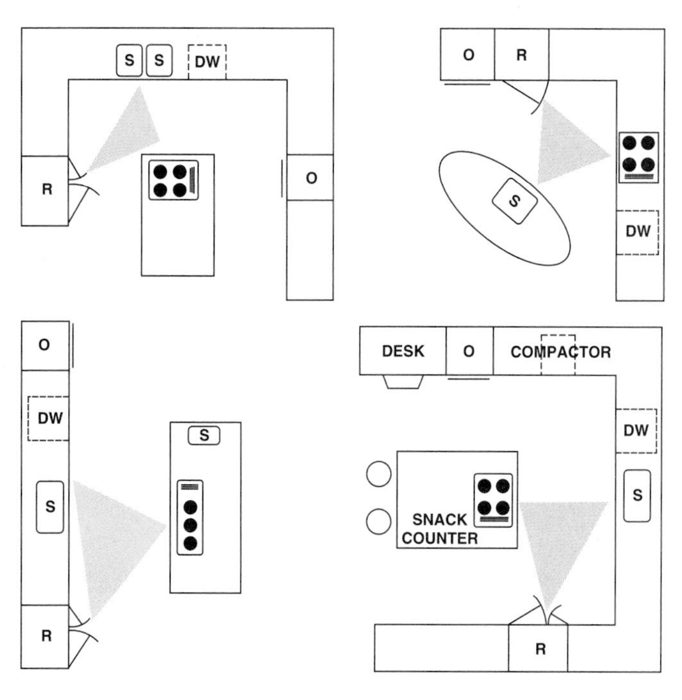

Fig. 10-13 ■ A family kitchen with a separate area for table and chairs. *Merrillat*

hooded vents. Allow at least 42″ (1067 mm) on all sides of an island. If used for eating, also add the depth of the chair or stool.

FAMILY KITCHEN The **family kitchen** is an open kitchen using any kitchen shape. The function of an open kitchen, however, is to provide a meeting place for the entire family—in addition to the usual kitchen services. A family kitchen often appears to have two parts in one room. The three food preparation work centers comprise one section. The dining area and family-room facilities comprise another section. See Fig. 10-13. Figure 10-14 shows several possible arrangements for family kitchens.

Family kitchens must be rather large to accommodate these facilities. An average size for a family kitchen is 225 sq. ft. (20 sq. m). Eating areas can be designed

Fig. 10-14 ■ Family kitchen arrangements.

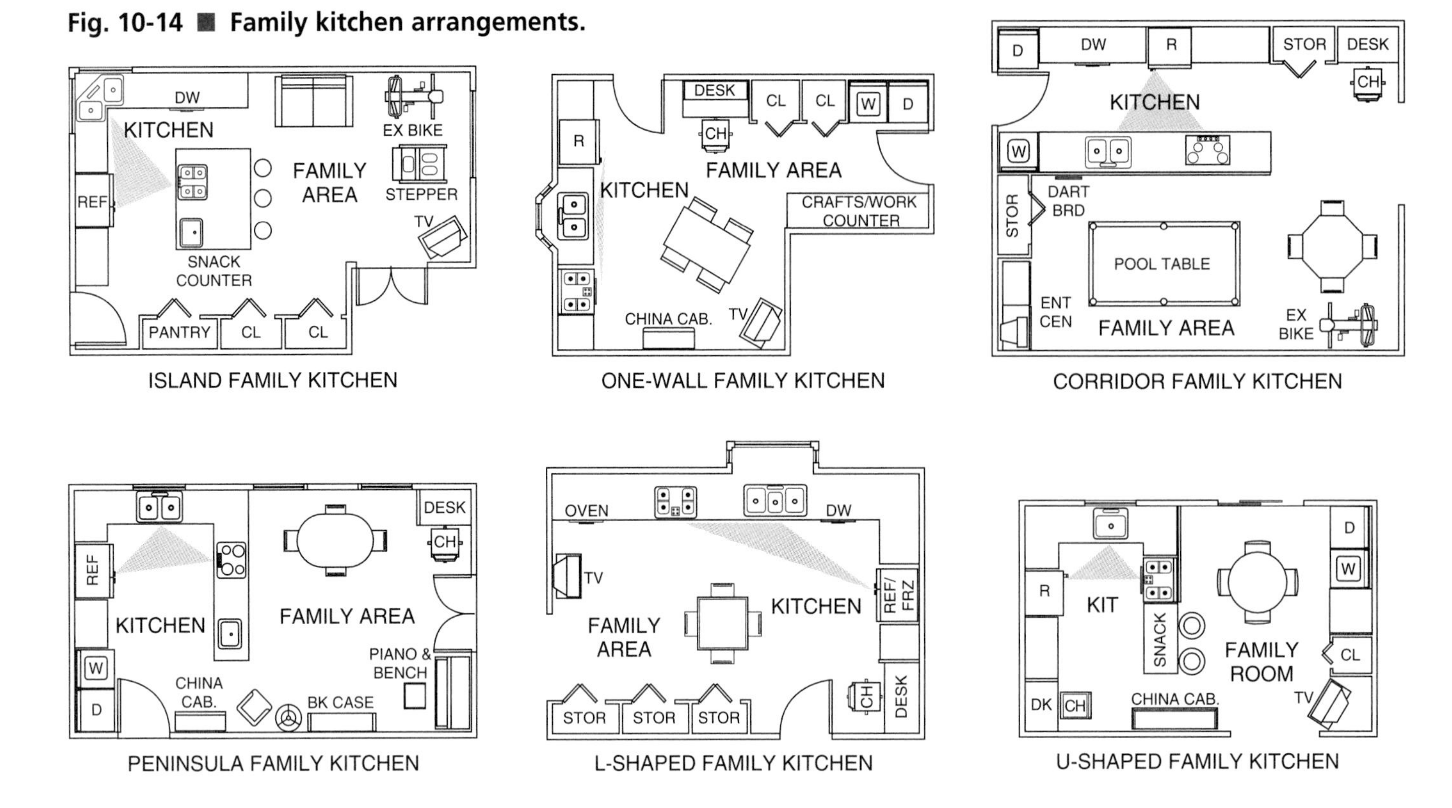

with either tables and chairs or with chairs and/or stools at a counter. When counters are used for eating, allow at least 12″ (305 mm) for knee space between the end of the counter and the face of the base cabinet.

Regardless of its shape, the kitchen is the core of the service area and should be located near the service entrance as well as near the waste-disposal area. The kitchen must be adjacent to eating areas, both indoors and outdoors. The children's play area should also be visible or easily accessible from the kitchen.

Decor

Kitchens cost more per square foot than any other room. Most of this cost relates to the selection of appliances, cabinetry, and fixtures. By selecting the least expensive models of appliances, hardware, and cabinetry, the same kitchen design can often be built for one-fourth the cost of a kitchen which contains the most expensive features.

Even though most kitchen appliances are produced in contemporary designs, some clients and designers prefer to decorate kitchens with a traditional style as a motif or theme. The cabinets, floors, walls, and accessory furniture would then be selected according to that chosen theme. Designing a totally harmonious kitchen is made easier by the wide variety of appliance sizes, colors, and styles.

Regardless of the style, the kitchen walls, floors, countertops, and cabinets should require a minimum amount of maintenance. See Fig. 10-15. Materials that are relatively maintenance-free include stainless steel, stain-resistant plastic, ceramic tile, washable wall coverings, washable paint, vinyl, molded and laminated plastic countertops, doors, drawers, and cabinet bases.

Options in kitchen design have broadened because of new synthetic and com-

Fig. 10-15 ■ Can you identify the features in this attractive kitchen that make it easy to maintain? *DAL-TILE*

posite materials and new construction methods for cabinets and countertops. Many kitchens now have what only the highest quality kitchens had a few years ago.

Size and Shape

When planning kitchens, cabinet sizes and spacing plus the size of appliances need to be considered to assure that adequate space is available for all elements of the design. See Figs. 10-16 and 10-17. Figure 10-18 shows typical sizes of common kitchen appliances.

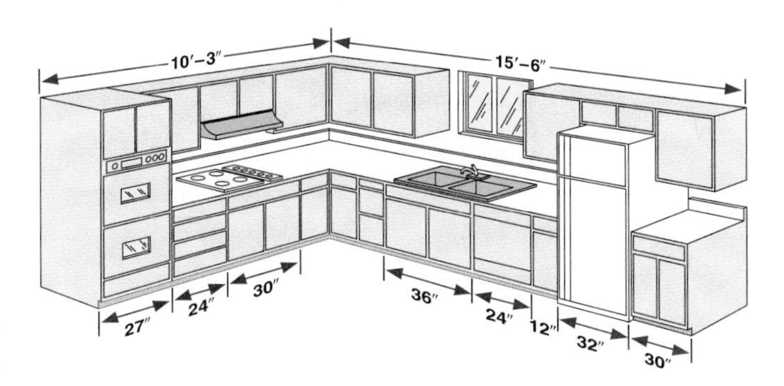

Fig. 10-16 ■ Standard horizontal dimensions used in kitchen design. *William Wagoner*

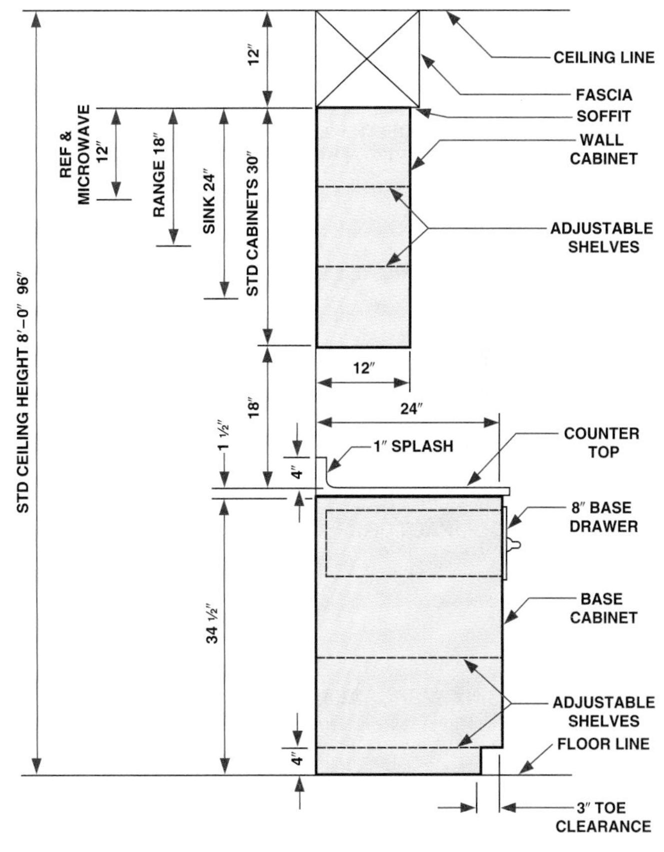

Fig. 10-17 ■ Standard dimensions for wall and base cabinets.

Fig. 10-18 ■ Common sizes of kitchen appliances.

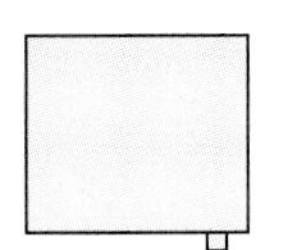

REFRIGERATORS
HT 56″ TO 66″
DEPTH 24″
WIDTH 24″ TO 32″

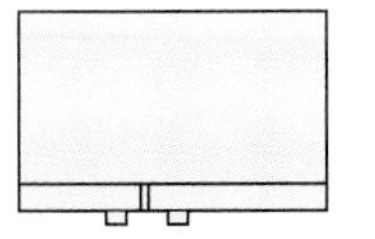

REF/FREEZERS
HT 56″ TO 66″
DEPTH 24″ 'TO 28″
WIDTH 30″ TO 42″

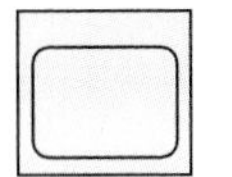

SINGLE SINKS
DEPTH 20″ TO 22″
WIDTH 24″ TO 30″

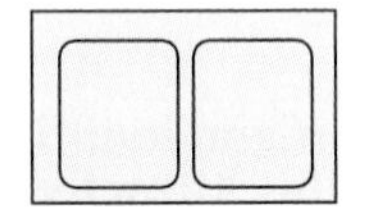

DOUBLE SINKS
DEPTH 20″ TO 22″
WIDTH 32″ TO 42″

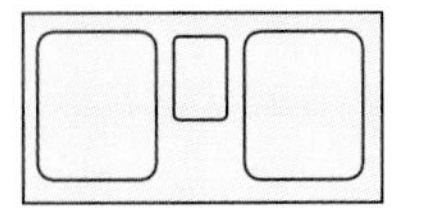

TRIPLE SINKS
DEPTH 20″ TO 22″
WIDTH 42″ TO 55″

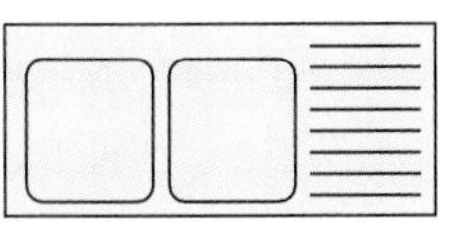

DBL SINKS/DRAIN BRDS
DEPTH 20″ TO 22″
WIDTH 50″ TO 60″

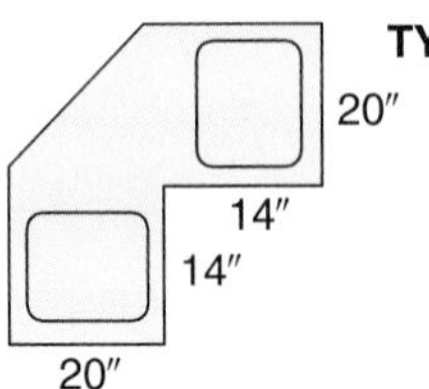

TYPICAL CORNER SINKS

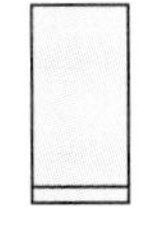

COMPACTORS
TYP. 12″ × 21″

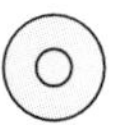

SINK DISPOSAL UNITS
TYP. HT 14″, 8″ DIAM

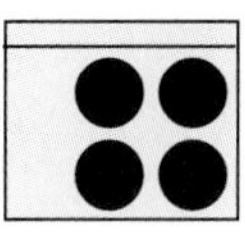

FREESTANDING RANGES
HT 36″
DEPTH 24″ TO 27″
WIDTH 20″ TO 40″

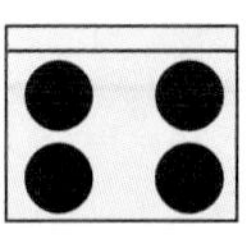

DROP-IN RANGES
TYP. 22″ × 30″

COOKTOPS/GRILLS
TYP. 21″ × 18″

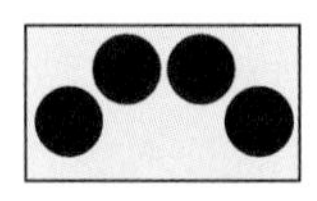

4-BURNER COOKTOPS
GAS/ELEC
TYP. 21″ × 26″

4-BURNER COOKTOPS/GRILLS
TYP. 21″ × 30″

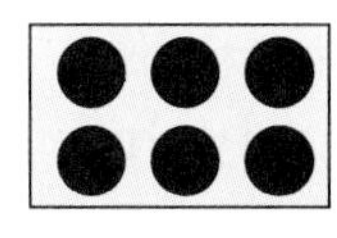

6-BURNER COOKTOPS
TYP. 21″ × 36″

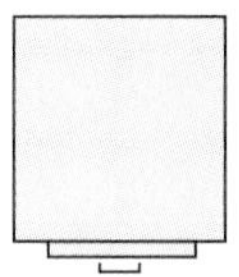

SINGLE OVENS
HT 30″ (DBL OVENS HT 50″ TO 70″)
DEPTH 24″
WIDTH 27″ TO 30″

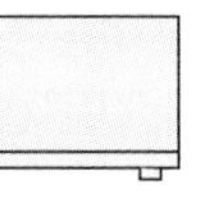

MICROWAVE OVENS
HT 18″ TO 20″
DEPTH 14″ TO 20″
WIDTH 20″ TO 30″

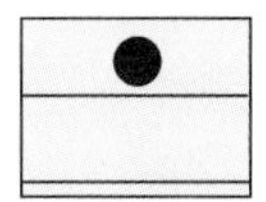

RANGE HOODS
HT VARIES
DEPTH 17″ TO 24″
WIDTH 30″ TO 72″

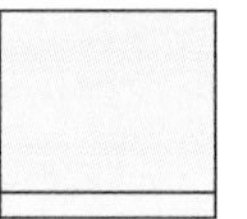

WASHERS/DRYERS
DEPTH 24″ TO 27″
WIDTH 24″ TO 29″

Kitchen Planning Guidelines

Remember the following guidelines for designing efficient kitchens.

FUNCTION

1. The food preparation centers include the storage and mixing center, the cooking center, and the cleanup center.
2. Each work area includes all necessary appliances and facilities.
3. Adequate storage facilities need to be provided throughout the kitchen.

LOCATION

4. Traffic lanes are clear of the work triangle.
5. The kitchen is located adjacent to the dining area.
6. The kitchen should be located near the children's play area.

SIZE AND SPACE

7. The work triangle measures no more than 22′ (6706 mm) or less than 12′ (3658 mm).
8. Lapboard heights are 26″ (660 mm).
9. Working heights for counters are 36″ (914 mm).
10. Working heights for tables are 30″ (762 mm).
11. Adequate counter space is provided for meal preparation.
12. Allow at least 12″ (305 mm) for knee space if counters are used as eating areas.
13. If space allows, include a pantry to store food staple quantities.
14. Allow at least 4′ (1219 mm) for aisle space between cabinets or appliances.
15. Allow at least 15″ (381 mm) on each side of an island cooktop for utensil storage.
16. Keep shelves within a reachable height (maximum height 84″).
17. Counter space is provided next to each appliance.

UTILITIES

18. An adequate number of electrical outlets are provided for each work center.
19. Shadowless and glareless light is provided and is concentrated on each work center.
20. Plumbing lines are planned for sink(s), icemaker, and any purified water system.
21. Provide adequate ventilation through overhead hoods, downdraft cooking-fume exhausts, circulating ceiling fans, and/or adequate heating/air-conditioning systems.

APPLIANCES

22. The oven and range are separated from the refrigerator by at least one cabinet and a continuous counter.
23. Cabinet and appliance locations are planned according to manufacturer's recommendations. Figures 10-16, 10-17, and 10-18 show standard dimensions.
24. Allow for door swing on all appliances.
25. The direction of door openings on appliances and cabinets should allow easy access from the work triangle. See Fig. 10-19.

Fig. 10-19 ■ Cabinet doors should not interfere with movement inside the work triangle.

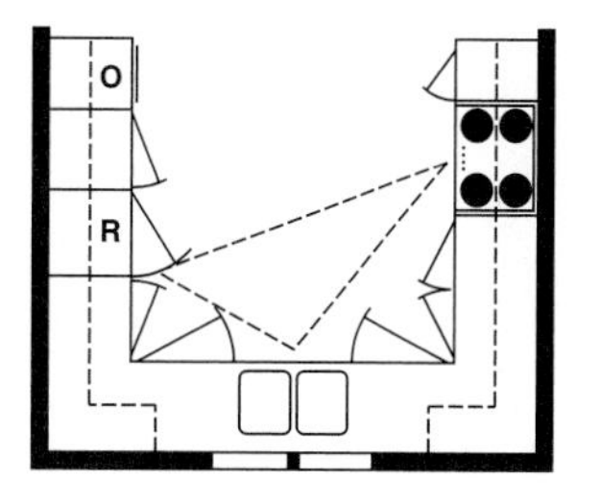

26. The refrigerator door should open toward the food preparation work space.
27. If a microwave oven is included, add a heat-resistant countertop between the cooktop and microwave.
28. Position the dishwasher next to the sink for easy loading.
29. The base cabinets, wall cabinets, and appliances create a consistent standard unit without gaps or awkward depressions or extensions.

Design considerations for persons with disabilities are provided in Chapter 13, Designing Floor Plans.

CHAPTER 10 Exercises

1. List the six types of kitchen shapes and give at least one advantage and one disadvantage of each.
2. Sketch a floor plan of one of the U-shaped kitchens shown in Fig. 10-4. Show the position of the dining area in relation to this kitchen, using the scale ½″ = 1′-0″.
3. Sketch a family kitchen using any of the six kitchen types in your design.
4. Sketch a floor plan of the kitchen in your own home. Prepare a revised sketch to show how you would propose to redesign this kitchen. Try to make the work triangle more efficient.
5. Sketch a floor plan of a kitchen you would include in a house of your own design, using the scale ½″ = 1′-0″.

6. Using a CAD system, draw one plan of any of the kitchen shapes shown in Fig. 10-14.
7. Calculate the space needed and plan a kitchen with a work triangle in a 14′ × 16′ peninsula kitchen and an L-shaped kitchen, 8′ × 14′.
8. Collect pictures of kitchens shown in magazines. Identify the kitchen type and find the work triangle in each. List good points and bad points of each kitchen design.

CHAPTER 11

General Service Areas

General service areas include utility rooms, garages and carports, workshops, and storage areas. Since a great number of different activities are related to these areas, they should be designed for great efficiency. Service areas should include facilities for the maintenance and servicing of the other areas of the home.

■ **General service areas are designed to serve many purposes.**
Whirlpool Corporation

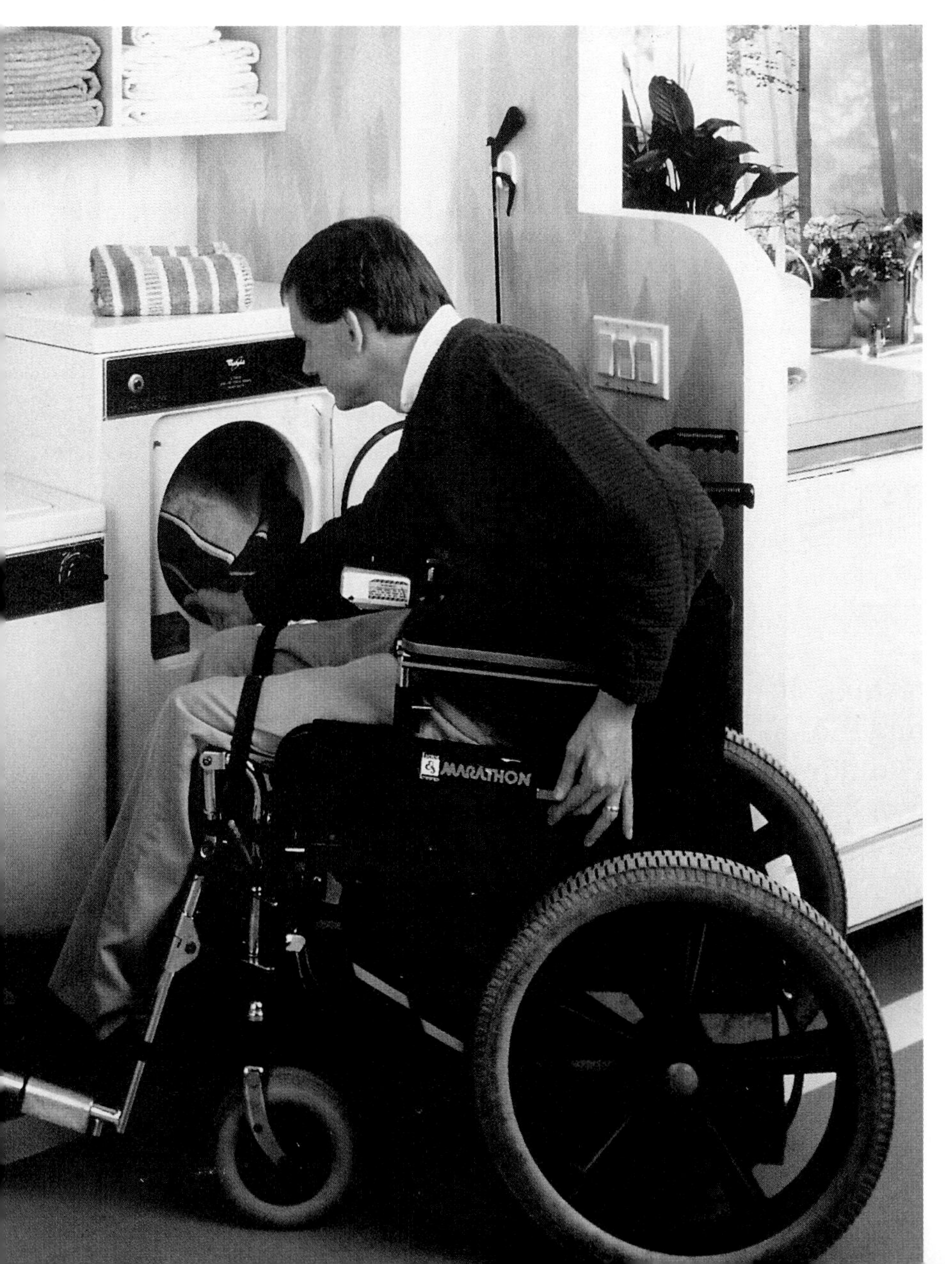

Objectives

In this chapter you will learn to:

- determine what kinds of equipment are included in a utility room.
- evaluate the best location for a utility room.
- sketch a garage and a carport.
- design storage facilities for a garage.
- calculate the area needed for garages and driveways.
- design and sketch an efficient and safe workshop area.
- design and sketch storage facilities.

Terms

carport
detached garage
dropleaf workbench
hand tools
integral garage
offset
peninsula workbench
power tools
utility room
ventilated shelving
wall closet

Utility Rooms

The **utility room** may include facilities for washing, drying, ironing, sewing, and storing household cleaning equipment. It may contain heating and air-conditioning equipment and/or even pantry shelves for storing groceries. Other names for this room are service room, all-purpose room, and laundry room.

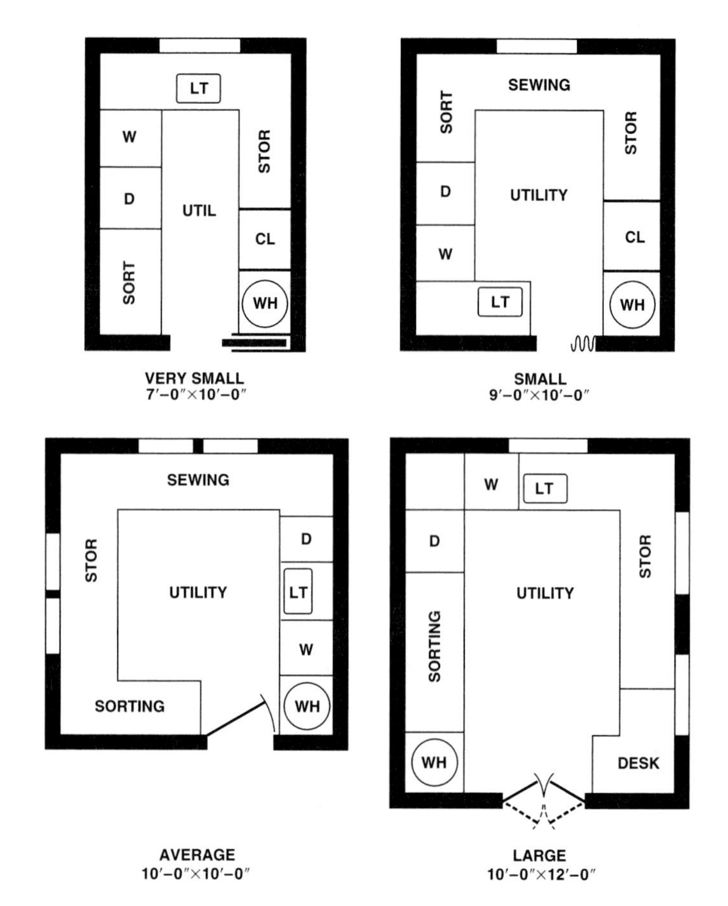

Fig. 11-1 ■ Arrange laundry equipment so that the process proceeds smoothly from one step to the next.

Function

The major function of the utility room is to serve as a laundry area. It may also accommodate water heating, water purification, heating, and air-conditioning equipment.

LAUNDRY To make laundry work as easy as possible, the appliances and working spaces in a laundry area should be located in the order in which they will be used. Such an arrangement will save time and effort during the steps in the process of laundering: receiving and preparing, washing, drying, ironing, storage, and often, sewing. See Fig. 11-1.

The first step in laundering—receiving and preparing the items—requires hampers or bins, as well as counters on which to collect and sort the articles. Storage facilities should be located nearby for laundry products such as detergents, bleaches, and stain removers.

The next step is the actual washing. It takes place in the area containing the washing machine (washer) and laundry tubs, trays, or sinks.

The equipment needed for the third step includes a dryer, indoor drying lines, and space to store clothespins. Dryers require either a 220-volt (220-V) outlet or gas access.

For the last step of the process, the required equipment consists of a counter for folding, an iron and ironing board, and a rack on which to hang finished ironing. Facilities for sewing and mending are often included. A sewing machine may be portable or it may fold into a counter or wall.

HEATING AND AIR CONDITIONING If the utility room is used for heating and air conditioning, additional space must be planned for the furnace, heating and air-conditioning ducts, water heater, and any related equipment such as humidifiers or air purifiers. See Fig. 11-2.

Location

A separate utility room is desirable because all laundry functions and maintenance equipment can be centered in one place. Space is not always available for a separate utility room, however. Laundry

Fig. 11-2 ■ Laundry, heating, cooling, and water purification facilities can all be located in the utility room.

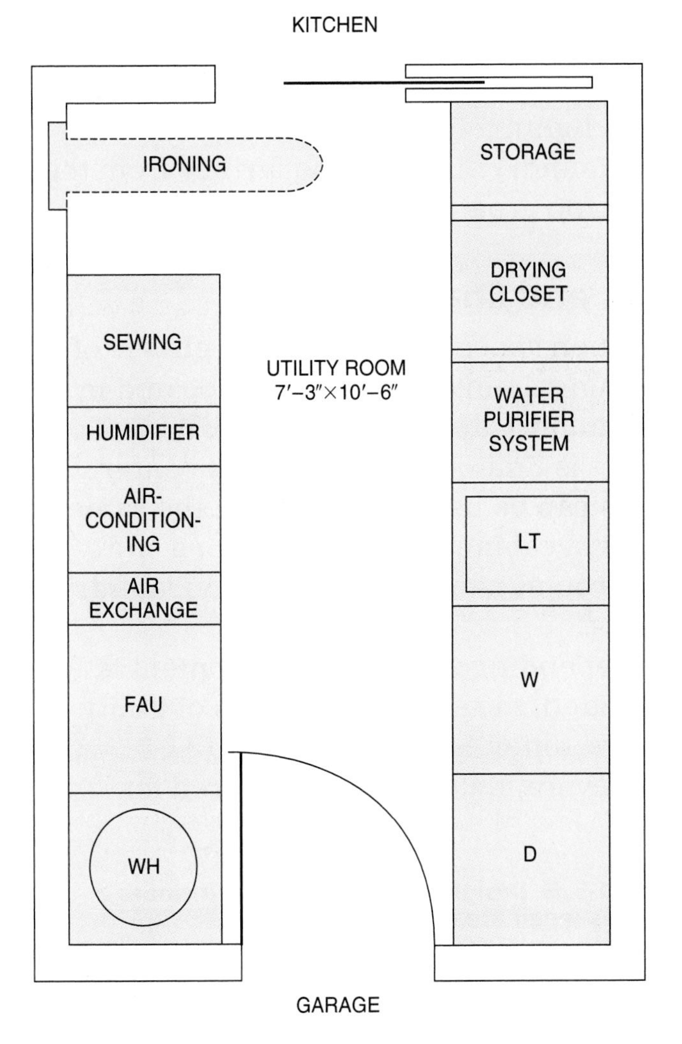

facilities may need to be located in some other area. See Fig. 11-3.

Placing the laundry appliances in or near the kitchen puts them in a central location and near a service entrance. Plumbing facilities are nearby, and some kitchen counters may be used for folding. However, these advantages may be offset by noise from the machines and odors

Fig. 11-3 ■ Laundry facilities can be located in other areas.

A. In the kitchen

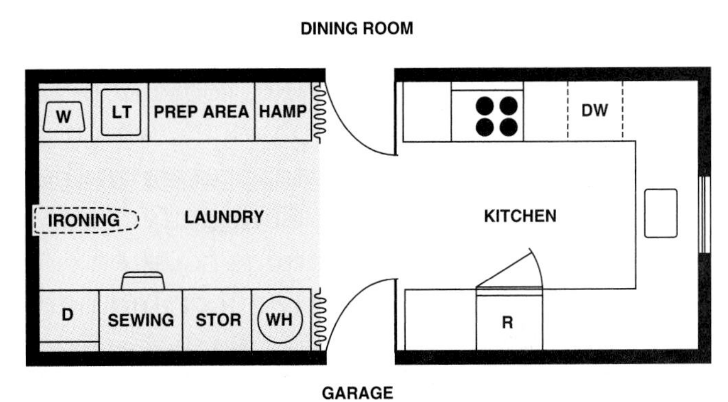

LAUNDRY IN SERVICE AREA

B. In closets

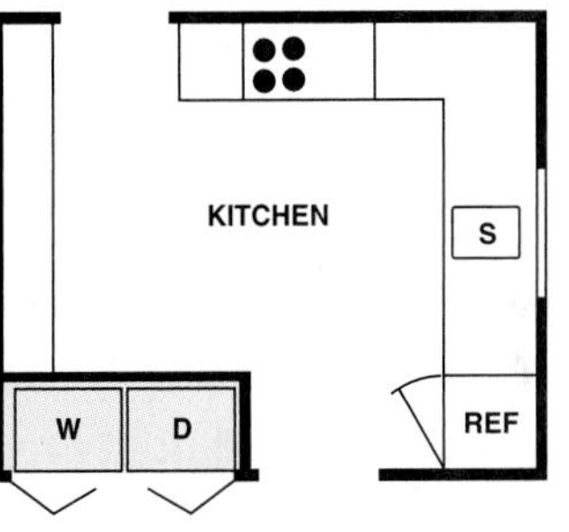

LAUNDRY IN CLOSET AREA

C. Near the sleeping area

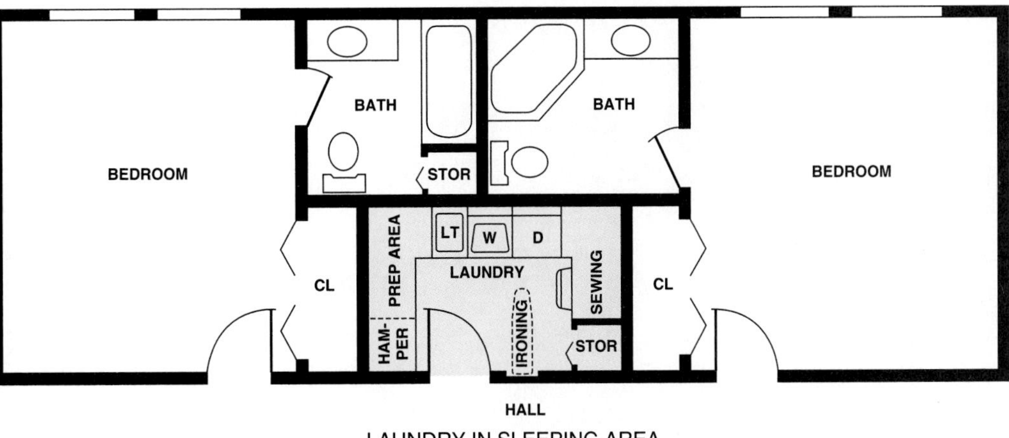

LAUNDRY IN SLEEPING AREA

from detergents, bleaches, and softeners near the kitchen. It's desirable also to keep laundering away from the preparation of food.

Other locations to consider for laundry appliances are less used areas such as the service porch, basement, garage, or carport. A closet or a family room may also work well for laundry facilities. Since clothing is changed in the sleeping area, laundry facilities in that area may be convenient.

Style and Decor

Style and decor in a utility room depend on the function of the appliances, which are themselves an important factor in the appearance of the room. Simplicity, straight lines, and continuous counter spaces produce an orderly effect and permit work to progress easily. Such features also make the room easy to clean. See Fig. 11-4.

An important part of the decor is the color of the paint used for walls and cabinet finishes. Colors should harmonize with the colors used on the appliances. All finishes should be washable. The walls may be lined with sound-absorbing tiles or wood paneling.

The lighting in a utility room should be 48″ (1219 mm) above any equipment used for washing, ironing, and sewing. However, lighting fixtures above work areas and laundry sinks can be farther from the worktop area.

Fig. 11-4 ■ A well-designed utility room. *Arnold & Brown*

Size and Shape

When space is available, all phases of clothing maintenance can be located in the laundry area. See Fig. 11-5. When space is limited, portable work center units can be used. For storage, the same sizes of cabinets used in kitchens and bathrooms can also work well in laundry areas.

Depending upon what equipment is included, the shapes and sizes of utility rooms differ. See Figs. 11-6 and 11-7. The average floor space required for

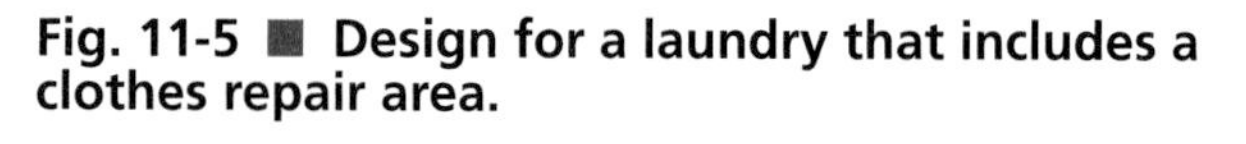

Fig. 11-5 ■ Design for a laundry that includes a clothes repair area.

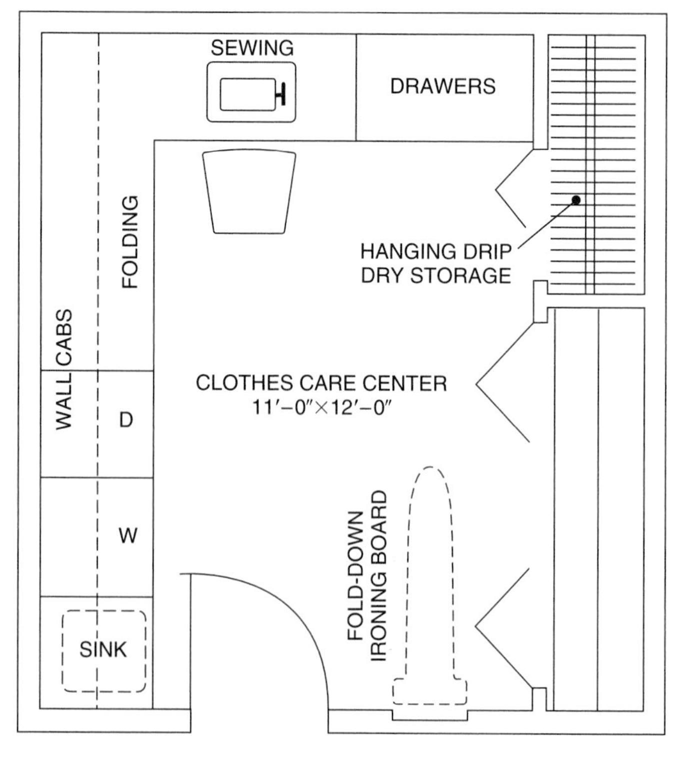

Fig. 11-6 ■ Typical utility room sizes and shapes.

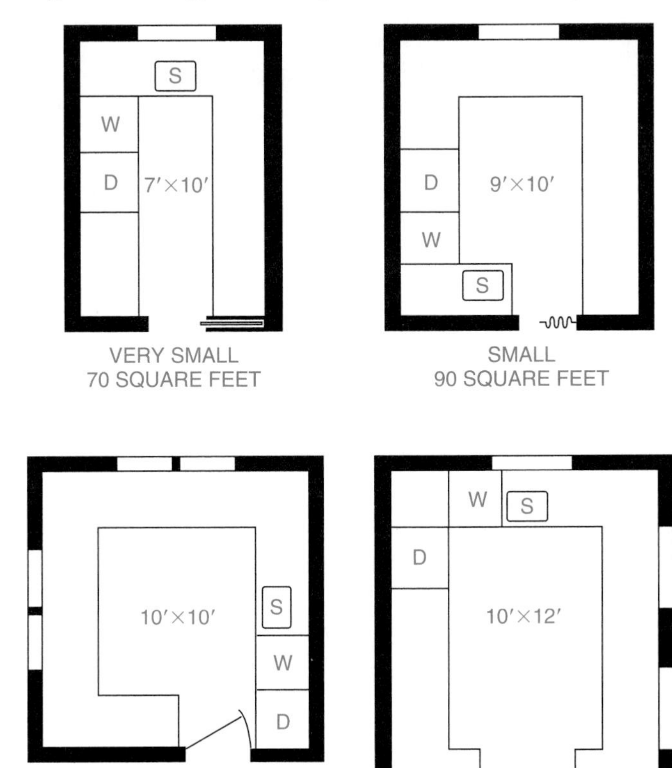

appliances, counters, and storage areas is 100 sq. ft. (9.3 m^2). However, this size may also vary according to the budget or needs of the household.

Garages and Carports

Areas for parking and storing vehicles—particularly automobiles—often make up a large percentage of the space available on a property. Therefore, the maximum utilization of space is important to consider when designing garages, carports, and driveways.

Function and Location

A garage is an enclosed structure designed primarily to shelter an automobile. It may be used for many secondary purposes—as a workshop, as a laundry room, or for storage space. A garage may be connected with the house, as an **integral garage**, or it may be a separate

Fig. 11-7 ■ Typical sizes of equipment placed in utility rooms.

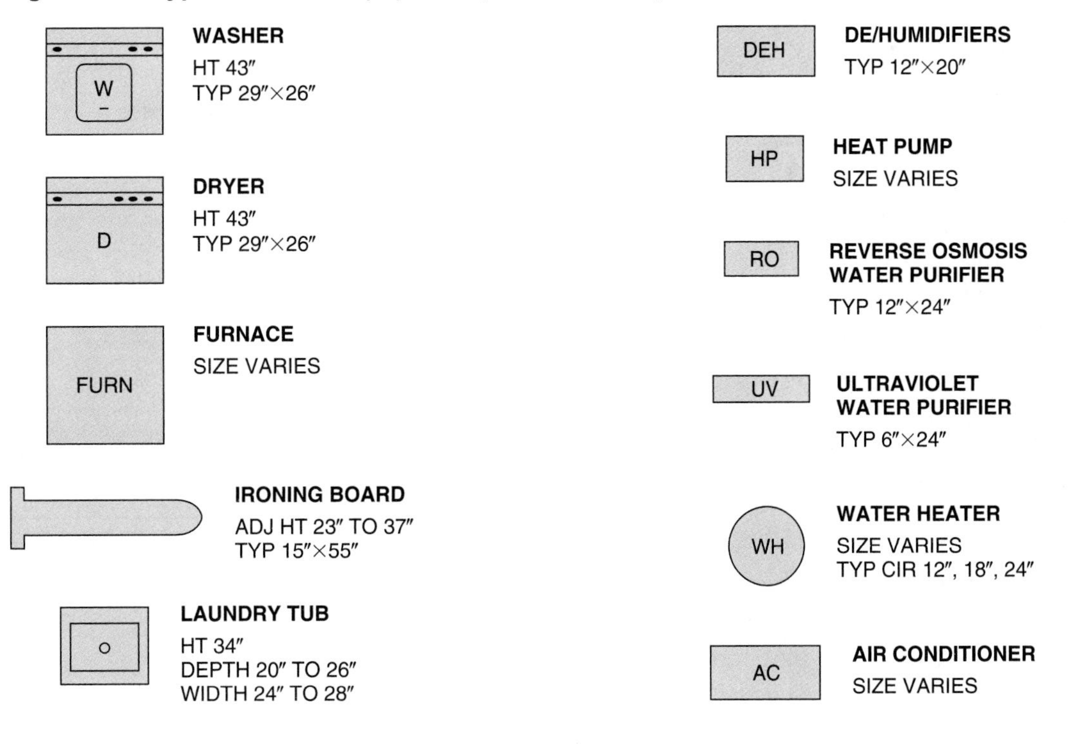

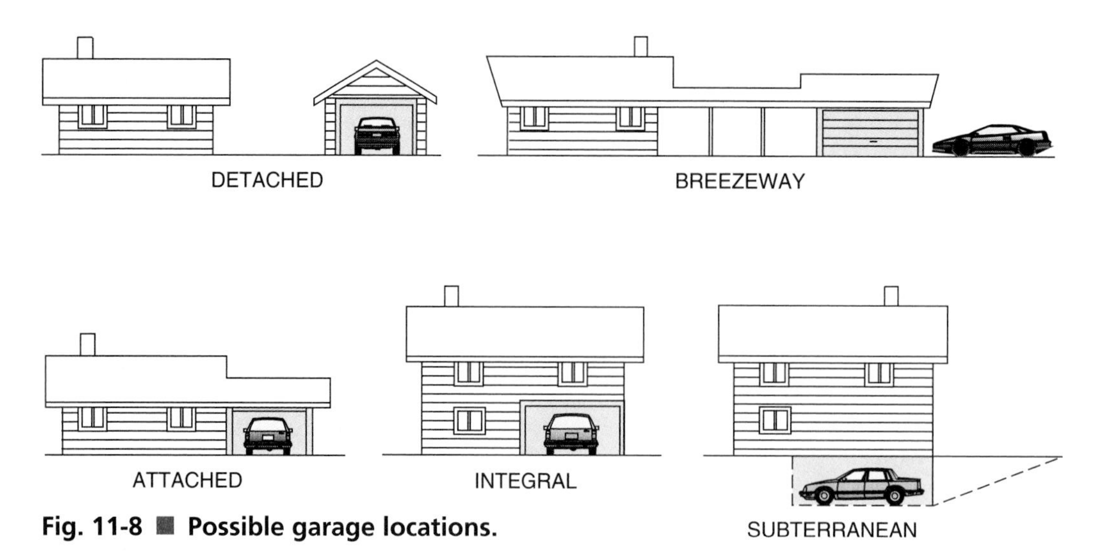

Fig. 11-8 ■ Possible garage locations.

building as a **detached garage**. In any case, there should be easy access from the garage to the service area of the house. Figure 11-8 shows several possible garage locations in relation to the house.

A covered walkway or breezeway from the garage or carport to the house should be provided if the garage is detached. Often a patio or porch is planned for this area to also integrate the detached garage with the house.

A utility room is often planned as a "mud room" transition from the garage to indoor areas. See Fig. 11-9.

A **carport** looks like a garage with one or more of the exterior walls removed. It may be completely separate from the house, or it may be built against the existing walls of the house or garage. See Fig. 11-10. Carports are most acceptable in mild climates where complete protection from cold weather is not needed. They offer protection primarily from sun and moisture.

Both the garage and the carport have distinct advantages. The garage is more secure and provides more shelter. However, carports lend themselves to open-planning techniques and are less expensive to build than garages.

Fig. 11-9 ■ Garage located off the utility room. Note that this plan provides access to a lavatory from the garage and from outside without passing through other rooms in the house. The shaded area in the garage represents the garage door in the open position.

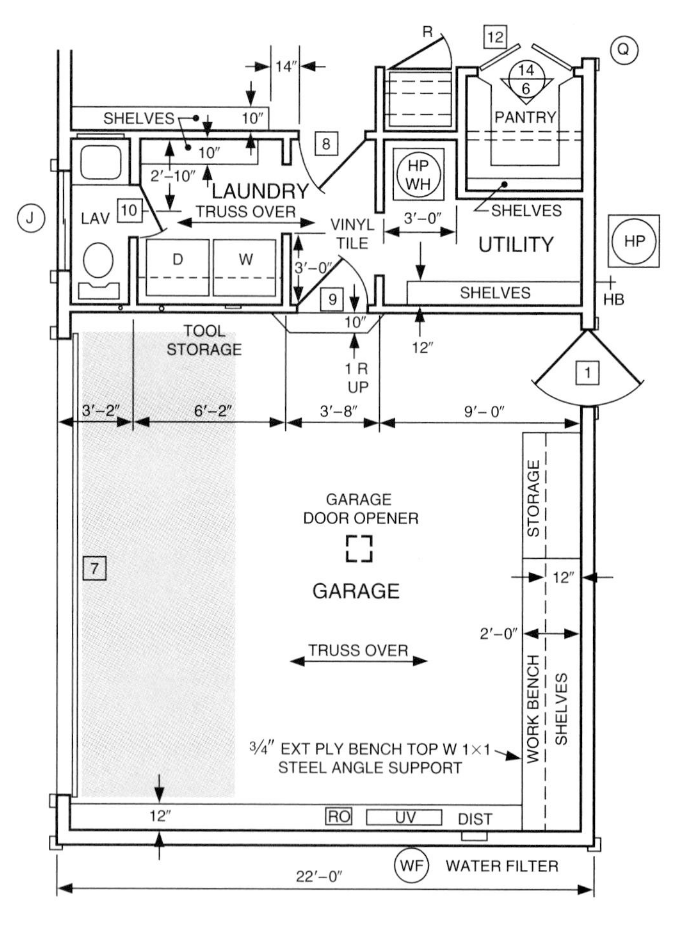

Fig. 11-10 ■ A carport located next to a garage. *Arnold & Brown*

Fig. 11-11 ■ Garage style integrated into house style. *Slattery & Root, Architects; E. Abraben, Architectural Photographer*

Decor

The lines of the garage or carport should be consistent with the major building lines and the architectural style of the house. See Fig. 11-11. The garage or carport must never appear to be an afterthought.

FLOOR The garage floor must be solid and easily maintained. A concrete slab 4″ (102 mm) thick and reinforced with welded wire mesh provides the best deck surface for a garage or carport. A vapor barrier of waterproof materials should be provided under the slab. The garage floor must have adequate drainage either to the outside or through drains located inside the garage. See Fig. 11-12.

DOORS The design of the garage door greatly affects the appearance of the house. Several types of garage doors are available: two-leaf swinging, overhead, four-leaf swinging, and sectional roll-up. See Fig. 11-13. Electronic devices are available for opening the door of the garage from the car.

For overhead doors, ceiling clearance must be planned to avoid interference between the opened door and light fixtures or other projections. Sectional roll-up doors require a 16″ (406 mm) clearance between the top of the door and the ceiling. Solid overhead doors require more clearance depending on the height of the door. Horizontal ceiling space must be at least 6″ (152 mm) above the height of the door. Refer again to Fig. 11-9.

Materials for garage doors include steel, aluminum, fiberglass, or wood—usually redwood or cedar. Metal doors are

Fig. 11-12 ■ Proper garage floor drainage is important.

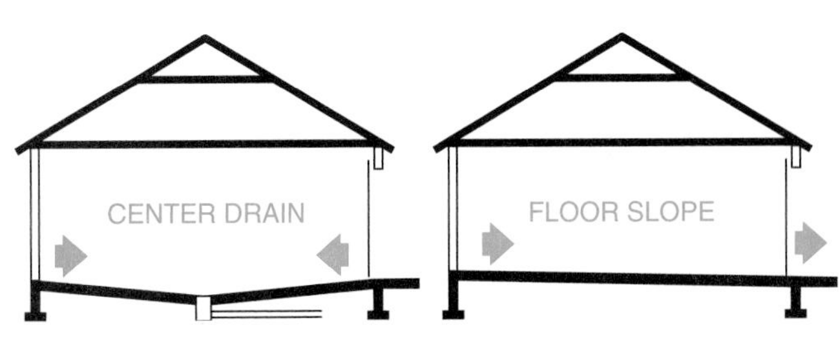

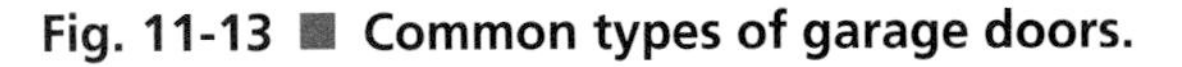

Fig. 11-13 ■ Common types of garage doors.

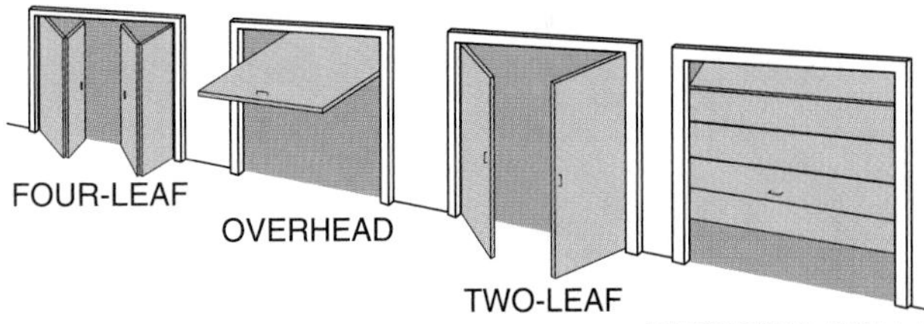

insulated or hollow. Solid overhead or roll-up doors are either manually operated or electronically controlled by a radio transmitter that activates a motor and lights. Garage doors of all types are available in a variety of patterns, styles, and sizes. They can be purchased with or without windows.

STORAGE DESIGN Most garages are also used for storage space. Storage facilities and even living space can be created in otherwise wasted space.

Cabinets should be elevated above the floor several inches to avoid moisture and to facilitate cleaning the garage floor. Garden-tool cabinets can be designed to open from the outside of the garage. The garage plan shown in Fig. 11-9 includes different types of storage facilities.

Size

The size of the garage depends upon the number of vehicles to be parked, plus space for any workshop facilities and storage. For example, storage space for bicycles, lawnmowers, and other lawn and landscape maintenance equipment must be considered as well as space for vehicles. Typical garage sizes are shown in Fig. 11-14.

Although some variations exist among manufacturers, standard garage door heights are 6′-6″, 6′-8″, 6′-9″, 7′-0″, 7′-6″, and 8′-0″. Standard garage door widths are 7′, 8′, 9′, 10′, 12′, 14′, 15′, 16′, 17′, 18′, and 20′.

Driveways

The main functions of a driveway are to provide access to all entrances and to the garage and to provide temporary parking space. See Fig. 11-15. However, a driveway can serve other purposes, too. A wide apron at the door of the garage can become a useful area—whether for car washing or for children's games. It can also enable cars to turn around without backing out onto a main street.

The driveway should be of brick, asphalt, or concrete construction. Concrete should be reinforced with welded-wire fabric to maintain rigidity and prevent cracking. Masonry pavers are often used over a concrete base. The driveway should be designed at least several feet wider than the track of a car which is approximately 5′-0″ (1524 mm). Slightly wider driveways of approximately 7′ to 9′ (2134 to 2743 mm) width are desirable for access and pedestrian traffic. To comfortably accommodate wheelchairs, a width of 10′ (3048 mm) may be needed. Sufficient space in the driveway should be provided for parking guests' cars. (Review Chapter 9, "Traffic Areas and Patterns," for more details concerning driveway configurations and sizes.)

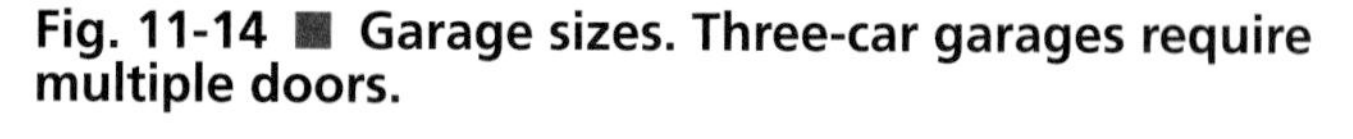
Fig. 11-14 ■ **Garage sizes. Three-car garages require multiple doors.**

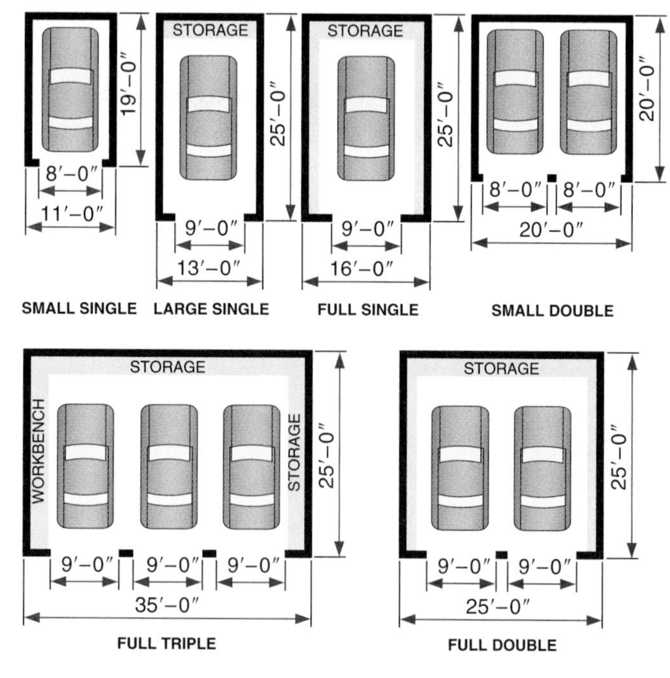

Fig. 11-15 ■ Positioning of cars on drives and aprons.

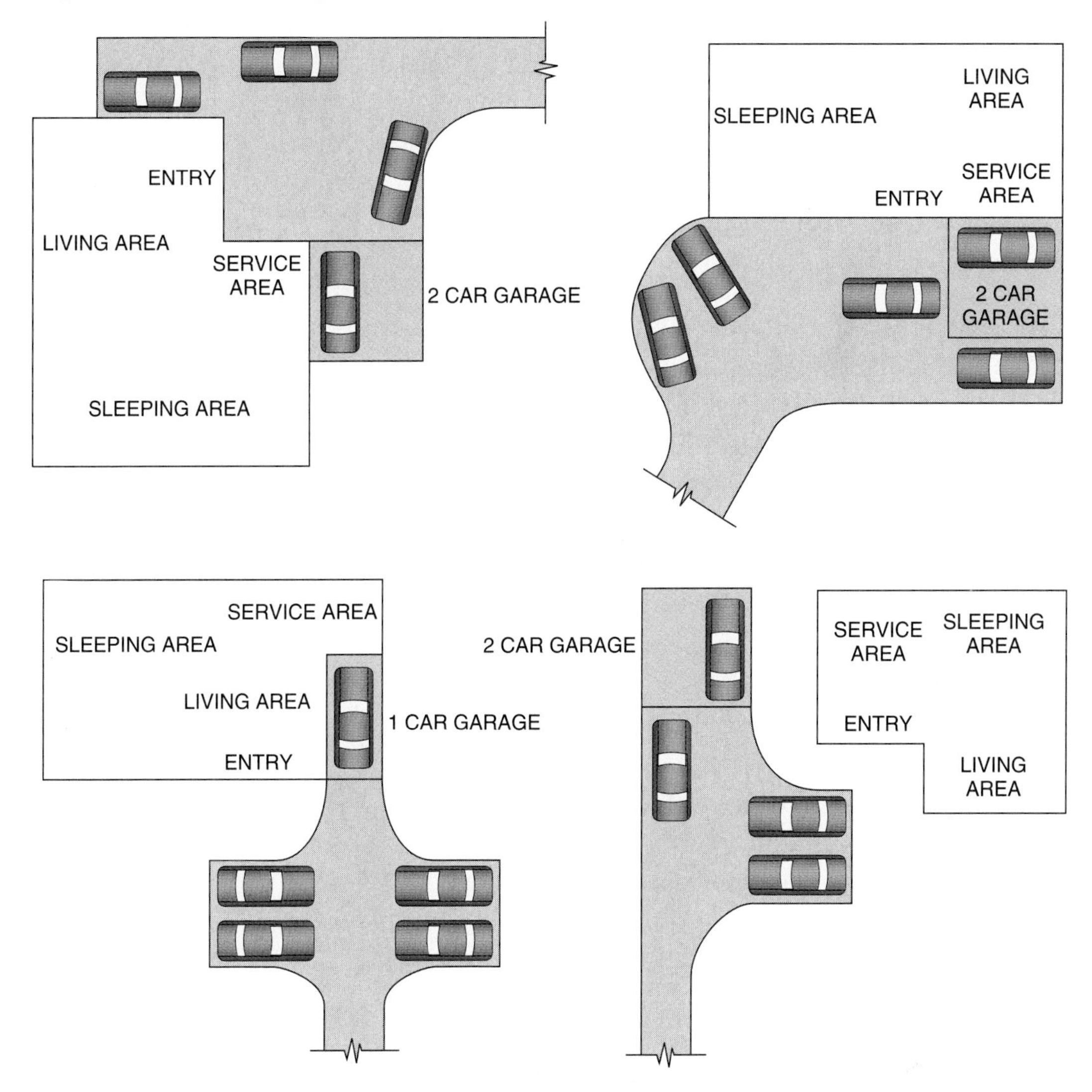

Workshops

The workshop is an area planned for working with equipment, tools, and materials.

Function and Location

A home workshop is designed for activities ranging from hobbies to home maintenance. See Fig. 11-16. As part of the service area, a workshop may be located in the garage, in the basement, in a separate room, or even in an adjacent building.

Workbench space, power tools, hand tools, and the storage areas should be

Fig. 11-16 ■ A home workshop designed for a variety of activities. *Lisanti, Inc.*

systematically planned to allow for the maximum amount of work space. Tools and equipment for working with large materials should be placed where the material can be handled easily. Any flammable finishing material, such as turpentine or oil-based paint, should be stored in metal cabinets.

WORKBENCH A workbench, usually with a vise for holding materials in place, is a major component of a home workshop area. The average workbench is 36″ (914 mm) high.

Workbenches are available to suit different needs. A movable workbench is appropriate for large projects. A **peninsula workbench** has three working surfaces with storage compartments on each of the three sides. A **dropleaf workbench** is excellent for work areas where a minimum amount of space is available. The side portions, or "drop leaves," can be extended for increased work space or folded down for storage in a small space.

HAND TOOLS Certain **hand tools** are necessary for any type of hobby or home-maintenance work. These basic tools include a claw hammer, carpenter's square, files, hand drills, screwdrivers, planes, pliers, chisels, scales, wrenches, saws, a brace and bit, mallets, and clamps.

Hand tools may be safely stored in cabinets that keep them dust-free, as shown in Fig. 11-16, or hung on appropriate hooks on *perforated hardboard*. Tools too small to be hung should be kept in drawers.

POWER TOOLS Electrically operated tools are called **power tools.** Those commonly used in home workshops include electric drills, saber saws, routers, band saws, circular saws, radial-arm saws, jointers, belt sanders, lathes, and drill presses.

In order to conserve motors, separate-drive motors can be used to drive more than one piece of power equipment. Separate 110V and 220V electrical circuits for lights as well as power tools should be planned for the home workshop area. Figure 11-17 shows work space clearances necessary for safe and efficient machine operation.

Multipurpose machines that can perform a variety of operations are convenient in a home workshop. Less equipment is necessary, and less space is required.

Decor

The work area should be as maintenance-free as possible. Glossy paint or tiles can retard the accumulation of shop dust on the walls. Exhaust fans help elimi-

Fig. 11-17 ■ Proper spacing for working safely with machinery.

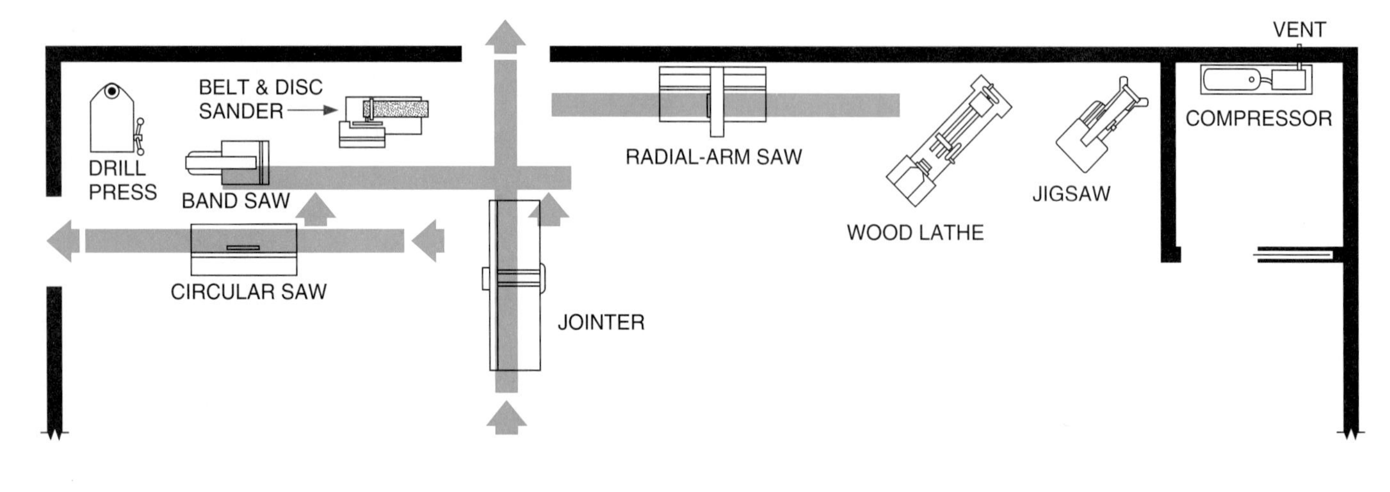

nate much of the dust and the gases produced in the shop. The shop floor should be concrete or linoleum. For safety, abrasive strips on floors around machines will eliminate the possibility of slipping.

Noise is a concern. Do not locate noisy equipment near the sleeping areas. Interior walls and ceilings should be soundproofed by offsetting studs and adding adequate insulation to produce a sound barrier. See Fig. 11-18.

Light and color are very important factors in designing the work area. Pastel colors, which reduce eyestrain, should be used for the general color scheme of the shop. Extremely light colors that produce glare and extremely dark colors that reduce effective illumination should be avoided. Choose colors that not only create a pleasant atmosphere in the shop but also help to provide the most efficient and safe working conditions.

Workshops should be well-lighted. General lighting should be provided in the shop to a high intensity level on machines and worktable tops. (Refer to Chapter 31 for electrical design.)

Size

The size of the work area depends on the size and number of tools, equipment, the workbench, and the storage facilities provided or anticipated. Plan the size of the work area for maximum expansion. At first, only a workbench and a few tools may occupy the area. If space is planned for the maximum amount of facilities, new equipment, when added, will fit appropriately into the basic plan. See Fig. 11-19. The designer must also anticipate the types and amounts of materials that will require storage space in the future and design the space accordingly.

Fig. 11-18 ■ Offset studs and insulation reduce the amount of noise that passes into other rooms.

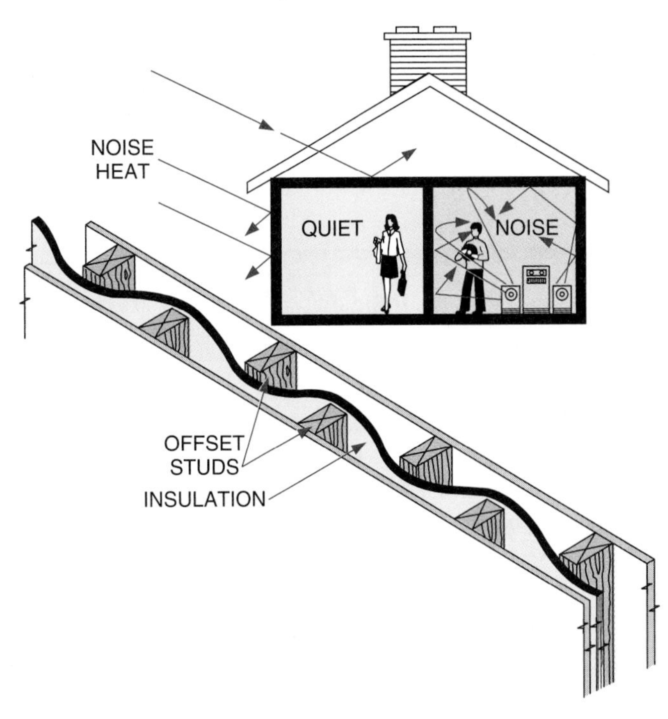

Storage Areas

Areas should be provided for general storage as well as for specific storage within each room.

Function and Types

Storage facilities—whether closets, cabinets, furniture, or room dividers—should be designed for convenient retrieval of stored articles. Those articles that are used daily or weekly should be stored in or near the room where they are needed. Those used only seasonally should be placed in more permanent storage areas. Areas that would otherwise be considered wasted space can become general storage areas. Parts of the basement, attic, or garage often fall into this category.

CLOSETS There are three basic types of closets: wardrobe, walk-in, and wall. A wardrobe closet is a shallow clothes closet built into the wall. The minimum depth for this closet is 24″ (610 mm). See Fig. 11-20.

Fig. 11-19 ■ Phased development of workshops.

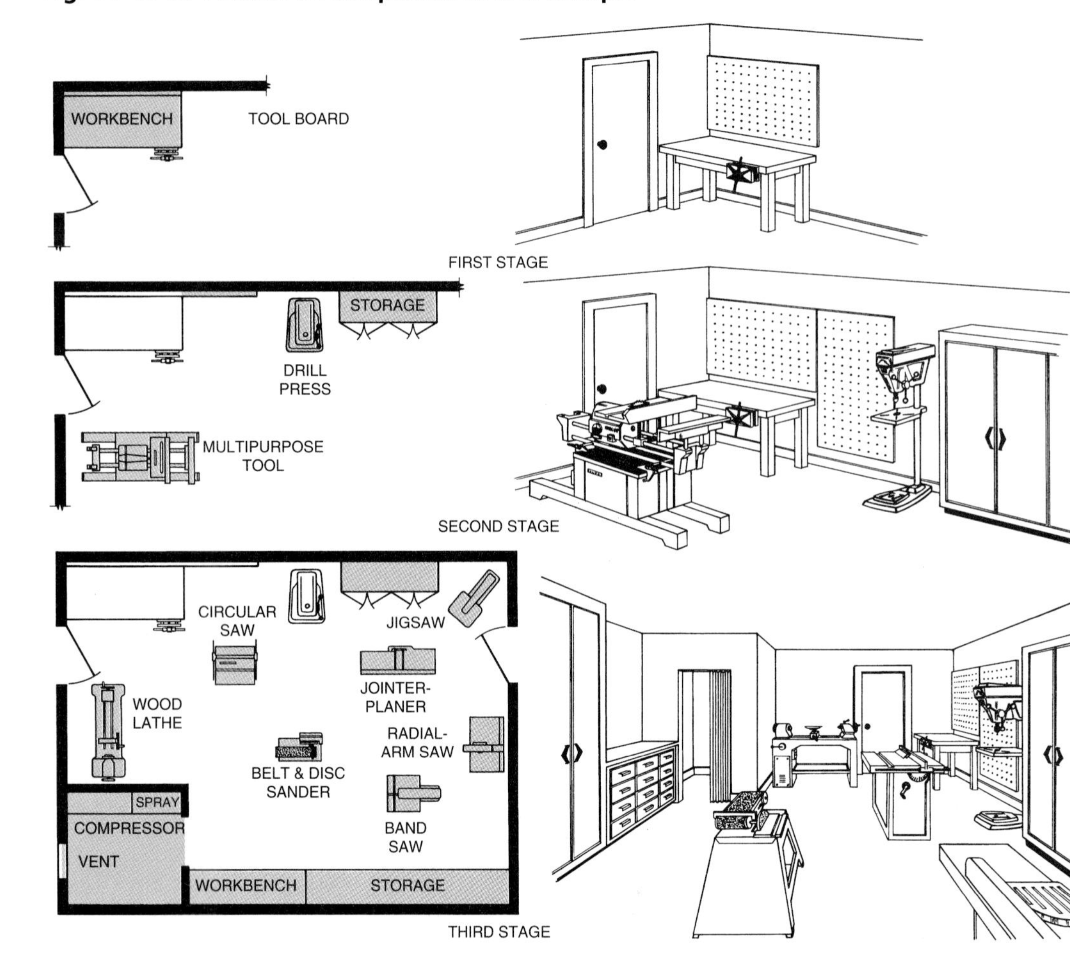

Fig. 11-20 ■ Wardrobe closet dimensions.

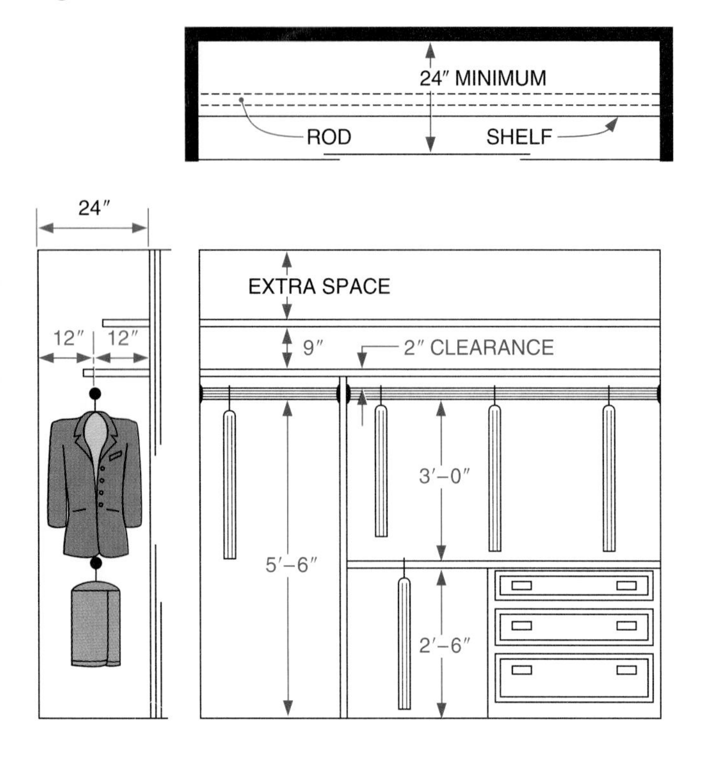

Depths of more than 30″ (762 mm) will make reaching the back of the closet difficult. Swinging or sliding doors should expose all parts of the closet that need to be within reach. A disadvantage of the wardrobe closet is the amount of wall space required for the doors.

A walk-in closet, as the name implies, is a closet large enough to enter and turn around. The area needed for this type of closet is equal to the amount of space needed to hang clothes plus space for a walkway. See Fig. 11-21. Although some space is wasted because of the walkway, the closet area takes up less wall space in the room. Only one closet door is needed.

A **wall closet** is a shallow closet in the wall for cupboards, shelves, and drawers.

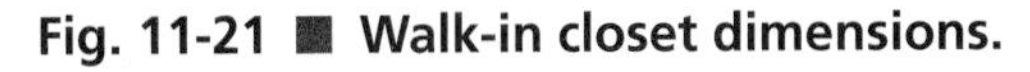
Fig. 11-21 ■ Walk-in closet dimensions.

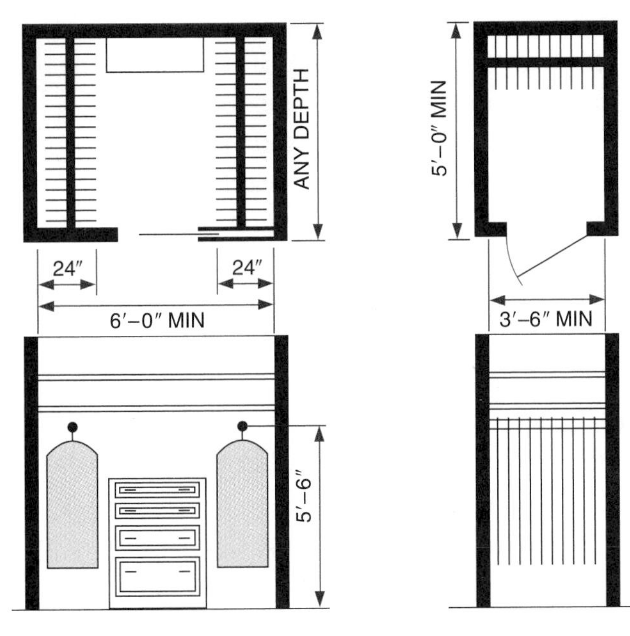

Fig. 11-22 ■ Built-in wall closet and drawer storage.
Closet Design Group

Wall closets are normally 18″ (457 mm) deep. This size provides access to all stored items without using an excessive amount of floor area. See Fig. 11-22.

Protruding closets that create an **offset** in a room should be avoided. By filling the entire wall of a room with closet space, a square or rectangular room can be designed without the use of offsets.

Doors on closets should be sufficiently wide to allow easy accessibility. Swing-out doors have the advantage of providing extra storage space on the back of the door. However, space must be allowed for the door swing. For this reason, sliding doors are usually preferred. All closets, except very shallow linen closets, should be provided with lighting.

FURNITURE AND BUILT-IN FEATURES Chests and dressers are freestanding pieces of furniture used for storage, generally in the bedroom. They are available in a variety of sizes and usually contain shelves and/or drawers.

Window seats are hollow, chest-like structures that are built-in below windows for persons to sit on. The hinged tops are often padded and can be raised to allow storage of items inside.

A room divider often doubles as a storage area. See Fig. 11-23. Room dividers often extend from the floor to the ceiling but may also be only several feet high. Many room dividers include shelves and drawers on both sides.

SHELVES Shelves for storage areas are available in a variety of sizes and materials including solid lumber, plywood, and hollow-core plywood. **Ventilated shelving** is made by welding steel rods together at ½″ or 1″ intervals. The rods are then coated with vinyl. Ventilated shelving is available in 9″, 12″, or 16″ depths and up to 12′ lengths.

Fig. 11-23 ■ Wall unit used for room dividers and storage. *Mirage Home Theatre*

Location

Different types of storage facilities are located in different areas of the home. Besides furniture, the most appropriate types of storage facilities for each room in the house are as follows:

Living room: room divider, wall cabinets, bookcases, window seats.

Dining area: room divider, closet.

Family room: built-in wall storage, window seats.

Recreation room: built-in wall storage.

Porches: storage under porch stairs.

Patios: sides of barbecue, storage shed.

Outside: storage areas built into the side of the house.

Halls: wall closets, ends of blind halls, bookshelves.

Entrance: room divider, closet.

Den: wall closet, bookcases.

Kitchen: wall and floor cabinets, room divider, wall closets.

Utility room: floor and wall cabinets.

Garage: cabinets built above the hood of a car, wall closets along sides, added construction on the outside of the garage.

Workshop: tool board, closets, cabinets.

Bedroom: closets, storage under, at foot, and at head of bed; cabinets; shelves.

Bathroom: cabinets, room dividers.

CHAPTER 11 Exercises

1. Design a utility room including a complete laundry facility within an area of 12′ × 12′ (144 sq. ft.). Show the location in relation to other areas of a house.
2. Using a CAD system, design a utility room for the house you are planning.

3. Design a full double garage and driveway for the house of your design. Include storage, laundry facilities, and a workbench. Identify the type of door you would use.
4. Using a CAD system, draw a plan of a garage and/or a workshop.
5. Design a work area for the house you are planning.
6. Add storage facilities to the house of your design.
7. Using a CAD system, draw a walk-in closet plan.

Sleeping Areas

CHAPTER
12

Approximately one-third of our time is spent in sleeping. Therefore, the sleeping area should be planned to provide facilities for maximum comfort and relaxation. The sleeping area should be located in a quiet part of the house and include both bedrooms and baths.

■ **The sleeping area occupies a major portion of a residence.** *Stock Market/ © Gary Chowanetz*

Objectives

In this chapter you will learn:

- **to plan and draw bedrooms for a sleeping area.**
- **to plan and draw baths appropriate to the size and arrangement of the floor plan.**
- **to design an efficient bath.**

Terms

central bath
compartment plan
fixtures
half-bath
lavatory
master bath
master bedroom
walk-in closet
wardrobe closet
water closet

Bedrooms

Houses are often categorized by the number of bedrooms. For example, a house may be described as a three-bedroom home or a four-bedroom home. A single person or couple with no children may require only a one-bedroom home. However, a minimum of two bedrooms is usually desirable, in order to provide one for guest use. Three-bedroom homes are most common, since they accommodate most families. See Fig. 12-1.

Function

The primary function of a bedroom is to provide facilities for sleeping. However, some bedrooms may also provide facilities for writing, reading, watching TV, listening to music, or relaxing. As with other rooms, the size and shape of each bedroom depends on the occupants, activities, and furniture designated for that room.

For babies and very young children, a bedroom may serve as a nursery, with a crib and related furniture and equipment. For older children, a bedroom may be a double room with twin beds and other furniture such as desks and entertainment equipment. A **master bedroom** for adults not only has a large bed or beds and other furniture, it usually also has an adjacent bath and, perhaps, a separate dressing room. See Fig. 12-2.

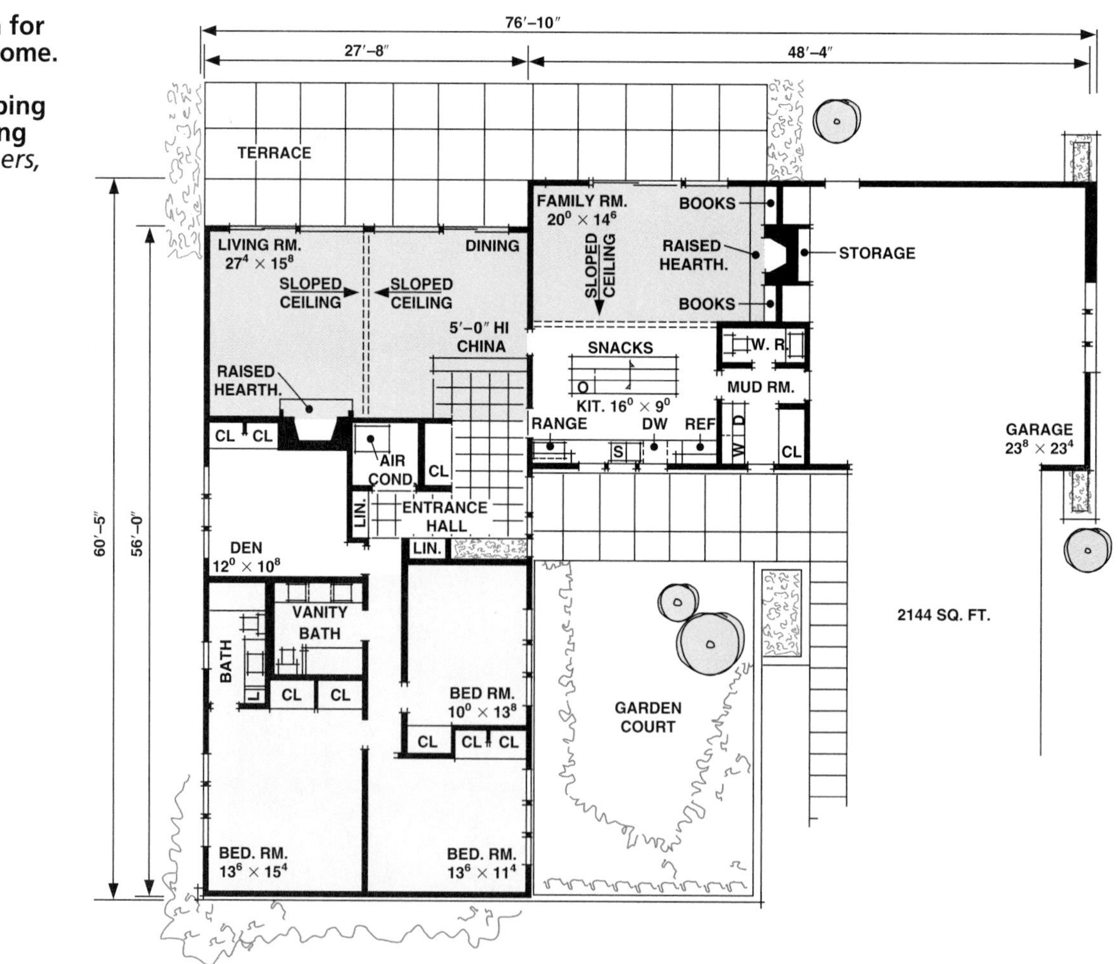

Fig. 12-1 ■ A plan for a three-bedroom home. Note the effective separation of sleeping areas from the living areas. *Home Planners, Inc.*

Fig. 12-2 ■ A master bedroom with an adjacent bath area. *Scholz Homes, Inc.*

Fig. 12-3 ■ Methods of minimizing noise.

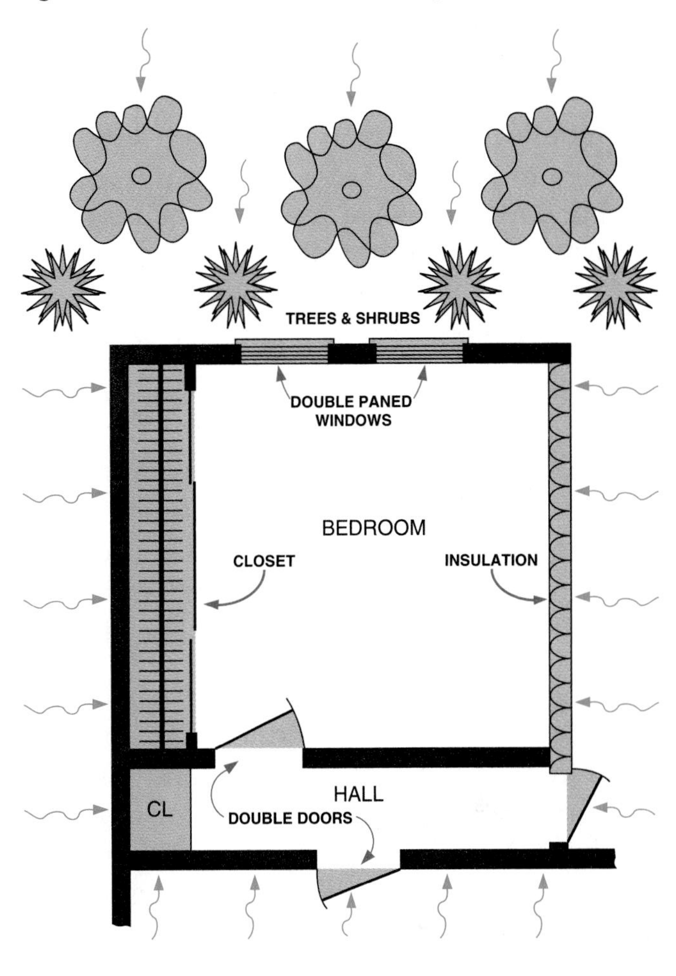

Location

For sleeping comfort and privacy, bedrooms should be grouped in a quiet part of the house, as far from the living area as possible. If a further separation is wanted between the master bedroom and children's or guest bedrooms, locating the master bedroom on a different level is recommended for multilevel dwellings. Regardless of the location, all bedrooms must have access to a hall from which a bath is also accessible.

GUIDELINES FOR NOISE CONTROL Since noise contributes to fatigue, location is particularly important for a restful bedroom area. To eliminate as much noise as possible, planned locations and well-selected materials can help accomplish this. See Fig. 12-3. The following guidelines are valuable for designing bedrooms that are quiet and restful:

1. The bedroom should be in the quiet part of the house, away from major street noises.
2. Air is a good insulator. Therefore, closets can be located to provide sound buffers. Clothing and other items stored in closets can also help muffle sound.
3. Carpeting or porous wall and ceiling panels absorb noises. Rooms above bedrooms should be carpeted.
4. Floor-to-ceiling draperies help to reduce noise.
5. Acoustical tile in the ceiling is effective in reducing noise.
6. Trees and shrubbery outside the bedroom help absorb sounds.
7. The use of double-glazed insulating glass for windows and sliding doors helps to reduce outside noise.
8. The windows of an air-conditioned room should be kept closed during hot weather. This eliminates noise and aids in keeping the bedroom free from dust and pollen.

9. In extreme cases when complete soundproofing is desired, the wall structure and materials may be designed to provide continuous sound insulation.
10. Placing rubber pads under appliances such as refrigerators, dishwashers, washers, and dryers often eliminates vibration and noise throughout the house.

WALL SPACE The bedroom entrance door, closet doors, and windows should be grouped to conserve wall space whenever possible. By minimizing the distance between doors and windows, the amount of usable wall space is expanded. Long stretches of wall space are best for efficient furniture placement.

DOORS Several types of doors may be used in bedrooms. Pocket, bypass (sliding), and bifold doors are used for closets. Swinging and pocket doors can be used for the entrance door. A swinging door should always swing into the bedroom against an adjacent wall, and not outward into the hall. See Fig. 12-4. The door connecting the bedroom with an outside deck, balcony, or patio should be French (glazed), swinging, or sliding glass to maximize light and ventilation.

Fig. 12-4 ■ Swinging bedroom doors should open into bedrooms toward a perpendicular wall, not the hall.

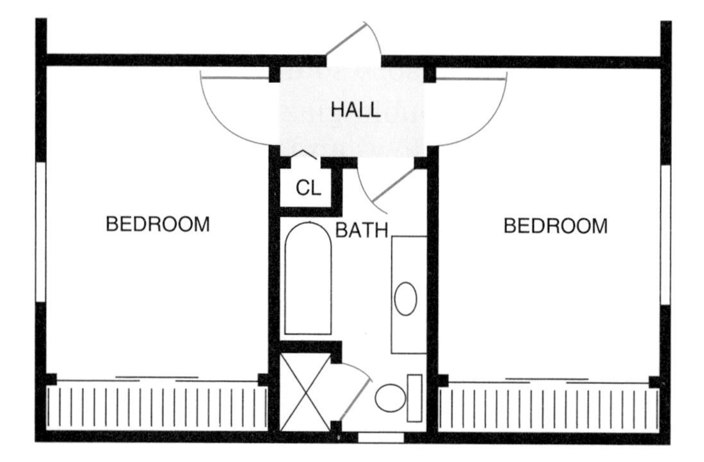

Standard door heights range from 6′-8″ to 8′-0″. Standard door widths for bedroom doors range from 2′-6″ to 3′-0″. A width of at least 2′-8″ is needed to allow passage of some furniture, and a width of 3′-0″ is needed for wheelchair entry.

WINDOWS AND VENTILATION Bedroom windows should be placed to provide air circulation, light, and sometimes solar heat. Similar to door placement, designers must also consider the efficient use of wall space for window placement. One method of conserving wall space for bedroom furniture is to use high windows. High, narrow-strip windows, called ribbon windows, provide space for furniture to be placed underneath. They also ensure some privacy for the bedroom. Many building or fire codes require an escape-size window in each room if no outside door exists.

Proper ventilation is necessary in bedrooms and is conducive to sound rest and sleep. When air-conditioning is available, the windows and doors may remain closed. Central air-conditioning and humidity control provide constant levels of temperature and humidity and are efficient methods of providing ventilation and air circulation.

Without air-conditioning, windows and doors must provide the ventilation. Bedrooms should have cross ventilation. However, the draft must not pass over the bed. See Fig. 12-5. High ribbon windows provide

Fig. 12-5 ■ Cross ventilation should not cause a draft over the bed.

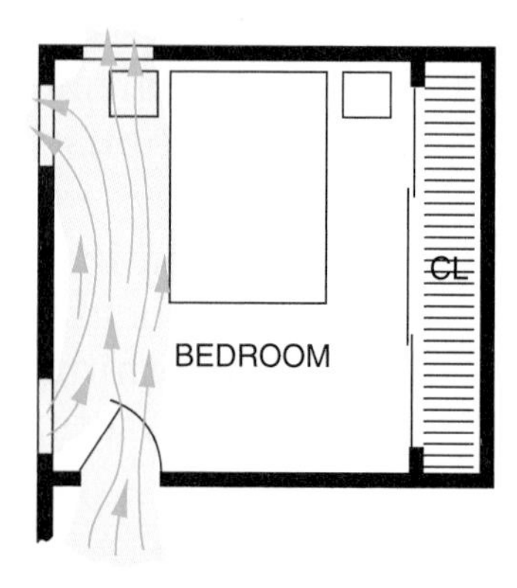

cross ventilation without causing a draft on the bed. Jalousie windows are also effective, since they direct air upward.

STORAGE SPACE Storage space placed in or adjacent to bedrooms is needed primarily for clothing and personal accessories. Furniture, closets, cabinets, and dressing rooms serve as storage facilities. These should be located within easy reach and should be easy to maintain.

Furniture storage space in the bedroom area is usually found in dressers, chests, vanities, and dressing tables. However, most storage space should be provided in the closets. See Fig. 12-6.

Closets should be a minimum of 24″ deep, allowing at least 6′ of clothes hanging rod space for each adult. **Walk-in**, or **wardrobe closets**, should be built-in. Built-in storage facilities should be designed to become part of a wall and thereby eliminate awkward offset closets that protrude into a room. Balancing an offset closet in one room with a closet in an adjacent bedroom helps solve this problem. In Fig. 12-1, this technique was used in the two smaller bedrooms.

Built-in storage *cabinets* also reduce the need for furniture that might take up valuable floor space. See Fig. 12-7.

DRESSING AREAS A dressing area is usually located adjacent to the master bedroom. It may be a separate room, an alcove, or part of the bedroom separated by a divider that also provides storage space. Figure 12-8 shows a dressing room, bath, balcony, and bedroom which comprises a master bedroom suite.

Fig. 12-6 ■ Storage areas for a variety of clothing articles. *Closet Design Group, Inc.*

Fig. 12-7 ■ The clothes closets and TV-audio center are both located behind sliding, mirrored doors. The plan for this room is shown in Fig. 12-8.

Fig. 12-8 ■ Master suite floor plan that includes a variety of areas.

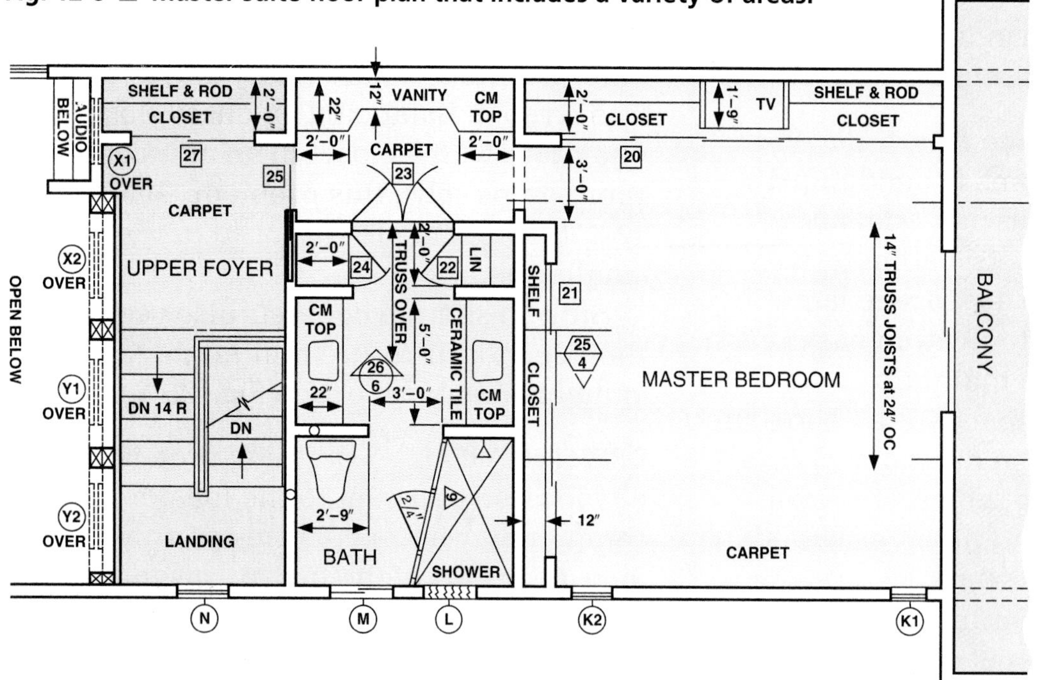

CHILDREN'S ROOMS Children's bedrooms and nurseries must be planned to be comfortable, quiet, and sufficiently flexible to allow change as the child grows and matures. See Fig. 12-9. For example, storage shelves and rods in closets should be adjustable so that they may be raised as the child becomes taller. Light switches should be placed low for small children and have a delay switch that allows the light to stay on for some time after the switch has been turned off.

Adequate facilities for study and hobby activities should be provided, such as a desk and worktable. Storage space for books, models, and athletic equipment is also desirable. Chalkboards and bulletin boards on the walls help make the child's room usable.

Grouping or dividing children's rooms into convertible double rooms is one method of designing for future change as children grow. Eventually the divided room can become a single large guest bedroom or study.

Fig. 12-9 ■ A well-planned child's bedroom. *Tony Stone Images/David Rigg*

Decor

In general, bedrooms should be decorated in quiet, restful tones. Matching or contrasting bedspreads, draperies, and carpets help accent the color scheme. Uncluttered furniture with simple lines also helps to develop a restful atmosphere in the bedroom.

Size and Shape

The type, size, and style of furniture to be included in the bedroom should be chosen before the size of the bedroom is established. In a preliminary design, the sizes and amount of furniture determine the size of the room and not the reverse. Because the bed or beds require the most space, the room size must provide more than enough space for the bed. A minimum-sized bedroom should accommodate at least a single bed, bedside table, and dresser. In contrast, a larger, master bedroom suite might include any of the following: a king and/or double bed, twin beds, bedside stands, chest of drawers, chaise and/or lounge chair, dressing area, and adjacent master bath. See Fig. 12-10.

The size of the furniture and the space between the furniture needs to be considered to determine the dimensions of the room. See Fig. 12-11. Using furniture templates, as for other rooms, is an effective method for designing bedroom sizes and shapes. With templates, the traffic patterns and the way the room functions can be foreseen.

Most bedrooms need to be a minimum of 100 sq. ft. An average size bedroom is between 100 sq. ft. and 200 sq. ft. Bedrooms over 200 sq. ft. are considered large.

Fig. 12-10 ■ A large master bedroom with well-coordinated furnishings. *Fran Murry Interiors*

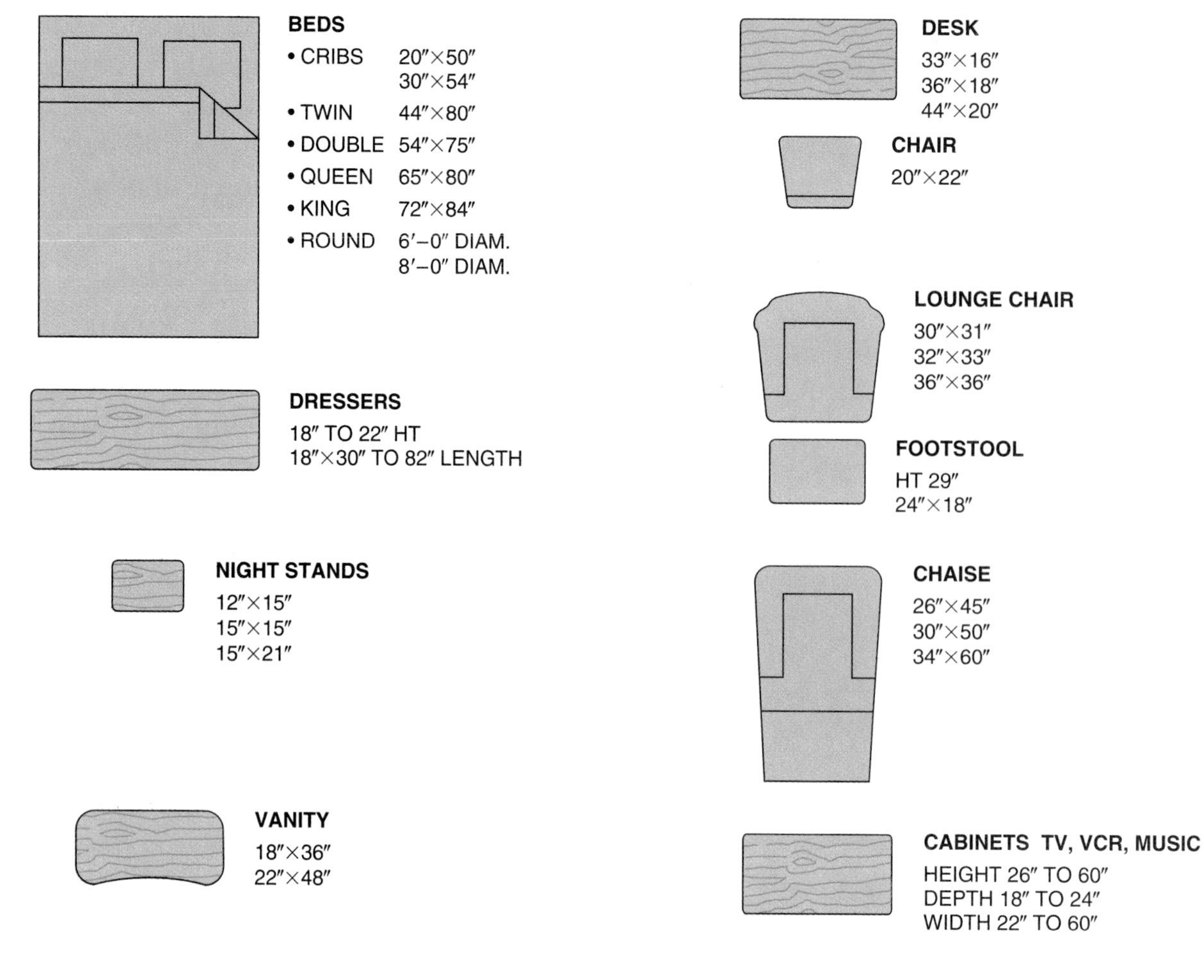

Fig. 12-11 ■ Typical sizes of bedroom furniture.

Baths

Baths (or bathrooms) must be planned to be functional, attractive, and easily maintained. They can vary widely in size depending on the space available and additional areas included.

Function

Designing a bath involves the appropriate placement of fixtures, cabinets, accessories, and plumbing lines. Adequate ventilation, heating, and lighting also needs to be planned. In addition to the normal functions, baths may also provide facilities for dressing, exercising, or laundering. Some may include a sauna and whirlpool bath. See Fig. 12-12.

FIXTURES The basic bath **fixtures** (items connected to plumbing lines) included in most baths are a **lavatory** (sink), a **water closet** (toilet fixture), and a bathtub or shower. The convenience of the bath depends upon the arrangement of these fixtures.

Lavatories, or sinks, are available in a wide variety of colors, materials, sizes, and shapes. They are manufactured with porcelain-covered steel or cast iron, stainless steel, brass, copper, or acrylic or other composite materials such as cultured marble. Sinks are either set into an opening, molded with the countertop into one piece, or made into a freestanding fixture on a pedestal. Sink areas should be well lighted and free of traffic, and include

Fig. 12-12 ■ Baths designed for many functions.

A. Complete bath including exercise area. *KCC.*

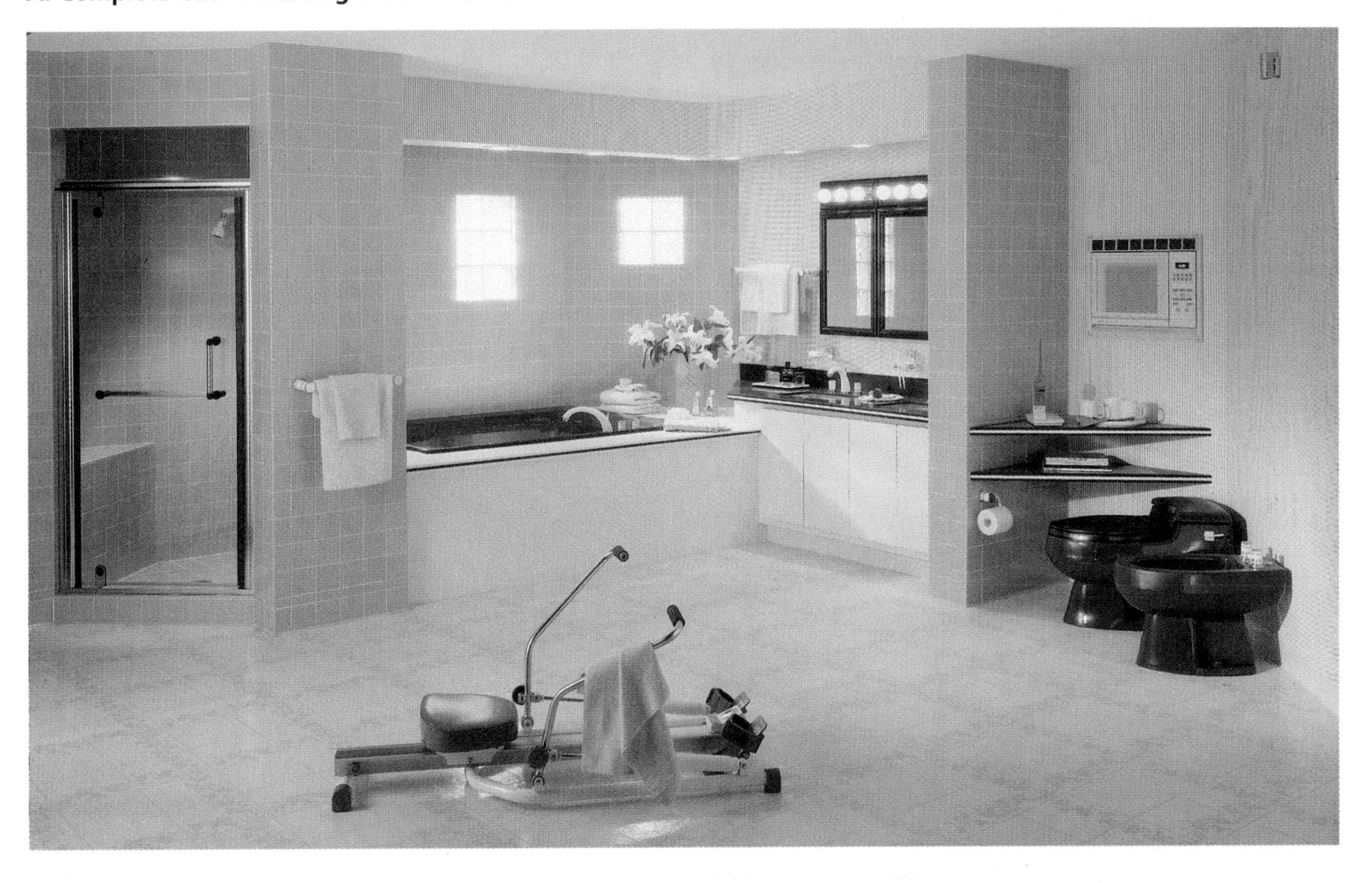

B. Plan that includes a whirlpool and a sauna.

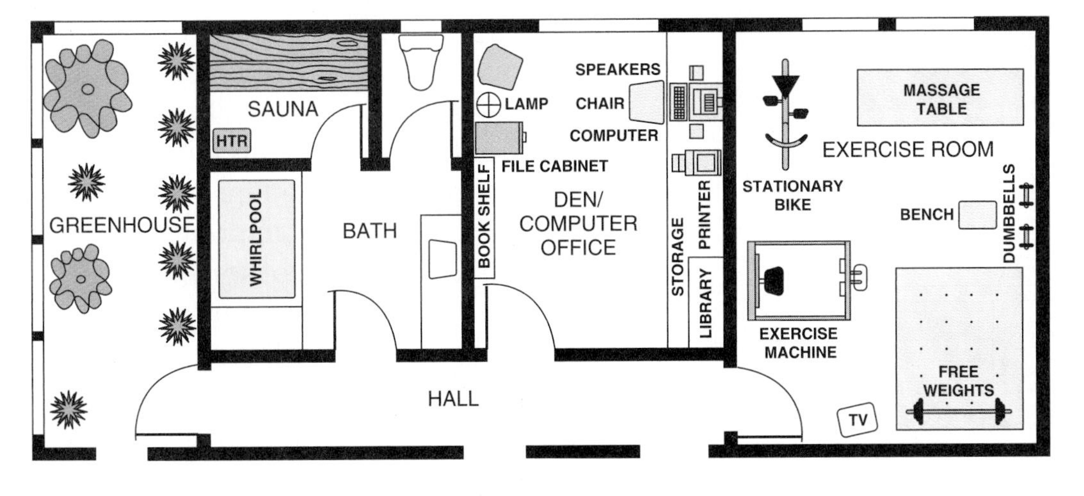

a mirror on the wall over the sink. Side mirrors provide additional angles for viewing. A comfortable sink height for most people is between 34″ and 36″. For wheelchair access, 36″ is usually required.

Water closets are available in either one-piece or two-piece models. One-piece models are either mounted on the floor or on a wall. Water closets need a minimum of 15″ distance from the center to a side wall or to other fixtures. For wheelchair access, a clear opening ranging from 2′-6″ to 4′-0″ is required in front and on the sides of a water closet. If space allows, the

Fig. 12-13 ■ Bathroom containing a tub-shower combination.

A. A shower curtain is used to close off the area.

B. Plan for the bathroom. Note the oval bathtub.

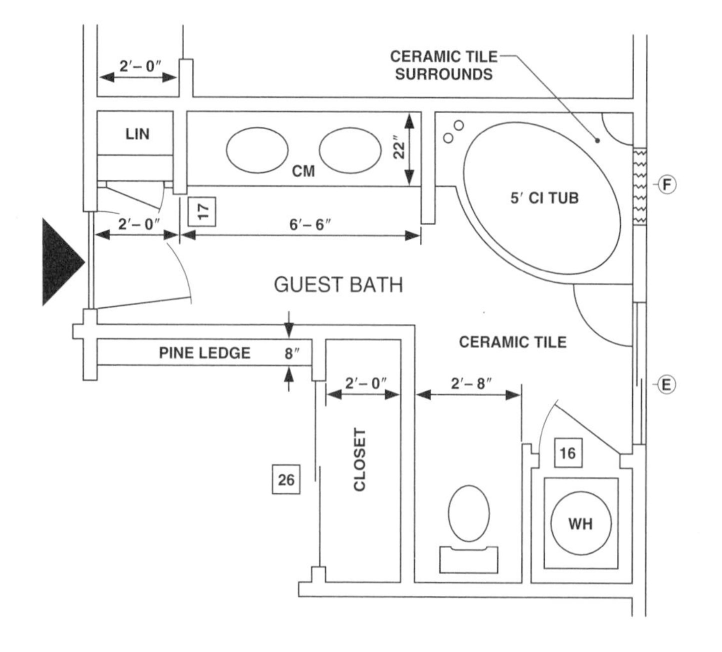

Fig. 12-14 ■ Open area bathtub without a shower. *Marvin Windows*

water closet should not be visible when the door to the bath is open.

The variety of tubs and showers available in squares, rectangles, or irregular shapes allow a great amount of flexibility in fixture arrangements. See Fig. 12-13. Many small or average-size baths include a showerhead on a tub wall. This tub-shower combination then requires an enclosure or a shower curtain rod should be installed.

Tubs without a showerhead can be located in open areas. See Fig. 12-14. Large tubs can include whirlpool outlets, jets, and controls for hydrotherapy. Special tubs with side openings that seal when closed are available for persons with physical disabilities. Grab bars installed beside the tub are simple, but effective safety features, particularly for older and disabled persons.

A separate shower stall is often preferred or substituted for a tub. One-piece prefabricated units are available made of materials such as fiberglass, acrylic, coated steel, or aluminum. Many shower stalls are constructed using ceramic tile, marble, or synthetic materials. Glass

Fig. 12-15 ■ Glass shower enclosure. *Century Shower Door Co.*

Fig. 12-16 ■ Design for a shower enclosure without doors.

shower walls, used for panels and doors, are made from shatterproof glass mounted between metal frames as shown in Fig. 12-15. Where space is available, showers can be designed without a door. See Fig. 12-16.

Some showerheads can now equalize hot and cold water pressure. For example, if cold water pressure is reduced during shower use, the hot water pressure is also automatically reduced to avoid scalding.

ACCESSORIES In addition to the three basic fixtures, many accessories help improve a bath's functions. Accessories range from various furnishings and plumbing devices to lights and heating devices. *Bath furnishings* include such items as a medicine cabinet, extra mirrors, a magnifying mirror, extra counter space, a dressing table, shelves for linen storage, and a clothes hamper. A bath designed for, or used in part by, children should include a low or tilt-down mirror, benches for reaching the lavatory, low towel racks, and shelves for bath toys.

Plumbing accessories such as a whirlpool bath or bidet might be installed. Figure 12-12 shows a bidet installed beside the water closet. Foot-pedal controls for water and single-control faucets are other types of accessories.

LAYOUT There are two basic types of bath layouts: the compartment plan and the open plan. In the **compartment plan**, partitions (sliding doors, glass dividers, louvers, or even plants) are used to divide the bath into compartments: the water closet area, the lavatory area, and the bathing area. See Fig. 12-17. In the open plan, all bath fixtures are completely or partially visible.

VENTILATION Baths should have either natural ventilation from a window or forced ventilation from an exhaust fan.

Care should be taken to place windows in a position where they will not cause a draft on the tub or interfere with privacy. A bath can be designed without windows. However, substitute sources of light and ventilation are then needed. This can be achieved by installing a light-ventilating fan combination which is controlled by a single switch.

LIGHTING Lighting should be relatively shadowless in the area used for grooming. See Fig. 12-18. Shadowless general lighting can be achieved with fluorescent tubes installed on the ceiling and covered with glass or plastic panels. Skylights can also help provide general illumination.

HEATING Heating in the bath is especially important. In addition to the conventional heating outlets, an electric heater or heat lamp is often used to provide instant heat. The source of heat should be placed under the window to eliminate drafts. All gas and oil heaters should be properly ventilated. Heat lamps should be controlled by a timer to avoid overheating.

Fig. 12-17 ■ Compartment bath with separate areas.

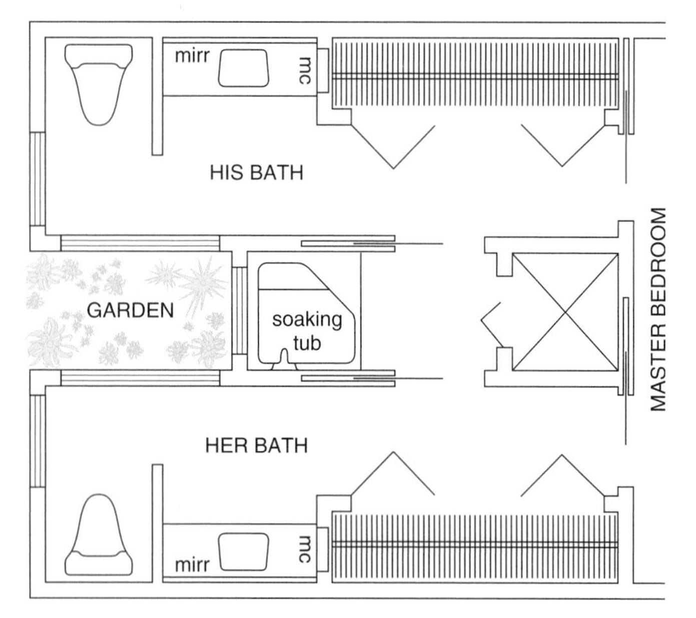

Location

Two factors in locating baths are positioning of plumbing lines and accessibility from other areas of the house.

PLUMBING LINES The plumbing lines that carry water to and from the fixtures should be concealed and minimized as much as possible. When two baths are located side by side, placing the fixtures back to back on opposite sides of the plumbing wall reduces the length of plumbing lines. See Fig. 12-19. In multiple-story dwellings, efficient use of plumbing lines can be accomplished if the baths are

Fig. 12-18 ■ Lighted side panels provide shadowless light in the sink area. *American-Standard*

Fig. 12-19 ■ Fixtures should be located to minimize the length and number of plumbing lines.

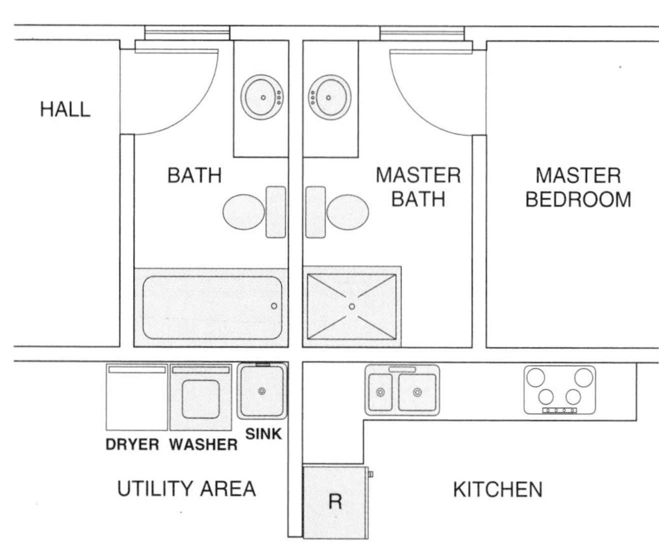

placed one directly above another. When a bath is placed on a second floor, a plumbing wall must be provided through the first floor for the soil and water pipes.

ACCESSIBILITY Ideally, a private bath should be located adjacent to each bedroom, as shown in Fig. 12-20. Often, this is not possible, and a **central bath** is designed to meet the needs of the entire family. See Fig. 12-21. The central or general bath should be in the sleeping area, accessible from all the bedrooms. A bath for general use plus a bath adjacent to the master bedroom is a desirable compromise. A **master bath** is accessible only from the master bedroom.

Bathing or showering facilities are usually not needed in the living or service areas. **Half-baths** containing only a lavatory and water closet are designed for these areas, unless a full bath is conveniently located nearby.

Fig. 12-20 ■ A bath for each bedroom is ideal.

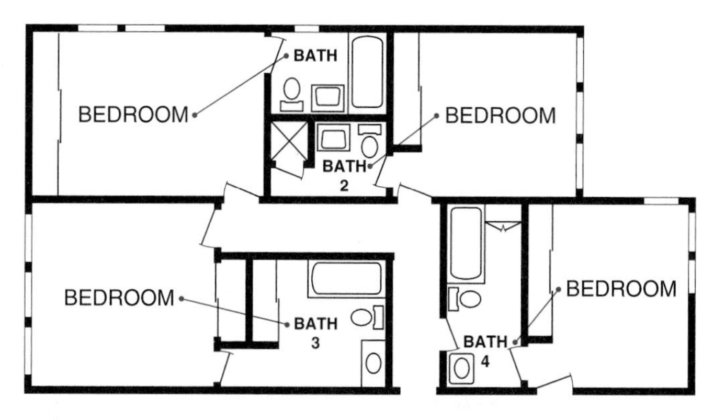

Fig. 12-21 ■ A central bath accessible from all bedrooms.

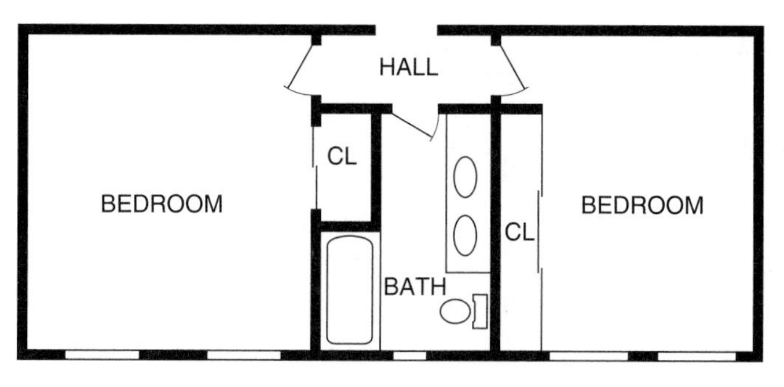

Decor

Today's baths need not be strictly functional and sterile in decor. They can be planned and furnished in a variety of styles. See Fig. 12-22. Baths should be decorated and designed to provide the maximum amount of light and color.

MATERIALS Materials used in the bath should be water-resistant, easily maintained, and easily sanitized. Tiles, linoleum, marble, plastic laminate, and glass are excellent materials for bath use. If wallpaper or wood paneling is used, it should be waterproof. If plaster or drywall construction is exposed, a gloss or semigloss paint should be used on the surface.

Fixtures are now available in a variety of colors, so that they can be coordinated or even matched with accessories. Matching countertops and cabinets are also available.

New materials and components enable the designer to plan baths with modular units that range from one-piece molded showers and tubs to entire bath modules. In these units, plumbing and electrical wiring are connected after the unit is installed.

Fig. 12-22 ■ Bath decor can be unique as well as attractive. *Du Pont CORIAN®*

Size and Shape

Bath sizes and shapes are influenced by the size and spacing of basic fixtures and accessories. The type of plan—whether compartmentalized or open—and the relationship to other rooms in the house also influences size and shape. Additional space may be required to accommodate wheelchairs. Figure 12-23 shows a variety of bath shapes and arrangements.

The minimum size bath which can include all three basic fixtures is 5′ × 8′. Sizes range from this minimum to luxury compartmentalized baths as shown in Fig. 12-17. Figure 12-24 shows the standard sizes of bath cabinets. Although cabinets may be custom-made to any size, choosing standard cabinet dimensions saves considerable time and expense. Figure 12-25 shows the various standard sizes of bath fixtures.

Fig. 12-23 ■ Common bath shapes and arrangements.

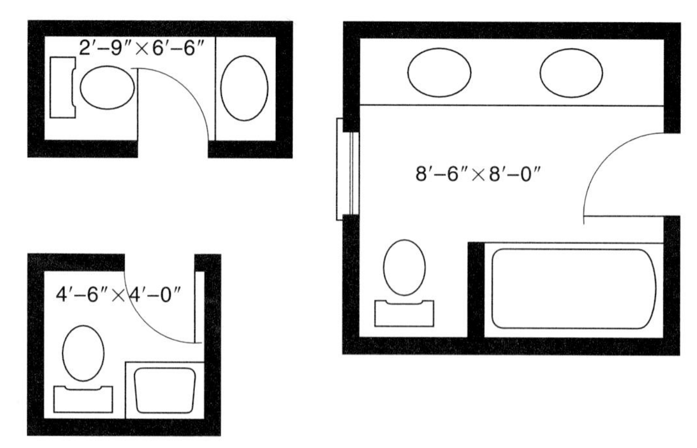

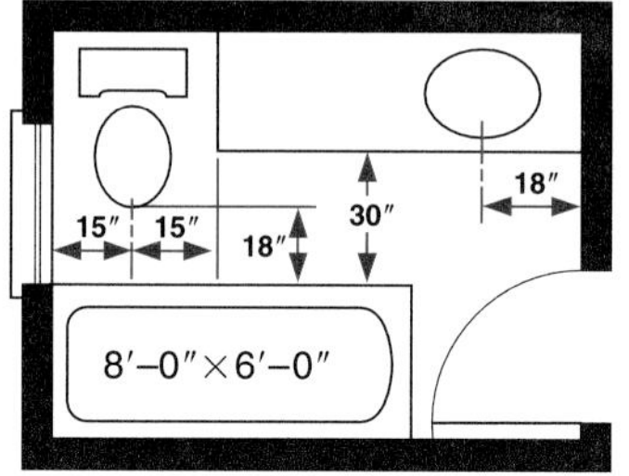

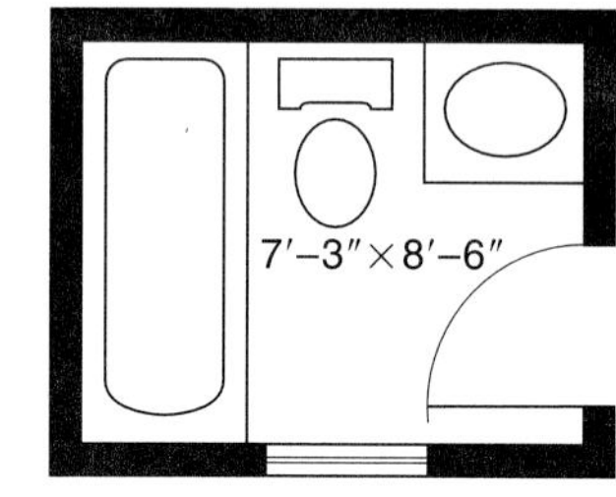

Fig. 12-24 ■ Standard sizes of common bath cabinets.

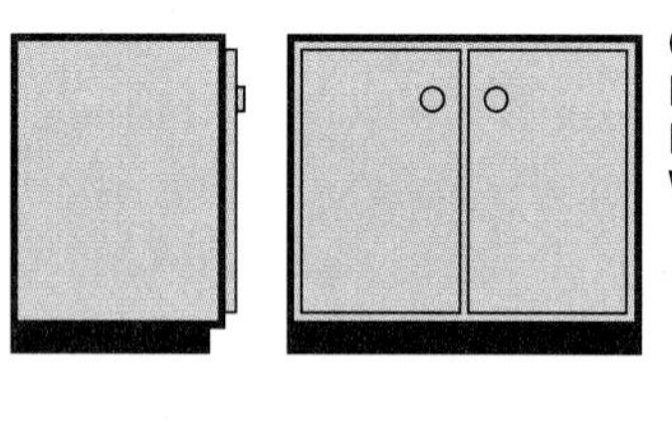

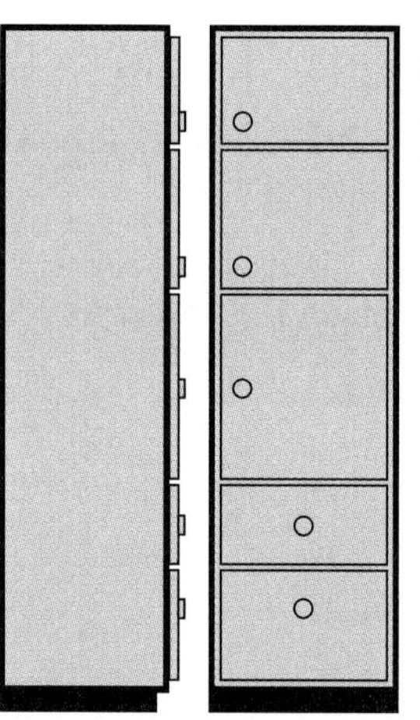

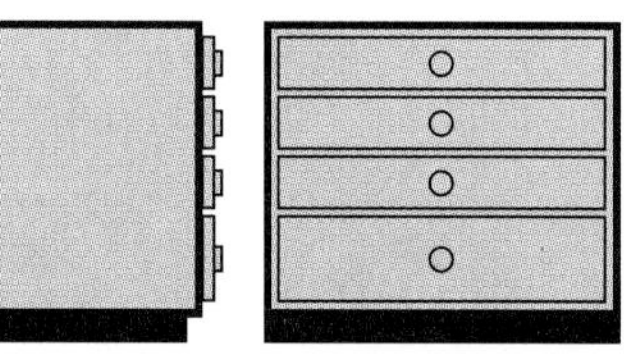

COMBINATION DRAWER & CABINET VANITIES
HT 30″
DEPTH 18″ TO 21″
WIDTH 12″ TO 60″

HAMPER VANITIES
HT 30″
DEPTH 18″ TO 21″
WIDTH 18″ TO 24″

BOWL VANITIES
HT 30″
DEPTH 18″ TO 21″
WIDTH 24″ TO 36″

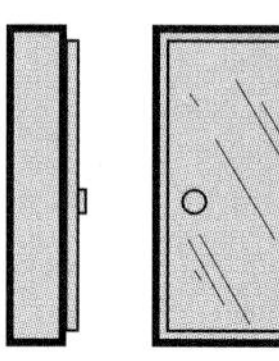

Fig. 12-25 ■ Standard sizes of bath fixtures.

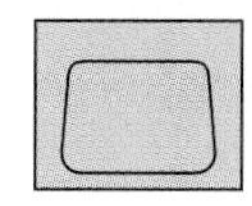

WALL-HUNG LAVATORIES
HT 31″ TO 33″
DEPTH 17″ TO 20″
WIDTH 16″ TO 28″

COUNTERTOP LAVATORIES
HT 31″ TO 33″
DEPTH 13″ TO 24″
WIDTH 19″ TO 24″
CIRCULAR 18″ TO 20″ DIAM.

PEDESTAL LAVATORIES
HT 31″ TO 33″
DEPTH 18″ TO 22″
WIDTH 18″ TO 30″

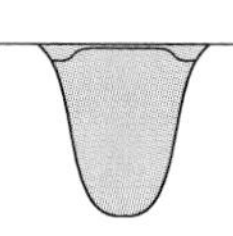

WALL-HUNG WATER CLOSETS
SEAT HT 14″ TO 15″
DEPTH 25″ TO 26″
WIDTH 14″ TO 16″

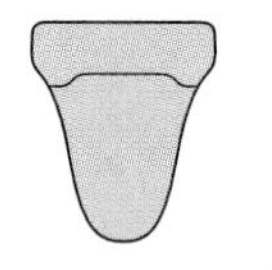

ONE-PIECE WATER CLOSETS
TANK HT 20″ TO 30″
SEAT HT 14″ TO 15″
DEPTH 28″ TO 30″
WIDTH 18″ TO 22″

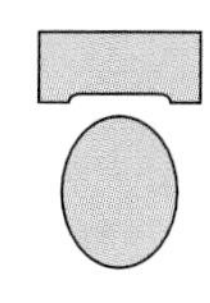

TWO-PIECE WATER CLOSETS
TANK HT 20″ TO 30″
SEAT HT 14″ TO 15″
DEPTH 26″ TO 30″
WIDTH 17″ TO 21″

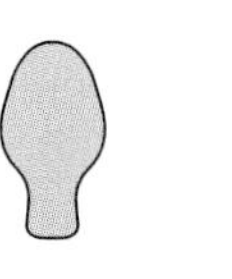

BIDETS
HT 14″ TO 15″
DEPTH 23″ TO 26″
WIDTH 14″ TO 16″

BATHTUBS
HT 14″ TO 22″
DEPTH 30″ TO 42″
LENGTH 54″ TO 72″

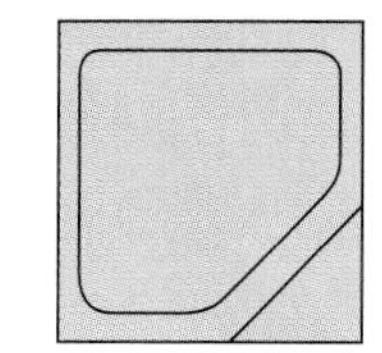

SQUARE BATHTUBS
HT 12″ TO 14″
DEPTH 37″ TO 60″
LENGTH 42″ TO 60″

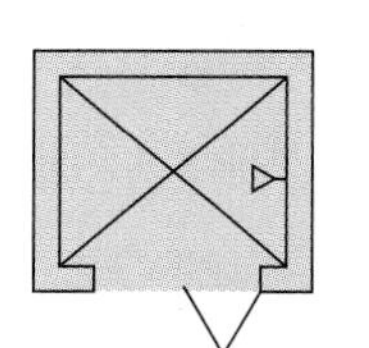

SHOWER STALLS
HT 84″
DEPTH 21″ TO 42″
LENGTH 21″ TO 72″

CHAPTER 12 Exercises

1. Design a bedroom, 100 sq. ft. in size, for a six-year-old child.
2. Design a bedroom, 150 sq. ft. in size, for a teenager.
3. Design a master bedroom with an adjoining bath that is 200 sq. ft. in size.
4. Using a CAD system, draw a bedroom shown in this chapter.
5. Plan the bedroom areas for the home you are designing.
6. Using dimensions provided in this chapter, calculate the minimum size of a bedroom that could accommodate a king size bed, built-in TV, a dresser, and a lounge chair.
7. Collect pictures of bedrooms you like. Identify good and bad design features.
8. Draw a plan with a master bath and a central bath.
9. Draw a plan of a bath you think is poorly designed. Then draw a plan for remodeling the bath to make it more functional.

10. Draw the plans for the bath areas in the home of your own design.
11. Using a CAD system, design and draw a bath that is 12′ × 8′.
12. Calculate the dimensions needed for a bath you design with: one cabinet vanity, two drawer-vanities, one hamper vanity, one wall cabinet, a countertop lavatory, a bathtub, and a one-piece water closet. (Use the standard sizes given in Figs. 12-24 and 12-25 for reference.)

PART
4

Basic Architectural Drawings

CHAPTER

13 Designing Floor Plans
14 Drawing Floor Plans
15 Designing Elevations
16 Drawing Elevations
17 Sectional, Detail, and Cabinetry Drawings
18 Site Development Plans

CHAPTER

13

Designing Floor Plans

Objectives

In this chapter you will learn to:

- gather information from a client that is needed to design an architectural project.
- analyze a building site.
- use the design process to prepare for drawing accurate and functional floor plans.
- create floor plan sketches.
- design floor plans to accommodate the needs of persons with physical impairments.

Terms

base map
conceptual design
easements
floor plans
idealized drawings
room template
setbacks
single-line drawing
site analysis
site-related drawing
situation statement
user analysis

The most commonly used architectural drawings are floor plans. Many examples are shown in Part Three. This chapter outlines procedures used in developing floor plan designs, beginning with a presentation of the steps in the design process. Guidelines for developing architectural designs to accommodate special needs are also included.

■ **Floor plans can be interesting as well as functional. This CAD floor plan shows furniture placement.** *David Karram*

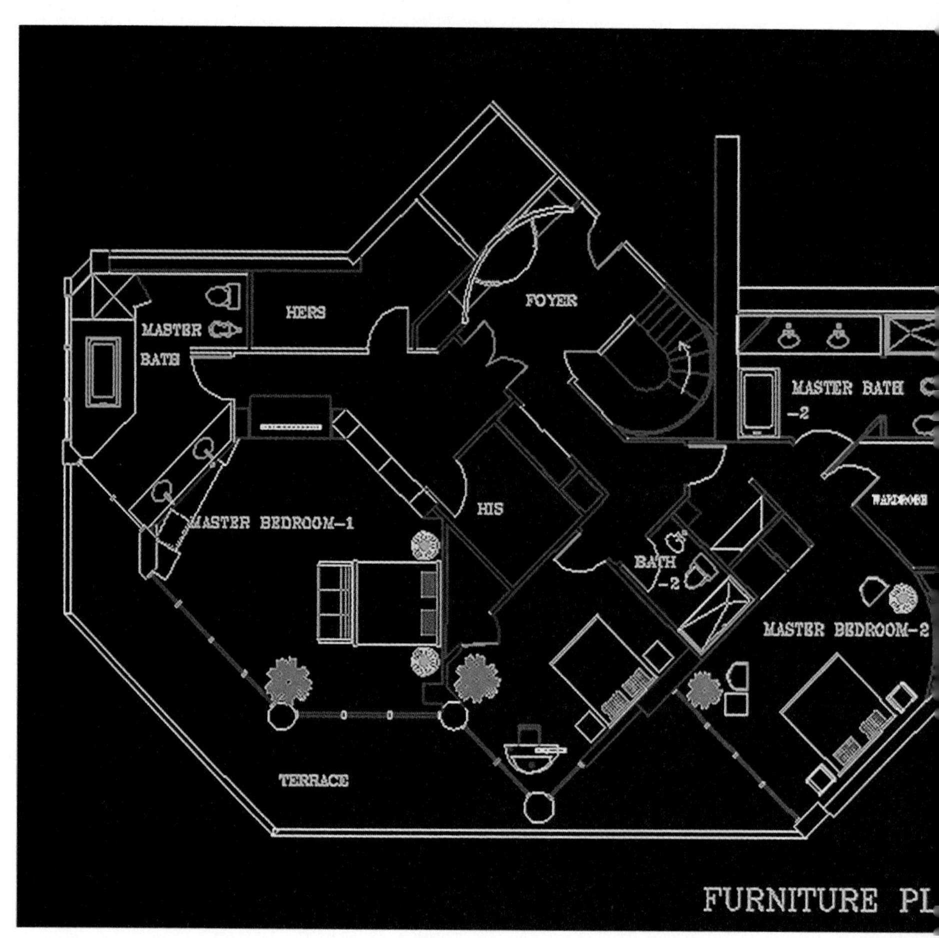

Floor Plan Development

Final **floor plans** contain more specific information about an architectural design than any other type of drawing. They include descriptions of locations, sizes, materials, and components contained in the design. Floor plans serve as a point of reference for other drawings in a set. For this reason the first phase of the architectural design process leads to the development of *basic floor plans*.

The Design Process

The architectural design process involves many personal, social, economic, and technical variables to create detailed working drawings. To effectively apply the principles and elements of design to an architectural project, established design sequences and procedures must be followed. This process is a logical sequence of thought and activities which begin with an inventory and analysis of the project. This process continues through the design of basic site and floor plans and proceeds through the completion of all working drawings. See Fig. 13-1.

Defining the Project

The first step in the design process is to define the project. An agreement established between a client and a designer involves the purpose, theme, scope of the project, budget, and schedules. This agreement is then translated into a **situation statement** as shown in Fig. 13-2. This statement identifies and records the client's major requirements and any special design requirements and problems. Any subsequent design drawings relate to this situation statement.

NEEDS AND WANTS The success of any design depends on how well the finished product meets the needs of the user. During the entire process, the needs of the residents and their "wants" as well must be kept clearly in mind. This includes physical and lifestyle considerations.

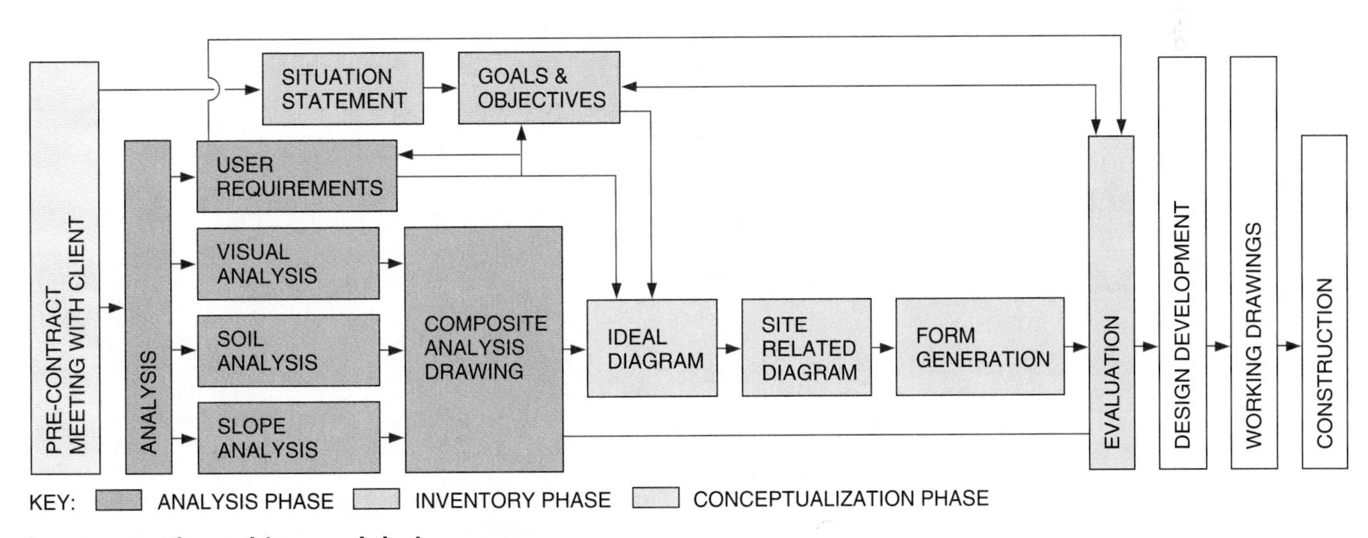

Fig. 13-1 ■ The architectural design process.

Situation

Mr. and Mrs. John Smith have acquired a five acre parcel of wooded land. They want the site developed as a residence and also for home office use. The Smith family of four have strong feelings for environmental preservation and are very fitness oriented. They want a high degree of privacy and need the capacity to entertain weekend guests. The total cost cannot exceed $60,000 more than the sale price of their existing home. Completion of construction (closing) must be no later than one year from the signing date of the design contract.

Fig. 13-2 ■ Design situation statement.

A *need* is an absolute requirement and must be implemented in the design. A *want* is a desirable feature, but not absolutely required. Wants can be compromised because of budget constraints, space, or code restrictions. An effective design must meet all of the client's needs and as many wants as possible.

The designer and the client prepare needs and wants lists. See Fig. 13-3. If the "wants" are listed in priority order, items on the list can be cut beginning at the bottom of the list and proceeding to the top, if necessary. In this way, what is wanted most is more likely to remain in the final design.

GOALS AND OBJECTIVES Once the client and designer agree on a situation statement and create a list of the client's needs and wants, major goals and specific objectives can be developed. This is an important step. Constant reference is made to these goals and objectives during all phases of the design process. They provide the focus of the project for the designer and become a basis for evaluating all aspects of the design. See Fig. 13-4. Once this is completed the analysis phase can begin.

Needs

- Living area open plan
- Contemporary design
- Home office or study which can be shared
- Living area fireplace
- Courtyard patio
- Pool with large living deck
- Boat dock
- Drive apron for parking
- Private MBR deck
- Privacy from road
- Three bedrooms including master suite
- Maximize view of mountains
- Dining facilities for 10 guests maximum

Wants

1. Jogging track
2. Shop area near garage
3. Badminton court
4. Minimum lawn area
5. Bridge connecting deck and dock
6. Whirlpool
7. Basketball court
8. Minimize tree removal
9. Maximize solar use
10. Separate home office for Mr. & Mrs. Smith
11. MBR fireplace
12. Greenhouse
13. Billiards room
14. Cabana near pool

Fig. 13-3 ■ Needs and wants list related to the situation statement.

Major goals

Design a contemporary residence for the existing site with good visual profiles, aesthetic appeal and emphasis on functional, non-destructive use of all site features, including maximum use of solar energy. Plan working, living, and recreation areas to conform to space and priority needs. Position all facilities so that all are not visible from one vantage point.

Objectives

1. Provide stimulating, casual, and open atmosphere.
2. Locate private and public areas to avoid user conflicts.
3. Position all facilities for minimum environmental impact and minimum maintenance.
4. Orient structures for maximum solar use.
5. Building areas to blend with existing site landform.
6. Relate interior living areas to exterior space.
7. Residence not to be completely visible from access road.
8. Plan circulation patterns for both vehicular and pedestrian traffic with parking area.
9. Plan facilities for badminton, swimming, basketball, jogging, and whirlpool.
10. Provide courtyard for seasonal use.
11. Interior and exterior dining facilities for maximum 10 guests.
12. Provide Mr. Smith with an office for evening and weekend use.
13. Provide Mrs. Smith with an accessible office area to meet clients daily.
14. Use natural contemporary lines and materials consistent with site.
15. Design a focal point fireplace for living area.
16. Design gradual realization for vehicular approaching traffic.
17. Keep total cost within limits established by clients.
18. Boat dock to have access from deck area.
19. Provide three bedrooms, including master suite.

Fig. 13-4 ■ Design goals and objectives based on needs and wants.

Analyzing the Project

After the project is defined, an organized and sequenced analysis must be made of user wants and needs, site features, soil conditions, the slope of the land, and the views. These separate analyses lead to the development of a comprehensive analysis.

USER ANALYSIS In a **user analysis,** each goal is further refined into descriptions of space elements, usage, size, and the relationships between areas. With a user analysis, a designer can break down each design element into manageable parts. To make evaluation, verification, and discussion easier, a chart is usually prepared. See Fig. 13-5.

The user analysis has great influence on the development of a design. No area should be omitted. If the user analysis is inadequate or contains erroneous information, the final design will not reflect the major goals and objectives of the project.

The user analysis is particularly important when designing a project that will be used by a person or persons with impairments. Design considerations and guidelines for developing designs to accommodate special needs are presented at the end of this chapter.

SITE ANALYSIS Any architectural project should be developed to take advantage of a site's positive features and minimize its negative features. Completing a **site analysis** not only helps the designer make proper design decisions, but also helps ensure appropriate land use.

Space elements	Primary users	Min. size	Notes and relationships
Living room	8–16 adults	16′ X 22′	Pool view–fireplace Access to foyer & dr.
Dining room	6–12 adults	14′ X 20′	Access to kit & lr.
Study #1	Mr. Smith	14′ X 20′	Private–quiet.
Study #2	Mrs. Smith	12′ X 14′	Private–client Accessible–joint office w/Mr. S ?
Entry	Family–guests	8′ X 10′	Visible from drive–baffle from street.
Parking	2 family cars 6 guest cars	9′ X 10′ stalls	Access to main entry.
Decks or terrace	Family–guests	6′ X 20′	Overlook pool–next to living area.
Courtyard	Family–guests	200 sq. ft.	For casual entertainment next to kit.
Kitchen	Family	12′ X 16′	Access to deck, lr, & laundry.
Garage	Family	20′ X 24′	Access to kit–convert to shop.
Service pickup	Service pers.	40 sq. ft.	Screen from living areas.
Master bedroom	2 adults	16′ X 24′	Morning sun–king bed–access to pool–suite w/ bath–quiet area.
Bedrooms	2 children	16′ X 18′	Plan for teen growth–away from living & master br.
Baths	Children & guests	2′–8′ X 10′	Access from children's rooms & guests.
Guest bedroom	Guests	12′ X 16′	Bath access–or convertible study ?
Site considerations	All	Entire site	Solve sitting water problem. Use rock formations & add foliage for visual appeal.
Solar considerations	All	Bldgs. & site	Use passive techniques–care in orientation of facilities.
Recreation facilities	All	Courts, pool	Orient w/sun & screen from residence.

Fig. 13-5 ■ User analysis.

Three types of site analyses are used to develop a final site analysis drawing: soil analysis, slope analysis, and visual analysis. Each of these three distinct factors affect the potential use of the different areas of the site. There are five phases in the development of a site analysis.

- *Phase 1—Development of a base map:*

A **base map** shows all fixed factors related to the site that must be accommodated in the site plan. It includes topographical features; the outline and location of property lines, adjacent streets, existing structures, walkways, paths, terraces, and utility lines; easements; setback limits; and the compass direction. **Easements** are right-of-ways across the land, as for utility lines. **Setbacks** are minimum distances structures must be located from property lines as set by the local government. (See Chapter 18 for setback details.)

Base maps are usually prepared to a scale of 1″ = 10′, 1″ = 20′, 1″ = 30′, or ⅛″ = 1′-0″. Many copies of this map will be used during the design process for analysis and development. See Fig. 13-6.

- *Phase 2—Soil analysis:*

Soil is composed of rocks, organic materials, water, and gases. Variations in the percentage of these ingredients determine the physical characteristics of the

Fig. 13-6 ■ Base map of property.

soil and its capacity to support the weight of structures. In general, coarse-grained soils, because of their drainage and bearing capacity, are preferred for buildings but not for plants. Conversely, fine-grained soils with high organic content are preferred for plants but not for buildings. The U.S. Department of Agriculture (the USDA) classifies four types of soil according to their quality for building a structure:

Type 1. *Excellent:* Coarse-grained soils—no clays, no organic matter.

Type 2. *Good to Fair:* Fine, sandy soils (minimum organic and clay content).

Type 3. *Poor:* Fine-grained silts and clays (moderate organic content).

Type 4. *Unsuitable:* Organic soils (high clay and peat content).

To prepare a soil analysis drawing, follow these steps:

1. Obtain a soil classification for the site from a county soil survey or from private borings.
2. Draw areas on the base map representing the different soil types, as shown in Fig. 13-7.
3. Note the bearing capacity and depth to bedrock for each soil category. This information is given in kilopounds (kips, or K) in the USDA survey book (1 K = 1000 lb. for 1 sq. ft. of soil area).
4. Provide a legend showing the categories of soil types and describe the soil

Fig. 13-7 ■ Soil analysis drawing.

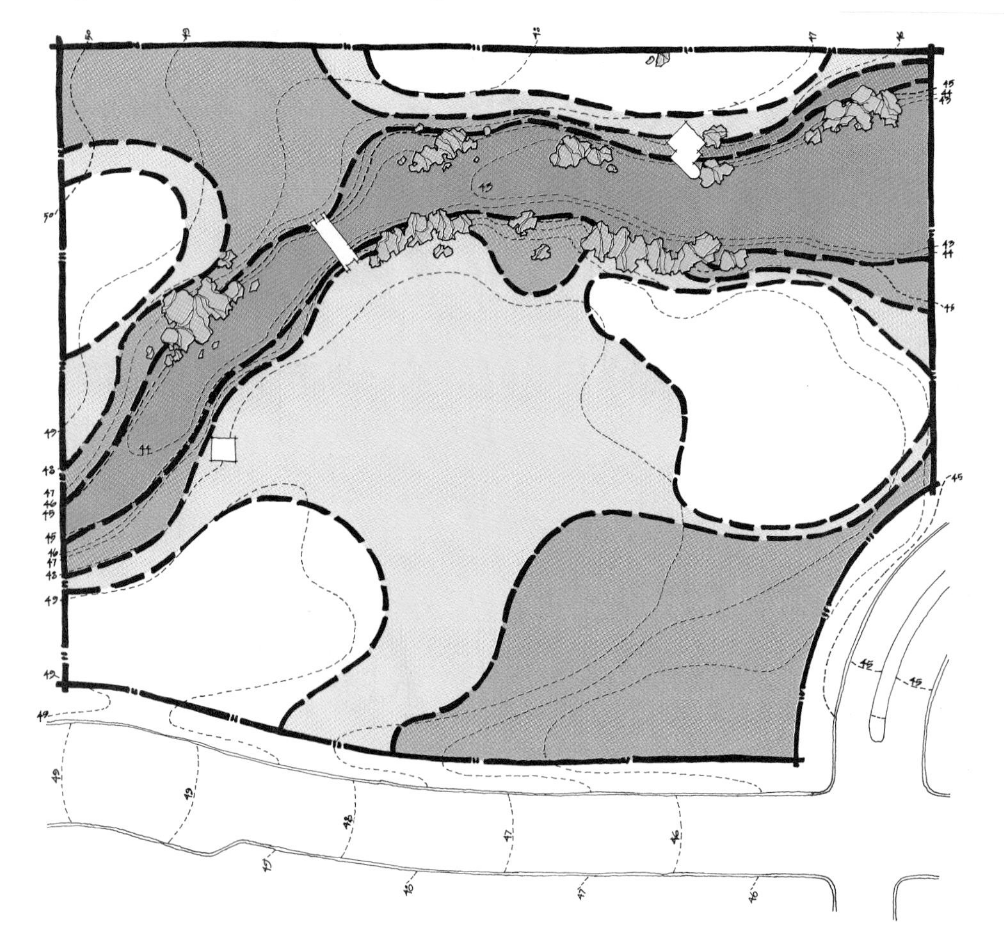

characteristics of each type. On the drawing, note where soil conditions can or cannot be used for development.

5 Color-code each soil capacity type in the legend to match the drawing.

- *Phase 3—Slope analysis:*

The slope of a particular site greatly affects the type of building that can or should be designed for it. The slope percentage also determines what locations are acceptable, preferred, difficult, or impossible for building. See Fig. 13-8. The cost of building may be greatly affected by excessive slope angles.

To complete a slope analysis drawing, refer to Fig. 13-9 and complete the following steps:

1 To the base map, add *contour lines* (lines connecting points that have the same elevation) derived from a U.S. Geological Survey (USGS) map of the area. If the site is very hilly, a surveyor may add more closely spaced contours to provide a more detailed description of the slope of the site. Existing contour lines are dashed, since the finished contour grade lines will later be drawn solid. (See Chapter 18 for more information on contour lines.)

2 Identify four classifications of slopes on the drawing:
   - 0% to 5%—excellent
   - 5% to 10%—good/fair
   - 10% to 25%—poor
   - over 25%—unsuitable

3 Identify each slope category using colors or tones to show the degree of development potential of each section. Generally, light colors are used for areas suitable for development and dark colors are used for less suitable areas.

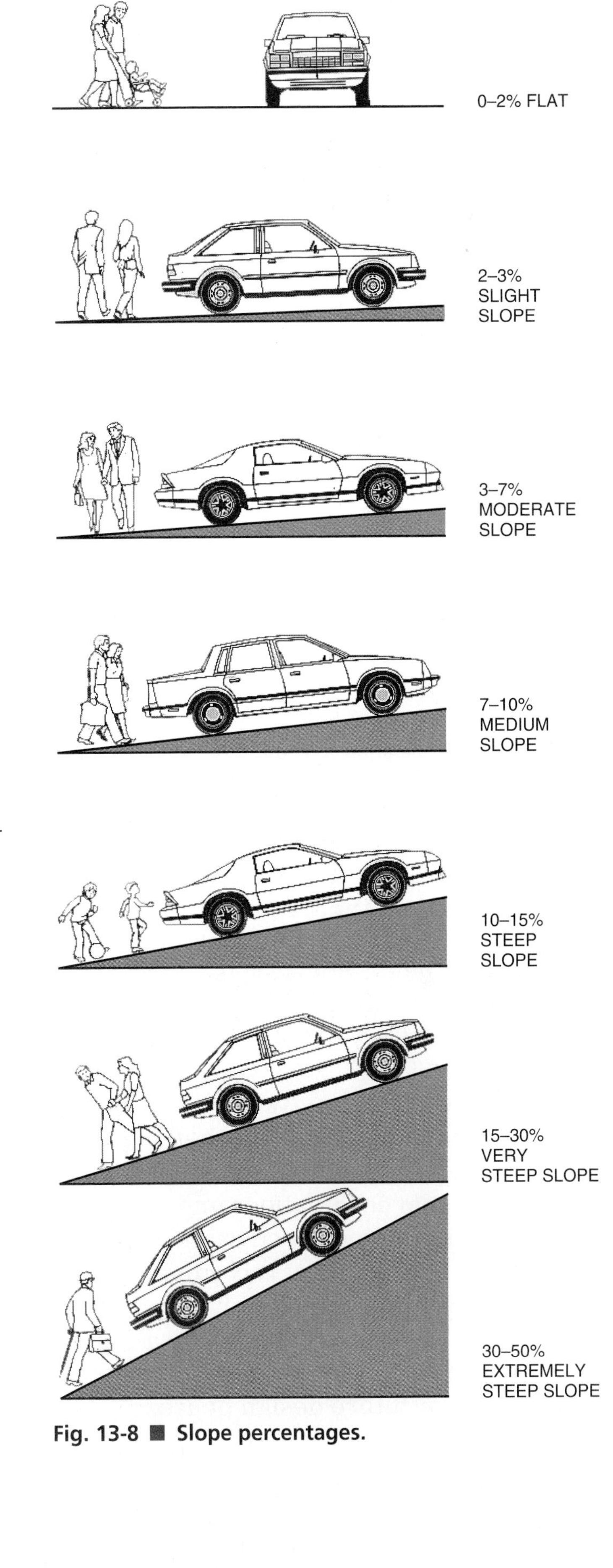

Fig. 13-8 ■ Slope percentages.

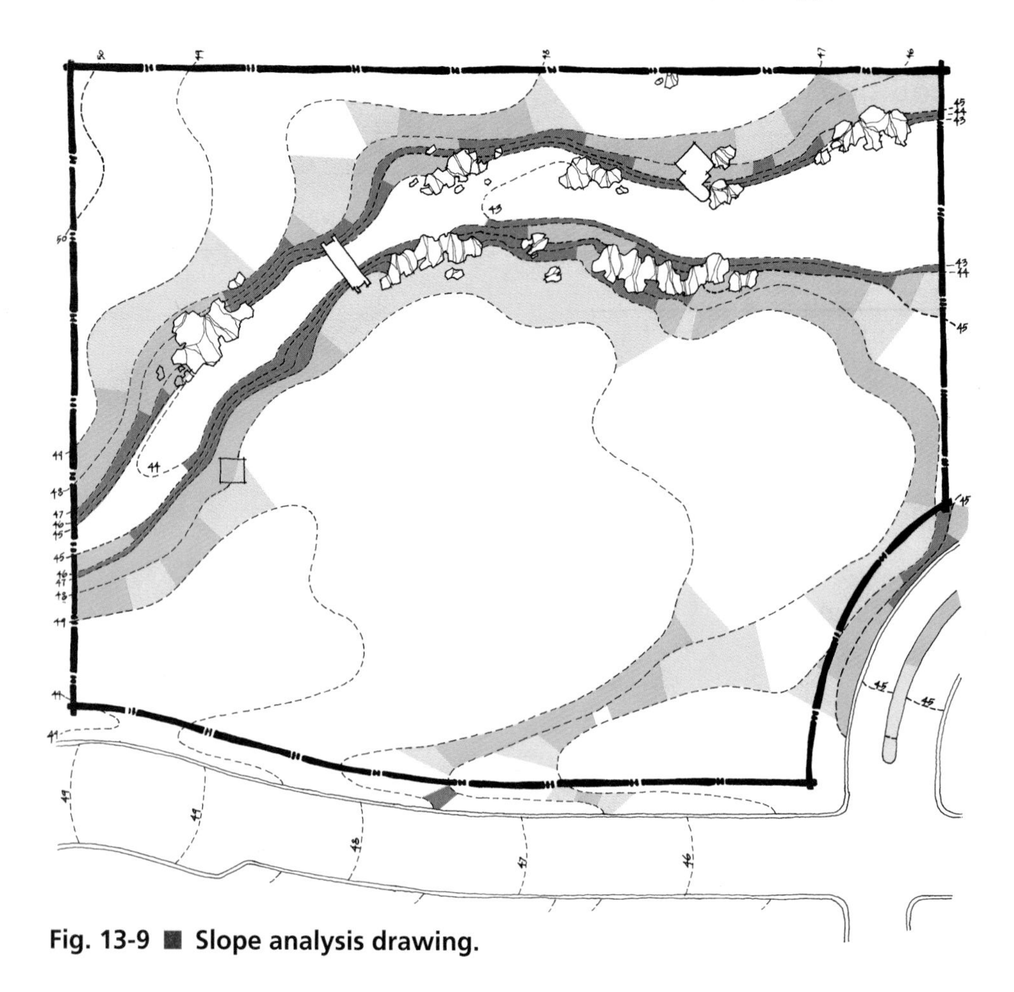

0–5%: EXCELLENT
- FLAT TO MODERATELY FLAT
- EASY TO DRAIN
- RESISTS EROSION
- REQ MIN GRADING
- IDEAL FOR RECREATION
- ALL TYPES ROADS FEASIBLE

5–10%: GOOD–FAIR
- SLOPING
- DRAINS EASILY
- MIN EROSION
- SOME GRADING REQ
- SLOPE RANGE EXCELLENT FOR BUILDING
- DRIVEWAYS SHOULD BE PLACED PARALLEL TO SLOPE

10–25%: POOR
- HILLY
- RAPID RUNOFF CREATES EROSION PROBLEMS
- STABILIZATION NECESSARY ON UNDEVELOPED SLOPES
- EXCESSIVE CUT & FILL NEEDED FOR STRUCTURES
- SINGLE-STORY STRUCTURES IMPRACTICAL
- ROADS REQUIRE REINFORCEMENT & BASE STRUCTURES

25%: UNSUITABLE
- SEVERE
- SERIOUS RUNOFF & EROSION PROBLEMS
- WILL NOT SUPPORT STRUCTURES WITHOUT MAJOR ALTERATIONS WHICH WOULD DESTROY THE ENVIRONMENT
- EXCELLENT FOR HIKING

Fig. 13-9 ■ Slope analysis drawing.

4 Provide a color-keyed legend of slope categories and note both potential and constraints for development for each slope category. Note assets and/or limitations for each slope category. Note erosion or drainage problems, if any, for each category.

- *Phase 4—Visual analysis:*

Designers analyze the aesthetic and environmental potential of a site visually. Because visual observations and aesthetic qualities are often subjective and elusive, an organized method of recording and analyzing is important. Refer to Fig. 13-10 and follow these steps to prepare a visual analysis drawing that can be used to provide input for future design phases:

1 On the base map, locate the direction of the best views from each important viewer position. Label the nature of each view, and rate it as good, fair, or poor. Make recommendations for the treatment of each view such as "enhance" or "screen."

2 Identify existing structures on the base map and describe their condition as good, fair, poor, unsound, or hazardous. Note suggestions to enhance, remove, or rehabilitate.

3 Draw the outline and location of all existing and significant plant material, such as large shrubs and trees. Label the type, and indicate the condition of each as good, fair, or poor. Also locate, draw, and indicate large stands of ground vegetation to be saved.

4 Identify any wildlife population and habitat areas to be saved. Indicate animal food and water sources.

5 With directional arrows, show the direction of prevailing winter winds. Also show the direction of prevailing summer breezes.

6 Find and label the source of any desirable fragrances and/or undesirable odors. For the latter, indicate possible solutions, such as minimizing with aromatic vegetation, screening, or removal of the source.

7 Locate and label exposed open space, semi-enclosed public space, and private space.

- *Phase 5—Composite analysis:*

Once the soil, slope, and visual analysis drawings are completed, this information is combined into a single composite analysis drawing. A composite analysis drawing is prepared to determine the best location zones (areas) for the placement of structures on the site. Location zones are judged and numbered for development potential:

1. Excellent
2. Good or fair
3. Poor
4. No development potential

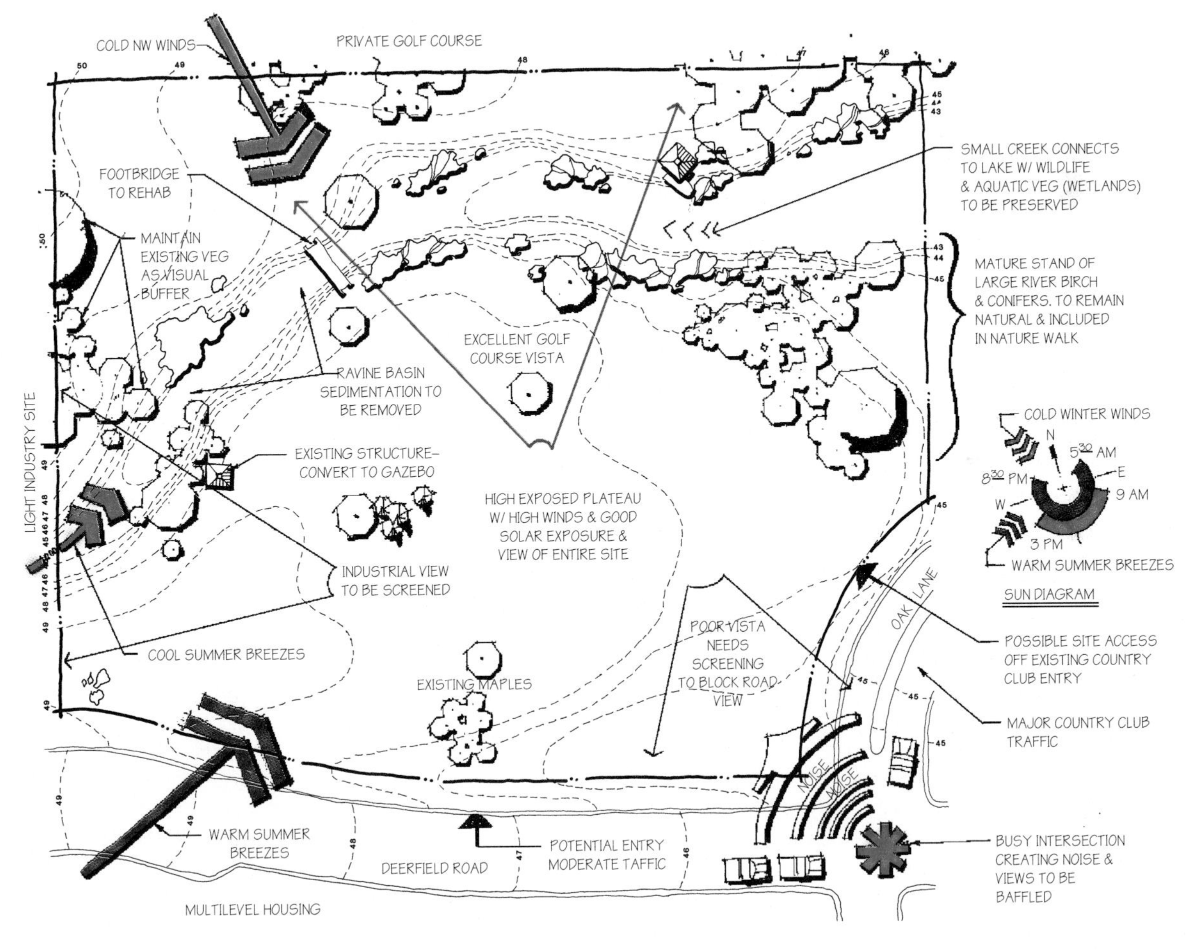

Fig. 13-10 ■ Visual analysis input.

To prepare a composite analysis drawing refer to Fig. 13-11 and follow these steps:

1 Place the soil analysis drawing directly over the slope analysis. Align the property lines with the base map and tape the base map to the drawing board.
2 Attach tracing paper over the slope and soil drawings and trace a line around each distinct area.
3 Determine which development zone each outlined area represents. One example would be if a 0 to 5 percent slope area overlaps with a coarse-grained soil area, the zone is labeled "1" (excellent potential). Another example would be a poor, clay soil area overlaps with a 20 percent slope area, the zone is labeled "3" (poor). Label each zone on an overlay drawing.
4 Place the overlay drawing over the visual analysis drawing and repeat the same outlining of areas covered in step 3 to complete the composite analysis drawing as shown in Fig. 13-11. Apply judgment concerning priorities when there is an overlapping area conflict.

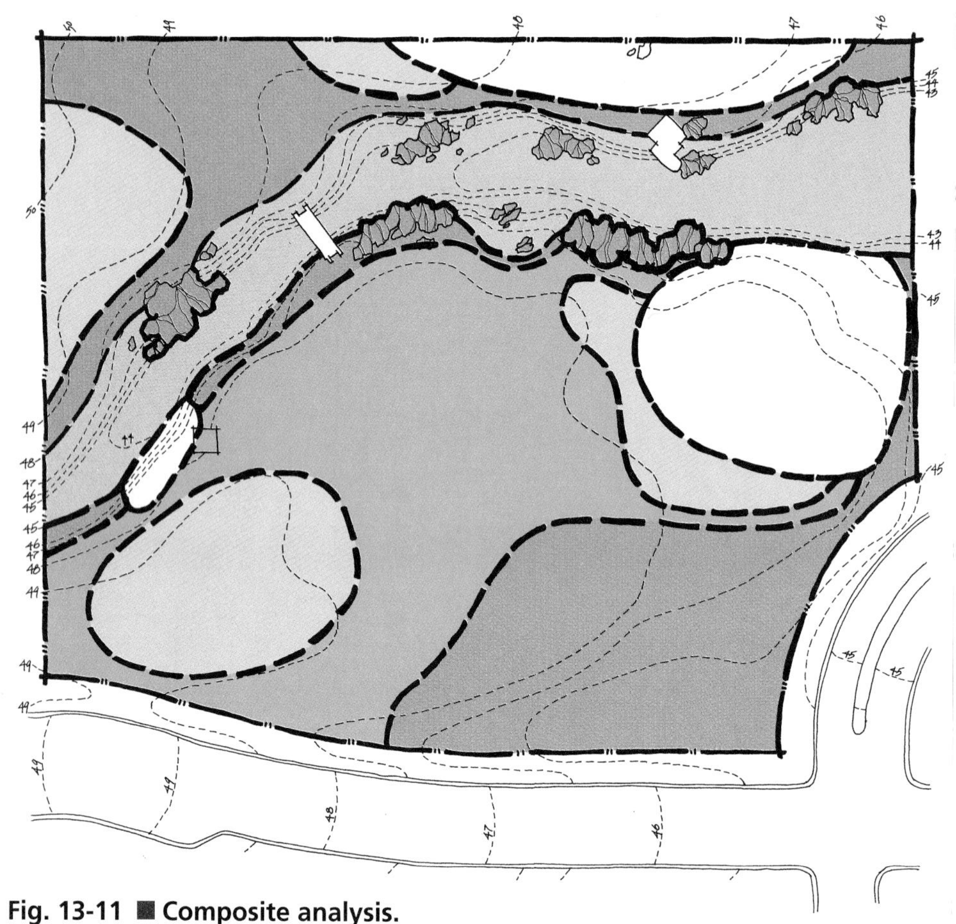

ZONE 1: EXCELLENT
- PRIME DEVELOPMENT LOCATION
- COMPACTABLE SOILS
- EXCELLENT DRAINAGE
- SLOPE 0–5%
- MINIMUM EROSION PROBLEMS
- ROAD & SERVICES EXCELLENT
- SOLAR ORIENTATION OF STRUCTURES POSSIBLE

ZONE 2: GOOD–FAIR
- GENTLE SLOPES 0–10%
- COMPACTABLE SOIL
- GOOD DRAINAGE
- EROSION PROBLEMS MINIMAL EXCEPT NEAR RAVINE
- PRIME AREA FOR SITING STRUCTURES

ZONE 3: RESTRICTED
- STEEP SLOPE 10–25%
- SOIL NOT SUITABLE FOR STRUCTURES
- HIGH RUNOFF RATES
- EROSION RISKS
- VEG REMOVAL COULD DAMAGE ECOTONE
- EXCELLENT FOR NATURAL USE FOR PATHS

ZONE 4: NO DEVELOPMENT
- ORGANIC SOIL
- POOR DRAINAGE
- GENERALLY UNDER WATER
- FOG POCKET DANGERS
- LEAVE UNDERDEVELOPED AND AS NATURAL AS POSSIBLE

ZONE 4A: NO DEVELOPMENT
- ESTABLISHED WOODED STANDS
- WILDLIFE HABITAT & FEEDING GROUNDS
- MUST BE SAVED AS IS

Fig. 13-11 ■ Composite analysis.

Developing a Conceptual Design

A **conceptual design** represents the best response to the information on the site analysis and in the user analysis chart. Two types of sketches are created to develop a conceptual design: idealized and site-related.

IDEALIZED DRAWINGS **Idealized drawings** or diagrams are a series of study sketches (usually drawn on inexpensive tracing paper, nicknamed "trash" or "bum wad"). These designate ideal spatial relationships of the user elements from the user analysis. Ideal diagrams are freehand, bubble-like sketches that show how the separate user elements fit together. The bubbles are not used to show the *sizes* of the areas, only their *spatial* relationships. Designers usually complete a number of studies or sketches. They do as many sketches as necessary until they achieve one study sketch that provides the best possible spatial relationship between the different elements. See Fig. 13-12.

SITE-RELATED DRAWINGS A **site-related drawing,** such as that shown in Fig. 13-13, is one that matches the idealized drawing to the site and introduces size requirements. The main effort at this stage of design is concentrated on "fitting" all the various elements of the user analysis onto the site, while maintaining the most ideal spatial relationships.

The relative scale (size) of each element is first introduced at this phase of the design process. The approximate position, size, and shape of each room, area, or feature are sketched on tracing paper placed over the composite site-analysis drawing. Now the floor-plan design and position begin to take physical form in relation to the land and the surroundings. Several site-related studies are usually completed which integrate the design with the site. At this stage, a designer needs to focus on

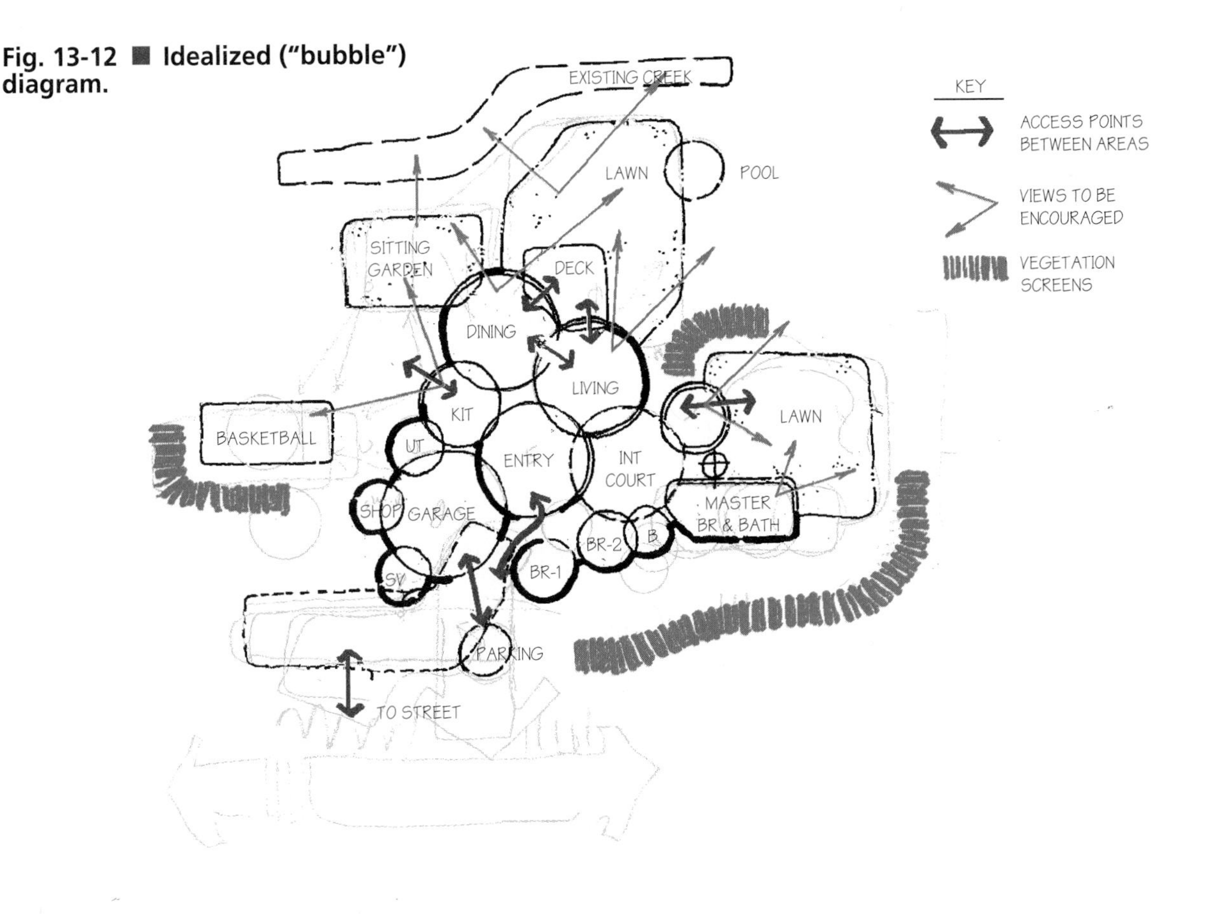

Fig. 13-12 ■ **Idealized ("bubble") diagram.**

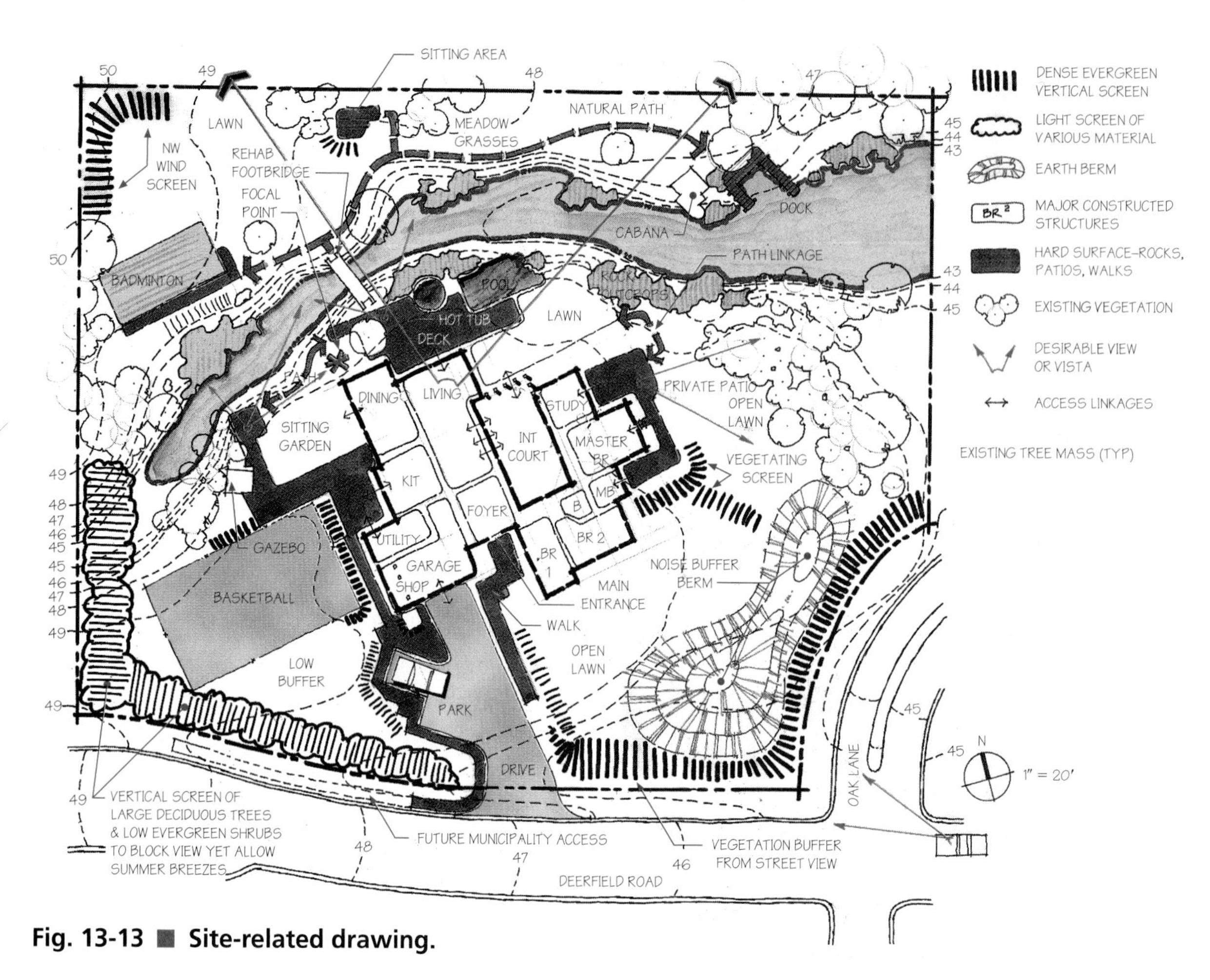

Fig. 13-13 ■ Site-related drawing.

the specific characteristics of the site—its constraints and opportunities.

From all the site-related diagrams and sketches, one is chosen that becomes the floor-plan conceptual design. The designer now begins to generate the form of the design. Drawings are refined into a loose graphic drawing for evaluation. See Fig. 13-14. This drawing is not to scale, but becomes a preliminary floor plan drawn with a straightedge.

Evaluating the Design

Evaluation is always needed to determine the degree of excellence of a design. Self-evaluation of a design is critical and necessary. This requires checking the quality to see if it measures up to the predetermined goals and objectives and to the user analysis requirements.

The conceptual design must be evaluated and necessary revisions made before beginning the final design-development phase. To redesign some elements at the conceptual design stage is easier, cheaper, and more time-efficient than later in the process.

Many details and sizes will not be determined at this point. However, the position of the structures and the relationship between the design elements should not change significantly after this evaluation is completed.

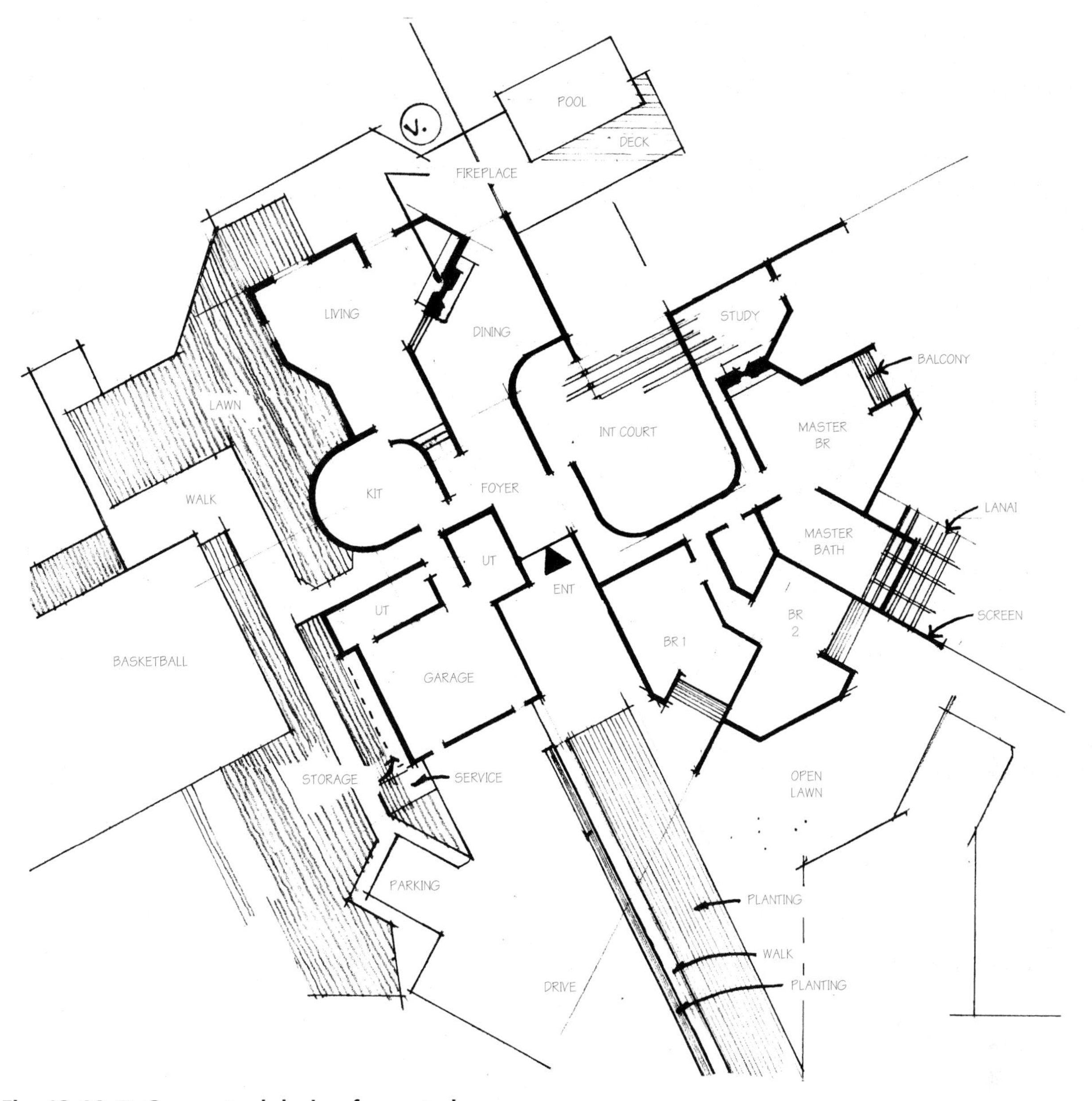

Fig. 13-14 ■ Conceptual design form study.

To evaluate a design, the contents of conceptual design must be compared with *each* specific goal and objective in the user analysis. If a goal has not been accomplished, then that part of the design must be altered to achieve the desired result. A well-developed design will contain very few discrepancies between the goals of the user analysis and the conceptual design.

Design Development

After the necessary changes have been made in the conceptual design as a result of evaluation and client feedback, the final design development phase begins. During this phase, details are added to the site-related diagram in progressive sketches. Sketches are redone until the outlines of the design parts fit together without overlapping and without awkward offsets.

Once the design is "smoothed out," a scaled, **single-line drawing** is prepared, as shown in Fig. 13-15. This drawing includes both floor plan and site features. After this drawing is completed, the designer can concentrate on refining interior building floor plans. See Fig. 13-16.

Functional Space Planning

Once a conceptual plan is developed, the exact space needed for each area needs to be finalized. Experienced designers can determine the relationships of areas and record design ideas through the use of sketches. Students and inexperienced designers should proceed through a process of determining final space requirements through the use of templates. The work done with templates, as presented in this chapter, can be performed manually or on a CAD system.

Planning Space for Rooms and Areas

FURNITURE AND EQUIPMENT Selecting the style, size, and amount of furniture needed for a room is the first step in determining a room's space requirements. Furniture should be chosen according to the needs of the residents—whether that means including a piano for someone interested in music or a large amount of bookcase space for an avid

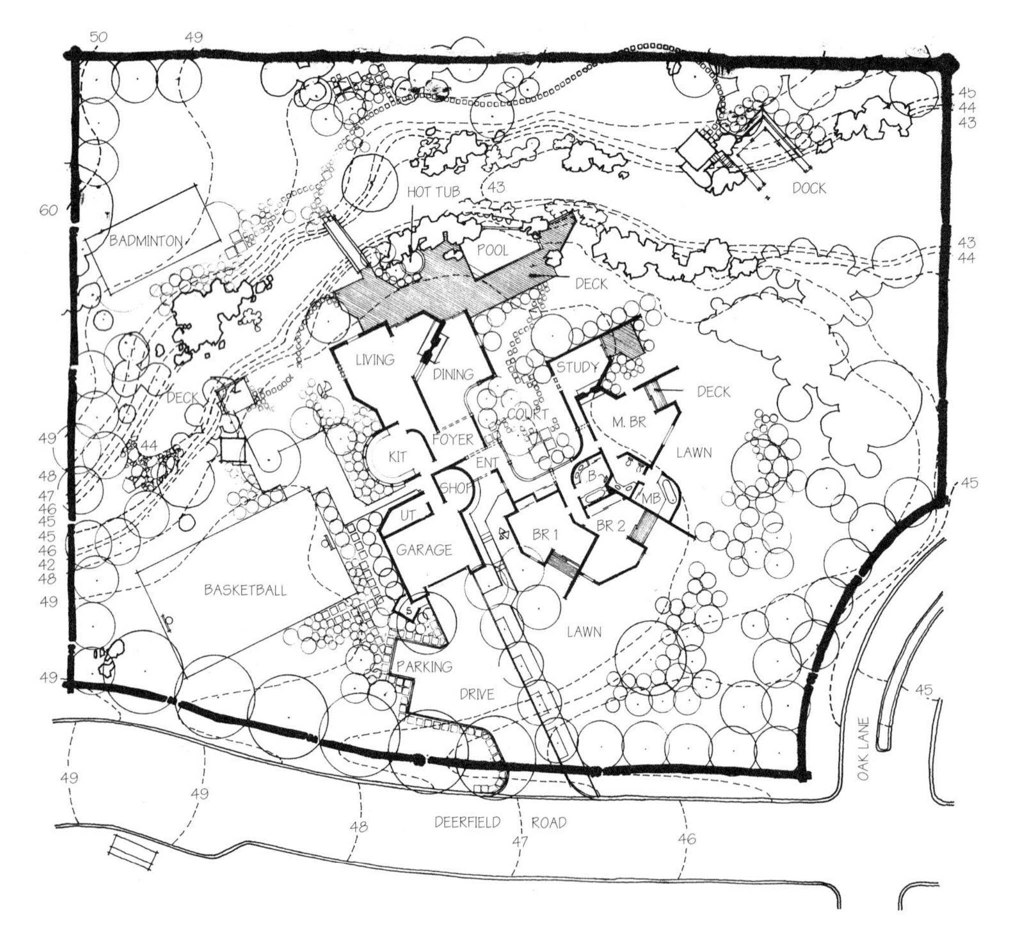

Fig. 13-15 ■ Scaled single line floor plan drawing.

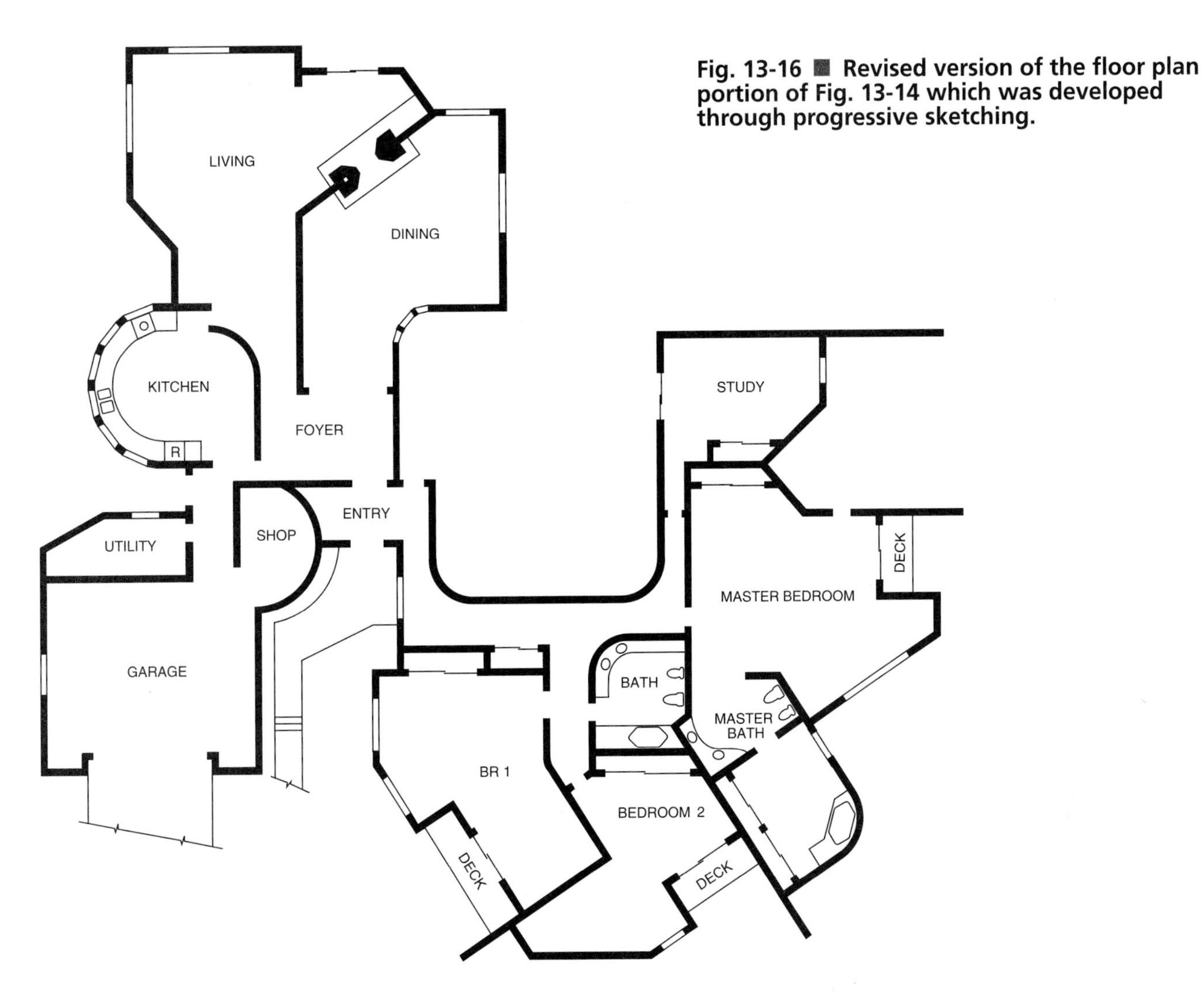

Fig. 13-16 ■ Revised version of the floor plan portion of Fig. 13-14 which was developed through progressive sketching.

reader. The artist, drafter, or engineer may require drafting furniture in the den or study. These individual pieces of furniture affect each room's specific size and shape.

After furniture dimensions are established, furniture templates can be made and arranged in functional patterns. As you learned in previous chapters, furniture templates are thin pieces of paper, cardboard, plastic, or metal that are used to determine exactly how much floor space each piece of furniture will occupy. See Fig. 13-17.

Fig. 13-17 ■ Furniture templates represent the width and length of pieces of furniture. *Art MacDillo's/Gary Skillestad*

Templates are always prepared to the scale that will be used in the final drawing of the house. The scale most frequently used on floor plans is 1/4″=1′-0″. Scales of 3/16″=1′-0″ and 1/8″=1′-0″ are sometimes used.

Wall-hung furniture, or any projection from furniture, even though it does not touch the floor, should be included as a template. This is necessary because the floor space under this furniture is not usable for any other purpose.

Figure 13-18 shows the use of furniture templates on a floor plan. Furniture templates are placed in the arrangement that will best fit the living pattern anticipated for the room. Space must be allowed for the free flow of traffic, as well as for opening and closing doors, drawers, and windows.

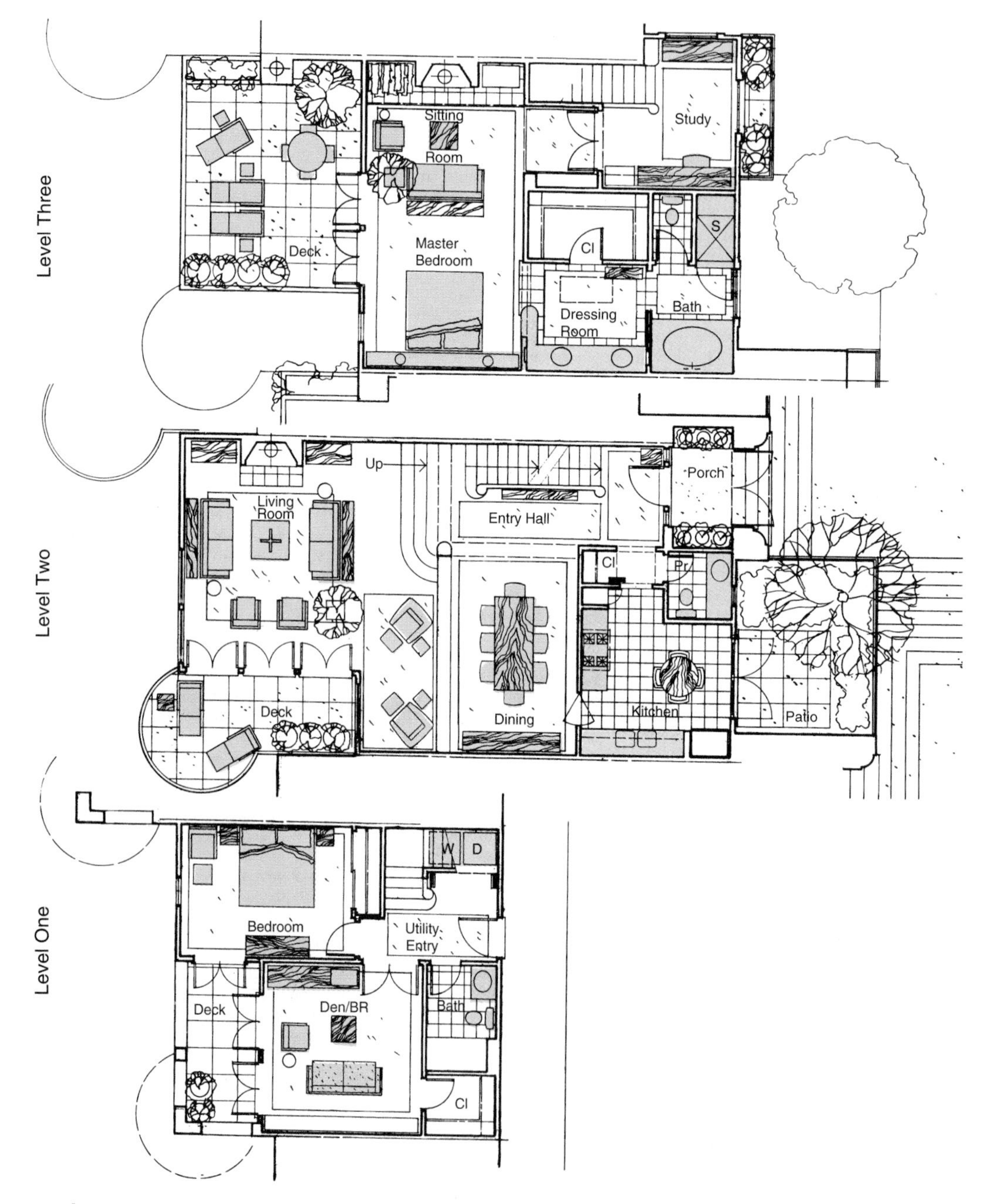

Fig. 13-18 ■ **Use of furniture templates to check room sizes and shapes.**

ROOM SIZES AND SHAPES After suitable furniture arrangements have been established, room dimensions can be determined by drawing an outline around the scaled furniture templates. Then a **room template** can be made by cutting around the outline of the room.

Since the cost of a home is largely determined by the size and number of rooms, room sizes must also be adjusted to conform to an acceptable price range. Figure 13-19 shows area sizes for each room in small, medium, and large dwellings. These areas represent only average sizes. Even where no financial restrictions exist, room sizes are limited by the functional requirements of the room. Just as a room can be too small, it can also be too large to function well for its intended purpose.

Visualizing the exact amount of real space that will be occupied by furniture or that should be allowed for traffic through a room is sometimes difficult. One device used to give a point of reference is a template of a human figure. See Fig. 13-20. This template will help you see how a person would move throughout the room. With a human figure template, you can check the appropriateness of furniture

Rooms	Typical room sizes (sq. ft.)		
	Small	Medium	Large
Living	200	300	600 +
Dining	120	200	300 +
Kitchen	120	160	250 +
Utility	40	60	100 +
Bedroom	150	180	250 +
Bath	40	80	120 +
Den/office	80	120	200 +
Family	180	250	300 +
Foyer	40	80	150 +
Porch	50	100	200 +
Closet	10	20	30 +
Walk-in Closet	30	50	100 +
Halls	3′ wide	3′–6″	4′–0″ +

Fig. 13-19 ■ Typical room sizes.

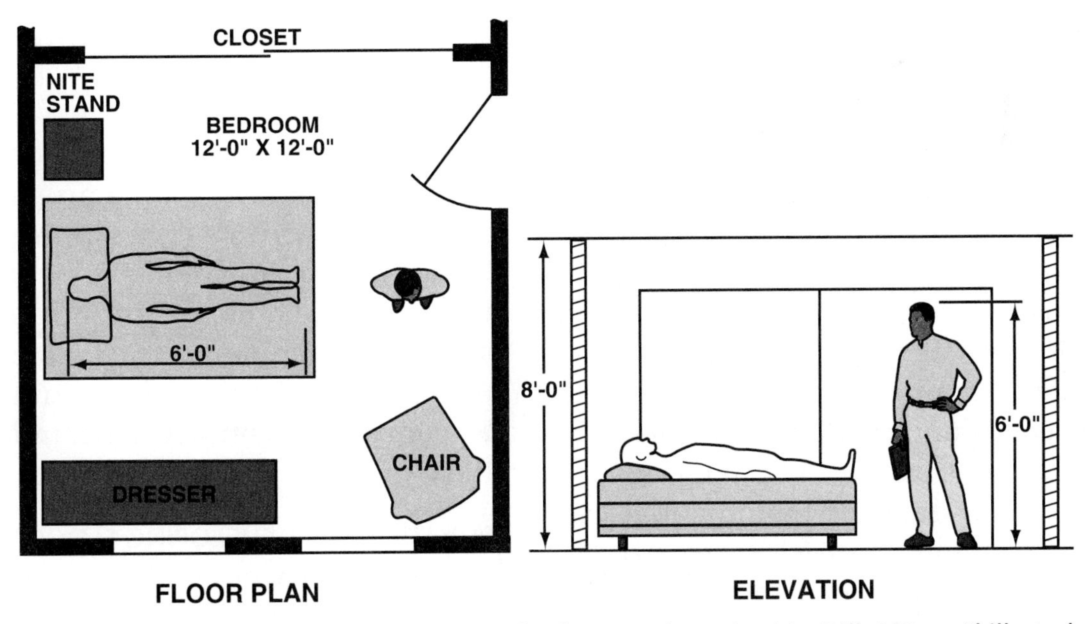

Fig. 13-20 ■ Use of human templates to check room sizes. *Art MacDillo's/Gary Skillestad*

size, number, placement, and the adequacy of traffic allowances.

Experienced designers may not always use furniture templates to establish room sizes, but they often use templates to check designs. Until students become completely familiar with furniture dimensions and the sizes of building materials, the procedures outlined in this chapter are recommended.

COMBINING AREAS INTO A FLOOR PLAN A residence or any other building is not just a series of separate rooms, but a combination of many activity areas. Rooms combine into areas and areas are positioned to create a total plan. Figure 13-21 shows the sequence of creating or evaluating space, starting with furniture needs through the development of a scaled sketch. As areas are combined, adjustments are made to allow space for such features as fireplaces, traffic flow, and storage space. Unlimited design variations may be possible within one defined area. The next step is to sketch these areas and rooms into a floor plan.

Floor Plan Sketches

A floor plan design sketch must satisfy all original goals and objectives. Once this is accomplished, many more sketch versions of the floor plan may be necessary. Think of the first sketch as only the beginning. Many sketches are usually necessary before a designer achieves an acceptable floor plan. Through successive sketches costly and unattractive offsets and indentations can be eliminated. By planning to use standard building materials and furnishings, many sizes are established. The exact positions and sizes of doors, windows, closets, and halls should be determined at this point. You may wish to consider both open and closed types of floor plans. Refer back to Chapter 7.

Further refinement of the design is done by resketching until a satisfactory design is reached. Except for very minor changes, making a series of sketches is always better than erasing and changing the original sketch. Many designers use tracing paper to trace the acceptable parts of the design and then add design improvements on the new sheet. This procedure provides the designer with a record of the total design process. Early sketches sometimes contain ideas and solutions to problems that might develop later in the final design process.

A final sketch should be prepared on grid paper to provide a more accurately scaled and detailed sketch. This sketch should include the exact positions of doors, windows, and partitions. It should also include the locations of shrubbery, trees, patios, walks, driveways, pools, and gardens. Once a final sketch is complete, three-dimensional conceptual CAD models may also be developed to aid in interpreting the conceptual design for the client and/or user.

After plans at this stage are approved by the client, contractor, and zoning and building departments, the preparation of working drawings can begin. Working drawings and documents are prepared to further refine the basic design concepts into very exact plan sets which can be used for bidding, budgeting, and construction purposes. You will learn more about these processes later in this text.

VARIATIONS IN DEVELOPING FLOOR PLANS The design process and sequence of preparing floor plans have been considered from the "inside-outside" point of

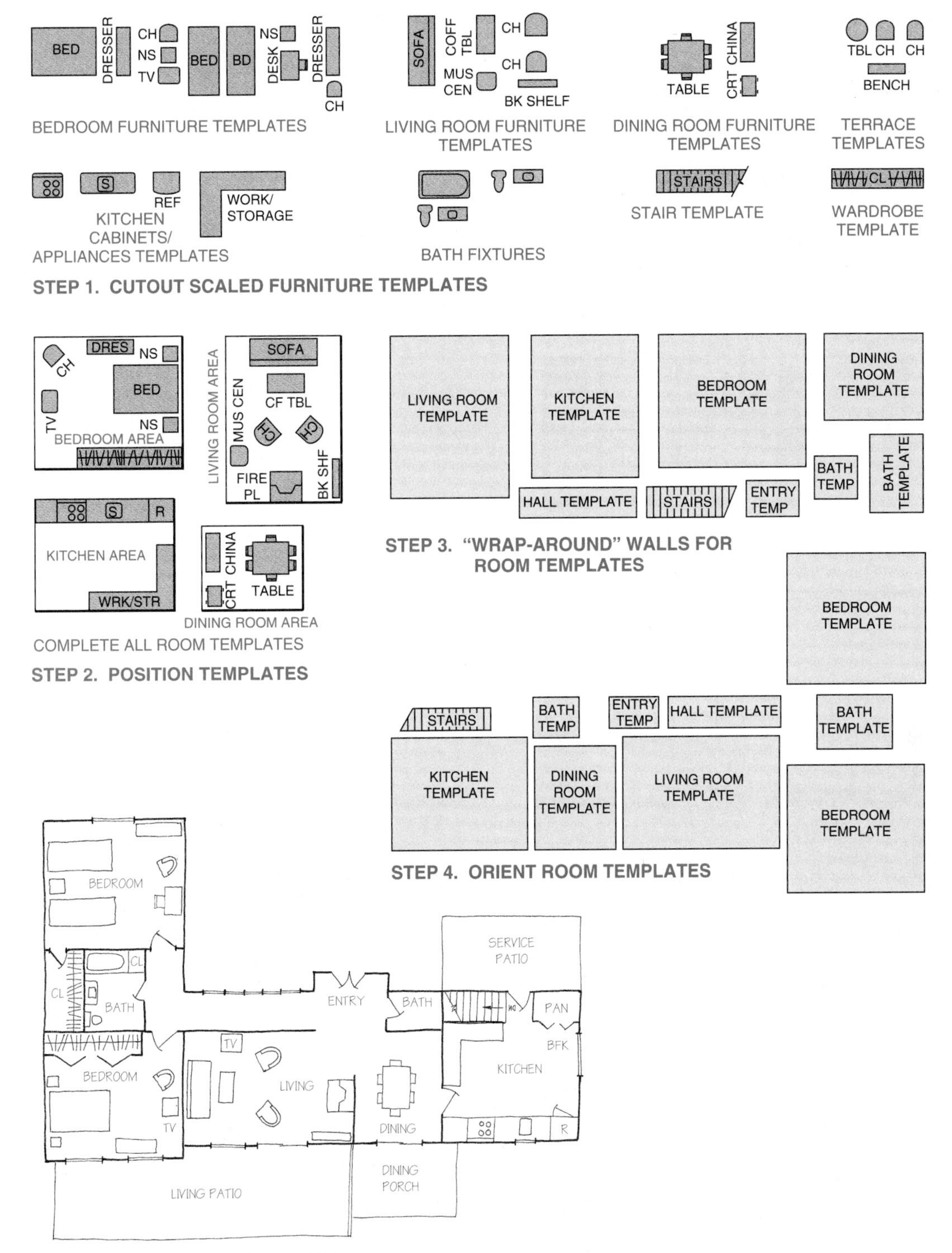

Fig. 13-21 ■ Steps in developing room sizes and relationships.

view. The needs of the inside areas determine the size and shape of the outside. However, some design situations require a plan to be developed within a given, predetermined outside area. Apartment, modular units, mobile, and manufactured home design fall into this category. See Fig. 13-22.

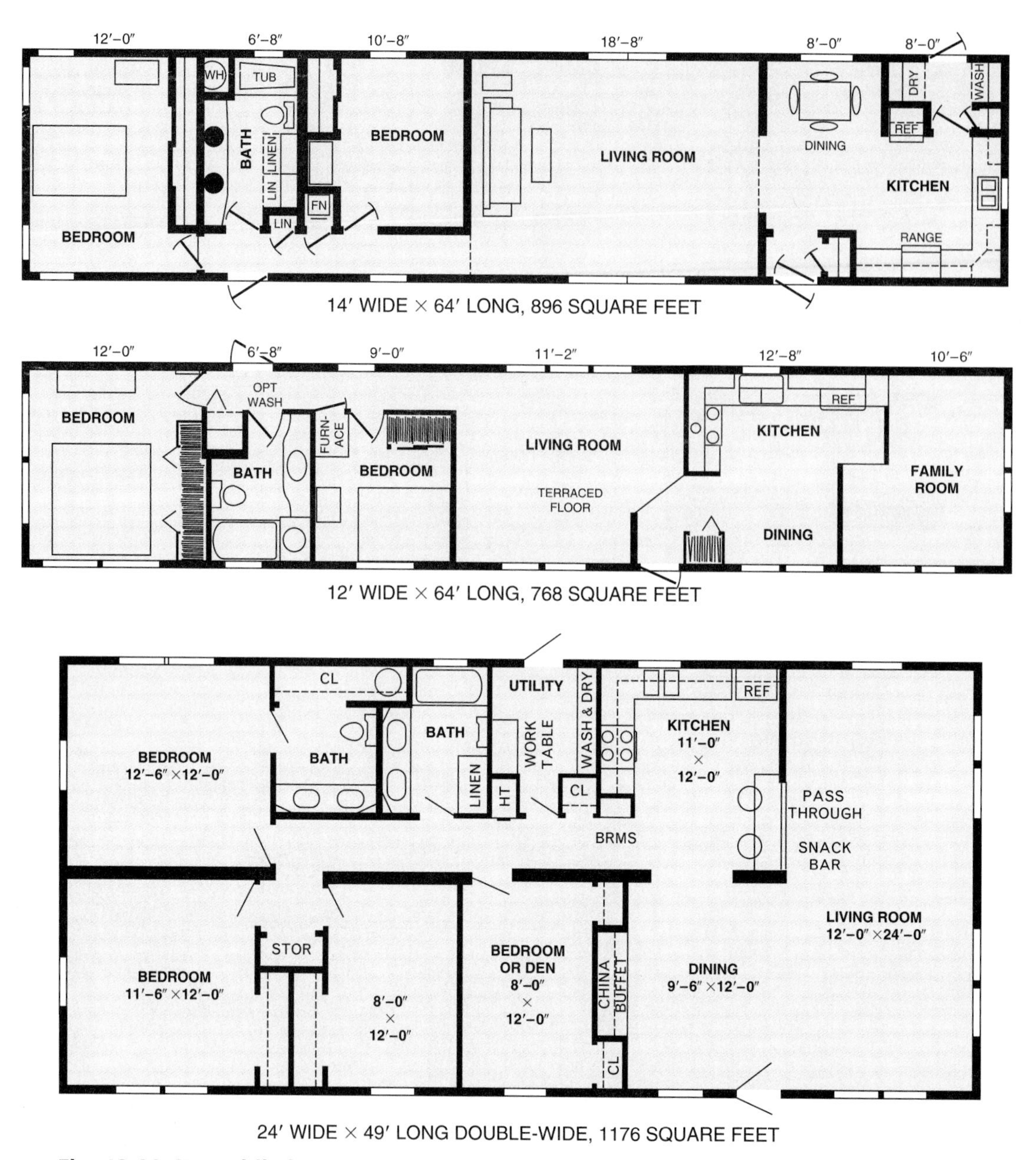

Fig. 13-22 ■ Mobile home designs must fit into a predefined area.

Because of limitations of time or money, it may be desirable to construct a house over a period of time. A house can be built in several steps. The basic part of the house can be constructed first. Then additional rooms (usually bedrooms) can be added in future years as the need develops.

When future expansion of the plan is anticipated, the complete floor plan should be drawn before the initial construction begins, even though the entire plan may not be complete at that time. If only part of the building is planned and built and a later addition is made, the addition will invariably look tacked on. This appearance can be avoided by designing the original floor plan for expansion. See Fig. 13-23.

Fig. 13-23 ■ Three-stage plan for expanding a floor plan. *Home Planners, Inc.*

Developing Plans to Accommodate Special Needs

Special design provisions must often be made to enable persons with physical impairments that restrict mobility, hearing, or vision to use areas and facilities within and around buildings. Design requirements for full accessibility in public buildings are found in building codes. Information for special residential designs can be obtained from federal, state, and local governmental agencies and some private organizations and companies. Check with your teacher, your school's media center, or the local library to find specific sources.

Following are two lists of some of the many design guidelines for planning buildings that can be fully, safely, and conveniently used by persons with impairments. One list applies to public buildings and the other to residences. However, the same general principles—and often the same dimensions—may apply to both. Always follow federal regulations and check with the state and local agencies in the area in which the structure will be built for regulations that apply to the development of plans for any specific project. Also, talk with persons who have impairments. Find out what kinds of problems they encounter. Then find ways to eliminate the problems by utilizing safe and acceptable design alternatives.

Public Buildings

All building codes contain design requirements for public buildings which prohibit the use of architectural barriers or infringements on the comfort and safety of persons with impairments.

OUTDOOR CONSIDERATIONS

1. A passenger loading zone at least 4′ × 20′ must be provided near an accessible entrance.
2. The minimum width of ramps is 3′-0″, with a maximum slope of 1:12 and a maximum rise of 2′-6″. Handrails must be provided for ramps longer than 6′-0″, and at least 6′-0″ of level area must be provided at the top of each ramp. Nonskid surfaces must be used on all ramps.
3. Parking facilities must be provided in the parking area nearest the building and marked with the international access symbol. This area must be out of the main traffic flow and connected to the building by a ramp if the level changes. Parking slots must have at least 4′-0″ of clearance on each side. Parking spaces for persons with impairments must be at least 8′-0″ wide and have an access area of at least 4′-0″ or 5′-0″ wide depending on local code. See Fig. 13-24.
4. Primarily for persons with visual impairments, walkways must be at least 4′-0″ wide, with no less than 6′-8″ headroom clearances. Walks must be level and ramps must be used when it is necessary to change level. Walks must be free of obstructions and be surfaced with nonskid material. At least 32″ must be allowed for a cane sweep width, as shown in Fig. 13-25.

ENTRANCES AND DOORS

5. At least one entrance must be accessible to wheelchair traffic and provide access to the entire building.
6. Doors must open at least 90° and be 2′-8″ (min.) to 3′-0″ wide, with threshold heights no more than ¾″ on exterior doors and ½″ on interior doors. Walls or objects can be no closer than 4′-0″ from the door hinge. See Fig. 13-26.

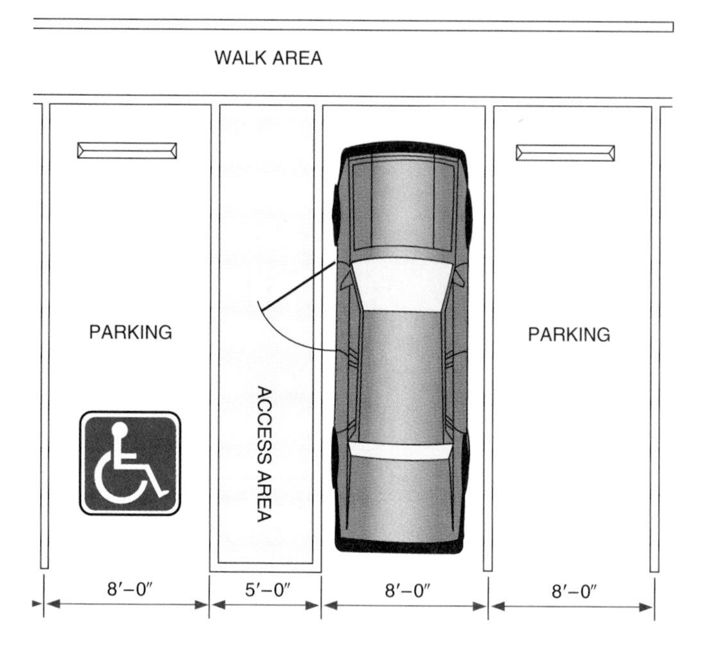

Fig. 13-24 ■ Parking space and access dimensions. Note the international access symbol.

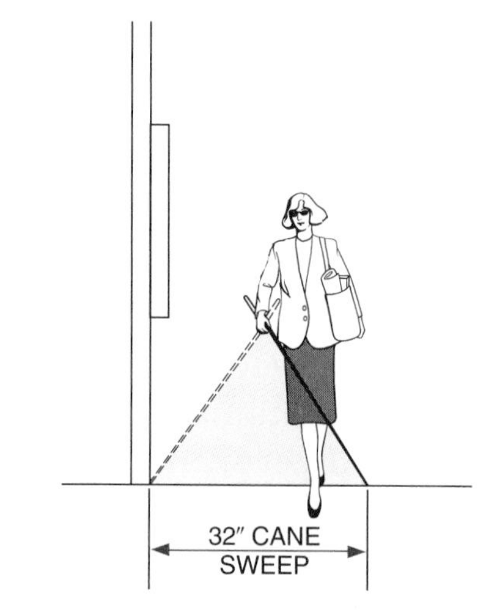

Fig. 13-25 ■ Typical cane sweep distance.

7. Doors should be provided with handles that can be opened with a closed fist.
8. Thresholds should not be higher than ½″ on swinging doors and ¾″ on sliding doors.
9. Door kickplates should cover the bottom 10″ of a door used by wheelchair users.

FLOORS AND PATHWAYS

10. Texture changes using raised strips, grooves, rough, or cushioned surfaces should be used to warn persons with visual impairments of an impending danger area, including ramp approaches.
11. Floors should have nonslip surfaces even when wet, or be covered with carpeting with a pile thickness of no more than ½″. Floors should have contrasting color borders to warn persons with limited vision of a solid object ahead.

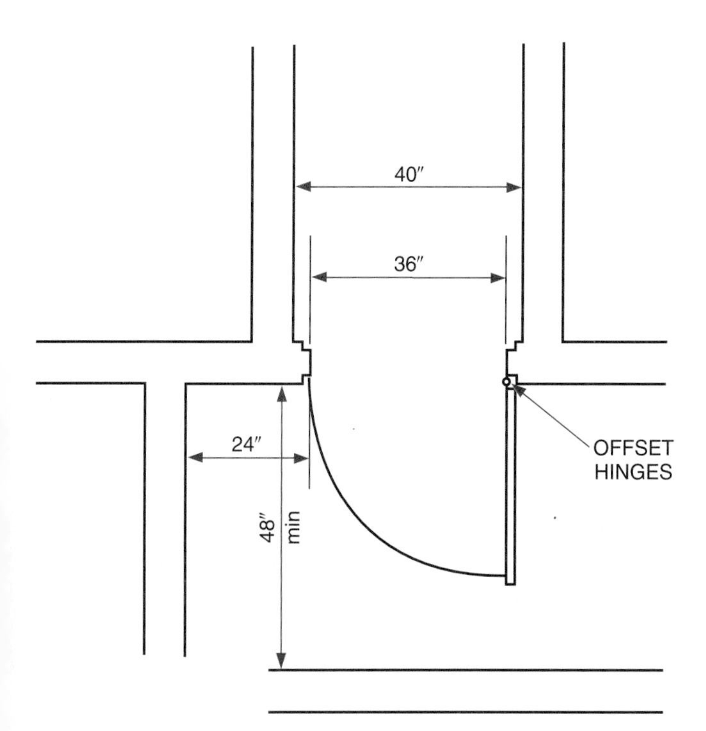

Fig. 13-26 ■ Door and wall clearance for persons with impairments.

INDOOR TRAFFIC AREAS

12. Hall widths must be at least 3′-0″, with 5′-0″ provided in all turning areas. Halls must provide access to all areas of the building without the need to pass through other rooms.
13. There should be at least three treads in a series of stairs. Treads and risers should be uniform and treads should have a minimum width of 11″ with round nosings. A landing should be planned for stair systems which contain more than 16 risers. Handrails must be at least 32″ above the floor and tread height. See Fig. 13-27.
14. At least one ramp (or elevator) must be provided as an alternative to stairs.
15. Wall projections, if located between 27″ and 80″ from the floor, cannot extend more than 4″ from a wall. Objects mounted below 27″ from the floor may project any amount. Freestanding objects between 27″ and 80″ may only project 12″ from their support, as shown in Fig. 13-28.

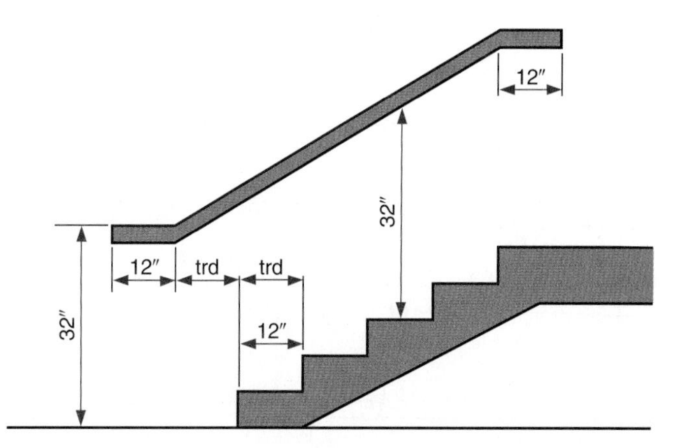

Fig. 13-27 ■ Handrail extensions and height requirements.

SIGNS, ALARMS, LIGHTING

16. Public phone amplifiers, eye-level warning lights to augment audio alarms, and high-frequency alarms should be provided for persons with hearing impairments.
17. Braille signs, level-change warning surfaces, and restrictions on wall protrusions over 4″ must be provided for persons with visual impairments.
18. Protruding signs must be at least 7′-6″ from the floor.
19. Emergency warning alarms should be both visible and audible.
20. Lighting should be free of glare or deep shadows.

LAVATORY FACILITIES

21. Lavatory facilities must include at least 2′-6″ × 4′-0″ clear floor areas. Water closet seat tops must be 1′-6″ from the floor. A 3′-0″ knee-room height must be provided under sinks and drinking fountains. See Fig. 13-29. Grab bars must be provided near water closets, sinks, and bath areas.
22. Sinks should be mounted no closer to the floor than 29″ and should extend a minimum of 17″ from the wall to provide adequate knee space, as shown in Fig. 13-30. A minimum floor space area of 2′-6″ × 4′-0″ must be provided around sinks for wheelchair access. See Fig. 13-31.

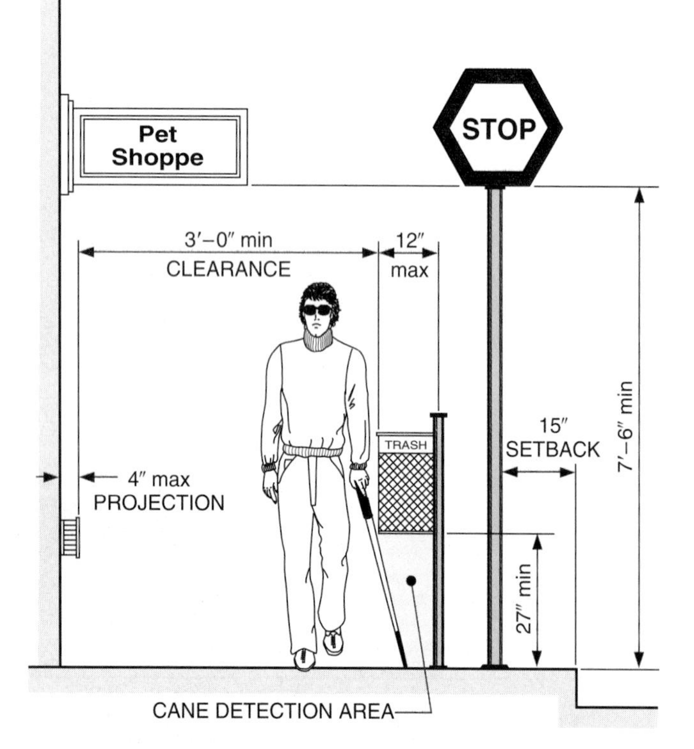

Fig. 13-28 ■ Projection limits of objects from walls.

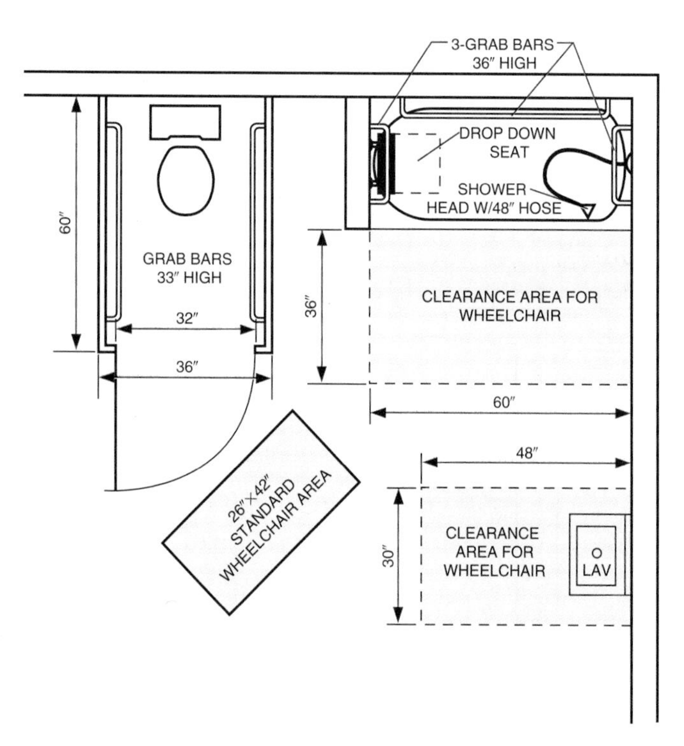

Fig. 13-29 ■ Lavatory clearances for use by persons with impairments.

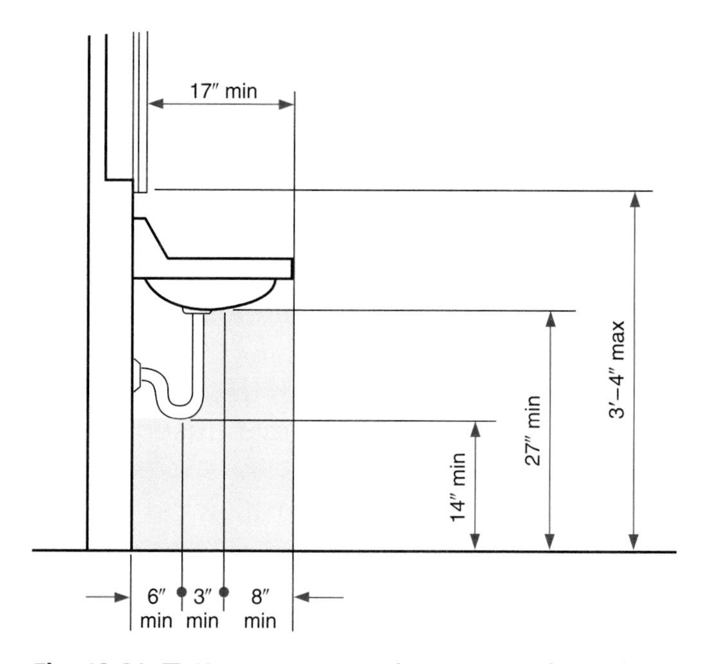

Fig. 13-30 ■ Knee-space requirements under sinks.

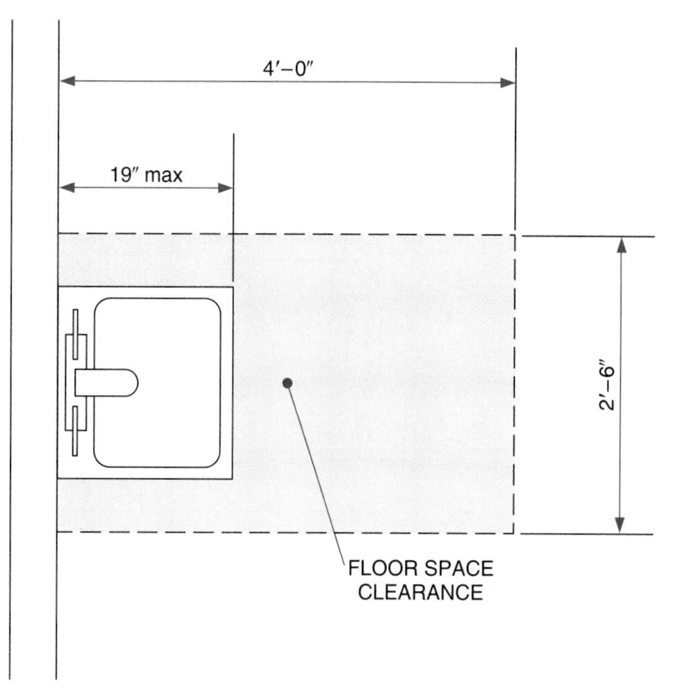

Fig. 13-31 ■ Floor-space clearances around sinks.

Residences

As for public buildings, a designer must follow building codes when designing a residence for a person with an impairment. Beyond that, working closely with the client is important. Except in extreme cases, persons with impairments have abilities as well. By identifying the specific needs and capabilities of the client, "overdesigning" can be avoided and a safe, convenient, and comfortable design can be achieved.

Design features that affect persons with mobility limitations usually relate to providing for wheelchair use. Since wheelchairs require the greatest amount of space, plans that accommodate wheelchairs will easily function for other design conditions. Figure 13-32 shows the dimensions and turning radius of a *standard* wheelchair. Wheelchairs do vary in size, however. When designing a home for a person who uses a wheelchair, be sure to find the exact dimensions of that wheelchair and plan accordingly.

Some general guidelines for designing residences are provided in the following list. Unless stated otherwise, design guidelines apply to wheelchair use.

OUTDOOR ENTRANCES/EXITS

1. To accommodate wheelchairs, the pathway or ramp to the entrance should be 36″ to 48″ wide and have a nonslip surface such as outdoor carpeting or sand paint.
2. If a ramp is used, a minimum landing platform of 5′ × 5′ should be located in front of the door. Plan the shape of the platform to accommodate the door swing and still allow easy access. For long ramps, more landings may be needed. A covering for protection from the weather is recommended. If room

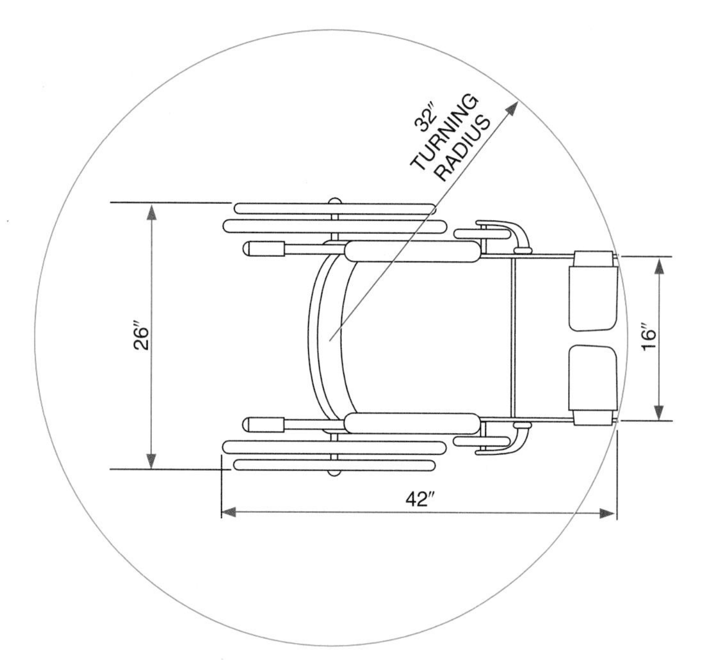

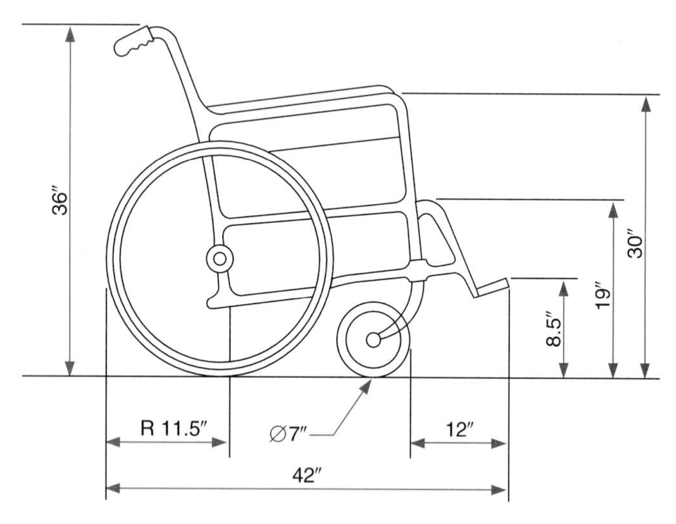

Fig. 13-32 ■ Standard wheelchair dimensions.

is unavailable for a ramp, consider planning for the use of a mechanical lift.

3. The vertical rise of a ramp should be 1:12. Handrails should be 32″ to 36″ high and extend 1′-0″ past the end of the ramp.
4. The height of the doorknob should be 36″ or less, with a door width of 36″ or more, if the wheelchair must move through at an angle. Threshold height should be ½″ or less.

INDOOR TRAFFIC AREAS AND FLOORS

5. Generally, levers on doors can be operated more easily than doorknobs.
6. Doorways need to be 32″ to 36″ wide. If a turn is required for a wheelchair to pass through a doorway, be sure the doorway is wide enough or provide extra turn space in front of the door.
7. For ease of use by persons in wheelchairs and persons with limited vision, doorways should not have raised thresholds.
8. Hallways must be 36″ to 48″ wide.
9. Hardwood floors or tiled surfaces are best. If carpeting is preferred, use carpet that has short, dense pile.

LIVING AREAS

10. Rooms should have 5′-0″ or more of clear area for turning a wheelchair.
11. Furniture planning and placement should allow adequate area for wheelchairs to move through the room. Persons with limited vision should have clear passageway through rooms and not be required to walk around articles of furniture.
12. Height of tables and work areas, such as desks, should be appropriate for wheelchairs.

KITCHENS

13. Allow adequate work area space for turning a wheelchair or for using crutches or a walker, usually 16 sq. ft. to 25 sq. ft. Important factors are the shape of the kitchen and the arrangement of appliances that create the work triangle.
14. Appliance cooking controls should be placed in front of the burners. Ovens should have side-hinged doors. All controls must be operable with a closed fist.

15. Braille control panels for persons who are blind or have extremely limited vision are available from some appliance manufacturers.
16. A clear 28″, 31″, or 36″ floor space should be provided under selected base cabinets or next to appliances.
17. Countertops should be 30″ to 33″ from the floor, with 27″ to 29″ for knee clearance. Cabinet pulls should be recessed.
18. Dishwashers, washers, and dryers should be front loading and have front controls.
19. Side-by-side refrigerator/freezers are most convenient.
20. Sinks need to be 34″ or less from the floor.

BATHS

21. Water closet seats should be 1′-6″ from the floor.
22. Sinks should not be higher than 34″ and should have adjoining counter space. They need to be open underneath. Exposed pipes should be insulated. A single faucet with a lever control is preferred.
23. The bottom of the mirror should be 40″ from the floor. (Consider the eye level of a person in a wheelchair.)
24. The top of the medicine cabinet should not be more than 50″ from the floor.
25. Space should be allowed near the bathtub to allow for transfer from a wheelchair. Reinforced grab bars should be installed.
26. A shower should be at least 5′ × 4′ in size, and the floor should have a nonslip surface. Reinforced grab bars should be installed. There should be no lip on the floor surface entrance to the shower.

BEDROOMS AND STORAGE AREAS

27. Bedrooms should be designed to allow wheelchair maneuverability and access to the bed, as shown in Fig. 13-33. The top surface of bedroom furniture should be a maximum of 34″ above the floor.

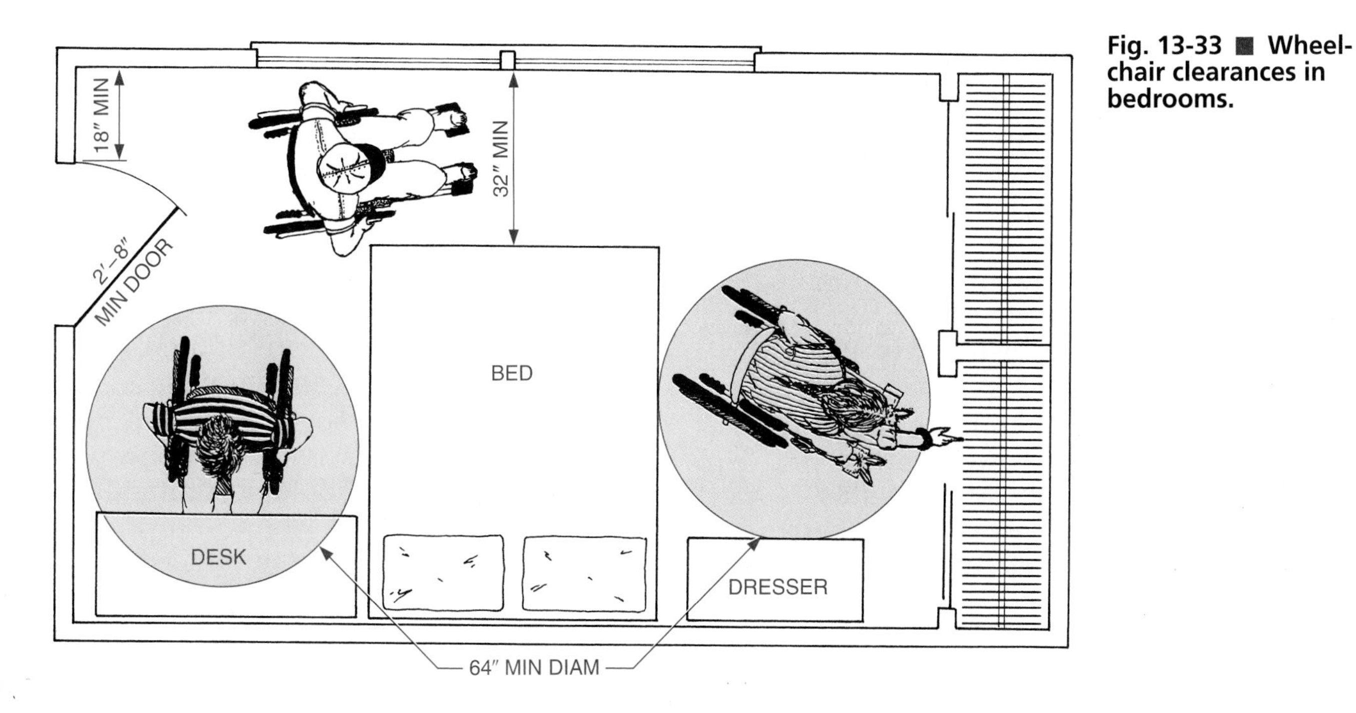

Fig. 13-33 ■ Wheelchair clearances in bedrooms.

Fig. 13-34 ■ Convenient reaching distances for persons in wheelchairs.

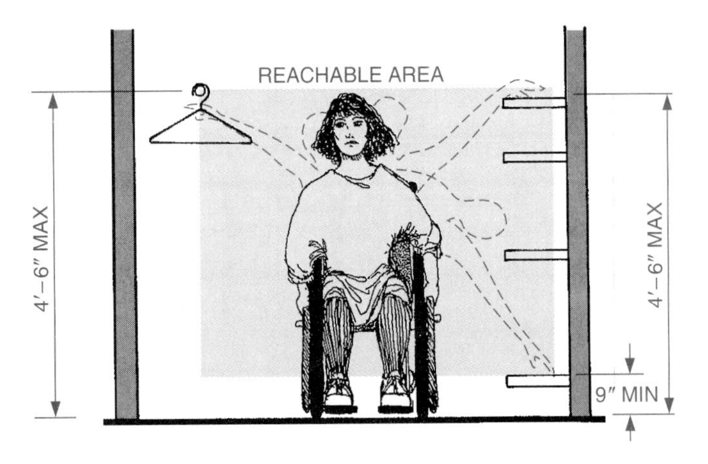

28. Sliding doors are preferred for closets. Hang rods should be 4'-6" or less from the floor.
29. Storage facilities should be designed for easy reach from a wheelchair. See Fig. 13-34.

ELECTRICAL CONSIDERATIONS

30. Switches and outlets should be 40" from the floor.
31. Lights or other visual cues (aids) should be connected to the telephone, doorbell, and other devices as needed for persons with hearing impairments.

CHAPTER 13 Exercises

1. List the design steps necessary to design a residence through the development of a conceptual design.

2. Prepare a situation statement and set goals and objectives for a house of your own design.
3. Explain how a composite analysis is prepared and used to create a plan of a design.
4. Prepare room templates and use them to make a functional arrangement for the living area, service area, and sleeping area of a house.

5. Arrange templates for a sleeping area, service area, and living area of your own design in a total plan.
6. Make room templates of each room in your own home. Rearrange these templates according to a remodeling plan, and make a sketch.

7. Make a list of furniture you would need for a home you will design. The list should include the number of pieces and size (width and length) of each piece of furniture.

8. Make a furniture template 1/4" = 1'-0" for each piece of furniture you will include in a home of your design.
9. Develop a floor plan for a family of four, including two small children and a person who uses a wheelchair.
10. Gather pictures of floor plans from real estate magazines and home planning catalogs. Evaluate them in terms of their space planning arrangements.
11. Choose a floor plan in Chapter 14 and redesign it in two ways. One plan should accommodate a person in a wheelchair and the other one a person who is visually impaired.

CAD Design Specialist

Developments in technology are revolutionizing the ways in which architects work. With computers, work can be done more quickly and with greater accuracy. Many career opportunities exist for people who can use computers to apply their architectural knowledge and skills.

Scott O'Brien

Scott O'Brien is an architect who specializes in CAD. On a recent design project for a public health facility, Scott took his architectural design team and some 3-D models he had developed in CAD to a virtual reality lab. There, with the help of powerful computers, the team members donned virtual reality helmets and "walked through" the facility they were designing. "We were able to 'move around' inside it and make some basic design decisions," says Scott.

Virtual reality helmets and software are quite specialized and too expensive for most architectural firms. There are alternatives, however. Scott's firm now has a relatively inexpensive virtual reality program for desktop computers. It enables designers to move around inside a building interactively on screen.

Virtual reality and 3-D modeling can often take the place of pictorial drawing. For example, Scott utilizes what he calls "material mapping" on his computer. He explains, "A building we were working on had copper shingles. We were able to scan a copper sample and then apply that 'material' to the 3-D model. We got a pretty good color rendering of it."

Many clients have difficulty understanding architectural drawings. With 3-D modeling, the design team can quickly show clients the future appearance of their new building from any angle or height. Any needed changes can be made early in the design process.

> "With 3-D models, you can make a lot of design decisions and incorporate the changes into the models quickly and easily."

Scott developed his expertise in computers gradually. His undergraduate degree is in mechanical engineering with a minor in computer science. As he studied for a master's degree in architecture, he had a job setting up and repairing computers and worked summers for an architecture firm which used CAD extensively. Describing how he became a CAD designer, Scott says, "I repaired computers and dealt with their problems, and at the same time I learned architecture."

Ted Mishima/Lord, Aeck & Sargent, Inc.

CHAPTER

14

Drawing Floor Plans

Objectives

In this chapter, you will learn to:

- use information on a scaled floor plan to draw a complete floor plan.
- name and explain the types of floor plans.
- use graphic symbols to communicate information on a floor plan.
- draw a floor plan according to a sequence of steps.
- draw dimensions that convey precise, accurate information for builders.

Terms

break line
dimension lines
multiple-level floor plans
object lines
overall dimensions
reflected ceiling plans
reversed plans
schedules (door/window)
subdimensions
symbols
working drawings

A complete floor plan is a scaled drawing of the outline and partitions of a building as seen if the building were cut (sectioned) horizontally about 4′ (1219 mm) above the floor line. See Fig. 14-1. There are many types of floor plans, ranging from very simple sketches to completely dimensioned and detailed floor-plan working drawings.

■ To achieve good results, architectural designs must be clearly communicated in drawings. Floor plans are key drawings in a set of working drawings. *Tom Knibbs/Karam-Princeton-Imhotep*

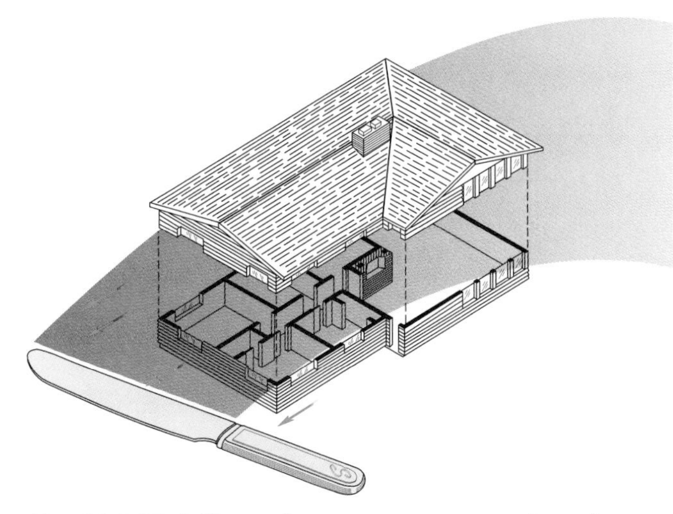

Fig. 14-1 ■ A floor plan represents a section view cut through a building.

Types of Floor Plans

Several types of floor plans are drawn. They are classified by their completeness or by the specialized information they convey.

As you have learned, preliminary plans are considered incomplete and consist of single-line drawings. These plans contain only overall room width and length—dimensions used primarily for sales purposes.

Drawings that contain all information needed to construct a structure are called **working drawings**. Completely dimensioned and accurately scaled floor plans are working drawings necessary for construction. Basic information included shows the size and position of all exterior walls, interior partitions, fireplaces, doors, windows, stairs, built-in furniture, appliances, cabinets, connecting walks, patios, lanais, or decks. Wall and surface construction materials are also shown.

The prime function of floor plans is to communicate information to building contractors. Complete working-drawing floor plans help prevent misunderstandings between designers and builders. Contractors should be able to correctly interpret working drawings without consultation.

Specialized floor plans are developed from basic working-drawing floor plans. For construction and installation of electrical, plumbing, and HVAC (heating, ventilating, air-conditioning) systems, separate specialized plans are drawn with specific symbols added. On very small projects, these symbols may all be included on the basic plan. However, this often makes the drawing too crowded and difficult to read. For most projects, electrical, HVAC, and plumbing plans are separate drawings. These are covered in detail in Chapters 31, 32, and 33.

Floor-Plan Symbols

On drawings, drafters use **symbols** to identify construction materials such as fixtures, doors, windows, stairs, and partitions. Using symbols saves time and space. Imagine trying to repeat a description every time that a material or component is used! It would be impossible.

Common symbols used on floor plans include symbols for walls, doors, windows, appliances, fixtures, sanitation facilities, and building materials. See Fig. 14-2. Floor-plan symbols for plumbing, heating, air conditioning, and electrical components are covered in later chapters. Architectural symbols are standardized. However, some variations of symbols are used in different parts of the country.

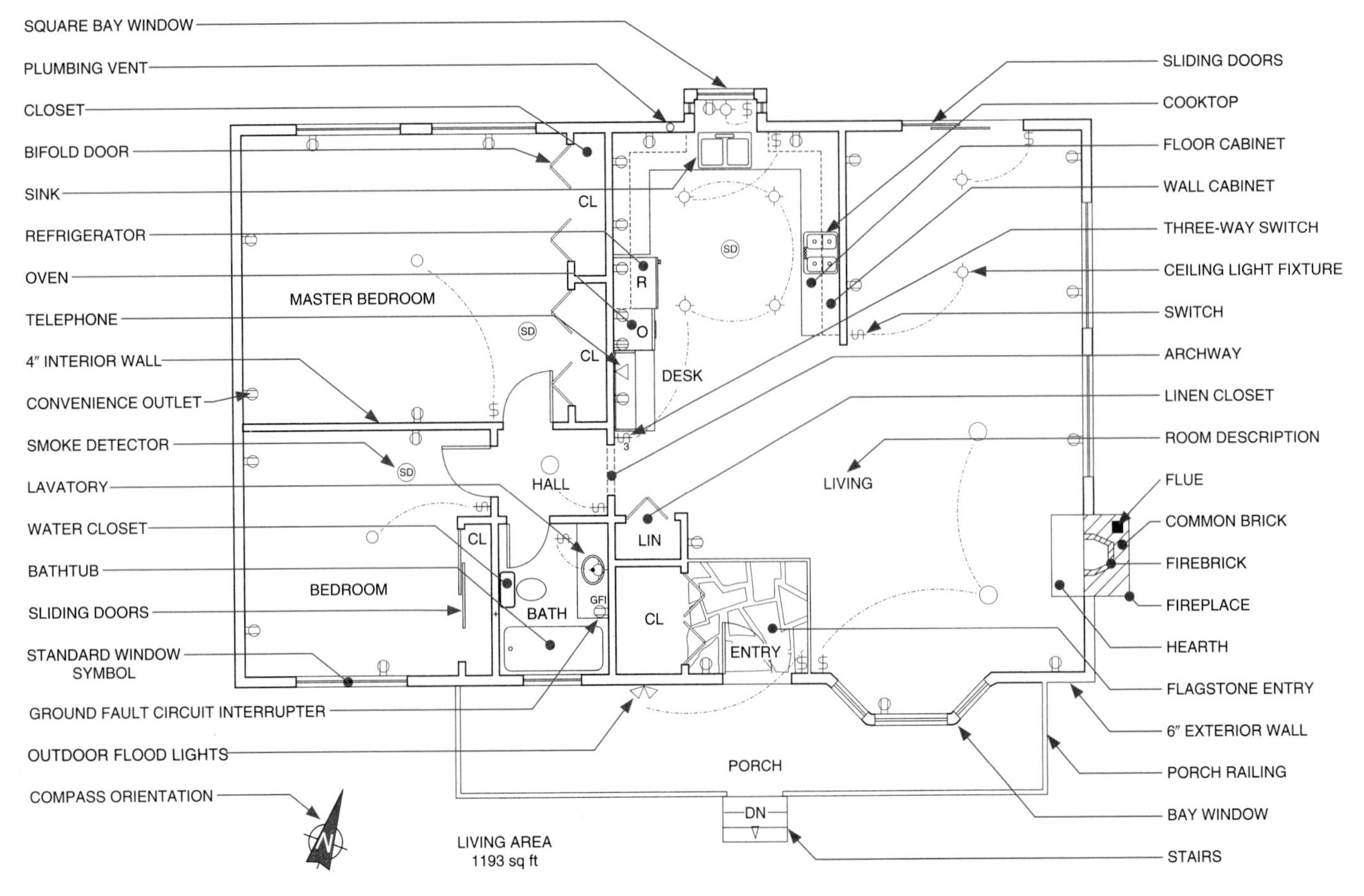

Fig. 14-2 ■ Examples of several floor-plan symbols.

Wall Symbols

Different types of wall construction are represented by different floor-plan wall symbols. On simple plans, walls are represented by single lines. However, on working-drawing floor plans, the actual scaled width of each wall is drawn. Figures 14-3 and 14-4 show methods of representing different types and variations of wall construction on floor plans.

Door Symbols

Floor-plan door symbols show the top view of a door and the width of each doorway. Door symbols usually show each door open 30° or 90° and connected to an arc that represents the door swing. Figure 14-5 illustrates a door symbol and methods of representing wall openings (or doorways) that do not include a door. Notice that the outline of the doorsill is added to all exterior door symbols on floor plans.

DOOR TYPES Interior doors, those located within interior partitions, and exterior doors, those that lead outdoors, are generally flush, paneled, or louvered. Interior flush doors have a hollow core covered with a thin wood, plastic, or metal veneer. The core of exterior flush doors is solid to provide strength and insulation, and to prevent warping from weather exposure. Panel doors are constructed from vertical stiles and horizontal rails. Thin panels of wood, plastic, metal, or glass are placed between the stiles and rails.

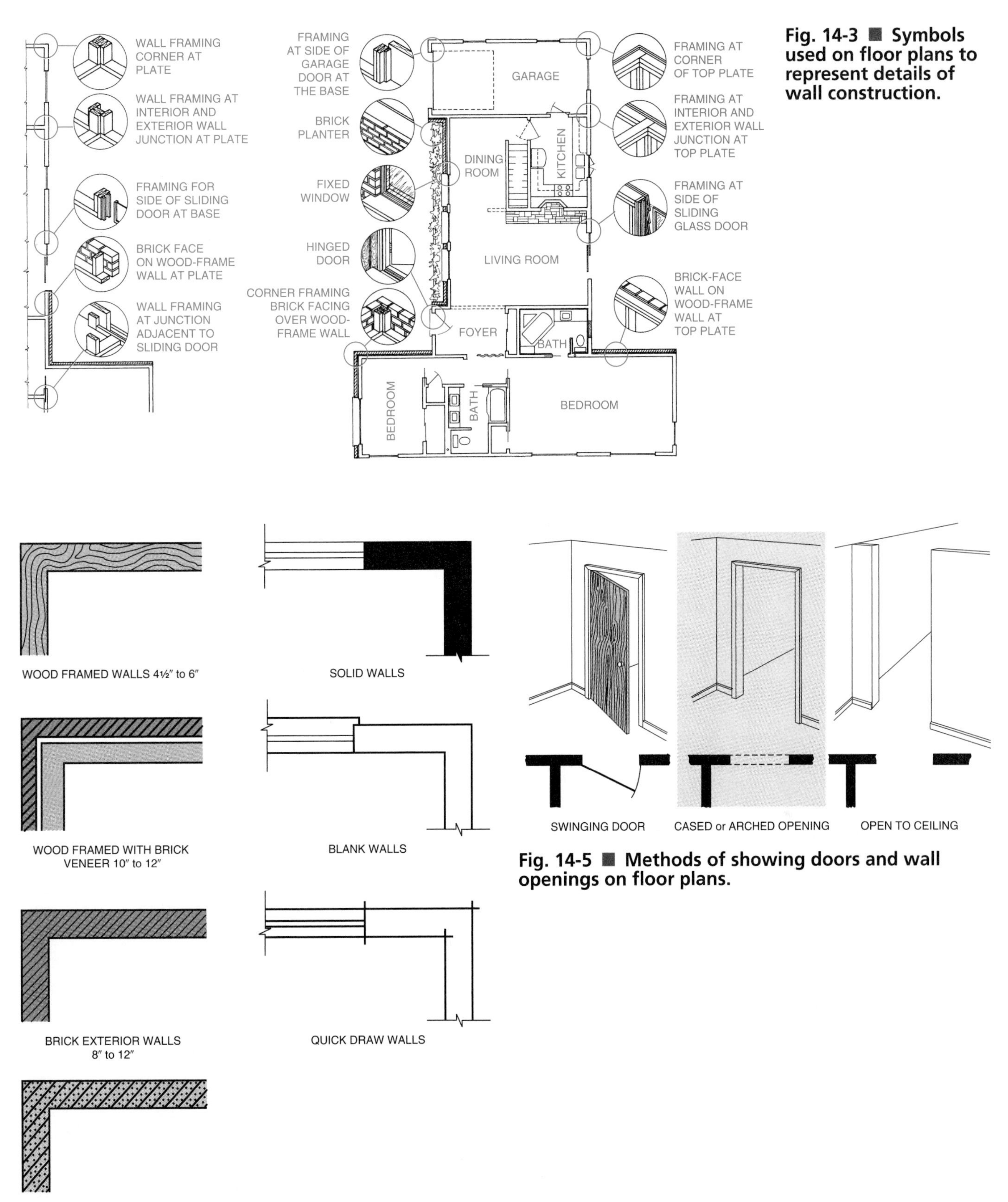

Fig. 14-3 ■ Symbols used on floor plans to represent details of wall construction.

Fig. 14-5 ■ Methods of showing doors and wall openings on floor plans.

Fig. 14-4 ■ Methods of showing types of wall construction.

DOOR STYLES Interior doors are manufactured in many different configurations to serve a variety of needs. Exterior doors are generally single- or double-swing doors for entrances and bypass sliding glass doors for patio or deck traffic. See Fig. 14-6. In addition to the normal two-unit sliders, many combinations of sliding and hinged doors are now available. See Fig. 14-7.

NAME	ABR	SYMBOL	ELEVATION	PICTORIAL
INTERIOR HINGED DOOR HOLLOW CORE	DR			
EXTERIOR HINGED DOOR SOLID CORE	DR			
BIPASSING SLIDING DOOR	BP SLDG DR			
DOUBLE FRENCH DOORS	DBL FR DR			
SLIDING POCKET DOOR	SLDG PK DR			
BIFOLDING DOORS	BI-FLD DR			
ARCH (CASED OPENINGS)	ARCH			
SECTIONAL ROLL-UP GARAGE DOOR	SEC RL UP GAR DR	SEC ROLL UP GAR DR		

Fig. 14-6 ■ Floor-plan symbols for commonly used doors.

Door styles are also indicated on door schedules and on elevation drawings that are cross-referenced with floor plans. **Schedules** are detailed lists that contain information such as size and type. Often

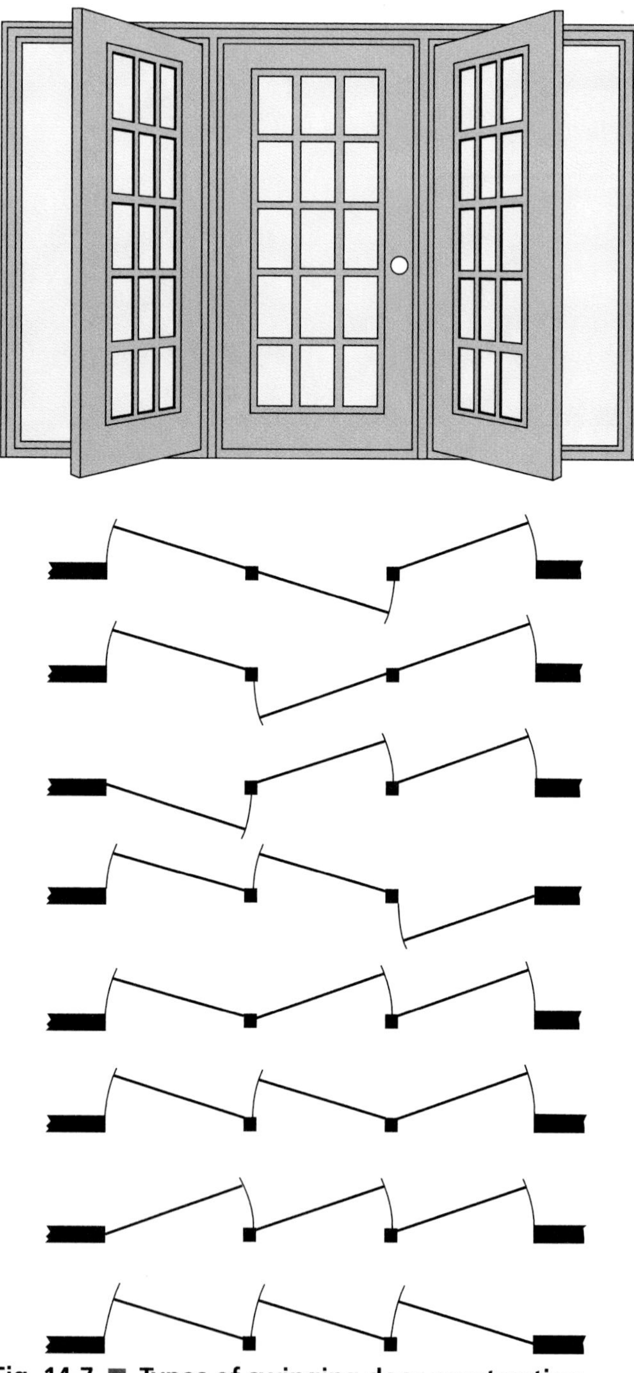

Fig. 14-7 ■ Types of swinging door construction. *Taylor Door Co.*

on a floor plan, a number in a square near the door symbol identifies the door style. See Fig. 14-8.

DOOR SIZES Different door types and styles are available in many size ranges for width, height, and thickness. See Fig. 14-9. If a door schedule is not used, door width and height dimensions are often shown directly on the door symbol. When this is done, the foot and inch dimensions are abbreviated, as shown in Fig. 14-10. Exact door framing information should always be determined from the manufacturer's data.

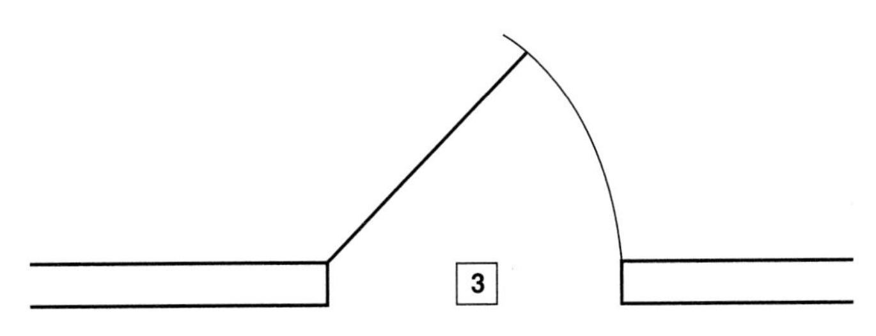

Fig. 14-8 ■ Use of door identification number keyed to door schedule.

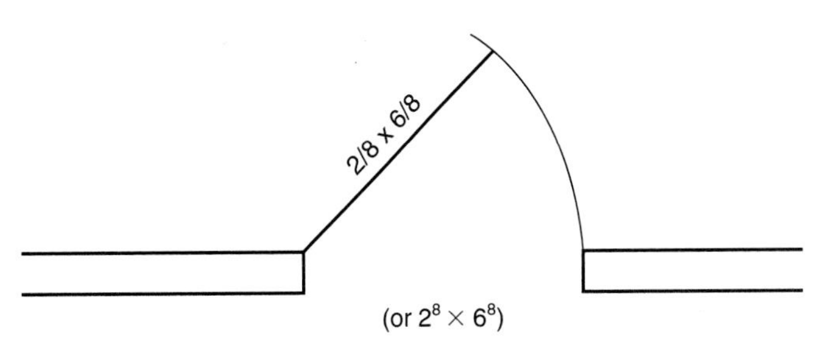

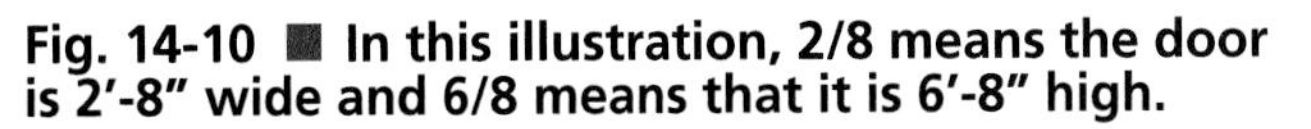

Fig. 14-10 ■ In this illustration, 2/8 means the door is 2′-8″ wide and 6/8 means that it is 6′-8″ high.

Fig. 14-9 Common Door Sizes for Light Construction.

Type	Width	Height
Exterior Swing	2/6 2/8 3/0 3/6 4/0	6/8 7/0 8/0
Sliding Glass:		
Single Unit	3/6 3/0 3/4 4/0 4/4	6/8 6/10 7/0 8/0
Double	5/0 6/0 8/0	6/8 6/10 7/0 8/0
3 Panel	7/0 8/0 9/0 12/0	6/8 6/10 7/0 8/0
4 Panel	10/0 12/0 16/0	6/8 6/10 7/0 8/0
Sidelights	1/0 1/2 1/4 1/6	6/8 7/0 8/0
Transom	2/6 2/8 3/0 3/6 4/0 6/0 8/0 9/0 10/0 12/0	1/0 1/2 1/6
Double French	3/0 4/0 5/0 6/0	6/8 7/0 8/0
Garage Roll-up	8/0 9/0 10/0 12/0 14/0 15/0 16/0 17/0 18/0 20/0	6/6 6/8 6/9 7/0 7/6 8/0
Interior Swing	2/4 2/6 2/8 3/0 3/6 4/0	6/8 7/0 8/0
Closet Slide Panel	1/0 1/2 1/4 1/6 1/8 1/10 2/0 2/4 2/6 2/8 3/0 3/6 4/0	6/8 7/0
Bifold	2/0 2/4 2/6 2/8 3/0	6/8 7/0
Pocket	2/6 2/8 2/10 3/0	6/8
Folding	4/0 6/0 8/0 12/0 16/0 20/0 24/0	6/8 7/0

Window Symbols

Floor-plan window symbols show the outline of the sash, glass position, and any mullions and muntins. See Fig. 14-11. Windows are often distinguished by the manner in which they open. For example, on casement windows, the direction of swing is indicated much like it is in a door symbol. On awning windows, the outline of the open window position is shown with dashed lines. On small scale drawings, often only the sash outline or glass position is shown. Figure 14-12 shows the symbols for common window types.

Only the width of windows is needed to draw window symbols on floor plans. Height dimensions are shown either on elevation drawings or stated in a window schedule which contains size, style, type, and manufacturer's information. Figure 14-13 shows common window sizes. Exact sizes for rough framing must always be secured from manufacturers' data.

On a floor plan, a letter in a circle is provided as a key or cross-reference to the window schedule. Refer again to Fig. 14-11.

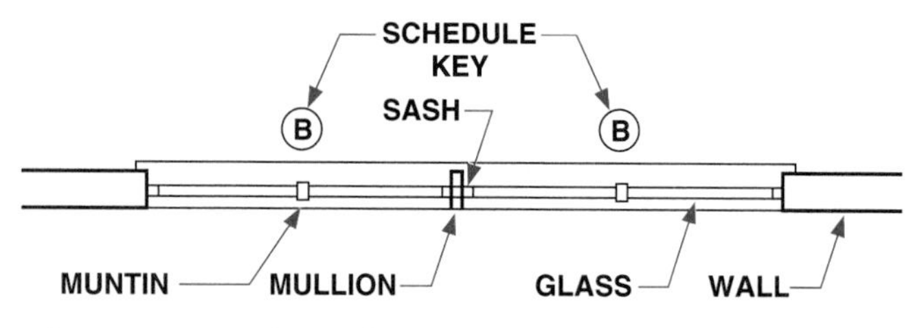

Fig. 14-11 ■ Floor-plan symbols for window components. Mullions are vertical strips between multiple window panes. Muntins are secondary framing strips used to separate and/or hold window panes within a window.

Appliance and Fixture Symbols

Figure 14-14 shows the plan and elevation symbols used for common appliances and fixtures. Overall width and length dimensions (as shown previously in Chapters 10, 11, and 12) can be helpful for drawing floor-plan symbols.

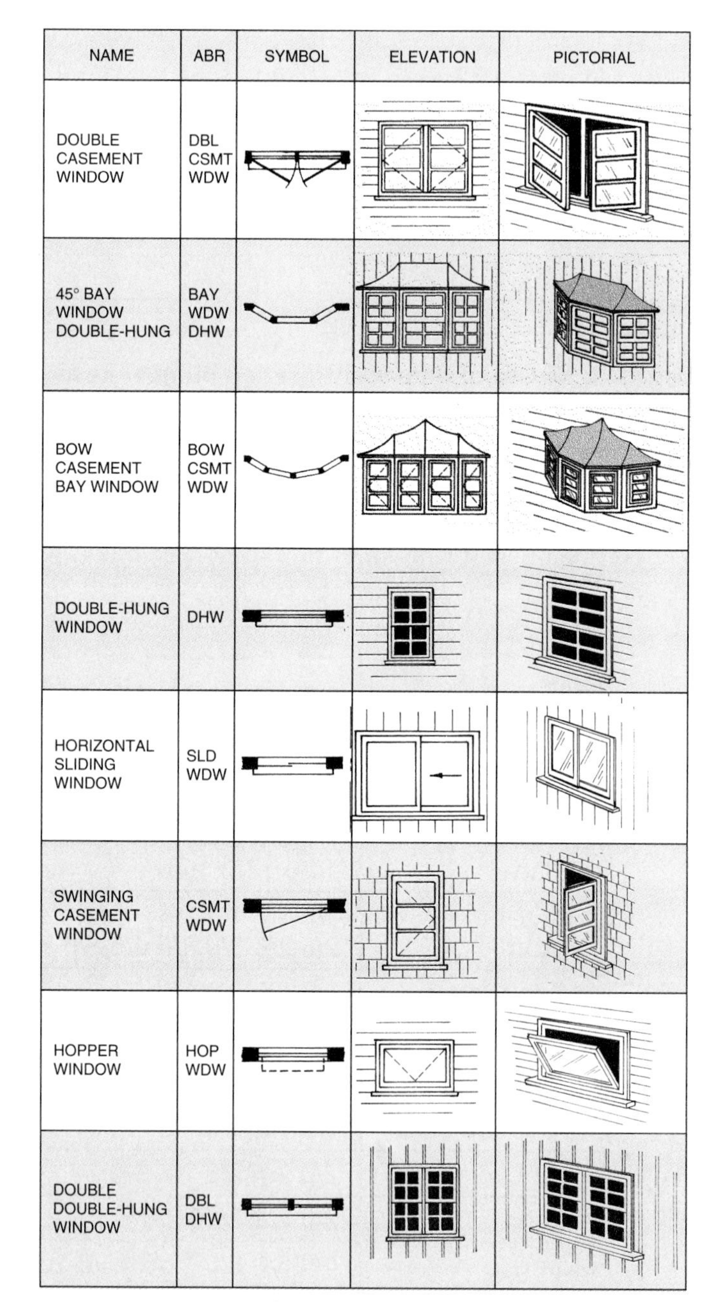

NAME	ABR	SYMBOL	ELEVATION	PICTORIAL
DOUBLE CASEMENT WINDOW	DBL CSMT WDW			
45° BAY WINDOW DOUBLE-HUNG	BAY WDW DHW			
BOW CASEMENT BAY WINDOW	BOW CSMT WDW			
DOUBLE-HUNG WINDOW	DHW			
HORIZONTAL SLIDING WINDOW	SLD WDW			
SWINGING CASEMENT WINDOW	CSMT WDW			
HOPPER WINDOW	HOP WDW			
DOUBLE DOUBLE-HUNG WINDOW	DBL DHW			

Fig. 14-12 ■ Floor-plan symbols for common window types.

Fig. 14-13 Common Window Sizes for Light Construction.

Type	Width	Height
Double-Hung	1/10 2/0 2/4 2/6 2/8 2/10 3/0 3/4 3/6 3/8 3/10 4/0 4/6	2/6 3/0 3/2 3/6 4/0 4/6 4/10 5/0 5/6 5/10 6/0
Horizontal Slider	3/8 4/8 5/8 6/0 6/6	2/10 3/6 4/2 4/10 5/6 6/2
Casement	1/6 2/0 3/0 3/4 4/0 6/0 8/0 10/0 12/0	1/0 2/0 3/0 3/4 3/6 4/0 4/6 5/0 5/4 5/6 6/0
Fixed	1/0 1/6 2/0 2/6 3/0 4/0 4/6 5/0 5/10 6/0	4/6 4/10 5/6 6/6 7/0 7/6 8/0
Hopper	2/0 2/8 3/6 4/0	1/4 1/6 1/8 2/0 4/0 6/0
Jalousie	1/8 2/0 2/6 3/0	3/0 4/0 5/0 6/0
Bay-Bow	4/0 6/0 8/0 10/0 12/0	3/0 3/4 4/0 5/0 6/0
Awning	2/0 2/6 2/8 3/0 3/4 4/0 5/4 6/0 6/8 8/0 10/0 12/0	1/6 2/0 3/0 3/6 4/0 5/0 6/0
Half-Elliptical	5/0 6/0 8/0	1/6 1/0 1/10
Half-Round Top	2/0 2/4 2/6 2/10 3/0 3/6 4/0 4/8 5/0 5/4 6/0 R	—
Quarter-Round	1/6 2/0 2/6 3/0 R	—

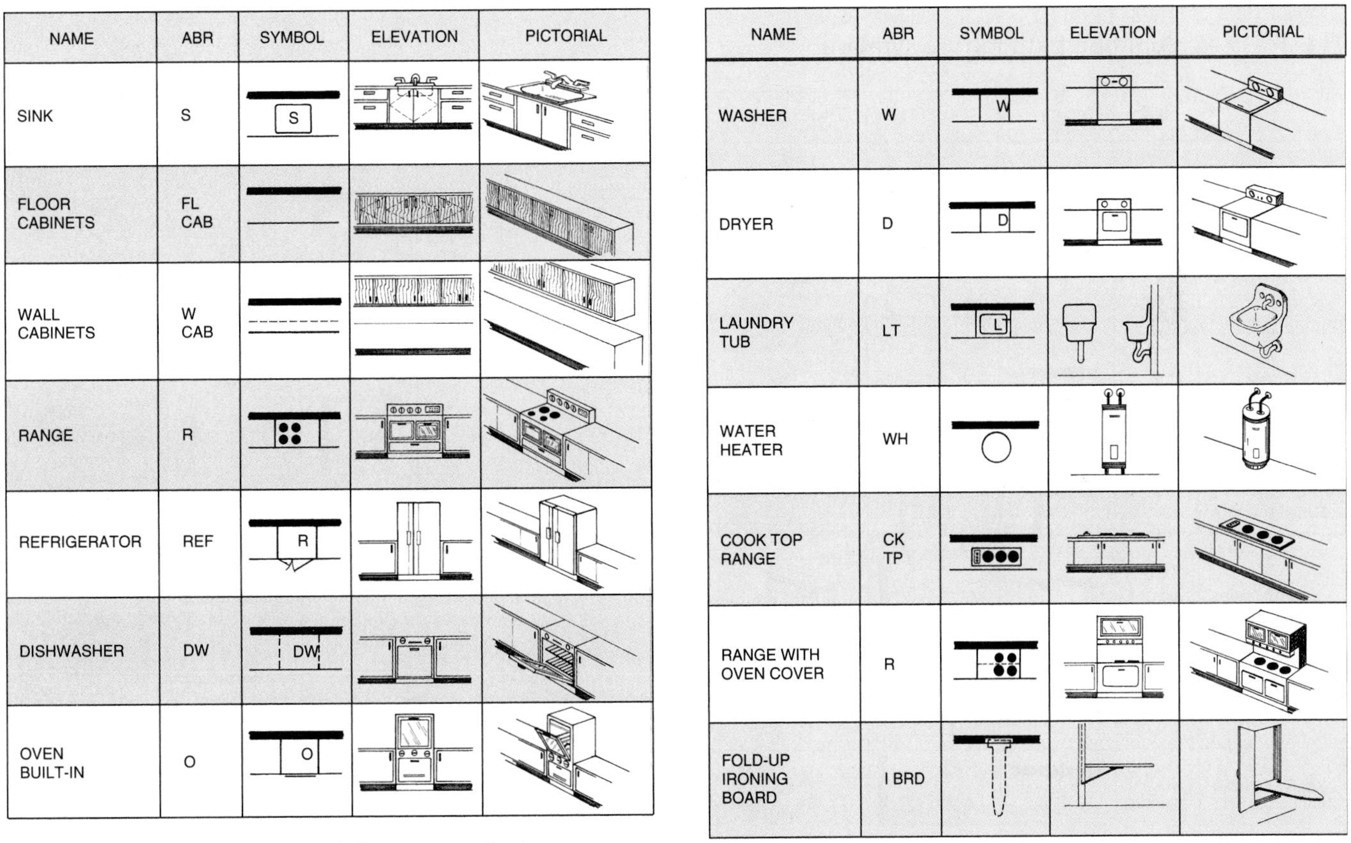

NAME	ABR	SYMBOL	ELEVATION	PICTORIAL
SINK	S	S		
FLOOR CABINETS	FL CAB			
WALL CABINETS	W CAB			
RANGE	R			
REFRIGERATOR	REF	R		
DISHWASHER	DW	DW		
OVEN BUILT-IN	O	O		

NAME	ABR	SYMBOL	ELEVATION	PICTORIAL
WASHER	W	W		
DRYER	D	D		
LAUNDRY TUB	LT	LT		
WATER HEATER	WH			
COOK TOP RANGE	CK TP			
RANGE WITH OVEN COVER	R			
FOLD-UP IRONING BOARD	I BRD			

Fig. 14-14 ■ Appliance and fixture symbols.

When preparing detail drawings, manufacturers' specifications must always be used. These show the exact dimensions of each unit plus the cutout dimensions for needed clearance. Appliance and fixture details are listed on schedules, as well as on floor plans. Schedules provide more information for purchasing, related cabinet design, and installation.

Bath Symbols

Figure 14-15 shows symbols for common bath fixtures. Symbols for freestanding units are usually drawn using a fixture template. Fixtures that align with cabinets must be carefully positioned. The type, style, and size specifications for each fixture must be taken from manufacturers' data to ensure proper fitting.

Furniture Symbols

Complete working-drawing floor plans do not usually include furniture symbols because they interfere with construction notes and dimensions. Furniture symbols are used mostly by interior designers on abbreviated floor plans to represent the width and length of each furniture piece. See Fig. 14-16. These symbols are either drawn with drafting instruments, furniture templates, or obtained from a computer software library. See Fig. 14-17.

Fig. 14-15 ■ Common bath fixture symbols.

NAME	ABR	SYMBOL	ELEVATION	PICTORIAL
LAVATORY FREESTANDING	LAV FR STN			
LAVATORY WALL HUNG	LAV WL HNG			
LAVATORY COUNTERTOP	LAV CNT TP			
LAVATORY CORNER	LAV COR			
WATER CLOSET ONE PIECE	WC 1 PC			
WATER CLOSET TWO PIECE	WC 2 PC			
WATER CLOSET WALL HUNG	WC WL HNG			

NAME	ABR	SYMBOL	ELEVATION	PICTORIAL
BATH TUB RECESSED	BT REC			
BATH TUB CORNER	BT COR			
BATH TUB ANGLE	BT ANG			
SHOWER HEAD	SH HD			
SHOWER SQUARE	SH SQ			
SHOWER CORNER	SH COR			

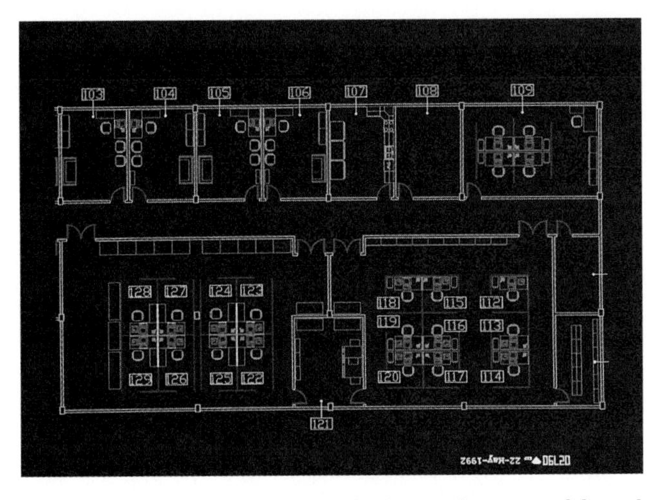

Fig. 14-16 ■ Furniture symbols used on an abbreviated floor plan.

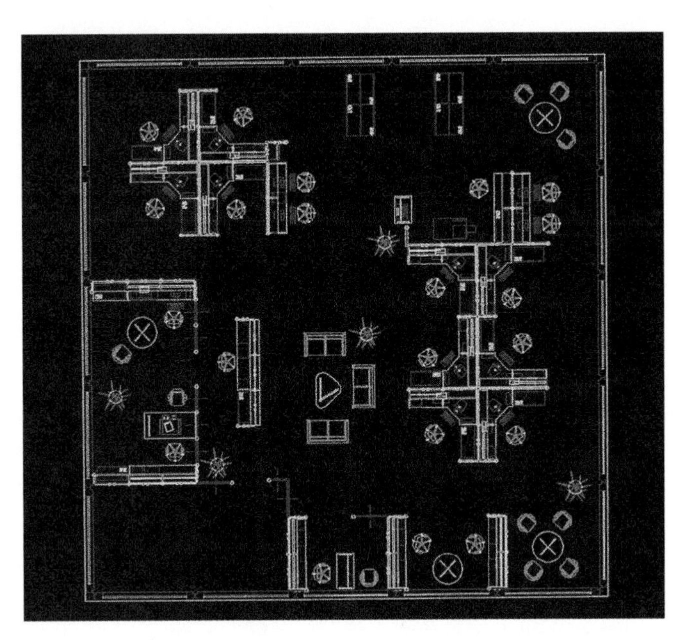

Fig. 14-17 ■ Use of furniture symbols from a computer software library.

Steps in Drawing Floor Plans

For maximum speed, accuracy, and clarity, the following steps should be observed in laying out and drawing floor plans. See Fig. 14-18, page 270.

1 Block in the overall dimensions of the house, and add the thickness of the outside walls with a hard pencil (4H).
2 Lay out the position of interior partitions with a 4H pencil.
3 Locate the position of doors and windows by centerline and by their widths (4H). Double check all dimensions. When establishing window dimensions, be sure the window square footage is at least 10 percent of the room's square footage.
4 Darken the **object lines** (visible lines), such as the main exterior walls and interior partitions, with an F pencil.
5 Add door and window symbols with a 2H pencil. Be sure to draw the door swing opening into perpendicular walls to provide the most convenient access. See Fig. 14-19.
6 Add symbols for stairwells, if applicable.
7 Erase extraneous layout lines if they are too heavy. If they are extremely light, they can remain.
8 Draw the outlines of kitchen and bath fixtures.
9 Add the symbols and sections for any masonry work, such as fireplaces and planters. Only the outline of the chimney, firebox, and hearth with material symbols are shown on floor plans. Fireplace construction is shown on detail drawings, as covered in Chapter 23.
10 Dimension the drawing as instructed later in this chapter.

Fig. 14-18 ■ Sequence of drawing floor plans.

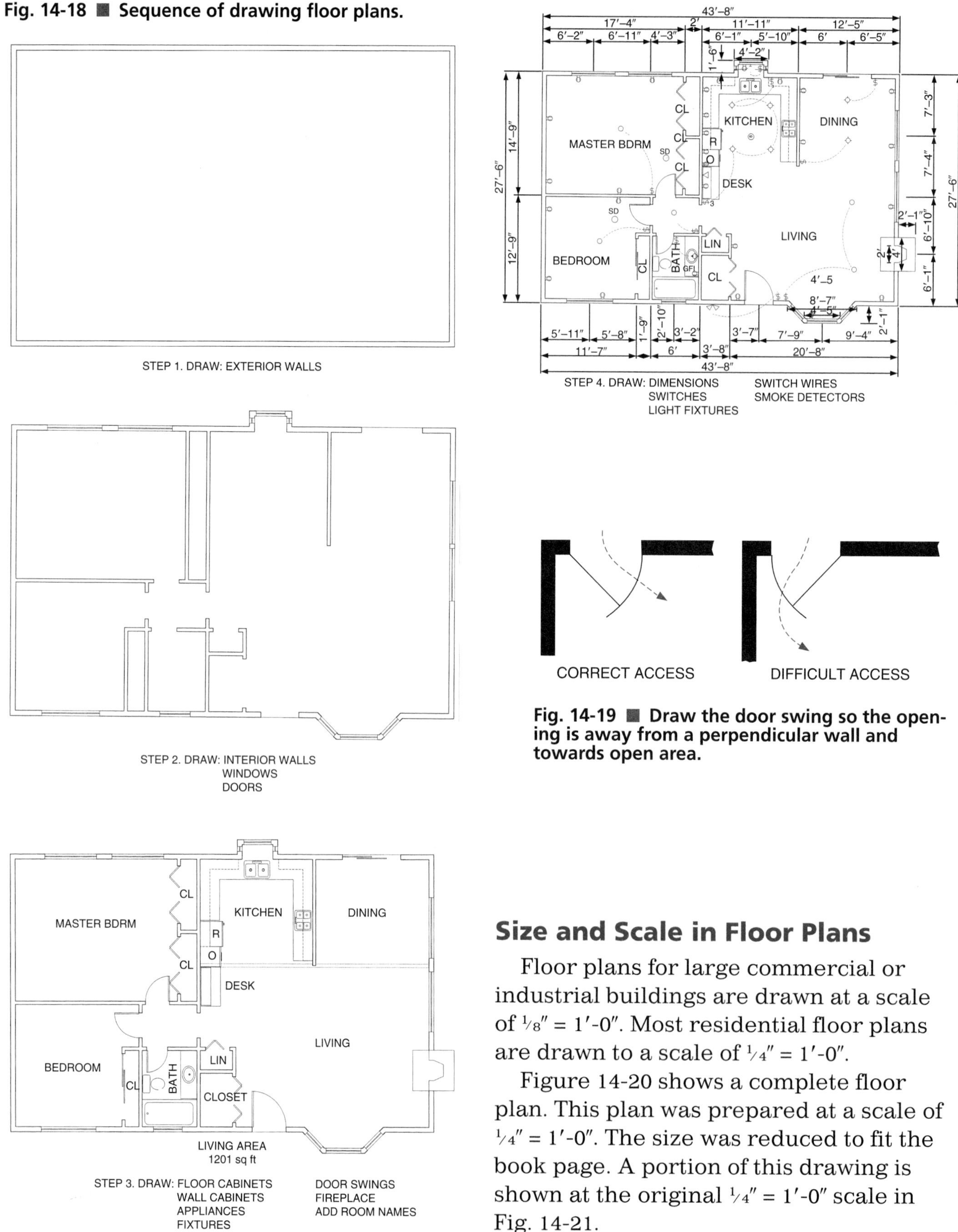

Fig. 14-19 ■ Draw the door swing so the opening is away from a perpendicular wall and towards open area.

Size and Scale in Floor Plans

Floor plans for large commercial or industrial buildings are drawn at a scale of 1/8″ = 1′-0″. Most residential floor plans are drawn to a scale of 1/4″ = 1′-0″.

Figure 14-20 shows a complete floor plan. This plan was prepared at a scale of 1/4″ = 1′-0″. The size was reduced to fit the book page. A portion of this drawing is shown at the original 1/4″ = 1′-0″ scale in Fig. 14-21.

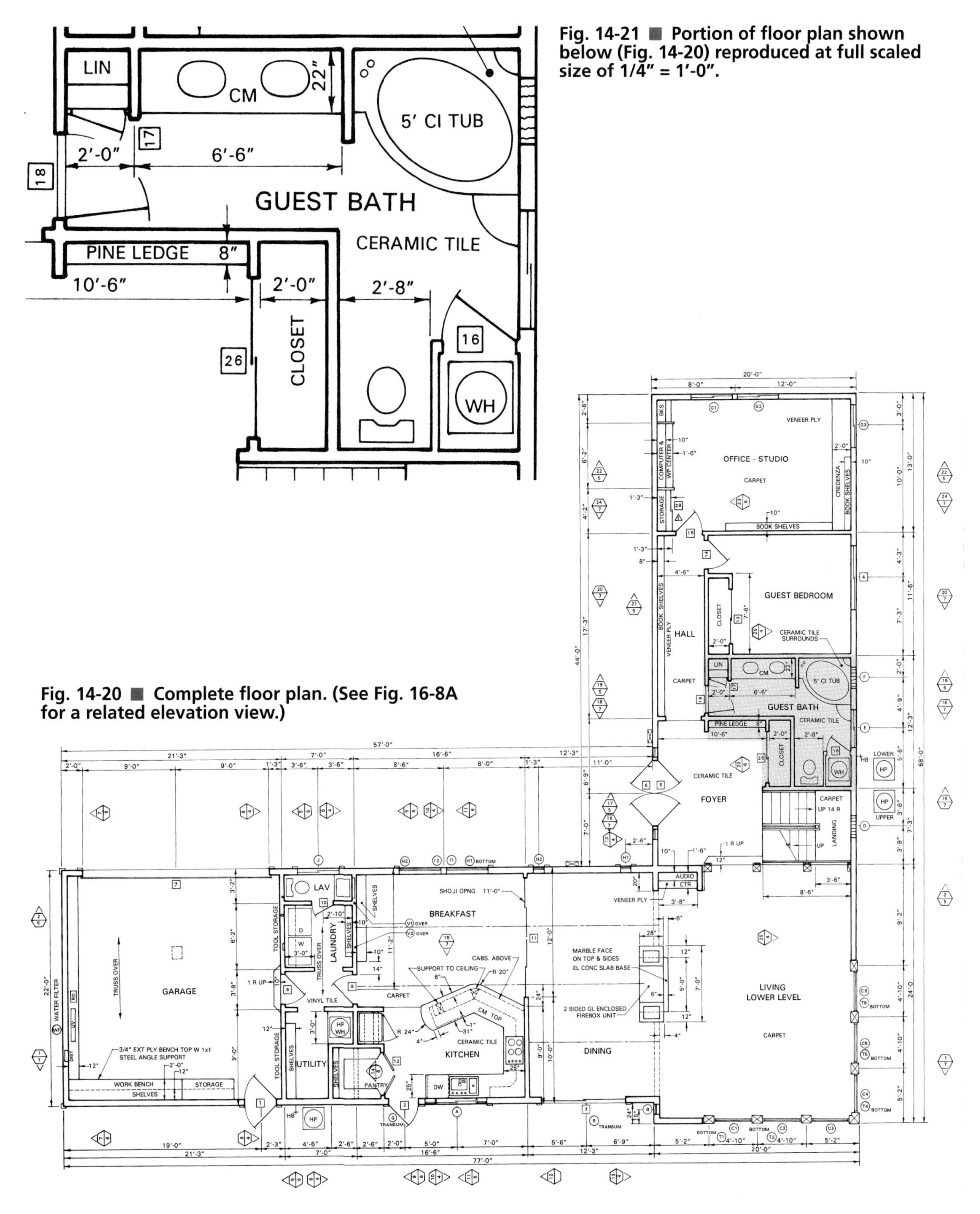

Fig. 14-21 ■ Portion of floor plan shown below (Fig. 14-20) reproduced at full scaled size of 1/4" = 1'-0".

Fig. 14-20 ■ Complete floor plan. (See Fig. 16-8A for a related elevation view.)

Multiple-Level Floor Plans

Drawing Separate Plans

Bilevel, two-story, one-and-one-half-story, and split-level homes require a separate floor plan for each level. The separate plans of **multiple-level floor plans** are prepared on tracing paper and drawn at the same scale as the first floor plan. The tracing paper needs to be placed directly over the first-floor plan to ensure alignment of exterior walls, partitions, and vertical features. Once the major outline has been traced, the first-floor plan is removed.

Alignment of features, such as stairwell openings, outside walls, plumbing walls, vents, and chimneys, is critical in preparing second-floor plans. Where no second floor exists over part of a first floor, the outline of the first-level roof is shown. See Fig. 14-22.

Figure 14-23 shows a typical second-floor plan of a one-and-one-half-story house. This drawing, in addition to revealing the second-floor plan, shows the outline of the roof as a single line, and the outline of the building as dotted lines under the roof.

Figure 14-24 shows a first-, second-, and third-floor plan of a three-level house. Visualize the position of each level by referring to the pictorial rendering of this house. Note that the three different levels of this plan do not stack evenly on top of each other. Upper floors are smaller in area. In drawing multiple floor plans with

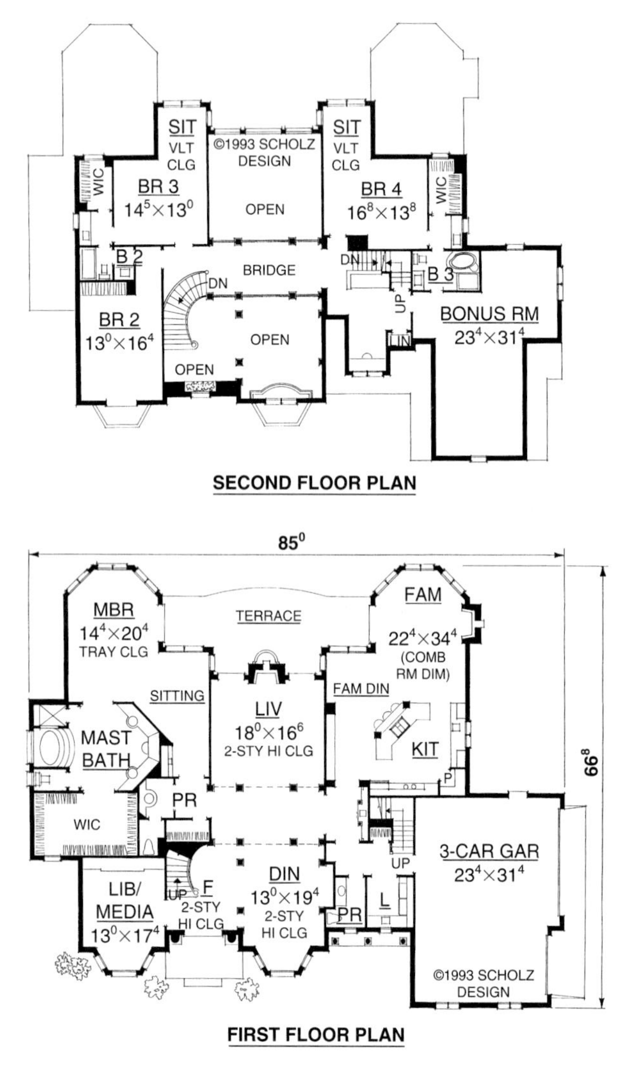

Fig. 14-22 ■ Aligned first- and second-floor plans. The rendering for this design is shown in Fig. 20-6.
Scholz Homes Inc.

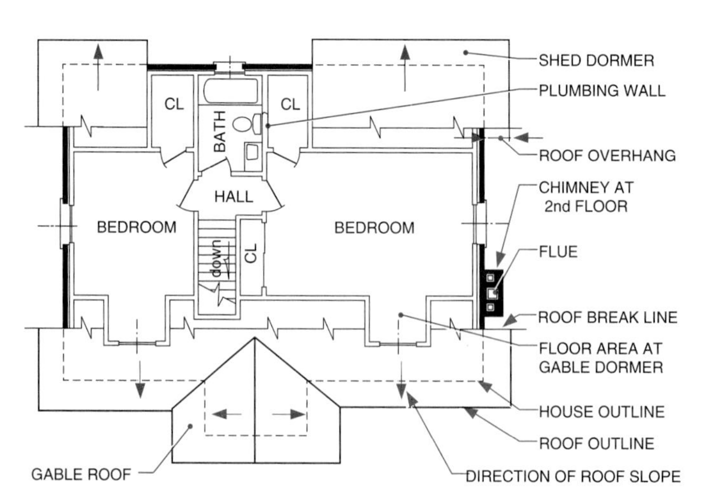

Fig. 14-23 ■ Second-floor plan for a one-and-one-half story house.

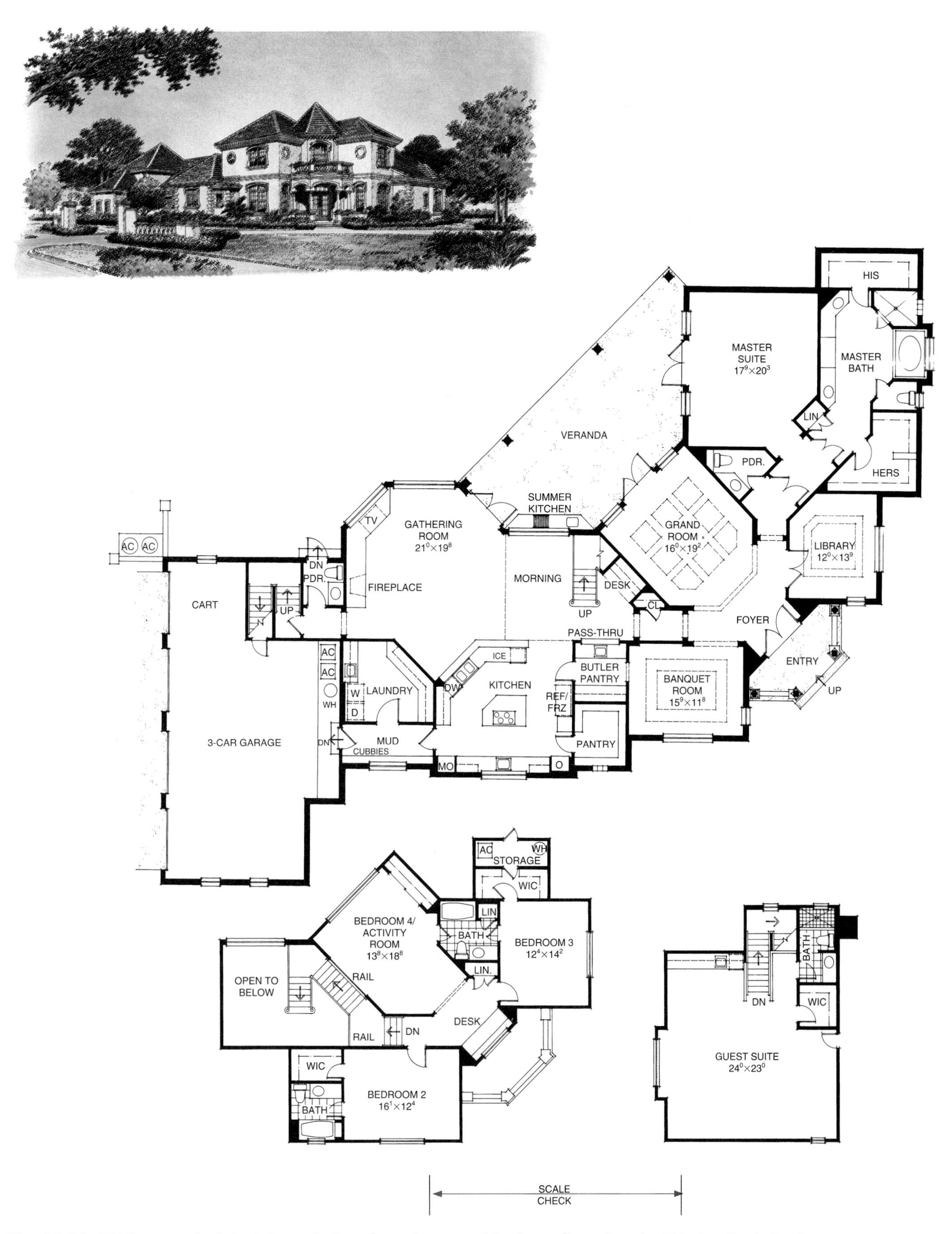

Fig. 14-24 ■ **Plans and pictorial rendering for a home with three floor levels.** *Winter Park Design*

a CAD computer program, the layering command places each plan on a separate layer which can be viewed separately or combined with other layers.

Calculating Dimensions for Stair Systems

Floor-plan stair symbols show the width and depth of each tread beginning at the plan level. A **break line** (see Fig. 4-14) is used to eliminate the need to draw every stair to the next level, either up or down. (See Chapter 9 for information on different types of stairs.)

When drawing multiple-level floor plans, stair systems must align on all levels. Also the number and width of treads (or steps) must be calculated and shown on all levels. To calculate stair tread width use the following formula:

$$\frac{\text{stair run (inches)}}{\text{number of treads}} = \text{tread width (inches)}$$

Example:
$144''/15 = 9.6''$

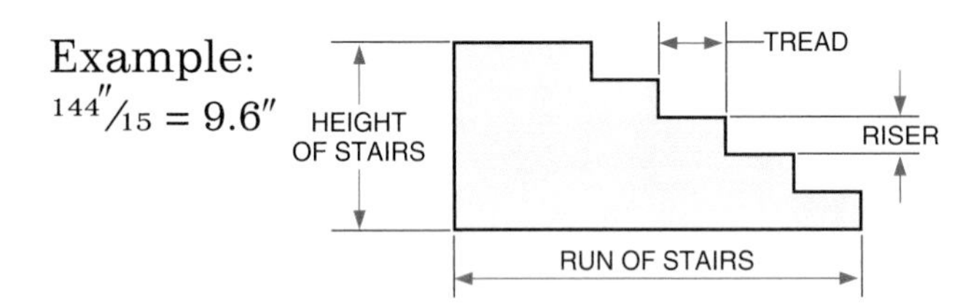

In addition to indicating the stair system, plans of second floors or higher levels must also show the outline of the stairwell opening. To determine the size of the stairwell opening, the complete stair system should be designed before the final floor plan is prepared.

Figures 14-25A through G show the sequence of steps necessary for determining the exact dimensions of a stair system.

FIGURE 14-25A

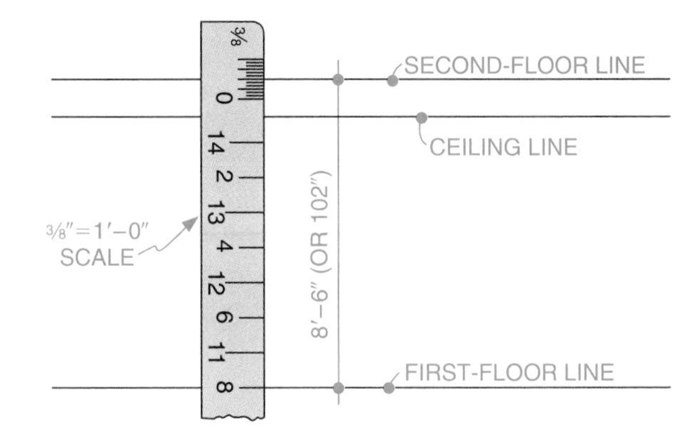

1 Lay out the distance from the first-floor level to the second-floor level exactly to scale. Convert this distance to inches and add the position of the ceiling line.

FIGURE 14-25B

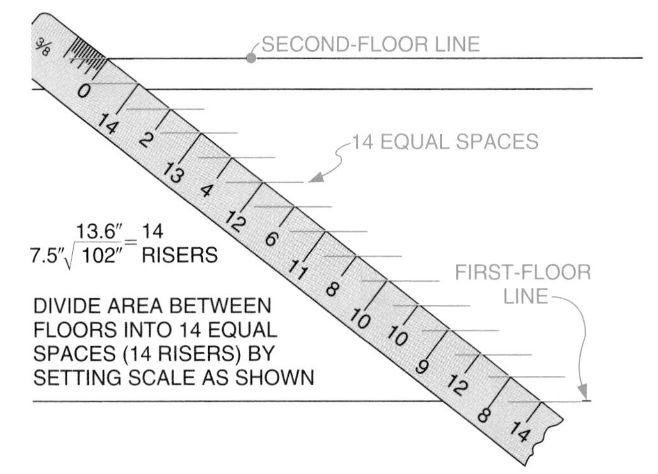

2 Determine the most desirable riser heights (7½″, or 190 mm, is average). Divide the number of inches (millimeters) between floor levels by the desired riser height to find the number of risers needed. Divide the area between the floors into spaces equaling the number of risers needed.

3 Extend the riser-division lines lightly for about an inch (25 mm). These lines represent the top of each tread.

FIGURE 14-25C

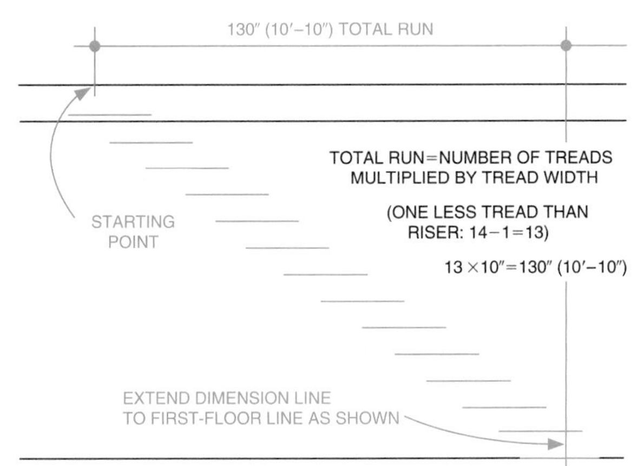

4 Determine the total length of the run. Lay out this distance from a starting point near the top riser line and measure the total run horizontally. Extend this line vertically to the first-floor line. The total run is the number of treads multiplied by the width or depth of each tread. There is always one less tread than riser because the floor level that is reached is the top or bottom tread.

FIGURE 14-25D

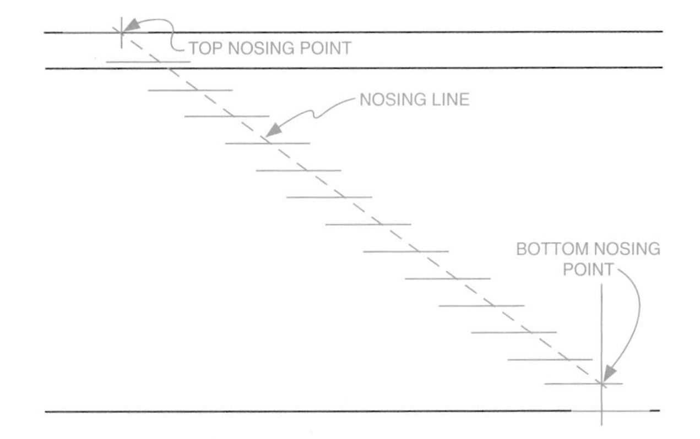

5 Locate the top and bottom nosing points. Mark the intersection between the starting point and the total run and the intersection between the end of the total run and the first riser line.

6 Draw the nosing line by connecting the bottom nosing point with the top nosing point.

FIGURE 14-25E

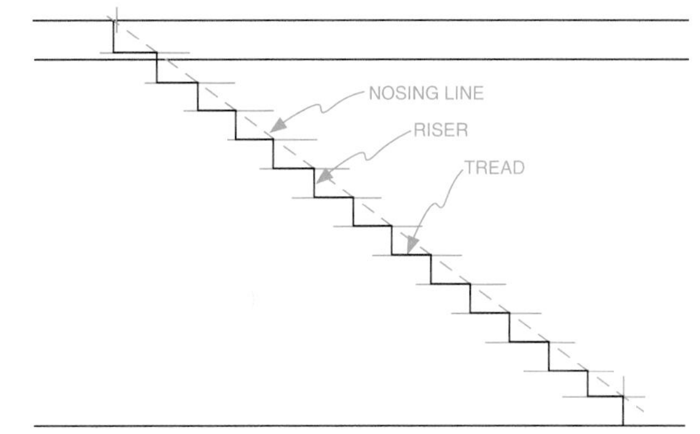

7 Draw vertical riser lines intersecting the nosing points and the light riser lines.

8 Make the tread lines and the riser lines heavy.

FIGURE 14-25F

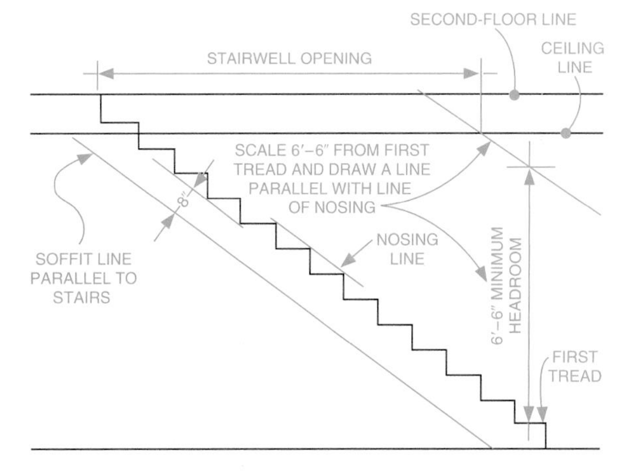

9 Draw the stringer line. Establish headroom clearances by drawing a parallel line 6′-6″ (1981 mm) above the nosing line. Establish the stairwell opening by cutting the joists where the headroom clearance line intersects the bottom of the joist.

10 Show the outline of the carriage or stringer assembly.

FIGURE 14-25G

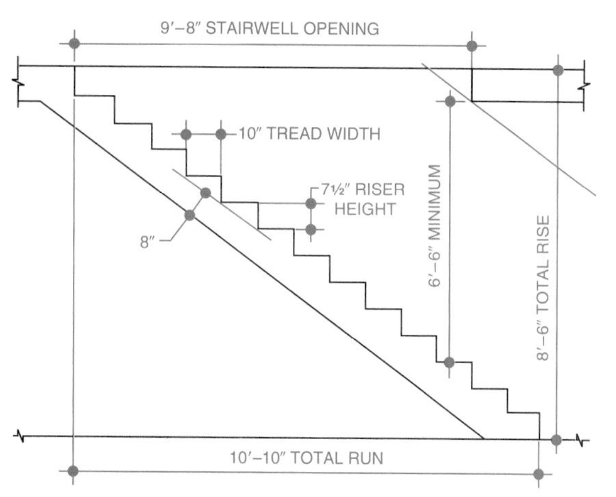

11 Erase all layout lines and make all object lines heavier.

12 Add dimensions to describe the length of the stairwell opening, the size of the tread widths, the riser height, the minimum headroom, the total rise, and the total run.

Reversed Plans

Floor plans offered as options in the development stage are often reversed to provide more plan choices. Reversing plans alters the appearance of a house and relocates rooms to avoid or take advantage of environmental and orientation factors and street locations. **Reversed plans** are accomplished by turning a plan over to provide a mirror image, as shown in Fig. 14-26. The plan can be traced in this position or a print created by feeding the drawing into a print machine with the front side facing down. On a finished plan, this should be done before lettering to avoid printing reversed letters. On a CAD system, a mirror command is used to produce a reversed plan.

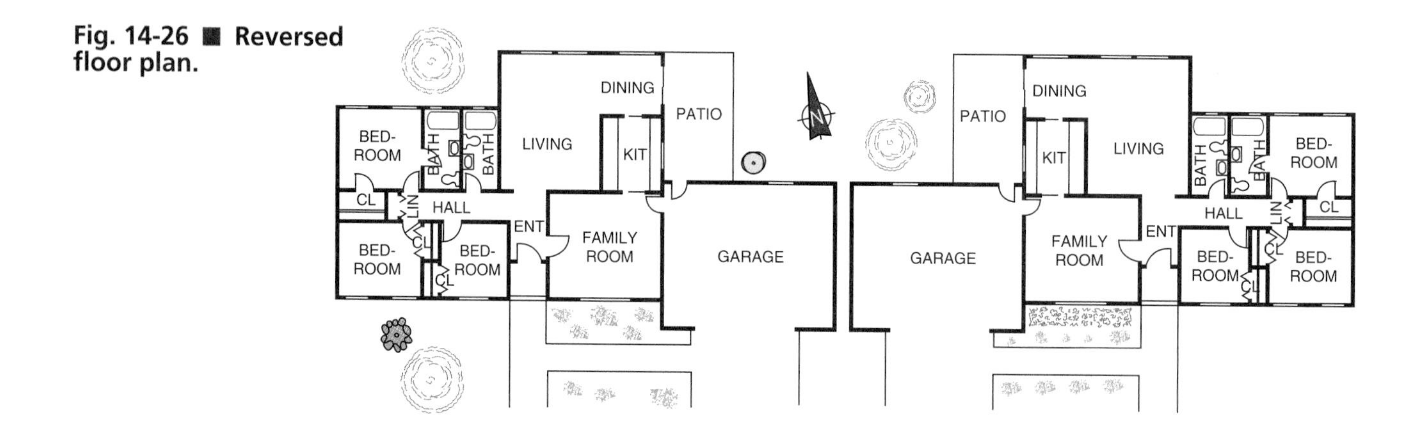

Fig. 14-26 ■ Reversed floor plan.

Reflected Ceiling Plans

Complex ceiling designs and multiple-lighting fixtures or levels often require the preparation of a **reflected ceiling plan.** See Fig. 14-27. These plans are drawn using a floor plan as a base. Floor plan walls and partitions are traced and symbols of ceiling features are drawn as the ceiling would be viewed if the floor were a mirror.

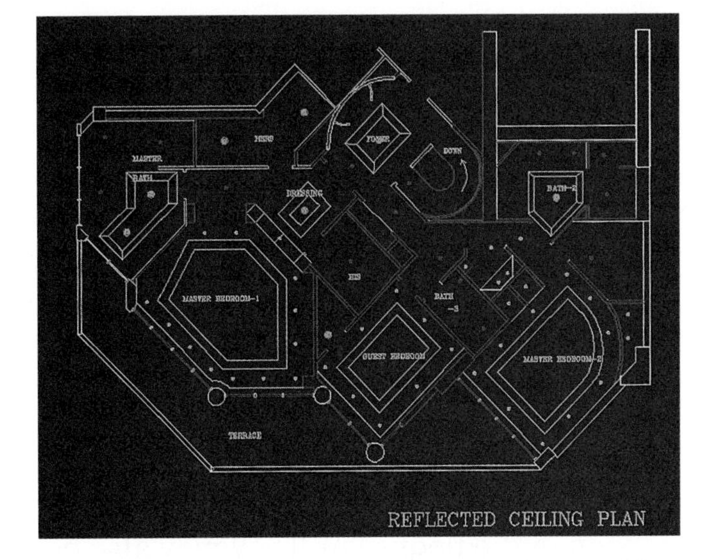

Fig. 14-27 ■ Computer-layered ceiling plan. *David Karram*

Floor-Plan Dimensioning*

A completely dimensioned drawing is necessary to complete any building exactly as designed. Dimensions on the floor plan show the builder the width and length of the building. They show the location of doors, windows, stairs, fireplaces, planters, and so forth. Just as symbols and notes show exactly what materials are to be used in the building, dimensions show the sizes of materials and exactly where they are located.

Drawing Dimensions

Architectural dimensioning practices differ among designers. Several of the most common methods of dimensioning floor plans are shown in Fig. 14-28.

Because a large building must be drawn on a relatively small sheet, a small scale

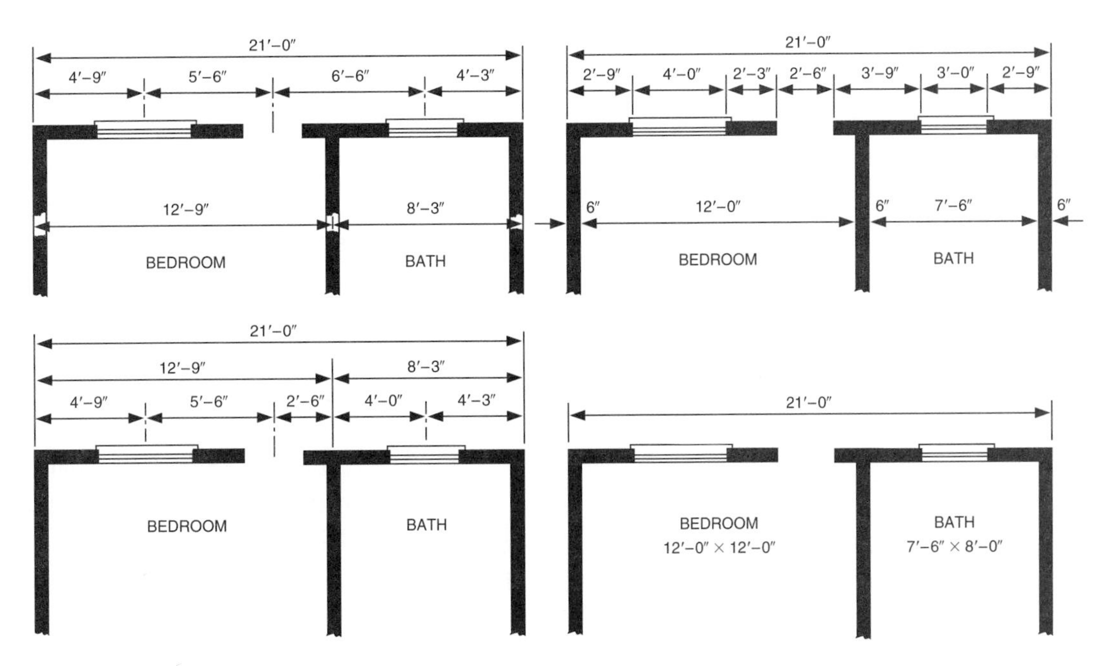

Fig. 14-28 ■ Methods of dimensioning frame construction floor plans.

* The floor plan dimensioning practices presented in this chapter adhere to the American Society of Mechanical Engineers (ASME) standards.

(¼″ = 1′0″ or ⅛″ = 1′0″) must be used. The use of such a small scale means that many dimensions must be crowded into a very small area. Therefore, only major dimensions such as the overall width and length of the building and of separate rooms, closets, halls, and wall thicknesses are shown on the floor plan. Dimensions too small to show directly on the floor plan are described either by a note on the floor plan or by separate, enlarged details.

Enlarged details, or detail drawings, are sometimes merely enlargements of a portion of the floor plan. They may also be section drawings cross-referenced on the floor plan. Separate drawings are usually necessary to communicate adequately dimensions for fireplaces, planters, built-in cabinets, door and window details, stairframing details, or any unusual construction methods.

Providing Dimensions

A floor plan must be completely dimensioned to ensure that the house will be constructed precisely as designed. Complete dimensions convey the exact wishes of the architect and owner to the builder. If adequate dimensions are not provided, the builder is placed in the position of a designer. A good builder is not expected to be a good designer. Supplying complete dimensions will eliminate the need for the builder to guess or interpret the size and position of the various features of this plan. All needed information about each room, closet, partition, door, and window is given.

A floor plan with only limited dimensions shows just the overall building dimensions and the width and length of each room. It is sufficient to summarize the relative sizes of the building and its rooms for the prospective owner, but insufficient for building purposes.

Guidelines for Dimensioning

Many construction mistakes result from errors made in architectural drawings. Most errors in architectural drawings are the result of mistakes in dimensioning. Dimensioning errors are therefore costly in time, efficiency, and money. Familiarization with the following guidelines for dimensioning floor plans will eliminate much confusion and error.

Math Connection

List of Guidelines

These guidelines are illustrated by the numbered arrows in Fig. 14-29.

1. Architectural **dimension lines** are unbroken lines with dimensions placed above the line. Dimension lines should be located no closer than ¼″ from object lines. Up to 1″ is preferred. Dimension lines should be spaced at least ¼″ between rows.
2. Foot and/or inch marks are normally used on architectural dimensions. Sometimes these marks are omitted and a dash or slash is used. For example 8-4 or 8/4 means 8′4″. If metric measures are used, the dimensions are always in millimeters. Therefore size unit notations are not needed. However, a note should be placed on the plan stating that all dimensions are in millimeters.
3. Dimensions over 1′ are expressed in feet and inches. Detail drawings often contain only inch dimensions regardless of size.

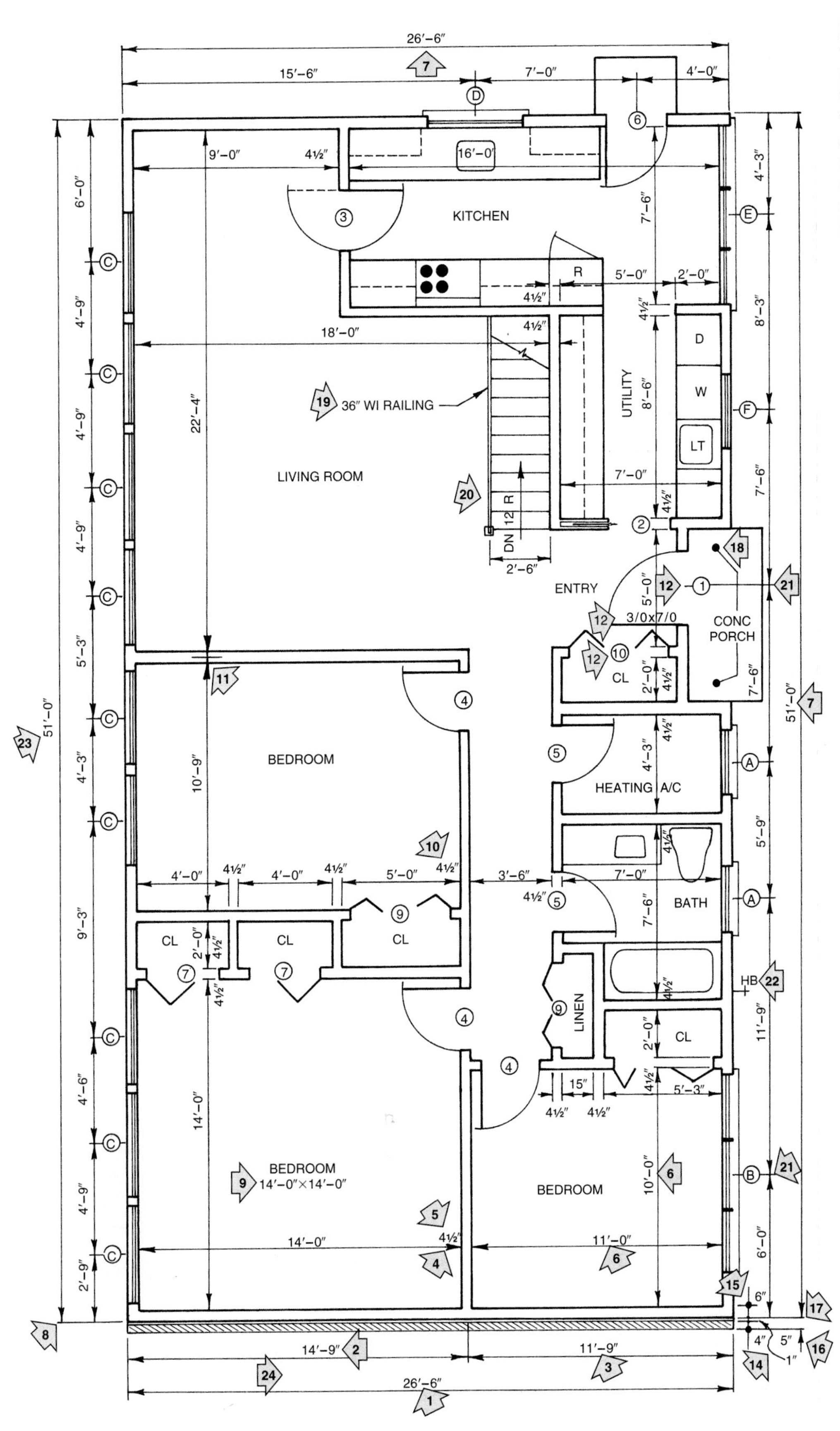

Fig. 14-29 ■ **Guidelines for dimensioning floor plans. Some arrows have the same number because they refer to different applications of the same guideline. Such arrows are printed in red. All single-reference arrows are blue.**

4. Dimensions less than 1′ are shown in inches.
5. A slash is often used with fractional dimensions to conserve vertical space.
6. Vertical dimensions should be placed to read from the right side of the drawing. Horizontal dimensions read from the bottom of the drawing.
7. **Overall dimensions** for the length and width of a building are placed outside other dimensions.
8. Line and arrowhead weights for architectural dimensioning are thin and dark as shown in Chapter 19. Arrowhead styles are optional. See Fig. 14-30.
9. Room sizes may be shown by stating width and length on abbreviated plans.
10. When the area to be dimensioned is small, numerals may be placed outside the extension lines.
11. Framed interior walls are dimensioned to the center of partitions.
12. Window and door sizes may be shown directly on the door or window symbol, or may be indexed to a door or window schedule with a reference callout.
13. Solid concrete walls are dimensioned from wall to wall, exclusive of wall coverings (not shown on 14-29; see Fig. 14-31).
14. Curved leaders are sometimes used to eliminate confusion with other dimension lines.
15. When areas are too small for arrowheads, dots may be used to indicate dimension limits.
16. The dimensions of brick or stone veneer must be added to the framing dimension. See Fig. 14-32.
17. When the space is small, arrowheads may be placed outside the extension lines.
18. A dot on the end of a leader refers to the entire area noted.
19. Dimensions that cannot be seen on the floor plan or those too small to place

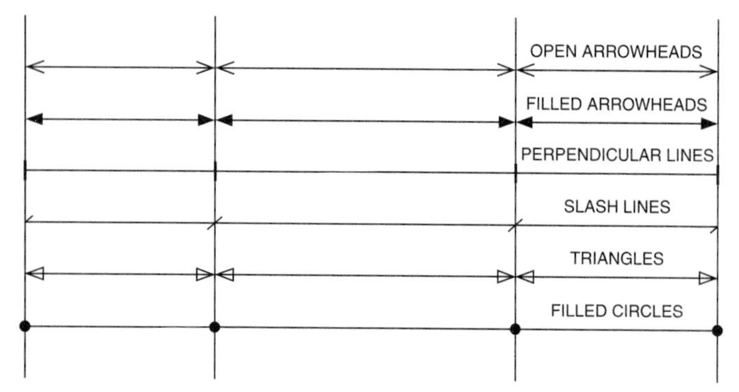

Fig. 14-30 ■ Arrowhead styles used on dimension lines.

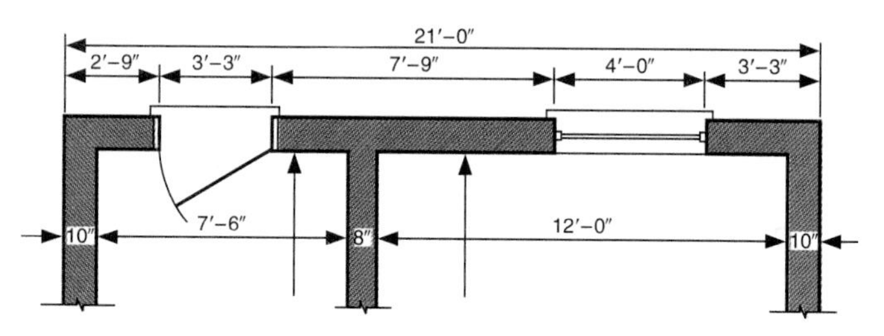

Fig. 14-31 ■ Methods of dimensioning concrete walls.

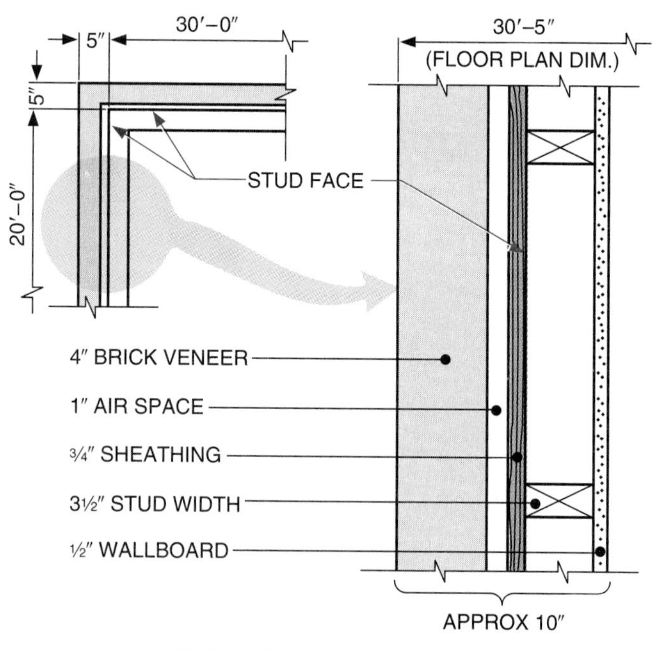

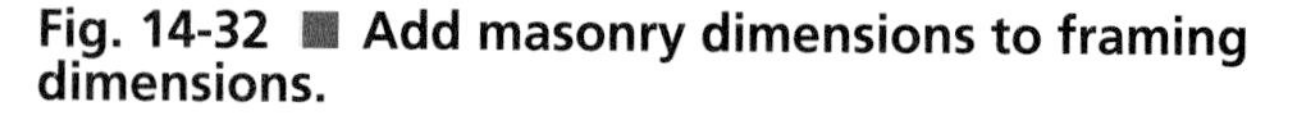

Fig. 14-32 ■ Add masonry dimensions to framing dimensions.

on the drawn object are noted with leaders for easier reading.

20. In dimensioning stairs, the number of risers is placed on a line with an arrow indicating the direction down (DN) or up (UP).
21. Windows, doors, pilasters, beams, construction members, and areaways are dimensioned to their centerlines. (Areaways are the sunken areas in front of basement doors and windows that allow light and air to reach the basement or crawl space.)
22. Use notes or abbreviations when symbols do not show clearly what is intended.
23. Architectural dimensions always refer to the actual size of the building regardless of the scale of the drawing. The building in Fig. 14-29 is 51′-0″ in length.
24. **Subdimensions** must add up to overall dimensions. For example: 14′9″ + 11′9″ = 26′6″. Most rows of dimensions include both feet and inches and may include fractional inches. In adding rows of mixed numbers such as these, add the inches separately, convert the inch total to feet and inches, and then re-add the foot total as follows:
    - Total number of inches ÷ 12 = feet and inch fractions
    - LCD = lowest common denominator in a series into which all denominators can be divided.

Example:	*To add:*	*Follow these steps:*	
	1′-7 7/8″	Step 1:	1′-7 14/16″
	2′-8 1/4″		2′-8 4/16″
	6′-10 9/16″		6′-10 9/16″
	11′-2 11/16″		9′-25 27/16″
		Step 2:	9′-26 11/16″
		Step 3:	11′-2 11/16″

The following guidelines are not illustrated in Fig. 14-29.

25. When framing dimensions alone are desirable, rooms are dimensioned by distances to the outside face of the studs in the partitions. See Fig. 14-33.
26. Since building materials vary somewhat in size, first establish the thickness of each component of the wall and partition, such as plaster, brick, or tile thicknesses. Add these thicknesses together to establish the total wall thickness. Common thicknesses of wall and partition materials are shown in Figs. 14-32 and 14-33.
27. The scale of each drawing can be noted in the title block. If separate drawings on a single drawing sheet are drawn at different scales, each drawing must be labeled with the appropriate scale.
28. An arrow showing the direction of north is placed on each floor plan unless the plan is prepared without any site identification. Various types of north arrows are shown in Fig. 14-34.

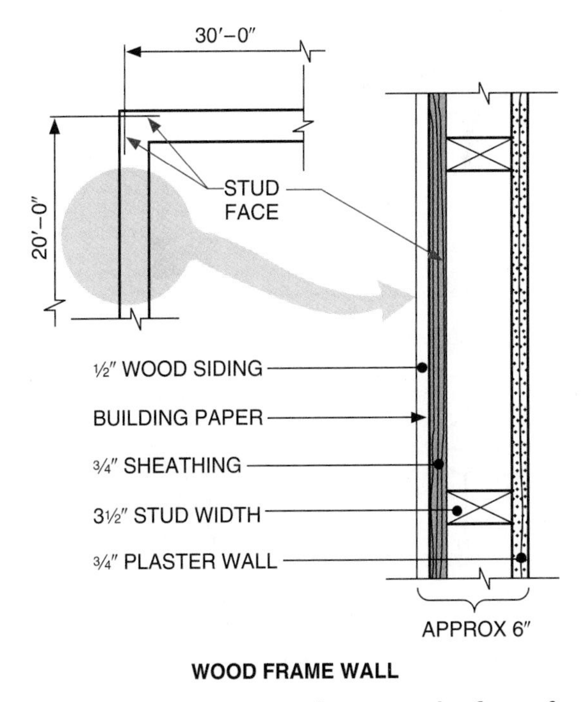

Fig. 14-33 ■ Dimensions shown to the face of studs.

Fig. 14-34 ■ Styles of north arrows. Designers and drafters may create their own arrow designs, but these must be precise and understandable.

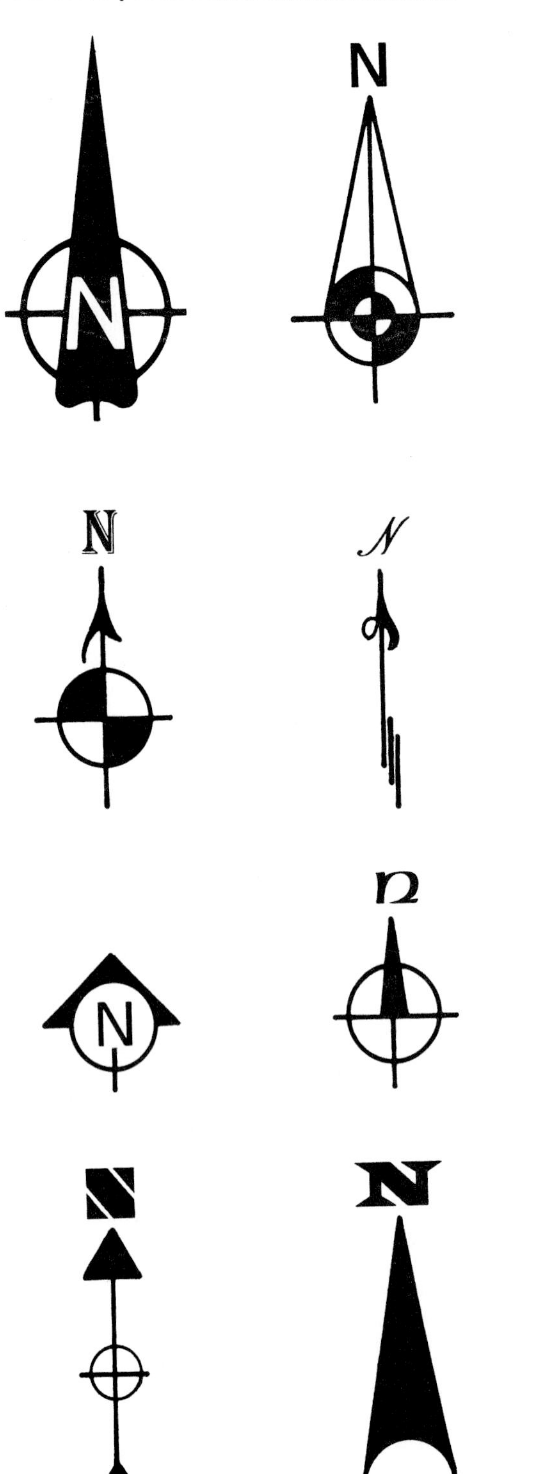

CAD Dimensioning

Floor plans can be dimensioned on a CAD system by placing the cursor or stylus at the end point of a line to be dimensioned. With the appropriate input, the computer will then measure and calculate this distance. When the cursor or stylus is placed on the location desired, the system will draw the dimension lines, arrowheads, and insert the dimension measurement. The drafter selects the dimensioning mode (customary, metric, decimal, fraction), arrowhead style, text size, and distance from the object to the dimension location. If scaled correctly, some software programs can dimension a floor plan automatically. (Refer back to Chapter 5.)

Metric Dimensioning

If a building is designed using metric sizes, all dimensions are shown in millimeters (mm), as shown in Fig. 14-35. Refer to Chapter 3 to review metric scales.

Modular Dimensions

An attempt should always be made to establish major dimensions to conform to 16″, 24″, and 48″ increments. However, if a building is designed to totally conform to modular grids, all dimensioning must conform to these standards as covered in Chapter 22.

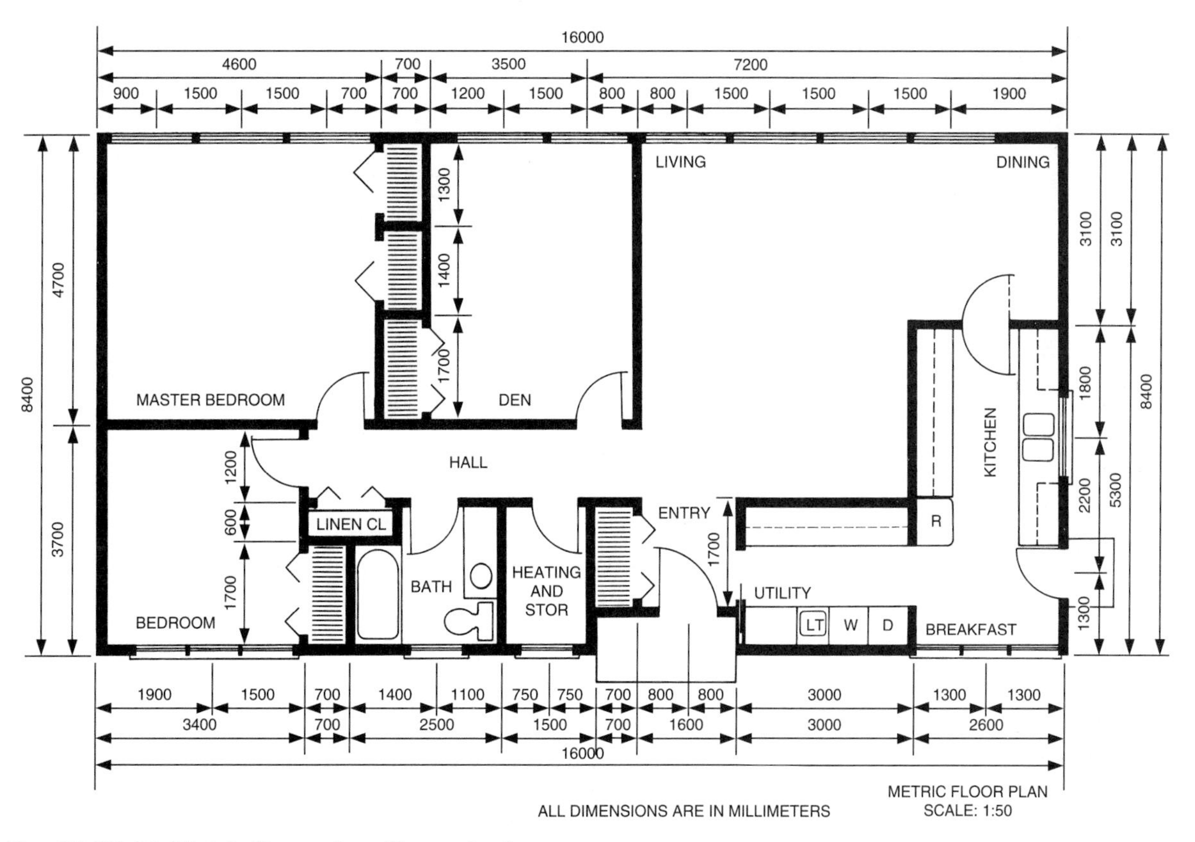

Fig. 14-35 ■ Metric floor-plan dimensioning.

CHAPTER 14 Exercises

1. Draw a complete floor plan, using a sketch of your own design as a guide, and using the scale ¼″ = 1′0″.
2. Complete Exercise 1 using a CAD system.
3. Dimension an original scaled floor plan that you have completed for a previous assignment.
4. Draw and dimension the floor plan of your own home.

5. Complete Exercise 3 using a CAD system.

CHAPTER

15

Designing Elevations

Objectives

In this chapter you will learn to:

- apply the principles and elements of design (Chapter 2) to creating elevation drawings.
- recognize different roof styles as options for roof design.
- select and design window styles in relation to elements of design and window functions.
- locate doors on an elevation design, considering style, size, and types of doors.

Terms

eave line
elevation drawings
fenestration
gable
ground line
overhang
pitch
ridge line
rise
run
slope (roof)

Elevation drawings, or elevations, show the vertical surfaces of a structure. Interior elevations, discussed later in Chapter 16, show interior walls. Exterior elevation drawings show the entire front, sides, and rear of a structure. In this chapter, you will learn how to apply the elements of design to creating the exterior form of a building, including the selection of roof, window, and door styles.

The term "elevation" also refers to the vertical surfaces themselves. Designing the elevations of a structure is only one part of the total design process. However, the elevation design reflects the part of the building that people see. The entire structure may be judged by the elevations.

■ **Attractive exterior elevations can make the whole structure more appealing.** *Georgia Pacific*

Relationship with the Floor Plan

Since a structure is designed from the inside out, the design of the floor plan normally precedes the design of the elevation. The complete design process requires a continual relationship between the elevation and the floor plan.

Flexibility is possible in the design of elevations, even in those designed from the same floor plan. Once the location of doors, windows, and chimneys has been established on the floor plan, the development of an attractive and functional elevation for the structure depends on various factors. Roof style, overhang, grade-line position, and the relationship of windows, doors, and chimneys to the building line must be considered. Choosing a desirable elevation design is not an automatic process that follows the floor-plan design, but a creative process that requires imagination.

The designer should keep in mind that only horizontal distances can be established on the floor plan. However, on an elevation, vertical heights, such as heights of windows, doors, and roofs, must also be shown. As these vertical heights are established, the appearance of the outside and the way that the heights affect the internal functions of the building must be considered.

Elements of Design and Elevations

Creating floor plans is a process of allocating interior space to meet functional needs. Designing elevations involves combining the elements of design to create functional and attractive building exteriors.

The principles and elements of design were defined in Chapter 2. In this chapter, the elements of design (line, form, space, color, light, materials, and texture) are applied to the creation of elevations. The total appearance of an elevation depends on the relationship among its component parts, such as surfaces, roofs, windows, doors, and chimneys. The balance of these parts, the emphasis placed on various components of the elevation, the texture of the surfaces, the light, the color, and the shadow patterns all greatly affect the general appearance of an elevation.

Lines

The lines of an elevation tend to create either a horizontal or vertical emphasis. The major horizontal lines of an elevation are **the ground line, eave line,** and **ridge line.** If these lines are accented, the emphasis will be placed on the horizontal. If the emphasis is placed on vertical lines such as corner posts and columns, the emphasis will be vertical. In general, a low building will usually appear longer and even lower if the design consists mostly of horizontal lines. The reverse is true for tall buildings with vertical lines.

Lines should be consistent. The lines of an elevation should appear to flow together as one integrated line pattern. Continuing a line through an elevation for a long distance is usually better than breaking the line and starting it again. Rhythm can be developed with lines, and lines can be repeated in various patterns.

When additions are made to an existing design, care must be taken to ensure that the lines of the addition are consistent with the established lines of the structure. The lines of the component parts of an elevation should relate to each other, and the overall shape should reflect the basic shape of the building.

Form and Space

Lines combine to produce form and create the geometric shape of an elevation. Elevation shapes should be balanced. The term "balance" refers to the symmetry of the elevation. Formal balance is used extensively in colonial and period styles of architecture. Informal balance is more widely used in contemporary residential architecture and in ranch and split-level styles. (Refer back to Chapter 2.)

Vertical space (elevation) emphasis, or accent, can be achieved by several different devices. An area may be accented by mass, color, or material.

In addition to the elements of design, the basic architectural style of a building needs to be considered when designing elevations. A building's style is more closely identified by the elevation design than by any other factor.

The type of building structure must also be compatible with the architectural style of the elevation. However, within basic styles of architecture, there is considerable flexibility in the type of structure. Figure 15-1 shows the basic types of structures. The elevation design can be changed to fit the building site and/or to create the appearance of several different architectural styles, as long as the basic building type is consistent with that style.

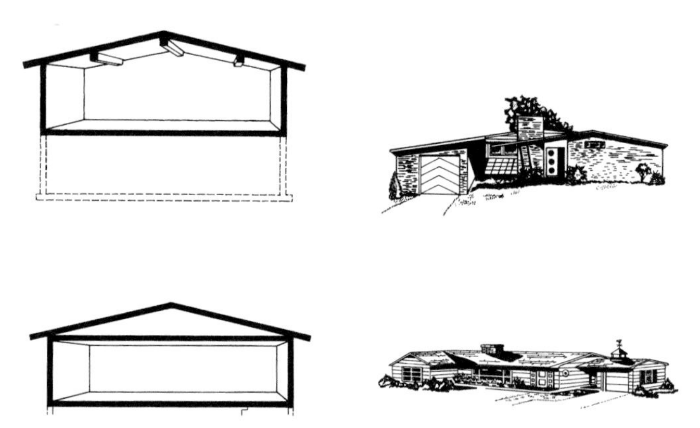

A. One-story.

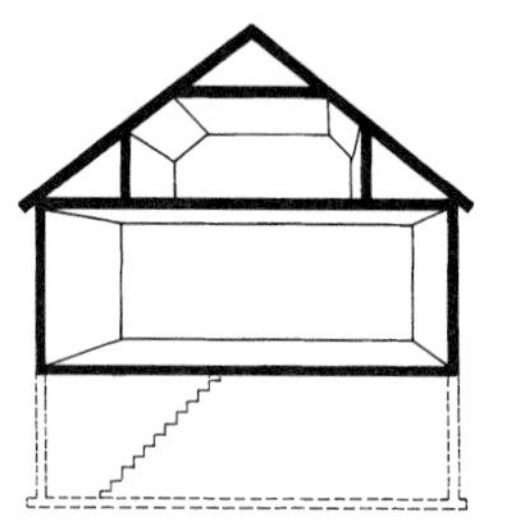

B. One-and-one-half-story.

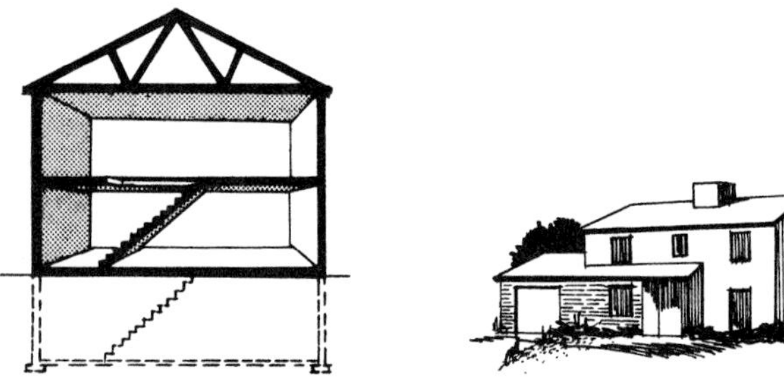

C. Two-story.

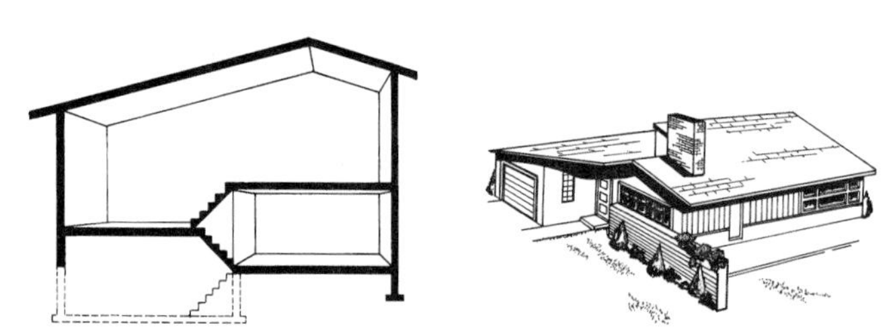

D. Split-level.

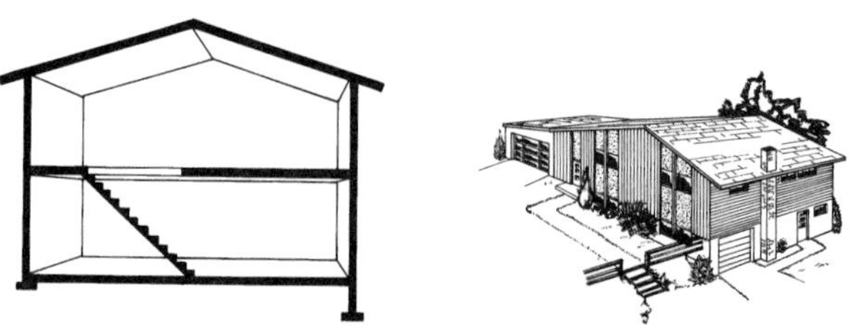

E. Bilevel.

Fig. 15-1 ■ Basic types of structures. *National Lumber Manufacturers Assoc.*

Elevations should appear as one integral and functional facade, rather than as a surface in which holes have been cut for windows and doors and other structural components. Doors, windows, and chimney lines should be part of a continuous pattern of the elevation and should not appear to exist alone. See Fig. 15-2.

Fig. 15-2 ■ Windows and doors should be related to the major lines of the structure.

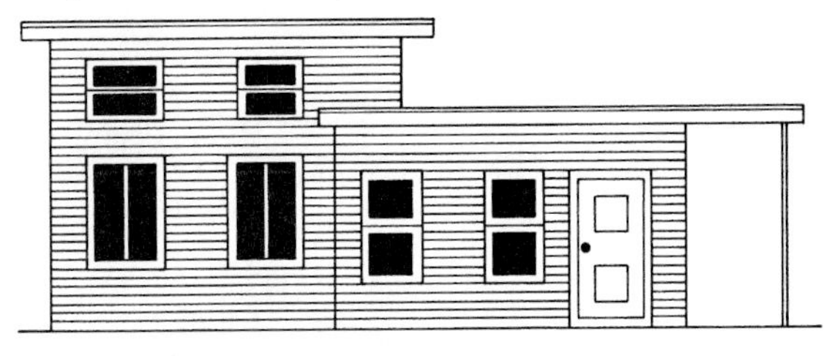

A. Unrelated.

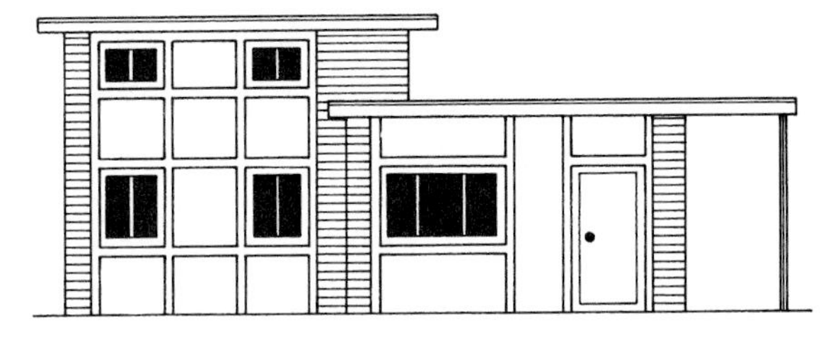

B. Related.

Light and Color

An elevation that is composed of all light areas or all dark areas tends to be uninteresting and neutral. Some balancing of light, shade, and color is desirable in most elevations. Shadow patterns can be created by depressing specific areas, using overhangs, texturing, and varying colors. Door and window trim, columns, battens (strips covering joints), and overhangs can be used to create most shadows.

Materials and Texture

An elevation may contain many kinds of materials, such as glass, wood, masonry, and ceramics. These must be carefully and tastefully balanced for the design to be effective. An elevation composed of too many similar materials is ineffective and neutral. Likewise, an elevation that uses too many different materials is equally objectionable. In choosing materials for elevations, designers should not mix horizontal and vertical siding or different types of masonry. If brick is the primary masonry, brick should be used throughout. It should not be mixed with stone. Similarly, it is not desirable to mix different types of brick or stone.

Design Sequence for Creating Elevations

The first step in elevation design is to choose an architectural style. (Refer to Chapter 1.) Then sketch the outline of an exterior wall showing the roof shape and the position of doors, windows, and other key features such as chimneys or dormers.

Next, create a series of progressive sketches to develop an elevation design. Experiment with different roof styles, door and window designs, siding materials for the exterior walls, overhangs, chimney shapes, roof materials, and trim variations. Sketches can also show various architectural styles derived from the same floor plan. See Fig. 15-3.

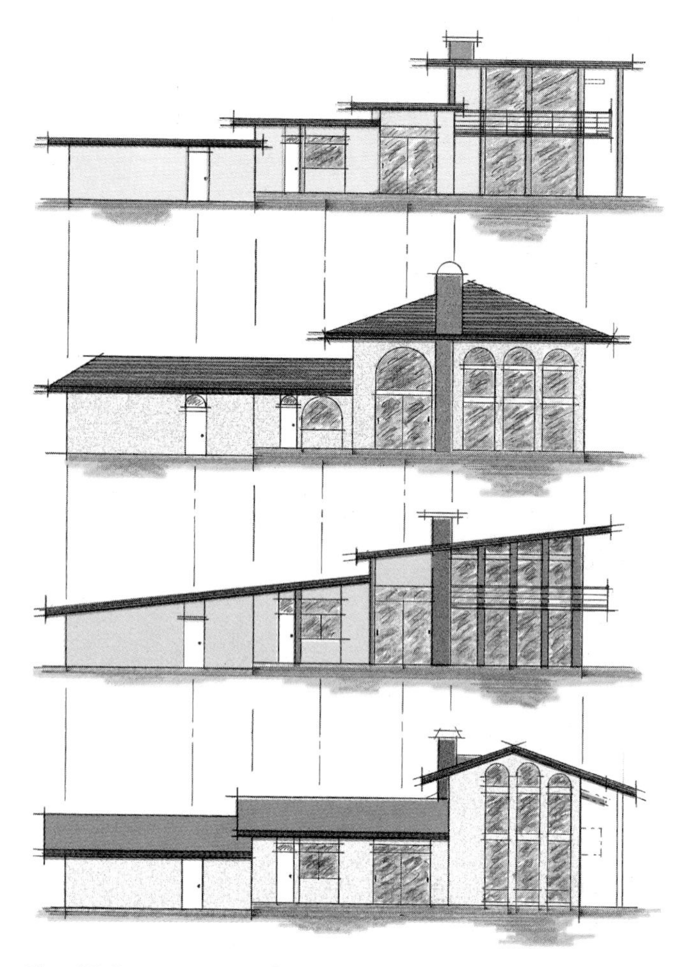

Fig. 15-3 ■ Progressive elevation line sketches using different architectural styles for the floor plan in Fig. 14-20.

Roof Style and Type

To design elevations, a designer needs to know roof styles and which style best matches the building's overall style. There are many styles of roofs. The gable, hip, flat and shed styles are the most popular. See Fig. 15-4. Other features which affect the appearance of the roof must also be considered. These include the size and shape of dormers, skylights, vents, chimneys, and cupolas. In addition to style, the overhang size and the roof pitch (angle) must be determined during the design process.

Roof framing plans are subsequently developed from the basic roof design that is developed. A roof plan shows the outline of the top view of a roof with solid ridge, valley, and chimney lines. Dashed lines represent the outline of the floor plan under the roof. Small arrows show the slope direction. Refer again to Fig. 15-4. (Detailed information on roof framing drawings is presented in Chapter 30.)

GABLE ROOFS A **gable** is the triangular end of a building. Roofs that fit over this area are gabled roofs, or, more simply, gable roofs. Gable roofs are the most common roof style because of their adaptability to a wide variety of architectural styles, from colonial to contemporary. They also drain and ventilate easily.

Variations of gable roofs include A-frames, winged, and pleated gables. A-frame roofs extend to the floor line, creating continuous ceilings and walls inside. Winged gable roofs are created by extending the ridge overhang further than the overhang at the corners. See Fig. 15-5, page 290. Pleated (folded plate) roofs consist of a series of aligned and connected small gable roofs. Refer again to Fig. 15-4.

HIP ROOFS Hip roofs provide eave-line protection around the entire perimeter of a building. The hip roof overhangs shade windows that would not be shaded at a gabled end. For this reason, hip roofs are very popular in warm climates. See Fig. 15-6, page 290. Another variation of the hip roof is the Dutch hip. A Dutch hip is created by extending the ridge outward to make a partial gable end at the top of the hip. Refer again to Fig. 15-4.

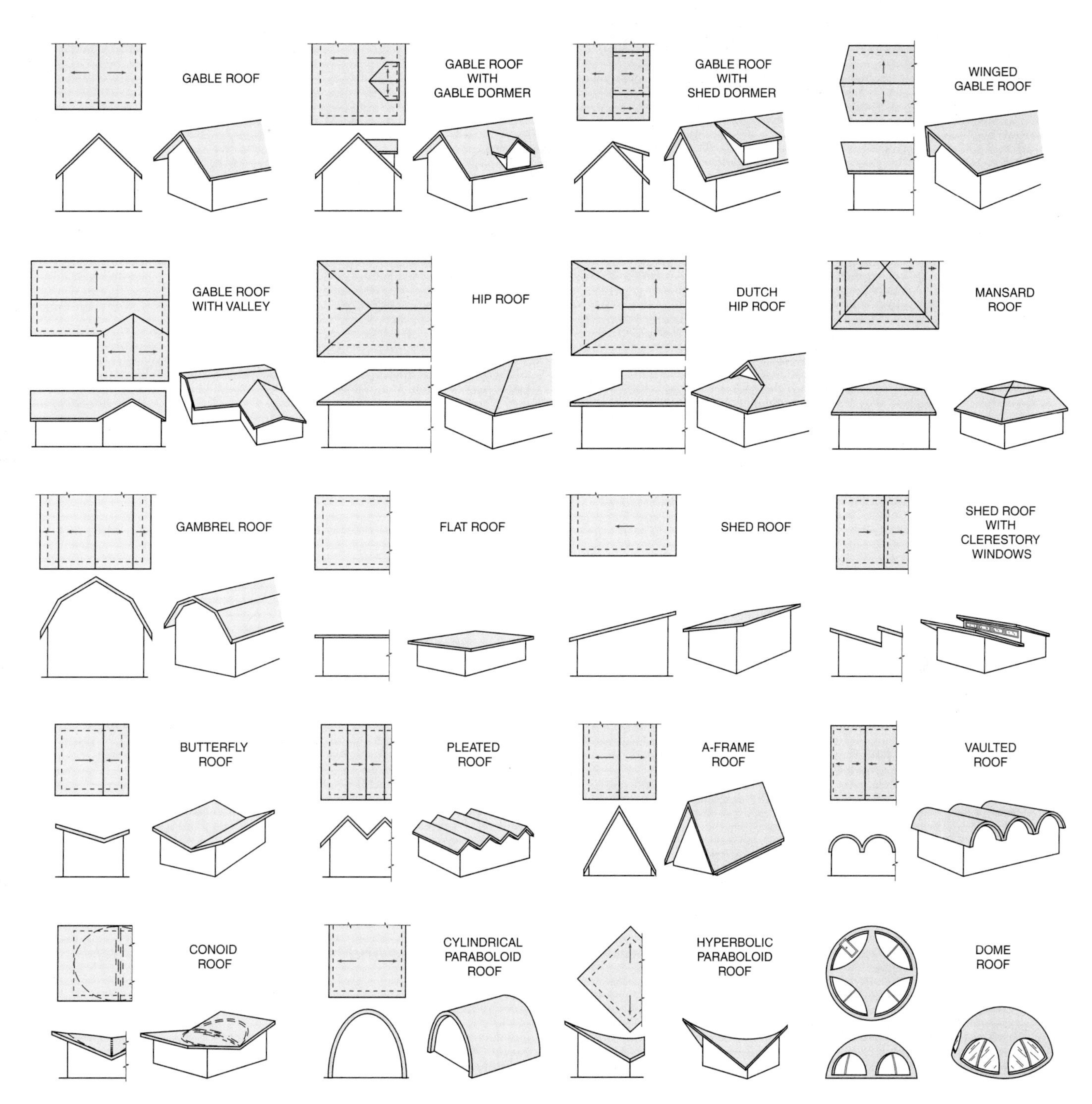

Fig. 15-4 ■ The roof plan and elevation views of many roof styles.

FLAT ROOFS When a low building silhouette is desirable, flat roofs are ideal. Flat roofs have a slight slope (⅛″ to ½″ per foot) for drainage, unless water is used as an insulator. They can also function as decks on multilevel structures. Flat roofs do not

Fig. 15-5 ■ Large winged gable roof supported by columns. *Two Creek Ranch, Fayetteville, TX—Lindi Shrovik, Coldwell Banker*

Fig. 15-6 ■ The house in this elevation drawing contains both gable and hip roofs. *John Henry, Architect; Jones Clayton, Construction*

have the structural advantage normally gained by rafters leaning on a ridge board. Heavier rafters (ceiling joists) are needed. Because of snow-load problems in cold climates, flat roofs are most popular in warm climates. See Fig. 15-7.

SHED ROOFS A shed roof is a flat roof that is slanted. If the down slope faces south, shed roofs are ideal for solar panels. When clerestory windows are added between offsetting shed roofs, light can be provided for the center of a building. Again, see Fig. 15-4. Many industrial buildings use multiple shed roofs and clerestory windows in a sawtooth pattern to maximize center light.

Fig. 15-7 ■ Flat roofs can also function as decks on multilevel structures. *Eagles Nest—Ponte Verdra Beach, FL; Robert C. Boward, Architect; E. Joyce Reesh, Arvida*

VAULTED ROOFS Vaulted roofs are curved panel roofs. They are composed of a series of manufactured curved panels which are erected side by side between two bearing walls. This arrangement allows for larger open areas since the curved construction is structurally stable.

BERMUDA ROOFS Bermuda roofs originated on the island of Bermuda in the Caribbean Sea. There the large fascia areas characteristic of the design are used to collect rainwater. This design effect is often used as a feature in other areas of the world. See Fig. 15-8.

DOME AND DOME-SHAPED ROOFS As you learned in Chapter 1, domes have been used in architecture for centuries. Dome roofs, like A-frames, provide both roof and walls in one structurally sound unit.

BUTTERFLY ROOFS Two shed roofs which slope to the center create a butterfly roof. This roof style allows for higher outside walls which can provide more light access.

GAMBREL ROOFS Gambrel roofs are double-pitched roofs. They are also known as barn roofs. Gambrel roofs are the distinguishing feature of Dutch colonial houses. They are used to create more headroom in one-and-one-half story homes.

MANSARD ROOFS Mansard roofs are double-pitched hip roofs with the outside constructed at a very steep pitch. This type of roof is used on French provincial homes. (Refer back to Fig. 1-9.)

Fig. 15-8 ■ The repetition of the large, Bermuda-like fascia and ridge caps is used as a major line emphasis in this house design. *Edna Carvin, Remax*

Because there is no need for internal support walls or columns, completely open floor space and flexible room sizes are possible.

The geodesic dome, developed by R. Buckminster Fuller, can be inexpensively mass produced at relatively low cost. Actually, geodesic domes are not true "domes." They are series of triangles which are combined to form hexagons and pentagons. See Fig. 15-9.

There are several restrictions to remember in designing dome structures. The use of domes for residential construction restricts the design to a predetermined area. Working with walls that are not plumb (exactly vertical) creates problems in fitting cabinets, fixtures, appliances, and furniture effectively into the design.

Fig. 15-9 ■ Epcot Center Geodesic Dome. Geodesic domes, as developed by Buckminster Fuller, can be built quite large because the weight is transmitted through the structural members to the circumference of the base. *James Eismont Photo*

NEW TECHNOLOGY AND ROOF STYLES The development of new building materials and methods in molded plywood, plastics, and reinforced concrete has led to the development of many shapes of roofs. The conoid, cylindrical parabolic, and hyperbolic parabolic roofs are among the most recent designs. Again, refer to Fig. 15-4.

ROOF OVERHANG The **overhang** is the portion of the roof that projects past the outside walls. Sufficient roof overhang should be provided to afford protection from the sun, rain, and snow. See Fig. 15-10. The length and angle of the overhang greatly affects a roof's appearance and ability to provide protection. Figure 15-11 shows

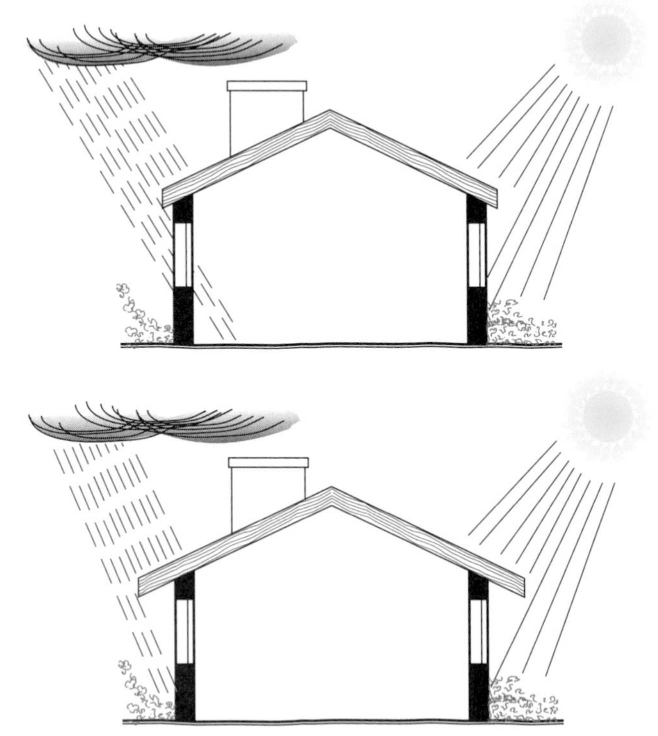

Fig. 15-10 ■ Effects of long and short overhang on sun and rain protection.

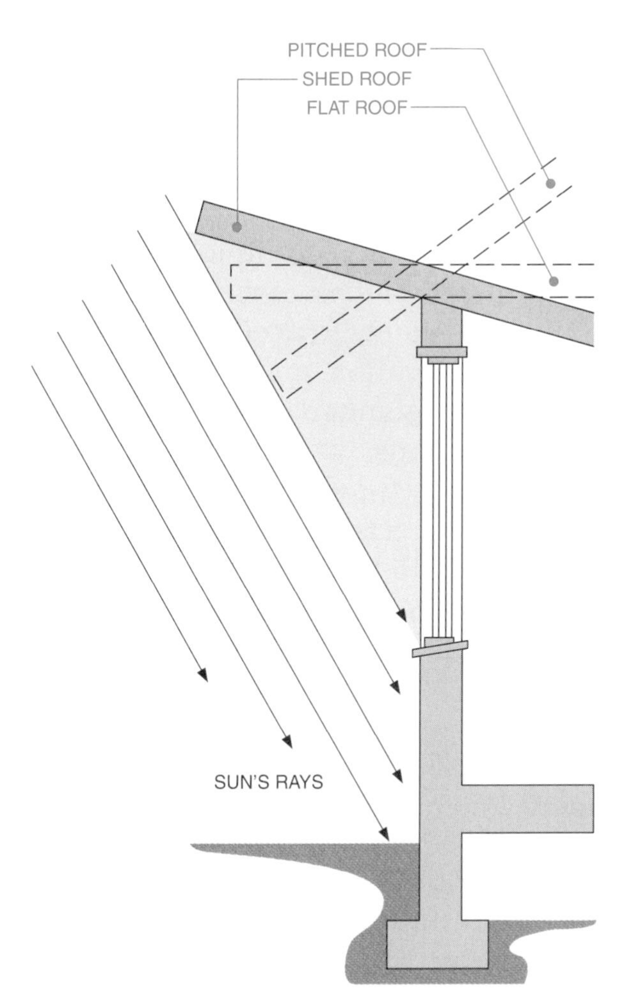

Fig. 15-11 ■ The angle of the roof determines how long the overhang needs to be.

Fig. 15-12 ■ The shape of the overhang may vary. *Western Wood Products*

that when the pitch is low, a larger overhang is needed to provide protection. However, if the overhang of a high-pitch roof is extended to equal the protection of the low-pitch overhang, it may block the view from the building. To provide protection and, at the same time, allow sufficient light to enter the windows, slatted overhangs may be used.

The fascia edge of an overhang does not always need to parallel the sides of a house. See Fig. 15-12. The amount of overhang is also determined by architectural style. Large gable end overhangs, such as the 9′ overhang shown in Fig. 15-13, must be supported by columns. Some gable

Fig. 15-13 ■ This large overhang provides protection to the lanais area below.

overhangs are enclosed to form a soffit as shown in the photo on page 284.

ROOF PITCH In designing roofs, the pitch or angle of the roof must be determined. The **slope** is the ratio of the rise over the run. The **run** is the horizontal distance between the ridge and the outside wall. The **rise** is the vertical distance between the top of the wall and the ridge. The run is always shown in units of 12. Therefore a slope of 6/12 means the roof rises 6″ for every 12″ of run. The **pitch** is the ratio of the rise over the span. Pitch is covered in more detail in Chapters 16 and 30.

Window Styles

Windows are designed and located to provide light, ventilation, a view, and—in some climates—heat. To accomplish these goals, the size, location, and shape of each window must be planned according to the following guidelines:

1. Relate window lines to the elevation shape to avoid a tacked-on look, as shown in Fig. 15-2A.
2. Plan window height to allow for furniture and built-in components that are placed near windows. See Fig. 15-14.
3. Plan window sizes to match available standard sizes. Refer back to Fig. 14-13.
4. Decide which windows need to open for ventilation and which should be fixed. See Fig. 15-15.

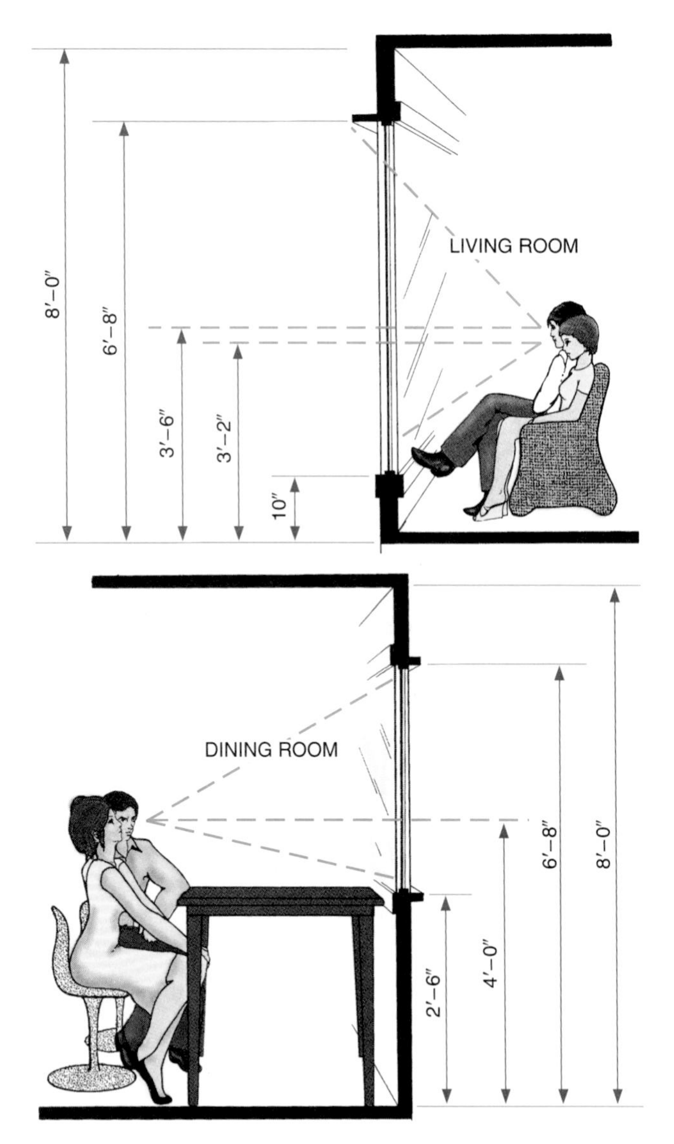

Fig. 15-14 ■ Position windows according to the functions of interior areas. *Small Homes Council*

Fig. 15-15 ■ Locating awning windows below fixed glass windows provides both ventilation and light. *Kolbe & Kolbe Inc.*

5. Be sure each window functions from the inside as required.
6. Position windows to access the best views. Avoid window placement that exposes undesirable views. Avoid mullions and muntins if they restrict views. See Fig. 15-16.
7. In warm climates, minimize the amount of window space on the south and maximize north-facing windows. Do the reverse in cold climates.
8. Keep the window style consistent with the architectural style of the house.
9. Where possible, align the tops of all windows and doors in each elevation.
10. If the building has more than one level, vertically align the sides of windows where possible.
11. Don't allow small areas between windows and other major features such as fireplaces. Design windows and major components to fill spaces.
12. If windows are to provide the entire light source during daylight hours, twenty percent of the room's floor area should be windowed. Ten percent is considered minimum.
13. Windows that provide ventilation should be located to capture prevailing breezes and provide the best air circulation.
14. If possible, locate windows on more than one wall in each room, to provide for the best distribution of light and ventilation.
15. **Fenestration** is the arrangement of windows or openings in a wall. Arrange fenestration patterns to conform to the elements of design. See Fig. 15-17.

Fig. 15-16 ■ Mullions and muntins may be used as decorative features. *Kolbe & Kolbe Inc.*

Fig. 15-17 ■ A well-designed elevation with window and door openings aligned. *Eagle Window and Door*

Fig. 15-18 ■ Interior view of common window types.

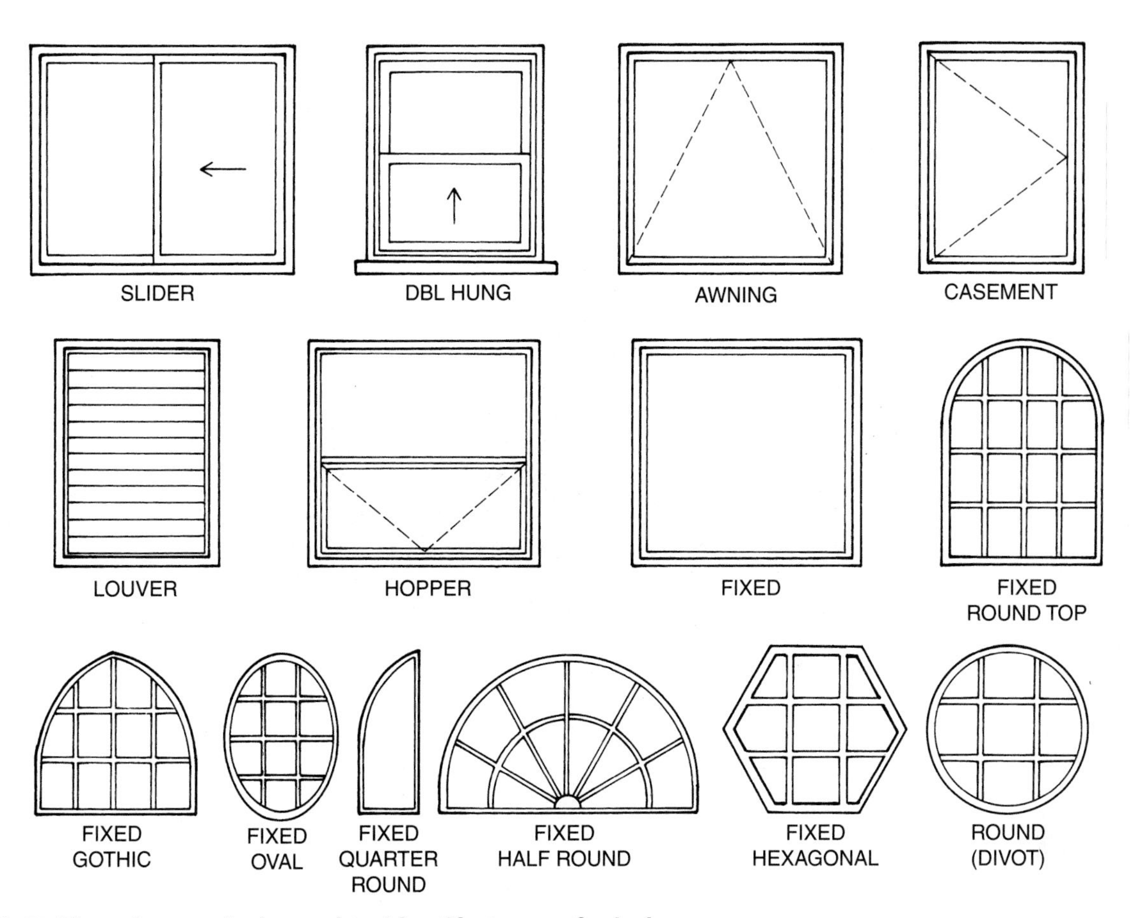

Fig. 15-19 ■ Elevation symbols used to identify types of windows.

WINDOW TYPES Windows slide, swing, pivot, or remain fixed. Choosing the right window type for each need requires a knowledge of the function and operation of each type. See Fig. 15-18. Figure 15-19 shows the elevation symbols for the most common window types.

Door Styles

The style, size, and location of doors do not have as great an effect on elevation design as windows. This is because of the limited options among door sizes and for door locations. Usually an elevation will either have only one door or will contain no doors at all. Nevertheless, the principles of placement relative to other elevation features is the same for doors as for windows. Refer back to Fig. 15-2.

Door types fall into three main categories: exterior, interior, and garage doors. Exterior doors provide security and visual privacy. Interior doors provide privacy and sound control between rooms. Figure 15-20 shows an elevation view of the most common types of interior and exterior doors.

Detailed information about doors is contained in a door schedule and cross-referenced to floor plans and/or elevations. Door framing information is presented in Chapter 29.

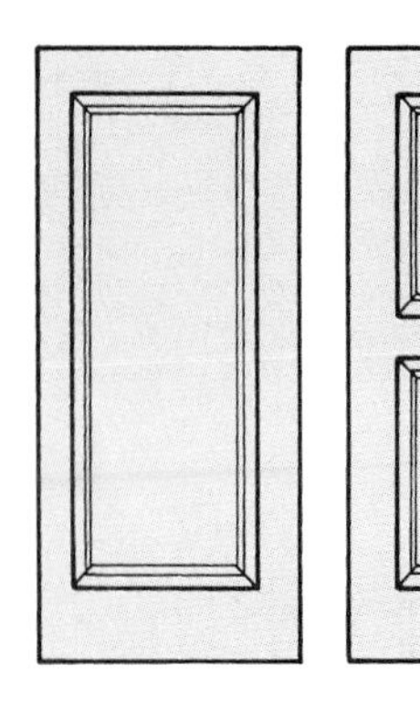

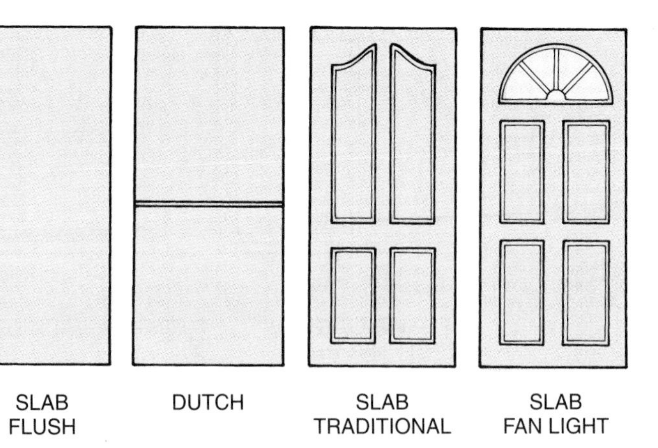

Fig. 15-20 ■ Common types of doors as represented on elevation drawings.

CHAPTER 15 Exercises

1. Sketch an elevation of your own design. Trace the elevation four times drawing in a flat roof, gable roof, shed roof, and butterfly roof. Choose the one you like best and the one that is most functional for your design. Explain why you made that choice.
2. Sketch the front elevation of your home or a home you like. Change the roof style, but keep it consistent with the major lines of the elevation. Move or change the doors and windows to improve the design.
3. Redesign the elevation from Exercise 2, moving the doors and windows and changing the materials. Be sure the door and window lines relate to the major lines of the building.

4. Complete Exercise 1 using a CAD system.
5. Collect pictures of roofs, windows, and doors that you particularly like. Try to identify house styles for which they are best suited.

CHAPTER 16

Drawing Elevations

The main features of the outside of a building are shown on elevation drawings. **Exterior elevation drawings** are orthographic representations of the exterior of a structure. These drawings are prepared to show the design, materials, dimensions, and final appearance of the structure's exterior components. In a building, these components include doors, windows, the surfaces of the sides, and the roof. Interior as well as exterior elevation drawings are projected from floor plans. Dimensions are used to show sizes, and elevation symbols are used to indicate various features on the drawings.

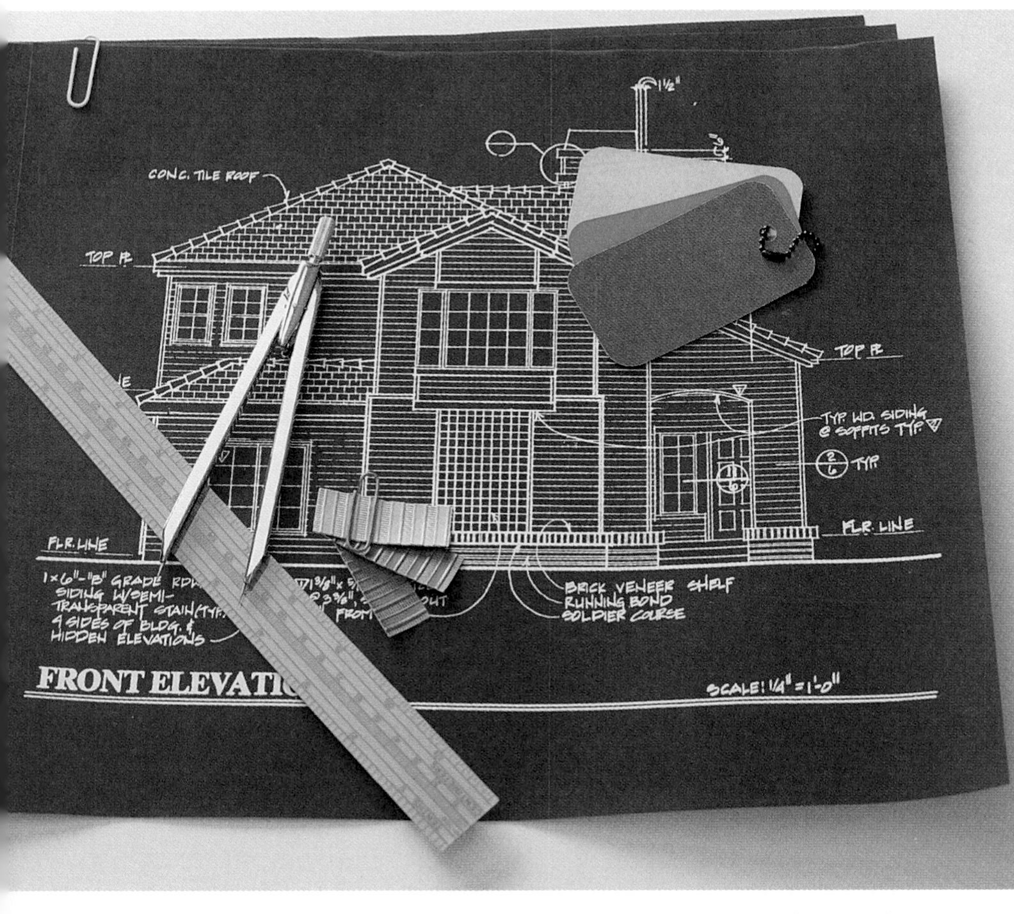

■ **An elevation drawing of a structure shows a direct view of one side or wall.** *Image Bank/G.K. & Vikki Hart*

Objectives

In this chapter you will learn:

- to follow steps to project elevations from a floor plan and complete an elevation drawing.
- to draw accurately scaled and dimensioned elevations.
- to mathematically establish the pitch of a roof.
- symbols used on elevations.
- pictorial drawing and rendering techniques to use on elevations.

Terms

auxiliary elevation
datum line
exterior elevation drawings
finished dimensions
framing dimensions
interior elevation drawings
orthographic projection
presentation drawings
profile drawings
slope diagram
span

Elevation Projection

In **orthographic** (multiview) **projection,** related views of an object are shown as if they were on a two-dimensional, flat plane. To visualize and understand orthographic projection, imagine a building surrounded by a transparent box, as shown in Fig. 16-1. If you draw the outline of the structure on the transparent planes that make up the box, you may create several orthographic views. For example, the front view is on the front plane, the side view on the side plane, and the top view on the top (horizontal) plane. If the planes of the top, bottom, and sides were hinged and swung out away from the box, as shown in Fig. 16-2, six views of the house would be created. Note how each view is positioned on an orthographic drawing. Study the position of each view as it relates to the front view. The right side is to the right of the front view. The left side is to the left, the top (roof) view is on the top, and the bottom view is on the bottom. The rear view is placed to the left of the left-side view, since, if this view were hinged around to the back, it would fall into this position.

Notice that the length of the front view, top (roof) view, and bottom view are exactly the same as the length of the rear view. Notice also that the heights and alignments of the front view, right side, left side, and rear view are the same. Memorize the position of these views and remember that the lengths of the front,

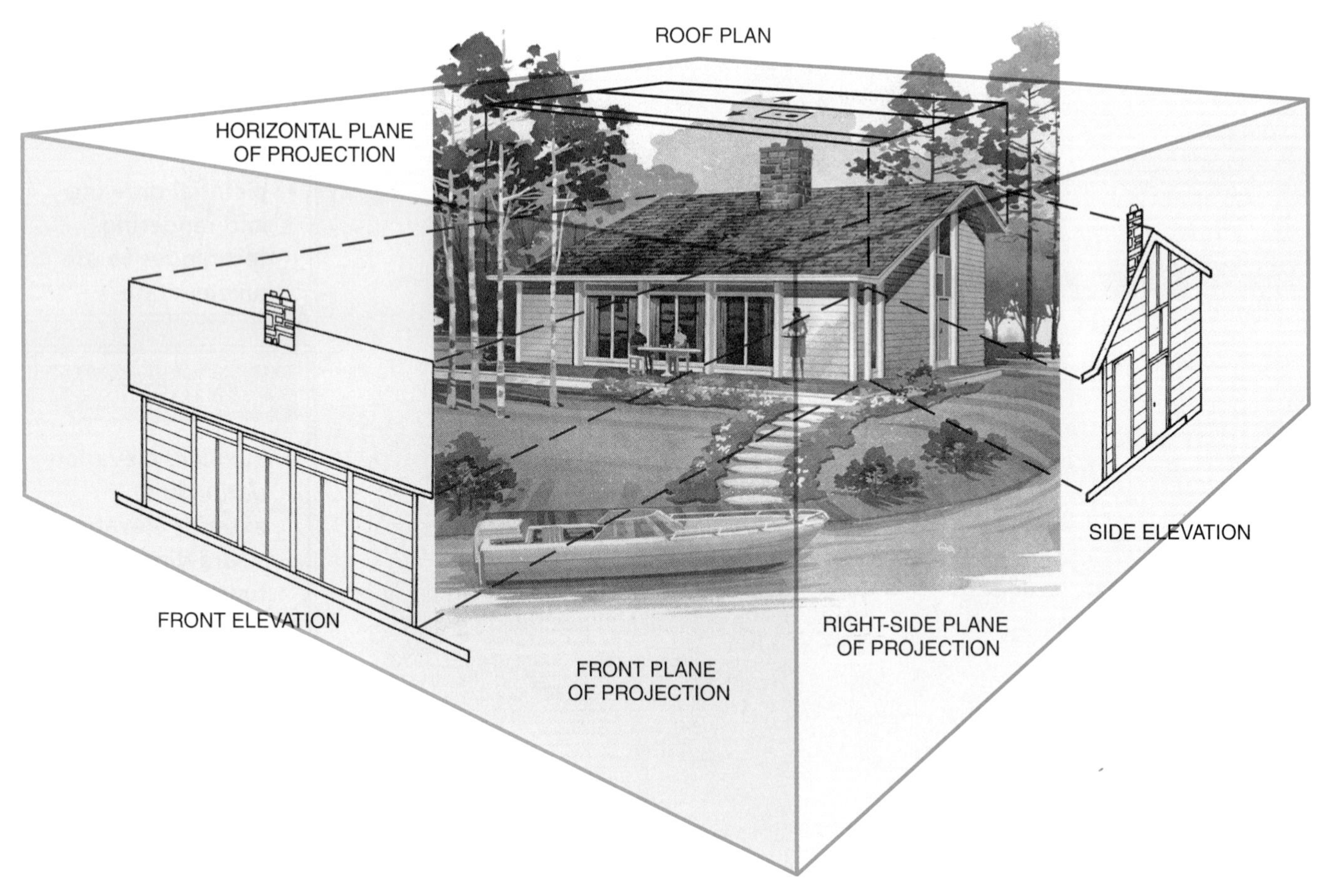

Fig. 16-1 ■ A projection box shows three planes of a building.

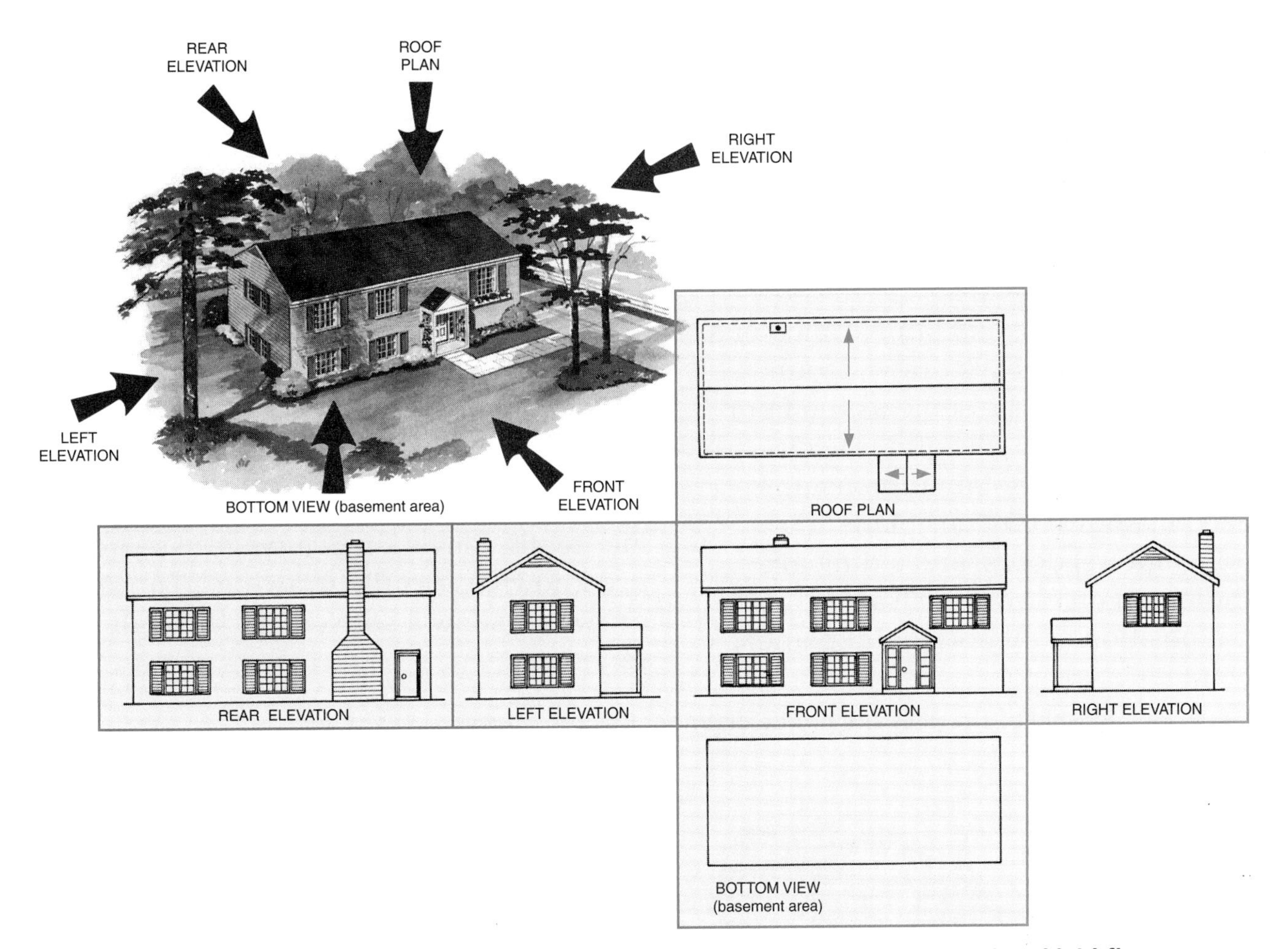

Fig. 16-2 ■ Six orthographic (multiview) views are shown when the box is opened and laid flat.

bottom, and top views are *always* the same. Similarly, the heights of the rear, left, front, and right side are *always* the same.

All six views are rarely used to depict architectural structures. Instead, only four elevations (sides) are usually shown. The top roof view is used to create floor plans. The roof plan is developed from the top view. The bottom view of a floor is not developed. Instead the foundation underneath the structure is described by a floor plan and elevation section.

Figure 16-3 shows how elevations are projected from the floor plan. The positions of the chimney, doors, windows,

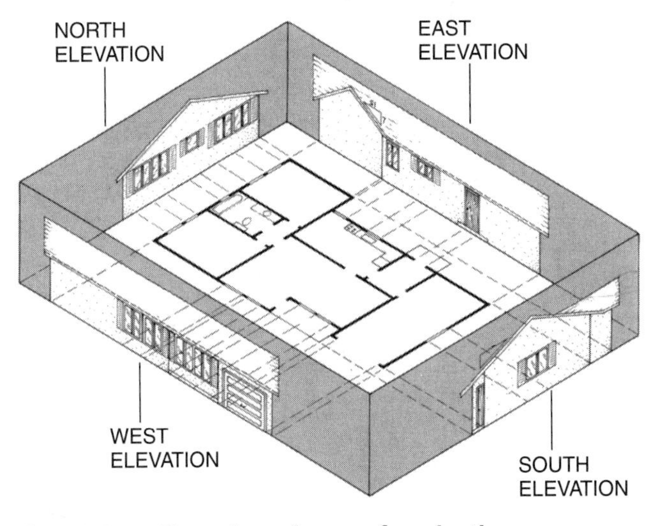

Fig. 16-3 ■ Elevation planes of projection.

overhang, and building corners are projected directly from the floor plan to the elevation plane.

Purposes of Projected Elevation Drawings

Elevation drawings are used to show the design of the finished appearance of a structure. Elevations are drawn to an exact scale, usually the same as the floor plan. For this reason, elevations accurately represent all height dimensions that are not shown on floor plans. The style of windows, doors, and siding are also indicated on elevation drawings. The vertical position of all horizontal planes, such as ground lines, floor lines, ceiling lines, deck or patio lines, and roof lines are only revealed on elevation drawings. Lines below the ground line such as foundation and footing lines are drawn by dashed lines.

Only through the use of elevation drawings can the vertical relationship of building components be visualized. For example, on the site plan in Fig. 16-4A, heights are shown by the contour lines. However, little height detail is apparent without the elevation drawing shown in Fig. 16-4B. Elevation drawings of a site are known as **profile drawings.** These drawings show a section cut through the terrain.

Fig. 16-4 ■ Relationship of site plan and profile drawing. *Bracken, Arrigoni, & Ross, Inc.*

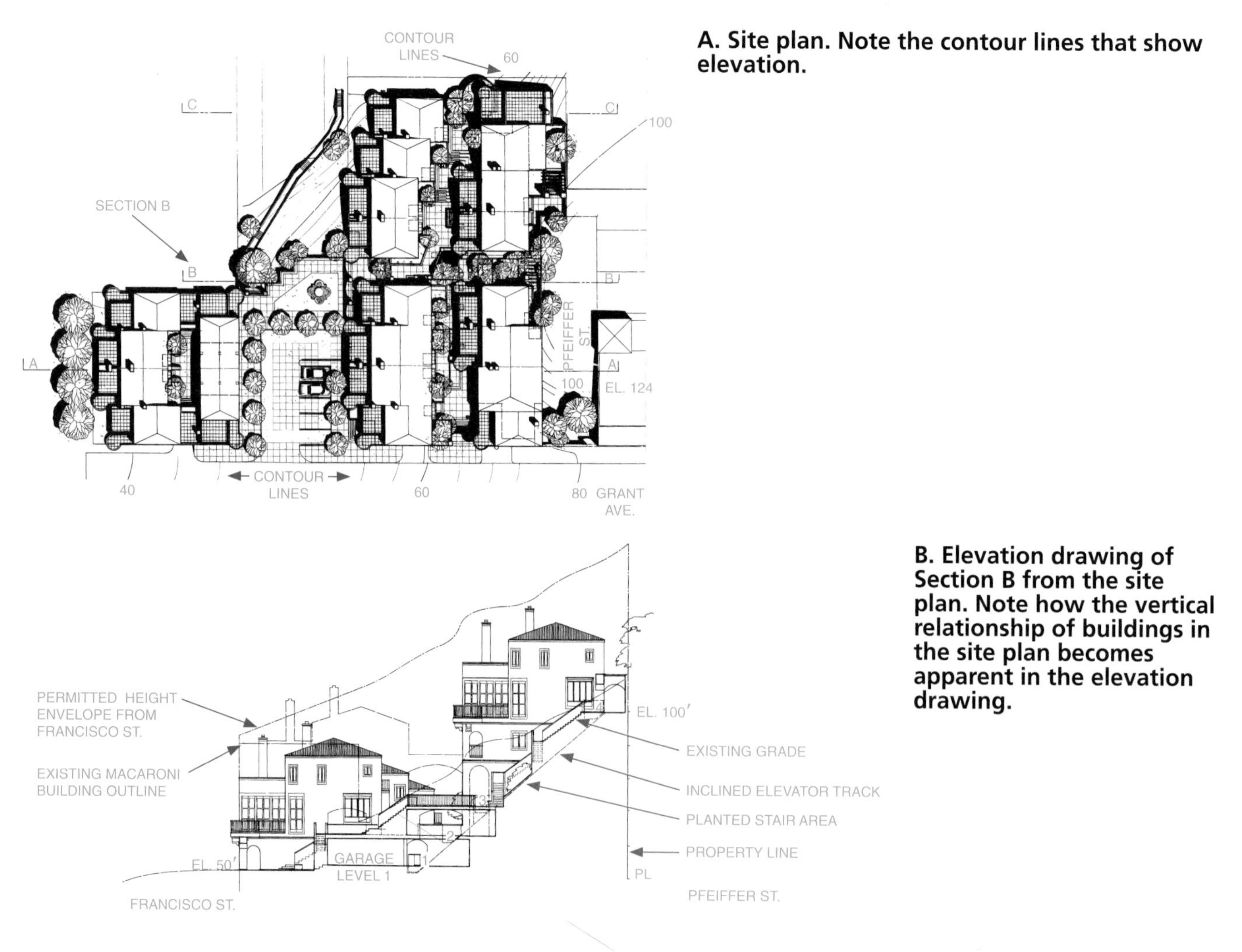

A. Site plan. Note the contour lines that show elevation.

B. Elevation drawing of Section B from the site plan. Note how the vertical relationship of buildings in the site plan becomes apparent in the elevation drawing.

Drawing Elevations from a Floor Plan

Think of an elevation as a drawing placed on a flat, vertical plane. Figure 16-5 shows how a vertical plane is related to and projected from a floor plan.

Orientation

Four elevations are normally projected by extending lines outward from each wall of the floor plan. When these elevations are classified according to their location, they are called the front, rear (or back), right, and left elevation. When these elevations are projected on the same drawing sheet, the rear elevation appears to be upside down and the right and left elevations appear to rest on their sides. See Fig. 16-6. Because of the large size of most combined floor plan and elevation drawings, and because of the need to show elevations as normally seen, the elevation drawing is rotated so each elevation can be drawn with the ground line on the bottom. See Fig. 16-7.

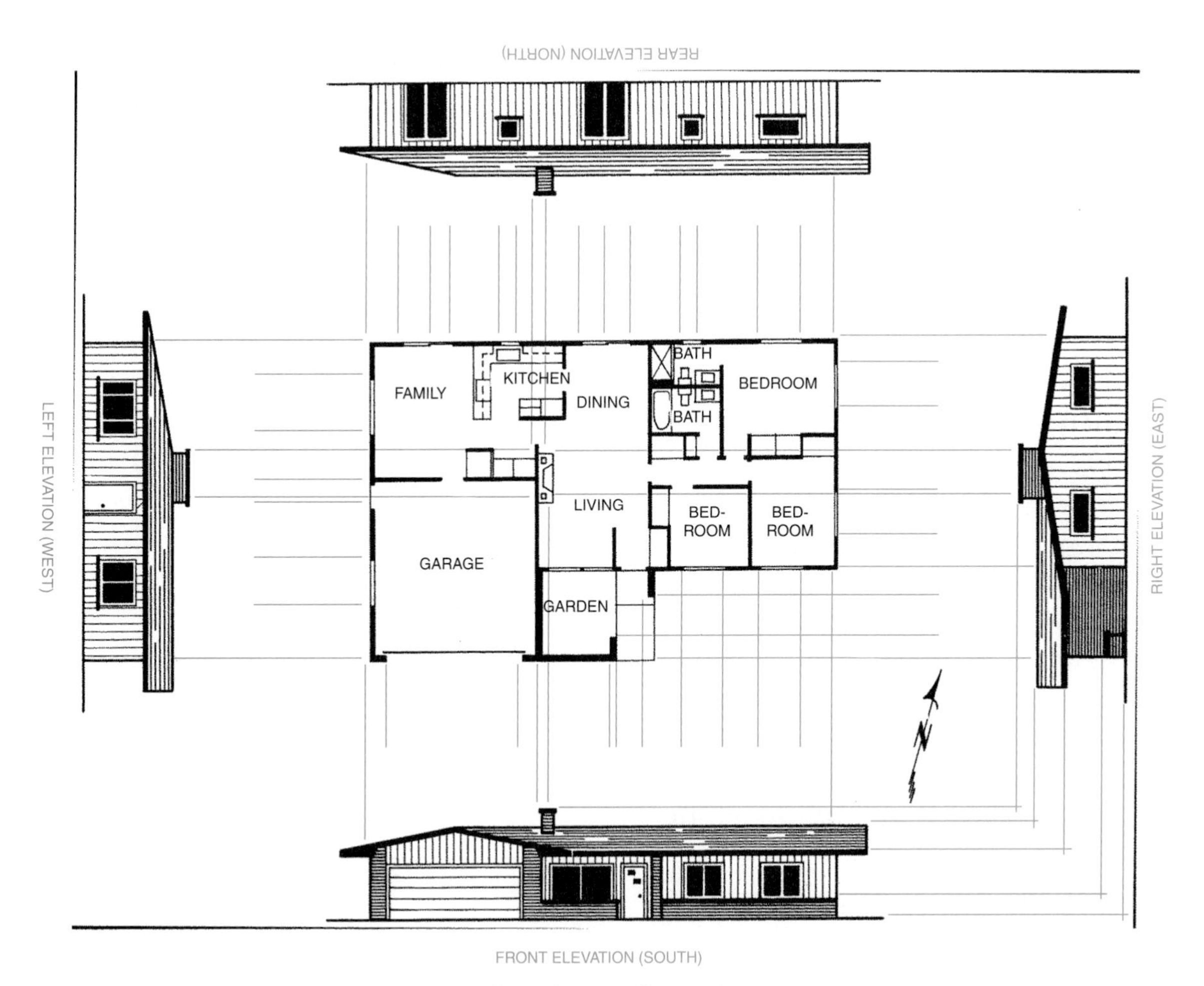

Fig. 16-5 ■ Projection of vertical elevation plans from a floor plan.

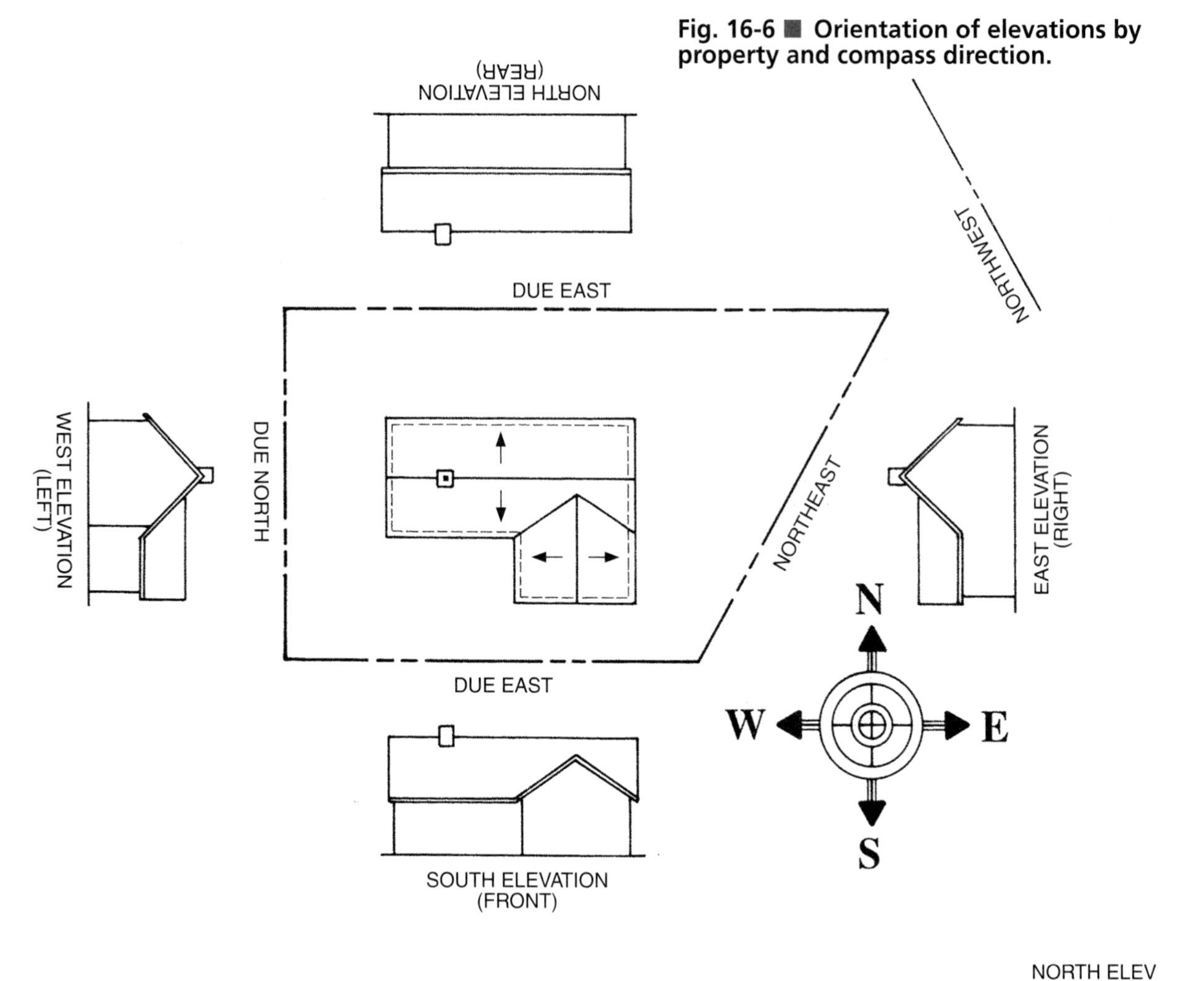

Fig. 16-6 ■ Orientation of elevations by property and compass direction.

The north, east, south, and west compass points are often used to describe and label elevation drawings. This method is preferred because it reduces the chance of elevation callout error. When this method is used, the north arrow on the floor plan or site plan is the key. For example, in Fig. 16-6, the rear elevation is facing north. Therefore, the rear elevation is also the north elevation. The front elevation is the south elevation, the left elevation is the west elevation, and the right elevation is the east elevation.

When elevations do not align exactly with the four major compass points, a split compass reading may be used. See Fig. 16-8.

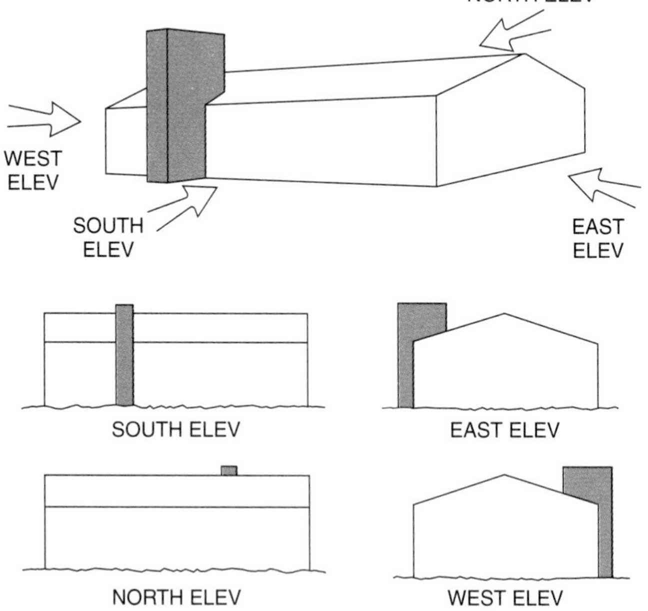

Fig. 16-7 ■ Elevations are drawn with the ground line at the bottom in a horizontal position.

Fig. 16-8 ■ Elevations for the house require split compass labeling.

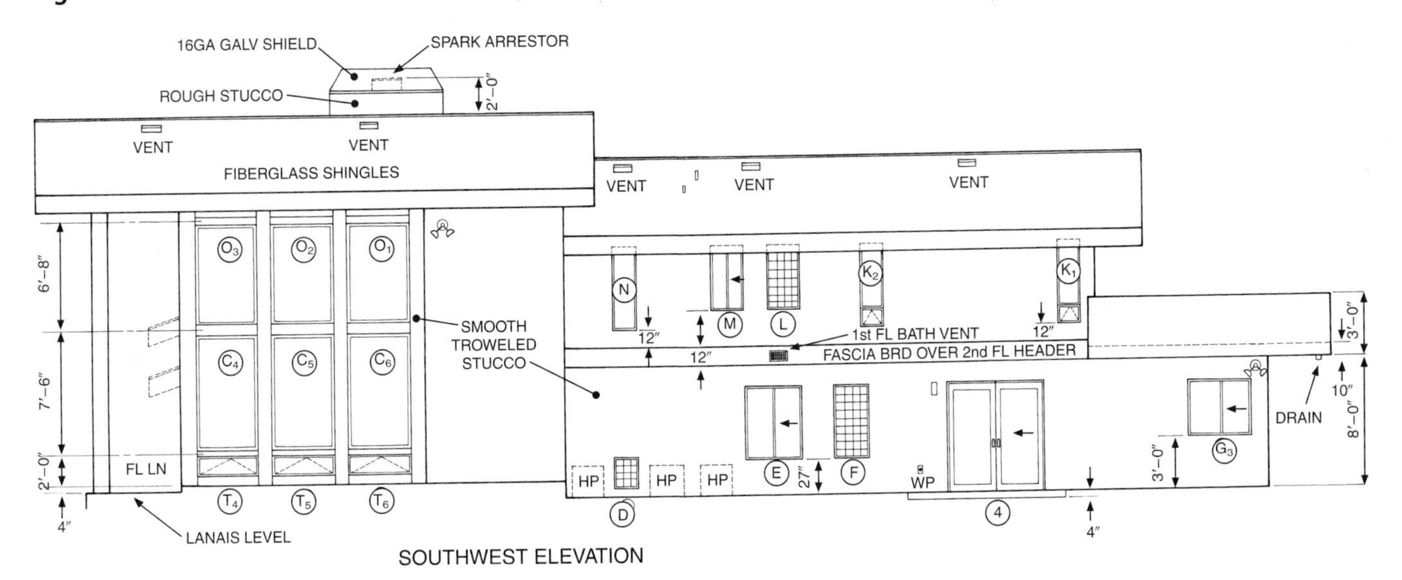

A. Southwest elevation.

B. Elevation under construction. *Diane Kingston Photo*

C. Compare features as shown on the drawing with the actual appearance in the finished structure. *Diane Kingston Photo*

Auxiliary Elevations

A floor plan may have walls at angles which deviate from the normal 90°. Then some lines and surfaces on the elevations may appear shortened because of the receding angles. An **auxiliary elevation** view may then be necessary to clarify the true size of the elevation. To project an auxiliary elevation, follow the same projection procedures as for other elevation drawings. See Fig. 16-9. When an auxiliary elevation is drawn, it is prepared in addition to—and does not replace—other standard elevation drawings.

Auxiliary elevations are also used to describe vertical design features of separate structures which cannot be shown on plan views. See Fig. 16-10.

Fig. 16-9 ■ Project the auxiliary elevation perpendicular to the wall of the floor plan from which you are projecting. *Home Planners, Inc.*

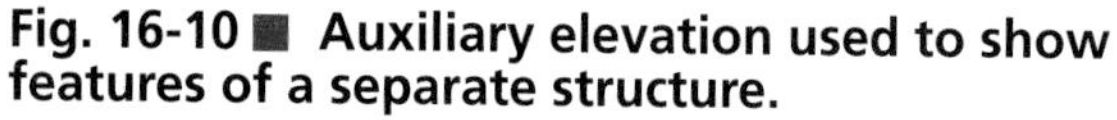

Fig. 16-10 ■ Auxiliary elevation used to show features of a separate structure.

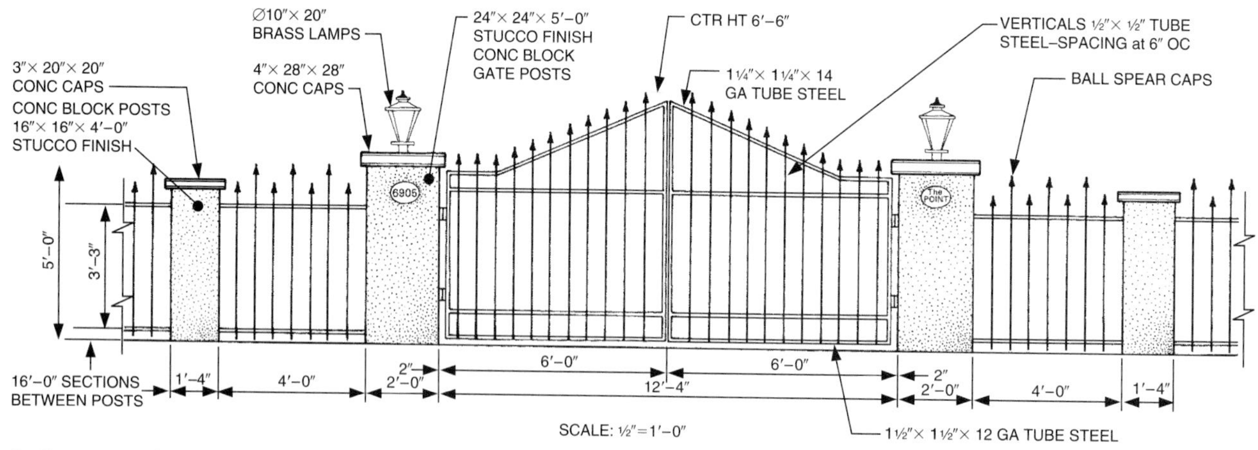

A. Fence and gate elevation.

B. Finished structure.

Elevations and Construction

Framing elevations show the position, type, and size of members needed for constructing the framework of a structure. When these are not prepared, builders rely solely on exterior and interior elevations for the height of framing members. This means that precise dimensions on the elevation drawings are crucial for accurate construction. Figure 16-8 shows an exterior elevation drawing that was drawn from the floor plan shown in Fig. 14-20. It also shows the related elevation under construction and in completed form. Observe and match the features of the elevation drawing with the framing details and appearance of the finished exterior. Look ahead also to Fig. 22-2 for another view of the same house.

Steps in Projecting Elevations

The major lines of an elevation drawing are derived by projecting vertical lines from the floor plan to the elevation drawing plane and measuring the position of horizontal lines from the ground line. To develop an elevation drawing which exactly reflects the features of a floor plan, refer to Fig. 16-11 and follow these steps:

1 Using the floor plan, *project the vertical lines* that represent the main lines of the building. These lines show the overall length or width of the building. They also show the width of doors, windows, and the major parts or offsets of the building.

When projecting an elevation on a CAD system, use the grid pick function to project the major lines from the floor plan to the elevation plane. During the drawing process, floor plans can be rotated 90° to position each elevation

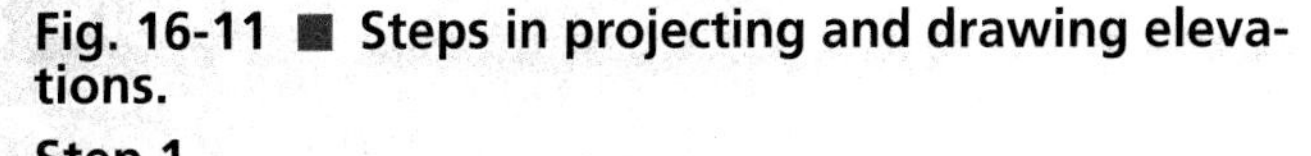

Fig. 16-11 ■ Steps in projecting and drawing elevations.

Step 1

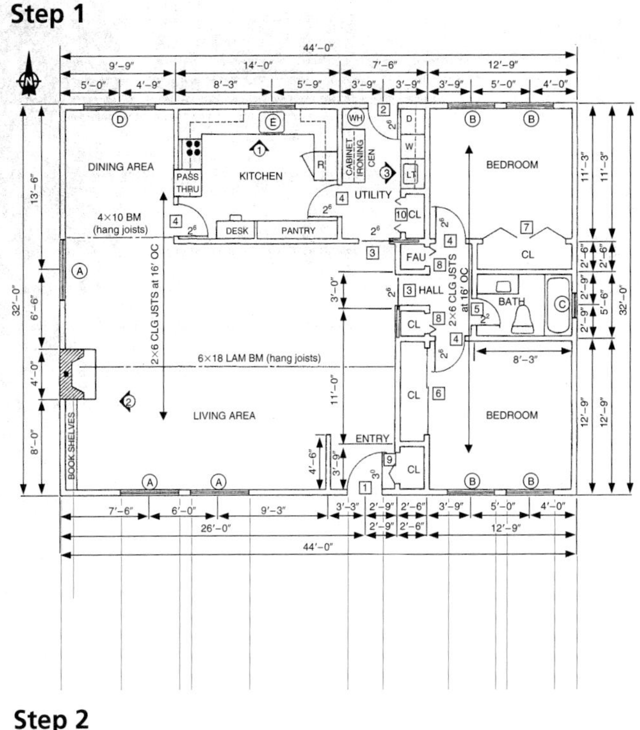

Step 2

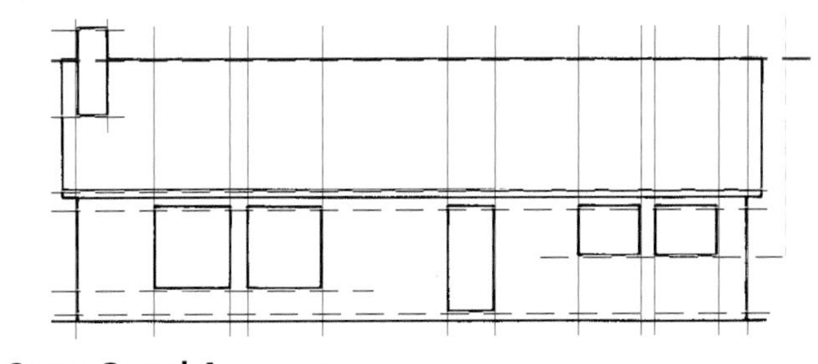

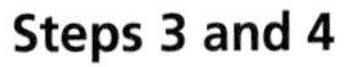

Steps 3 and 4

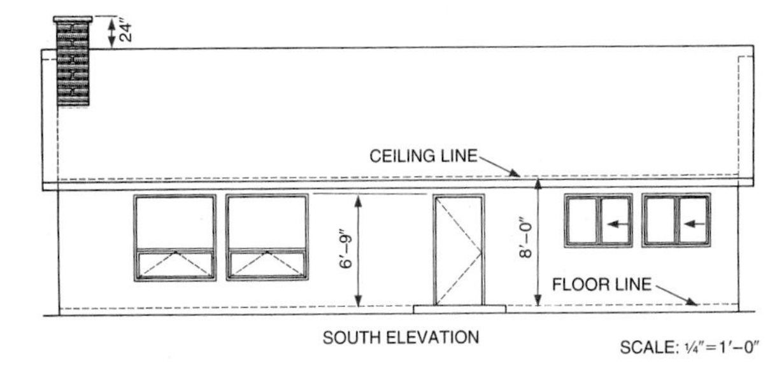

Steps 5 and 6

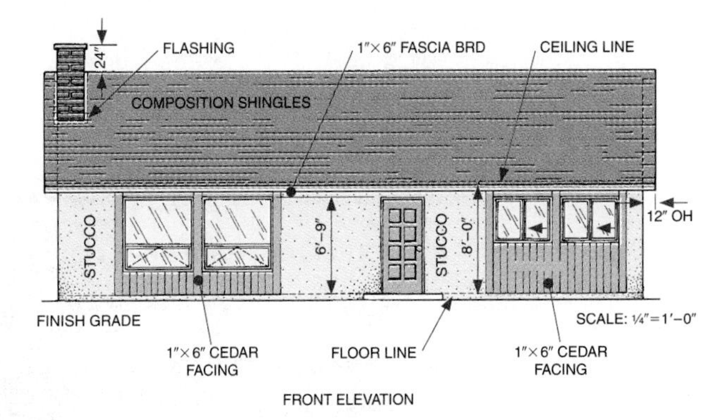

with the ground line on the bottom during the drawing process. Some architectural software can create elevations from floor plans if height dimensions are input.

2 *Measure and project horizontal lines* that represent the height of the ground line, footing, doors, tops and bottoms of windows, chimney, siding, breaks, planters, and other key features. To eliminate the repetition of measuring each of these lines for each elevation, a sheet showing the scaled lines is often prepared. See Fig. 16-12.

3 *Complete the basic elevation.* First, develop the roof elevation projection. In order to determine the height of the eave and ridge line, the roof slope (angle) must be established. On a high-slope roof, there is a greater distance between the ridge line and the eave line than on a low-slope roof. See Fig. 16-13. Pitch is the angle of the roof described in terms of the ratio of the rise over the span (rise/span) described

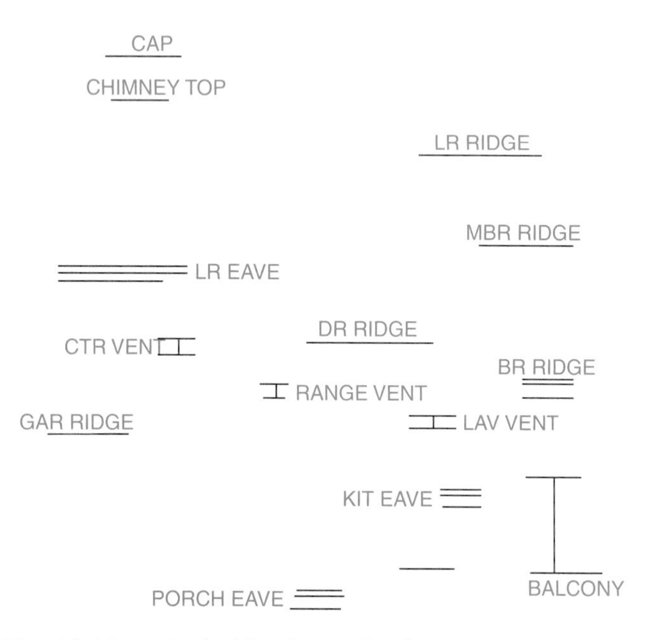

Fig. 16-12 ■ Scaled horizontal reference lines can be transferred to all elevations.

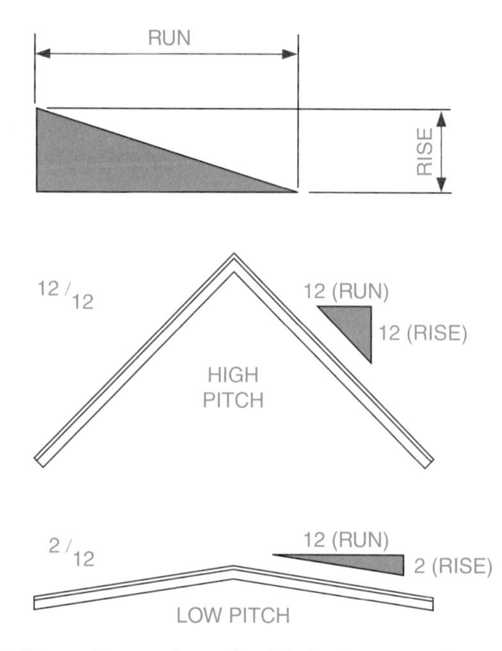

Fig. 16-13 ■ Examples of a high-slope roof and a low-slope roof.

in terms of the ratio of the rise over the run (rise/span). Span is the horizontal distance covered by a roof. Rise is the vertical distance. The run is always expressed in units of 12. The span is the run doubled. Therefore the span is always expressed in units of 24.

After the pitch is established, a **slope diagram** must be drawn on the elevation, as shown in Fig. 16-14. The slope diagram is developed on the working drawings by the drafter. The carpenter must work with the pitch fraction (ratio) to determine the angle of the rafters from a pitch angle table, so the ends of the rafters can be correctly cut. Double the run to find the span. The **span** is the distance between the supports of the roof. It is a constant of 24. Place the rise over the span (24) and reduce if necessary. This fraction is used by the carpenter to determine the rafter angle in degrees.

A roof elevation can be projected from a roof plan. See Fig. 16-15. Note that the end of every eave, and every

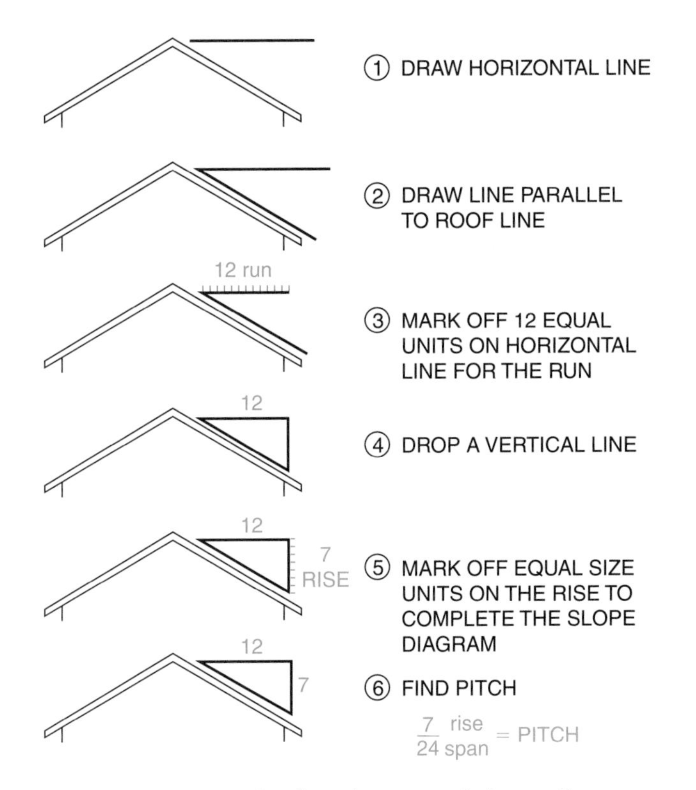

Fig. 16-14 ■ Steps in drawing a roof slope diagram.

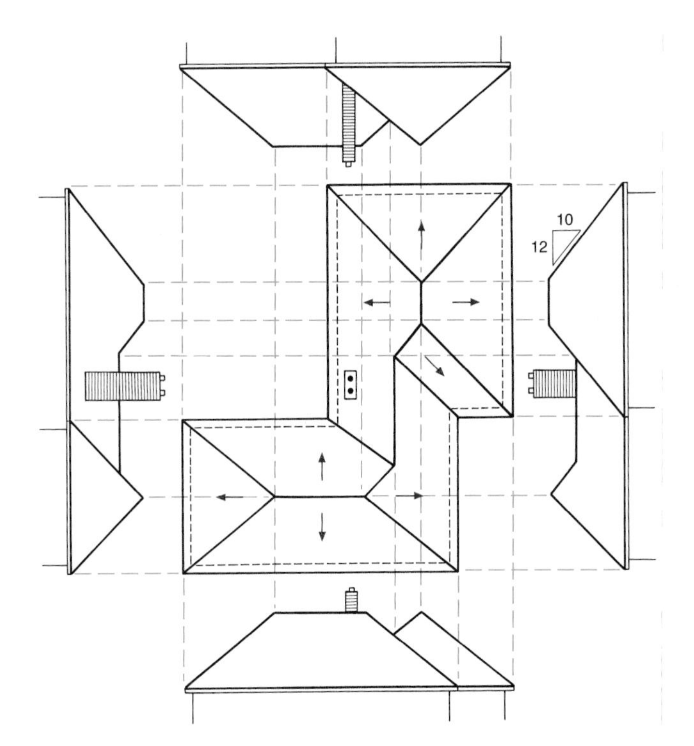

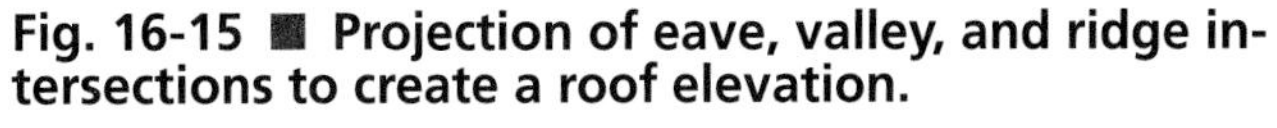

Fig. 16-15 ■ Projection of eave, valley, and ridge intersections to create a roof elevation.

valley and ridge intersection is projected at a right angle to the plan view outline. Figure 16-16 shows a comparison of roof dimensions between elevation views of common roof types.

4 *Establish the intersection of all vertical and horizontal lines,* including the eave and ridge line. These represent the outline of all features to be shown on the elevation. After they are established, darken the lines to identify the position of each.

5 *Add details and symbols,* such as indicating door and window trim, mullions, muntins, siding, and roofing materials.

6 *Add final dimensions, labels, and notes.*

Many different elevation styles can be projected from one floor plan. The roof style, pitch, overhang, grade level, windows, chimney, and doors can all be manipulated to create different effects.

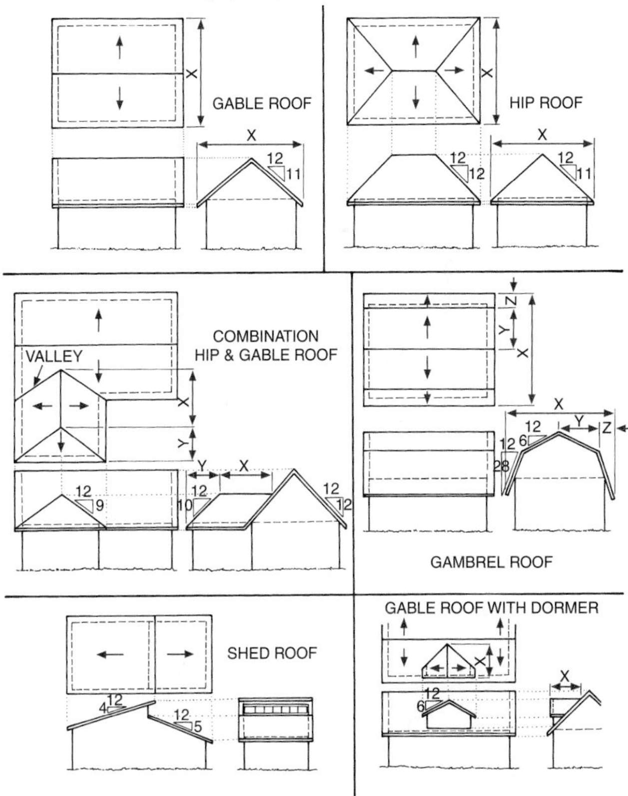

Fig. 16-16 ■ Note how the X, Y, and Z dimensions on each roof plan are the same as those on the corresponding elevation drawing.

Elevation Symbols

Symbols are needed to clarify and simplify elevation drawings. They help to describe the basic features of an elevation. Some symbols identify door and window styles and positions. Standardized patterns of dots, lines, and shapes show the types of building materials used on exterior walls. These kinds of symbols on an elevation make the drawing appear more realistic. See Fig. 16-17.

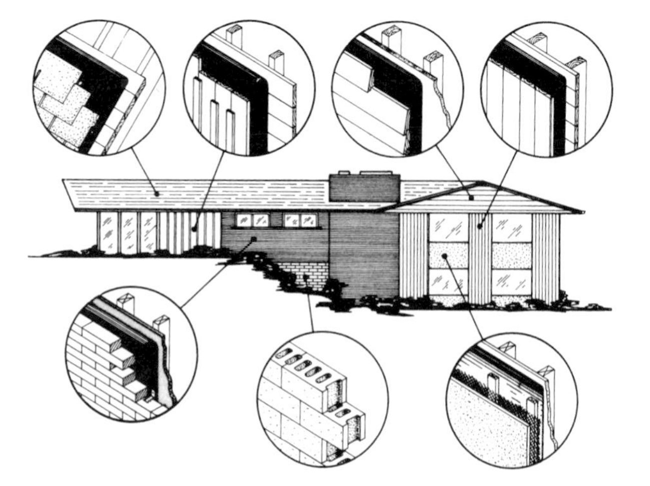

Fig. 16-17 ■ Relationship between material symbols used on an elevation and the actual material as it is used in construction.

Material Symbols

Most standard architectural symbols resemble the material they represent. However, in many cases the symbol does not show the exact appearance of the material. For example, the symbol for brick does not include all the lines shown in a pictorial drawing. Refer again to Fig. 16-17. Representing brick on an elevation drawing exactly as it appears is a long, laborious, and unnecessary process. Many elevation symbols appear as if the material were viewed from a distance.

When using a CAD system, elevation symbols such as doors and windows can be stored in symbol libraries. For example, the material symbols function can be used to add siding material symbols on elevation surfaces. See Fig. 16-18.

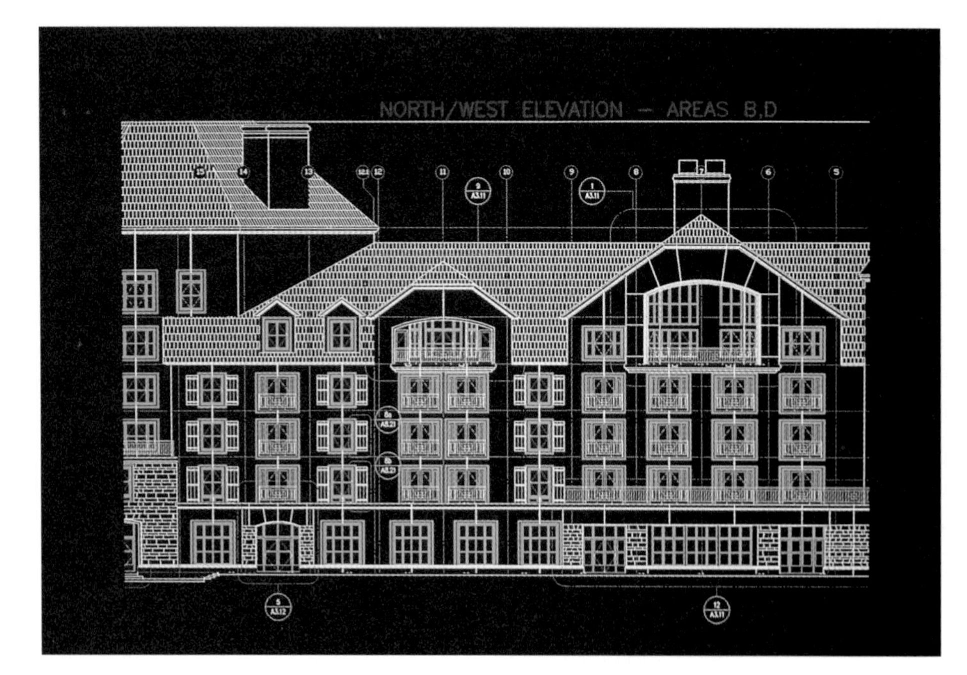

Fig. 16-18 ■ Material symbols on a computer-generated elevation. *Autodesk, Inc.*

Window Symbols

The position and style of windows greatly affects the appearance of elevations. Windows are, therefore, drawn on the elevation with as much detail as the scale of the drawing permits. Parts of windows that should be shown on all elevation drawings include the sill, sash, mullions, and muntins, if any. See Fig. 16-19.

Fig. 16-19 ■ Elevation window symbols.

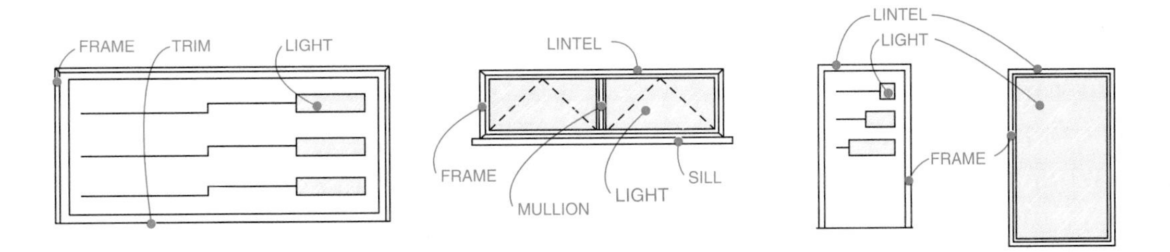

A. Typical elevation symbols for fixed windows.

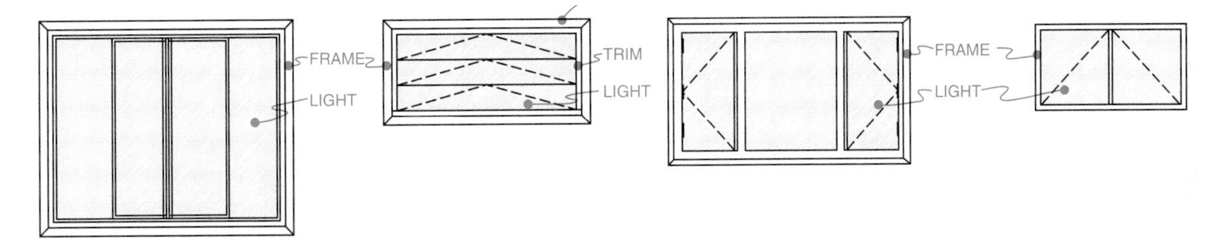

B. Methods of showing sliding, awning, and casement windows on elevations.

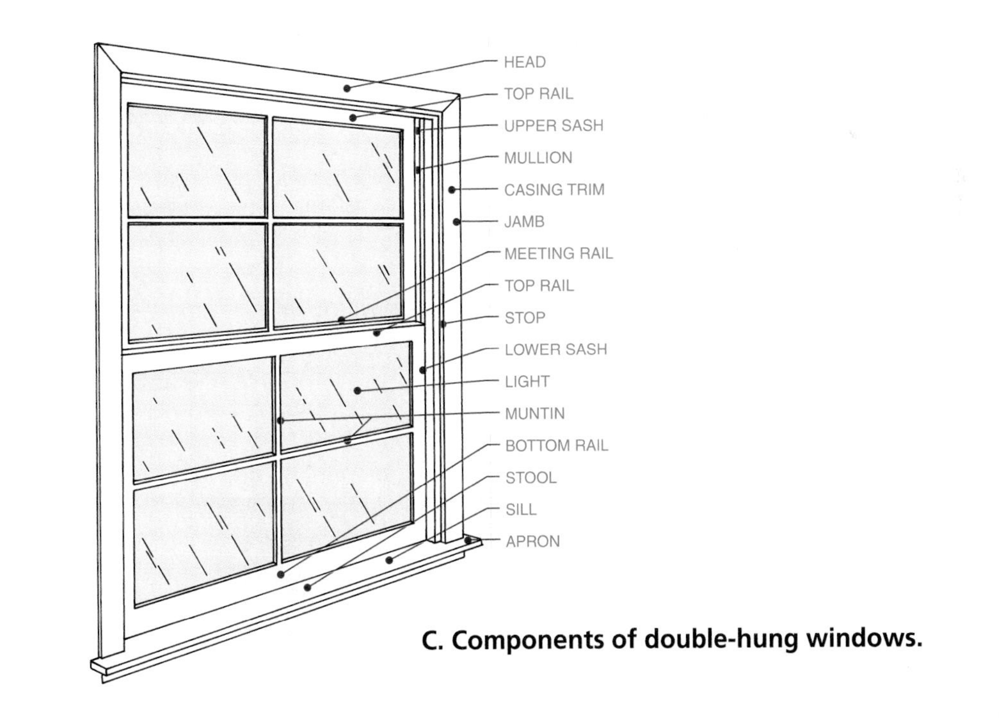

C. Components of double-hung windows.

In addition to showing the parts of a window, it is also necessary to show the direction of the hinge for casement and awning windows. Dotted lines are used on elevation drawings, as show in Fig. 16-20. The point of the dashed line shows the part of the window to which the hinge is attached.

Many different styles of windows are available. Refer back to Fig. 15-18. These illustrations show the normal amount of detail used in drawing windows on elevations. An alternative method of showing window styles on elevation drawings is used to include one window detail for each style on the plan drawn to a larger scale. See Fig. 16-21. When the elevation drawing is prepared, the size and outlined position of the window are shown with a letter or number to refer to a detail drawing. This detail drawing is also indexed to a window schedule which contains complete purchasing, framing, and installation data for each window. (See Chapter 35 for examples of schedules.)

Door Symbols

Doors are shown on elevation drawings by methods similar to those used for illustrating window styles and positions. They are either drawn completely, if the scale permits, or shown in abbreviated form. Sometimes the outline is indexed to a door schedule. See Fig. 16-22. The com-

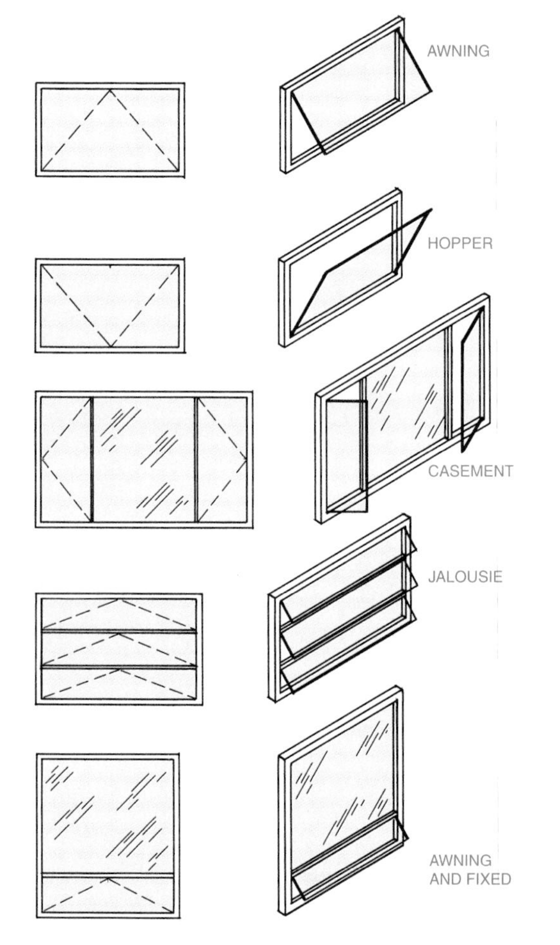

Fig. 16-20 ■ Hinge placement is indicated by the intersection of dashed lines.

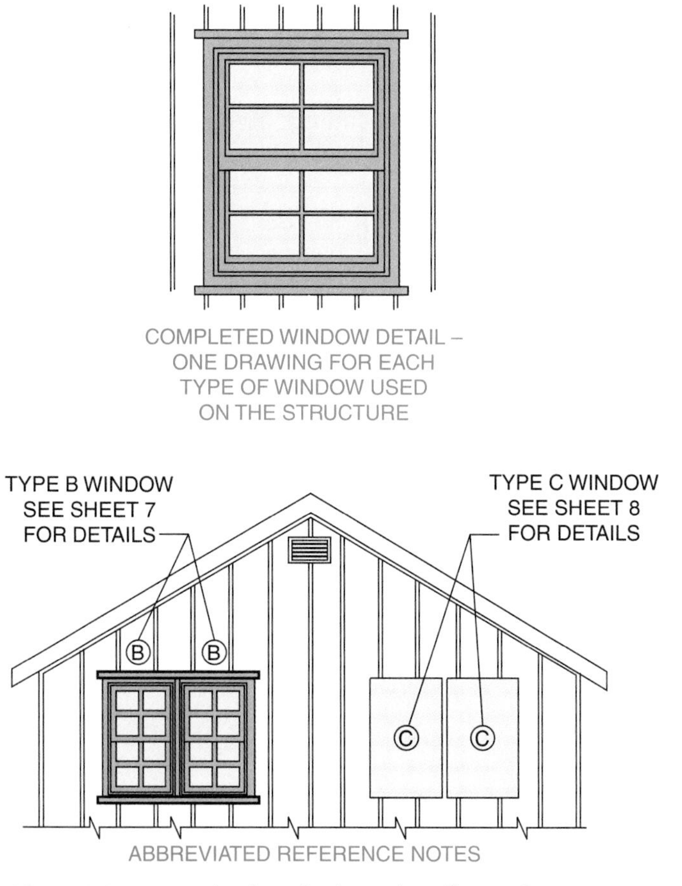

Fig. 16-21 ■ A single window detail may be prepared to show a window style.

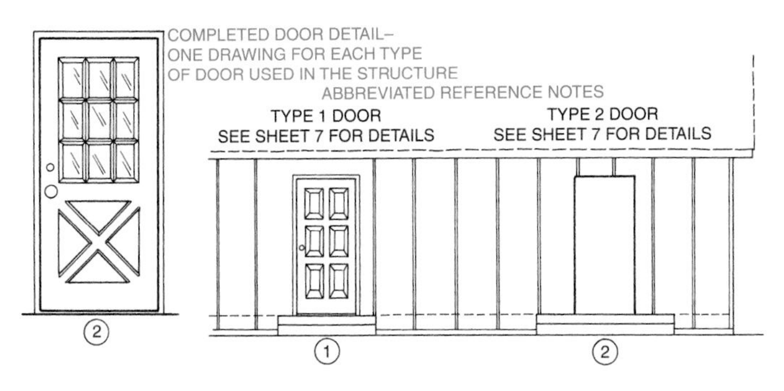

Fig. 16-22 ■ Use of code index to indicate door details.

plete drawing of the door, whether shown on the elevation drawing sheet or as a detail drawing on a separate sheet, should show the division of panels and lights, sill, jamb, and head-trim details.

Many exterior door styles are available. Refer back to Fig. 15-20. The total relationship of the door and trim to the entire elevation cannot be seen unless the door trim is also shown. Exterior doors are normally larger than interior doors. Exterior doors must provide access for larger amounts of traffic and large enough to permit the movement of furniture. They must also be thick enough to provide adequate safety, insulation, and sound barriers. Refer back to Fig. 14-9 for common door sizes.

Interior Elevations

As exterior elevations illustrate the outside walls, **interior elevation drawings** are necessary to show the design of interior walls (vertical planes). Because of the need to show cabinet height and counter arrangement detail, interior wall elevations are most often prepared for kitchen and bathroom walls. See Fig. 16-23. An interior wall elevation shows the appearance of the wall as viewed from the center of the room.

A coding system is used to identify the walls on the floor plans for which interior elevations have been prepared. The code

Fig. 16-23 ■ Elevations are made of vertical interior areas.

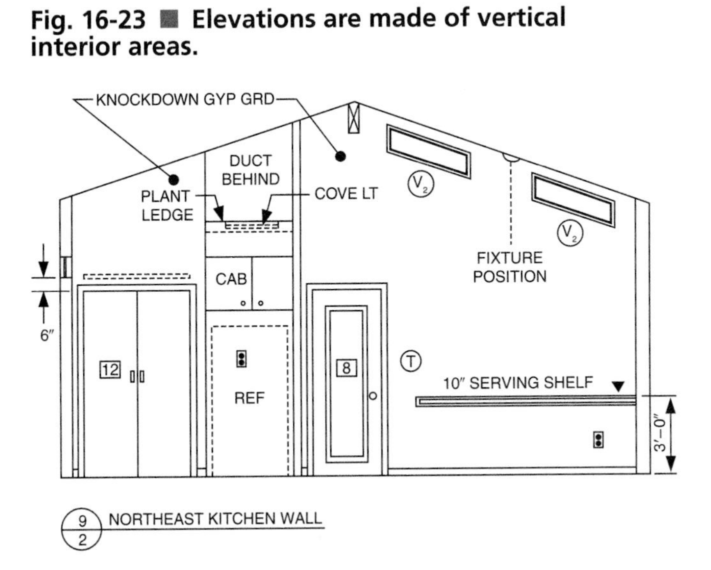

A. Kitchen elevation drawing.

B. Completed construction of the same area. *Diane Kingston Photo*

symbol tells the direction of the view, the elevation detail number, and the page or sheet number. See Fig. 16-24. If only a few interior elevations are prepared, then the title of the room and the compass direction of the wall are the only identification needed.

Steps in Drawing an Interior Elevation

The following steps in drawing an interior elevation are outlined in Fig. 16-25:

1 Provide a drawing of the floor plan.
2 Project lines perpendicular from the floor plan outline at each corner.

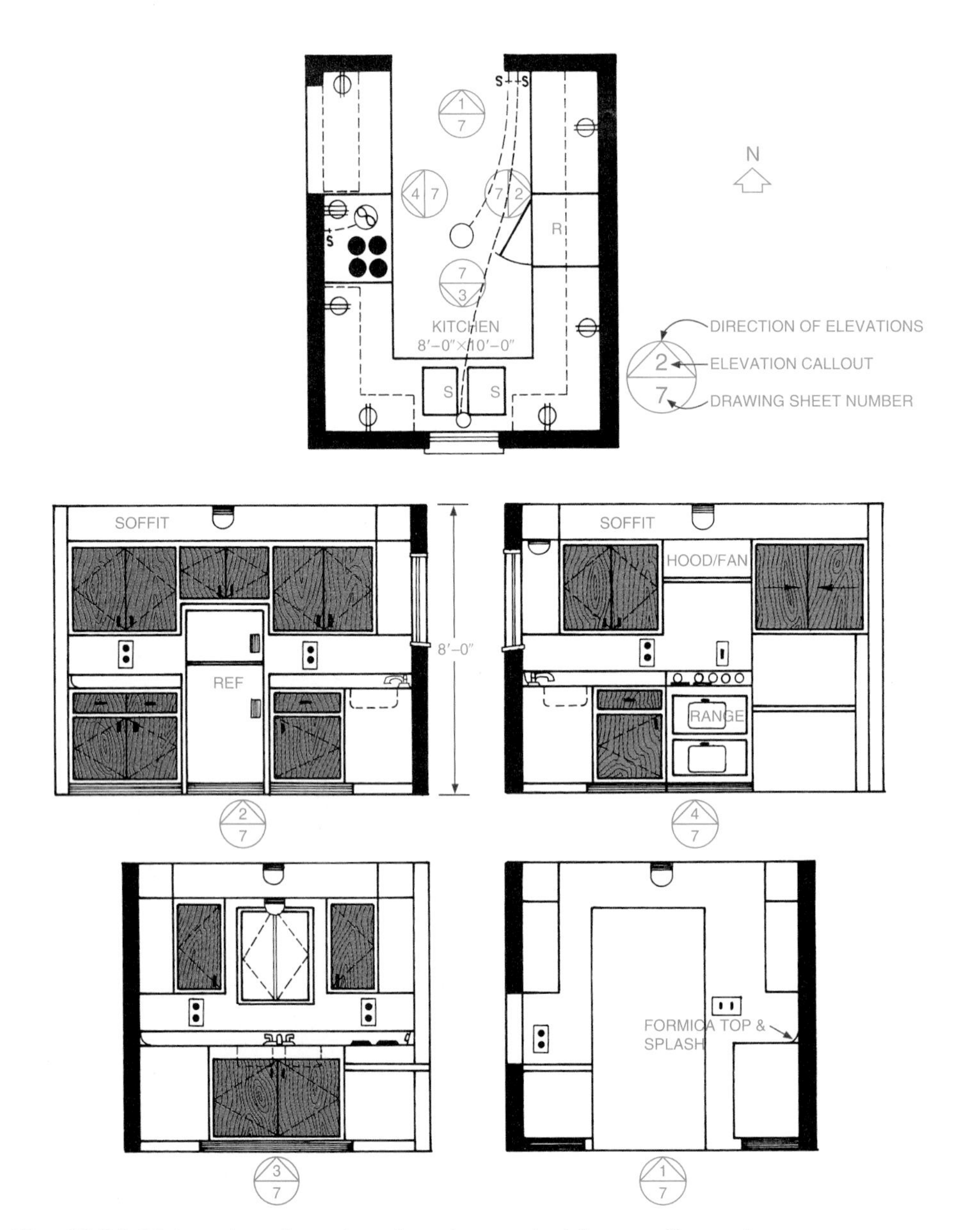

Fig. 16-24 ■ Interior elevation drawing coded from a floor plan.

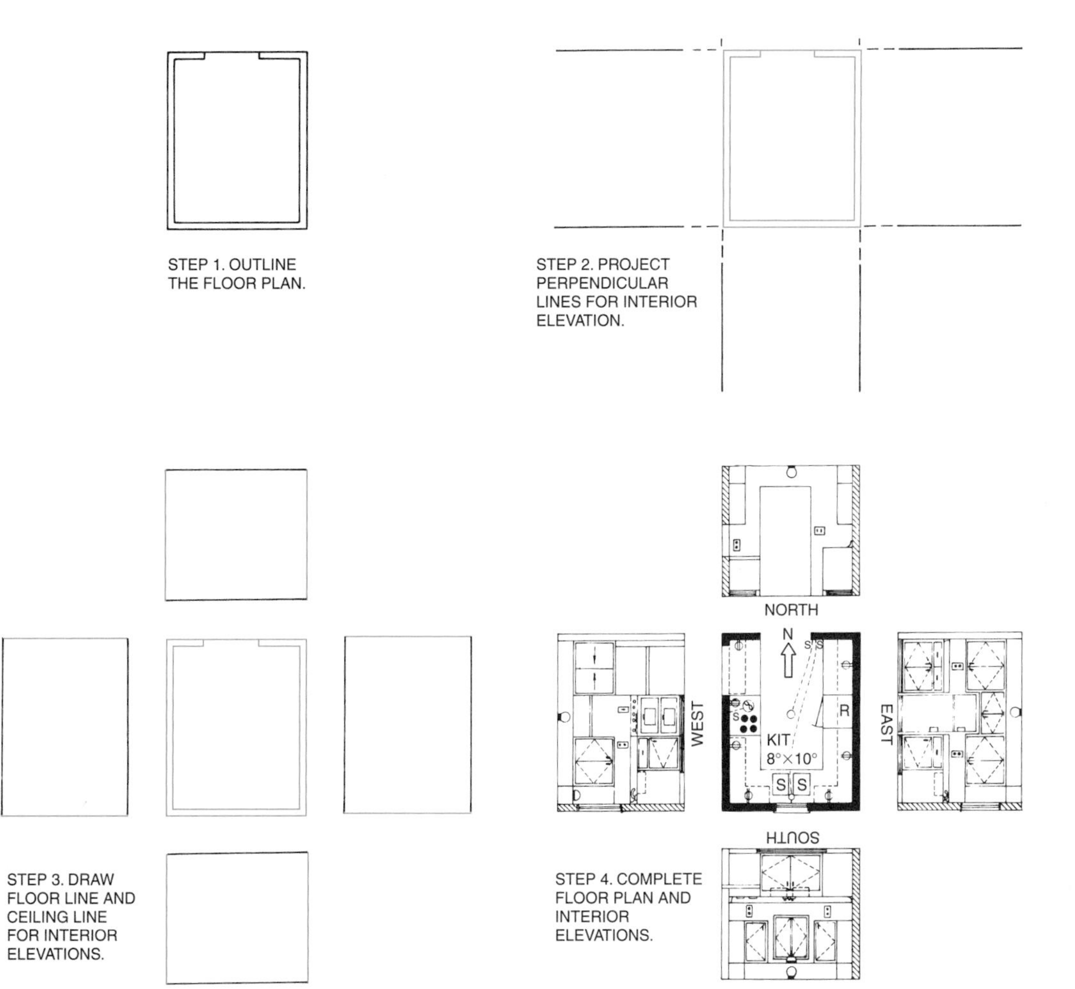

Fig. 16-25 ■ Sequence of drawing interior elevations.

3 Add ceiling lines and floor lines to give each wall its specified height.
4 Project lines directly from the key locations on the floor plan to each elevation drawing. This includes cabinets, appliances, windows, and other details.

Projecting the interior elevation in this manner is appropriate for accurate drawing, but results in an elevation drawn on its side or upside down. Therefore, interior elevation drawings, like exterior elevations, are not left in the position as they were originally projected from the floor plans. Interior elevations are repositioned so that each floor line appears on the bottom as a room would normally be viewed.

Once the features of the wall are projected to the elevation from the floor plan, dimensions, instructional notes, and additional features can be added to the drawing. See Fig. 16-26.

Interior elevations provide a great amount of detail: the height of all cabinets, shelving, ledges, railings, wall lamps, fixtures, valances, mirrors, chair rails, electrical outlets, switches, landings, and stair profiles. Elevation drawings also include wall surface treatment labeling. See Fig. 16-27. Using a common floor line and ceiling line for several elevations eliminates much layout work. In some situations, an interior elevation can span

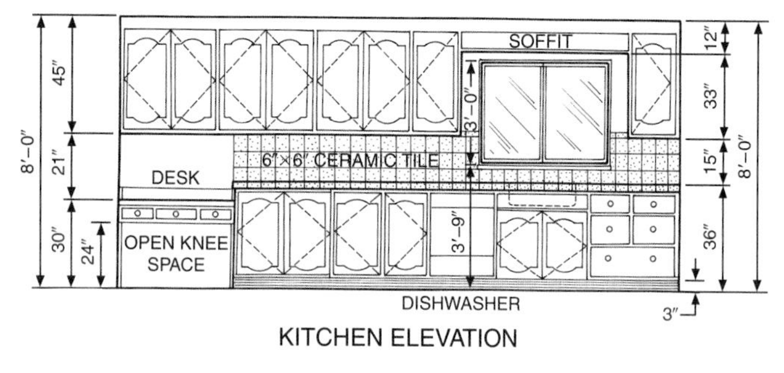

Fig. 16-26 ■ Complete interior drawing.

several levels, as shown in Fig. 16-28. (Check the floor plan in Fig. 14-20 for the sources of elevations shown in Figs. 16-23, 16-27, and 16-28.)

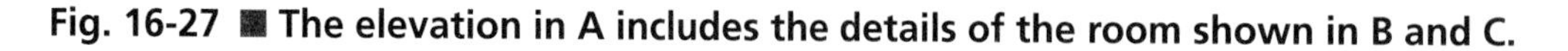

Fig. 16-27 ■ The elevation in A includes the details of the room shown in B and C.

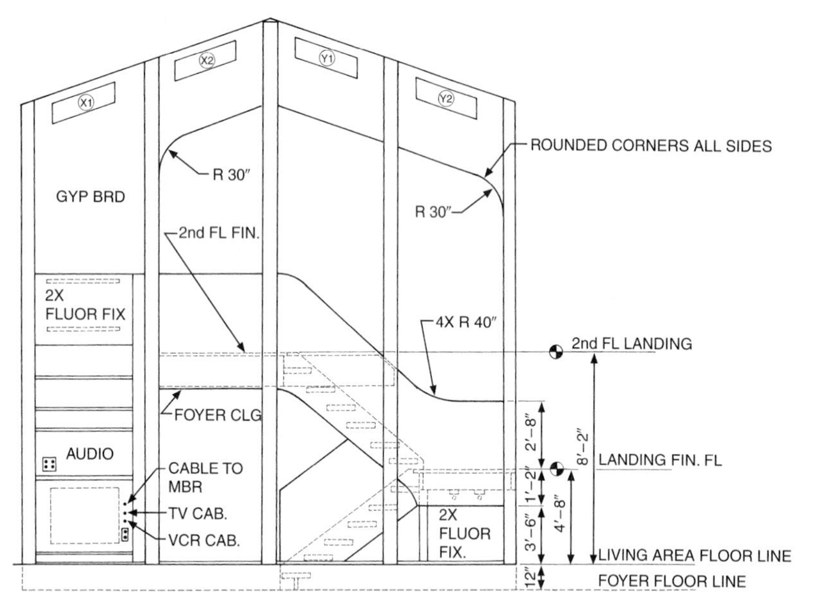

A. This type of interior elevation can be prepared for each outside wall.

B. Elevation under construction. *James Eismont Photo*

C. Area after construction was completed. *James Eismont Photo*

Fig. 16-28 ■ Two-level interior elevation drawing.

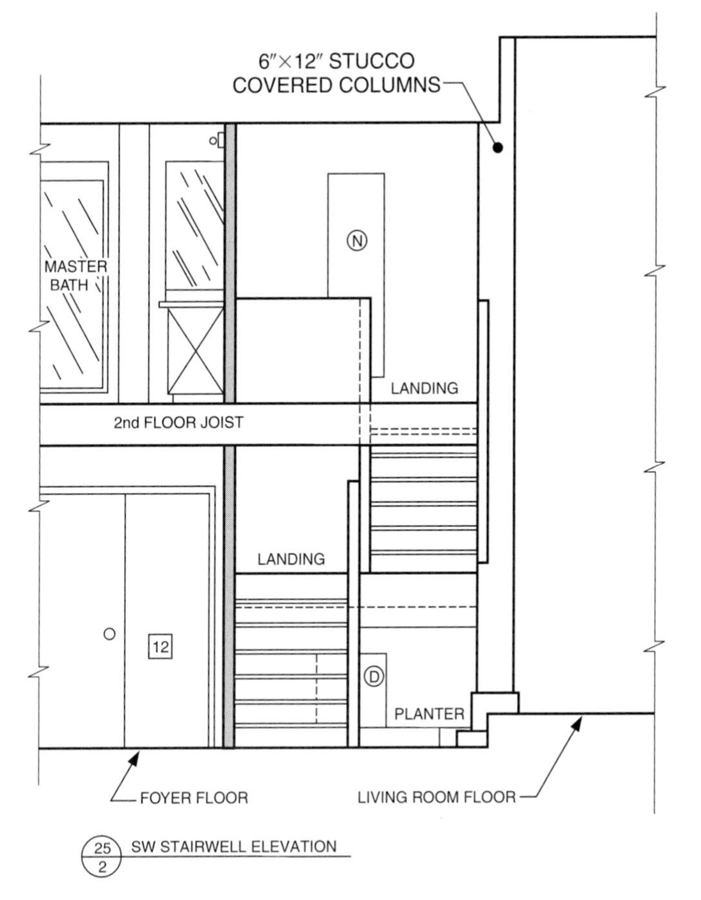

A. This elevation includes the stairwell, foyer, and two bath walls.

B. Area after construction was completed. *James Eismont Photo*

Elevation Dimensioning

The vertical (height) dimensions are as important on elevation drawings as horizontal (width and length) dimensions are on floor plans. Many dimensions on elevation drawings show the vertical distance from a datum line. The **datum line** is a reference that remains constant. Sea level is commonly used as the datum or basic reference for many drawings. However, any given line can be conveniently used as a base or datum line for vertical reference.

Dimensions on elevation drawings show height above the ground line. They also show the vertical distance from the floor line to the ceiling and roof ridge and eave lines, and to the tops of chimneys, doors, and windows. Distances below the ground line are shown by dotted lines.

Standards and Guidelines for Dimensioning

Elevation dimensions must conform to basic standards to ensure consistency of interpretation. The arrows on the elevation drawing in Fig. 16-29 show the applications of the following guidelines for elevation dimensioning:

1. Vertical elevation dimensions should be read from the right side of the drawing.
2. Levels to be dimensioned should be labeled with a note, abbreviation, or term.

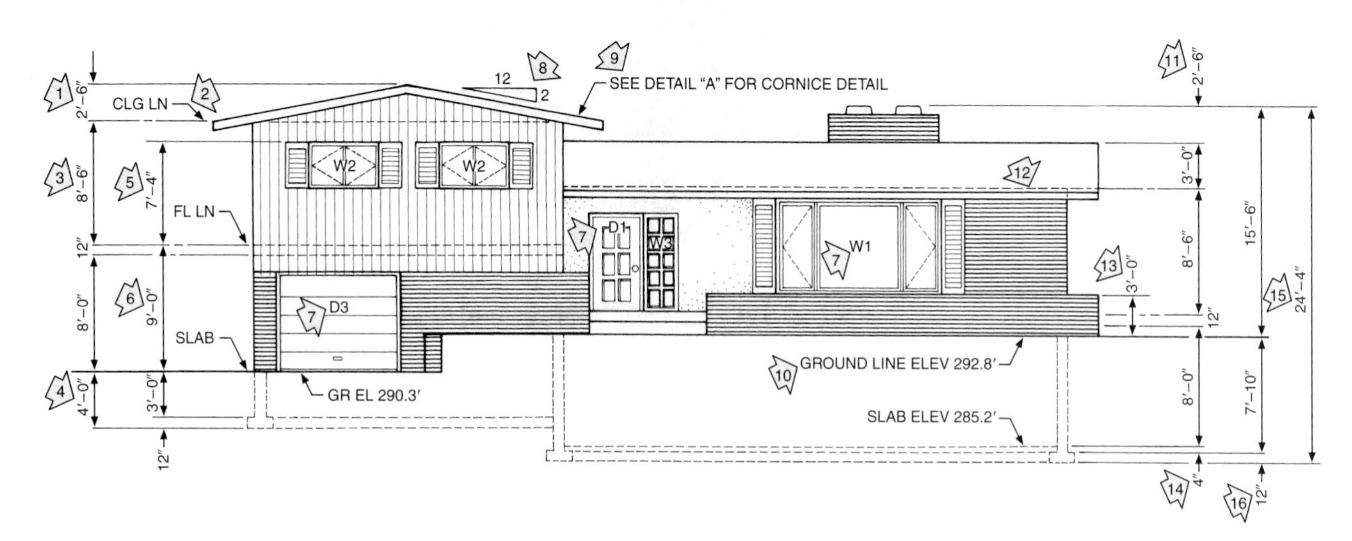

Fig. 16-29 ■ Arrows indicate portions of the drawing referred to in the guidelines for dimensioning.

3. Room heights are shown by dimensioning from the floor line to the ceiling line.
4. The depth of footings is dimensioned from the ground line.
5. Heights of windows and doors are dimensioned from the floor line to the top of the windows or doors.
6. Elevation dimensions show only vertical distances (height). Horizontal distances (length and width) are shown on floor plans.
7. Windows and doors may be indexed by a code or symbol to a door or window schedule, if the style of the windows and doors are not shown on the elevation drawing. See also Fig. 16-30.
8. The slope of the roof is shown by indicating the rise over the run.
9. Dimensions for small, complex, or obscure areas should be indexed to a separate detail.
10. Ground-line elevations are expressed as heights above a datum point (for example, sea level).
11. Heights of chimneys above the ridge line are dimensioned.
12. Floor and ceiling lines are shown with hidden lines.
13. Heights of planters, fences, and walls are dimensioned from the ground line.
14. Thicknesses of slabs are dimensioned.
15. Overall height dimensions are placed on the outside of subdimensions.
16. Thicknesses of footings are dimensioned.
17. Refer to Fig. 16-31. When the level to be dimensioned is obscure or extremely close to other dimensions, use an elevation line symbol and label the level line.
18. Datum must be identified with a note, if not part of the elevation drawing.

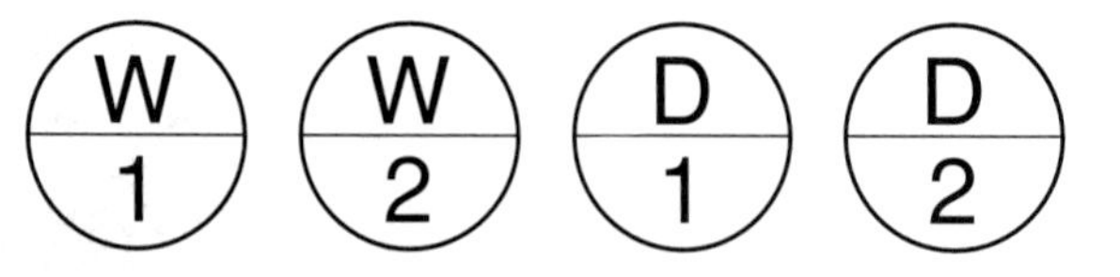

Fig. 16-30 ■ Code used on an elevation drawing to refer to a door or window schedule.

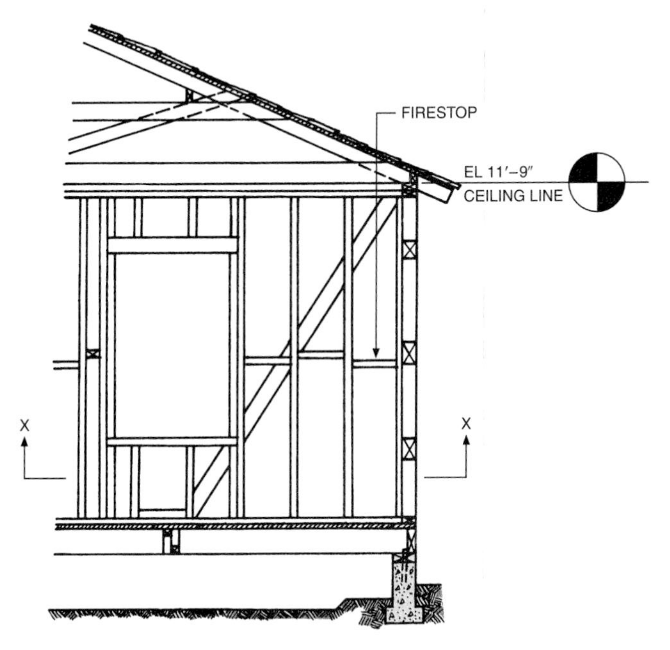

Fig. 16-31 ■ Use of an elevation line symbol.

Types of Dimensions

Two types of elevation dimensions are used: framing dimensions and finished dimensions. **Framing dimensions** show the actual distances between framing members. This is the most common method and is preferred by most builders. To avoid an accumulation of measuring errors, framing member dimensions are often dimensioned to their centers.

Finished dimensions show the actual vertical distances between finished features, such as from the finished floor to finished ceiling levels. These two types of dimensions should not be alternately used on the same drawing, unless the exception is clearly noted. A note on each drawing or set of drawings should indicate which method is used.

Interior vs. Exterior Dimensioning

Features that are shared by an interior and exterior wall, such as doors and windows, are usually shown on both interior and exterior elevations, but dimensioned only on the exterior elevation drawings. Many other features or components—such as cabinets, shelving, counters, ledges, railings, wall lamps, switches and receptacles—can only be dimensioned on interior elevations. See Fig. 16-32.

Landscape on Elevation Drawings

Creating a Realistic Drawing

Elevation drawings, although accurate in every detail, do not show exactly how the building will appear when it is complete and landscaped. The reason is that elevation drawings do not show the position of trees, shrubbery, and other landscape features that would be part of the total elevation design. Adding these landscape features to the elevation drawing creates a more realistic drawing or rendering of the house.

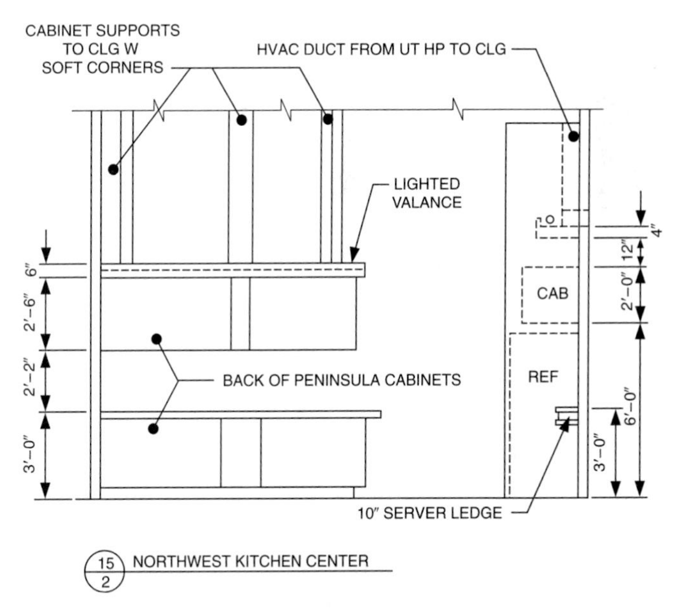

Fig. 16-32 ■ Typical dimensions used on an interior elevation drawing.

Fig. 16-33 ■ Adding landscape features to an elevation give it a more realistic appearance.

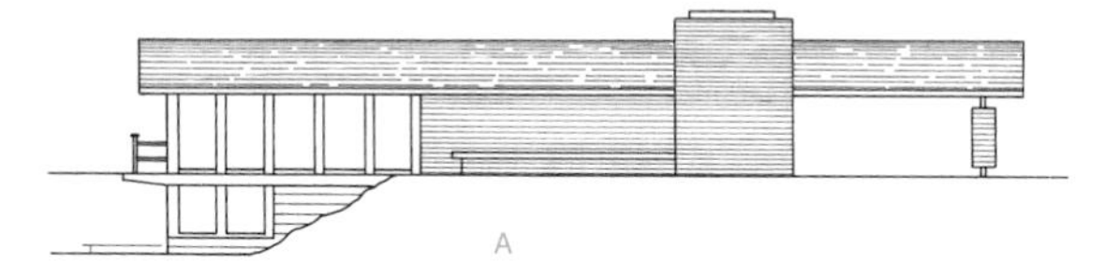

A. Before landscape rendering.

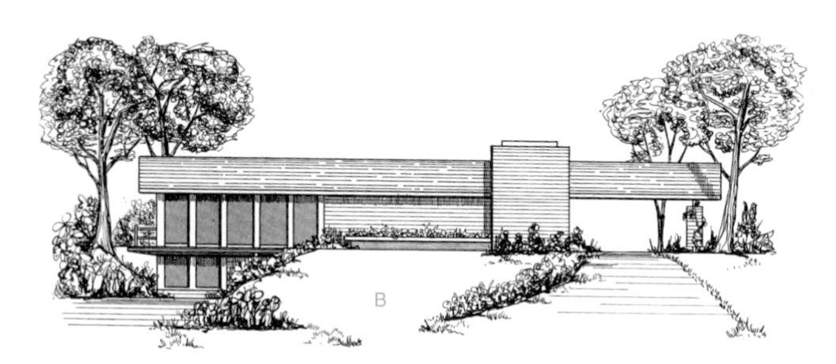

B. After landscape rendering.

Figure 16-33 shows some of the advantages of adding rendered landscape features to an elevation drawing. The elevation shown in Fig. 16-33A, when dimensioned, would be adequate for construction purposes. However, the illustration shown in Fig. 16-33B more closely resembles the final appearance of the house. Dimensions and hidden lines are omitted.

When landscape features are added to elevations, the drawings function as **presentation drawings.** These presentation drawings are prepared only to interpret the final appearance of the building primarily to clients. They are not used for construction purposes. See Fig. 16-34.

Fig. 16-34 ■ An elevation rendering can be used to show clients the expected appearance of the structure after construction has been completed. *Jenkins & Chin Shue Incorporated*

Steps for Making Landscaped Elevations

An elevation drawing is converted into a landscape elevation drawing in several basic steps, as shown in Fig. 16-35. After material symbols are placed on the elevation, the positions of trees and shrubs are added. The elevation lines within the outlines of the trees and shrubs are erased, and details are added. Finally, shade lines are added to trees, windows, roof overhangs, chimneys, and other major projections of the house.

Fig. 16-35 ■ Sequence of adding surface treatments and landscape features to an elevation.

Drawing and Rendering Techniques

The addition of landscape features should not hide the basic lines of the house. If many trees or shrubs are placed in front of the house, it is best to draw them in their winter state. Different drawing techniques may be used. See Fig. 16-36. The drafter should use the medium that best suits the elevation drawing to be rendered. (See Chapters 18, 19, and 20 for information about making pictorial drawings and landscape renderings.)

Fig. 16-36 ■ Methods of adding landscape features to elevation drawings.

A. Line drawing.

B. Wash drawing.

CHAPTER 16 Exercises

1. Project the front, rear right, and left elevations of a floor plan of your own design. Add elevation symbols.
2. Draw a kitchen wall elevation of a kitchen of your design.
3. Sketch and dimension the front elevation of your home or another home that you are familiar with.
4. Using a CAD system develop a library of symbols for elevations to use in drawing your own design. Note or list which symbols are already included in the CAD program.
5. Project and sketch or draw the front elevation suggested in one of the pictorial drawings in Chapter 19.

6. Copy five of the trees, shrubs, or plants as practice for creating a landscape rendering. Then create one with shadows to an elevation of your own design.
7. Complete Exercises 1 and 2 using a CAD system.

CHAPTER

17

Sectional, Detail, and Cabinetry Drawings

Objectives

In this chapter you will learn to:

- describe types of sectional drawings.
- communicate views of sections based on a cutting plane.
- draw sections, using correct codes and proper dimensioning.
- evaluate when a detail sectional drawing is needed.
- read and prepare detail drawings.
- design and prepare cabinet drawings.

Terms

break-out sectional drawings
cabinet coding system
cutting plane
cutting-plane line
detail sections
full section
horizontal wall sections
longitudinal section
removed section
sectional drawings
transverse section
vertical wall sections

Architectural sections are drawings that are important in the design process. They show details not visible on floor plans or elevations. Some architectural sections contain symbols, reference codes, and dimensions to indicate construction information. Other sections may only be sketches to compare heights, while still others may be pictorial drawings. Sections are used to show the exact details of construction. The ability to prepare technical architectural drawings depends on a thorough understanding of sectional drawings.

■ **Detail sections are used to show the construction of cabinets and other built-in features.** *© Gary Buss/FPG International Corp.*

Sectional Drawings

Sectional drawings reveal the internal construction of an object. An architectural sectional drawing (or an architectural section) that is prepared for the entire structure is called a **full section.** A sectional drawing that shows only specific parts of a building is a **detail section.** The size and complexity of the parts usually determine whether a full section and/or detail sections are needed. Because they provide all information needed for construction, architectural sections are most often used as working drawings. Sections are also useful as design concept drawings and presentation drawings.

Design Concept Sections

Architectural designs that involve multiple levels and variations in height often require the preparation of full-section sketches (in addition to floor plans) during the design process. Sections are often needed to clarify the elevation relationships of ground, footing, floor, and roof lines.

Presentation Sections

When a drawing is needed to show the general appearance of both the interior and exterior of a building in one drawing, a pictorial section is often completed. This is usually done using one-point perspective methods. See Fig. 17-1.

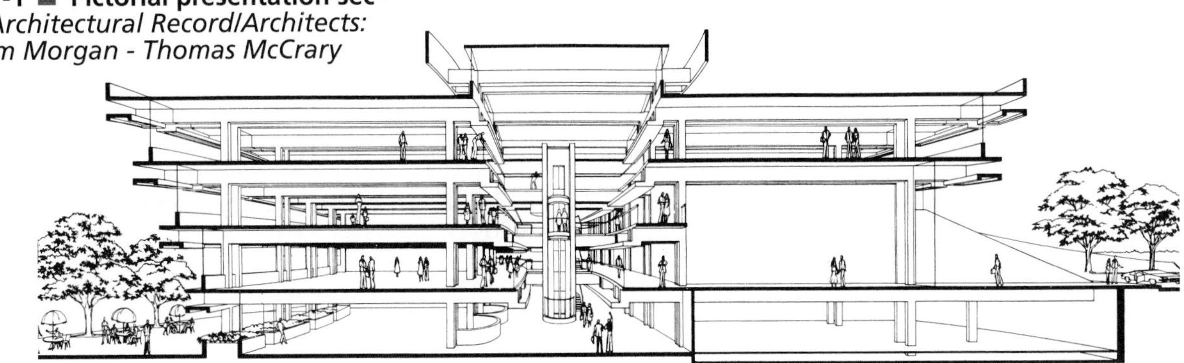

Fig. 17-1 ■ Pictorial presentation section. *Architectural Record/Architects: William Morgan - Thomas McCrary*

Full Sections

In full-section drawings, entire buildings are drawn as if they were cut in half. The purpose of full sections is to convey how a building is constructed from the foundation through the roof.

The Cutting Plane

A **cutting plane** is an imaginary plane that passes through a building. The position of a cutting plane is shown by the **cutting-plane line** that is drawn as a long heavy line with two dashes. Sectional drawings may show either transverse or longitudinal sections. A **transverse section** shows a cutting plane across the shorter or minor axis of a building, as shown in Fig. 17-2. A **longitudinal section** shows a

Fig. 17-2 ■ A transverse section and its cutting plane.

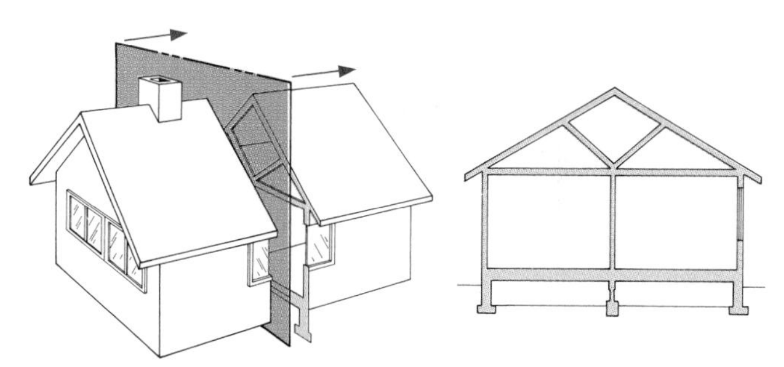

Fig. 17-3 ■ A longitudinal section.

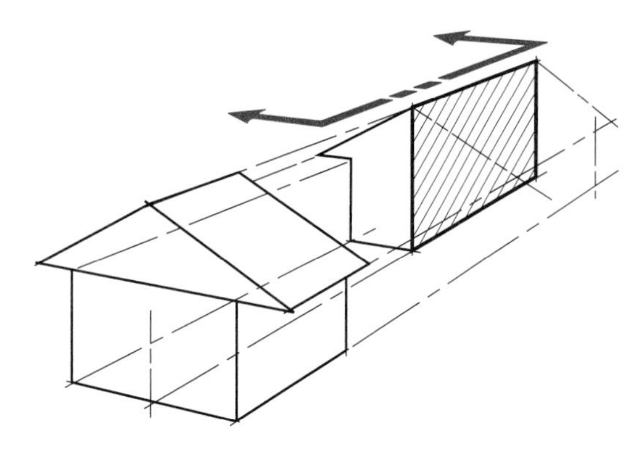

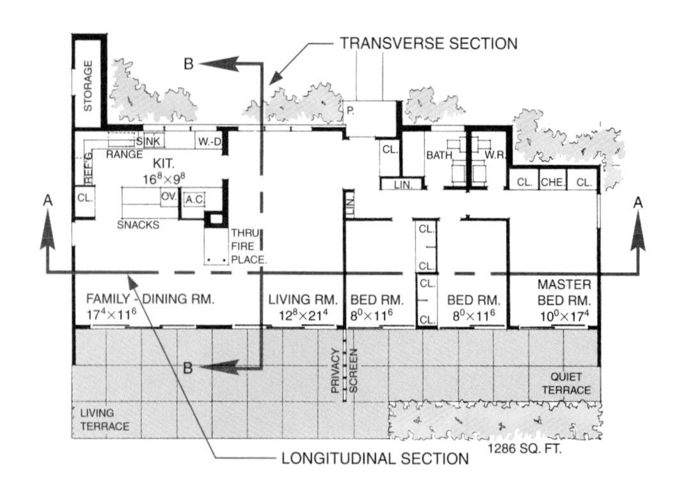

Fig. 17-4 ■ Both transverse and longitudinal cutting-plane lines are shown on this floor plan.

cutting plane along the length or major axis of a building. See Fig. 17-3. Notice that although sections convey information about the inner construction, both transverse and longitudinal full sections share the same external outlines as elevation drawings.

The cutting-plane line is usually placed on a floor plan to tell which part is drawn as a section. The arrows at the ends of the line tell the direction of the view. See Fig. 17-4.

Because the cutting-plane line can easily interfere with dimensions, notes, and details, only the extreme ends of the cutting-plane line are indicated on most architectural drawings. The part of the line that is omitted on the drawing is then assumed to be a straight line between the ends of the cutting-plane line. When a cutting-plane line must be offset to show a different area, the offsetting corners of the line are drawn as they appear in Fig. 17-5. An offset cutting plane is often used to include sections of different walls on one sectional drawing.

Symbols

When sections are referenced to another drawing, the symbol shown in Fig. 17-6 should be used. This referencing method is the same method introduced in Chapter 4. Figure 17-7 shows a full section of a residence. and the portion of a floor plan from which it is referenced.

A floor plan is a small scale horizontal section. Therefore, many floor-plan symbols are also used in plan detail drawings.

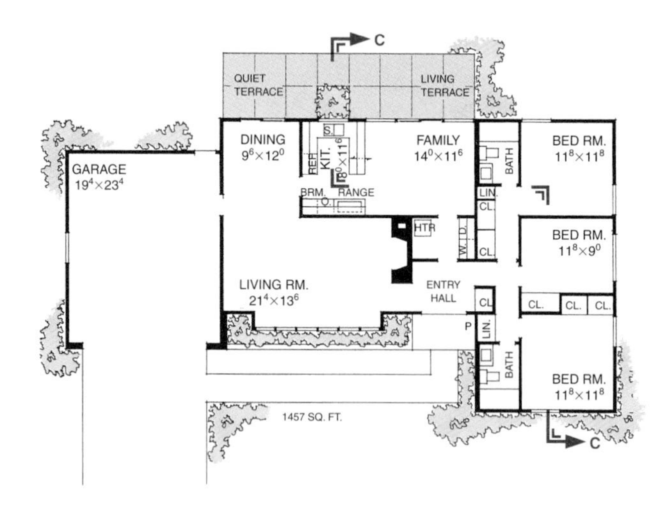

Fig. 17-5 ■ An offset cutting-plane line. *Home Planners, Inc.*

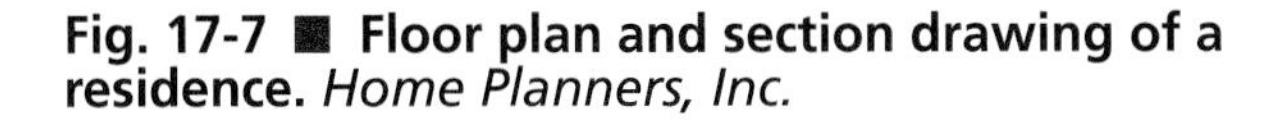

Fig. 17-7 ■ Floor plan and section drawing of a residence. *Home Planners, Inc.*

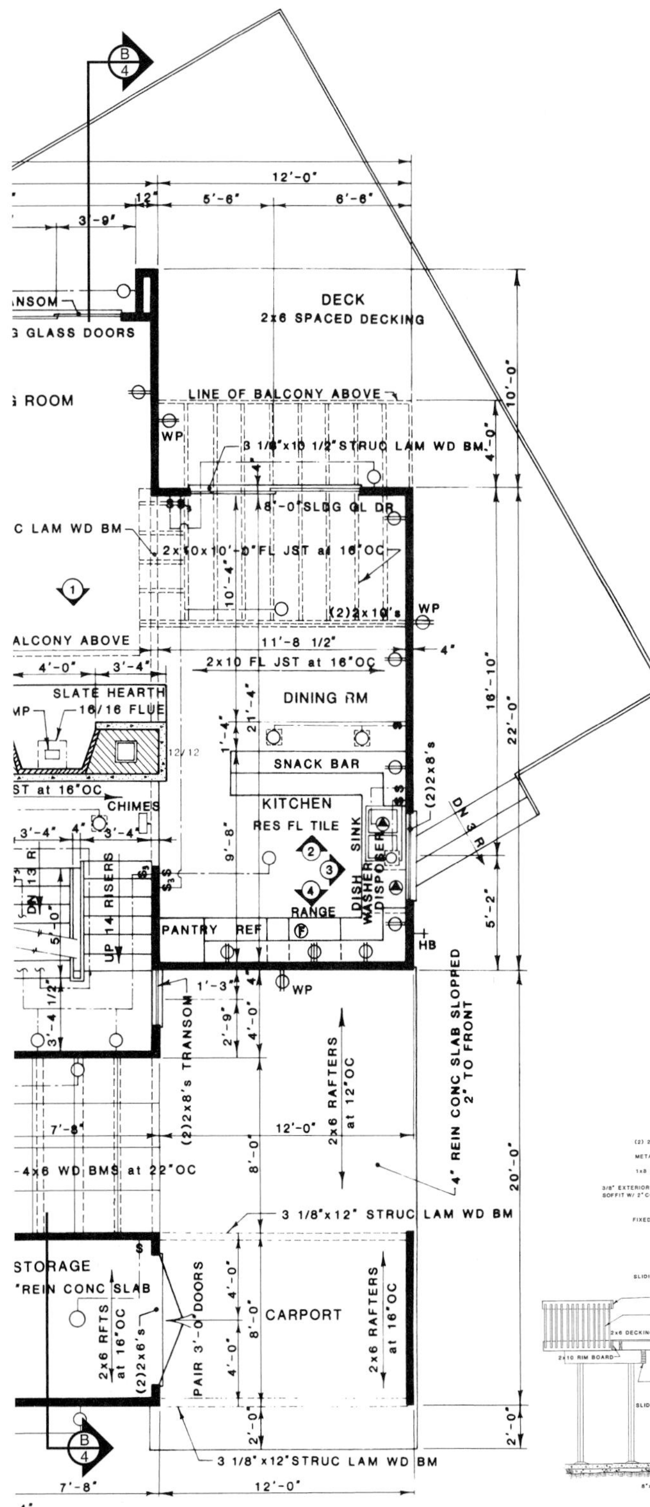

A. Locate cutting-plane line B-4 on this floor plan.

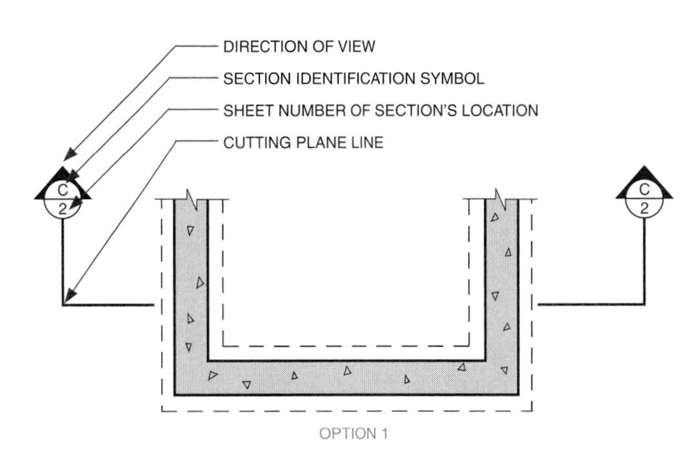

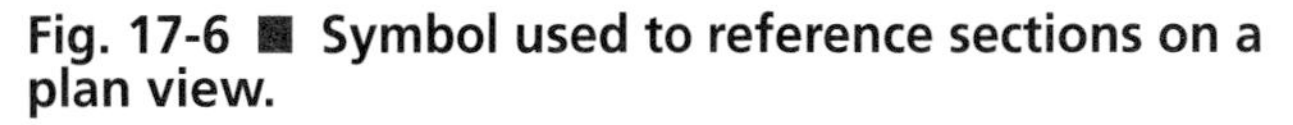

Fig. 17-6 ■ Symbol used to reference sections on a plan view.

Many *section-lining symbols* appear realistic, as though materials were cut through. Others are simplified in order to save time on the drawing board.

A building material is only sectioned if the cutting-plane line passes through it. The outline of all other materials visible behind the plane of projection must also be drawn in the proper position and scale. Symbols for building materials are shown in Fig. 17-8.

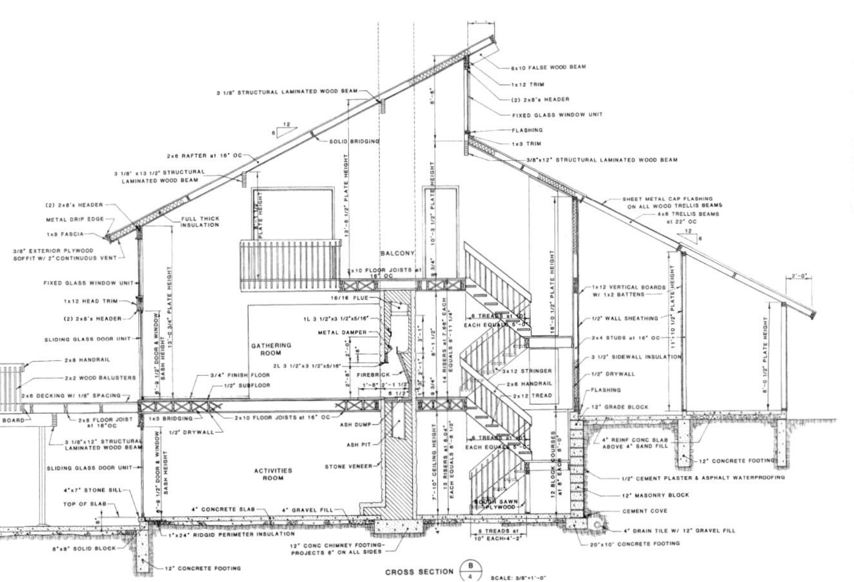

B. Full section referenced by cutting-plane line B-4 on the floor plan. *Homeplanners, Inc.*

NAME	ABBRV	SECTION SYMBOL	ELEVATION	NAME	ABBRV	SECTION SYMBOL	ELEVATION
COMMON BRICK	COM BRK			WELDED WIRE MESH	WWM		
FACE BRICK	FC BRK			FABRIC	FAB		
FIREBRICK	FRB			LIQUID	LQD		
GLASS	GL			COMPOSITION SHINGLE	COMP SH		
GLASS BLOCK	GL BLK			RIDGID INSULATION SOLID	RDG INS		
STRUCTURAL GLASS	STRUC GL			LOOSE-FILL INSULATION	LF INS		
FROSTED GLASS	FRST GL			QUILT INSULATION	QLT INS		
STEEL	STL			SOUND INSULATION	SND INS		
CAST IRON	CST IR			CORK INSULATION	CRK INS		
BRASS & BRONZE	BRS BRZ			SHEET METAL (FLASHING)	SHT MTL FLASH		
ALUMINUM	AL			REINFORCING STEEL BARS	REBAR		

Fig. 17-8 ■ Section symbols for common building materials.

Scale

Because full sections show construction methods used in the entire building, they are drawn to a relatively small scale (¼″ = 1′-0″). However, a scale that is too small often makes the drawing and interpretation of minute details extremely difficult. One method of maintaining a larger scale for large sections is to insert break lines. See Fig. 17-9. Break lines indicate that much of the repetitive portion of the building was removed from the drawing. With break lines, a very large area can be drawn at a readable scale and still fit on one sheet.

Another method is to draw detail sections of the removed portions of the building separate from the full-section drawing. **Removed sections** drawn at a larger scale are used to clarify small details.

NAME	ABBRV	SECTION SYMBOL	ELEVATION	NAME	ABBRV	SECTION SYMBOL	ELEVATION
EARTH	E			CUT STONE, ASHLAR	CT STN ASH		
ROCK	RK			CUT STONE, ROUGH	CT STN RGH		
SAND	SD			MARBLE	MARB		
GRAVEL	GV			FLAGSTONE	FLG ST		
CINDERS	CIN			CUT SLATE	CT SLT		
AGGREGATE	AGR			RANDOM RUBBLE	RND RUB		
CONCRETE	CONC			LIMESTONE	LM ST		
CEMENT	CEM			CERAMIC TILE	CER TL		
TERAZZO CONCRETE	TER CONC			TERRA-COTTA TILE	TC TL		
CONCRETE BLOCK	CONC BLK			STRUCTURAL CLAY TILE	ST CL TL		
CAST BLOCK	CST BLK			TILE SMALL SCALE	TL		
CINDER BLOCK	CIN BLK			GLAZE FACE HOLLOW TILE	GLZ FAC HOL TL		
TERRA-COTTA BLOCK LARGE SCALE	TC BLK			TERRA-COTTA BLOCK SMALL SCALE	TC BLK		

Fig. 17-8 ■ (continued)

Sectional Dimensions

Full sections expose the size and shape of building materials and components not revealed on floor plans and elevations. These sections are an excellent place on which to locate many detail dimensions. Full-section dimensions primarily show specific elevations, distances, and the exact size of building materials. See Fig. 17-10. The guidelines for dimensioning elevation drawings apply also to full-elevation sections and to detail sections. (See Chapter 16.)

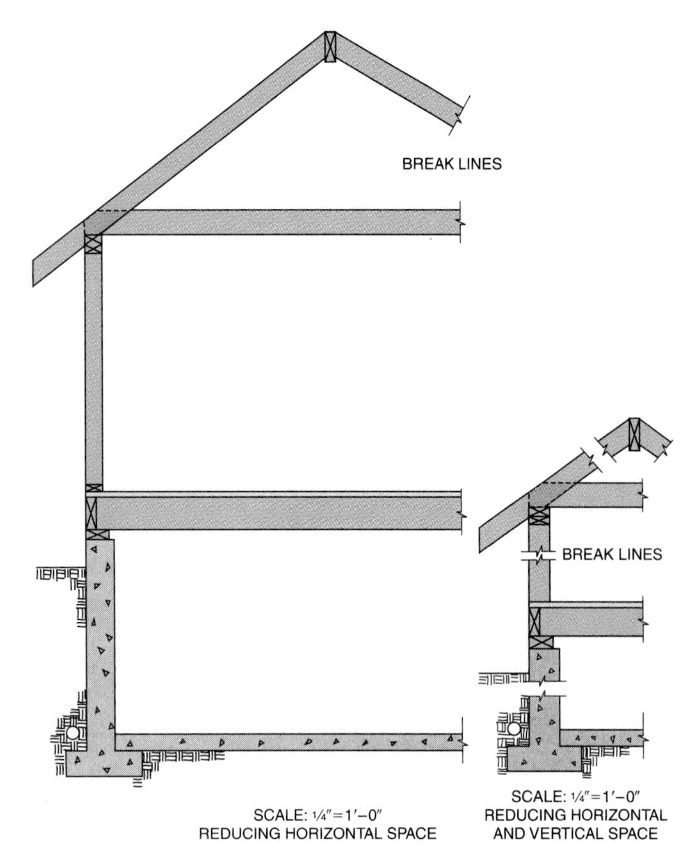

Fig. 17-9 ■ Using break lines reduces the drawing area required. Drawings may also be done at a larger scale.

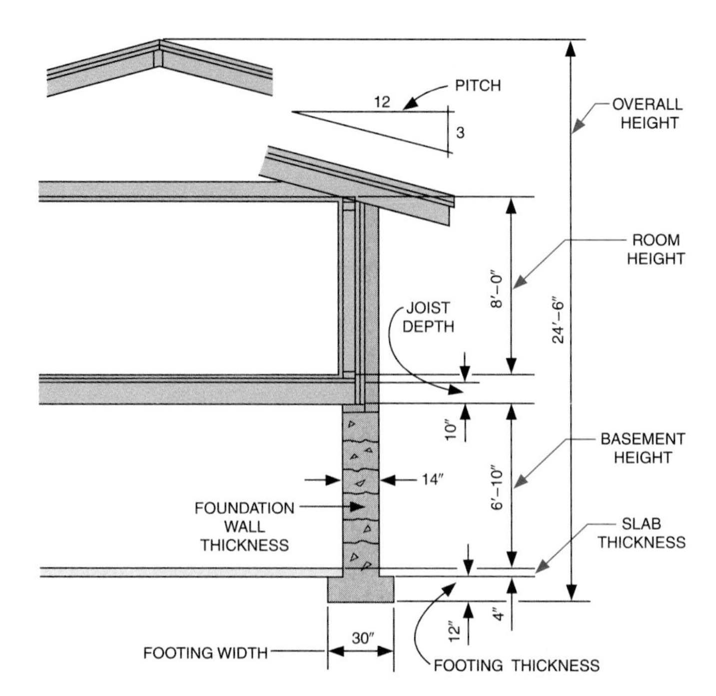

Fig. 17-10 ■ This drawing includes some of the more important dimensions that can be placed on full sections.

Steps in Drawing Full Sections

In drawing full sections, the architect "constructs" the framework of a house on paper. Figure 17-11 shows the progressive steps in the layout and drawing of a section of the side edge of a house.

1 Lightly draw the finished floor line approximately at the middle of the drawing sheet.
2 Measure the thickness of the subfloor and of the joist and draw lines representing these under the floor line.
3 From the floor line, measure up and draw the ceiling line.
4 Measure down from the floor line to establish the top of the basement slab

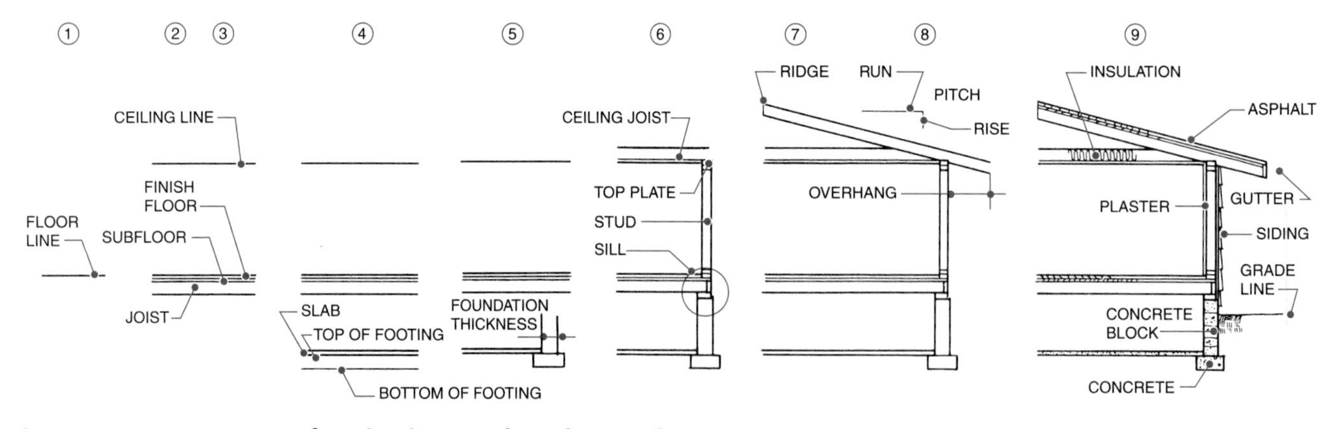

Fig. 17-11 ■ Sequence of projecting an elevation section.

and footing line, and draw in the thickness of the footing.

5 Draw two vertical lines representing the thickness of the foundation and the footing.

6 Construct the sill detail and show the alignment of the studs and top plate.

7 Measure the overhang from the stud line and draw the roof pitch by projecting from the top plate on the angle determined by dividing the rise by the span.

8 Establish the ridge point by measuring the distance from the outside wall horizontally to the center of the ridge line. This is usually at the center of the structure.

9 Add details and symbols representing siding and interior finish.

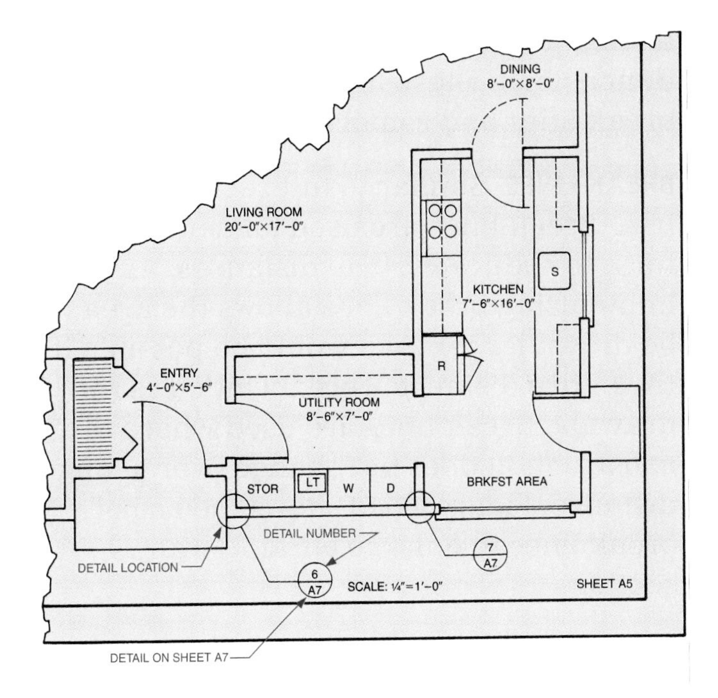

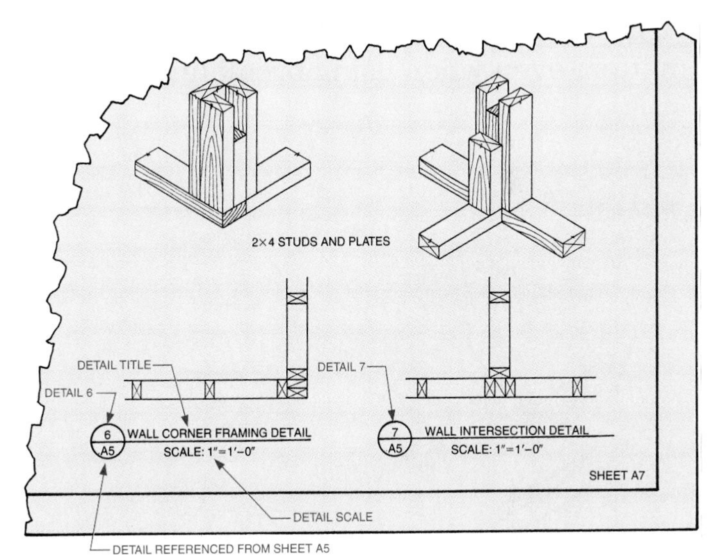

Fig. 17-12 ■ Symbols for cross-referencing detail drawings.

Detail Sections

Because full sections are usually drawn to a small scale, many small parts are difficult to interpret. To reveal the exact position and size of many of these small parts, enlarged detailed sections are prepared. Detail sections clarify any construction feature that could not be described on the basic floor plans, elevations, or full sections. Detail sections may be prepared on a vertical (elevation) plane or a horizontal (plan) plane. Like full sections, detail sections are keyed to a plan or elevation view. See Fig. 17-12.

Vertical Wall Sections

Vertical wall sections show exposed construction members on a vertical plane. They are prepared for exterior and interior walls.

EXTERIOR WALLS Elevation drawings do not reveal construction details because of their small scale and because many details are hidden by siding materials. Several other methods are used to produce exterior wall section drawings large enough to show construction details and dimensions. These include the use of break lines and removed sections.

BREAK LINES Similar to full-section drawings, break lines are used on detail sections to reduce vertical distances. As you know, using break lines allows the area to be drawn larger than would be possible if the entire distance were included in the drawing. Break lines are used where the construction does not change over a long distance. Figure 17-13 shows the use of break lines to enlarge a frame-wall section.

REMOVED SECTIONS Sometimes it is impossible to draw an entire wall section to a large enough scale to show needed information, even when using break lines. In these cases, a removed section is drawn at a larger scale, separate from the original location. Removed sections are frequently drawn for the ridge, cornice, sill, footing, and beam areas.

Cornice sections are used to show the relationship between the outside wall, top plate, and rafter construction. See Fig. 17-14. Some cornice sections show gutter details.

Sill sections, as shown in Fig. 17-15, show how the foundation supports and intersects with the floor system and the outside wall.

A *footing section* is needed to show the width and height of the footing, the type of material used, and the position of the foundation wall on the footing. See Fig. 17-16 shows several footing details and the pictorial interpretation of each type.

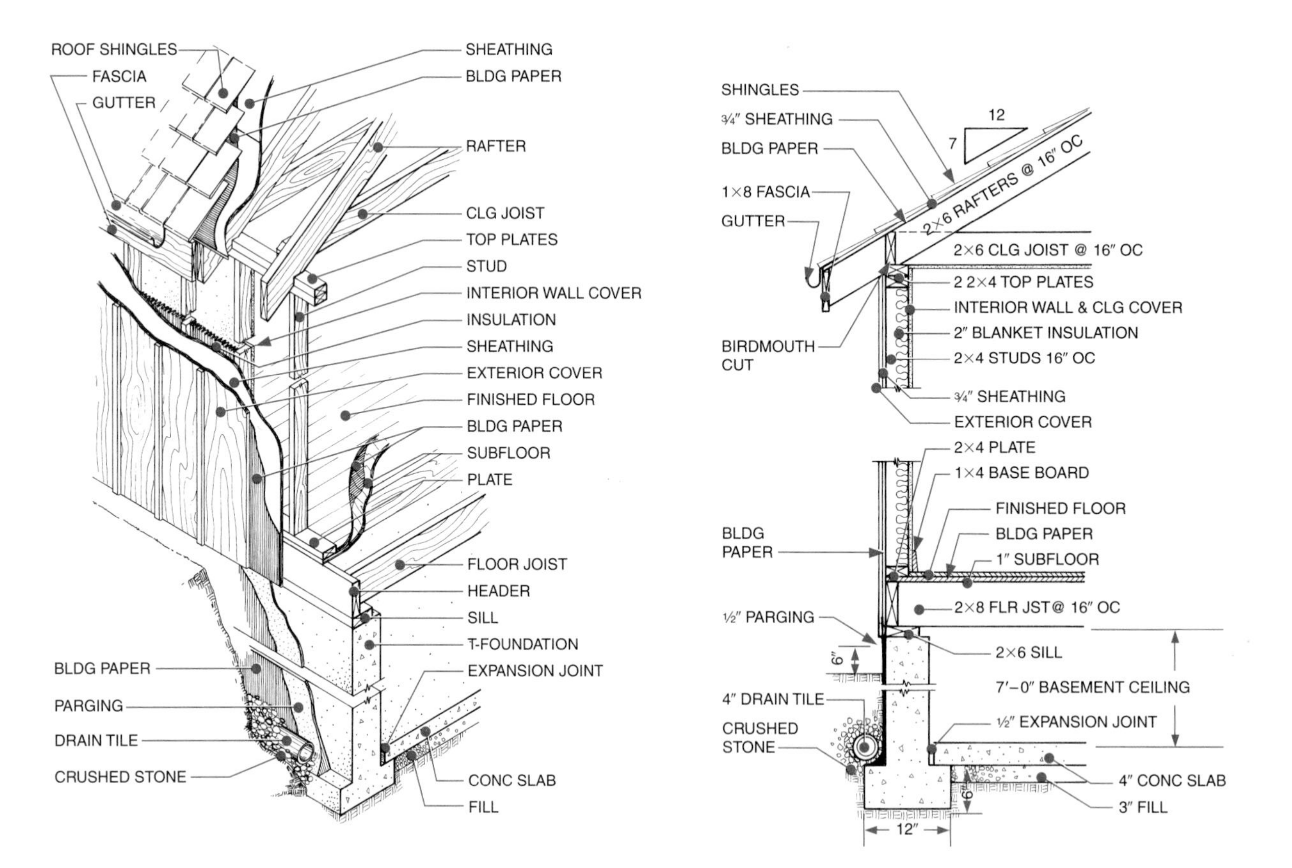

Fig. 17-13 ■ Use of break lines on pictorial and orthographic detail sections.

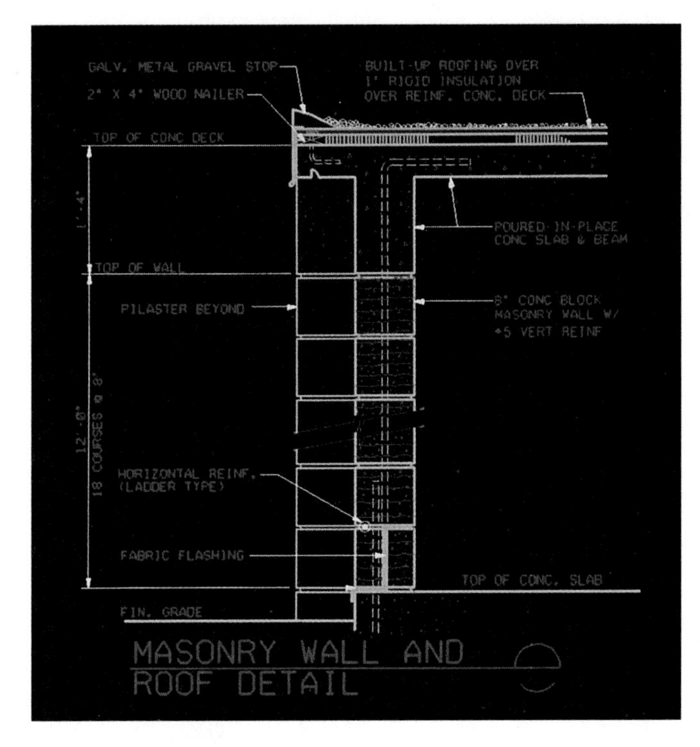

Fig. 17-14 ■ Cornice section prepared on a CAD system. *Autodesk, Inc.*

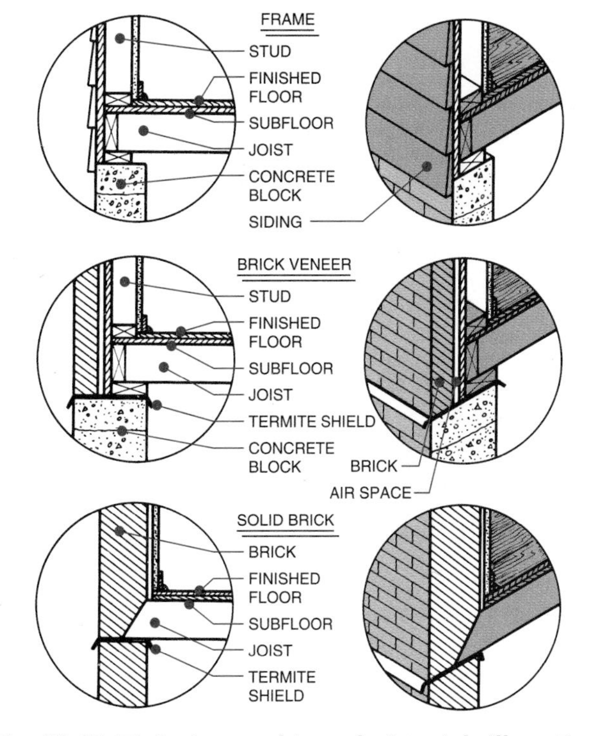

Fig. 17-15 ■ Orthographic and pictorial sill sections.

Beam details are necessary to show how the joists are supported by beams and how the columns or foundation walls support the beams. As for all sections, the position of the cutting-plane line is extremely important. Figure 17-17 on page 332 shows two possible positions of the cutting plane. If the cutting-plane line is placed parallel to the beam, you see a cross section of the joist, as shown in drawing A. If the cutting-plane line is placed perpendicular to the beam, you see a cross section of the beam, as shown in drawing B.

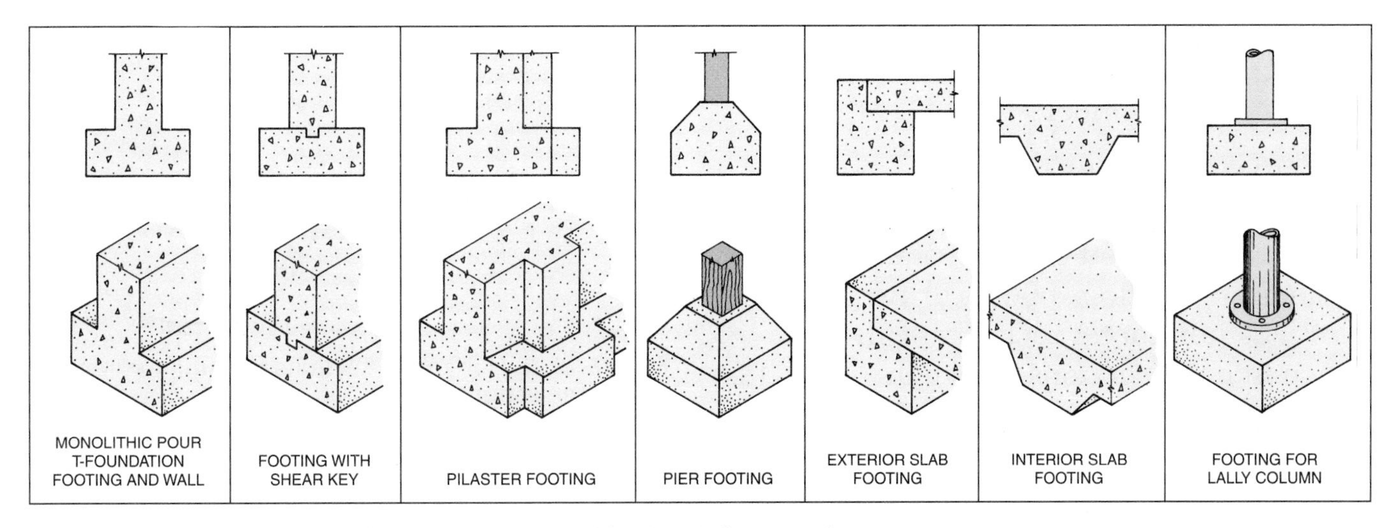

Fig. 17-16 ■ Orthographic footing sections with pictorial comparisons.

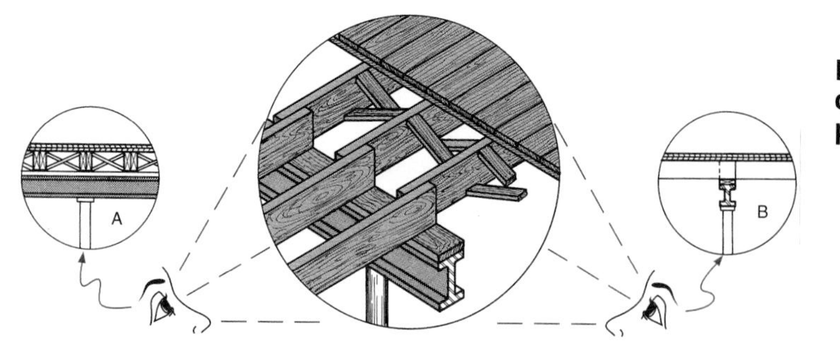

Fig. 17-17 ■ What is seen in a beam detail drawing depends on the position of the cutting plane.

INTERIOR WALLS To illustrate the methods of constructing inside partitions, sections are often drawn of interior walls at the base and at the ceiling. A base section shows how the wall-finishing materials are attached to the studs and how the intersection between the floor and wall is constructed. A crown section shows the intersection between the ceiling and the wall and how the finish construction materials of the wall and ceiling are related. Sections may also be prepared for stair and fireplace details.

Horizontal Wall Sections

Horizontal wall sections of exterior and interior walls are drawn to clarify how walls are constructed. Walls in these sections are similar to those on floor plans, but are drawn at a larger scale in order to show a horizontal sectional view of each construction member.

EXTERIOR WALLS Although a floor plan is a horizontal section, many construction details are omitted because of the small scale used. If a floor plan is drawn exactly as a true horizontal section, it will look much like the sections shown in Fig. 17-18. However, more information is usually needed to describe the exact construction of corners, intersections, and window and door frames. Larger horizontal sections of these are prepared. See Fig. 17-19. Such

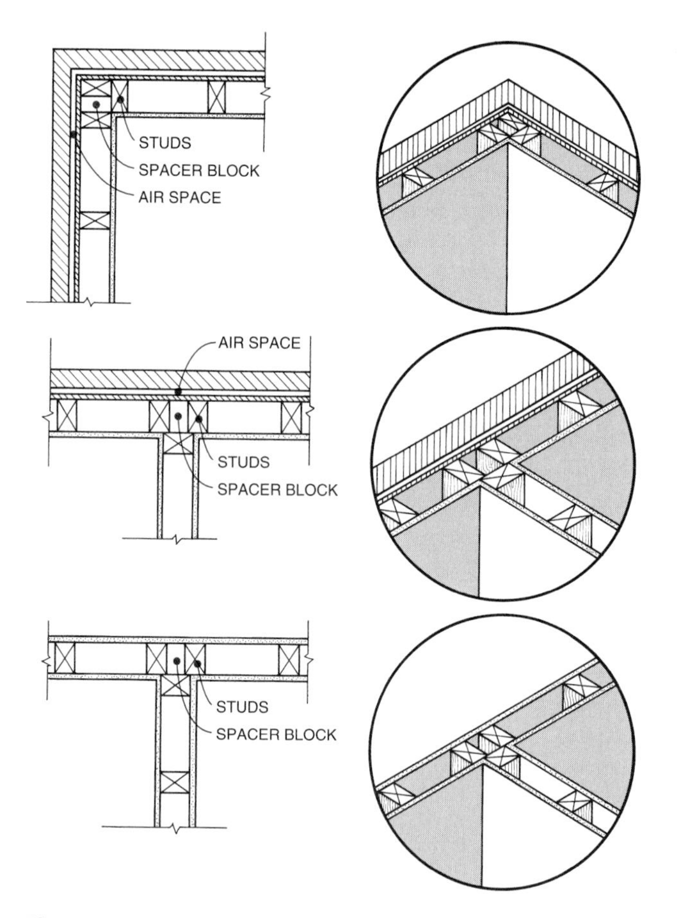

Fig. 17-18 ■ Horizontal sections through exterior wall intersections.

Fig. 17-19 ■ Corner of the living room shown in the floor plan in Fig. 14-20.

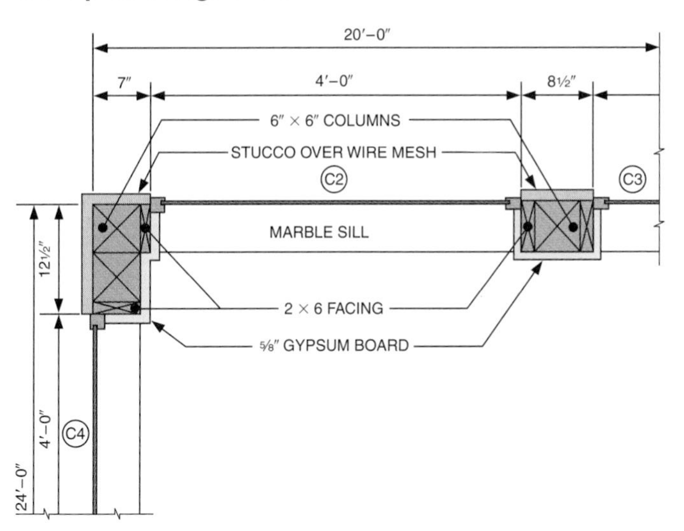

A. Horizontal wall section detail.

B. Area under construction.

C. Completed area.

drawings are known as stud layout details. Stud layouts are covered in detail in Chapter 29.

INTERIOR WALLS Typical horizontal sections of interior wall intersections indicate construction methods. For example, horizontal sections are needed to show the inside and outside corner construction of a paneled wall and how paneled joints and other building joints are constructed. See Fig. 17-20. Horizontal sections are also very effective for illustrating various methods of attaching building materials.

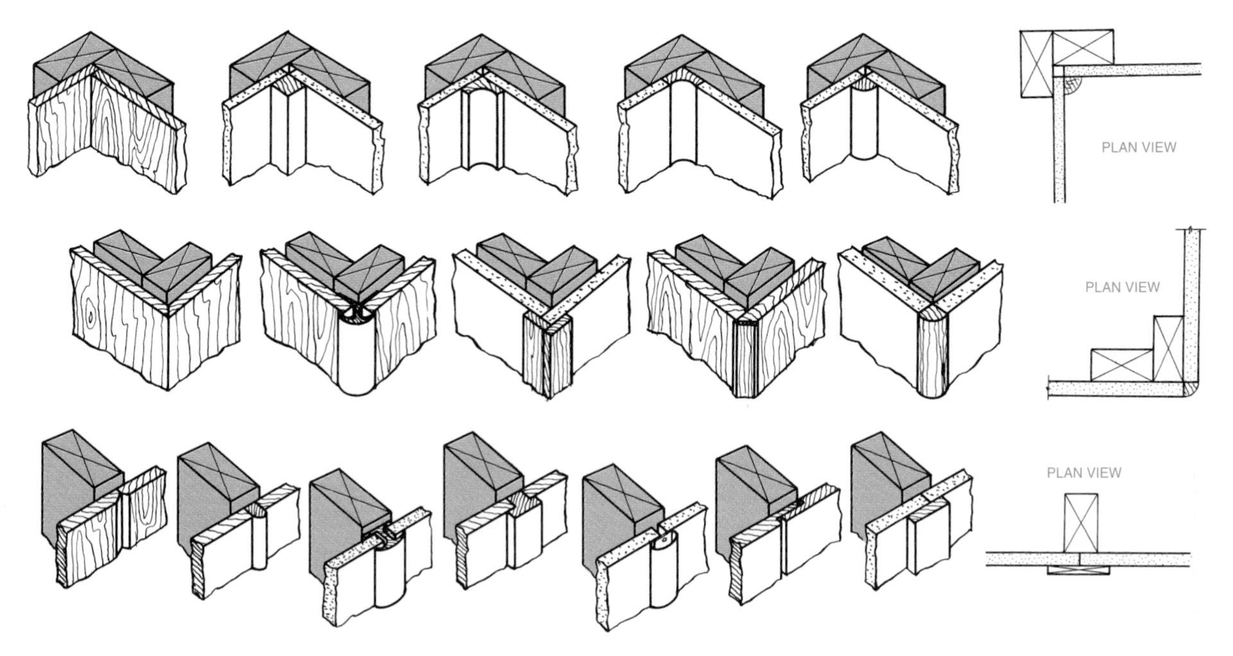

Fig. 17-20 ■ Sections showing interior panel-wall intersections.

Window Sections

Because window construction is hidden, sectional drawings are needed to show construction details. Figure 17-21 shows both horizontal and vertical window construction areas that are commonly sectioned. These include head, sill, and jamb construction. Window manufacturers generally use pictorial sections to illustrate key features and methods of installation. See Fig. 17-22.

Head and sill sections are vertical sections. Preparing head and sill sections on the same drawing is possible only when a small scale is used. If a larger scale is needed, the head and sill must either be drawn independently or break lines must be used. Figure 17-23 shows the relationship between the cutting-plane line and the head and sill sections. The circled areas in Fig. 17-24 show the areas that are removed when a separate head and sill section is prepared.

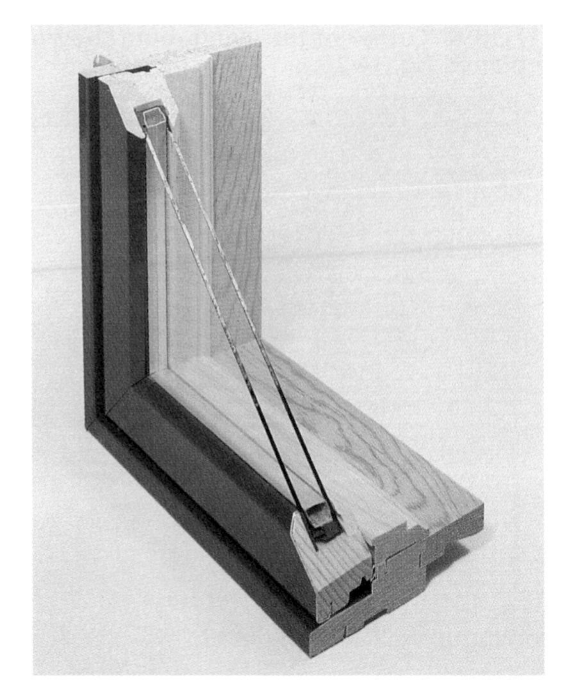

Fig. 17-22 ■ Pictorial section showing window construction. *Kolbe and Kolbe*

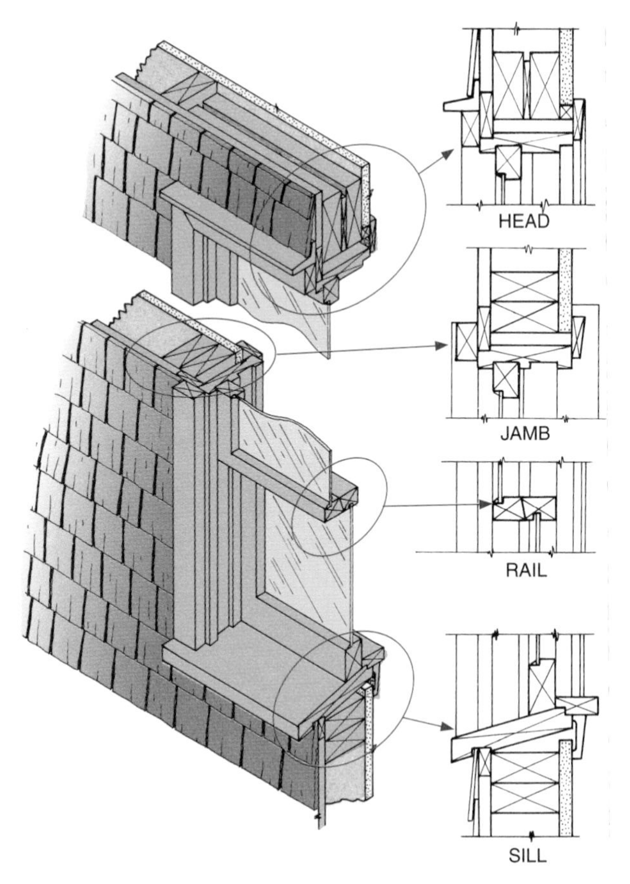

Fig. 17-21 ■ Window areas commonly sectioned.

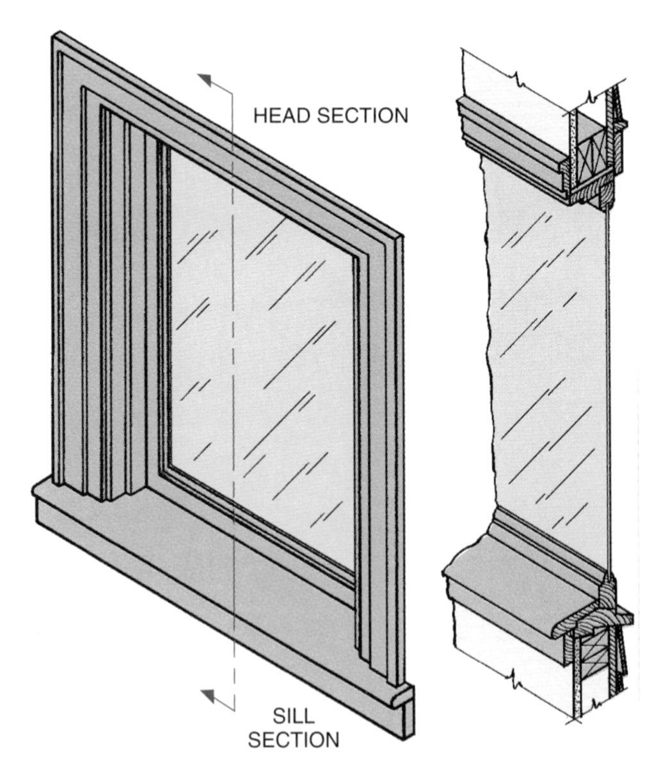

Fig. 17-23 ■ The head and sill sections are in the vertical plane.

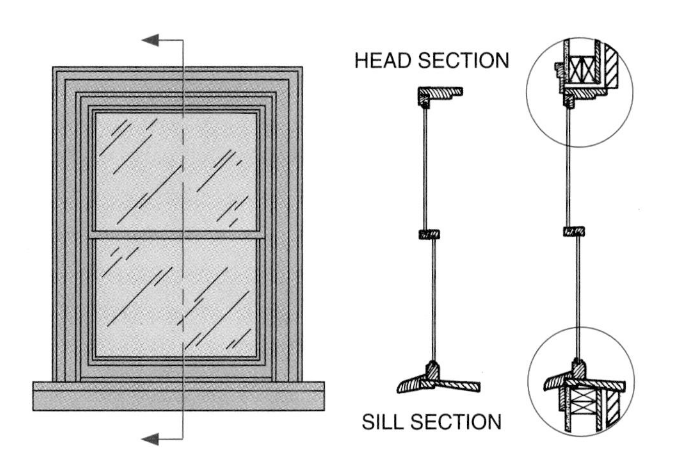

Fig. 17-24 ■ Projection of window head and sill sections.

When a cutting-plane line is extended horizontally across the entire window, the resulting sections are known as jamb sections. Jamb details (the horizontal section) are projected from the window-elevation drawing. See Fig. 17-25. The construction of both jambs is usually the same, with the right jamb drawing being the reverse of the left. Only one jamb detail is normally drawn. The builder then interprets one jamb as the reverse of the other.

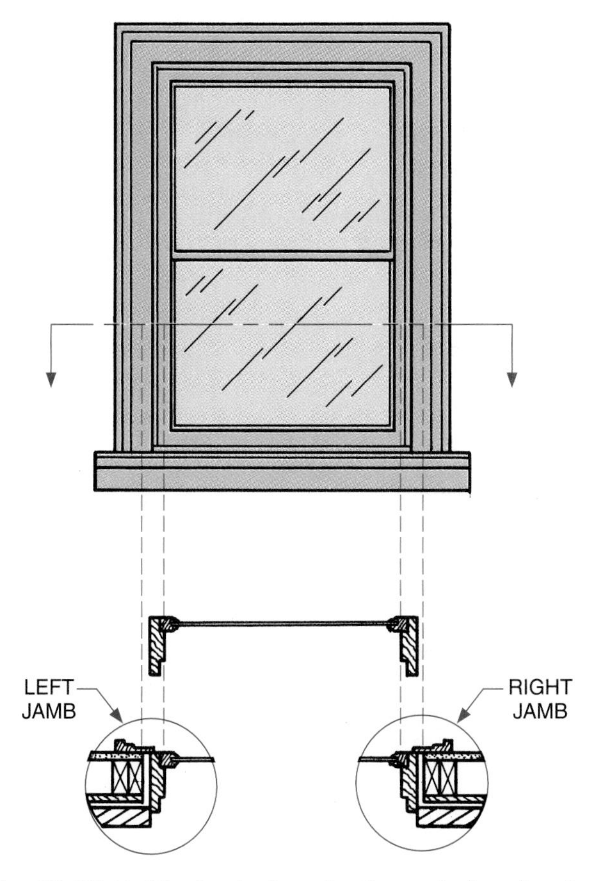

Fig. 17-25 ■ Method of projecting window jamb sections.

Door Sections

A horizontal section of all doors is indicated on a floor plan. However, a floor plan does not include sufficient detail for installation. Similar to window construction sections, sill, head, and jamb sections are necessary to show door construction.

When a cutting-plane line is extended vertically through the sill and head, a section as shown in Fig. 17-26 is revealed. These sections are often too small to show the desired degree of detail necessary for construction. A removed section is drawn at a larger scale to show head or sill sections. See Fig. 17-27.

Since doors are normally not as wide as they are high, an adequate jamb detail can be projected without the use of break lines or removed sections. See Fig. 17-28.

Sectional drawings of the framing details of the door sill, head, and jamb, exclusive of the door and door frame

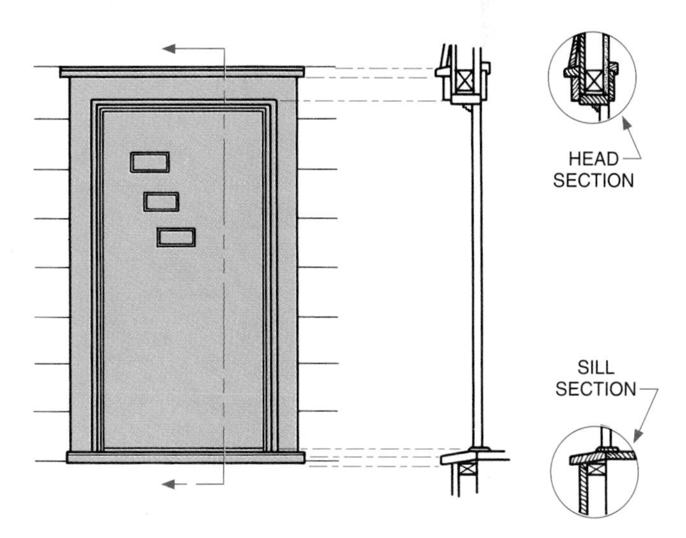

Fig. 17-26 ■ Projection of door head and sill sections.

assembly, are often prepared for framing purposes. In such drawings, the framing section is drawn separately, with the door frame and door removed. Usually, however, door sections are prepared with the framing, trim, and door in their proper locations. See Fig. 17-29. Drawings of garage doors and industrial-size doors are usually prepared with sections of the brackets and apparatus necessary to house the door assembly.

In addition to vertical and horizontal sections, break-out sectional drawings are prepared to show the internal construction of components. In Fig. 17-30, this method is used to show door construction and other types of internal construction.

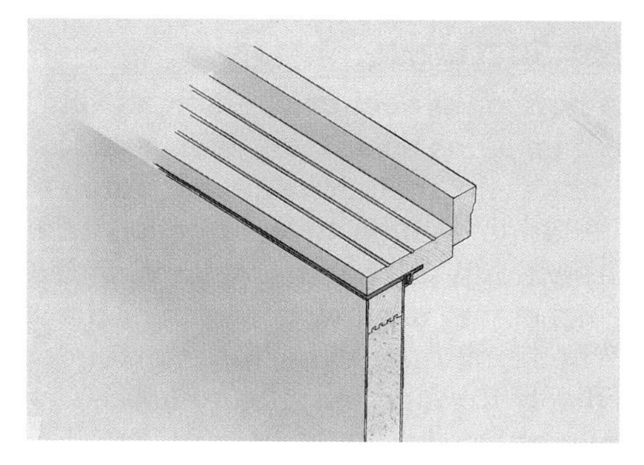

Fig. 17-27 ■ Removed head section of a door. *Kolbe and Kolbe*

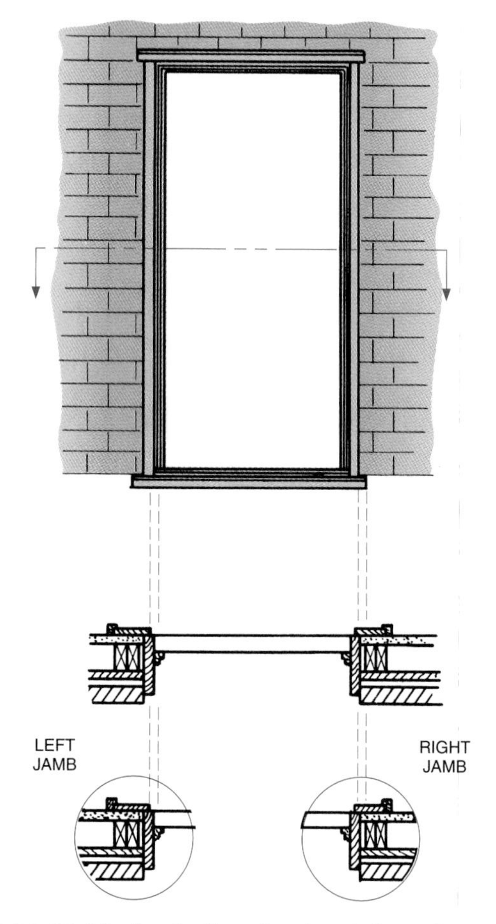

Fig. 17-28 ■ Method of projecting left and right jamb sections from the door elevation drawing.

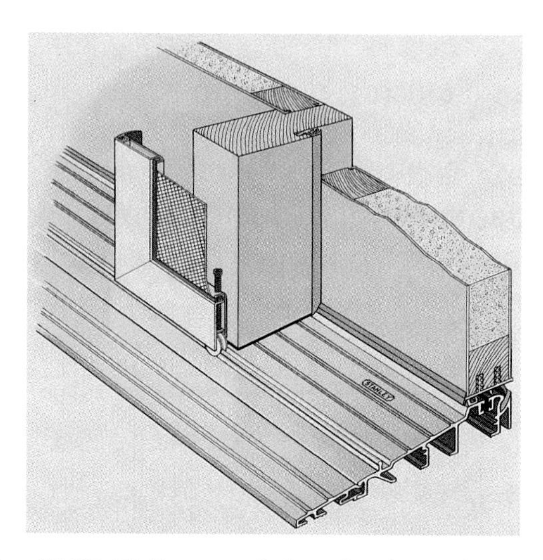

Fig. 17-29 ■ Removed door-jamb section, including all related components. *Kolbe and Kolbe*

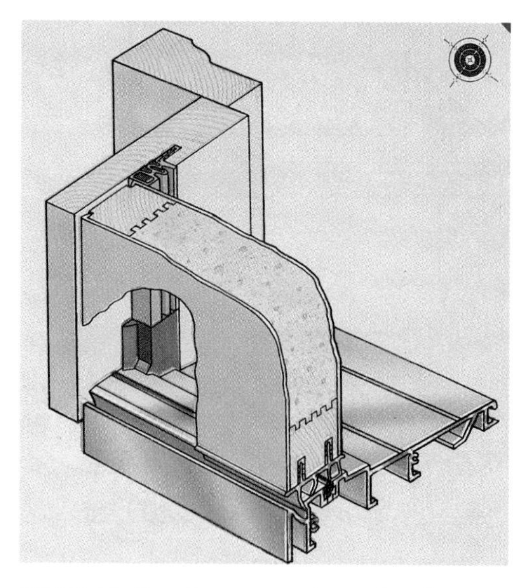

Fig. 17-30 ■ Break-out section of a door assembly. *Kolbe and Kolbe*

Cabinetry and Built-in Component Drawings

Basic architectural plans are prepared at a scale too small to show the exact size and construction details of many cabinets and built-in components. On simple designs, details are often explained with a note or reference to a manufacturer's product. On larger, more complex designs, separate details are provided in pictorial, plan, elevation, and/or sectional drawings.

Cabinet Construction and Types

Cabinets are either custom-built (made to order) or manufactured using modular sizes. In either case, the quality of the finished product depends on the materials, joints, hardware (hinges, pulls, latches), finish, and the accuracy of construction and installation. The quality of components range widely from economy units, which use the least expensive materials and methods, to premium components, which resemble fine furniture.

Cabinets are either wall-hung or positioned on the floor (base cabinets). See Fig. 17-31. Numerous styles and sizes are

Fig. 17-31A ■ Common dimensions of kitchen cabinets.

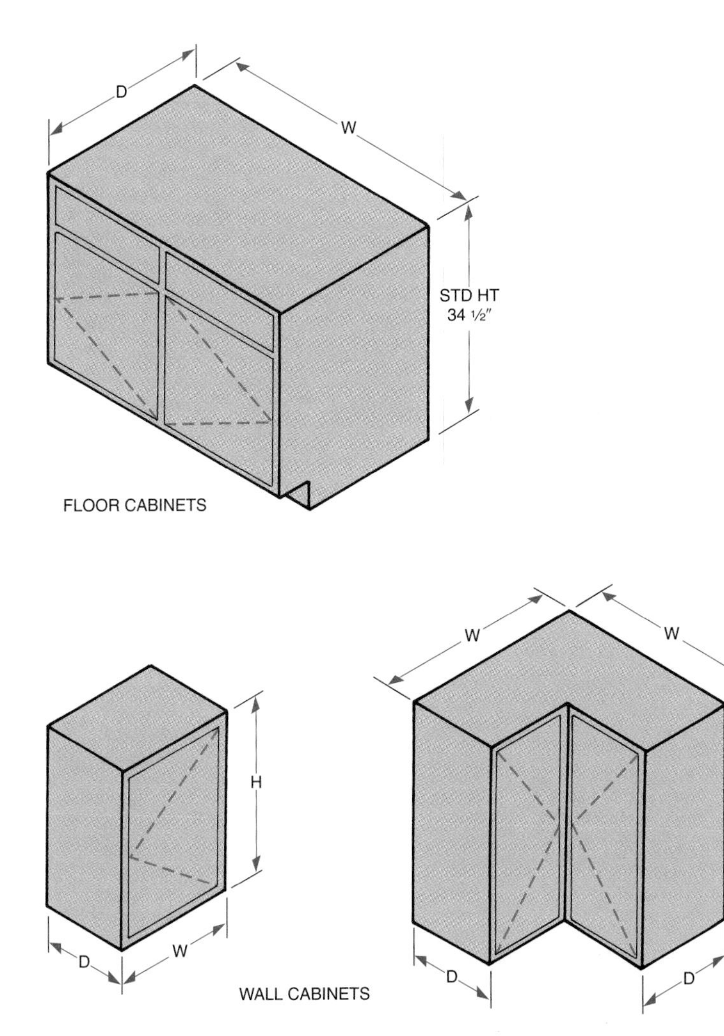

Fig. 17-31B ■ Kitchen cabinet schedules.

Sym	Amt	Depth	Height	Width
Base cabinets				
A	1	24″	34 1/2″	36″
B	2	24″	34 1/2″	36″
C	1	24″	34 1/2″	21″
D	1	24″	34 1/2″	48″
E	1	24″	34 1/2″	9″
Wall cabinets				
1	1	12″	36″	36″
2	1	12″	18″	42″
3	2	12″	36″	12″
4	2	12″	36″	24″
5	1	12″	36″	30″
6	1	12″	36″	18″
7	1	12″	18″	30″
8	1	12″	30″	21″

available, as shown in Fig. 17-32. Most modular base cabinet sizes are standardized at 34½″ in height and 24″ deep. Modular wall cabinets are usually 12″ deep. Custom-built wall cabinets are usually 15″ deep and floor cabinets are 26″ deep. Cabinets are manufactured for baths, kitchens, or laundry rooms.

Materials used in the construction of cabinets and built-in components include hardwood, softwood, stranded lumber, laminates, ceramic tile, marble, and synthetic materials. See Fig. 17-33. Though not always seen, the types of joints and fasteners used to attach parts has an important effect on the overall appearance and durability of cabinets. For example, simple butt joints are typically used on economy cabinets, while joints such as dovetails and mortise and tenons are used in premium units. See Fig. 17-34.

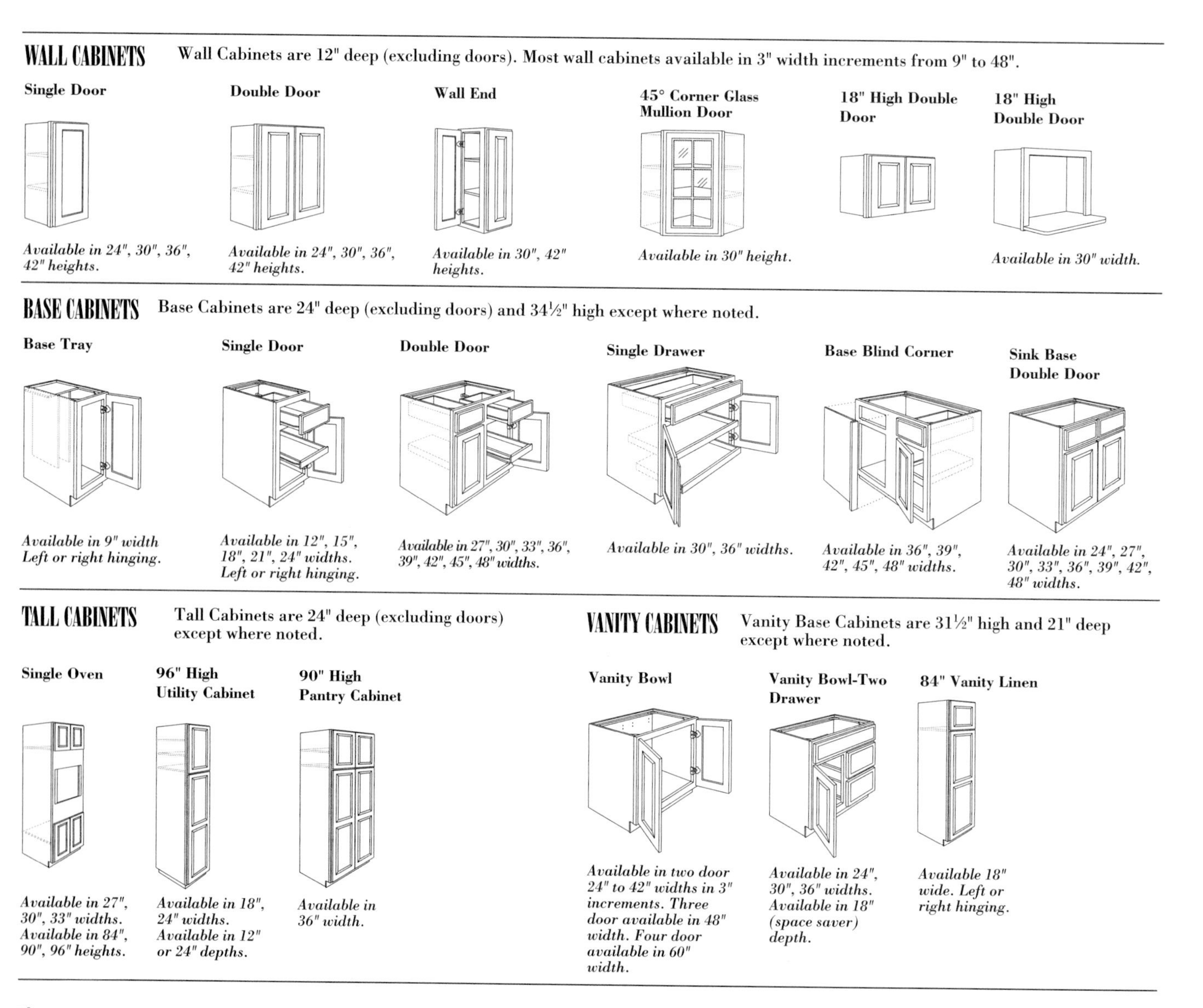

Fig. 17-32 ■ **Standard modular dimensions of common cabinets.** ***Merillat Industries, Inc.***

Fig. 17-33 ■ Most manufactured cabinets are used in kitchens. *Merillat Industries, Inc.*

Built-in Components

Because of the intricate details involved, only the outlines of most built-in components are drawn on floor plans and elevation drawings. Separate, large-scale dimensioned details are used to show construction and/or installation information.

Most built-in component designs include precise joinery, hidden hinges, roller sliding parts, and special hardware. Manufactured units for built-in products are usually prefabricated in modular units. The surrounding finished carpentry must be designed to blend with the unit.

Fig. 17-34 ■ Types of joints used in the construction of cabinet framework and drawers.

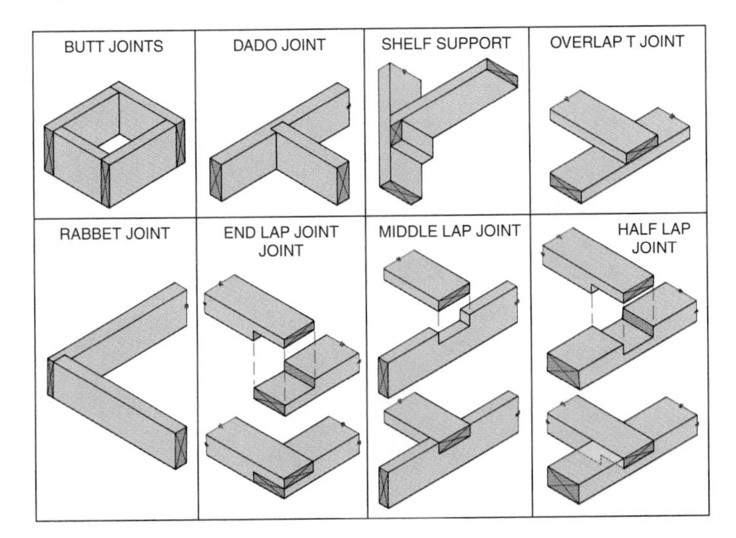

A. Joints used in economy cabinets.

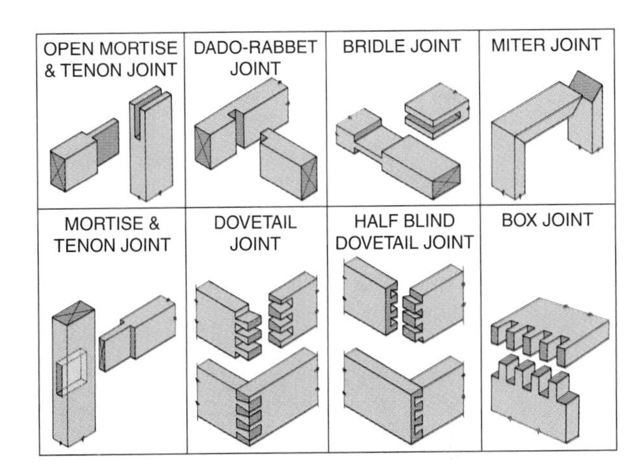

B. Joints used in custom cabinets.

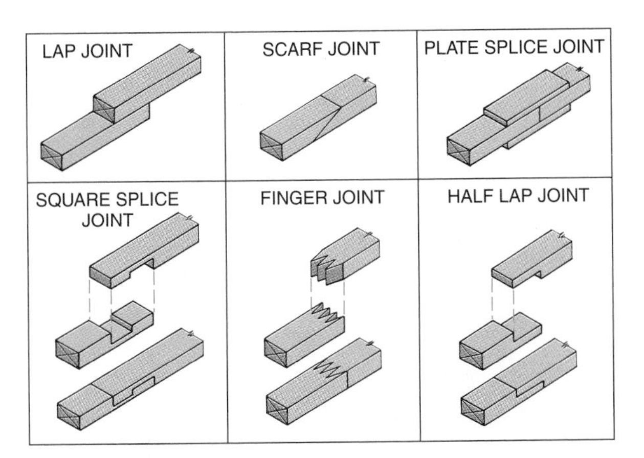

C. Typical end joints.

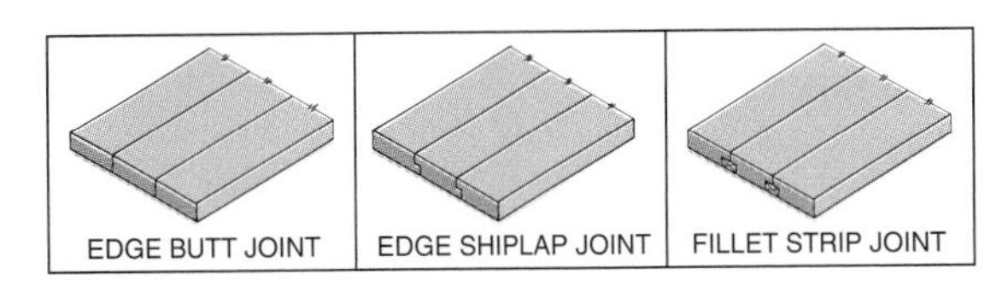

D. Typical edge joints.

This requires framing and detail drawings. When prefabricated units are not used, special framing drawings of the walls, shelves, fascia, and soffits are needed. See Fig. 17-35. Built-ins of simpler design, such as shelves, mantels, and planters are usually built on site. These may only require an interior elevation plus floor plan notes and dimensions. See Fig. 17-36.

Fig. 17-35 ■ Design and construction of built-in entertainment center.

½″ PLY COVER
36″ FLUORESCENT FIXTURE
ALIGN W RAILING WALL TOP
DINING ROOM CEILING LINE
SMOOTH FIN GYP BRD
DR OUTLET
TO AIR EXCHANGE
6″
2′–9″
2×4 FRAMING
5½″
2nd FLOOR LINE
36″× ¾″ × ¾″ MINI LIGHT STRIP
19″
13″
1′–1½″
ALL SHELVES ¾″ VENEER PLY W ½″ × 1½″ PINE NOSING
12″ RECORDS
1′–3″
CD'S
10″
1′–11″
AUDIO COMPONENT LEDGE
23″
1½″
TUBE WELL
26″
3′–6″
¾″ T&G SUBFLOOR
12″
SECTION A-A

A. Detail drawing.

B. Area under construction.

C. Completed area.

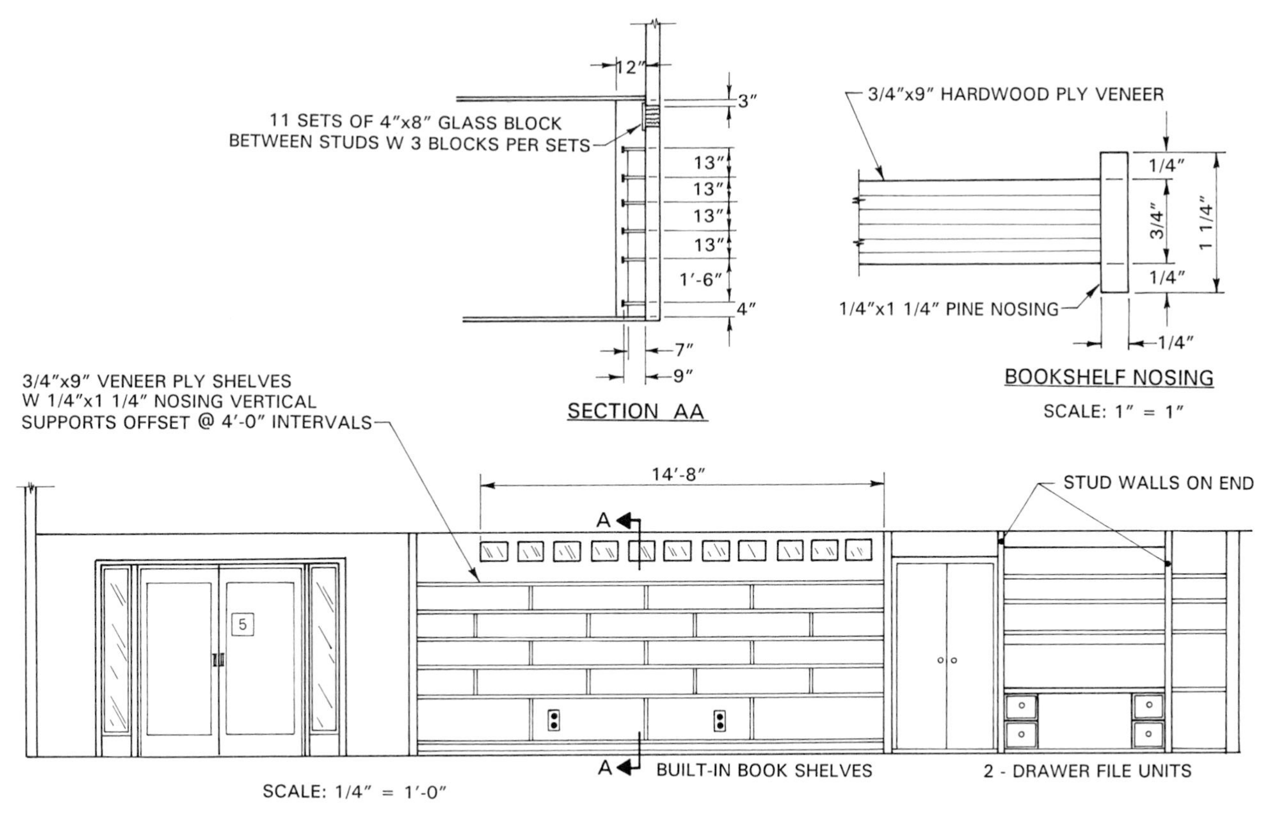

A. Elevation drawing.

B. Completed shelves.

Fig. 17-36 ■ Design and construction of bookshelves.

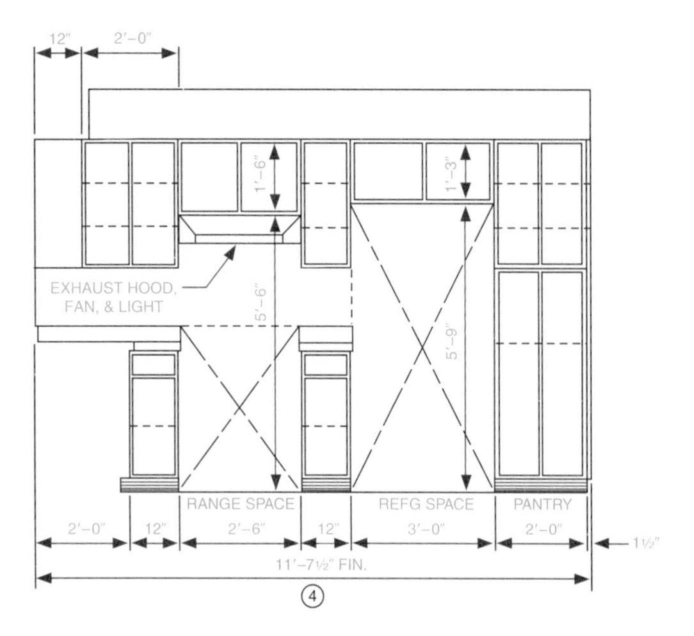

Fig. 17-37 ■ Dimensioning of space reserved for manufactured components.

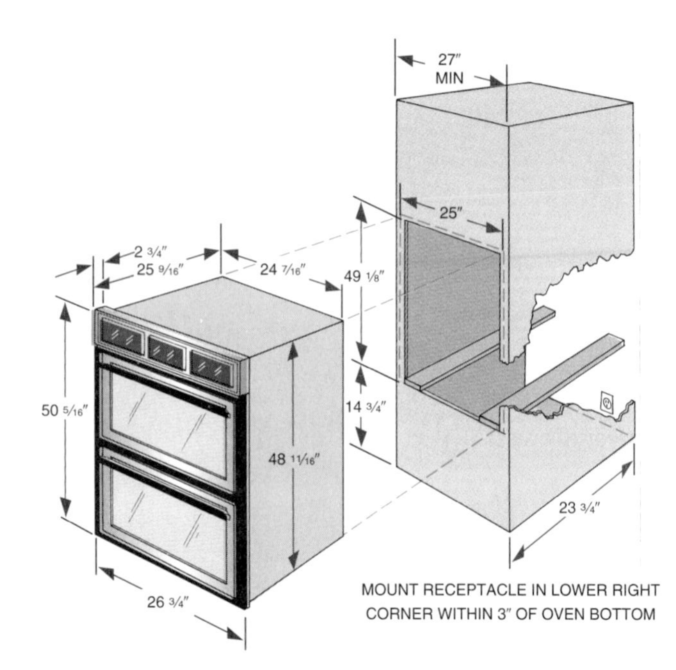

Fig. 17-38 ■ Positioning drawing showing the dimensions for the component and its housing.

Dimensioning Cabinetry and Built-ins

When cabinets or built-ins are custom-made or built on site, normal dimensioning practices are used. When factory-produced components are used, they must be precisely positioned with other components, cabinets, and/or framing. Appliances and plumbing fixtures are designed to stand alone, slide into a space between cabinets, or "drop into" (fit within) a countertop space. In all these cases, adequate space must be dimensioned on plans and elevations or detailed to ensure proper fitting. See. Fig. 17-37. Some cabinetry must be built to accommodate all component dimensions (width, height, and depth). A drawing that shows the dimensions of the component and housing is then prepared for accurate positioning. See Fig. 17-38. Positioning drawings are often included in the manufacturer's specifications.

Usually many sizes of models of a manufactured product are available. An alphabetical dimensioning system is used, like the one charted in Fig. 17-39 to show model size differences.

When unusual design features are used, drawings of the surrounding framing must be detailed. See Fig. 17-40. This is necessary whether the components are site-built or factory-built. Normally cabinets are factory-built and assembled on site.

Fig. 17-39 ■ An alphabetical dimensioning system allows for model size differences.

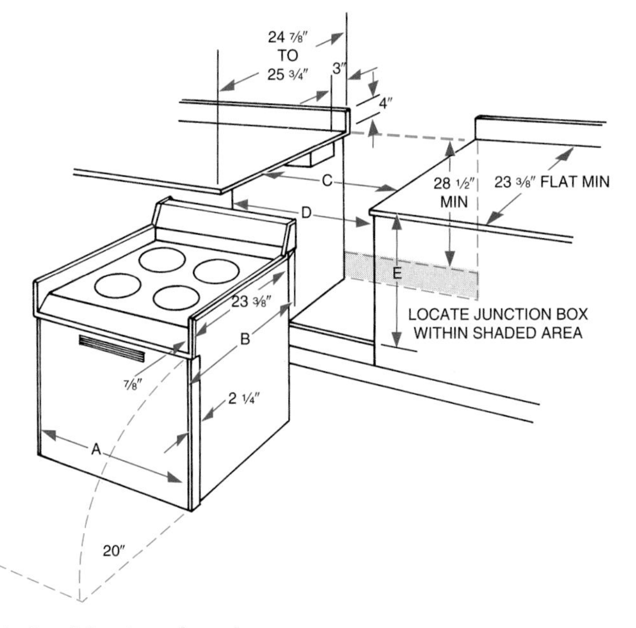

A. Positioning drawing.

Fig. 17-39B Cabinet and cutout dimensions.

Model	A	B	C	D	E
1	26 3/4″	24 1/4″	26″	27″	26 1/16″
2	26 3/4″	24 1/2″	26″	27″	28 7/16″
3	24 3/8″	24 3/4″	22 1/2″ min.	27″	34 1/2″
4	44 1/2″	24 3/4″	22 1/2″ min.	27″	14 3/4″
5	28 1/8″	22″	24″ min.	24″	32 1/2″
6	28 1/8″	24 3/4″	24″ min.	27″	32 1/2″
7	49 5/8″	24 3/4″	24″ min.	27″	13 1/4″
8	47 1/4″	24 3/4″	24″ min.	27″	15 5/8″

Fig. 17-40 ■ Design and construction of kitchen cabinet area. (continued next page)

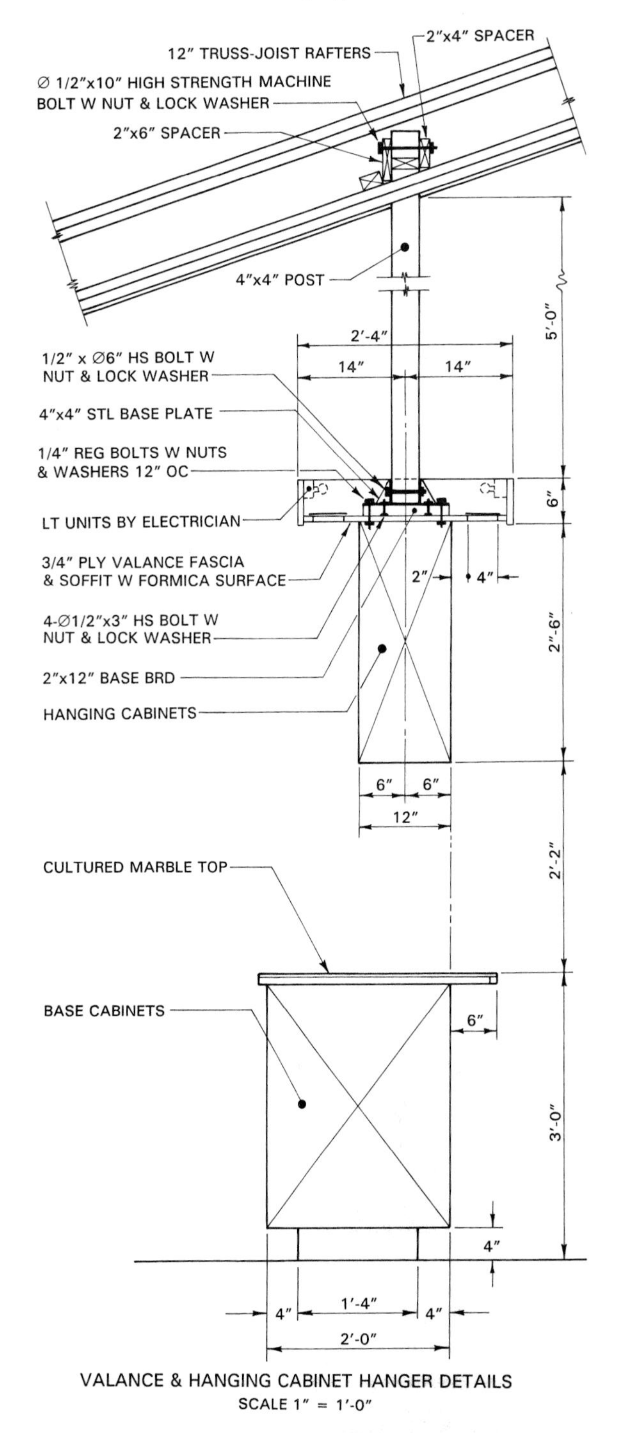

A. Detail drawing.

B. Completed area.

A shortcut method of dimensioning cabinets is the use of a **cabinet coding system.** The manufacturer's code number for standard modular units is shown on the cabinet outlines drawn on a floor plan. See Fig. 17-41. This system is also used for custom-made cabinets by referencing the code number to a detailed drawing of each cabinet.

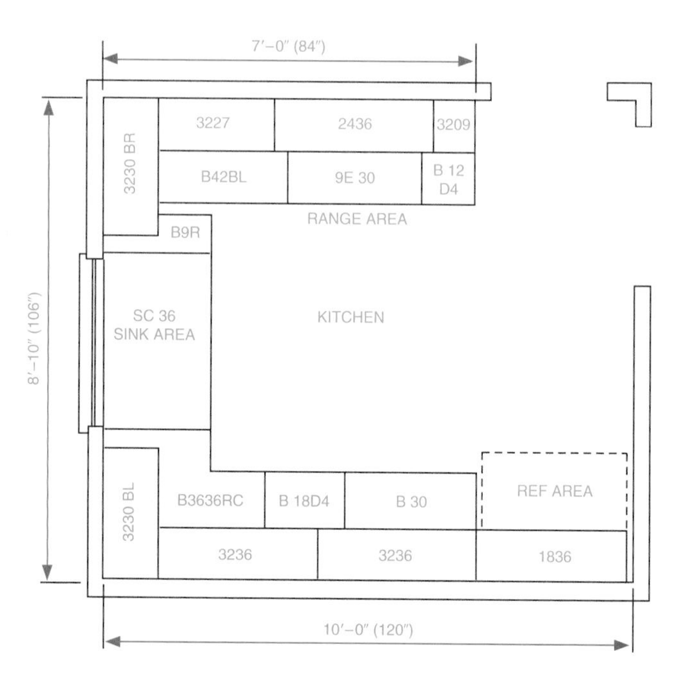

Fig. 17-41 ■ A cabinet coding system may be used to locate modular cabinet units on a floor plan.

CHAPTER 17 Exercises

1. Draw a full section of a house you have designed.
2. Draw a section through the view shown in Fig. 17-10, revolving the cutting-plane line 90°.

3. Draw a head, jamb, and sill section of a typical window and a typical door of the house you have designed.

4. Draw a sill, cornice, and footing section of the house you have designed.

5. Complete Exercises 2 through 4 using a CAD system.
6. Name the two methods of cabinet construction and describe what affects their quality.
7. List materials used in cabinet construction.

8. Draw a coded cabinet plan for the kitchen and bath of the house you are designing.
9. Complete Exercise 8 using a CAD system.

CHAPTER

18

Site Development Plans

Landscape architecture is primarily concerned with the use of space and the integration of landform, site character, and architecture. Achieving this goes beyond simply planting trees and shrubs. It involves the development of the entire site. Site development is an integral part of the design process. The design sequence presented in Chapter 13 should be followed carefully when developing a site plan. A site design should provide proper orientation and use of natural features of the site.

Site plans describe the characteristics of the land and the relationship of all structures to the site. The outline and dimensions of all constructed features (buildings, driveways, etc.) and their exact position on the site are shown on site plans. Also included are the shape of the landform as well as the locations and types of plants, such as trees. Specialized site plans include survey plans, plot plans, plats, landscape plans, and renderings. These are discussed in this chapter. Various features of these plans are often combined into one composite site plan.

■ **A good landscape design integrates a structure with its surroundings and makes the site more useful and attractive.** *Alan Goldstein Photography*

Objectives

In this chapter, you will learn to:

- identify the major elements used in site design.
- understand the role and uses of zoning ordinances in the design process.
- draw survey, plat, and plot plans.
- understand the polar coordinate system and its application to site plans.
- design, draw, and render landscape plans and elevations.

Terms

building envelope
building permit
contour lines
density
landscape plans
phasing
planting schedules
plat
plot plans
setbacks
survey
zoning ordinance

Site Analysis

Completing a site analysis is the first step in producing an acceptable site design. The design should meet the needs of the user, as well as protect and enhance the environment. Future inhabitants of the site must also be considered. Both environmental and human-related elements that influence development and design are analyzed in the site analysis. See Figs. 18-1 and 18-2. The surrounding area of a site should also be considered to determine future plans. For example, commercial zoning changes may affect plans for constructing a residence.

Suitability Levels

Environmental and human-related elements must be analyzed to determine the level of the site's suitability for development and building.

- *High suitability*: Many favorable conditions exist to make this area relatively inexpensive to develop with a minimum amount of environmental impact.
- *Moderate suitability:* Some special design and construction measures will be needed to modify this land and preserve the environment.
- *Low suitability:* Conditions exist that place serious restrictions on building in this area. For example, either environmental damage will result if the site is disturbed, or high construction costs will be needed to avoid damage to ecosystems or to protect the public.
- *Not suitable*: Disturbance or impact to this area will cause significant environmental damage or adversely impact the public safety and welfare. Costs to develop the area are excessive.

- Slope
- Soils
- Vegetation
- Wildlife and factors related to habitat, especially if rare and endangered species inhabit the site
- Hydrology:
  - Surface water
  - Flood hazard
  - Groundwater
  - Wetlands (FWW/Tidal)
- Climate (regional) and microclimate (specific to site)
- Geology
- Visual character

Fig. 18-1 ■ Environmental elements influencing site design.

- Existing site, street layout, and topography
- Existing land use and zoning
- Historical significance and preservation
- Available utilities
- External factors
  - Noise
  - Site accessibility
- Demographics (population characteristics)
- Socioeconomic forecasts

Fig. 18-2 ■ Human-related elements influencing site design.

Suitability levels may apply to the entire site or to selected zones of a site. Refer back to the composite analysis shown in Fig. 13-11.

Zoning Ordinances

Before beginning the design concept, all local *zoning ordinances* must first be thoroughly checked. **Zoning ordinances** are laws or regulations designed to provide safety and convenience for the public and to preserve or improve the environment.

Local building codes must also be checked prior to beginning a design. Redesigning or redrawing plans, if they are in conflict with laws, is very time-consuming and costly. Working drawings should not be started until the basic site plan is approved by the local zoning authorities.

Specific zoning ordinances may differ among communities. However, zoning categories are very similar. For example, zoning laws specify the type of occupancy, population density (number of persons in an area), land use, and the building type allowed in each zone of a community. Maps, usually called *tax maps*, are found in each community's building department.

Most codes divide municipalities (cities) into residential, commercial, and industrial zones. Residential zones are divided into single-family dwellings, multiple-family dwellings (duplex, triplex, and quadraplex), and apartments which include units for five or more families. Commercial zones include offices, retail stores, and medical facilities. Industrial zones include factories, warehouses, or any facility requiring the movement and/or storage of large vehicles or equipment. Where appropriate, separate zones are established for hazardous areas, such as those where the danger of flooding exists.

Structural Types

The types of structures allowed in each community are related to the designated use of each zone. A community committee often regulates these ordinances. Zoning ordinances may be intended to maintain a degree of architectural consistency within a given area. Ordinances of this type may restrict or allow only specific styles, periods, materials, landscaping, heights, colors, or sizes.

Maximum and sometimes minimum building sizes are often specified to control the use of space, traffic, and the impact on the environment. For example, a single-family residential zone in one community may require a minimum size home of 2000 sq. ft. and a maximum of 5000 sq. ft. Another community may require a minimum of 3000 sq. ft. with an unlimited maximum area. In addition to the amount of square footage allowed, some codes specify maximum building width and length with reference to street frontage and property lines.

All codes now include maximum building heights to allow neighbors maximum access to views, air circulation, and sunlight. A maximum height of 35 ft. is often used for residential zones. See Fig. 18-3.

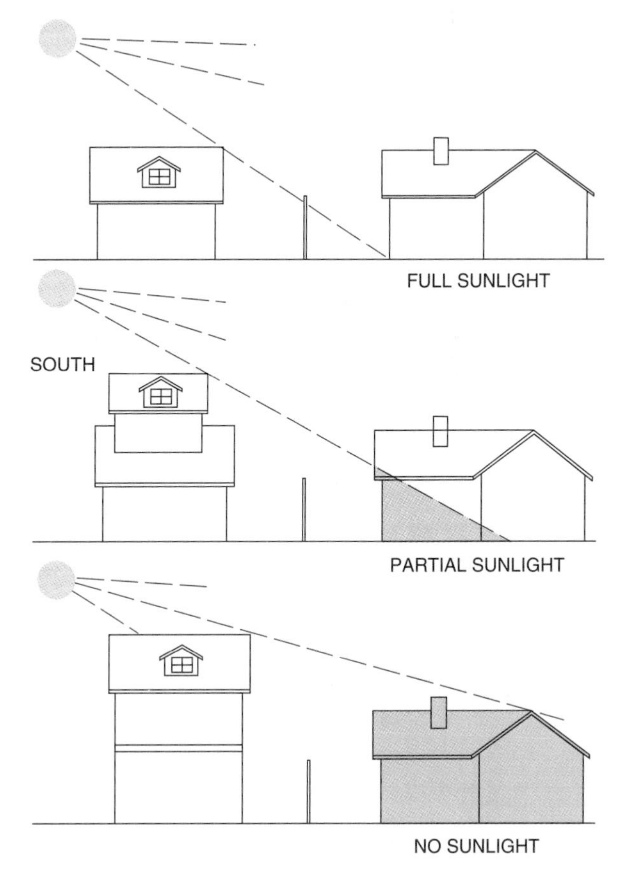

Fig. 18-3 ■ Many codes restrict blockage of sunlight on adjacent buildings.

Many newer codes now include limits on the amount of space used in upper floors. This pyramid principle, called the *daylight plane,* is required to allow more light to reach adjacent properties located on the north side. See Fig. 18-4. The impact of shadowing is becoming more critical as building areas become more densely built. Using constructed models or computer-drawn models, shadow patterns are studied and checked during different seasons and times.

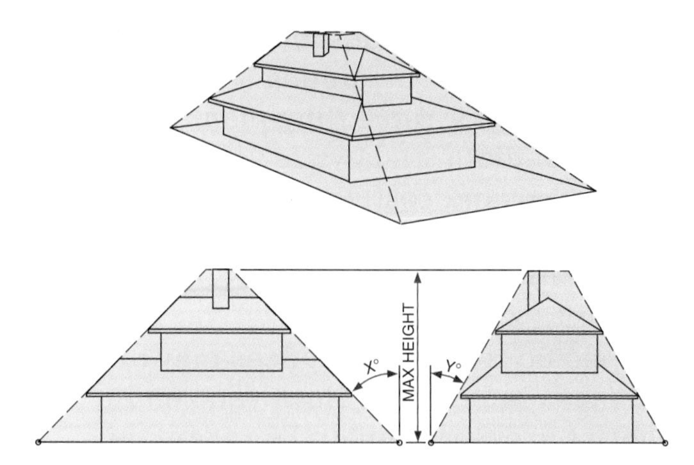

Fig. 18-4 ■ Giving the house a pyramid shape by making the second story smaller than the first level allows more light to reach adjacent buildings.

Land Coverage and Setbacks

Regardless of the square footage of a building, laws may also restrict the percentage of property allowed for building. This is done to preserve as much *green space* as possible. Percentages vary from one community to another, usually from 25% to 40%. For example, a small site, 50′ × 120′, with a coverage of 40% can contain a maximum size building of 2400 sq. ft. See Fig. 18-5.

Zoning laws include codes that restrict the distance permitted from any building to the property lines. These distances are known as **setbacks**. Refer again to Fig. 18-5. Some codes require that a structure be placed no closer than 5′ from a prop-

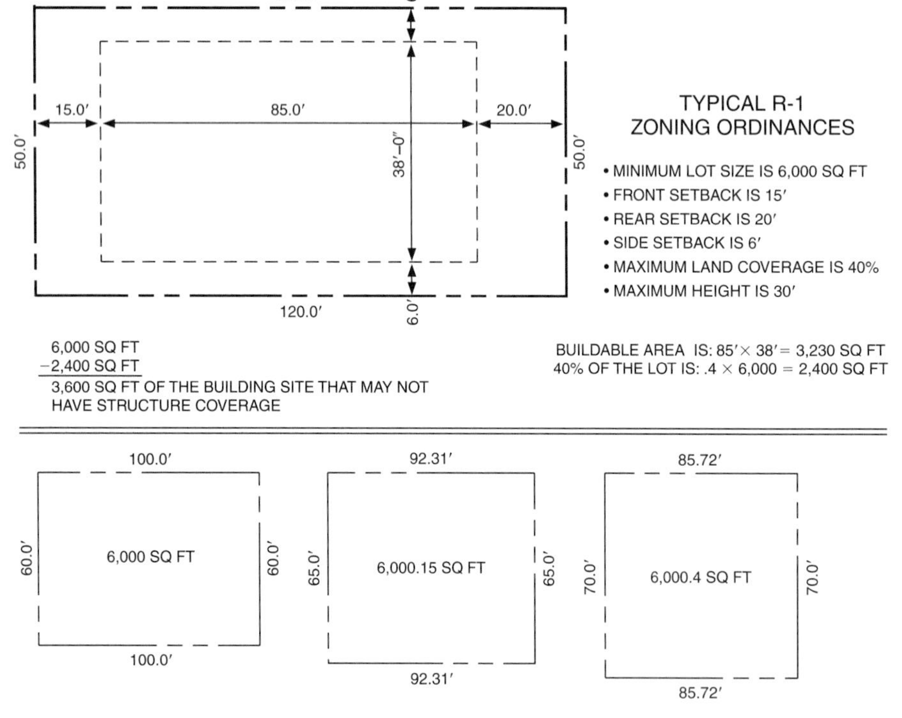

Fig. 18-5 ■ Example of zoning ordinances applied to a building lot.

erty line. Other codes may require 10′ or more. Setback distances may be different between front, side, and rear property lines.

Setback lines are drawn within and parallel to property lines to indicate the acceptable building area for a lot. Setbacks and other zoning dimensions also vary for pools, garages, corner lots, hillside sites, easements, parking areas, and walks. See Fig. 18-6.

Typical single-family-dwelling zoning ordinances (often called R-1) limit the front, rear, and side setbacks. They also limit the lot size and maximum land coverage. Figure 18-5 shows an example of how to calculate the **building envelope** (building area) for a 50′ × 120′ lot.

Figure 18-7 shows a lot with a building and accessory envelope which is determined as follows:

1. Draw setback lines for front yard, rear yard, and side yard setbacks. Also add accessory (other buildings and features) setbacks.
2. The remaining area (71′ × 66′) shows the allowable building envelope (4,686.0 sq. ft.).
3. A typical R-15 code states the maximum gross floor area can be no more

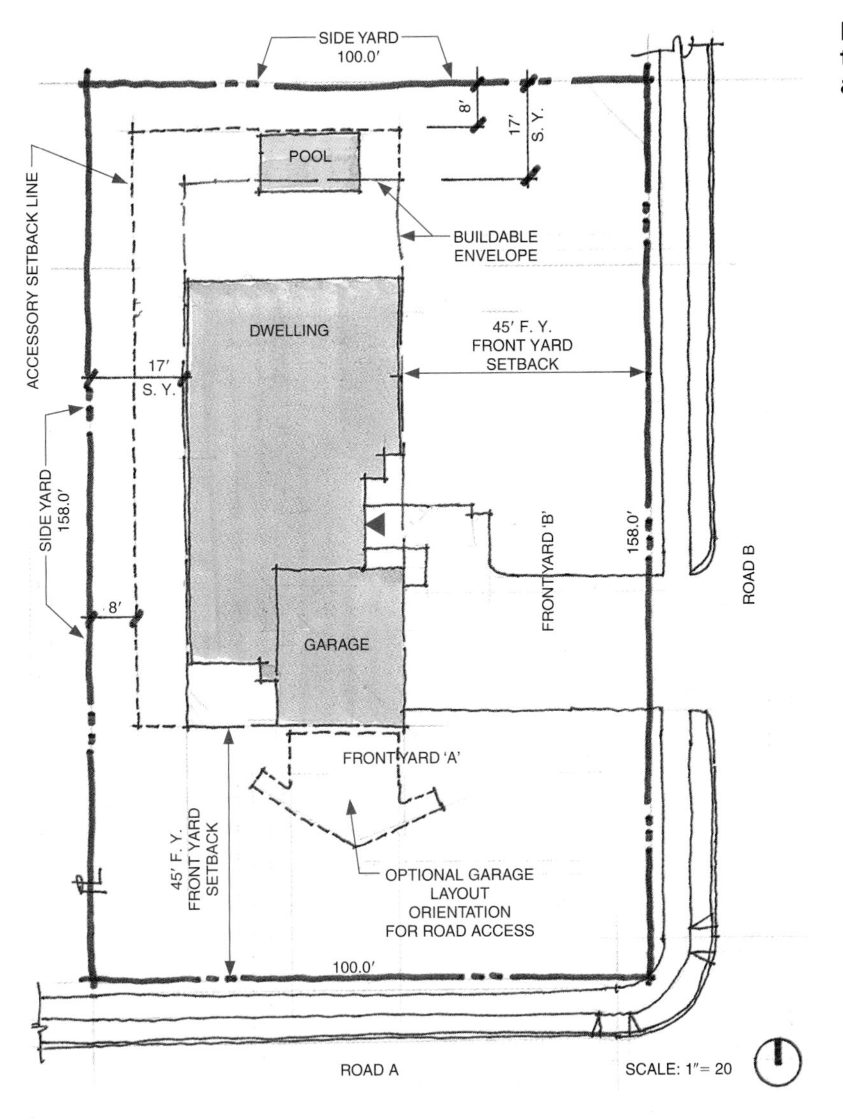

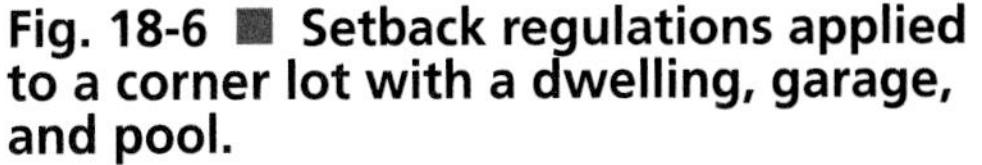

Fig. 18-6 ■ Setback regulations applied to a corner lot with a dwelling, garage, and pool.

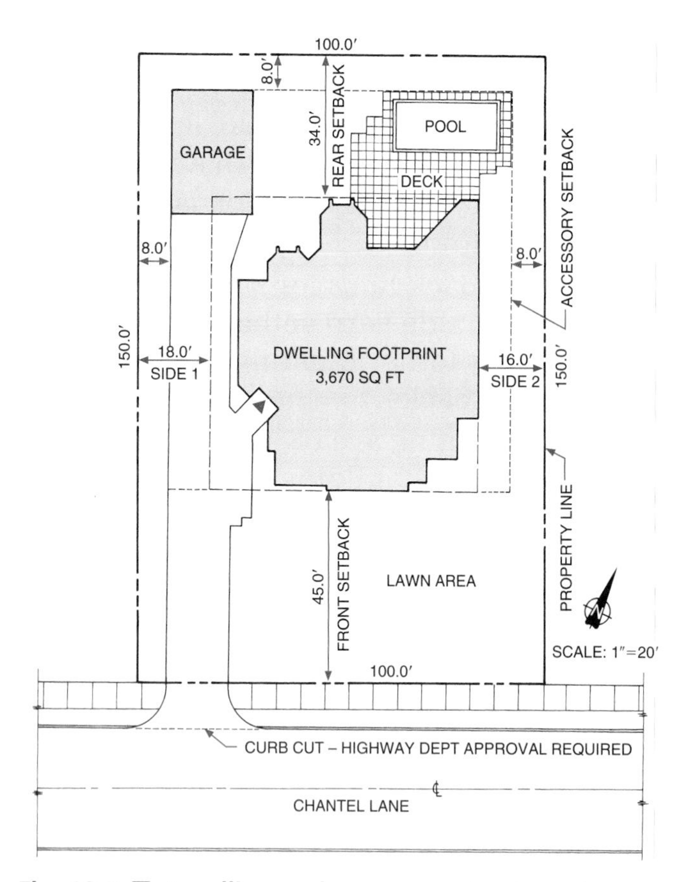

Fig. 18-7 ■ Dwelling and accessory structures locations on a lot zoned R-15, using required setbacks.

than 25% of the lot area. Thus 0.25 × 15,000 sq. ft. = 3,750 sq. ft. This is the maximum building envelope (or footprint) of the structure.

4. Note that the entire building envelope cannot be used. To calculate the amount that can be used:

   Actual gross floor area = 3,750 sq. ft.

   Allowable building envelope = 4,686 sq. ft.

   4,680 - 3,750 = 936 must remain open in the building area.

5. Thus 80% (3,750 ÷ 4,686 = 0.80) of the allowable building envelope can be covered by the house. This does not include decks, patios, garages, or other buildings.

Setback, usage, land coverage, structural type, and height restrictions must all be combined to determine the location, size, and height of structure allowed for a particular lot. See Fig. 18-8. It is important to check with the local zoning department

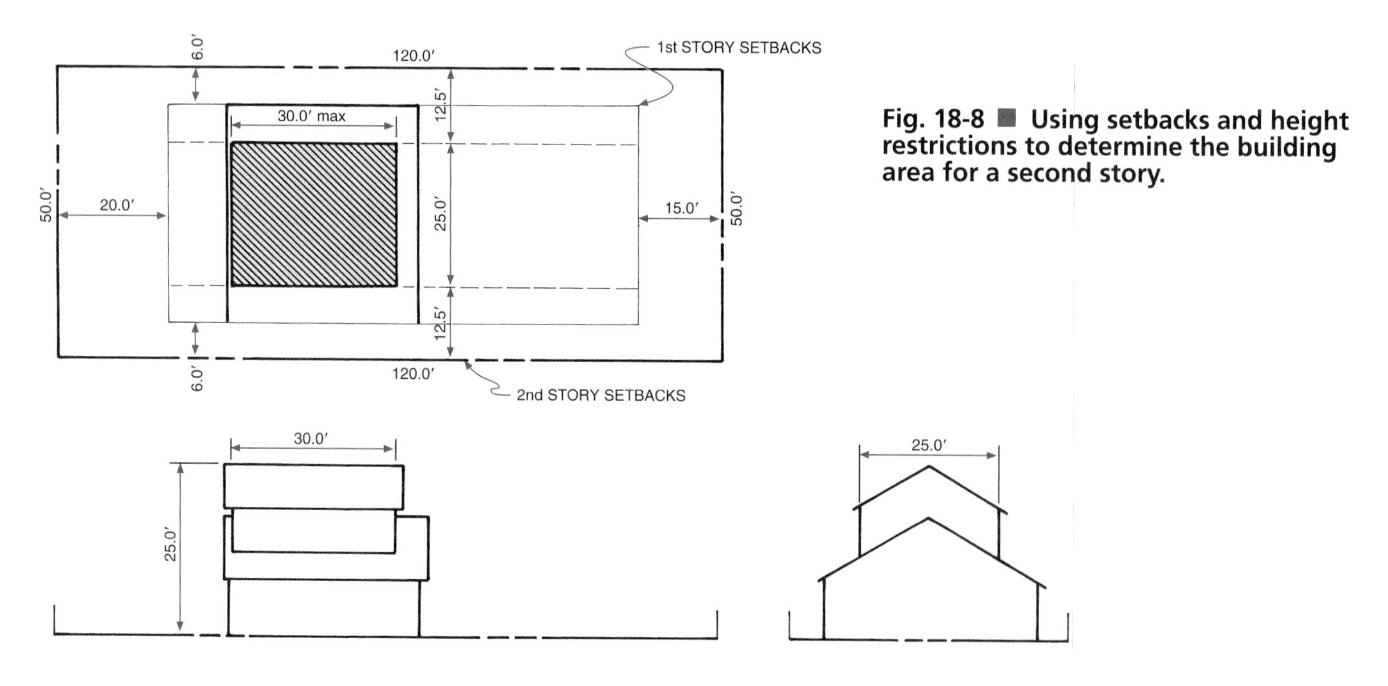

Fig. 18-8 ■ Using setbacks and height restrictions to determine the building area for a second story.

to determine the exact requirements for setbacks. Setback dimensions are measured from the property line to the wall line, eave line, projecting fireplace, or cantilevered second story of the structure, depending on local setback codes.

Density Zoning

To create a good living environment and prevent overbuilding in an area, architects and builders who plan and build multiple-home developments must conform to density zoning laws. **Density,** in architectural terms, is the relationship of the number of residential structures and people to a given amount of space. The density of an area is the number of people or families per acre or square mile. For example, a town may have a density of 3 families per acre or 1920 families per square mile.

The *average density* of an area is the ratio of all inhabitants to a specific geographic area. Density patterns may vary greatly within different parts of one area. Some parts of that area may be crowded, while other parts are less populated. In other patterns, the population may be evenly distributed. Density planning for an area must be based upon the maximum number of people who will occupy the area, regardless of the patterns. See Fig. 18-9.

To prevent overcrowding, local zoning ordinances may restrict the size of each building. This automatically restricts the number of families allowed to occupy a specific area. This approach also spreads the density patterns equally. As another approach, zoning laws may encourage clustering many residents into fewer structures, such as high-rise apartments or town houses (attached houses) with larger open public areas. In higher

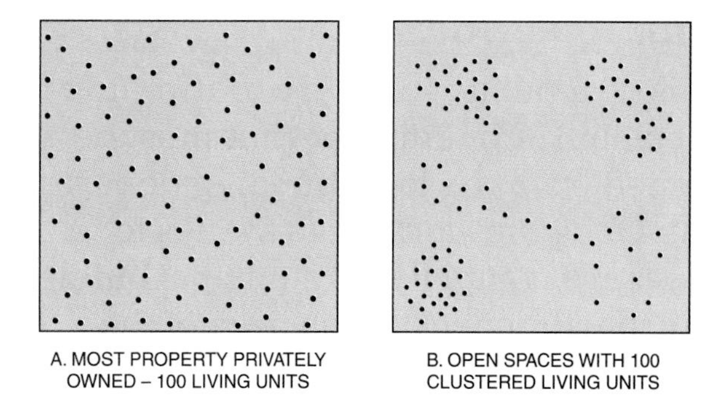

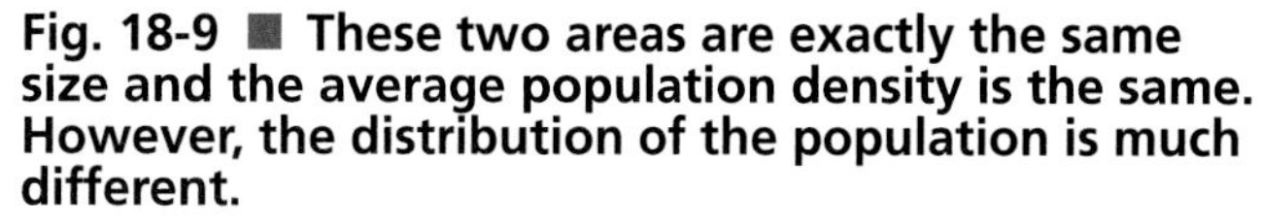

Fig. 18-9 ■ These two areas are exactly the same size and the average population density is the same. However, the distribution of the population is much different.

density developments, smaller size lots can be used more efficiently by eliminating one side yard (zero lot line-ZLL) and reducing the front, rear, and other side yard. See Fig. 18-10.

A combination of plans involves zoning different parts of an area for single-family residences, town houses, and high-rise apartments. The amount of space planned for each type of structure depends on the average density desired.

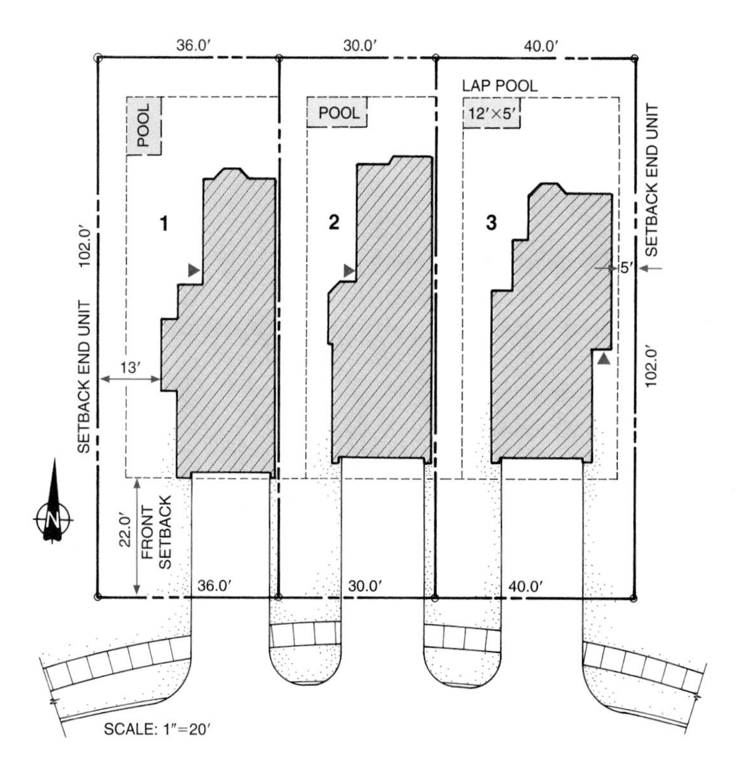

Fig. 18-10 ■ Use of zero lot lines in a high-density development.

Building Permits

Before a public structure or dwelling can be built, a **building permit** must be obtained from the local building department. Once the working drawings for a project are complete, a municipal building inspector checks each area of the design. The design must be in compliance with all existing codes and ordinances. If the drawings and specifications meet the code requirements, a building permit will be issued. See Fig. 18-11. By carefully checking all local and regional code requirements before finalizing the design and preparing working drawings, revisions can be avoided.

In addition to local building departments, the administration of local codes may be co-regulated and/or controlled by other local or governmental agencies such as city planning commissions, air pollution control districts, fire departments, public health departments, water pollution control boards, and perhaps even art and design commissions or historical preservation societies. If a federal building is involved, the Department of Housing and Urban Development (HUD), Department of Health, Education and Welfare (HEW), Federal Housing Authority (FHA), or other agencies may also be involved in the approval process.

Zoning ordinances and some building codes do contain allowances for exceptions to the law. If a building cannot be designed or sited to conform to all local laws, builders can request a *variance* from the building department. In making this request, the builder must show that the exemption will not harm or inconvenience

PERMIT

Page 1 of 1

Permit #: 970012 Type: Residence Issued: 1-3-- - by: JP
Job Location: 603 Issue Loc:
Lot: 8 Subdiv: Lake Estates
Parcel: 8A
Owner: Kingston Elev: 6' Fl Map: A4
Project: 95 K Seawall Datum
Job Description: Residence & Pool
Applicant Name: J R Smith Type: Skeleton Frame
Applied Date: 1-2- -- Appl Oper: 4
Contract Phone: 364 8752 Inspector Area: 6 Work w/o Permit Fee: 50
Contractor Name: T. Jones Cert Nbr: 336
Business Name: Capitol Builders Septic Tank: Dwg A1

Setbacks Front: 50 Left: 35 Right: 35 Rear: 35
FCC Code: 329
Square Footage: 3560 Rate: TBD Job Value: 195,000
Number of Units: 1 Floors: 2 Buildings: 1 + Dock
ROW: 24' ℄ RD Zoning: PUD Map No: 00617
Minimum Floor Elevation: 6' Seawall Datum Residential/~~Commercial~~

H. Mitchell *1/3/--*
Building Official or Authorized Signature Date

Fig. 18-11 ■ A building permit, such as this one, must be posted on each building site. *Lake County Florida Building Department*

neighbors, the community, or the environment in any way. Variances are often requested for setbacks and building sizes and types.

Survey Plans

A **survey** is a drawing showing the exact size, shape, and levels of a property. When prepared by a licensed surveyor, a survey is used as a legal document to establish property rights. It is filed with the deed to the property. The lot survey includes the length of each boundary, tree locations, utility lines, corner elevations, contour of the land, and position of streams, rivers, roads or streets. It also lists the owner's name and the owners or titles of adjacent lots. A survey drawing must be accurate and must also include a complete written description of the lot.

Establishing Dimensions

Surveying a site involves locating points (coordinates) on the earth's surface. To indicate and connect these points on a drawing, the *polar coordinate system* is used. In this system, a fixed, true north-south reference line, called a *meridian*, is established. The direction of a line on the survey drawing is given in relation to the meridian as the line's *bearing*, an angle toward east or west.

The exact plan and shape of a lot is shown by property lines. Property line dimensions are shown directly on the line by length and angle. Angles are dimensioned using either the American system or the azimuth system. See Fig. 18-12. In the American system a compass is divided into four quadrants: NE, SE, SW, and NW. Angular dimensions are shown by noting the degrees, minutes, and seconds from either N, E, S, or W and toward another direction. There are 360 degrees (°) in a circle, 60 minutes (') in a degree, and 60 seconds (") in a minute. Thus a 45° line in the northeast quadrant is dimensioned N 45° 00' 00" E. This means the line is 45° from north, heading toward east.

In the azimuth system, each line is dimensioned as an angle, reading clockwise from the north meridian, from 0° to 360°. These angular lines are drawn by aligning the 0° or 360° line on a protractor with the north meridian line. Place the center of the protractor at the intersection of the meridian and the east-west line. The degree is located on the circumference of the protractor and connected with a line to the protractor center, as on Fig. 18-12.

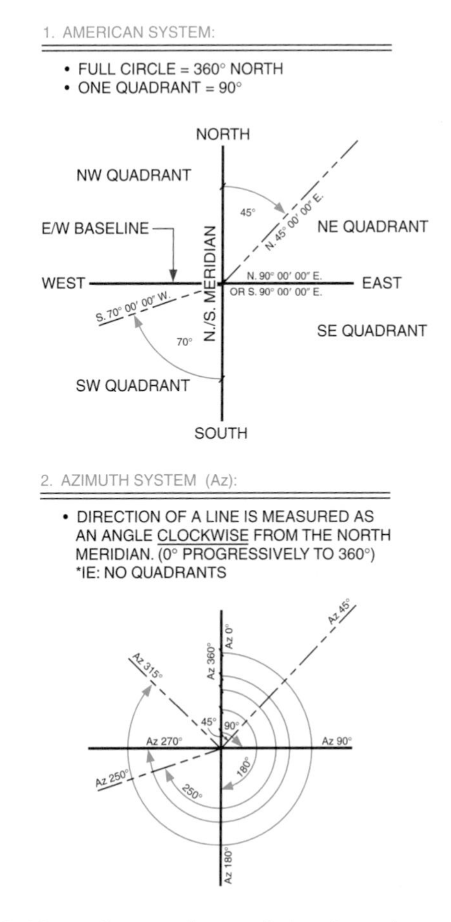

Fig. 18-12 ■ Comparison of the American system and the azimuth system for dimensioning property lines.

ELEVATIONS (HEIGHTS) The height of any point on a site is dimensioned from a fixed elevation point called a *datum*. The universal datum elevation is sea level. On small sites a fixed point on a road, sidewalk, or seawall may be used as datum. Datum is always zero and all elevation measurements are made up or down from the datum.

Several types of drawings and lines are used to show elevation distances and shapes on survey drawings. These include profile drawings (land sections), elevation point notations, and contour lines.

CONTOUR LINES On maps or drawings that need to show terrain, **contour lines** connect points on the land surface that are the same elevation above datum. See Fig. 18-13.

The vertical distance between contour lines is called the *contour interval*. The interval can be any convenient distance, but is usually an increment of 5′. Contour intervals of 5′, 10′, 15′, and 20′ are common on large surveys. The use of smaller intervals (1′ or 2′) gives a more accurate description of the slope and shape of the terrain than does the use of larger intervals. Note the different levels of detail between the 2′ and the 10′ interval of the same terrain as shown in Figs. 18-14A and 18-14B. The size of the interval depends on the scale of the drawing and on the size of the area to be shown. Large contour intervals are normally used on large regional maps.

Fig. 18-13 ■ Contour lines may be thought of as imaginary cuts made horizontally through the terrain at regular intervals.

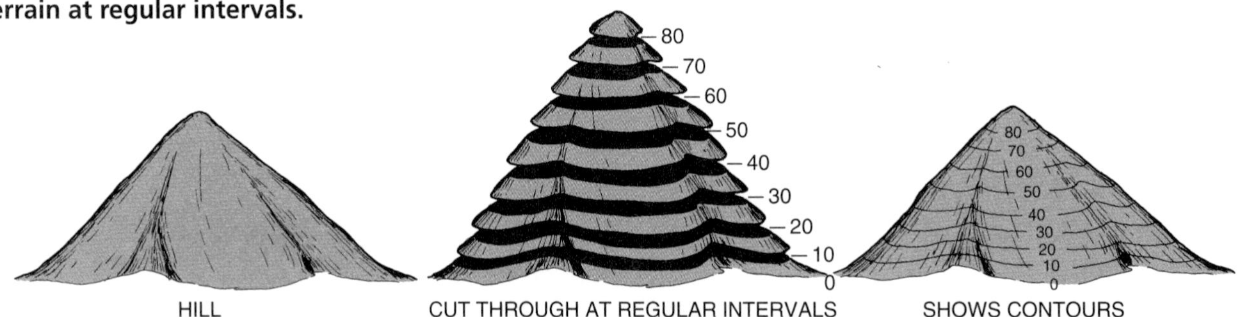

Fig. 18-14 ■ The choice of contour interval determines how much detail is shown on the drawing.

A. Large intervals show only the general slope.

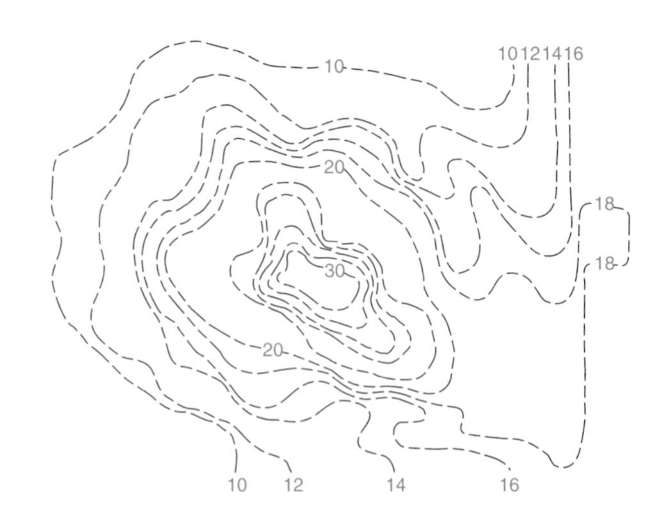

B. Small intervals reveal more terrain details. The irregular lines used here indicate a very rugged surface.

Contour lines that are very close together indicate a steep slope. Lines spaced far apart indicate a gradual slope. See Fig. 18-15.

Elevation notations are important. On a contour drawing, a hill may appear the same as a valley until the notations are added to the contour interval lines. Lighter contour lines without notations may be used to show intermediate contour intervals. See Fig. 18-16.

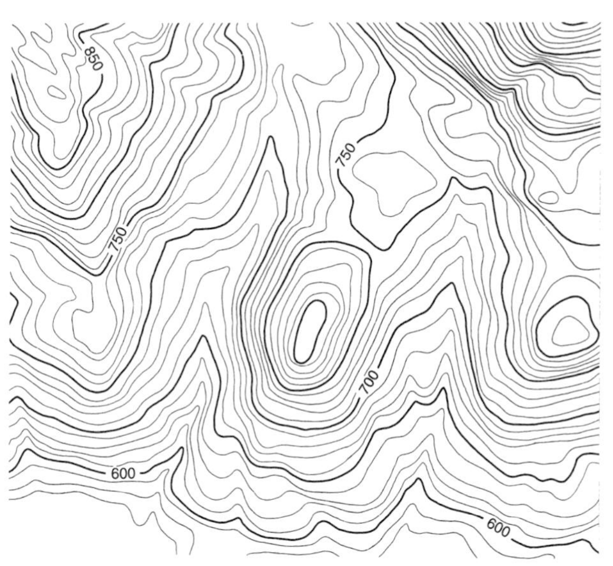

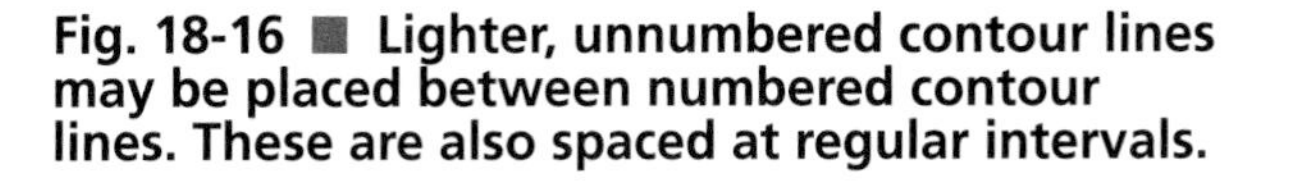

Fig. 18-16 ■ Lighter, unnumbered contour lines may be placed between numbered contour lines. These are also spaced at regular intervals.

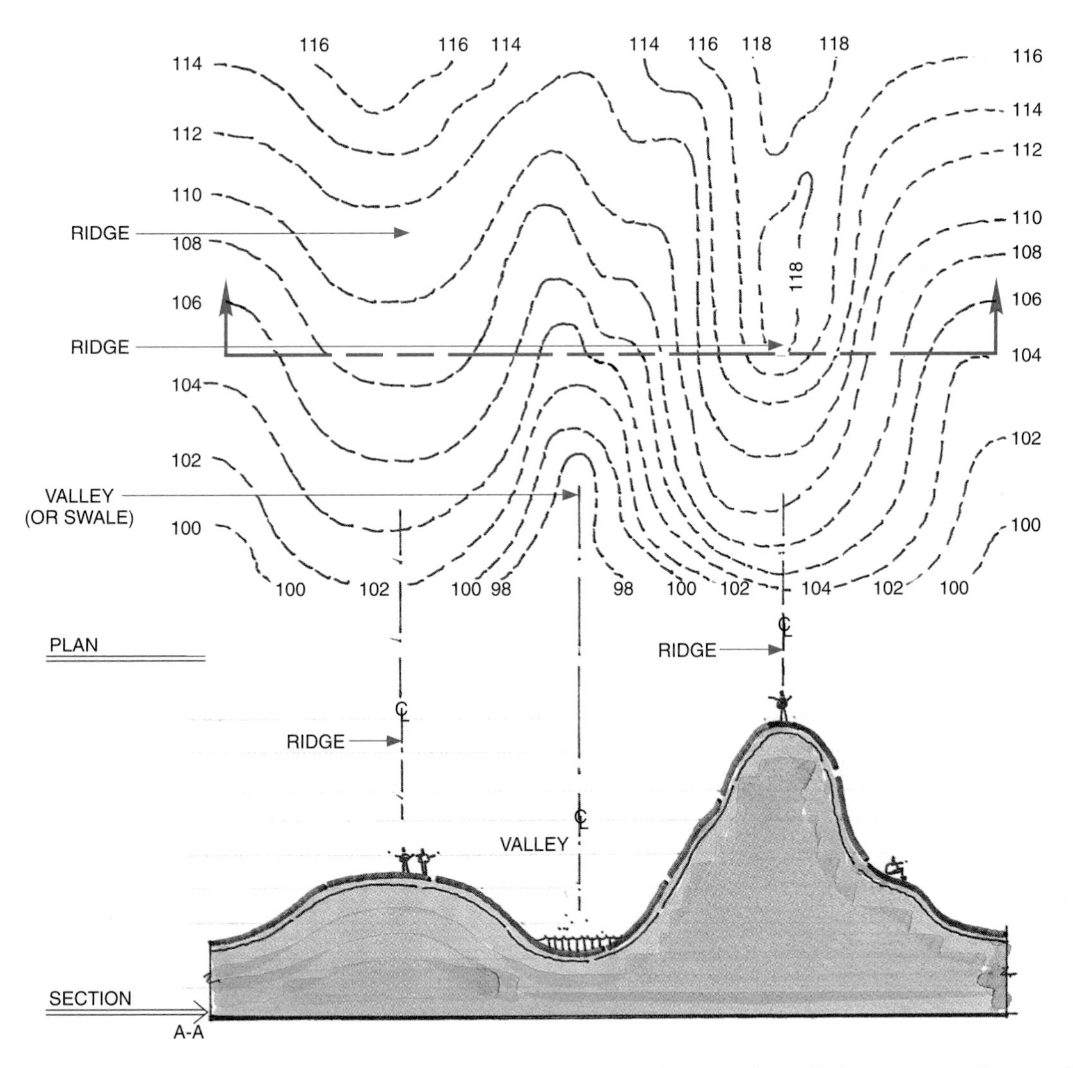

Fig. 18-15 ■ Profile (section) through a terrain. Note how the spacing of the contour lines reflects the degree of slope of the land.

For large, flat projects such as parking lots and airport tarmacs (runway areas, etc.), contour lines may not be used because the slope differential is less than a foot. Only the elevations of selected points are placed on the drawing. An arrow is drawn beside each point to show slope direction.

Symbols on Survey Drawings

Symbols are used extensively to describe the features of the terrain. Some symbols, such as tree symbols, resemble the appearance of a feature. Most survey symbols are graphic representations. See Figs. 18-17 and 18-18.

Guidelines for Drawing Surveys

The numbered arrows in Fig. 18-19, page 359, correspond to the following guidelines for preparing survey drawings:

1. Record the elevation above the datum of the lot at each corner.
2. Represent the size and location of streams and rivers by wavy lines (blue lines on geographical surveys).

NAME	ABBREV	SYMBOL	NAME	ABBREV	SYMBOL	NAME	ABBREV	SYMBOL	NAME	ABBREV	SYMBOL
TREES	TR		CULTIVATED AREA	CULT		BUILDINGS	BLDGS		LARGE RAPIDS	LRG RP	
GROUND COVER	GRD CV		WATER	WT		SCHOOL	SCH		WASH	WSH	
BUSHES SHRUBS	BSH SH		WELL	W		CHURCH	CH		LARGE WATERFALL	LRG WT FL	
OPEN WOODLAND	OP WDL		PROPERTY LINE	PR LN		CEMETARY	CEM	† CEM	BOUNDARY, U.S. LAND SURVEY TOWNSHIP	BND US LD SUR TWN	
MARSH	MRS		SURVEYED CONTOUR LINE	SURV CON LN	75	POWER TRANSMISSION LINE	PW TR LN		BOUNDARY, TOWNSHIP APPROXIMATED	BND TWN	
DENSE FOREST	DN FR		ESTIMATED CONTOUR	EST CON		GENERAL LINE LABEL TYPE	GN LN	oil line	BOUNDARY, SECTION LINE U.S. LAND SURVEY	BND SEC LN US LD SUR	
SPACED TREES	SP TR		FENCE	FN		BOUNDARY, STATE	BND ST		BOUNDARY, SECTION LINE APPROXIMATED	BND SEC LN	
TALL GRASS	TL GRS		RAILROAD TRACKS	RR TRK		BOUNDARY, COUNTY	BND CNTY		BOUNDARY, TOWNSHIP NOT U.S. LAND SURVEY	BND TWN	
LARGE STONES	LRG ST		PAVED ROAD	PV RD		BOUNDARY, TOWN	BND TWN		INDICATION CORNER SECTION	COR SEC	
SAND	SND		UNPAVED ROAD	UNPV RD		BOUNDARY, CITY INCORPORATED	BND CTY		U.S. MINERAL OR LOCATION MONUMENT	U.S. MIN MON	▲
GRAVEL	GRV		POWER LINE	POW LN		BOUNDARY, NATIONAL OR STATE RESERVATION	BND NAT OR ST RES		DEPRESSION CONTOURS	DEP CONT	
WATER LINE	WT LN		HARD-SURFACE HEAVY DUTY ROAD – FOUR OR MORE LANES	HRD SUR HY DTY RD		BOUNDARY, SMALL AREAS: PARKS, AIRPORTS, ETC	BND		FILL	FL	
GAS LINE	G LN		HARD-SURFACE HEAVY DUTY ROAD – 2 OR 3 LANES	HRD SUR HY DTY RD		LEVEE	LEV		CUT	CT	
SANITARY SEWER	SAN SW		IMPROVED LIGHT DUTY ROAD	IMP LT DTY RD		RIVER	RV		LAKE, INTERMITTENT	LK INT	
SEWER TILE	SW TL		TRAIL UNIMPROVED DIRT ROAD	TRL UNIM DRT RD		STREAM PERENNIAL	ST PER		LAKE, DRY	LK DRY	
PROPERTY CORNER WITH ELEVATION	PROP CR EL	EL 70.5	ROAD UNDER CONSTRUCTION	RD CONST		STREAM INTERMITTENT	ST INT		SPRING	SP	
SPOT ELEVATION	SP EL	+ 78.8	BRIDGE OVER ROAD	BRG OV RD		STREAM DISAPPEARING	ST DIS		PILINGS	PLG	
WATER ELEVATION	WT EL	80	ROAD OVERPASS	RD OVP		SMALL RAPIDS	SM RP		SWAMP	SWP	
BENCH MARKS WITH ELEVATIONS	BM/EL	BM X 84.2 BM △ 84.2	ROAD UNDERPASS	RD UNP		SMALL WATERFALL	SM WT FL		SHORELINE	SH LN	

Fig. 18-17 ■ Common symbols used on survey drawings.

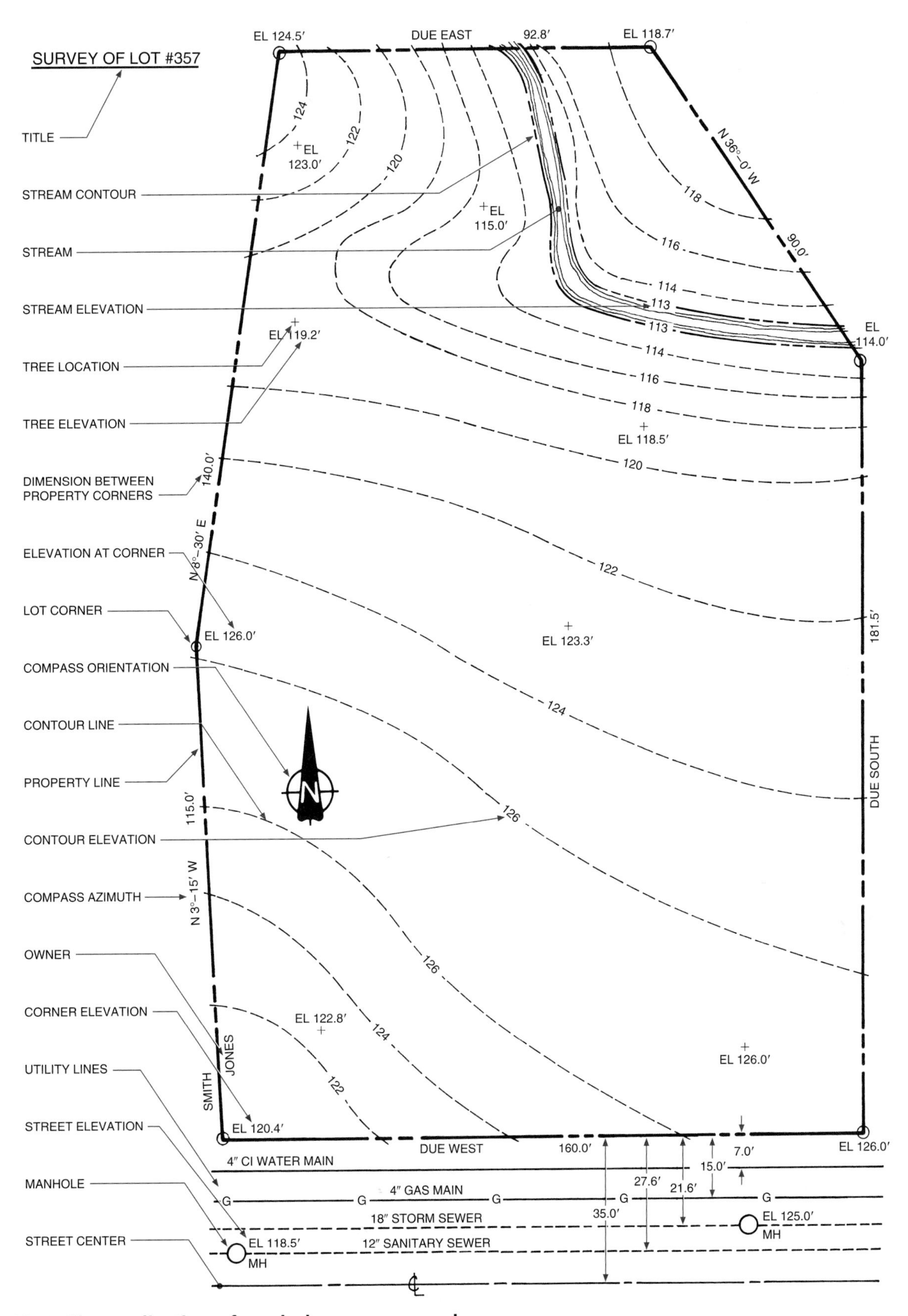

Fig. 18-18 ■ The application of symbols on a survey plan.

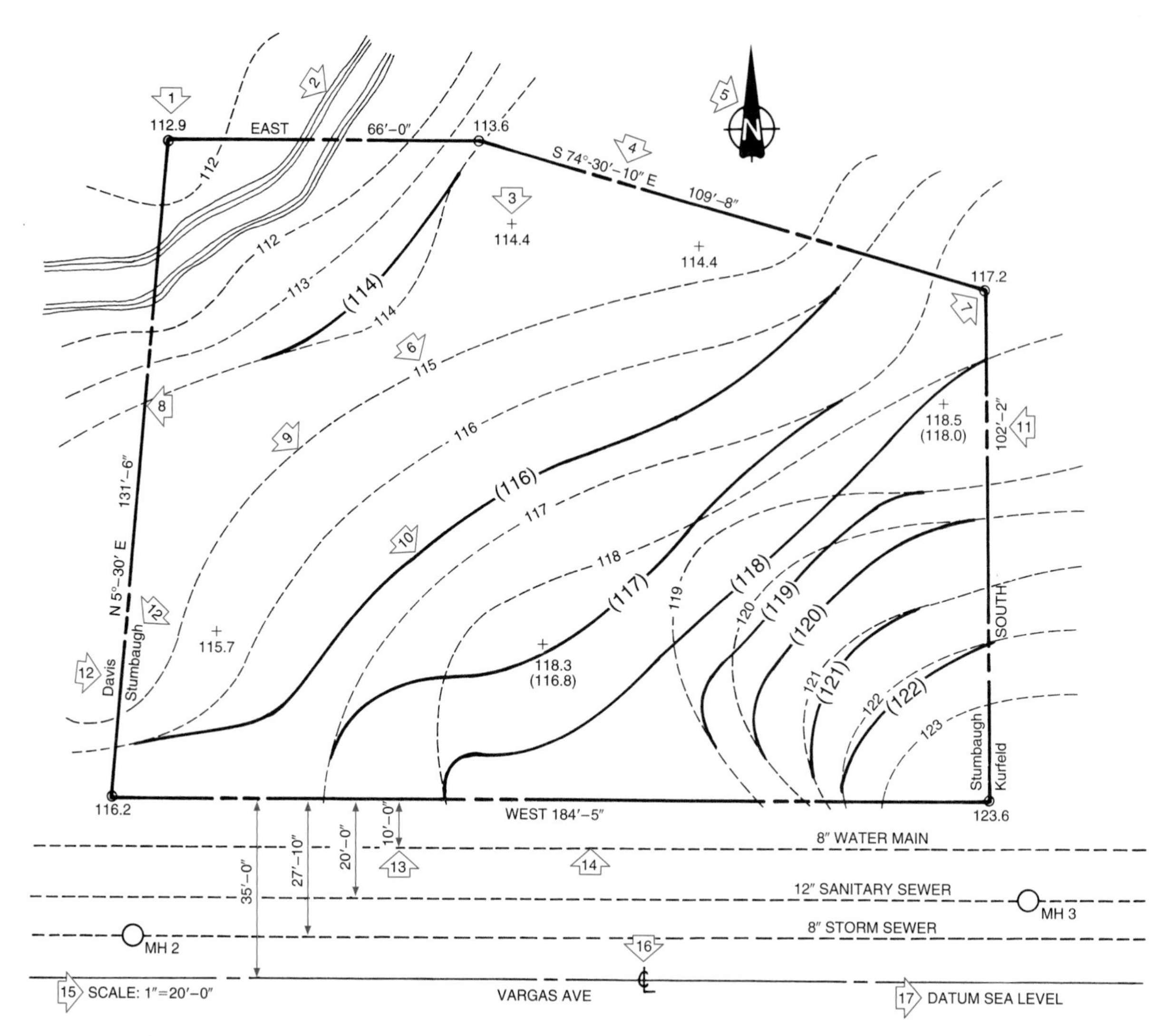

Fig. 18-19 ■ Example survey drawing. Refer to the guidelines in the text.

3. Use a cross to show the position of existing trees. The elevation at the base of the trunk is shown on some drawings.
4. Indicate the compass direction of each property line by degrees, minutes, and seconds.
5. Use a north arrow to show compass direction.
6. Break contour lines to insert the height of each contour above the datum.
7. Show lot corners by small circles or overlapping property lines.
8. Draw the property line symbol by using a heavy line with two dashes repeated throughout.
9. Show elevations above the datum by contour lines.
10. Show any proposed change in grade line by contour lines. Use dotted contour lines to show original grade and solid contour lines to show the new proposed grading levels.
11. Show lot dimensions directly on the property line. The dimension on each line indicates the distance between property corners.

12. Give the names of owners of adjacent lots outside the property line. The name of the owner of the site is shown inside the property line.
13. Dimension the distance from the property line to all utility lines.
14. Show the position of utility lines by dotted lines. Utility lines are labeled according to their function.
15. Draw surveys with an engineer's scale. Dimensions are shown as feet and decimal parts of a foot (for example, 6.5′). Common scales for surveys are 1″ = 10′ and 1″ = 20′.
16. Show existing streets and roads either by centerlines or by curb or surface outlines.
17. Indicate the datum level used as reference for the survey.

Geographical Survey Maps

United States Geographical Survey Maps (USGS) are similar to property surveys except they cover extremely large areas. The entire world is divided into geographical survey regions. However, not all regions have been surveyed and mapped.

Geographical survey maps show the general contour of the area, natural features of the terrain, and structures. See Fig. 18-20. Computer programs are used to create pictorial contour drawings as shown in Fig. 18-21. This is done using X-Y coordinate input data and the corresponding datum level (Z) for each coordinate.

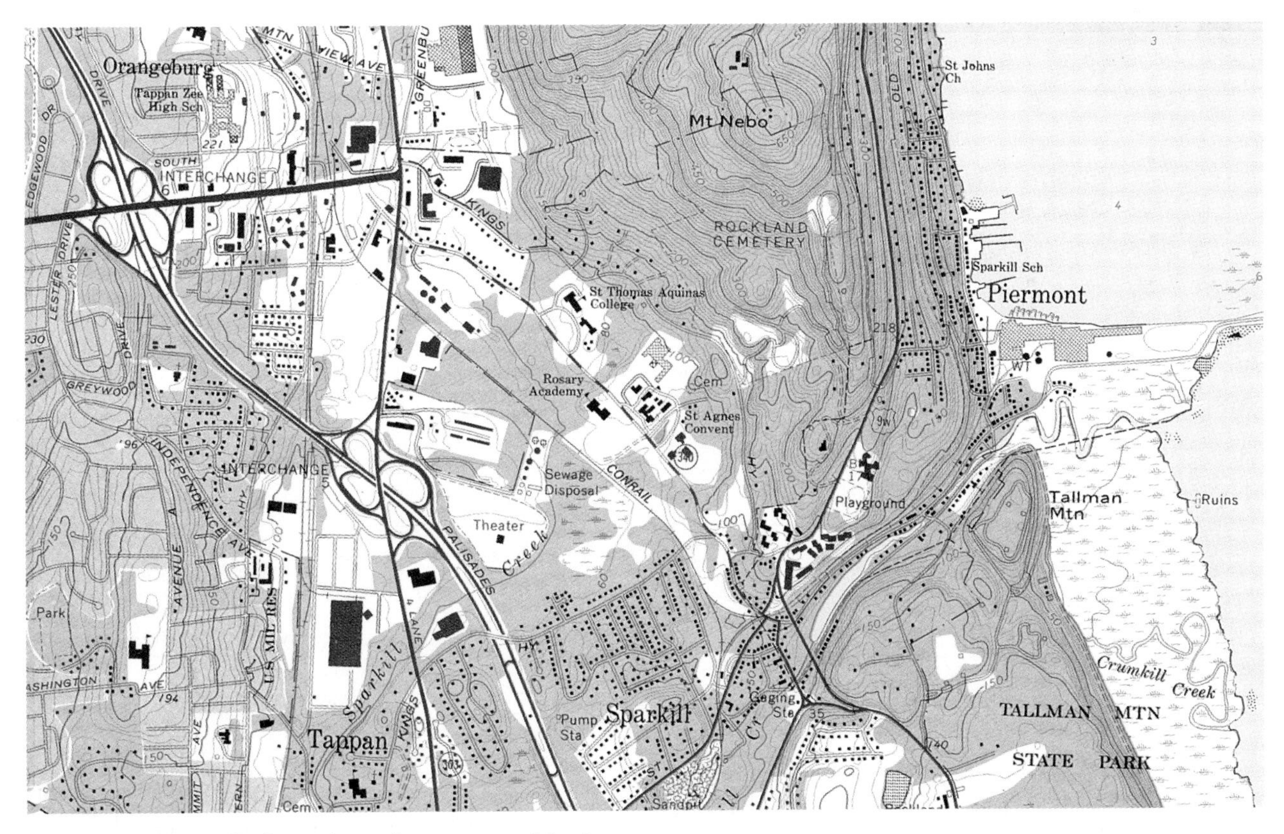

Fig. 18-20 ■ Typical portion of a geographical survey map.

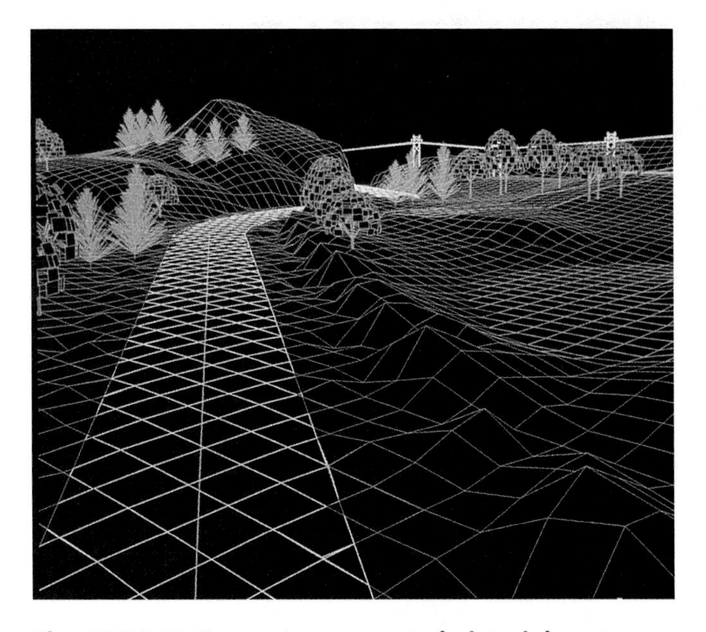

Fig. 18-21 ■ Computer-generated pictorial contour drawing. *Autodesk, Inc.*

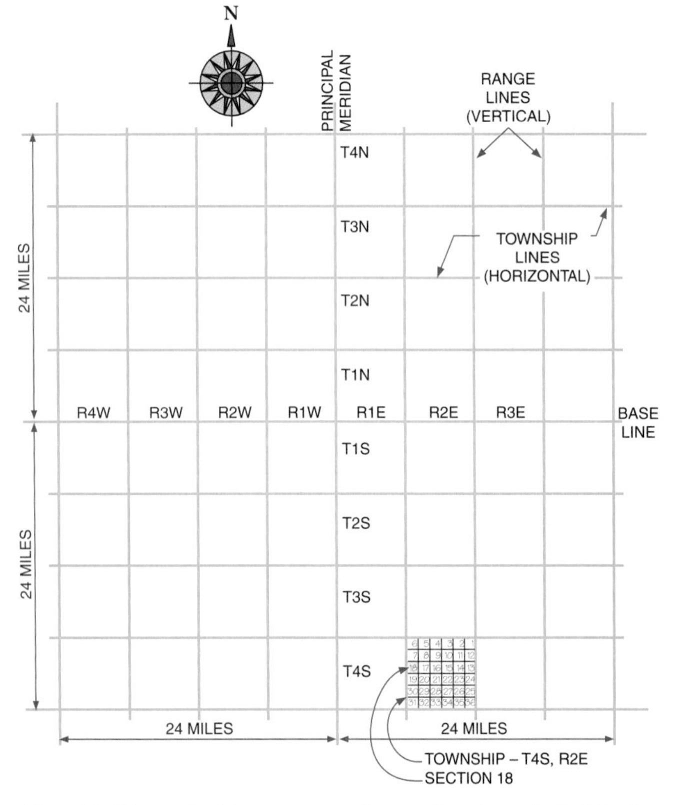

Fig. 18-22 ■ A 48-square-mile region is subdivided into four 24-square-mile quadrants, that are further subdivided into twenty-four 6-square-mile townships. Each township contains thirty-six 1-square-mile sections.

Plats

A **plat** is a survey (map, chart, or plan) of multiple connected properties. Plats are legal descriptions of a land site and are identified by plat name, section, township, county, and state. A plat is part of a geographical survey region which is divided into areas that contain further subdivisions. See Fig. 18-22.

Plats may include the compass bearing (direction) of the plat area, dimensions of each property line, and the position of all roads, utility lines, and easements. Some show only lot shapes, as in Fig. 18-23. Others identify lots by numbers which refer to a more detailed survey. See Fig. 18-24. Plats are prepared for residential developments, industrial parks, urban developments, and shopping complexes.

Plot Plans

Plot plans are used to show the size and shape of a building site and the location and size of all buildings on that site. The position and size of walks, drives, pools,

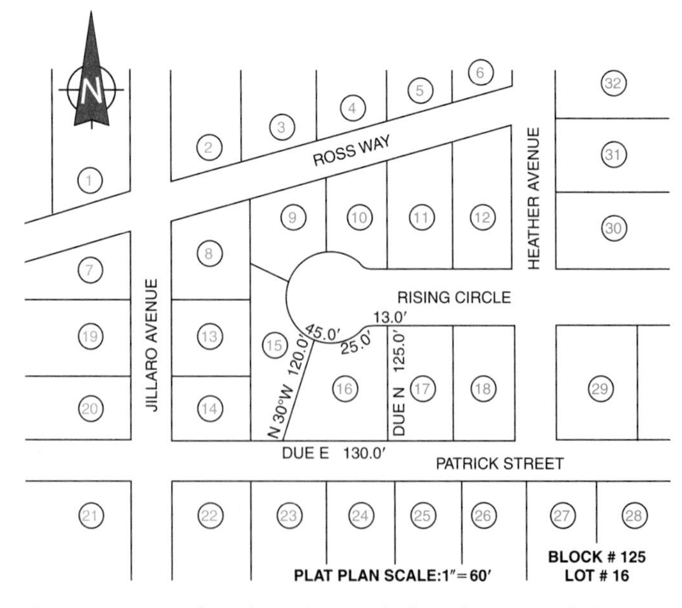

Fig. 18-23 ■ Plat showing only lot shape and size.

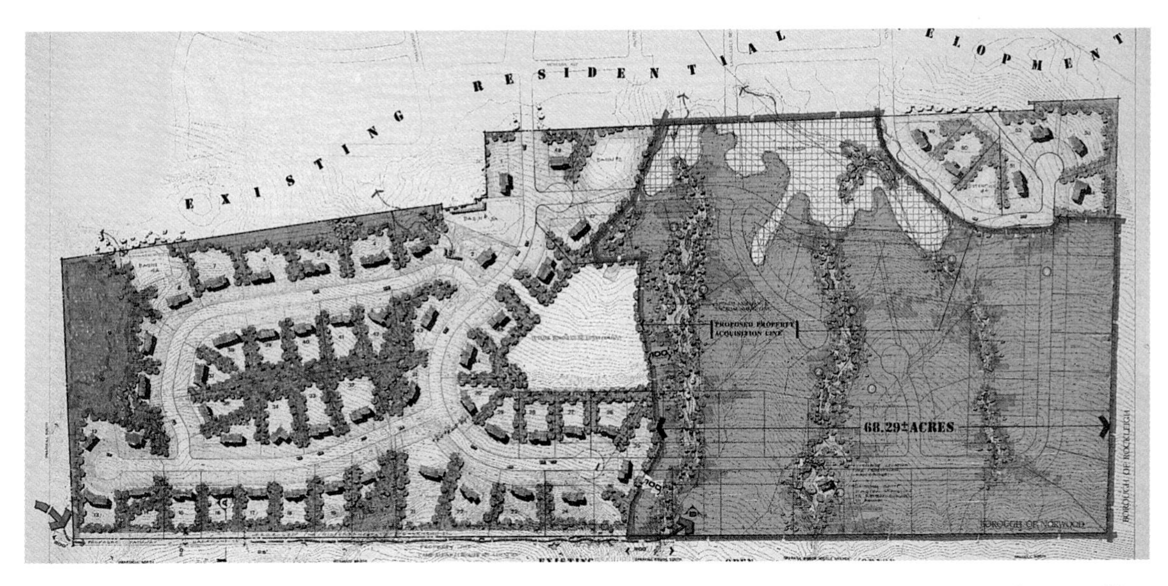

Fig. 18-24 ■ Portion of a rendered plat used for presentation purposes. *Norwood East Hill Tract, Bergan County, NJ.*

streams, patios, and courts are also shown. Compass orientation of the lot is given, and contour lines are sometimes shown. Plot plans may also include details showing site construction features. See Fig. 18-25. If a separate survey and/or a landscape plan is prepared, contour lines, utility lines, and planting details are usually omitted from a plot plan.

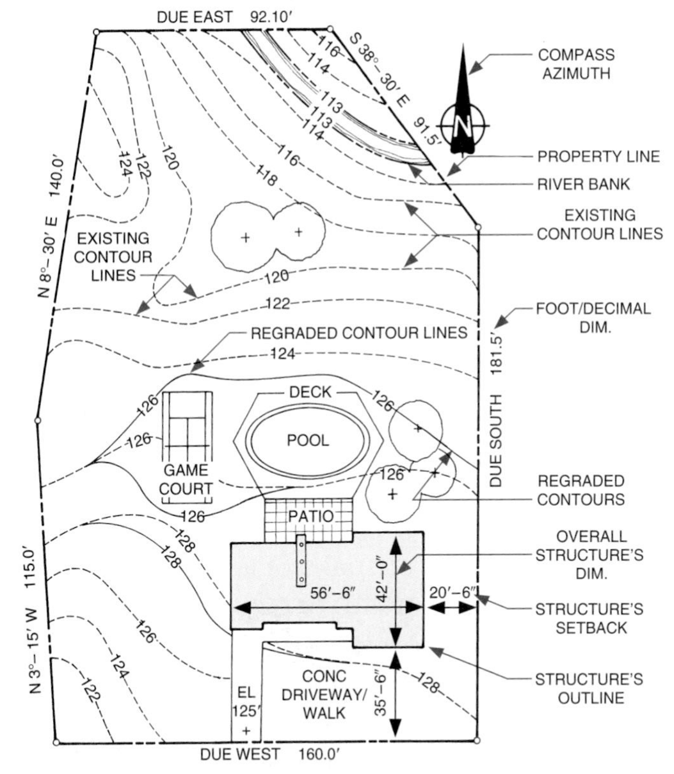

Fig. 18-25 ■ A plot plan. Note the various symbols used.

Guidelines for Drawing Plot Plans

When plot plans are prepared, the features indicated by the numbered arrows in Fig. 18-26 should be drawn according to these guidelines:

1. Draw the outline of the main structure on the lot. Crosshatching is optional.
2. Draw the outlines of other buildings on the lot.
3. Show overall building dimensions.
4. Locate each building by dimensioning perpendicularly from the property line to the closest point on a building. On curved or slanted property lines, dimension to points of tangency (touching but not intersecting), as shown here. The property line shows the legal limits of the lot.
5. Show the position and size of driveways.
6. Show the location and size of walks.
7. Indicate elevations of key surfaces such as floors, ground line, patios, driveways, and courts.
8. Outline and show the symbol for surface material used on patios and terraces.
9. Label streets adjacent to the site.

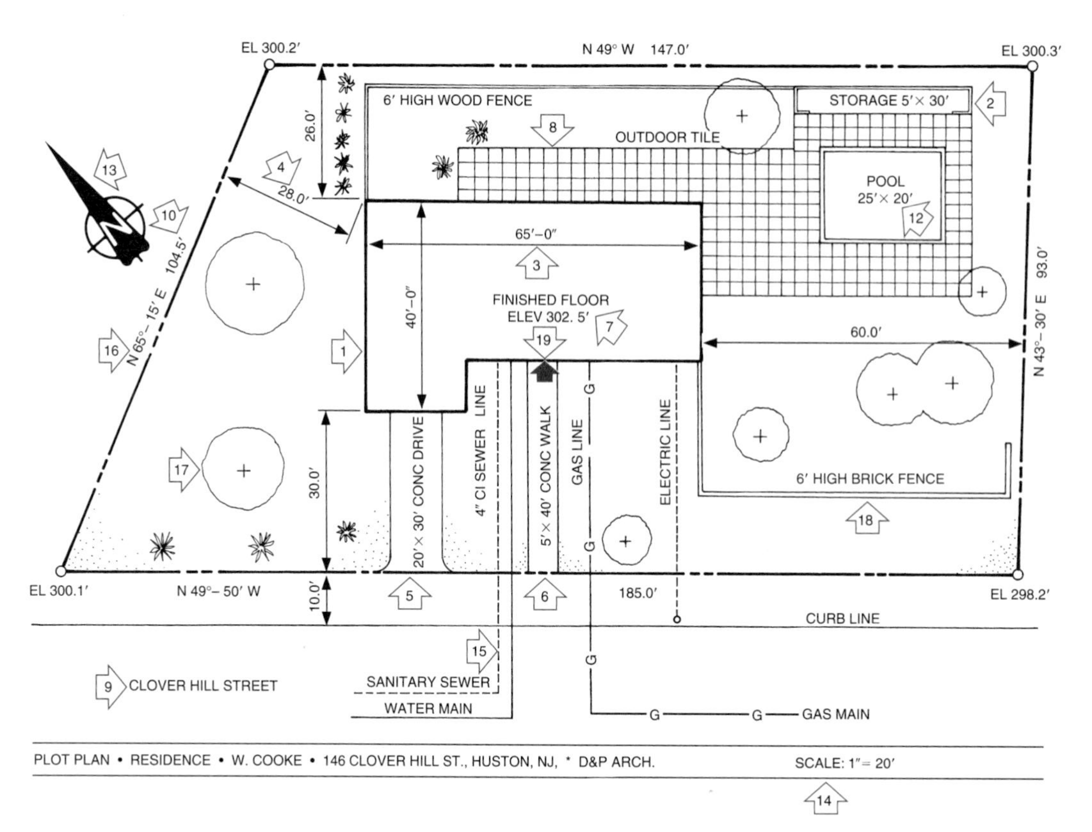

Fig. 18-26 ■ Example plot plan. Refer to guidelines in the text.

10. Place overall lot dimensions either on extension lines outside the property line or near the property line.
11. Show the size and location of constructed recreation areas, such as tennis courts. (None on this plan.)
12. Show the size and location of pools, ponds, or other bodies of water.
13. Indicate the compass orientation of the lot with a north arrow.
14. Use a decimal (civil engineer's) scale, such as 1" = 10', or 1" = 20', for preparing plot plans.
15. Show the position of utility lines.
16. Include the compass direction with the perimeter dimensions for each property line.
17. Show trunk base location and coverage of all major trees. Trunk diameter may also be noted.
18. Label and dimension all landscape construction features and auxiliary structures.
19. Identify the location of entrances. Arrows are most often used. See Fig. 18-27.

In addition, if a septic system is used, draw and dimension the location and minimum distance allowed from system components to the nearest building. See Fig. 18-28, pages 364-365.

Fig. 18-27 ■ Entrance location symbols.

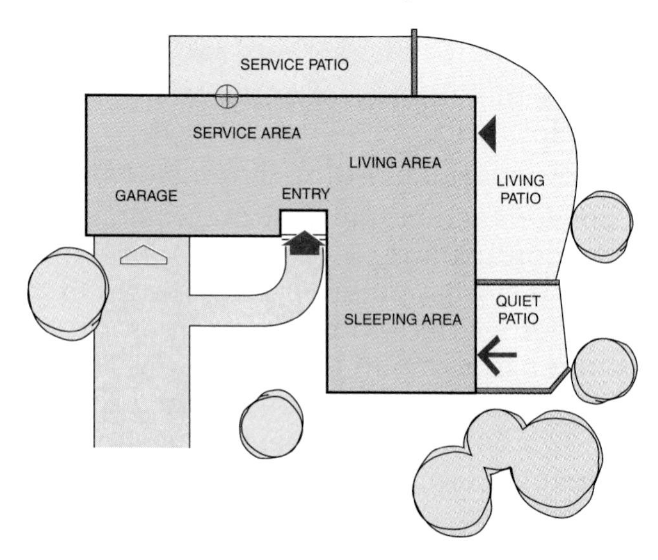

Variations

Although plot plans should be prepared according to the standards shown in Fig. 18-26, many optional features may also be included in plot plans. For example, a plot plan may show only the outline of the building, or it may include shading, outlines of the roof intersections, or cross-hatching. Sometimes the interior partitions of a building are drawn to reveal connections between outside living areas and the inside rooms. Refer again to Fig. 18-28.

Landscape Plans

Landscape plans are drawings that show the types and locations of vegetation. They may also show contour changes and the position of buildings. Such features are necessary to make the placement of the vegetation meaningful. Symbols are used on landscape plans to show the position of trees, shrubbery, flowers, vegetable gardens, hedges, and ground cover. See Fig. 18-29.

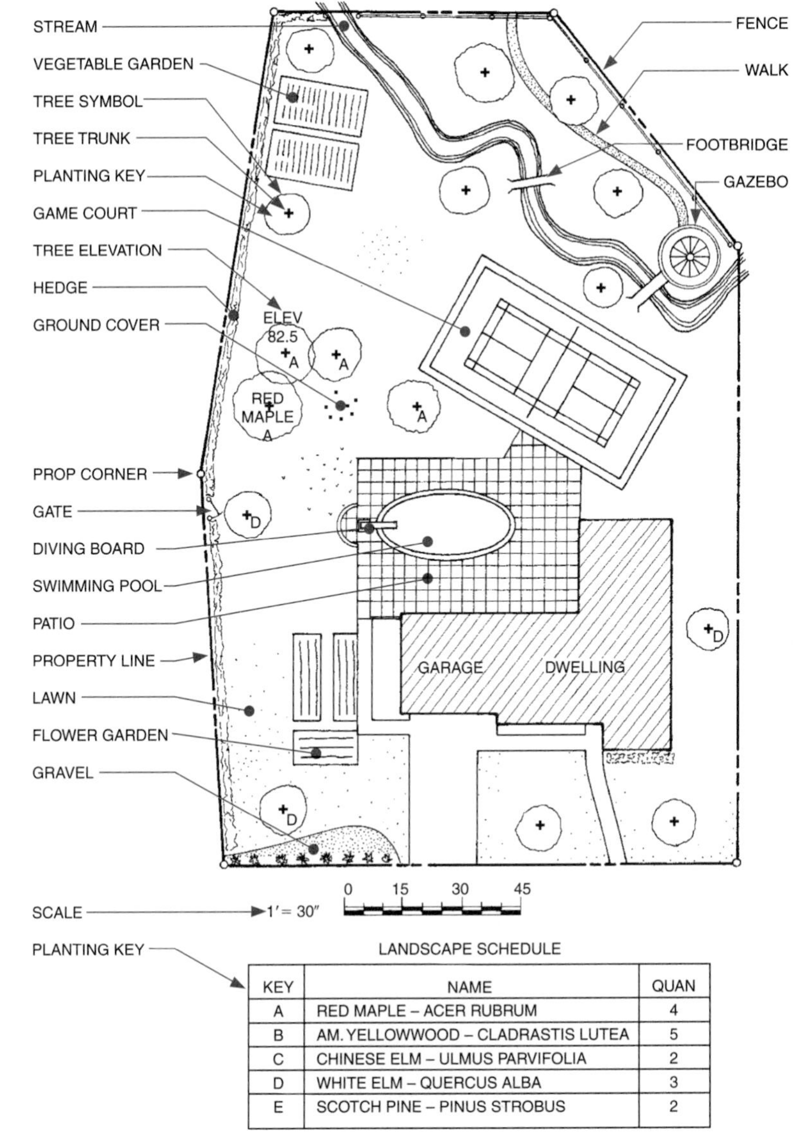

KEY	NAME	QUAN
A	RED MAPLE – ACER RUBRUM	4
B	AM. YELLOWWOOD – CLADRASTIS LUTEA	5
C	CHINESE ELM – ULMUS PARVIFOLIA	2
D	WHITE ELM – QUERCUS ALBA	3
E	SCOTCH PINE – PINUS STROBUS	2

Fig. 18-29 ■ Common symbols used on landscape drawings.

Fig. 18-28 ■ Note the changes made to this site by comparing the plot plan (A) with the original aerial photograph (B).

A. Detailed plot plan. This plan is related to the floor plan shown in Fig. 14-20. Refer back to Fig. 1-21B to see the completed structure.

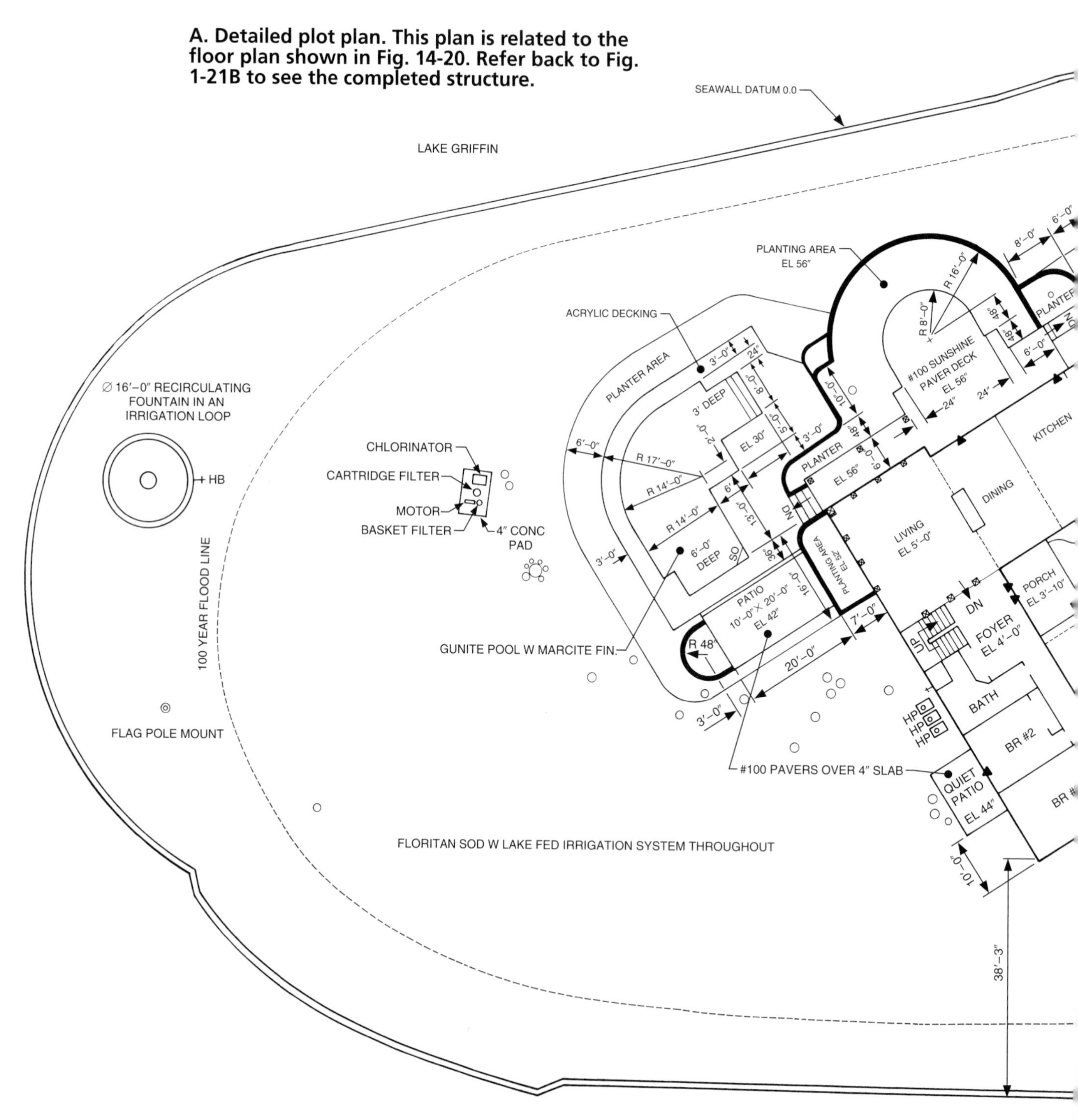

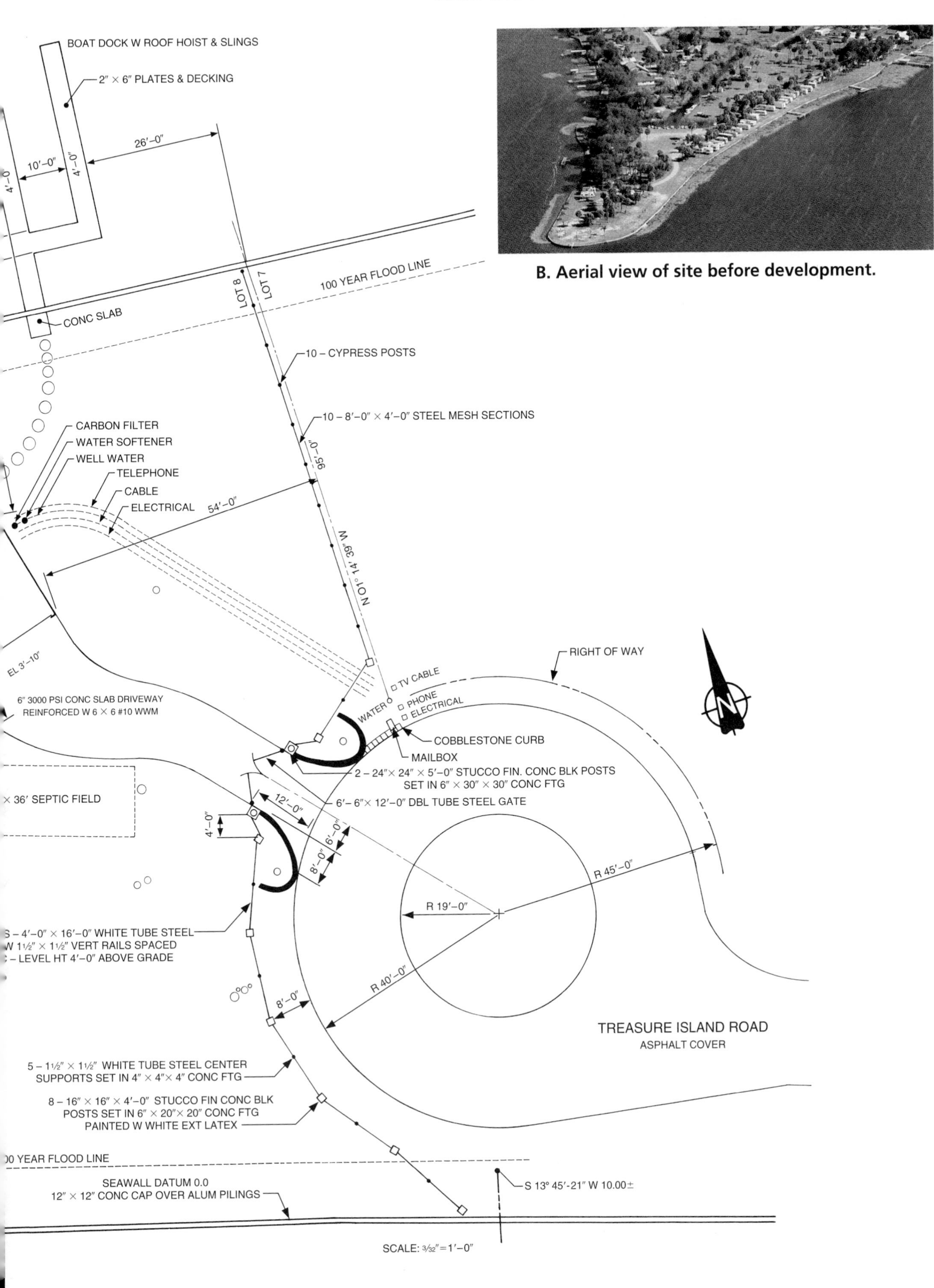

B. Aerial view of site before development.

Fig. 18-30 ■ Example landscape plan. Refer to the guidelines in the text.

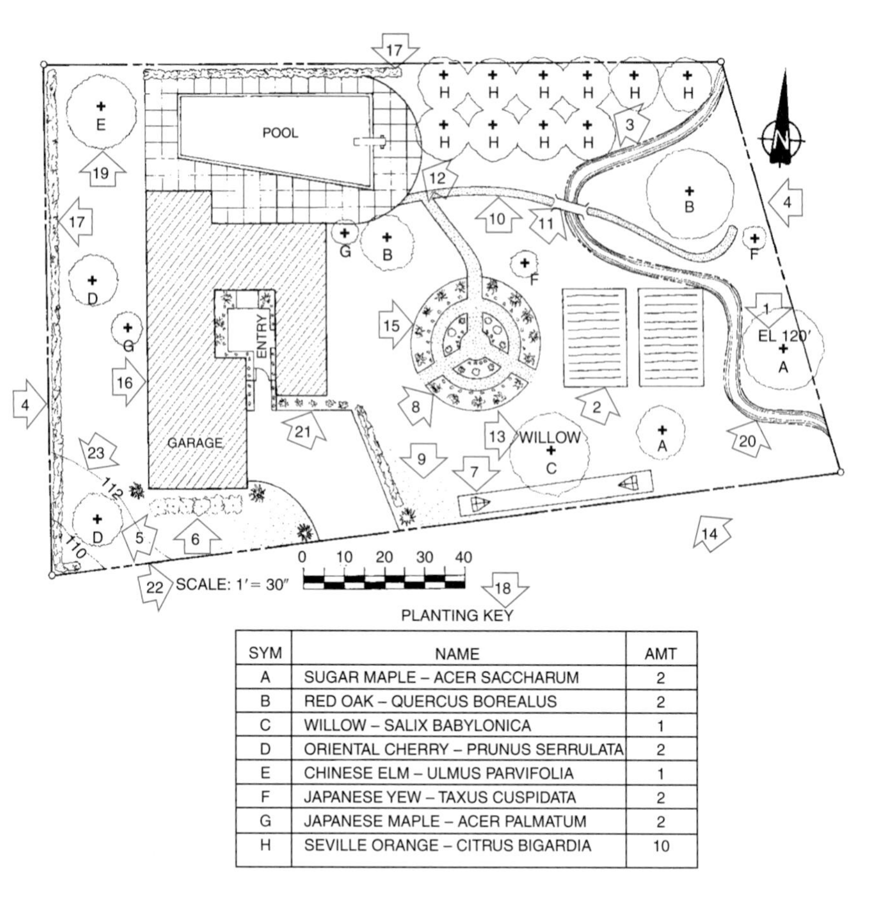

SYM	NAME	AMT
A	SUGAR MAPLE – ACER SACCHARUM	2
B	RED OAK – QUERCUS BOREALUS	2
C	WILLOW – SALIX BABYLONICA	1
D	ORIENTAL CHERRY – PRUNUS SERRULATA	2
E	CHINESE ELM – ULMUS PARVIFOLIA	1
F	JAPANESE YEW – TAXUS CUSPIDATA	2
G	JAPANESE MAPLE – ACER PALMATUM	2
H	SEVILLE ORANGE – CITRUS BIGARDIA	10

Guidelines for Drawing Landscape Plans

The following guidelines for drawing landscape plans are illustrated by the numbered arrows in Fig. 18-30.

1. Tree data including elevation, botanical name, limb coverage, and trunk diameter is shown on the drawing or keyed to a tree schedule.
2. Vegetable gardens are shown by outlining the planting furrows.
3. Orchards are shown by outlining each tree in the pattern.
4. The property line is shown to define the limits of the lot.
5. Trees are located to provide shade and windbreaks and to balance the decor of the site.
6. Shrubbery is used to provide privacy, define boundaries, outline walks, conceal foundation walls, and balance irregular contours.
7. The outlines of constructed recreation areas are shown.
8. Flower gardens are shown by the outline of their shapes.
9. Lawns are shown by small, sparsely placed dots.
10. The outlines of all walks and planned paths are shown.
11. Conventional map symbols are used for small bridges.
12. The outline and surface covering of all patios and terraces are indicated.
13. The name of each shrub and perennial flower bed is either labeled on or beside the symbol or indexed to a planting schedule.

14. All landscaping should enhance the function and appearance of the site.
15. Flowers should be located to provide maximum beauty and ease of maintenance. Other plantings, such as ground cover, should be indicated, as shown in Fig. 18-31.
16. Buildings are outlined, crosshatched, or shaded. In some cases, the outline of the floor plan is shown in abbreviated form. This helps to show the relationship of the outside to the inside living areas.
17. Hedges are used as screening devices to provide privacy, to divide areas, to control traffic, or to serve as windbreaks.
18. If labeling is impractical, each tree or shrub is indexed to a planting schedule. (Refer again to guideline 13.)
19. A tree is shown by drawing an outline of the area covered by its branches. This symbol varies from a perfect circle to irregular lines representing the appearance of branches. A plus sign (+) indicates the location of the trunk.
20. Water is indicated by irregular parallel lines.
21. Shrubbery in front of the house should be low so it does not interfere with site traffic patterns or with views from windows.
22. An engineer's scale that divides the foot into decimal parts is used to prepare landscape plans. This is the measure used by surveyors, civil engineers, and landscape architects.
23. Contour lines are drawn and labeled.

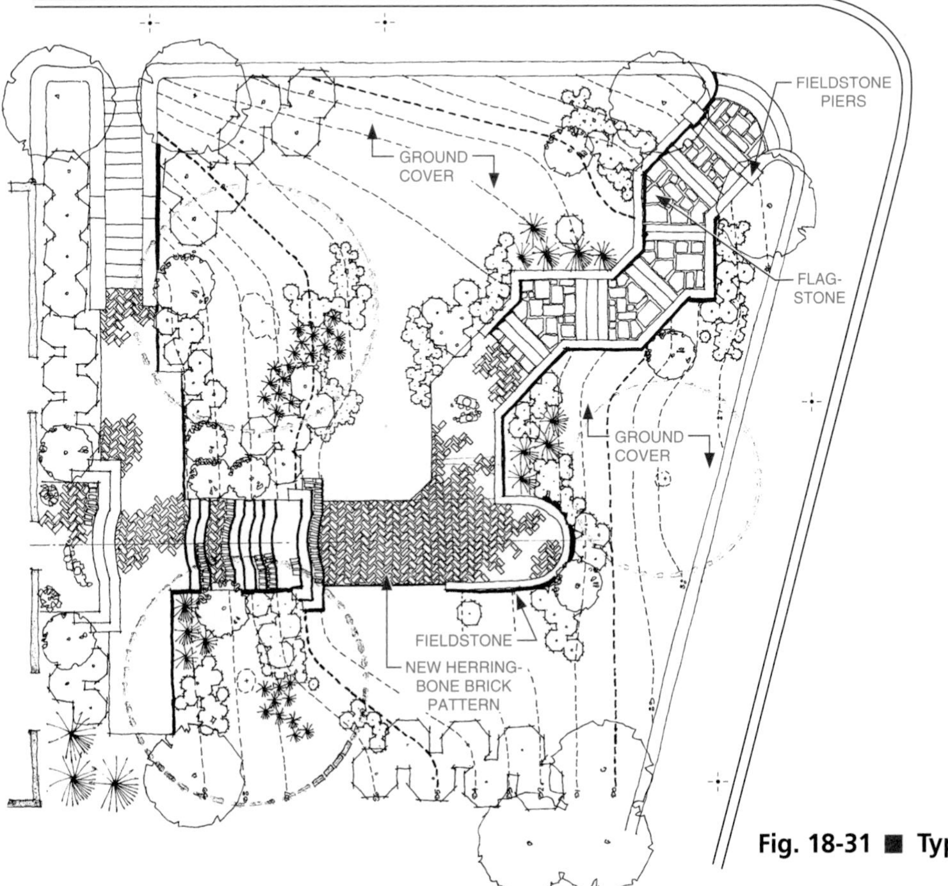

Fig. 18-31 ■ Typical landscape plan.

Phasing

The complete landscaping of a site may be prolonged over several years because of a lack of time or money. **Phasing** spreads the project over a larger period of time. Parts of the plan are completed at different times.

When a landscape plan is phased, the total plan is drawn. Then different shades or colors are used to identify the items that will be planted in the first month or year, and in successive months or years.

Landscape Rendering

Landscape renderings are used by landscape architects, architects, planners, and environmental professionals to realistically communicate the scale and visual character of a project. They visually convey the integration of the architecture and the site. Proper landscape design is accomplished by using a variety of plants, pavings, water, topography, scale, form, color, and texture.

Rendering Media

Landscape drawings require the use of many different types of inks, markers, pens, pencils, paints, and papers. You will learn more about these and techniques for using them in Chapter 20.

If rendering techniques are to be added directly to a print, a heavy blackline or brownline print, not a blueline print, should be used. Blueline prints make a poor rendering medium if defined shapes and lines are to be maintained. Blue lines do not stand out when other colors are added.

Plan Rendering

Plan views are the most common form of site illustration because plans best define the overall scope of a project or development. A rendered landscape plan can be used to describe the landscaping of large parcels or small residential sites. A landscape rendering can instantly tell the viewer whether the site is a hilly, wooded area or a grassy knoll. A rendered landscape plan typically shows trees, shrubs, ground cover, grass, walks, driveways, curbs, steps, pools, patios, rock outcrops, walls, and bodies of water.

The amount of rendering and intensity of detail should be consistent with the scale of the drawing. To describe a large site development, a minimum amount of rendering and detail is often adequate because of the small scale used. See Fig. 18-31.

A large site at the scale of 1″ = 200′ can show trees and existing vegetation in simple groups or masses. A small residential plan at a scale of $\frac{1}{4}$″ = 1′-0″, can show a considerable amount of detail. Individual trees and even branches can be shown, as well as textured paving and ground cover. Elements such as pools, arbors, shrubs, and decks can also show some amount of texture.

Before rendering a full landscape plan, the individual elements should be mastered. The following illustrations show a progression of rendering techniques designed to build specific skills before rendering a complete plan. In these illustrations note how the image changes when color, value, texture, and shadow all come together.

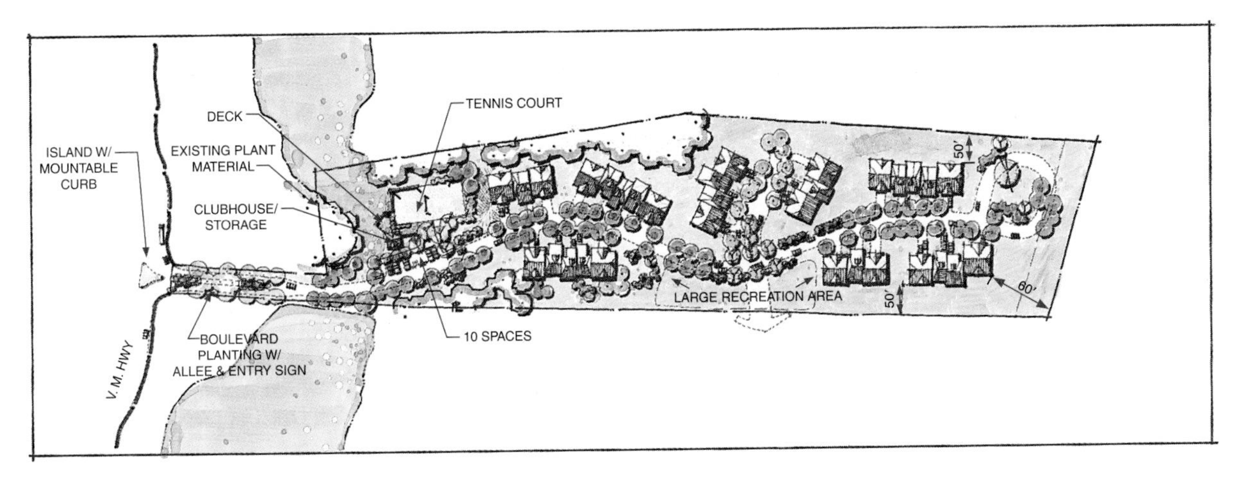

Fig. 18-32 ■ **Development plan with landscape features rendered in color.**

PAVING Figure 18-33 shows three steps in plan rendering of different paving types. Practice these steps with successive sheets of tracing paper until you are satisfied.

TREES AND SHRUBS In a plan view, tree branches cover and hide plant material, ground cover, and some structural areas located below. Because of their shape, the angle of the sun, and resulting shadows, treetops in a plan view contain a mixture of light, shade, and shadow. These are rendered by adding value, color, and texture. See Fig. 18-34. Also note that larger trees cast larger shadows, and conifers (pine trees) cast shadows of a different shape. Deciduous trees cast different shadows in summer than in winter.

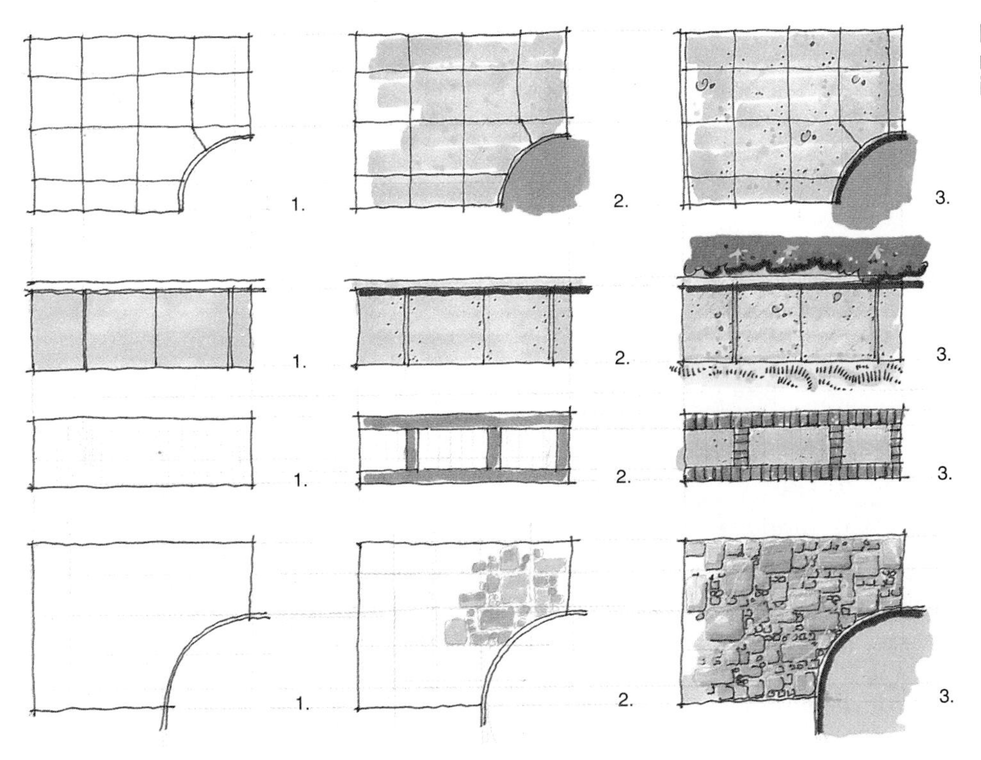

Fig. 18-33 ■ **Techniques for rendering pavement surfaces in plan view.**

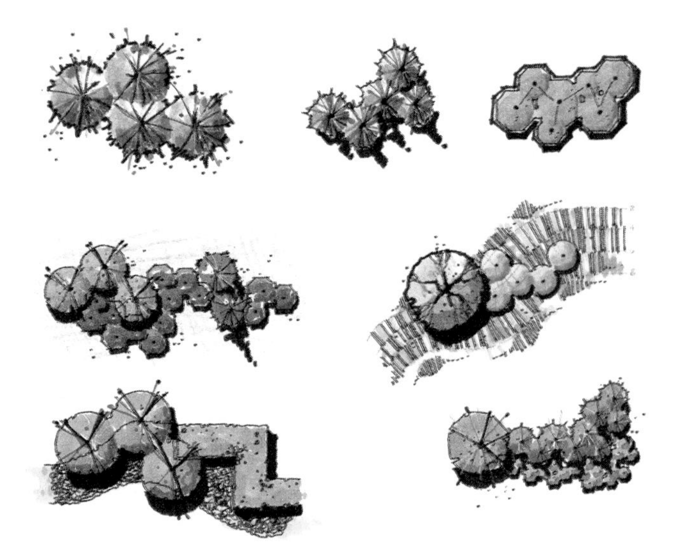

Fig. 18-34 ■ Plan view of rendered trees and shrubs.

WATER The rendering of water is challenging because water tends to be monochromatic (one color) and appears flat. Many aspects of the appearance of water result from its reflective qualities, which are difficult to render. Several techniques can be used to illustrate the reflections and movement of pool and stream water. See Fig. 18-35.

PEOPLE, VEHICLES, AND OTHER OBJECTS Including people, vehicles (especially automobiles) and other familiar objects in a rendering helps establish the size of the project and its various components. The key to effectively rendering people and

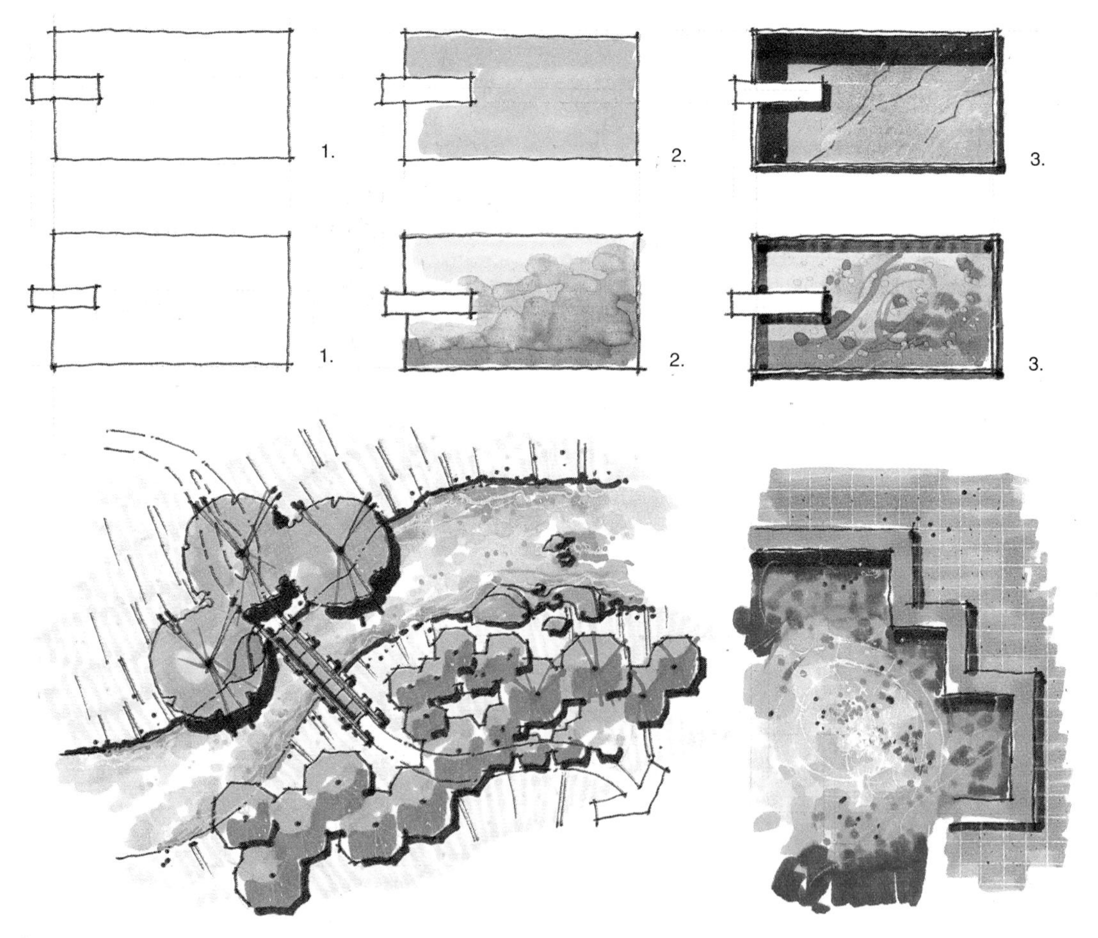

Fig. 18-35 ■ Techniques for rendering water surfaces in plan views.

objects is the proper use of scale and shadow. Everything in the rendering must be drawn at the same scale. In plan views, shadows are also very important. See Fig. 18-36.

RENDERING A COMPLETE LANDSCAPE PLAN

The techniques just described were applied in rendering the complete landscape plan shown in Fig. 18-37. To prepare a complete landscape rendering, begin by determining the color scheme on a separate but similar print or tracing. Then render higher elements, such as trees, arbors, and trellises for plants. Next, render lower elements, such as shrubs, decks, and ground cover. Render people, vehicles, and other objects last. Finally, add texture and shadows.

Fig. 18-36 ■ In a plan view, people and familiar objects are drawn with shadows to give the viewer a sense of scale.

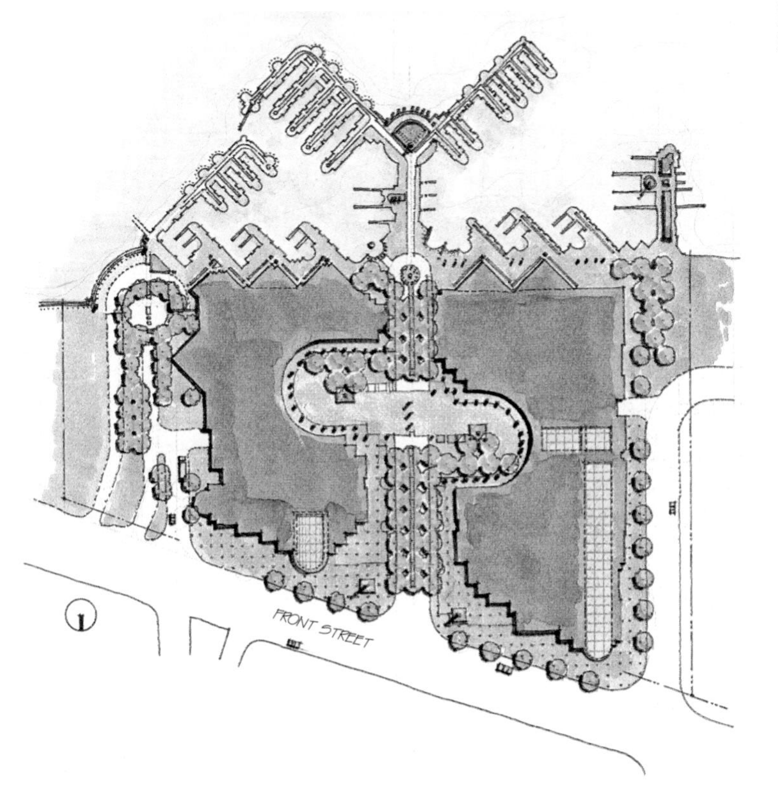

Fig. 18-37 ■ Landscape plan rendering of a park and marina.

Elevation Rendering

Plan views may be best for showing all the features of a site, but elevation drawings are also needed to show landscape features as realistically viewed from eye level. Landscape elevation renderings combine drawings of trees, shrubs, walls, topography, vehicles, and people, with the structure to produce a realistic view of the site. These drawings are used by design professionals to reveal form and scale of the project. The addition of color, value, texture, shades, and shadows provide depth to elevation drawings. Rendered elevation drawings are effective tools for sales and client-approval presentations because they help viewers perceive spatial depth and the final appearance of the project. See Fig. 18-38.

Fig. 18-38 ■ Rendered elevation drawing of a natural waterfall.

TREES AND SHRUBS Vegetation sizes, shapes, colors, and textures vary greatly. These differences must be considered in rendering. See Fig. 18-39. Trees and shrubs are located and rendered to show them at full maturity even though they will be planted at much smaller sizes. Trees especially should help define space without totally blocking desirable natural views outside and eye-level views from inside structures. See Fig. 18-40.

Fig. 18-39 ■ Rendering techniques used to show different types of trees on elevation drawings.

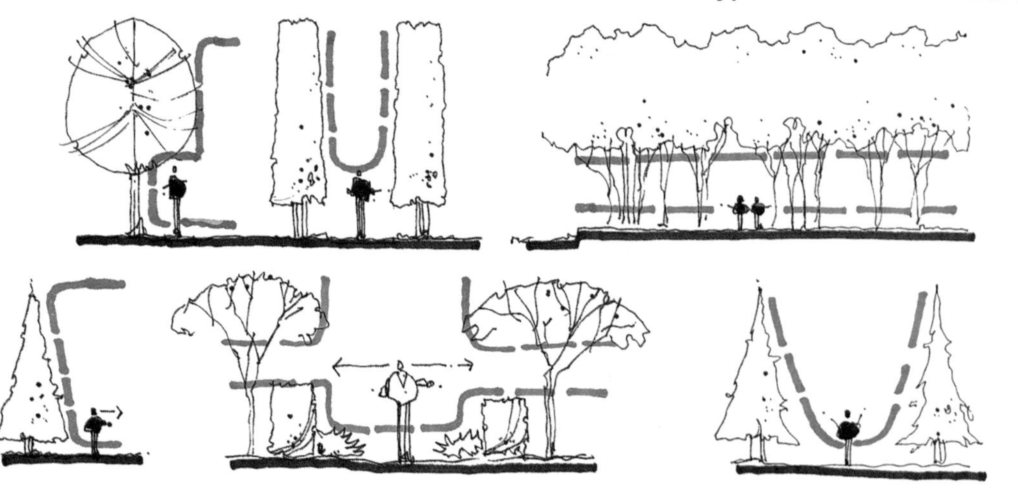

Fig. 18-40 ■ Open space and views must be maintained when selecting and locating trees.

Where dense vegetation is used as a baffle, outlining or rendering the winter form of deciduous trees and shrubs is recommended.

WATER Water in elevations is found in the form of natural waterfalls, constructed waterfalls, and fountains. Water as a focal point in an elevation should be vibrant and alive. This is accomplished through the use of color and texture, with a proper mix of current flow and froth (bubbling). See Fig. 18-41. Also, refer back to Fig. 18-38.

Fig. 18-41 ■ Techniques for rendering water in elevations.

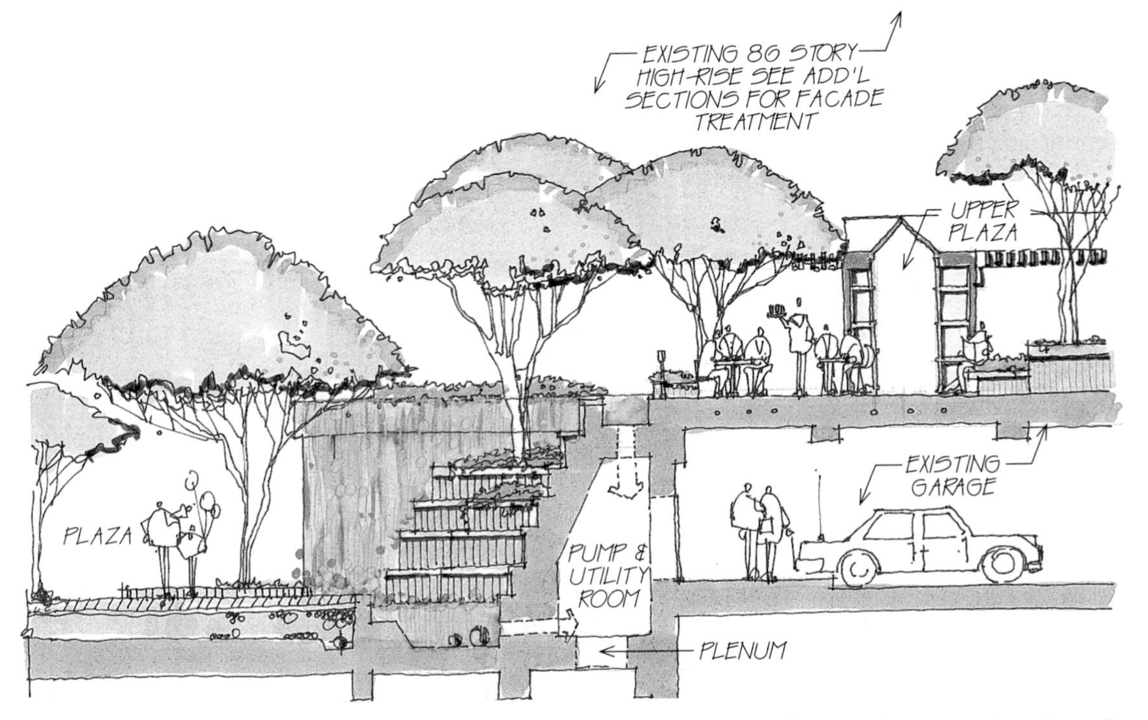

A. Rendering of a constructed waterfall designed as a transition feature between levels and as a focal point. Note that the relatively gentle flow of water is shown with the use of solid color.

B. The rendering of a fountain requires a more open and sparse use of color and more use of bubbling action.

PEOPLE AND FAMILIAR OBJECTS As with complete plan renderings, adding people and familiar objects to an elevation rendering adds both realism and scale. Refer again to Fig. 18-41A.

RENDERING WALLS Retaining walls are a common element in most multi-level landscape designs. These are constructed from wood timbers, brick, stone, and cast-concrete forms. Effective renderings duplicate the wall size, color, joints, texture, shades, and shadows. See Fig. 18-42.

Fig. 18-42 ■ Techniques for rendering wall elevations.

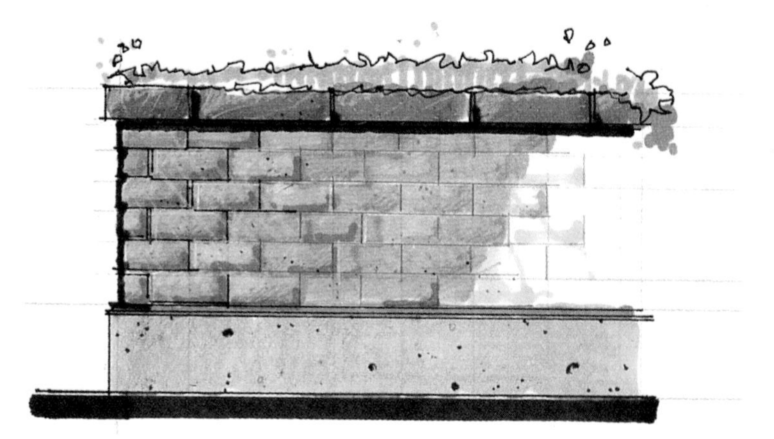

A. A masonry wall.

B. Lattice wall design including plant material placement. Note how the landscaping is rendered to show the vegetation and boulder placement without hiding the basic elements of the wall design.

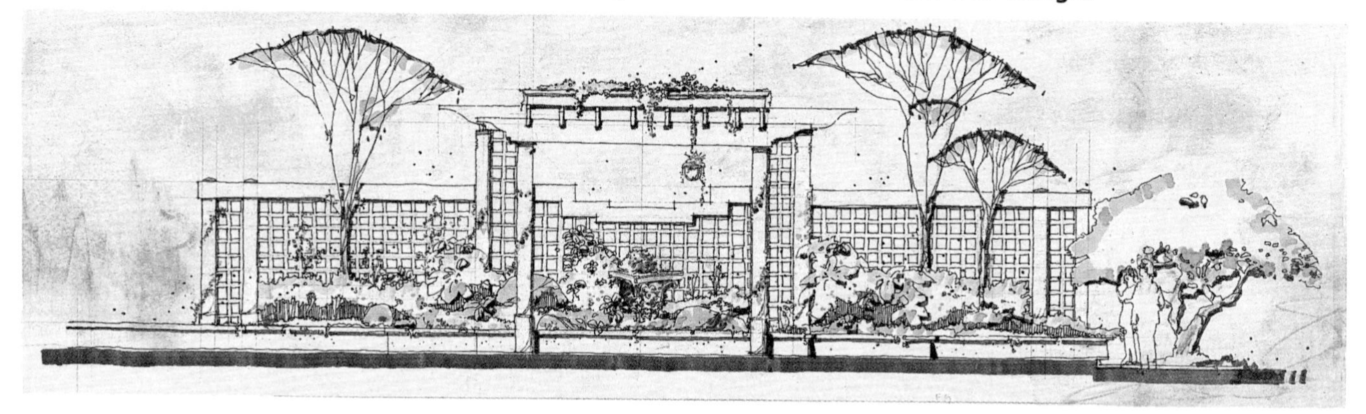

RENDERING ELEVATION SECTIONS When a design contains multiple levels, a rendered section is often the best way to describe the relationship of the levels. Adding vegetation, people, and automobiles also provides a sense of scale and proportion. Refer back to Fig. 18-41A.

RENDERING BUILDING ELEVATIONS When adding landscape features to a building elevation, care must be taken to not hide the major lines of the building. First add color, texture, and shadows to the building. Then begin rendering the foreground and work toward the background. See Fig. 18-43.

Fig. 18-43 ■ Rendered building elevation. Note that planned plant materials will not obstruct views from inside.

Site Details and Schedules

Site development involves the design of landscape features, such as plants, with structural elements. Details, usually sections or profiles, are needed to ensure that construction is completed as designed. See Fig. 18-44. For example, planting instructions are detailed to ensure that the installation of major plant materials meets horticultural planting standards. See Fig. 18-45.

Planting schedules are prepared and indexed with numbers corresponding to a landscape plan. These schedules function as a guide for the purchase and placement of each size and species of plant material. See Fig. 18-46.

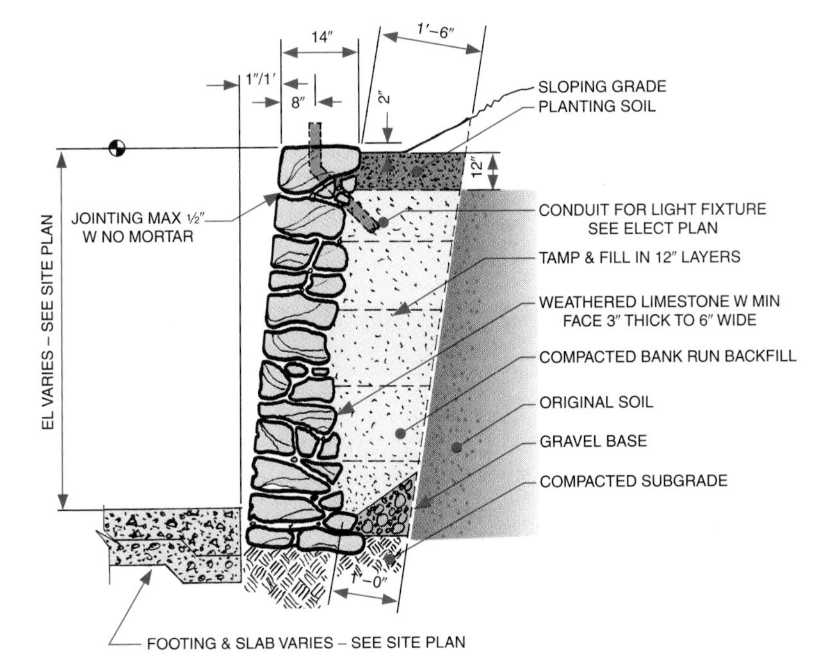

Fig. 18-44 ■ Structural details are prepared for the construction of landscape features.

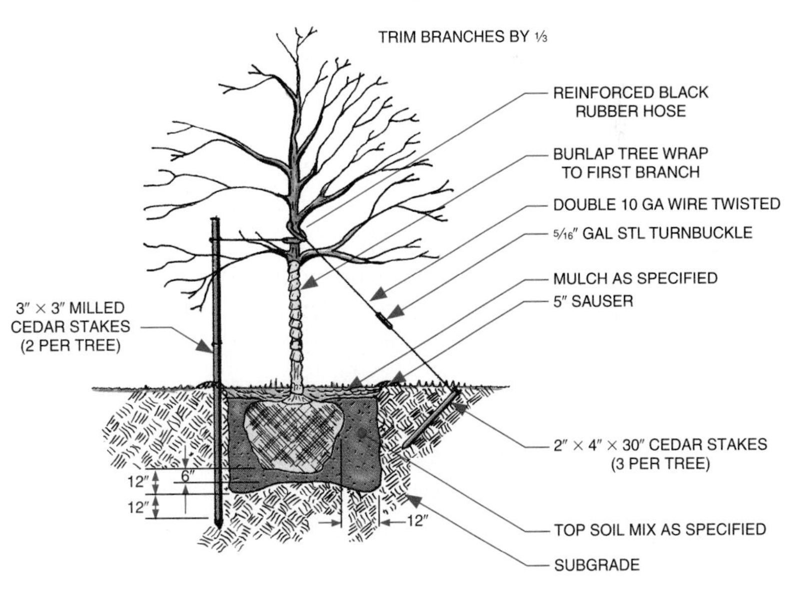

Fig. 18-45 ■ Plant material details ensure that proper planting techniques are used.

KEY	BOTANICAL NAME	COMMON NAME	QTY.	SIZE
As	Alnus serrulata	Smooth Alder	1	12–18″ #1 cont.
Aa	Aronia arbutifolia	Red Chokeberry	4	12–18″ b/b
Bh	Baccharus halimifolia	Groundsel Tree	4	2 qt.
Co	Cephalanthus occidentalis	Buttonbush	14	12–18″ b/b
Cl	Clethra alnifolia	Sweet Pepper Bush	3	18–24″ b/b
Cr	Cornus amomum	Silky Dogwood	6	18–24″ b/b
Cor	Cornus sericca 'Ruby'	R.R. Osier Dogwood	1	18–24″ b/b
Hib	Hibiscus moschentos	Marsh Hibiscus	3	1 qt. pot
If	Iva frutenscens	High-Tide Bush	39	18–24″ stab.
Jun	Juneus rocmerianus	Black Needle Rush	35	1 qt. pot
Kos	Kosteletzkya virginica	Seashore Mallow	5	1 qt. pot
Le	Leersia oryzoides	Rice Cutgrass	3	1-3/4″ peat pot
Lc	Limonium Carolinianum	Sea Lavender	65	1 qt. pots

HEPLER ASSOCIATES, P. C., Landscape Architects & Planners

Fig. 18-46 ■ A planting schedule.

CHAPTER 18 Exercises

1. Identify and discuss the environmental and human-related influences that affect site design.
2. Describe the zoning daylight plane ordinances for second-story setbacks. List why these are important.
3. Draw the setback and building area for a lot 130′ × 65′ according to the zoning requirements in any of the figures in this chapter. Determine the maximum size of a house for that site.

4. Describe the zoning laws and a density pattern you would prefer for an area in which you wish to locate a house of your design.
5. Using a CAD system, locate a house on a 60′ × 120′ lot. Setbacks are 5′ on the sides and 20′ front and rear.
6. Prepare a survey plan for a home you are designing.
7. Determine the bearing of property lines for a lot in your area using both the azimuth and American systems. Estimate the contour lines and include them in a sketch of this property.
8. Study a plot plan in this chapter and identify the highest and lowest levels above datum. What changes and/or structures would you recommend for this site?
9. Sketch a plat of your neighborhood using roads or streets as the outer boundaries.

10. Draw a plot plan of a house you are designing.

11. Draw a plot plan of a property in your area using a CAD system.
12. Draw and render a landscape plan for a house and property of your own design.
13. Redesign, draw, and render a landscape plan for a property in your area.

Landscape Architect

The landscape architect applies artistic and scientific principles to the design and management of the natural and built environments. Doing this requires a diversity of design, scientific, and cultural knowledge such as human behavior, settlement patterns, natural resources, grading, and landscape planting. Much of the work requires detail site analysis and master planning. The objective is to create and maintain meaningful and culturally enriching human places. A background in natural sciences, architectural design, and graphic communication is important.

> "Landscape architects should be able to look at the landscape in multiple ways and have a good sense of why the land expresses itself the way it does."

Ignacio San Martin

Ignacio San Martin is a professor of landscape architecture and urban planning at Arizona State University in Tempe, Arizona. In his teaching, he stresses the importance of respect for the land, for the ecology, for the cultural roots of the area, and for the people who will inhabit the land.

Ignacio has a bachelor's degree in environmental geology and two master's degrees—one in landscape architecture, the other in urban design. Before coming to Arizona to teach, he practiced landscape architecture and urban planning in California. He notes that the different regions present different challenges to the landscape architect. In California, site hazards such as flooding, earthquakes, landslides, and fire were important considerations in site planning and analysis. In Arizona, the dry climate presents new challenges, but also provides multiple opportunities. "We try to create landscapes in the southwest which use native plants, not only because they don't need much water, but because doing this helps preserve the character of the southwest landscape."

There is also a very strong Native American heritage in Arizona. "This cultural influence," says Ignacio, "can be expressed through the use of materials, space enclosures, texture of walls, and color."

Ignacio believes that "Landscape architects must be concerned with understanding the regional integrity of the land, its ecology, and the cultural roots of the area while creating new landscape designs that reflect modern living."

PART 5

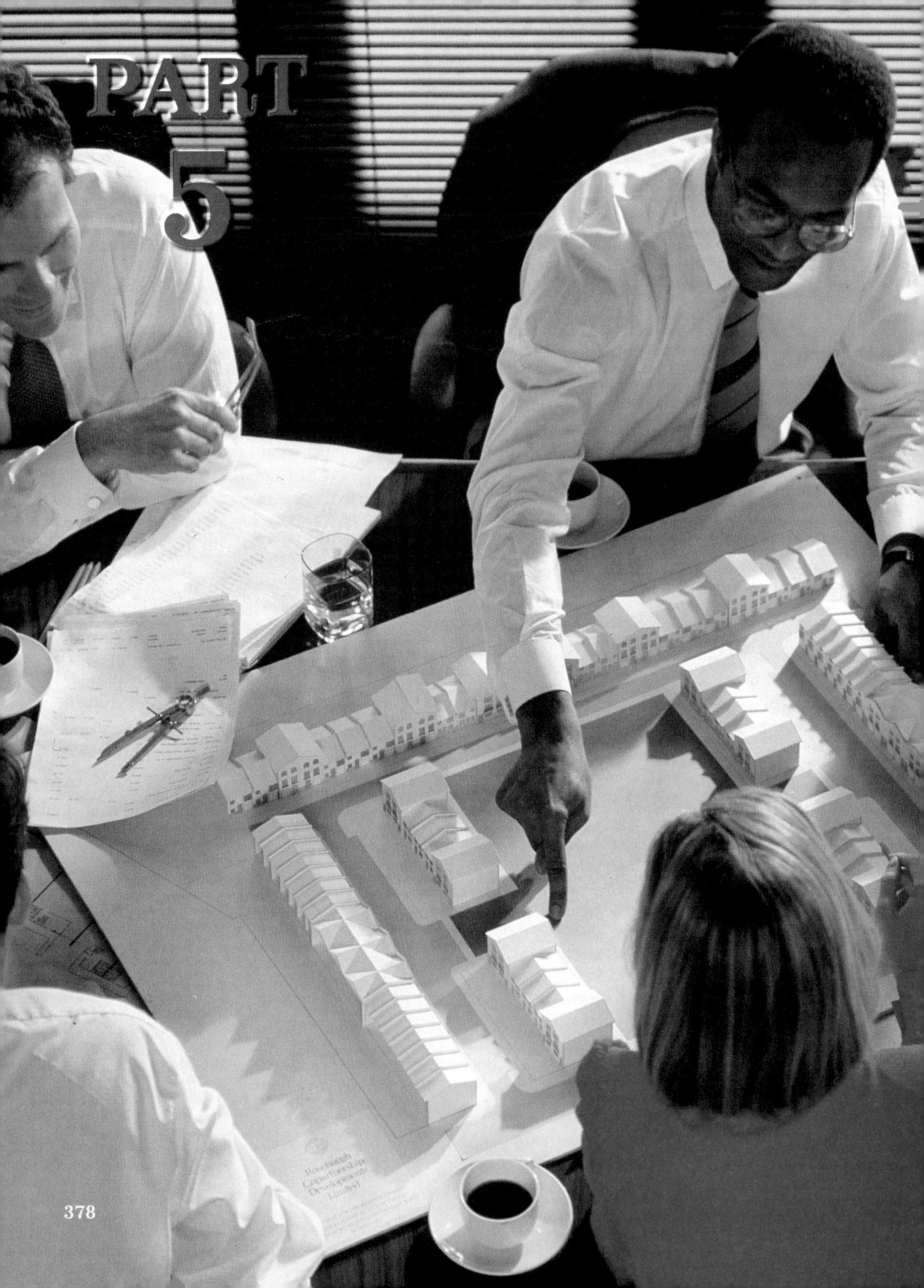

Presentation Methods

CHAPTER

19 Pictorial Drawings
20 Architectural Renderings
21 Architectural Models

CHAPTER

19

Pictorial Drawings

Objectives

In this chapter, you will learn:

- to differentiate between isometric, oblique, and perspective drawings.
- geometric principles involved in projecting lines (from a given point or at a constant angle) to create 3D images.
- to apply principles of perspective drawing to create interior and exterior pictorial drawings.
- projection methods for drawing pictorials.

Terms

horizon line
isometric drawings
oblique drawings
one-point perspective
perspective drawings
picture plane
station point
three-point perspective
two-point perspective
vanishing point

Pictorial drawings are picturelike drawings. Unlike elevations that reveal only one side of an object, pictorials show several sides of an object in one drawing. Pictorial drawings have lines that recede, creating an illusion of depth.

■ **A pictorial drawing is a three-dimensional drawing that provides a realistic view of a structure.** © *Kurt Alan Williams*

Types of Pictorial Projection

Three types of pictorials used in architectural drawing are oblique, isometric, and perspective. See Fig. 19-1.

Oblique and isometric drawings are created by drawing horizontal lines parallel. An **oblique drawing** is created by adding an angled side to an elevation drawing. The receding lines of this side are drawn at a very low angle of 10° to 15°. (Within a particular drawing, the angle stays constant.) The receding lines in an **isometric drawing** are projected at a constant angle of 30° from the horizontal. Because the angle of the projected lines remains constant, all receding lines are parallel. Thus oblique and isometric drawings are created by *parallel angle projection.*

Objects on oblique and isometric drawings are shown true scale. Because of the large size, especially length, of most architectural structures, isometric and oblique drawings usually result in visual distortion of the receding areas and produce an unrealistic appearance. Architectural isometric drawings are primarily used for smaller details and for small objects. See Fig. 19-2. Most pictorial drawings of entire buildings are prepared in perspective form, which is more realistic to the human eye.

Fig. 19-1 ■ Compare the types of pictorial drawings.

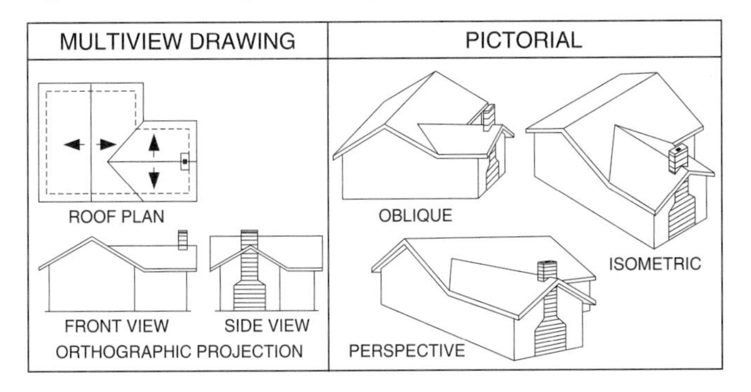

Fig. 19-2 ■ Isometric drawings are used for construction detailing.

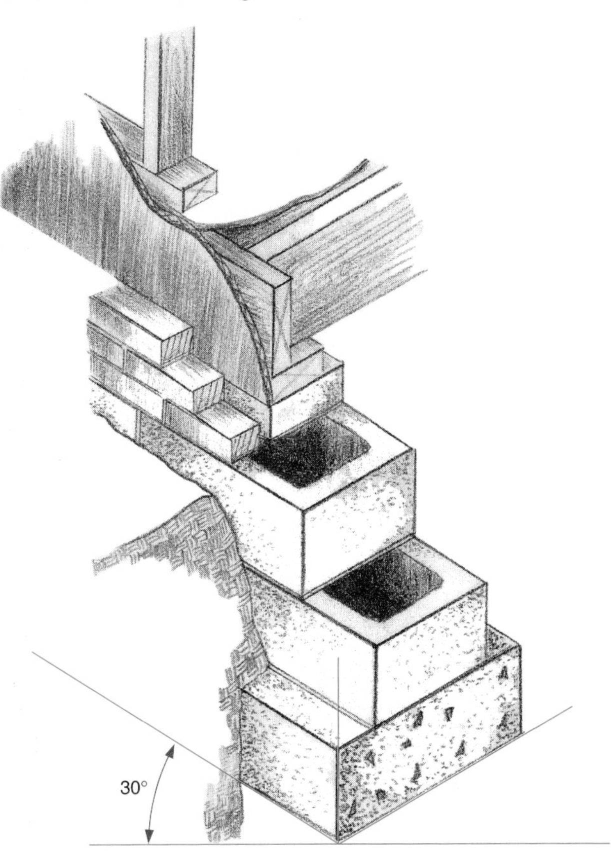

Perspective Drawings

In **perspective drawings,** receding lines of a building appear to meet. They are not drawn parallel. A perspective drawing, more than any other kind of drawing, most closely resembles the way people actually see an image. If you look down railroad tracks, the parallel tracks appear to come together and vanish at a point on the distant horizon. Similarly, horizontal lines on a perspective drawing appear to meet at a distant point. See Fig. 19-3. The point at which these lines seem to meet and disappear is known as the **vanishing point.** When lines are drawn as if they meet at a vanishing point or project out from that distant point, a perspective drawing is created.

Fig. 19-3 ■ On a perspective drawing, horizontal lines appear to meet at the vanishing point.
Autodesk, Inc.

On perspective drawings, a **horizon line** is established. This line is considered the observer's eye level. The location of the observer is the **station point.** The vanishing points in a perspective drawing are always placed on the horizon line. If the horizon line is placed through a building, the building will appear to be at eye level (Fig. 19-4A). If the horizon line is placed low or below a building, the building will appear as if you were looking up at it (Fig. 19-4B). If the horizon line is placed above a building, it will appear to be below your line of sight (Fig. 19-4C). Objects placed close to the horizon line—either on it, just above it, or just below it—are less distorted than objects placed a greater distance from the horizon.

We are accustomed to seeing areas decrease in depth as they recede from our point of vision. That is why the sides of an isometric drawing (prepared with the true dimensions of a building) appear distorted. To make perspective drawings appear more realistic, the actual lengths of the receding side lines of the building are shortened. Because perspective

Fig. 19-4 ■ Effects of horizon line placement. A. Through a building.
Jenkins & Chin Shue Incorporated

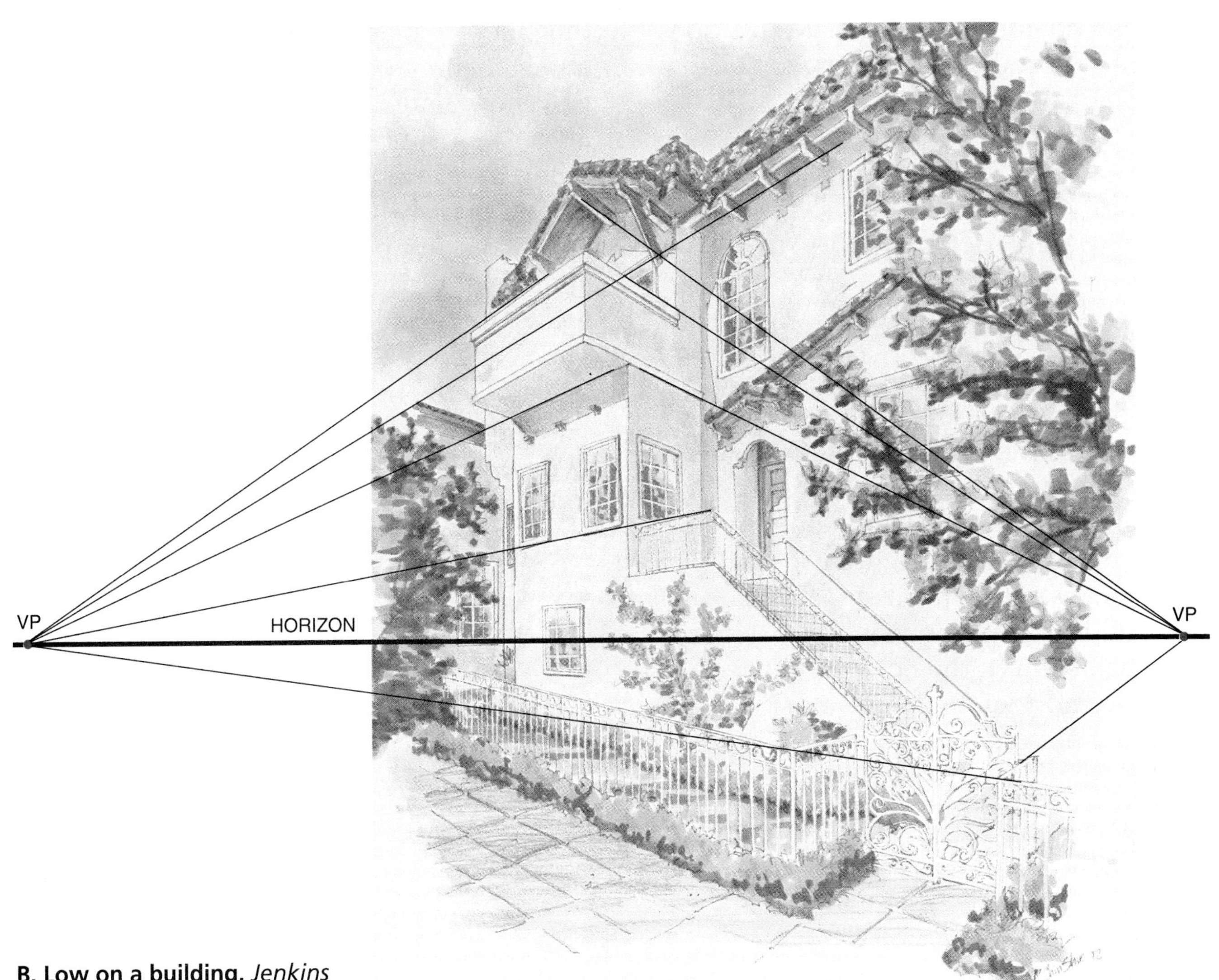

B. Low on a building. *Jenkins & Chin Shue Incorporated*

C. Above a building. *Turner Lechner & Romero Architects*

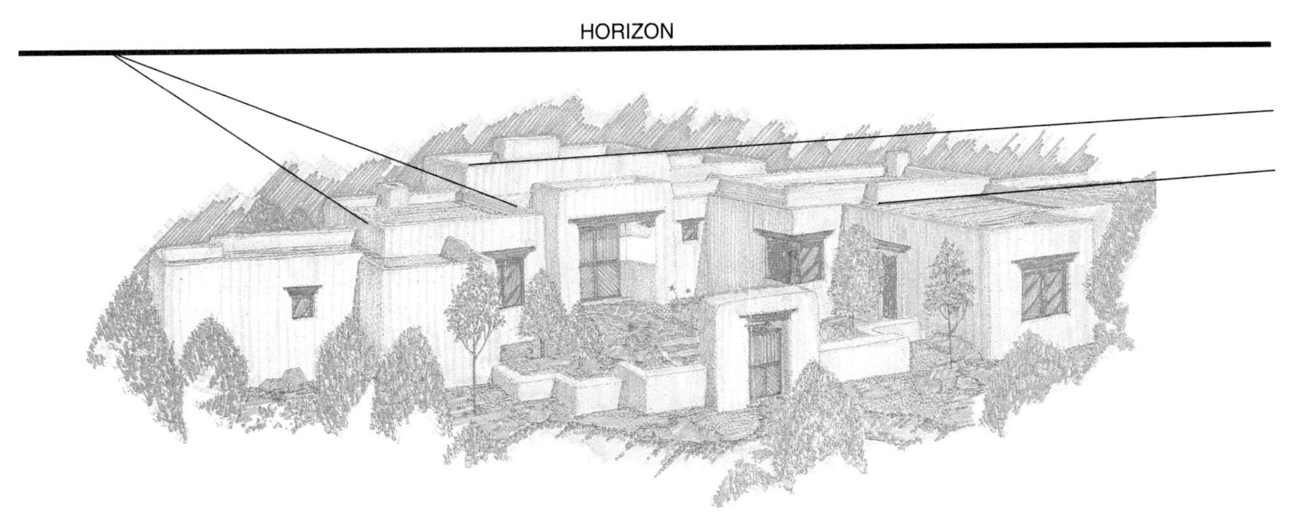

drawings do not reveal the true size and shape of the building, they are not used as working drawings.

Projection of Exterior Perspective Drawings

One-Point Perspective

A **one-point perspective** is a drawing in which the front view is drawn to its true scale and all receding sides are projected to a single vanishing point located on the horizon line. If the vanishing point is placed directly to the right or to the left of the object, with the horizon passing through the object, only one side (left or right) will show. If the object is placed above the horizon line and vanishing point, the bottom of the object will show. If the object is placed below the horizon line and vanishing point, the top of the object will show. Vanishing points need not always fall outside the building outline. When they are located within, only the frontal plane will show. See Fig. 19-5.

A one-point perspective drawing is relatively simple to create. The front view is drawn to the exact scale of the building. The corners of the front view are then projected to one vanishing point.

Steps in Drawing a One-Point Perspective

Follow these steps in drawing or sketching a one-point perspective. See Fig. 19-6.

1 Draw the horizon line and mark the position of the vanishing point. With the

Fig. 19-5 ■ Vanishing point located within the building outline.
Jenkins & Chin Shue Incorporated

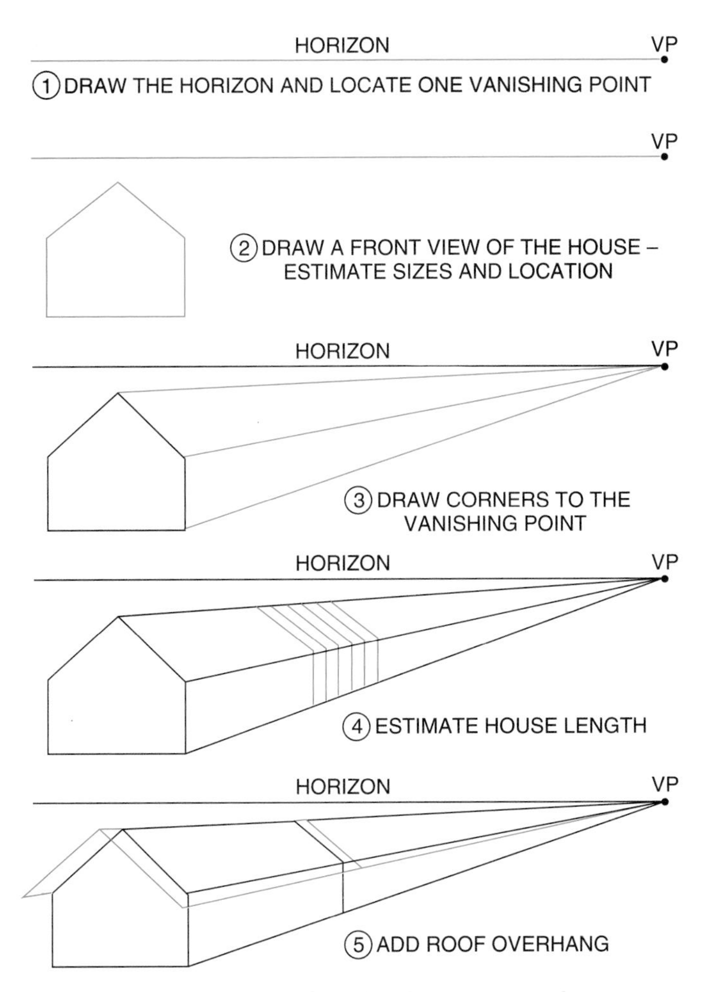

Fig. 19-6 ■ Sequence of projecting a one-point perspective drawing.

vanishing point to the left, you see the left side of the building. With the vanishing point to the right, you see the right side of the building. If the vanishing point is to the rear, you see only the front of the building, provided the horizon line extends through the building.

2 Draw the front view of the building to a convenient scale.

3 Project all visible corners of the front view to the vanishing point.

4 Estimate the length of the house. Draw lines parallel with the vertical lines of the front view to indicate the back of the building.

5 Make all object lines heavy, such as roof overhang. Erase the projection lines leading to the vanishing point.

One-point perspective drawings are also extensively used to prepare landscape or multi-building complex drawings.

Two-Point Perspective

A **two-point perspective** drawing is one in which the receding sides are projected to two vanishing points, one on opposite ends of the horizon line. See Fig. 19-7. In a two-point perspective, no sides are drawn exactly to scale. All sides recede to vanishing points. Therefore the only true-length line, the one that is to scale, on a two-point perspective is the vertical line in the corner of the building from which both sides are projected.

When the vanishing points are placed close together on the horizon line, considerable distortion results because of the acute receding angles, as shown in Fig. 19-8. When the vanishing points are placed farther apart, the drawing looks more realistic. Placing the drawn object closer to the horizon also helps create a more realistic appearance.

One vanishing point is often placed farther from the station point than the other vanishing point. This placement allows one side of the building to recede at a

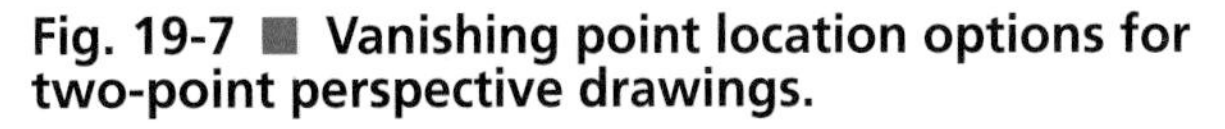

Fig. 19-7 ■ Vanishing point location options for two-point perspective drawings.

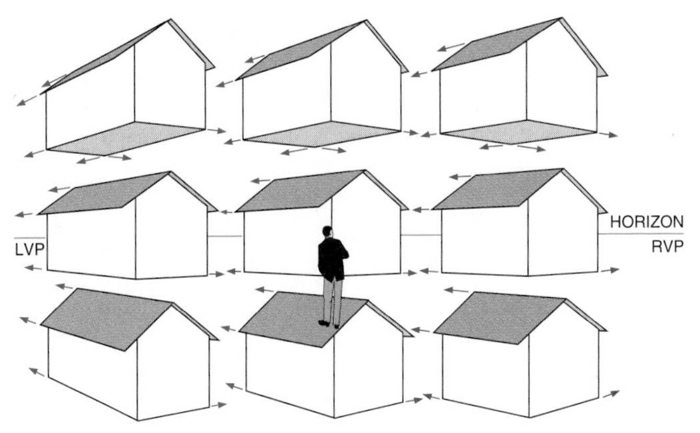

Fig. 19-8 ■ The distance between the vanishing points affects the receding angles.

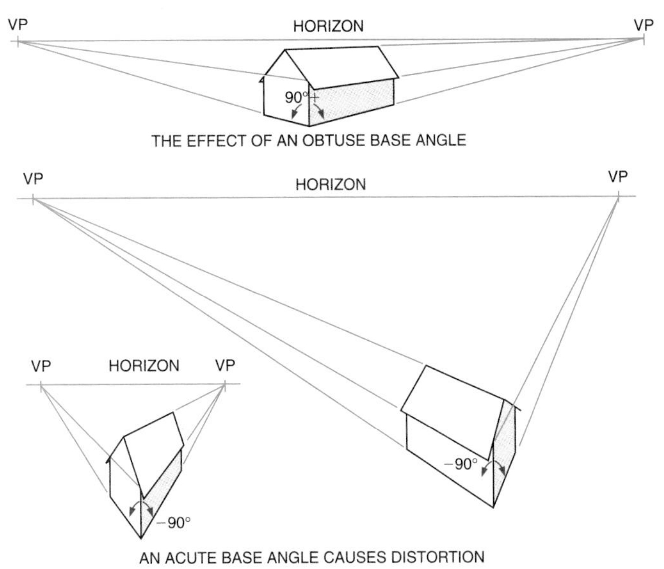

sharper angle than the other. See Fig. 19-9. Figure 19-10 shows the relationship between the station point, vanishing points, object, and an imaginary picture plane. A **picture plane** is an imaginary plane between the station point and the object upon which a perspective view is visualized by the observer. In drawing or sketching a simple two-point perspective, the steps outlined in Fig. 19-11 can be followed.

Three-Point Perspective

Three-point perspective drawings are used to overcome the height distortion of tall buildings. In a one- or two-story building, the vertical lines recede so slightly that, for practical purposes, they are drawn parallel. However, the top or bottom of extremely tall buildings appears

Fig. 19-9 ■ Compare how the side placement of the vanishing points affects the appearance of the building as shown.

A. The left vanishing point placed closer to the true-length line than the right vanishing point. *Jenkins & Chin Shue Incorporated*

B. The right vanishing point placed closer to the true-length line than the left vanishing point. *Jenkins & Chin Shue Incorporated*

Fig. 19-10 ■ Visualization of a two-point perspective shape on a picture plane.

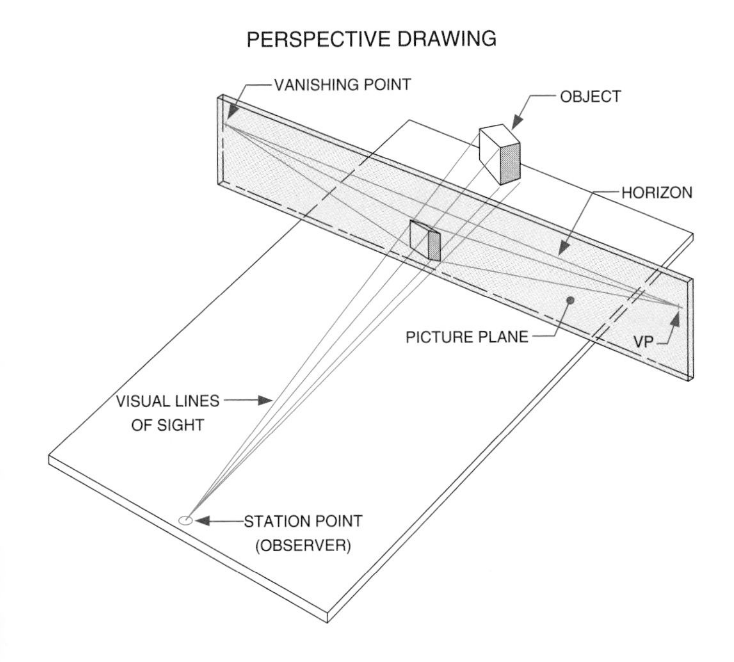

smaller than the area nearest the viewer. A third vanishing point may be added to provide the desired recession. See Fig. 19-12. The farther away the third vanishing point is placed from the object, the less the distortion. If the lower vanishing point is placed so far below or above the horizon that the angles are hardly distinguishable, then the advantage of a three-point perspective may be lost because the vertical lines are almost parallel.

Projection Devices

There are four basic devices used for projecting building lines to vanishing points.

- Use a straightedge held against a pin placed at the vanishing point.
- Use an underlay grid with preprinted lines converging on vanishing points.

Fig. 19-11 ■ Steps in drawing a simple two-point perspective.

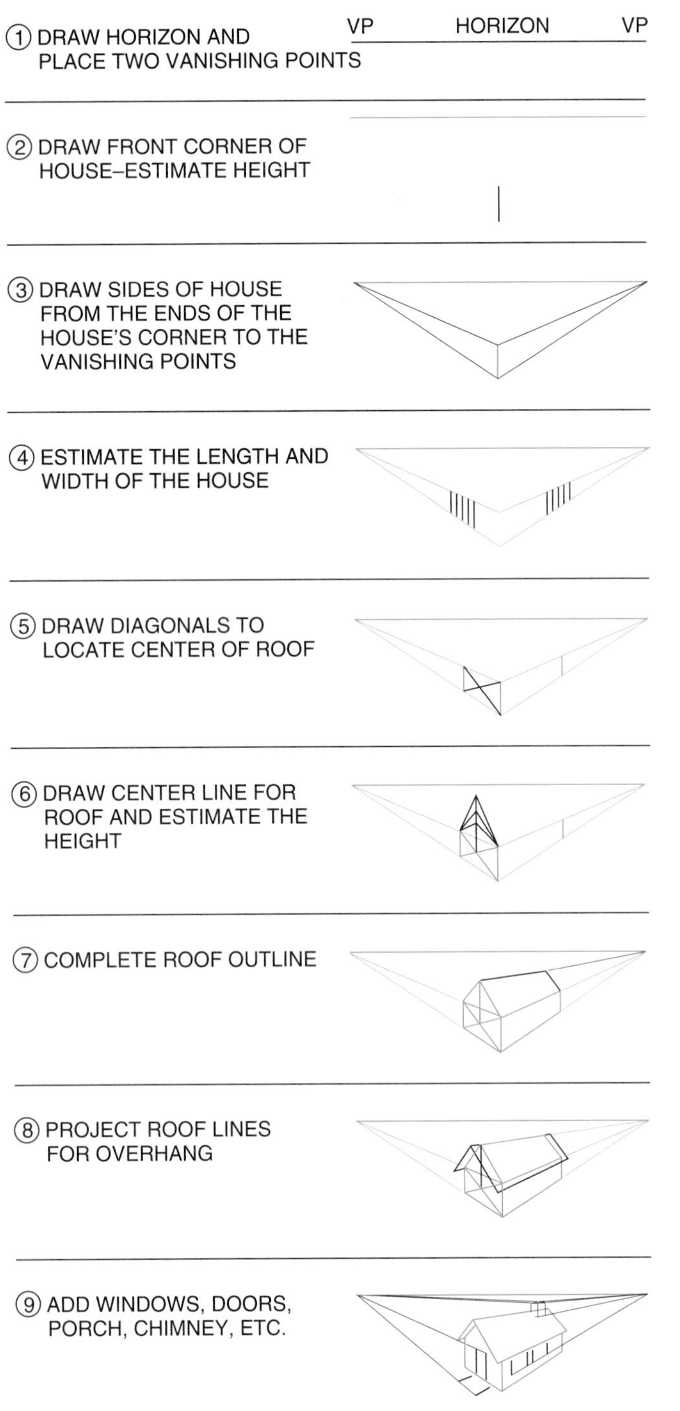

(Gridsheets are available with vanishing points established at different distances. Grids are produced for one-point, two-point, and three-point perspective patterns.)

Fig. 19-12 ■ Comparison of one-, two-, and three-point perspective drawings.

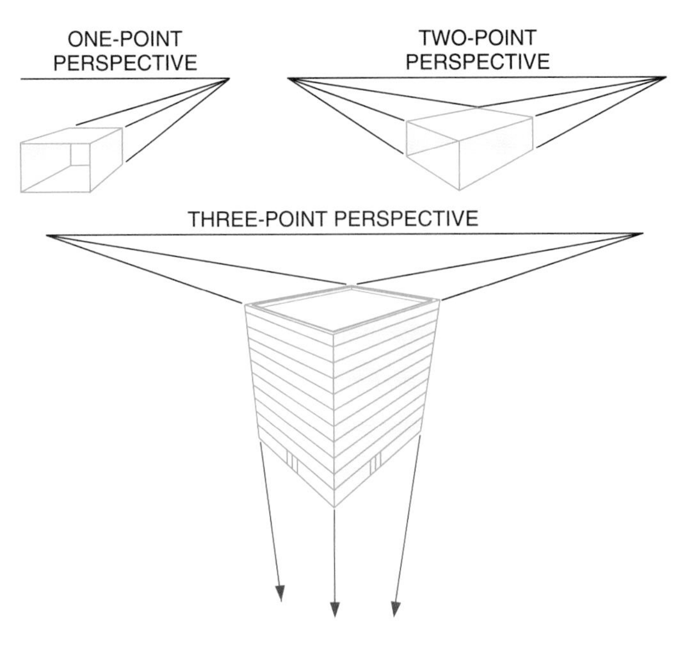

- Use special T-square devices that automatically lock into perspective angles.
- The fastest-growing method is to use 3D CAD systems as covered in Chapter 5. Figure 19-3 shows an example of a perspective drawing created on a 3D CAD system.

Projection of Interior Perspective Drawings

A pictorial drawing of the interior of a building may be an isometric drawing, a one-point perspective, or a two-point perspective drawing. Pictorial drawings may be prepared for an entire floor plan. More commonly, however, a pictorial drawing is prepared for a partial view of a single room or to show a particular interior detail.

Isometric Drawings

Isometric drawings are only satisfactory for small interior areas. See Fig. 19-13. Because receding isometric lines (at 30° angles) are always parallel, these draw-

Fig. 19-13 ■ Interior isometric drawing.

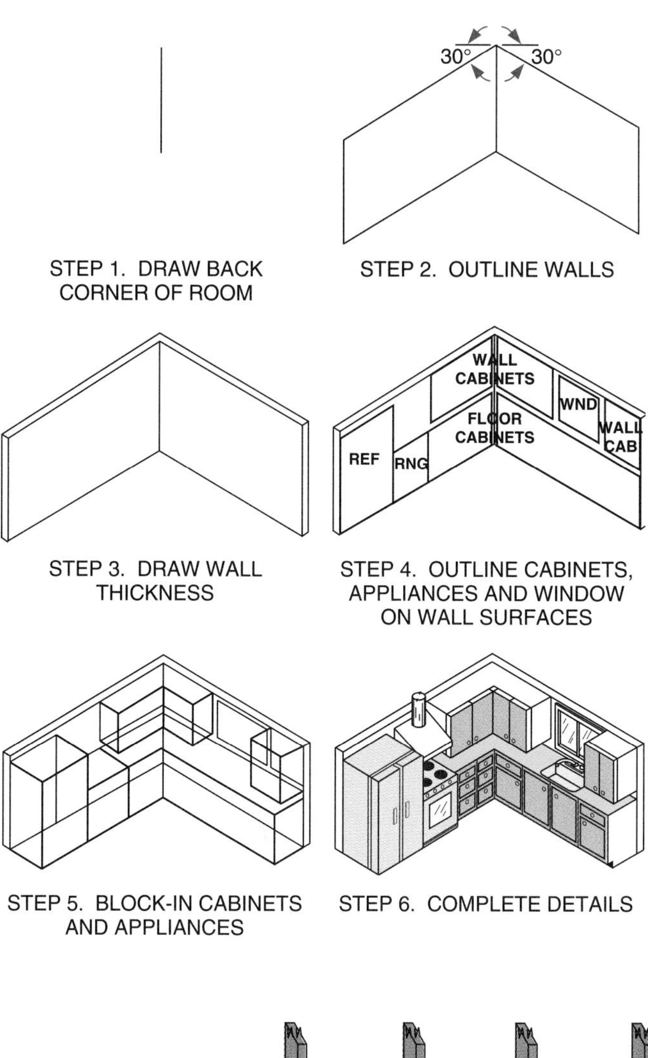

ings may be prepared to an exact scale. Isometric drawings of large interior areas are usually not desirable because they lack receding perspective lines and therefore appear distorted. Isometric or oblique (Fig. 19-14) drawings are, however, excellent ways to convey interior construction details that may not be understood from floor plans or interior elevation drawings.

One-Point Perspective

A one-point perspective of the interior of a room is a drawing in which all the intersections between walls, floors, ceilings, and furniture are projected to one vanishing point. Drawing a one-point perspective of the interior of a room is similar to drawing the inside of a box with the front of the box removed. In a one-point interior perspective, walls perpendicular to the plane of projection, such as the back wall, are drawn to scale. The vanishing point on the horizon line is then

Fig. 19-14 ■ Oblique interior construction detail.

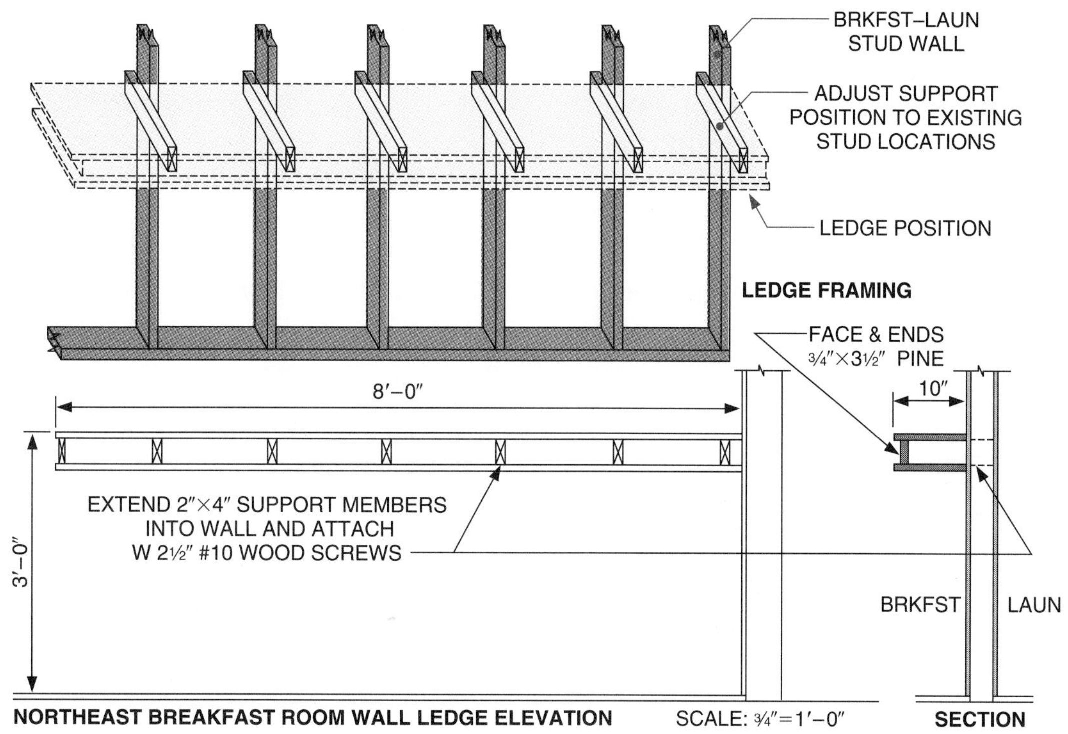

Fig. 19-15 ■ Projection of interior lines to a single vanishing point.

placed somewhere on this wall (actually behind this wall). See Fig. 19-15. The points where this wall intersects the ceiling and floor are then projected from the vanishing point to form the intersection between the side walls and the ceiling and the side walls and the floor.

VERTICAL PLACEMENT Vertical placement of the vanishing point affects the appearance of a drawing. See Fig. 19-16. If the

Fig. 19-16 ■ Effects of vanishing point locations on drawing appearance.

A. Above the horizon.

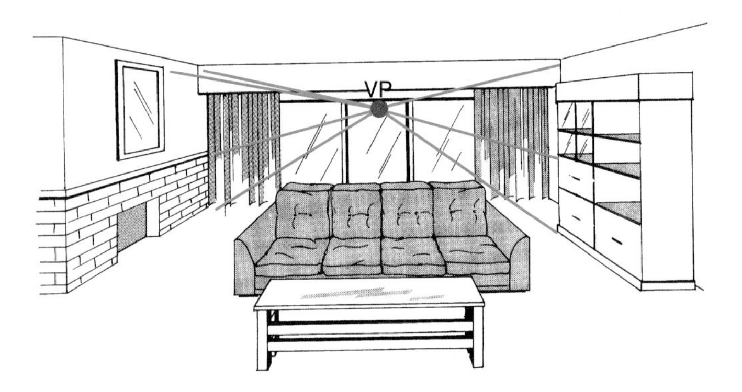

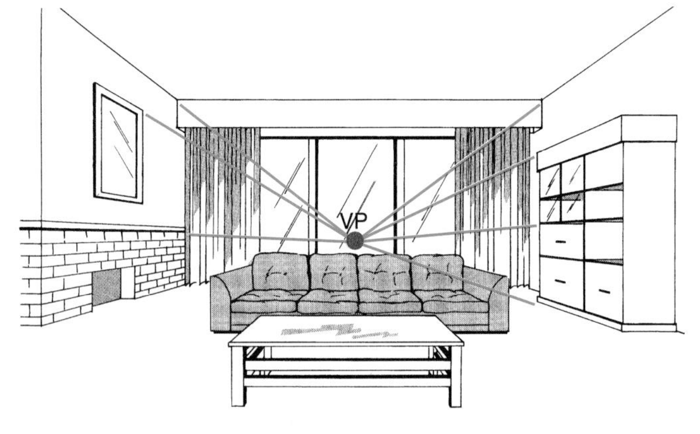

B. Center of the horizon.

C. Below the horizon.

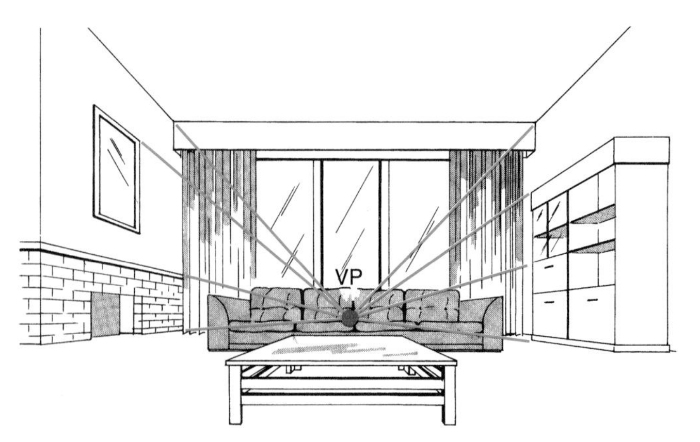

vanishing point is placed high, very little of the ceiling will show in the projection, but much of the floor area will be revealed. If the vanishing point is placed near the center of the back wall, an equal amount of ceiling and floor will show. If the vanishing point is placed low on the wall, much of the ceiling but very little of the floor will be shown. Since the horizon line and the vanishing point are at your eye level, you can see that the position of the vanishing point affects the angle from which you view the object.

HORIZONTAL PLACEMENT Moving the vanishing point from right to left on the back wall has an effect on the view of the side walls. This effect is the same for interior walls and enclosed exterior walls. If the vanishing point is placed toward the left, more of the right wall will be revealed. See Fig. 19-17A. Conversely, if the vanishing point is placed near the right side, more of the left wall will be revealed in the projection, as shown in Fig. 19-17B. If the vanishing point is placed in the center, an equal amount of right wall and left wall will be shown. When one wall should dominate, place the vanishing point on the extreme end of the opposite wall.

When projecting wall offsets and furniture, always block in the overall size of the item to form a perspective view. The steps are shown in Fig. 19-18A, B, and C. The details of furniture or closets or even of persons can then be completed within this blocked-in cube or series of cubes.

Fig. 19-17 ■ Effects of side placement of the vanishing point.
A. Left side. *Jenkins & Chin Shue Incorporated*

Fig. 19-17 ■ Continued
B. Right side. *Addison Estates, Boca Raton, FL*

Fig. 19-18 ■ Steps in developing an interior one-point perspective.

A. First step.

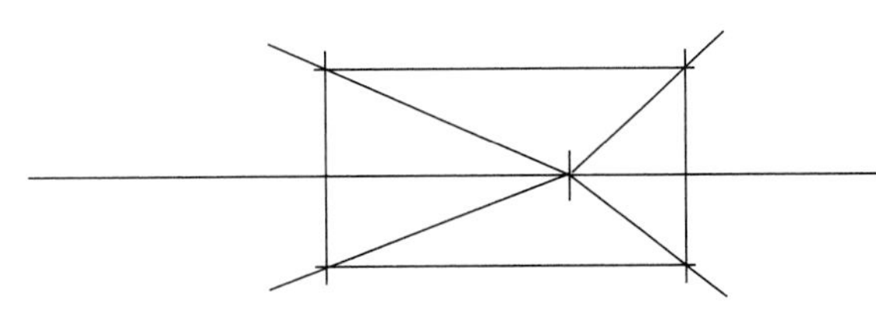

B. Second step.

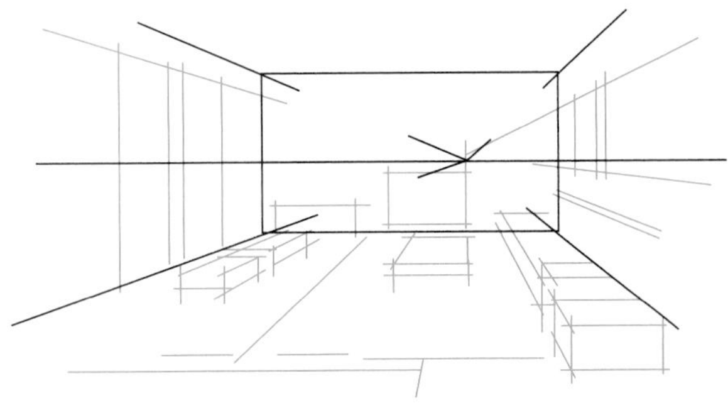

C. Final steps.

Two-Point Perspective

Two-point perspectives are normally prepared to show the final design and decor of two walls of a room. The vertical true-length or base line on an interior two-point perspective is similar to the base line on an exterior two-point perspective. On an interior drawing, this line may be a corner of a room, an article of furniture, or other vertical line. Not only are the walls projected to the vanishing points in a two-point perspective, but each object in the room is also projected to the vanishing points. The sequence of steps in drawing two-point interior perspectives is shown in Fig. 19-19A, B, and C.

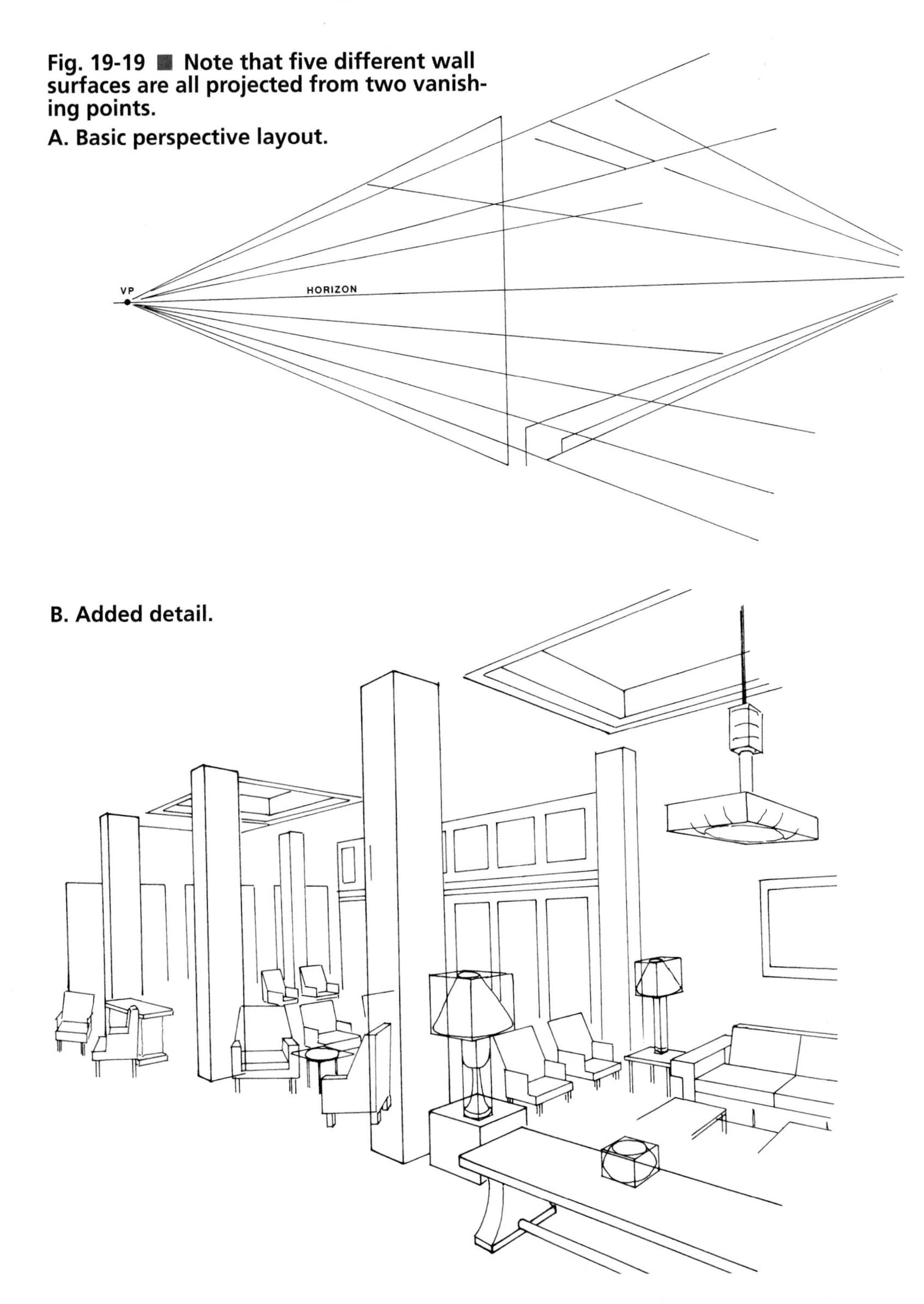

Fig. 19-19 ■ Note that five different wall surfaces are all projected from two vanishing points.
A. Basic perspective layout.

B. Added detail.

Fig. 19-19 ■ Continued
C.Finished rendering. *Jenkins & Chin Shue Incorporated*

Pictorial grids are used to eliminate the need to draw projection lines from the vanishing points and can thereby speed up the drawing process. However, the proper grid must be selected to provide a correctly placed horizon line and vanishingpoint position that will reveal the room at the desired angle. See Fig. 19-20.

Fig. 19-20 ■ **Use of a perspective grid for two-point perspective drawing.**

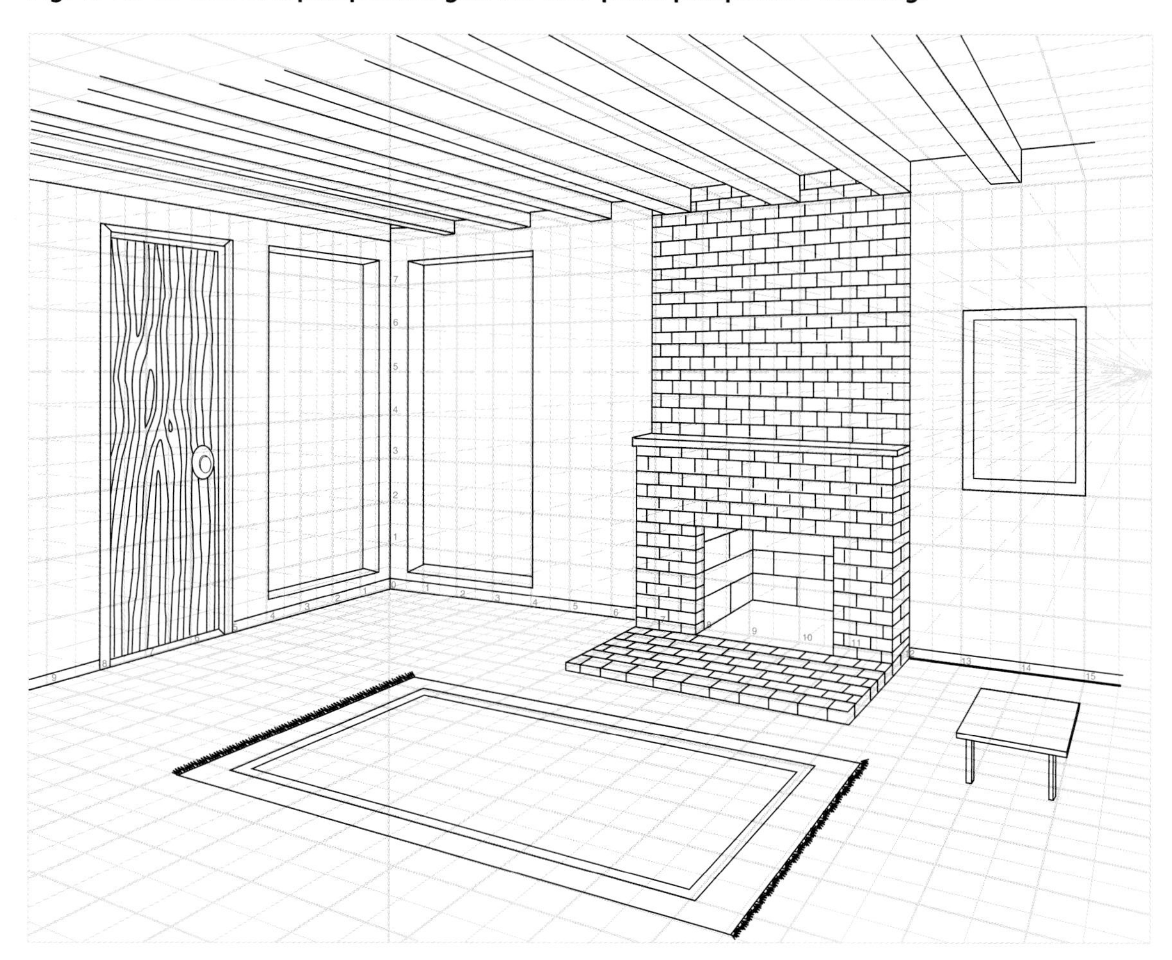

CHAPTER 19 Exercises

1. Draw a two-point perspective of a building of your own design.
2. Draw a one-point and a two-point perspective of your own home.
3. Sketch a three-point perspective of the tallest building in your community.
4. Prepare a one-point interior perspective of your own room.
5. Prepare a one-point perspective of a room in the house of your own design.
6. Draw a one-point interior perspective of a classroom. Prepare one drawing to show more of the ceiling and left wall. Prepare another drawing to show more of the floor and right wall.

7. Complete Exercises 1 through 6 using a CAD system.

CHAPTER

20

Architectural Renderings

Objectives

In this chapter, you will learn to:

- recognize the wide selection of media available for renderings.
- evaluate when to use which media to achieve an artistic effect.
- add realism to drawings by the use of shading, shadows, texture, entourage, and landscapes.
- follow the correct sequence for preparing a rendering.

Terms

acrylics
entourage
pastels
render
wash drawing

Because pictorial drawings are three-dimensional, their shape resembles a realistic view of a building. Our eyes see more than shape though. We see color, texture, shades, shadows, people, and landscape features. In a rendering, these features are added to a pictorial drawing. To **render** a drawing is to make the drawing appear more realistic—whether it is a plan, an elevation drawing, or a perspective drawing. Drawings are rendered by adding realistic texture and establishing shade and shadow patterns. This may be done using a variety of media.

■ Renderings of a design help people visualize the completed project. This attractive rendering was prepared using watercolors. *Tangerine Bay Club—Tangerine Development Corp.*

Choosing Media for Rendering

A rendering may utilize only one medium, or several media may be combined to create various images. Media used to render drawings include: pencils, charcoal, ink, watercolors, felt markers, **pastels** (light-colored, water-based drawing medium), oil paint, and **acrylics** (water-based permanent paints).

Pencil Renderings

Soft pencils (Bs) are effective media for rendering architectural pictorials. Changes in the weight and density of lines create many tones. See Fig. 20-1. Variations in the spacing of pencil lines and in the pressure of the pencil can create values and different contrasting effects. Smudge blending to add tone is accomplished by rubbing a finger over soft penciled areas, as shown in Fig. 20-2. For more sketch-like effects, charcoal pencils, which are extremely soft, can be used.

Pencil renderings are popular because shading and texture can easily be added to penciled pictorial outlines. Moreover, pencil renderings are easy to reproduce with the remainder of the drawing set. Colored or pastel pencils can be used to make the various colors and values of building surfaces appear very realistic.

Fig. 20-1 ■ A pencil rendering. Note the various techniques used to create different effects. *Home Planners, Inc.*

Fig. 20-2 ■ Pencil smudge shading.

Ink Renderings

Since ink lines cannot be blended, the distance between lines is controlled to create the appearance of texture, light, shade, and density. Figure 20-3 shows ink line patterns as rendering techniques. Ink lines and strokes placed close together produce dark effects. Farther apart, they create lighter effects. See Fig. 20-4.

Watercolor and Wash Drawings

The use of wash drawings or watercolors is a fast and effective method of adding realism to pictorial drawings. When only black and gray tones are used, this type of rendering is called a **wash drawing.** When color is added, these drawings are known as watercolors. Watercolor paints blend to create a variety of attractive color and gradation effects. Therefore watercolors are used extensively for presentations and for advertising. Perhaps the most effective use of watercolor techniques is for the pictorial combination of landscape settings with structures. See Fig. 20-5.

Oil and Acrylic Renderings

Architectural renderings in oil paints or acrylics are more time-consuming and expensive than any other medium. For this reason, they are rarely prepared, except as works of art for display.

Felt-Marker Renderings

The use of felt (or felt-tip) markers is a fast way of adding color to pictorial drawings. Stroke lines do not blend easily, so this method is usually restricted to adding patches of color to existing drawings.

Fig. 20-3 ■ Common ink line strokes used on renderings.

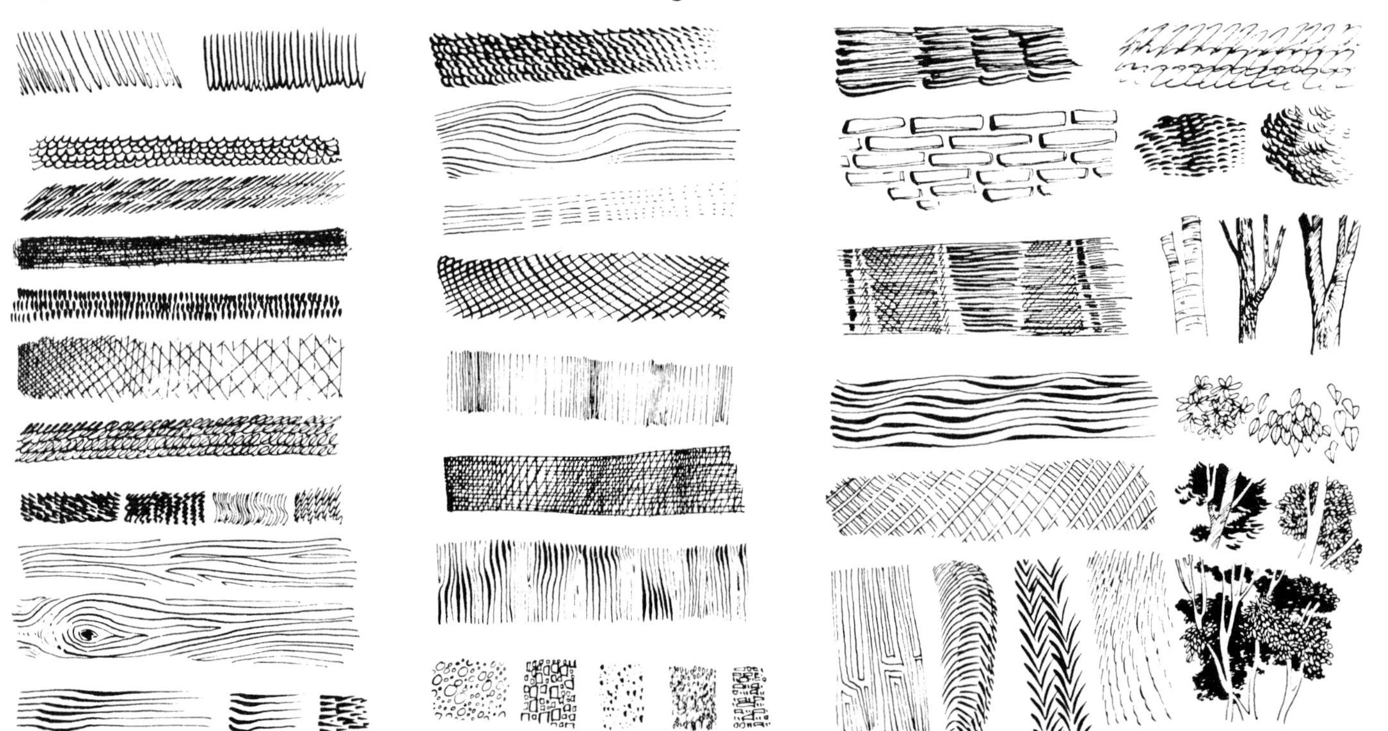

Fig. 20-4 ■ An ink rendering. Note the various line techniques used and how they affect the appearance of the structure and its surroundings. *Albert Lorenz—Rapidograph*

Fig. 20-5 ■ Watercolor renderings can be used effectively to show structures within the landscape setting. *Amelia Island Estates*

Pressure-Sensitive Overlays

A popular technique that is used to convert a perspective drawing into a rendering is to apply a pressure-sensitive overlay to the drawing. Preprinted pressure-sensitive screens are used to add tones, texture, and shadow. These effects are created by variations in the distance between lines or dots, in the width of lines, and in the blending of lines. Pressure-sensitive overlays can be used to create gray tones or solid black areas for contrast or for light and shadow patterns.

Media Combinations

Most architectural renderings include a variety of media, depending upon the cost and the emphasis desired. For example, in Fig. 20-6 the same exterior view is rendered in watercolors with and without ink lines. Note the difference in appearance and in the amount of detail presented in the two renderings. (The floor plan for this house is shown in Fig. 14-22.)

Interior design features are frequently rendered using watercolor over line work to reveal color and texture of materials. See Fig. 20-7. Other media normally used in combination with pencil or ink line drawings include airbrush, pastels, and felt markers.

Another method of preparing pictorial renderings is to use colored or gray illustration board. When this method is used, all lines, shades, or tints are added with a variety of white watercolor and/or acrylic paint. Notice how the black, white, and gray tones in Fig. 20-8 emphasize texture and depth.

The effectiveness of any medium is related to how skillfully it is used to create realistic textures, shades, shadows, and landscape features.

Fig. 20-6 ■ Different media can be combined to produce different effects.

A. Line work is emphasized with fine detail done in ink.
Howard Associates—Scholz Design Inc.

B. Softer wash techniques (without ink) reveal less detail.
Howard Associates—Scholz Design Inc.

Fig. 20-7 ■ Combining media to prepare an interior rendering.

A. A pencil layout. *Jenkins & Chin Shue Incorporated*

B. Rendering completed using pastel pencils and watercolor. *Jenkins & Chin Shue Incorporated*

Fig. 20-8 ■ Rendering done on gray stock. *John Henry, Architect*

Showing the Effects of Light

Light Source and Shade

When shading a building, consider the location of the sun or other light source. Areas exposed to the light source should appear lighter. Areas not exposed to the light source should be shaded or darkened. See Fig. 20-9. When an object with sharp corners is exposed to a light source, one side may be extremely light and the other side of the object extremely dark. However, objects and buildings often have areas that are round (cylindrical). These areas change gradually from dark to light. A gradual shading from extremely dark to extremely light must be made.

Shadow

To determine which areas of a building should be drawn darker to indicate shadowing, the angle of the sun must first be established. Once the angle of the sun is established, then all shading should be consistent with the direction and angle of the shadows, as shown in Fig. 20-10. In addition to the light source, consider the building outline and site contours when drawing shadows on buildings. Note the connection between the light source angle, building outlines, and site contours. Also, shadowing is often used to reveal hidden features such as overhang depth, building depressions, offsets, and extensions. See Fig. 20-11.

Fig. 20-9 ■ Shading methods.

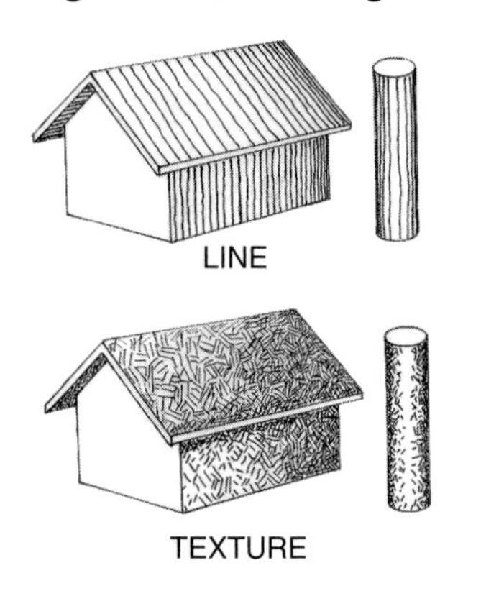

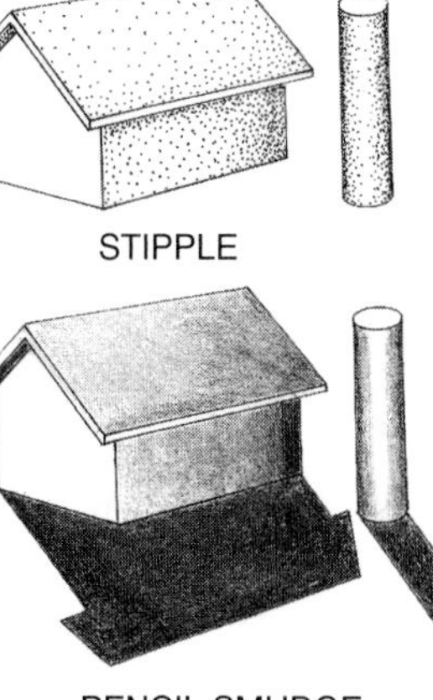

Fig. 20-10 ■ Shadows depend on the direction and angle of the light source.

A. Shadowing effect when a light source is located at the middle left.

B. Shadow patterns shift as light source moves from upper to middle left.

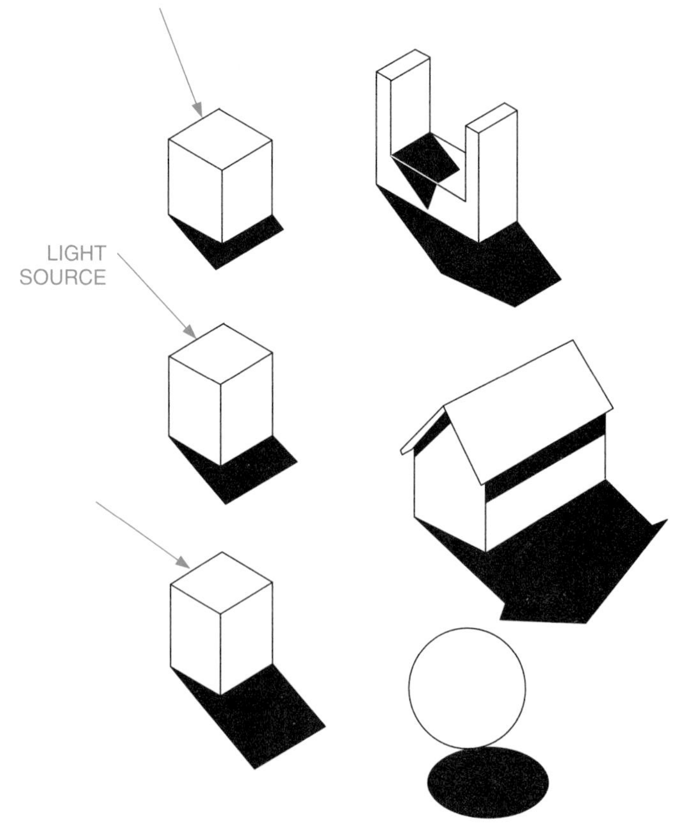

Fig. 20-11 ■ The shape of the shadow is related to the building outlines and site contours as well as the light source angle.

A. Drawing a shadow based on the light source direction.

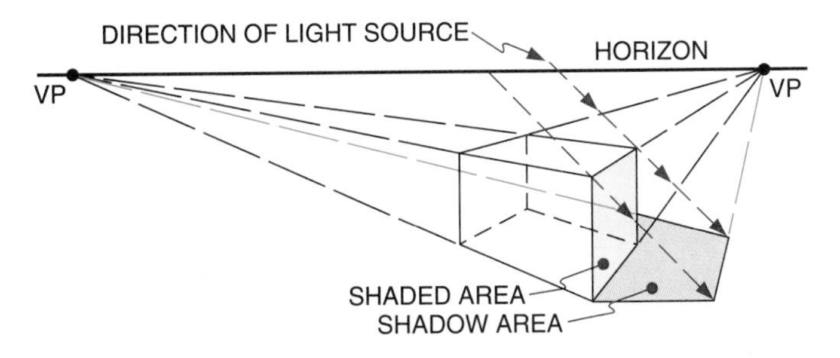

B. Shadow outlines reveal shapes and parts of the structure on object drawn.

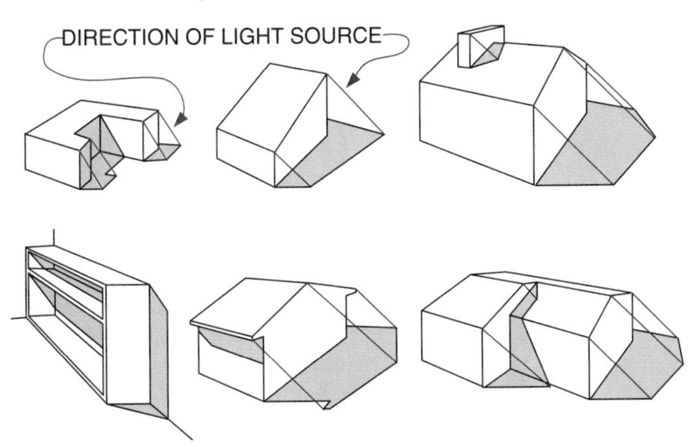

Various techniques can be used to indicate shadows. Figure 20-12 shows how depth and shadows are rendered on windows. Some windows show reflected light, and others are drawn to reveal the room behind, as though the window were open. Keep in mind that most windows look dark during the day because the inside of the building is darker than the outside. To produce realistic-looking windows, dark colors or even black surfaces are often used. See Fig. 20-13.

Texture

Giving texture to an architectural drawing means making building materials appear as smooth or as rough as they actually are. Smooth surfaces are very reflective and hence are very light. Only a few reflection lines are usually necessary to illustrate smoothness of surfaces such as aluminum, glass, and painted surfaces. For rough surfaces, the material can often

Fig. 20-12 ■ Use of shadows in rendering windows.

Fig. 20-13 ■ Rendering using blue sky and foliage reflections on windows. Note that the portion of the windows in a shadow are darker blue. *Jenkins & Chin Shue Incorporated*

Fig. 20-14 ■ Compare the presentations of textures in these renderings.

A. Using black and white media. *Jenkins & Chin Shue Incorporated*

B. Using color media. *Jenkins & Chin Shue Incorporated*

be shown by shading and texturing. A variety of texturing methods and rendering techniques for exterior materials used on siding and roofs can be used. See Fig. 20-14. Notice how the shading of the texture depends on the light source. Also note how the shadows from trees and overhangs are incorporated into the texture and values to add more realism to the drawing. Refer back to Fig. 20-1 and note ways in which texture and shadow are shown using pencil.

Rendering "stand-alone" features such as fences, walls, and chimneys requires close attention to independent shadow and shade effect on texture. See Fig. 20-15.

Entourage

Entourage is the people or objects that are part of a building's surroundings used to enhance the size, distance, and reality of renderings. Sketches of people—sitting, standing, and walking—are often necessary to show the relative size of a building and to put the total drawing in proper proportion. Since people should not interfere with the view of the building, architects frequently draw people in outline or in extremely simple form. In a drawing, people may indicate pedestrian traffic patterns, as well as provide a feeling of perspective and depth. Automobiles are

Fig. 20-15 ■ Shadow and shade effects to indicate texture.

A. Pencil renderings of "stand-alone" features.

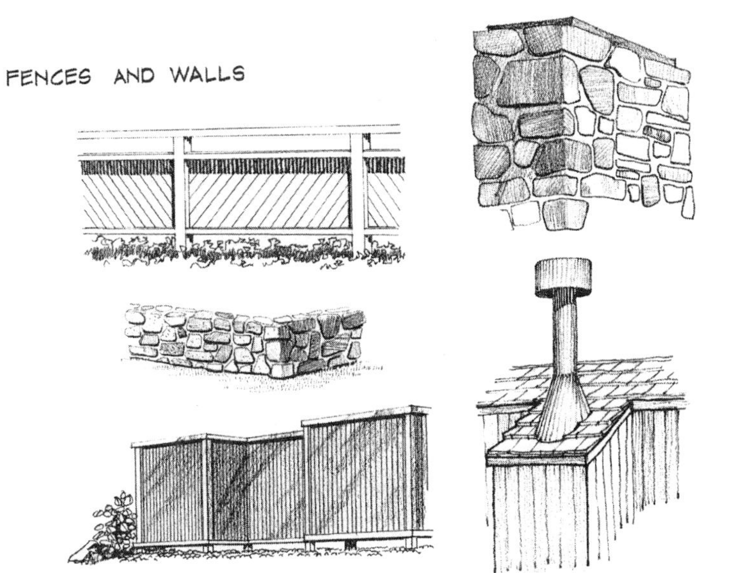

B. Ink rendering of stone texture.

Fig. 20-16 ■ Samples of entourage drawings available for use in renderings.

also added to architectural renderings to indicate relative size and to give a greater feeling for external traffic patterns.

People, boats, and automobile outlines in a variety of settings, angles, and scales are available on pressure-sensitive sheets or in traceable entourage publications. See Fig. 20-16. Entourage figures are often photo reduced or enlarged to fit within a drawing. Figures closer to the vanishing point are progressively smaller than figures near the station point. See Fig. 20-17.

Computer 3D CAD programs now offer libraries on disk. These libraries are similar to entourage publications and provide a wide variety of drawings, including people in walking, sitting, and standing positions. See Fig. 20-18.

Fig. 20-17 ■ Size differences of people indicate distances from the station point. *Jenkins & Chin Shue Incorporated*

Fig. 20-18 ■ People depicted in a 3D CAD rendering. *Autodesk, Inc.*

Landscape

In rendering, adding landscape features—whether to pictorial or elevation drawings—involves drawing trees, shrubs, ground cover, and drive and walkway surfaces. Trees should be placed so as not to block out the view of the buildings. See Fig. 20-19.

Rendering landscape pictorial drawings also involves using the same techniques to render accessory structures, water features, retaining walls, trellises for vines or other plants, and many other features. See Fig. 20-20. Also, refer back to Chapter 18.

Fig. 20-19 ■ Note the drawing and rendering techniques used to show landscape features. *Jenkins & Chin Shue Incorporated*

Fig. 20-20 ■ Note how the many and varied landscape features are presented in this rendering. *Tangerine Bay Club—Tangerine Development Corp.*

Steps in Preparing a Rendering

To prepare a pictorial rendering, proceed in the following sequence:

1. Block in with single lines the projection of the perspective.
2. Sketch the outline of building materials in preparation for rendering. This work can be done with a soft pencil or inking pen. Establish the light source and sketch shadows and shading. Darken windows, door areas, and the underside of roof overhangs.
3. Add texture to the building materials. For example, show the position of each brick with a chisel-point pencil. Leave the mortar space white and lighten the pressure for the areas that are in direct sunlight.
4. Complete the rendering by emphasizing light and dark areas and establishing more visible contrasts of light and dark shadow patterns.

CHAPTER 20 Exercises

1. Render a perspective drawing of your own house.
2. Render a perspective drawing of a house of your own design.
3. Render a perspective sketch of your school. Choose your own medium: pencil, pen and ink, watercolors, pastels, or airbrush.
4. Complete Exercises 1, 2, or 3 using a CAD system.
5. Collect illustrations that could be adapted for use on renderings: drawings of people, cars in different sizes and positions, landscapes, plants, etc.

Architectural Illustrator

Architectural illustrators prepare renderings used for presentations. They need a knowledge of architectural drawing (both manual and CAD), commercial art techniques, 3D computer modeling, and often photography and photo retouching.

> "Interior renderings make you feel that you are looking at a drawing or photograph of a finished room, except the room doesn't really exist yet."

Mark Englund

Visualizing a finished structure from a two-dimensional drawing can be difficult. Mark Englund, an architectural illustrator, helps people do just that. He prepares renderings that look like drawings or photographs of finished rooms and structures *before* the rooms or structures are built. Some of his house renderings have been seen in the Associated Press as "House of the Week" in newspapers across the country. Others have appeared in publications such as *House Beautiful* and *Better Homes and Gardens* magazines.

Hand-drawn illustration is Mark's first love. Most of the homes in magazines are shown as rendered architectural perspectives. Mark prepares his renderings in two layers using pen and ink and watercolors. The top layer is an ink drawing of the home done on clear acetate. The second layer is watercolor paint on paper. "There will always be a fondness for the old techniques," says Mark, "but illustrations and renderings can be done faster and easier on a computer. I think more and more artists will not resist art on a computer, but will actually start to prefer it."

Working on a computer, the illustrator still has the flexibility to highlight or minimize selected features. Mark is adapting his manual skills by touching up renderings on the computer. He also works on printouts. In his office, drafters create perspective drawings at CAD stations. Mark embellishes and "humanizes" the printouts. Computer drawings are being improved, however, and soon this may not be necessary.

"The whole industry is changing rapidly with advancements in computer technology," says Mark. "I would suggest that anyone taking architecture or art classes right now learn CAD drawing and 3D modeling. I think the two—computers and art—need to be merged. I believe that is the future of architectural illustration."

Photo by Mark Englund

CHAPTER

21

Architectural Models

Objectives

In this chapter, you will learn to:

- describe architectural models made for design study purposes.
- explain the differences between presentation and design study models.
- tell what input is needed to create a computer model.
- construct an architectural model.

Terms

basic layout model
contour-interval model
design study model
detailed model
interior design model
presentation model
solid form model
structural model

Architectural models are three-dimensional representations of a complete design. Models are made to scale and can be viewed from any angle or distance. They may be a constructed replica of a building, or they may be computer images.

The two basic types of models are design study models and presentation models. To check basic design ideas or construction methods during the design process, a **design study model** is made. A **presentation model** is used for sales purposes because people can understand a design more easily by looking at a model than they can by looking at drawings.

■ This model was used during the design and construction of the completed structure behind it. *James Eismont Photo*

Design Study Models

Design study models are helpful during the design process. They are used to check the form of a structure, verify the basic layout, clarify construction methods, or show interior design options. They can also be used to finalize the orientation of a structure on a site. Models of this type are used to study sun angles and show patterns at different times and on different days. See Fig. 21-1.

Solid Form Models

Before final dimensions are applied to a design, a **solid form model** like the one in Fig. 21-2 is often made. It is used to check the overall proportions of a building. Solid form models which contain no details can help to study the size and relationship of building clusters. Solid form models are made from Styrofoam™, balsa, clay, or soap. As the name implies, they are solid all the way through.

Basic Layout Models

Design study models also provide a means to check the overall layout and function of a design. First the preliminary floor plan and elevations must be available. Then **basic layout models** are made. These models are constructed to the same scale as the floor plans and elevations, usually ⅛″ = 1′-0″ or ¼″ = 1′-0″. They are not finely detailed because they will probably be revised and altered many times. See Fig. 21-3.

Fig. 21-1 ■ Model used to study sun angles and shadow patterns.

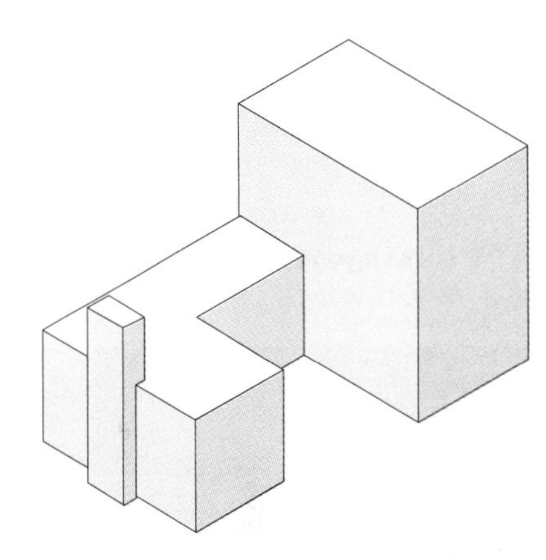

Fig. 21-2 ■ Solid form model.

Fig. 21-3 ■ Design options can be studied with the help of a basic layout model. This model shows a flat roof option for the same basic design shown in Fig. 21-1.

Structural Models

Only the structural members of a building are shown on a **structural model**. Builders use these models to check unique structural methods or to study framing options. These checks are especially important if many houses with the same structural design are to be built.

Building a model of this type is also a good way to learn framing methods.

Structural models are usually built to a scale of 1″ = 1′-0″, ¾″ = 1′-0″, or ½″ = 1′-0″. Smaller scales ⅛″ = 1′-0″ or ¼″ = 1′-0″ are difficult to handle.

Structural members are cut to scale or purchased from model stock. Some parts are assembled into panels by placing them directly over a wall-framing drawing or over elevation drawings. The panels and other components are then assembled with glue and pins as shown in Fig. 21-4. A structural model is built in the same sequence as a full-size house.

Fig. 21-4 ■ Assembled construction model. *NRI, A McGraw-Hill Company*

Interior Design Models

An **interior design model** is used to show individual room designs. These can be shown effectively with one-room models. One-room models usually include the floor, ceiling, and three walls. One wall remains open for viewing, like a dollhouse. One-room models are usually built to a scale of 1″ = 1′-0″ or 1½″ = 1′-0″. Scaled human figures, decor, and furniture add to the realism of interior design models. See Fig. 21-5.

Presentation Models

Most presentation models are used to promote the sale of a building, land parcels, or community development projects. **Presentation models** replicate (copy) the actual appearance of the real project in as much detail as the scale allows. A presentation model, like the large high-rise building in Fig. 21-6, usually includes the building site.

Fig. 21-5 ■ An interior room design model with the ceiling removed for viewing. *Helene Norrman*

Fig. 21-6 ■ Presentation model.
Auto-Trol Technology

Contour-Interval Models

Presentation models are frequently used to show the landform around a building. A **contour-interval model** represents the shape and slope of a site. In this model, the thickness of each layer is equal (in scale) to the different levels of the land. The shape of each layer is the same shape as the contour lines on a survey drawing. See Fig. 21-7.

To develop smooth contours, small posts are used to represent the contour height. These posts are spaced throughout the site. Flexible screens are then laid on the posts and covered with papier-mâché which can then be smoothed. See Fig. 21-8.

Fig. 21-8 ■ Contour-interval model construction.

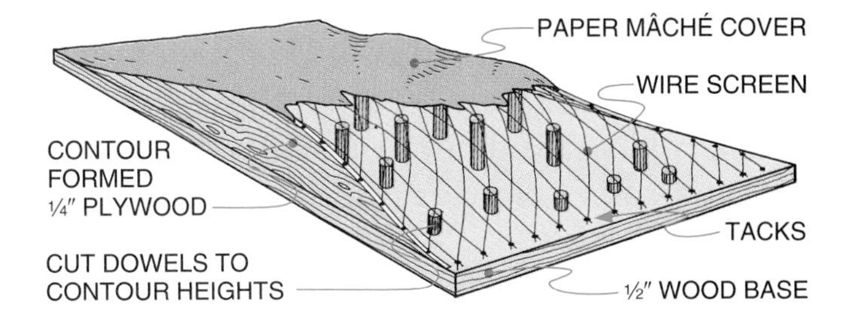

Fig. 21-7 ■ Landform shown with contour interval modeling. *Edward Aueber, UK2 Architects Inc.*

Detailed Models

City and housing tract developers are the largest users of presentation models. To show building relationships, developers often surround a **detailed model** with solid form models of adjacent and nearby buildings, as shown in Fig. 21-9. Constructed models of this type are often used to create a subject for photographs. A photograph can be retouched later to add or eliminate details and features.

Housing developers use landform models with small-scale solid form models to show specific lot locations and their relationship to other features. The placement of people, cars, and trees on these models adds realism and better defines the scale of the project. See Fig. 21-10.

Steps in Constructing a Model

An accurate and realistic model is constructed to a precise scale with careful attention to detail. Many materials and items used to build models, such as those in Fig. 21-11, may not be available in the scale needed. These must then be constructed by the model maker. Professional model makers may use manufactured parts and/or fabricate others.

Fig. 21-9 ■ Detailed and solid form models are used to show the relationship of the planned structure with surrounding buildings. This model was created with DataCAD 6.0, Photoshop™ 3.0, 3D Studio version 4.0. *Savannah Courthouse Annex computer models created by David Karram and Todd Groves*

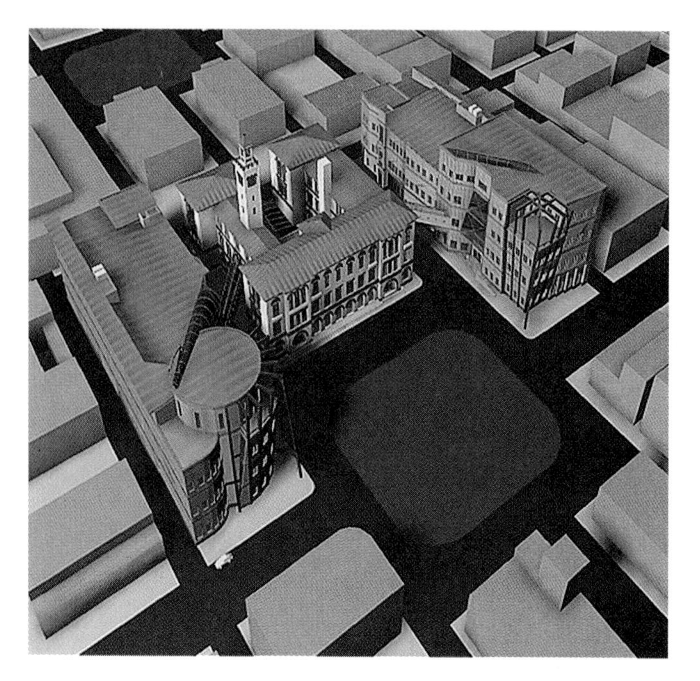

Fig. 21-10 ■ Housing development model. *The Plantation at Leesburg*

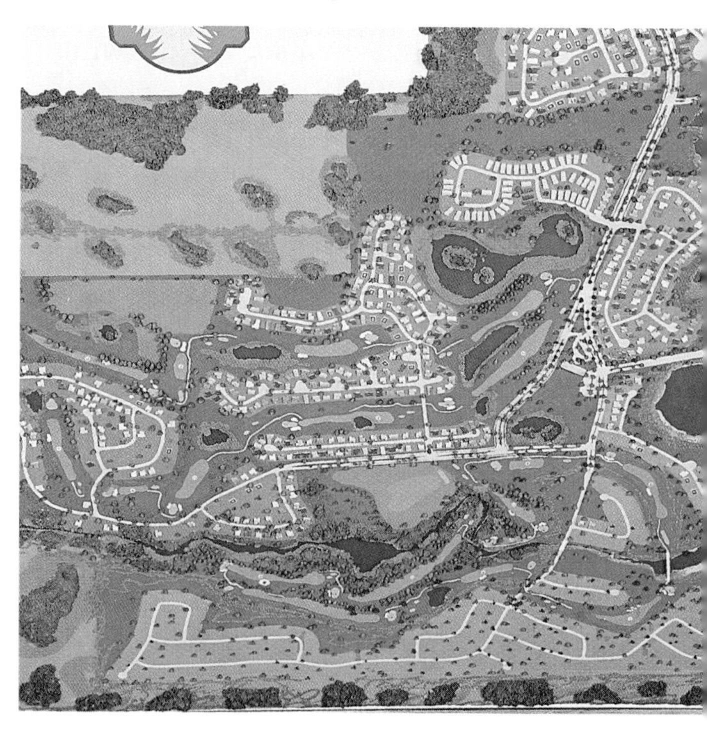

Fig. 21-11 Supplies and methods for constructing models.

Part	Model Materials	Methods of Construction
• STRUCTURE		
base	plywood; particleboard	Cut to maximum, but convenient, size for structure and site.
walls	softwood; cardboard; acrylic; matboard; foamboard; Styrofoam™; wallpaper; fabric; plywood	Cut walls to exact dimensions of elevations. Allow for overlapping of joints at corners. Wall thickness must be to scale.
floors	flocking for carpet; printed paper of floor type; thin wood veneer; vinyl scraps	Paint floor area with slow-drying colored enamel and apply flock. Remove excess when dry. With paper, glue in place. With wood veneer, rule black lines for strip effect and glue in place.
windows and doors	purchased, premade strips of wood or plastic; thin, clear acetate; acrylic	Glue premade windows and doors in place. Cut strips for casing, sill, and window frames, and glue in place. Draw the windows and doors directly on the walls.
roofs	thin, stiff cardboard or wood; paint; colored sand; wood pieces for shingles; premade/printed roof coverings; sandpaper	Cut out roof patterns and assemble. For roof coverings, glue on sand, wood pieces for shingles, or preprinted roof coverings.
• BUILDING MATERIALS		
siding materials	scored sheets of balsa wood; foamboard; preprinted paper patterns; wood strips for board and batten and horizontal siding	Glue or paint siding materials to model walls.
stucco	spackle; plaster of paris; sandpaper; sand and thick paint	Mix and dab on with a brush leaving a rough texture.
brick and stone	printed paper; embossed plastic sheets; thin softwood	Glue paper in place. Cut grooves in wood, and paint color of bricks or stone.
wood paneling	printed paper; thin veneer wood; molded plastic sheets	Glue paper or plastic sheet in place. With veneer wood, rule lines for strip effect.

continued

Fig. 21-11 *Continued*

Part	Model Materials	Methods of Construction
• FURNISHINGS		
furniture, appliances, fixtures	commercial models; doll furniture (to scale); cardboard; softwood; clay; Styrofoam™; soap; fabric; paint	Purchase commercial model furniture or carve/sculpt to shape. Paint or flock for a finish.
fireplaces	(Refer to brick and stone materials.)	Carve fireplace, and simulate finish.
• BUILDING SITE		
topography	wood base; wire screen; papier-mâché; fine gravel	Build up sloped areas with sticks and wire screen. Place papier-mâché over wire, and glue gravel for soil effect.
geologic features, terrain, water	stones; sand; colored gravel and sand	Glue small rocks for boulders. Paint high-gloss blue paint for water.
• LANDSCAPE		
grass	green paint and flock	Paint grass area and apply flock. Remove excess when dry.
trees and bushes	sponges; lichen; small twigs	Grind up sponges and paint shades of green. Glue pieces to twigs for trees and bushes. Lichen may be purchased in model stores in bulk or as model trees.
wood fences	wood strips	Glue together to form a fence.
masonry fences	(Use same materials as for masonry walls.)	Form the same as walls to fence size.
gazebo	wood strips	Assemble from a working drawing.
• MISCELLANEOUS		
automobiles	commercial models; toys; clay; soap; Styrofoam™	Purchase or shape from soft materials and paint.
people	commercial models; toys; clay; soap; Styrofoam™	Purchase or shape from soft materials and paint.
swimming pools	wood strips; blue paint or paper; sheet glass; acrylic	Outline pool with glued wood strips. Paint or glue paper for water and attach clear glass. Ripple acrylic surface while drying for ripple effect.

Methods of model construction vary depending upon the material and the amount of detail required. The following procedures represent a typical sequence for constructing solid wall models.

1 *Floor plan base.* Attach a print of all floor plans to a sheet of rigid foamboard or Styrofoam™ sheet with rubber cement or use a pre-glued sheet, as shown in Fig. 21-12. Allow some space on all sides.
2 *Wall construction.* Glue a print of all exterior elevations to a sheet of foamboard. This foamboard should be the same scaled thickness as the outside walls indicated on the floor plan. Cut the foamboard to create a wall for each elevation. See Fig. 21-13.
3 Repeat this procedure for all interior partitions. If interior elevations are not available, at least the outline of each partition must be drawn and cut from foamboard. Usually all windows and doors are cut out of interior and exterior walls, as shown in Fig. 21-14. An alternative, particularly for design study models, is to simulate them using paint or a suitable material.
4 The next step is to attach window trim and acetate. See Fig. 21-15. Doors may

Fig. 21-12 ■ Glue floor plan print to foamboard.

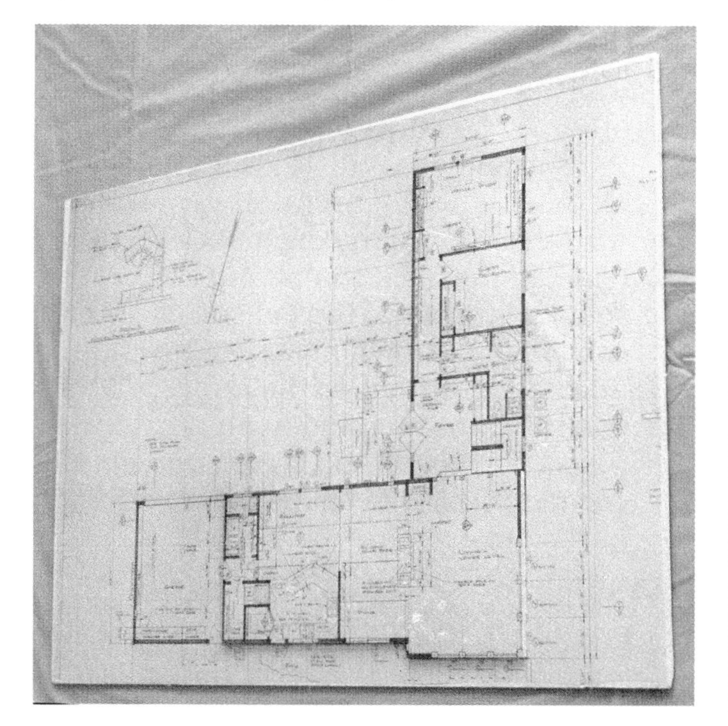

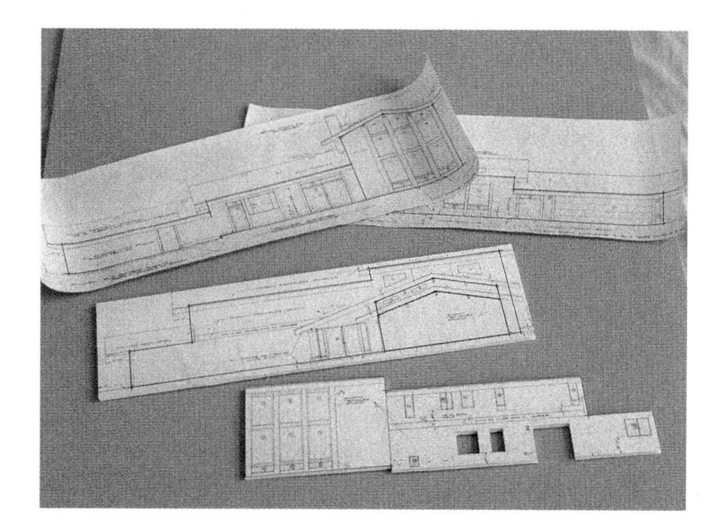

Fig. 21-13 ■ Glue elevations to foamboard and cut out walls. *James Eismont Photo*

Fig. 21-14 ■ Cut out door and window openings. *James Eismont Photo*

Fig. 21-15 ■ Attach window trim and acetate.

A. Using transparent tape to attach acetate. *James Eismont Photo*

B. Assembly of window trim.

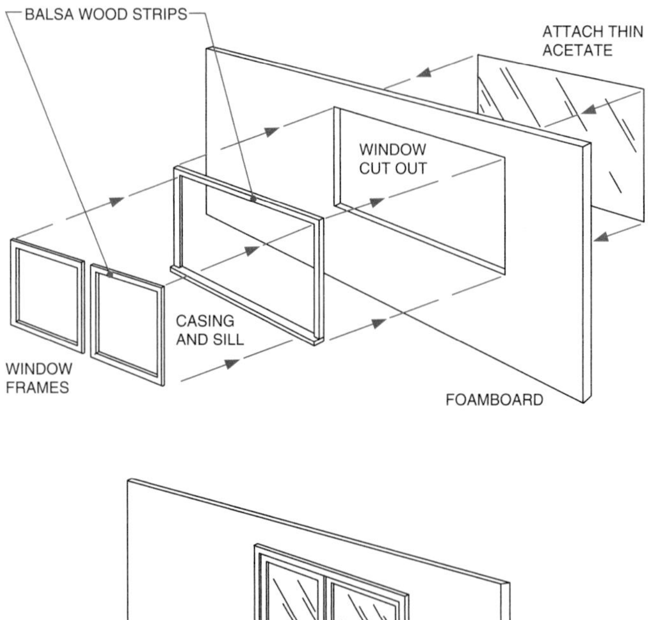

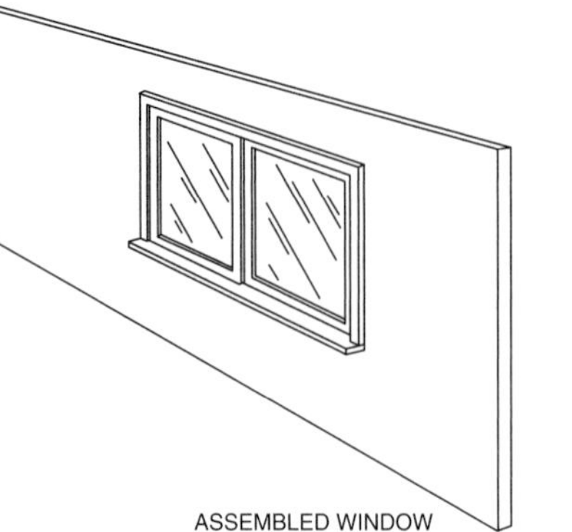

Fig. 21-16 ■ Windows may be darkened if not removed. *James Eismont Photo*

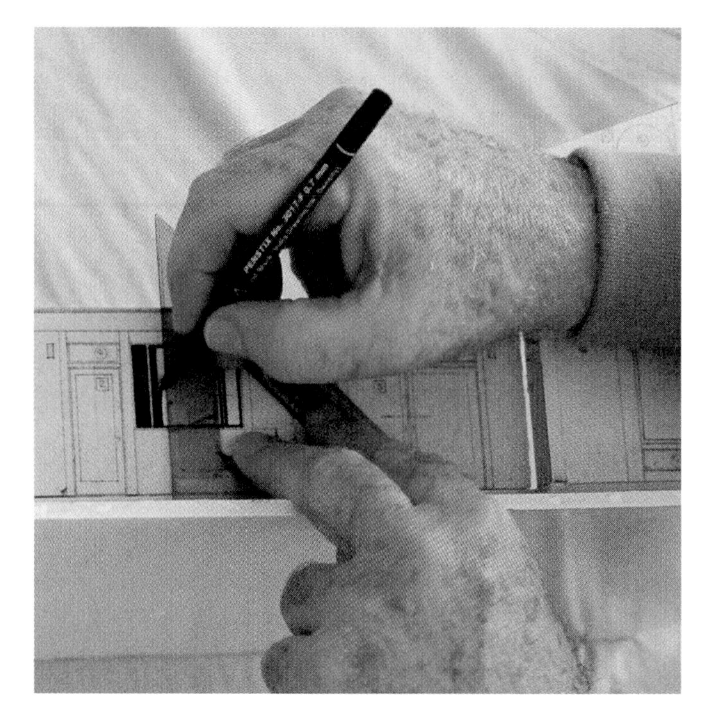

be cut from thin, stiff cardboard and hinged with transparent tape. If windows and doors are not cut out, they could just be shaded black or gray for effect, as in Fig. 21-16, or the selected material should be glued in place.

5 *Wall attachment.* Glue the exterior walls to the floor plan base and to each other at the corners. See Fig. 21-17. Fit the corners carefully so that the overall outside wall lengths align properly with the floor plan corners. Use a 90° triangle to check the plumb and squareness of the tops of corner intersections, as shown in Fig. 21-18. Glue interior partitions to the appropriate partition lines indicated on the floor plans. See Fig. 21-19. Glue intersecting corners to outside walls and to other partitions.

Fig. 21-17 ■ Attach exterior walls to the base. *James Eismont Photo*

Fig. 21-18 ■ Check walls for plumb and squareness. *James Eismont Photo*

Fig. 21-19 ■ Attach interior walls to the base and exterior walls. *James Eismont Photo*

6 *Wall finishing.* Paint interior walls and floors with white or lightly tinted tempera paint. Apply texture or simulated (imitative) coverings to represent floor tiles, masonry surfaces, fireplaces, and chimneys. Add siding materials or textures to exterior surfaces. See Fig. 21-20. Spraying the finished surfaces with fixative will help surface treatments adhere better.

7 *Cabinetry and fixtures.* Make solid 3D forms to represent built-ins. These may include kitchen and bath cabinet fixtures, fireplaces, chimneys, and bookshelves. If desired, add color to cabinets, countertops, or fixtures.

Fig. 21-20 ■ Add surface texture to exterior walls.
James Eismont Photo

8 *Roof construction.* Roofs should be constructed separately so that they can easily be removed to reveal the interior. Flat roofs can be constructed directly from a roof plan. The pitched roof in Fig. 21-21 must be cut to align with the pitch indicated on an elevation drawing. Construct the roof in panels to represent continuous flat surfaces. Then glue the panels together. Other roof components and simulated roof coverings can then be added. Construct the chimney and glue it to the top of the roof to align with the interior chimney. This is done to allow the part of the chimney above the roof to be removed along with the roof.

9 *Outdoor areas.* After the structural part of the model is complete, remove the extra base material from the floor plan. Proceeding outward from the house, add outdoor areas, such as patios, pools, ponds, decks, or lanais. Include all the features that are within the property's perimeter.

10 *Landscape features.* A complete architectural model includes landscaping and details of the site. Construct walkways, driveways, and steps to connect different levels of the property. Retaining walls might need to be added to separate parts of the site. Add landscape features such as trees, fountains, shrubbery, and ground cover. Cars, people, and furniture can be included to help define the scale and the areas.

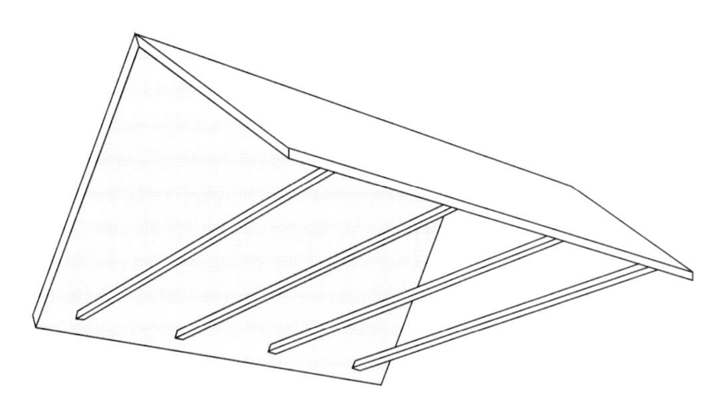

Fig. 21-21 ■ Construct roofs separately for easy removal. Shown here is gable roof construction.

Computer Modeling

Many types of 3D computer models are used in the design process as described in Chapter 5. Computer-generated models can be used to check the structural stability, orientation, and pictorial appearance of the design. Refer again to Fig. 21-9. Some elements of the design are often changed after the model is studied.

A 3D CAD system can eliminate the need for a physically constructed model. A computer-generated model, however, remains a two-dimensional display on the monitor. One of its main advantages is that it can easily be changed and updated. Both formats have advantages.

CHAPTER 21 Exercises

1. What are two basic purposes of architectural models?
2. Describe four types of design study models and explain their functions.
3. What features does a presentation model usually include?
4. List the steps for constructing a model.
5. Construct a basic layout model of a house shown in Chapter 1.

6. Construct a model of the house you designed.
7. Complete a computer model of the house you designed.

PART 6

Foundations and Construction Systems

CHAPTER
22 Principles of Construction
23 Foundations and Fireplace Structures
24 Wood-Frame Systems
25 Masonry and Concrete Systems
26 Steel and Reinforced-Concrete Systems
27 Disaster Prevention Design

CHAPTER 22

Principles of Construction

Objectives

In this chapter, you will learn to:

- name and define physical forces that act on a building.
- describe the factors that determine the strength of structural components.
- draw a modular floor plan, elevation, and detail drawing.

Terms

bearing-wall structures
building load
cantilever
compression force
dead load
deflection
lateral load
live load
modular components
module
prebuilt home
prebuilt module
prefabricated home
shear force
skeleton-frame structures
tension force
torsion force

New construction materials and new methods of using conventional materials provide designers with great flexibility in construction design. Stronger buildings can now be erected with lighter and fewer materials.

Although the basic principles for preparing construction drawings are the same for all types of construction, the use of symbols, conventions, and terms changes drastically from system to system. Construction systems are broadly divided into four material groups: wood, steel, masonry, and concrete. Specific drawings for these groups will be discussed in detail in subsequent chapters. Most contemporary designs include a combination of these systems and materials. Sometimes parts are preassembled and brought to the building site. Information about construction using prefabricated components is included in this chapter.

Buildings designed according to factory standards and sizes can be prebuilt in sections and assembled quickly on the site.
Robert Bennett/FPG International Corp.

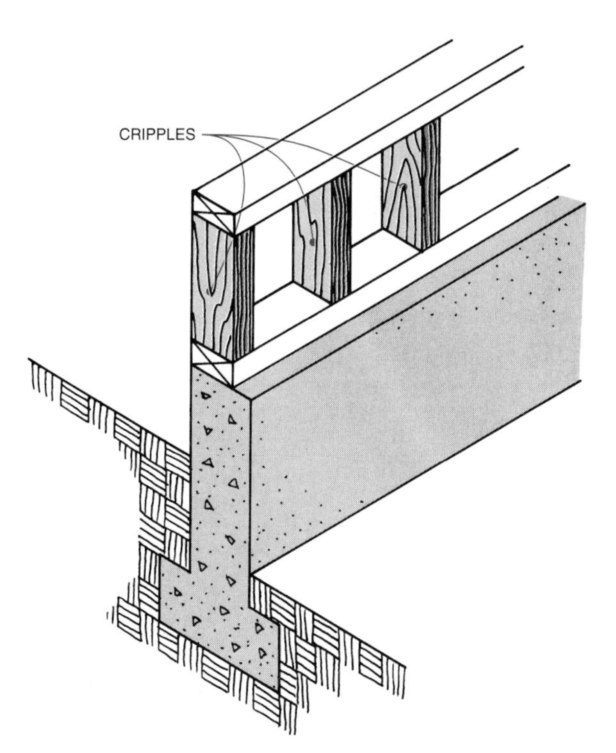

Fig. 23-18 ■ Cripples raise floor height without raising the foundation height.

spaced closer than the normal 16″ on center. ("On Center" [OC] means the spacing is measured from the center of one member to the center of the next.) Shear stress plywood must be used over cripples to overcome lateral stresses.

Slab Foundations

Slab foundations, or "slabs," are made of reinforced concrete. They are either monolithic (one piece) or separate pieces. In a monolithic slab, the slab floor and footing are poured as one piece. See Fig. 23-19. Rebars, wire mesh, plumbing line risers, waterproof membrane, electrical conduits (tubes for wires), HVAC (heating, ventilating, and air conditioning) ducts (if in the floor) must all be securely in place on a compacted soil base before pouring. See Figs. 23-20 and 23-21.

Fig. 23-20 ■ Area prepared for pouring a monolithic slab foundation.

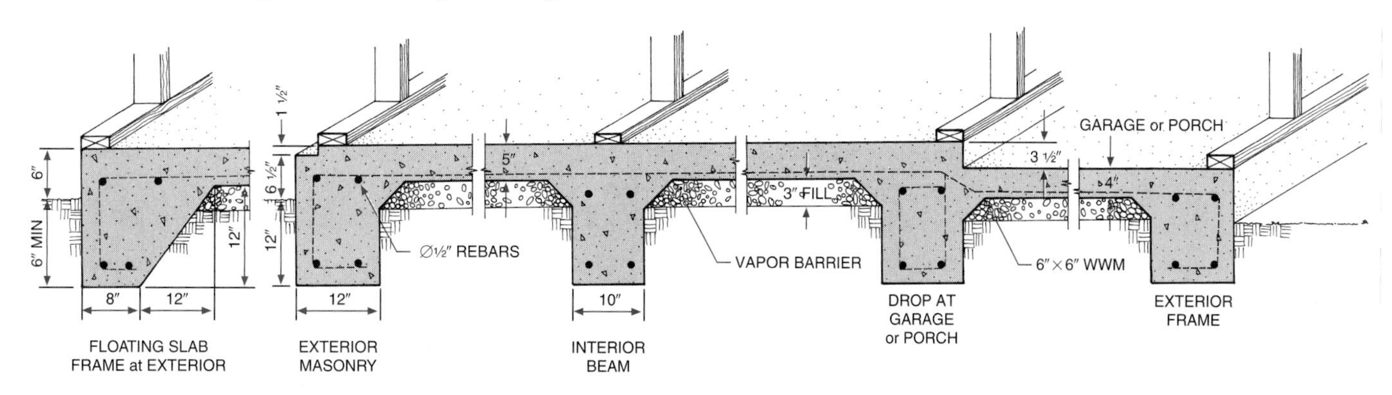

Fig. 23-19 ■ Types of monolithic slab foundations.

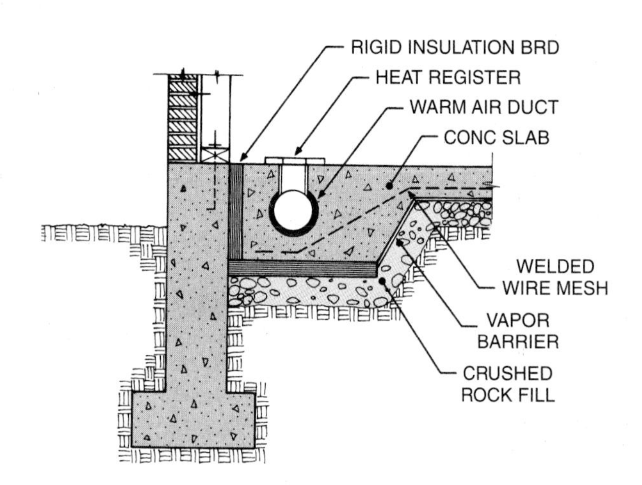

Fig. 23-21 ■ HVAC duct embedded in a slab.

Fig. 23-22 ■ Two-piece slab.

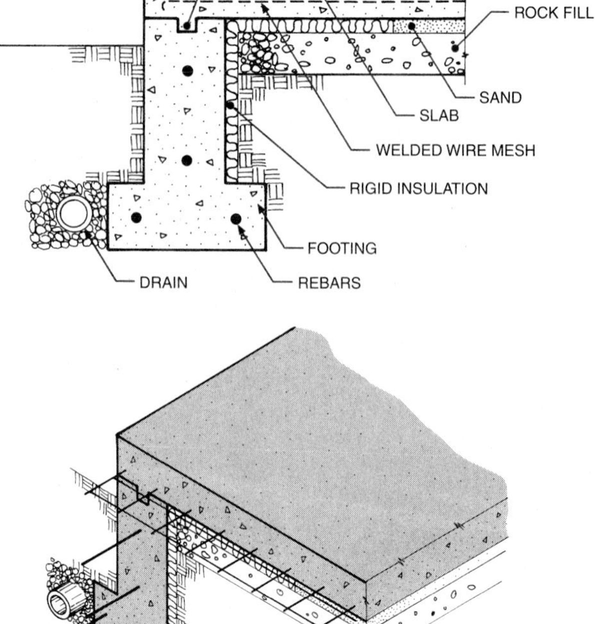

In slab foundations that are not monolithic, the footings are poured separately from the floor slabs, as shown in Fig. 23-22. Figure 23-23 shows a variety of slab floor,

Fig. 23-23 ■ Slab foundation details.

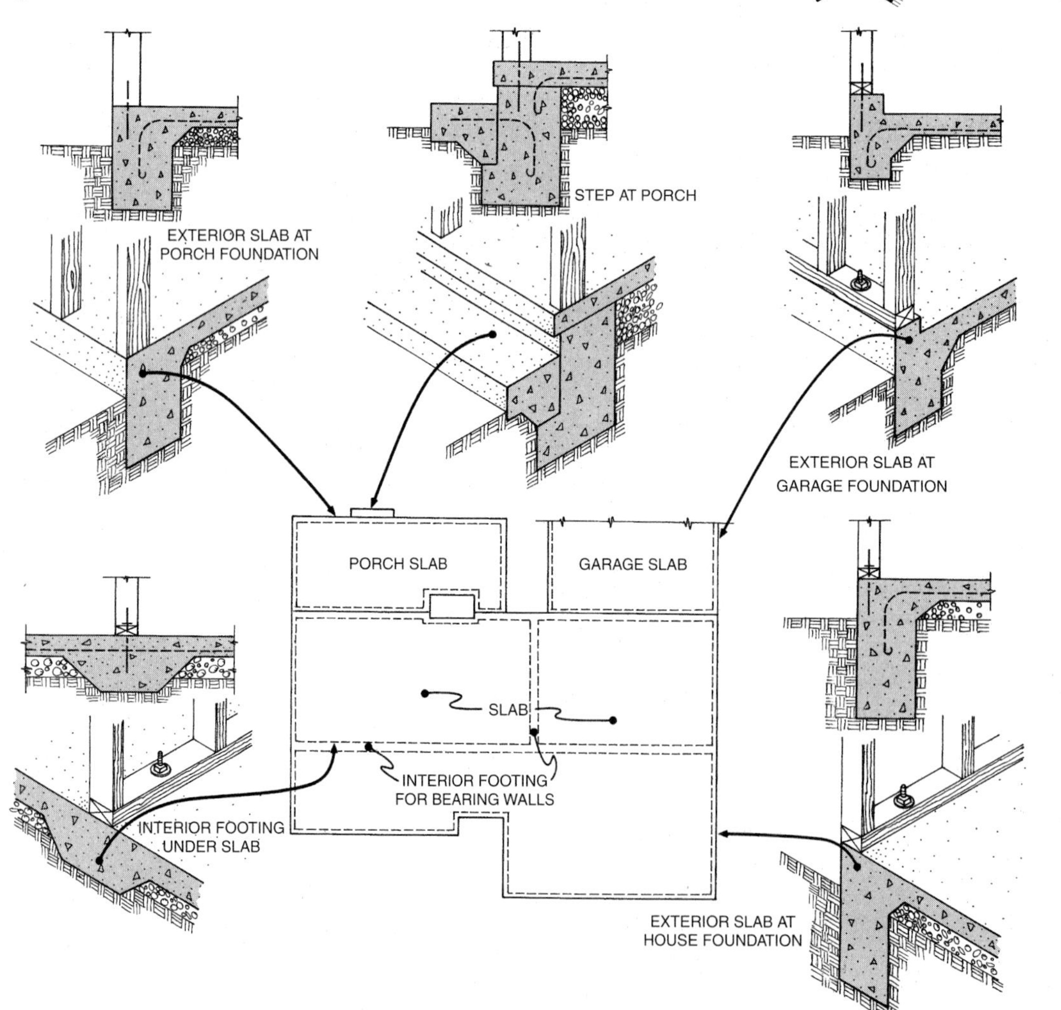

wall, and footing details used in different foundation situations. Where concrete or concrete block is used separately for foundation walls, the floor slab and footing must be secured to the block with rebars. See Fig. 23-24.

Building codes specify the minimum distance from the top of a slab to the grade line, usually 6″ to 8″. Since slabs lose heat around the perimeter adequate insulation is important.

Pier-and-Column Foundations

Pier-and-column foundations consist of individual footings (piers) upon which posts and columns are placed. Posts and columns are vertical members used to support floor systems. See Fig. 23-25.

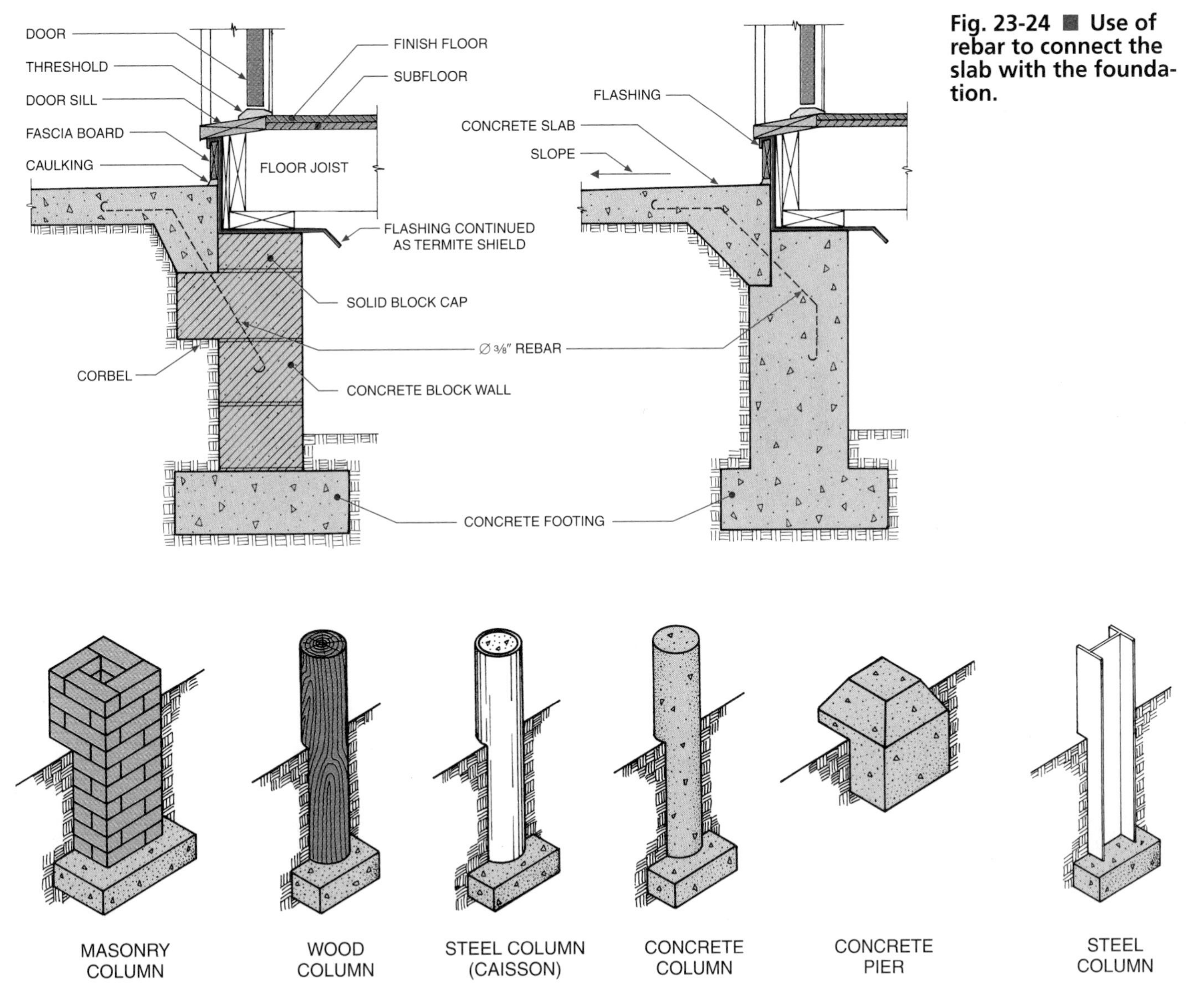

Fig. 23-24 ■ Use of rebar to connect the slab with the foundation.

Fig. 23-25 ■ Materials used to construct piers and columns.

Individual footings are known as piers. Piers may be sloped or stepped in order to spread the load of the structure with a minimal amount of added volume. See Fig. 23-26.

Posts and columns are vertical members that support girders and beams and transmit their weight and the weight of floors to the footings. See Fig. 23-27. The terms post and column are often used interchangeably. Generally, short vertical supports are called posts and longer vertical members are known as columns. Posts are usually made of wood, and columns are usually steel or masonry.

Piers and posts or columns may be used as the sole support of the structure. They also may be used in conjunction with foundation walls to provide intermediate support for horizontal members. See Fig. 23-28. Fewer materials and less labor are needed for pier-and-column foundations, but they are seldom used in basements because they occupy needed open space.

BEAMS AND GIRDERS Beams are horizontal structural members that support a load. Girders are large beams that are the major horizontal support members for a floor system. In common practice, the terms beam and girder are often used interchangeably. However, technically, girders are members that are supported by piers and columns and secured to

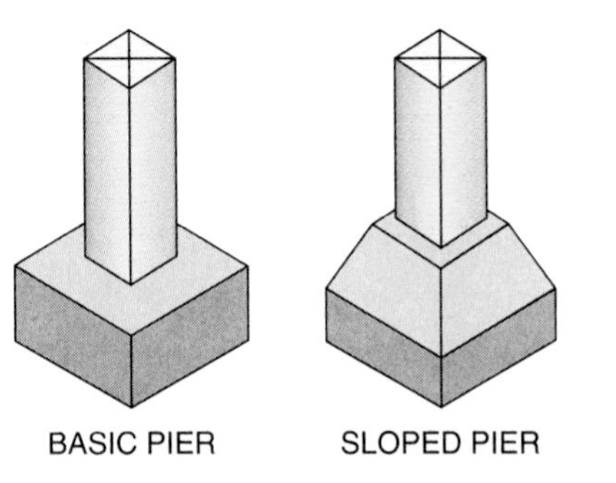

Fig. 23-26 ■ Basic shapes of piers.

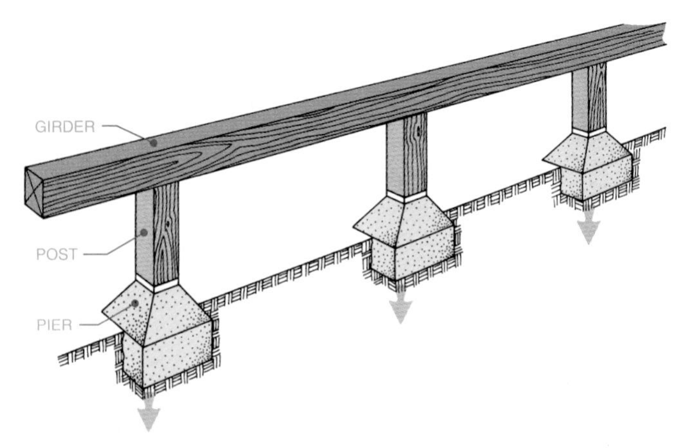

Fig. 23-27 ■ Posts transmit the weight above them to the footings.

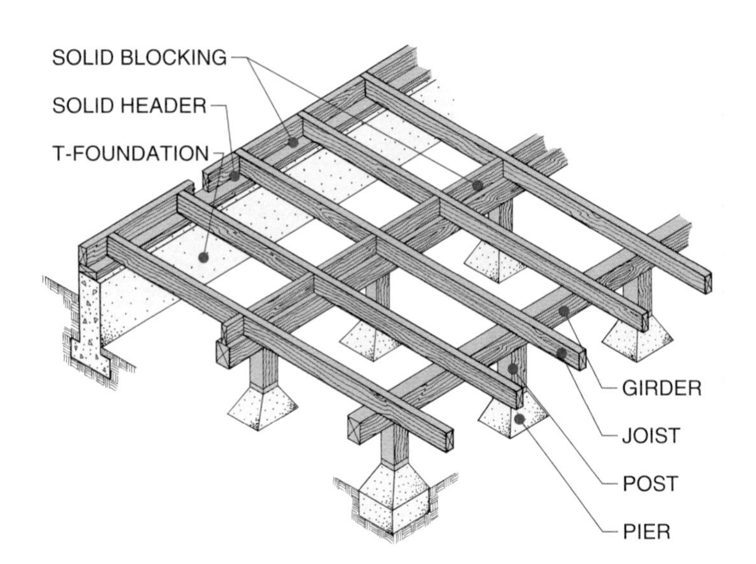

JOIST SPAN FOR STANDARD GRADE WOOD

JOIST SIZE	JOIST SPACING	JOIST SPAN
2 × 6	12″ 16″ 24″	10′ 9′ 7′–6″
2 × 8	12″ 16″ 24″	13′ 12′ 10′–6″
2 × 10	12″ 16″ 24″	16′ 15′ 12′
2 × 12	12″ 16″ 24″	20′ 18′ 15′
2 × 14	12″ 16″ 24″	23′ 21′ 17′

GIRDER SPAN FOR STANDARD GRADE WOOD

GIRDER SIZE	SUPPORTING WALLS	NO WALL SUPPORT
4 × 4	3′–6″ 3′–0″	4′–0″ 3′–6″
4 × 6	5′–6″ 4′–6″	6′–6″ 5′–6″
4 × 8	7′–0″ 6′–0″	8′–6″ 7′–6″

Fig. 23-28 ■ Piers and columns used to provide intermediate support to girders.

foundation walls. See Fig. 23-29. Beams are placed horizontally at right angles on or between girders and are supported by them.

Girder sizes are closely regulated by building codes. The allowable span of the girder depends on the size of the girder itself. A decrease in the size of a girder means that the span must be decreased. This is done by adding additional pier-and-column supports underneath the girder.

Most wood girders for residential construction are built up from 2 × 6s, 2 × 8s, or 2 × 10s that are spiked or nailed together. Steel beams or girders can perform the same function as wood beams or girders, but steel members can span larger distances than wood members of the same size.

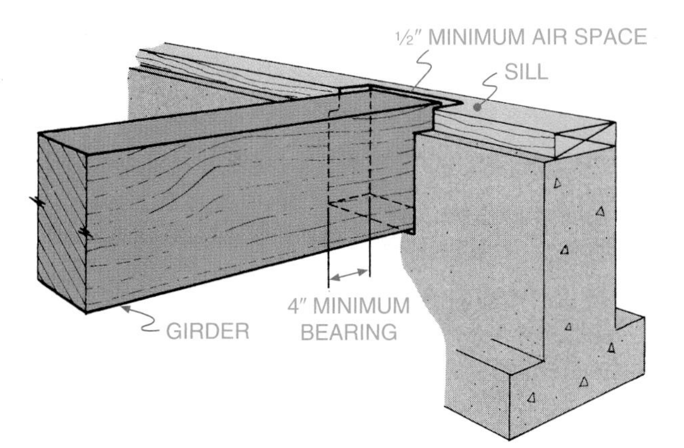

Fig. 23-29 ■ Pocket intersection for a girder.

JOISTS Joists are the parts of the floor system that are placed perpendicular to the girders. See Fig. 23-30. Joists span either from girder to girder or from girder to the foundation wall. The ends of the joists butt against a header or extend to the end of the sill. Bridging is placed between joists. Bridging consists of smaller structural members fastened between the joists to add stability and keep spacing consistent.

PILES When piers are driven into supporting soil or bedrock, without a separate footing, they are known as piles. Masonry, wood, steel, and concrete are used for piles, as shown in Fig. 23-31. Piles are used to support large structures. They are driven deep into the soil to support structures on sites where stable soil conditions do not exist near the surface.

Fig. 23-30 ■ Joists provide support for the floor and rest on girders.

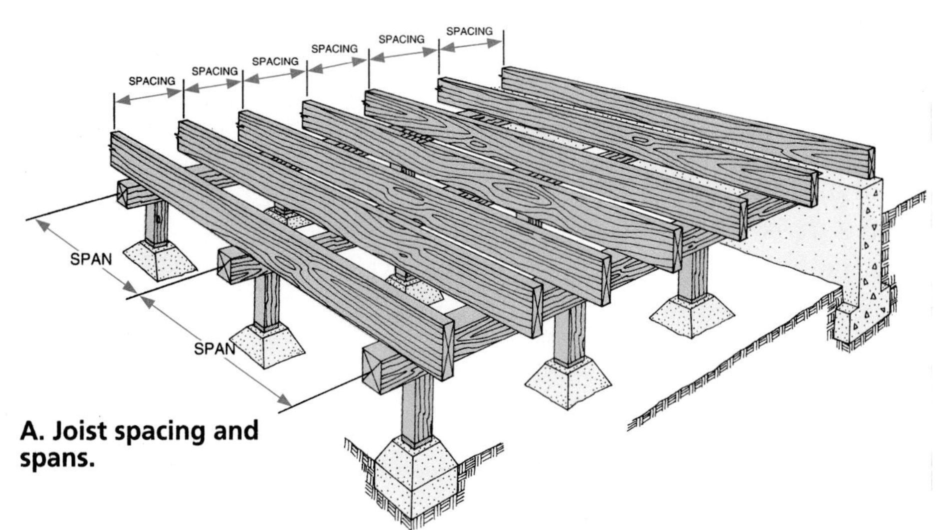

A. Joist spacing and spans.

TYPICAL JOIST DIMENSIONS

JOIST SIZE	JOIST SPACING	JOIST SPAN
2 × 6	12″ 16″ 24″	10′–0″ 9′–0″ 7′–6″
2 × 8	12″ 16″ 24″	13′–0″ 12′–0″ 10′–6″
2 × 10	12″ 16″ 24″	16′–0″ 15′–0″ 12′–0″
2 × 12	12″ 16″ 24″	20′–0″ 18′–0″ 15′–0″
2 × 14	12″ 16″ 24″	23′–0″ 21′–0″ 17′–0″

B. Typical floor joist dimensions.

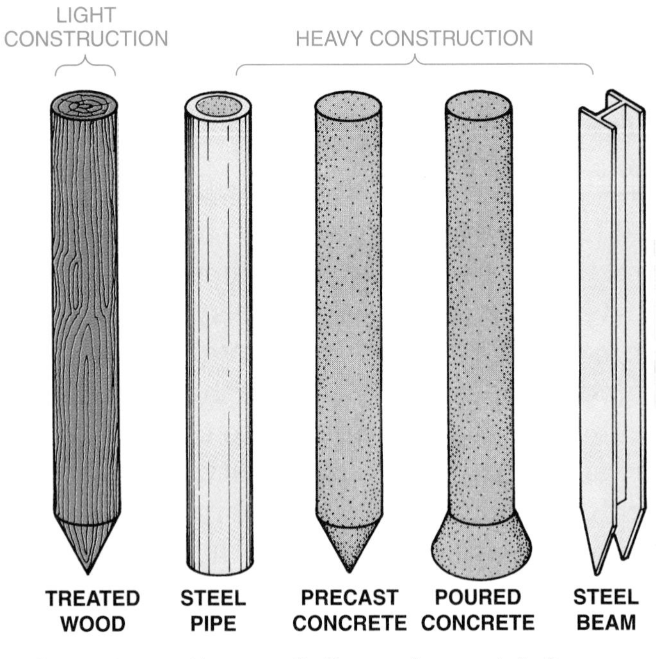

Fig. 23-31 ■ Shapes of piles and materials from which they are made.

Piles support building loads in several ways: by friction with the soil, with self-contained footings as shown in Fig. 23-32, or through contact with bedrock. Although bedrock may support the compression load of a building, sufficient soil bearing capacity and/or horizontal ties must be used to prevent piles from drifting out of position.

Permanent Wood Foundations

Permanent wood foundations are constructed similarly to wood frame walls in other parts of a building. See Fig. 23-33. There is one important difference: the plywood and lumber components of permanent wood foundation walls are pressure-treated with wood preservatives that become chemically bonded in the wood. This process permanently protects the foundation from fungi, termites, and other causes of decay. The lumber species used is highly stress resistant.

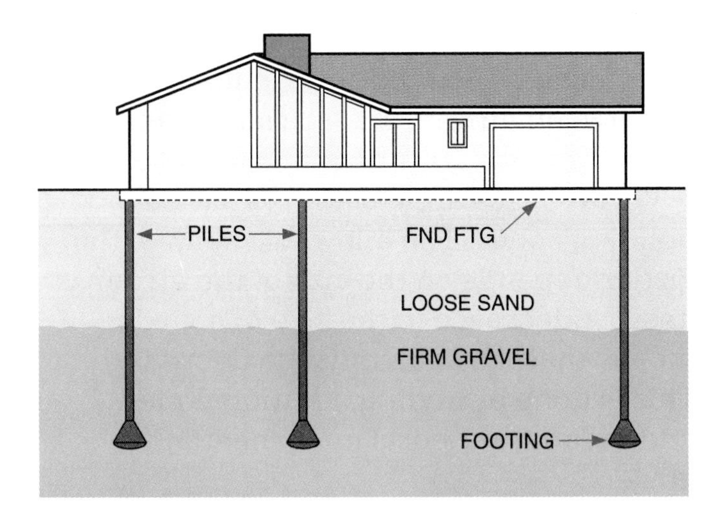

Fig. 23-32 ■ Deep piles are used to support buildings when footings cannot be stabilized in loose soil near the surface.

Permanent wood foundation walls are engineered to absorb and distribute loads and stresses that frequently crack and split other types of foundations. Another advantage of permanent wood foundation design is that it prevents the types of moisture problems that typically plague

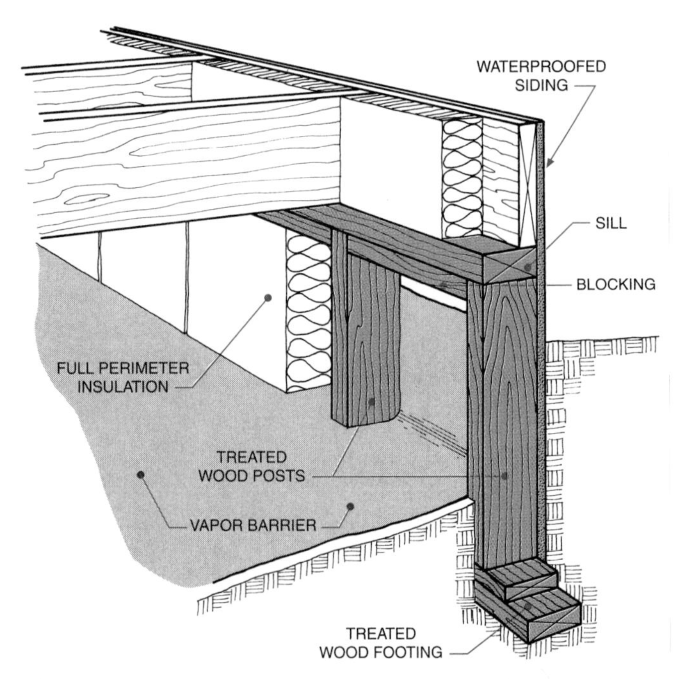

Fig. 23-33 ■ Permanent wood foundation members.

conventional basements. The design incorporates moisture deflection and diversion features, such as vapor barriers, horizontally along the ground below the floor and vertically along the outside of the foundation. See Fig. 23-34. Figure 23-35 shows a pictorial drawing of a wood foundation system.

Fireplaces

Because masonry fireplaces and chimneys are exceptionally heavy, they cannot be supported by the normal building footings. Special provisions must be made. The design of the fireplace influences the type of foundation that is needed. In designing fireplaces, the style, type, support, framing, size, materials, components, ratio of the opening and firebox, and the height of the chimney must all be considered. Fireplaces are classified by: fuel type, type of opening, construction, and architectural style (contemporary, Spanish, colonial, etc.).

Fuel Types

Fireplaces burn either wood, natural gas, or synthetic materials.

WOOD Oxygen is the vital ingredient needed for effective wood burning and proper functioning of the fireplace. Since warm air rises, air in the room is drawn into the fireplace, supplying the fire with needed oxygen. Since cold air continually replaces rising warm air, much of the heat produced by many fireplaces goes up the chimney. To reduce this heat loss and redirect some of the heat back inside, warm-air outlets that balance the inlet of cold-air are effective. Using outlets of this type allows heat to reenter the room,

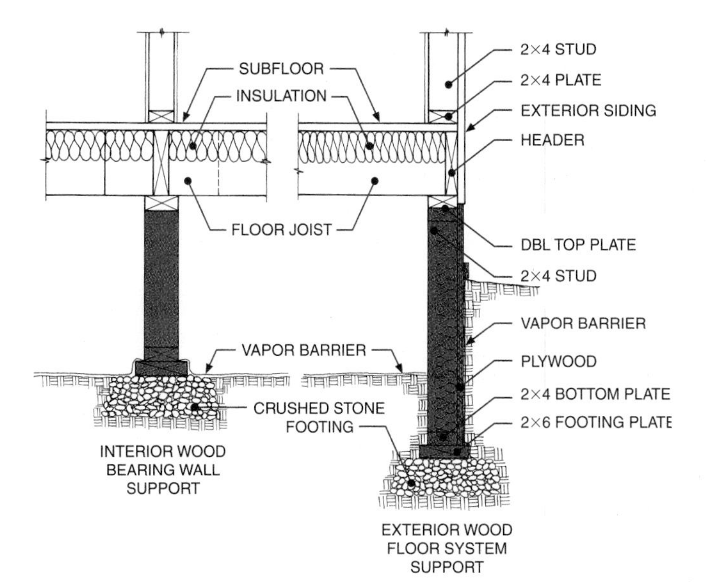

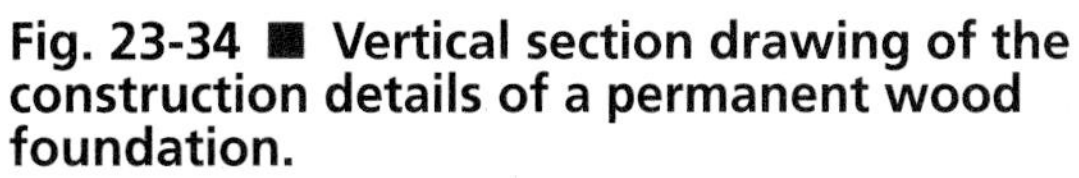

Fig. 23-34 ■ Vertical section drawing of the construction details of a permanent wood foundation.

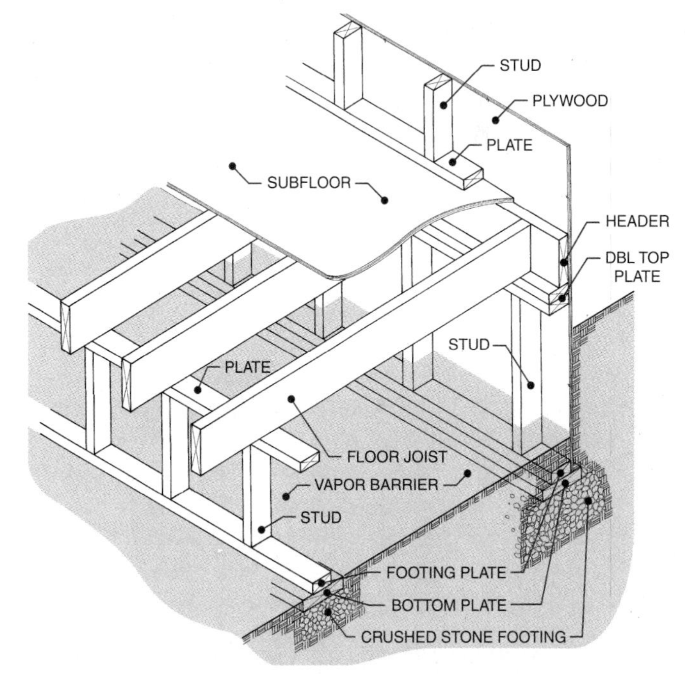

Fig. 23-35 ■ Details of a wood foundation system.

while smoke, debris, and toxic fumes are directed outside through the chimney.

NATURAL GAS A gas fireplace offers maximum operating convenience and warm-air circulation. Concealed circulating fans keep air moving around the firebox and expel heated air into the room. Gas fireplaces have automatic pilot lights that are

Fig. 23-36 ■ Types of fireplace openings.

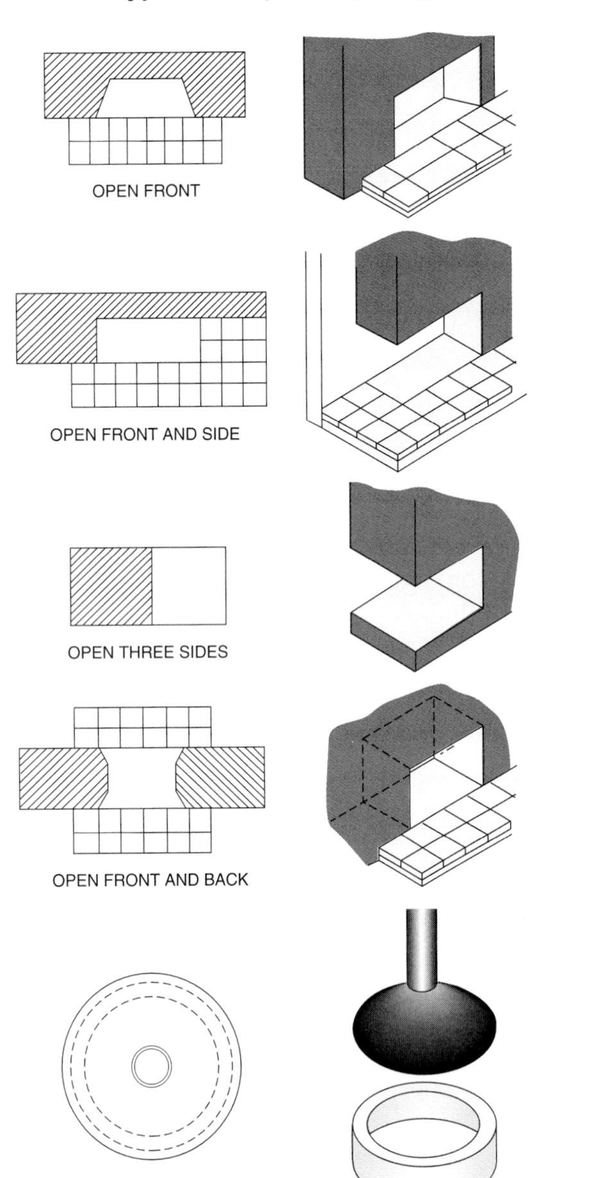

quickly turned off or on. Units can also be thermostatically controlled. “Logs” used in gas fireplaces resemble wood in appearance but are made of noncombustible ceramic or masonry materials.

Requiring neither heavy masonry and foundations nor front hearth, prefabricated gas fireplaces can rest on any type of flooring with no limitations on enclosure size or trim. Gas fireplaces require no direct vertical flue. Therefore the area

Fig. 23-37 ■ A flush (single-faced) fireplace. *Majestic*

Fig. 23-38 ■ A two-sided or corner fireplace appropriate for use in L-shaped areas. *Heatilator, Inc.*

above the firebox need not be a chimney flue. Refer back to the photo on page 440.

SYNTHETIC MATERIALS Some fireplaces are designed to burn synthetic materials, such as gelled alcohol. These fireplaces require no venting to the outside, no hearth, and no special flooring.

Types of Fireplace Openings

Fireplaces are divided into five basic types of openings: flush, two-sided, three-sided, double-flush or see-through, and open on all sides. See Figs. 23-36 through 23-41.

Fig. 23-39 ■ A three-sided or peninsula fireplace can be used as a separation between rooms. When positioned on a straight wall, the fire can be seen from any position in either room. *Majestic*

Fig. 23-40 ■ A see-through fireplace is a variation of the flush fireplace. This design allows one fireplace to be enjoyed in two different rooms simultaneously. *Heatilator, Inc.*

Fig. 23-41 ■ A design concept sketch of a fireplace that is open on all sides. These open-pit fireplaces are rarely installed in residences. They require a large hood and exhaust fumes are difficult to control. Glass enclosures may be used. *John Henry, Architect*

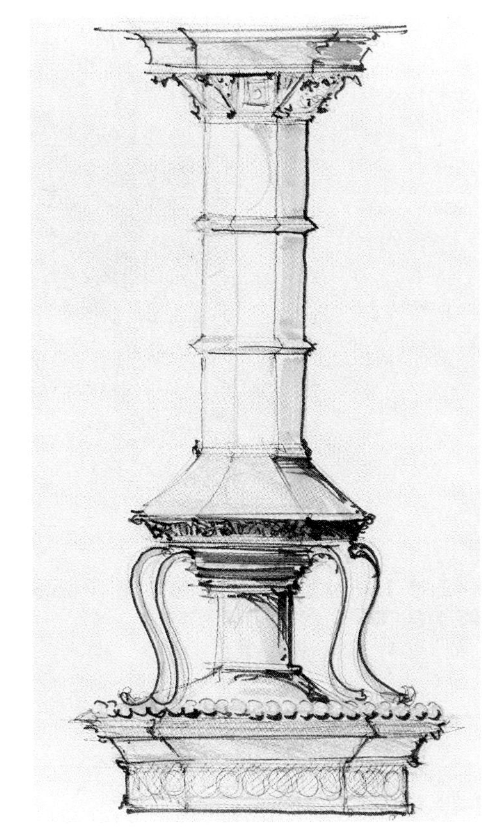

Fireplace Construction

Today, fireplaces are rarely constructed completely on site. Most fireplaces consist of manufactured components around which framing or masonry walls are placed. Fireplace components are designed to produce fire and to provide for safety, convenience, and efficiency. See Fig. 23-42.

FIRE-PRODUCING COMPONENTS A firebox to support the fire, a damper to control air flow, and a flue system to exhaust fumes are necessary to create and maintain combustion in most fireplaces. A **firebox** or **fire chamber,** contains and supports a fire while reflecting heat and exhausting smoke through a flue system. Fire chambers may be built on site, but most are factory-built. See Fig. 23-43. The floor of the firebox or fire chamber (also called the internal hearth) and the wall surfaces that are in direct contact with the fire are covered with firebrick, which is laid in fire-repellent mortar known as fireclay.

If the depth of the firebox is excessive, only a small percentage of heat will be reflected into the room. Smoke may

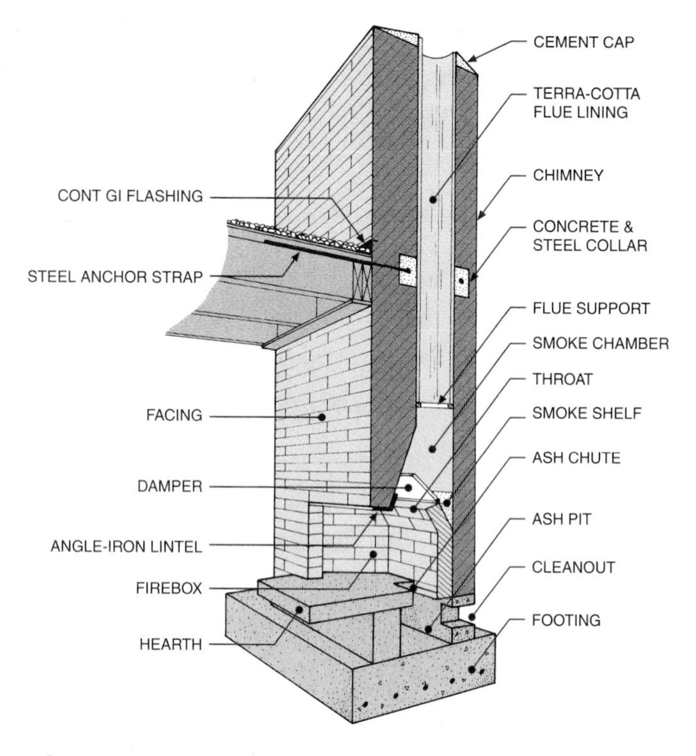

Fig. 23-42 ■ Major components of a fireplace and chimney structure.

escape into the room if the depth is too shallow. The size and proportion of the firebox is critical to the effective operation of a fireplace. Therefore, care must be taken to use the most efficient ratio (usually 1:10) of flue size to firebox size. See Fig. 23-44.

The **flue** is the opening in a chimney through which smoke passes. A **damper** is a door that separates the firebox from the flue area. When the fireplace is in use, this door is opened to allow the upward flow of hot air to create a draft that expels smoke and gas from the firebox to the flue. When the fireplace is not in use, the damper is closed to prevent downdrafts from the flue to the firebox. Part of a damper system is a smoke shelf. When the damper is open, it deflects cold downdrafts into the rising warm air currents.

Fig. 23-43 ■ Types of fireboxes.

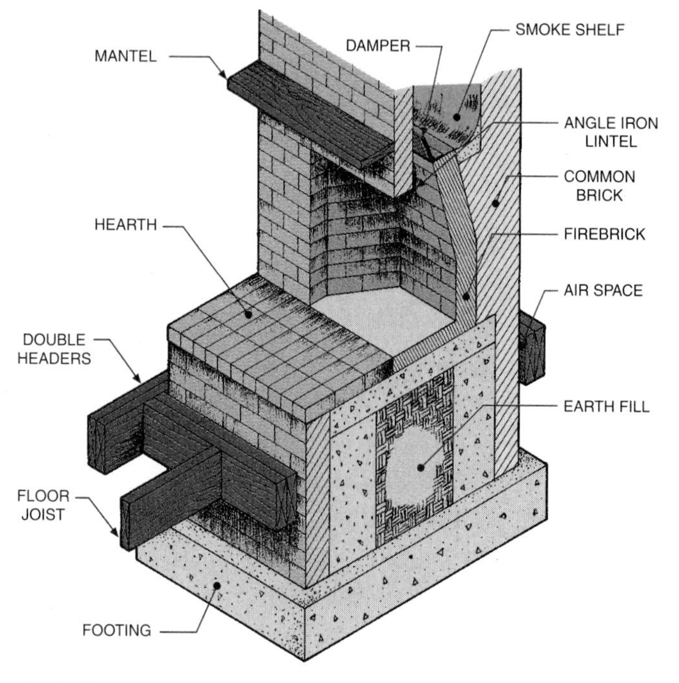

A. Built on site.

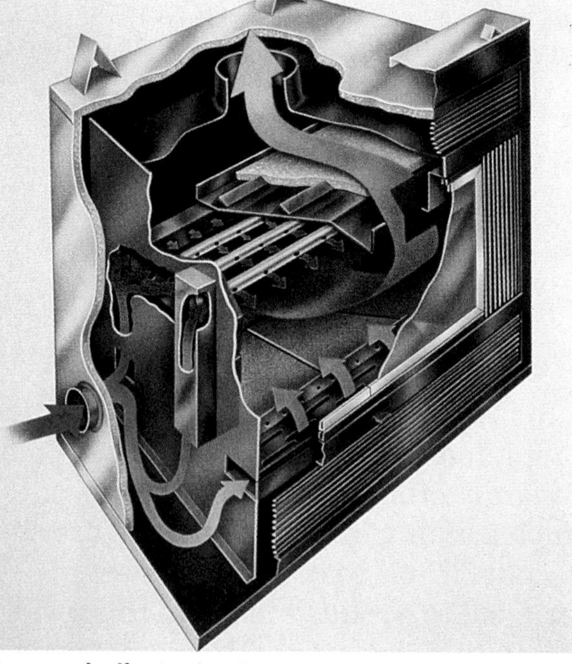

B. Factory-built. *Majestic*

Designing for adequate warm-air rise (called draw) is critical for proper fireplace functioning. Inadequate draw, either from an undersized flue or from improper chimney placement, can result in smoke leaking into the room. The cross-section area of a flue should be a minimum of one-tenth the area of the fireplace opening. One flue is necessary for each fireplace or furnace, but multiple flues can extend vertically through one chimney if properly offset, as shown in Fig. 23-45A. Often masonry caps or uneven flue projections prevent downdrafts. See Fig. 23-45B.

CHIMNEYS Chimneys extend from the footing through the roof of a house. The chimney extends above the roof line to provide a better draft for drawing the smoke and to eliminate the possibility of sparks igniting the roof. Spark arrestors are required by many building codes. See Fig. 23-46.

The minimum required height of a chimney above the roof line varies somewhat among local building codes. In most areas the minimum distance is 2′ (610 mm) if the chimney is closer than 15′ to

Fig. 23-44 ■ Computing proportions when designing an efficiently operating fireplace.

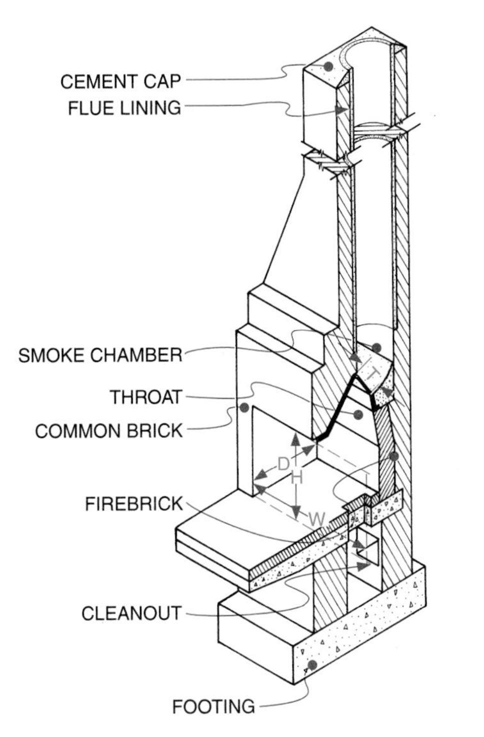

A. Key dimensional areas. Substitute dimensions from the chart in part B to determine the correct proportions.

Fig. 23-44B ■ Dimensional fireplace proportions.

Fireplace dimensions (*in inches*)		**Recommended flue sizes** (*in inches*)					
		Fireplace Width W	**Rectangular Flues**			**Round Flues**	
W	24 to 84		Nominal or Outside Dimension	Inside Dimension	Effective Area	Inside Diameter	Effective Area
H	⅔ to ¾ W	24	8½ × 8½	7¼ × 7¼	41°″	8	50.3°″
D	½ to ⅔ H (16 to 24 (Rec) for Coal; 18 to 24 (Rec) for Wood)	30 to 34	8½ × 13	7 × 11½	70°″	10	78.54°″
FLUE (*effective area*)	⅛ WH for unlined flue 1/10 WH for rectangular lining 1/12 WH for circular lining	36 to 44 46 to 56	13 × 13 13 × 18	11¼ × 11¼ 11¼ × 6¼	99°″ 156°″	12 15	113.0°″ 176.7°″
T (*area*)	5/4 to 3/2 Flue area	58 to 68	18 × 18	15¾ × 5¾	195°″	18	254.4°″
T (*width*)	3″ minimum to 4½″ minimum	70 to 84	20 × 24	17 × 21	278°″	22	380.13°″

Fig. 23-45 ■ Positioning flues in chimneys.

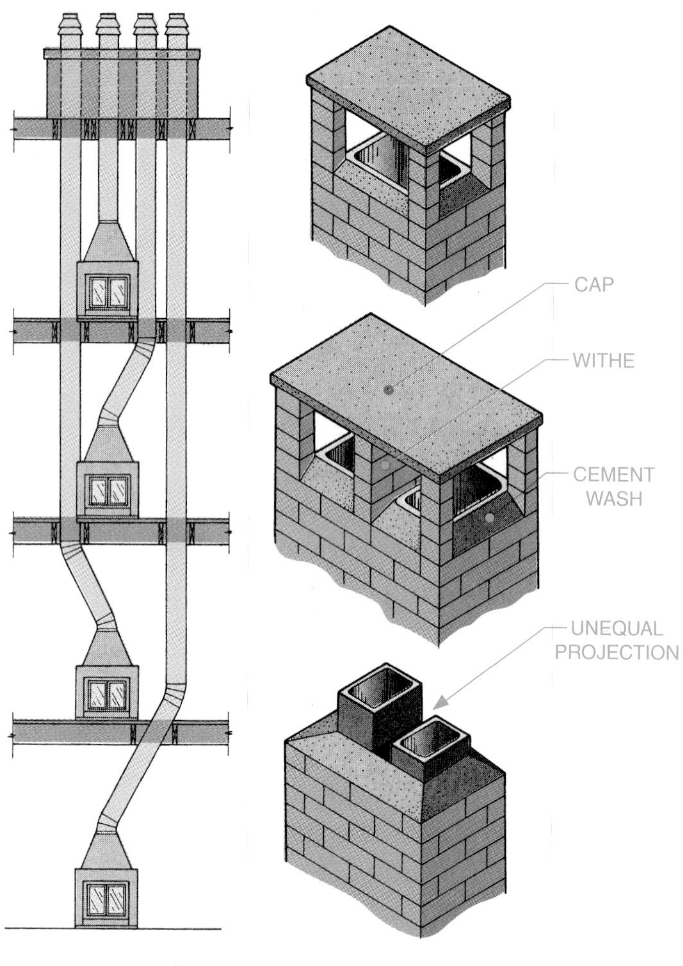

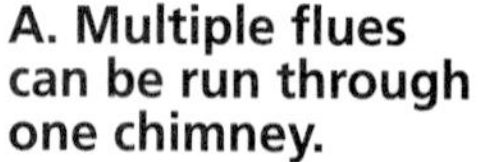

A. Multiple flues can be run through one chimney.

B. Chimney designs for downdraft control.

the nearest ridge. See Fig. 23-47. The footing must be of sufficient size to support the entire weight of the chimney.

SAFETY COMPONENTS Some components are not necessary for fire maintenance, but are necessary to prevent flames from spreading outside the firebox. These are covered specifically in building codes.

- **Hearth.** Most codes require an extended hearth to cover a floor area 16″ beyond the firebox face and 6″ to each side of the firebox. Hearths may be elevated or flush with the floor level.

Fig. 23-46 ■ A chimney with a spark arrestor and sheet-metal enclosure to prevent downdrafts.

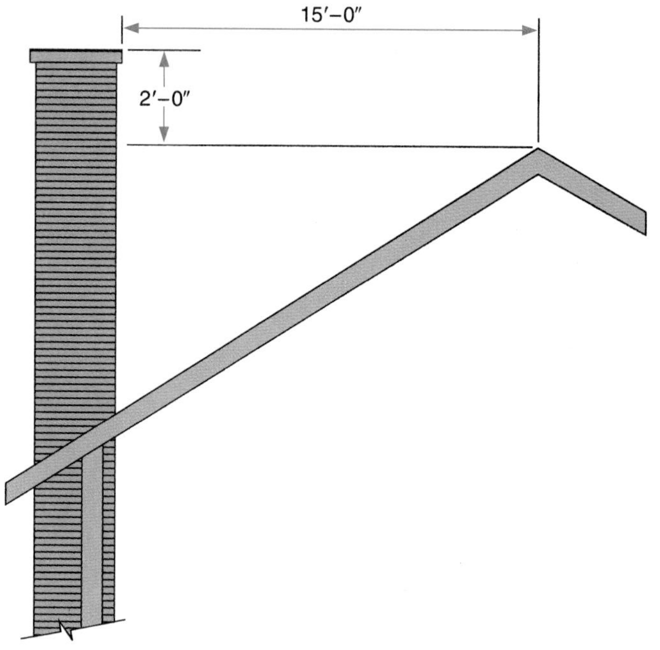

Fig. 23-47 ■ Common chimney code requirements.

- **Materials.** Noncombustible materials must be used in fireplace components that will either be in contact with flames or be excessively heated during operation. Firebrick and fire mortar must be used in the firebox. Generally all masonry materials such as concrete, brick, stone, tile, or marble are safe for use outside the firebox. Wood products are not acceptable. Flues must be made of heat-reflecting materials such as terra-cotta, or a 3″ minimum air space must be provided between metal flues and wood framework. An airspace or fire-resistant wallboard or insulating material must be placed between any wood members and a firebox.
- **Structural Support.** During the structural design phase, provisions must be made so that the heavy loads of the fireplace and chimney assemblies rest on a solid footing. A solid reinforced concrete footing is most often used for residential construction. Extra footing depth must be provided under the fireplace area. Refer back to Fig. 23-42. Wood-framed fireplaces with manufactured fireboxes and sheet metal flues may require a shallow footing of 6″. A masonry fireplace and chimney may require a footing depth of 12″ to 24″, depending on the size of the chimney structure. Footings should extend at least 6″ beyond the fireplace outline on all sides.
- **Safety Screens.** To prevent sparks from projecting outside the firebox, a safety screen should cover the fireplace opening. Wire mesh screen can be opened to load the firebox.

CONVENIENCE COMPONENTS Some items are not necessary for fire production or safety but do add to the convenience and efficiency of using a fireplace.

- **Glass Enclosures.** As shown in Fig. 23-38 and others, glass enclosures allow the flames to be seen while preventing smoke or sparks from escaping. When glass enclosures are closed, most heat from the fire is transmitted to the room through vents in the firebox.
- **Ash Pit.** Except for fireplaces on a slab construction, a metal trap door can be placed on the inner hearth. Cold ashes are then dumped through this door to a metal container below. This container has a cleanout door for the removal of ashes.
- **Blowers and Remote Outlets.** To project heat into a room beyond normal air movement, blowers can be used. They may direct air into the room or to remote outlets. See Fig. 23-48.

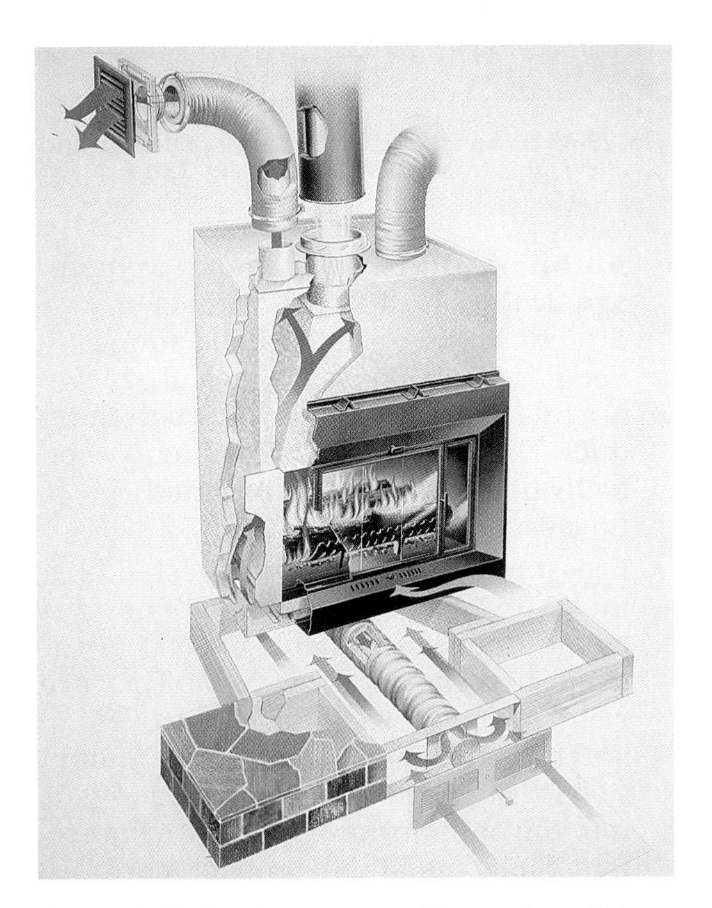

Fig. 23-48 ■ Fireplace system with remote outlets.
Olympic Fireplaces

Fig. 23-49 ■ An air duct connected to a firebox and run through a frame wall. See also Fig. 23-51.

- **Air Intake Ducts.** To add more oxygen to a firebox, an air duct connected directly to the outside can be added. See Fig. 23-49. Some systems use blowers to accelerate air flow through the duct. The use of an outside duct is especially helpful in tightly insulated houses.
- **Freestanding Fireplaces.** Freestanding metal fireplaces constructed of heavy-gauge steel are available in a variety of shapes, as shown in Fig. 23-50. They are relatively light wood-burning stoves and therefore need no concrete foundation for support. A stovepipe leading into the chimney provides the exhaust flue. Since metal units reflect more heat than masonry, metal fireplaces are much more heat efficient, especially if centrally located. For safety, fire-resistant materials such as concrete, brick, stone, or tile must be used beneath and around these fireplaces.

Fig. 23-50 ■ Freestanding fireplace designs. *Majestic Fireplaces/Art MacDillo's/Gary Skillestad*

Fireplace Detail Drawings

A horizontal section through a fireplace firebox is drawn on floor plan drawings. The outline of the fireplace opening and the chimney design are shown on interior elevation drawings. However, floor plan and elevation scales are not large enough to show the amount of detail needed to build a fireplace system. Enlarged sections are used to show construction details for masonry fireplaces. For fireplaces that include manufactured components in framed walls, enlarged orthographic views and/or sectional framing drawings are necessary to show construction details. See Fig. 23-51.

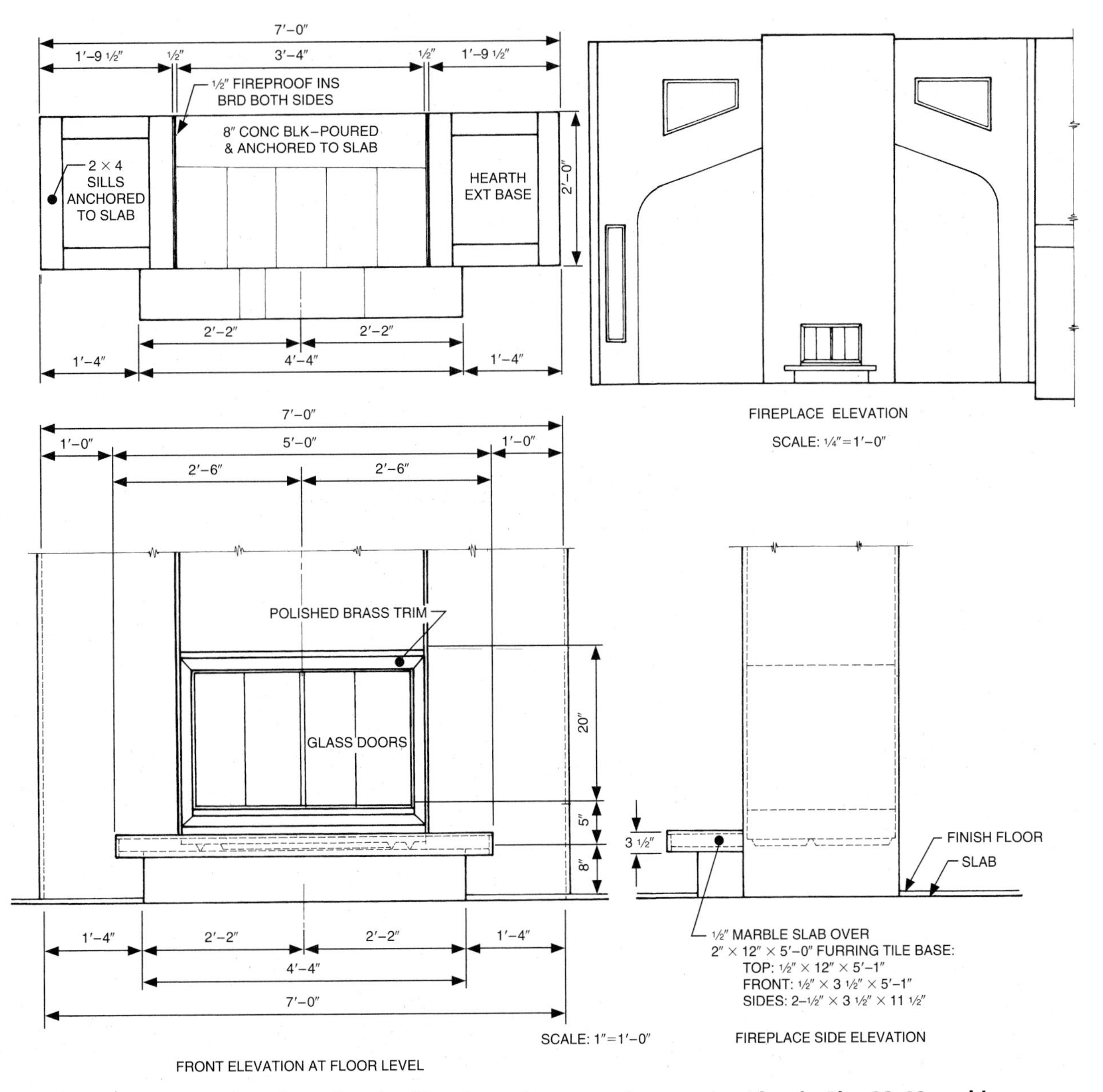

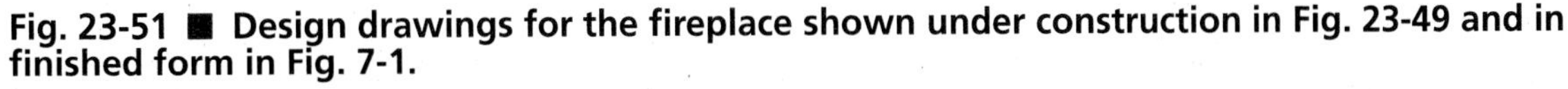
Fig. 23-51 ■ Design drawings for the fireplace shown under construction in Fig. 23-49 and in finished form in Fig. 7-1.

Floor and roof framing drawings related to fireplaces and chimney construction are covered in more detail in Chapters 28 and 30.

Foundation Drawings

Before learning to draw foundation plans, an understanding of foundation layout and excavation is necessary.

Layout

Establishing the exact position of a building on a lot requires locating each building corner. This is done by measuring the distance from property lines to building corners on the plot plan.

If the positions of only two building corners are shown on the plot plan, all others can be plotted by turning angles with a transit. Right angles (90°) can be plotted by using the 3.4.5 unit method. See Fig. 23-52. The Pythagorean Theorem may be used to determine if the measurements are correct.

Pythagorean Theorem:

Square of the hypotenuse of a right triangle = the sum of the square of the two sides

$$c^2 = a^2 + b^2$$

Example:

$$c^2 = a^2 + b^2$$

$$\sqrt{c^2} = \sqrt{a^2 + b^2}$$

$$c = \sqrt{32^2 + 24^2}$$

$$c = \sqrt{1024 + 576}$$

$$c = \sqrt{1600}$$

$$c = 40''$$

If only one point is dimensioned, the azimuth of one side needs to be known to accurately lay out the building. (You may wish to refer back to Chapter 18.) For rectangular areas the accuracy of the layout can be checked by measuring across the diagonals. The diagonal distances will be equal if the measurements are true. Once each corner is located, string attached to batter boards is used to identify the corners during excavation, as shown in Fig. 23-52.

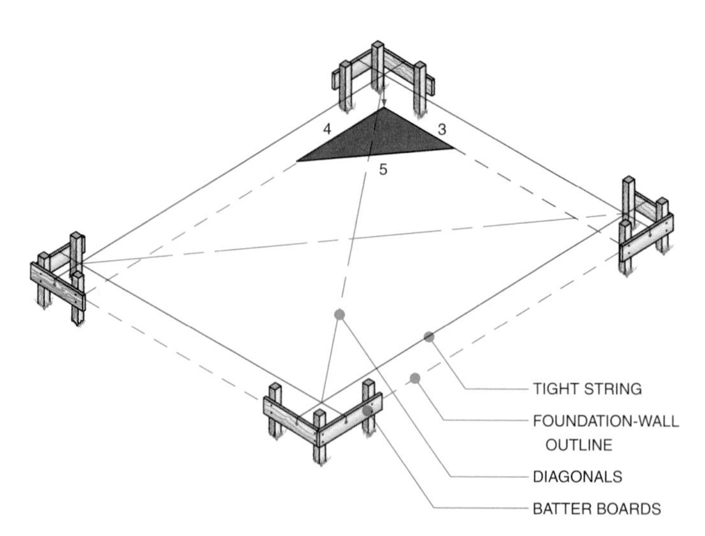

Fig. 23-52 ■ Building layout, using the 3.4.5 unit method, batter boards, and line.

Excavations and Forming

Foundation plans should clearly show whether areas of a foundation are to be completely excavated for a basement, partly excavated for a crawl space, or unexcavated. The depth of the excavation should also be indicated on the elevation drawings. If a basement is planned, the entire excavation for the basement is dug before the footings are poured. If there is to be no basement, a trench excavation is made. To calculate the volume of an excavation use the following formula:

$$\text{Formula: cu. yds.} = \frac{W' \times L' \times D'}{27}$$

Example:

(T-foundation wall)

$$\text{cu. yds.} = \frac{.5' \times 40' \times 1.25'}{27} = \frac{25}{27} = .93 \text{ cu. yd.}$$

(footing)

$$\text{cu. yds.} = \frac{2' \times 40' \times .75'}{27} = \frac{60}{27} = 2.22 \text{ cu. yd.}$$

$$\text{Total} = 3.15 \text{ cu. yd.}$$

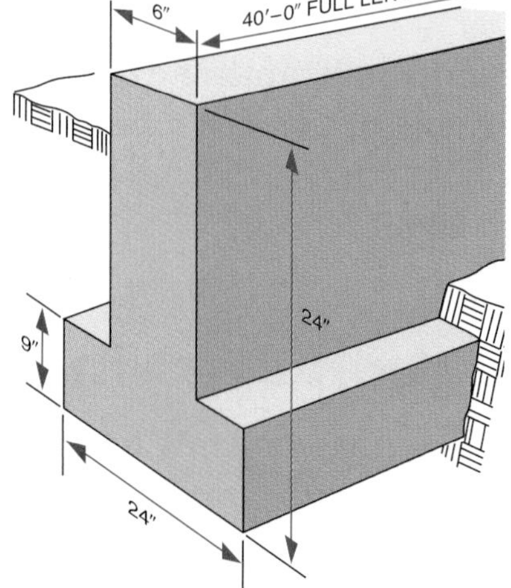

Although slab foundations may only require trench footings, all organic soil material must be removed and replaced with non-organic, compactable soil, as shown in Fig. 23-53. Excavation must always be made at least 6″ below the frost line. All HVAC, electrical, or plumbing lines which are to be embedded in or pass through a poured area must be installed when the form for the foundation is constructed. See Fig. 23-54.

Some foundation plans include the position of plumbing lines if lines are to pass under or through a poured area. Usually, though, the position of plumbing lines is read from wall and fixture positions on either a floor plan or plumbing plan.

Foundation Plans

Foundation plan drawings are floor plans drawn at the foundation level. The same plan symbols, such as those for brick and concrete, show construction materials. Positions of footings and piers under the foundation are shown with hidden lines on a foundation plan. See Fig. 23-55.

Fig. 23-53 ■ Preparing a site for a foundation includes replacing organic soil with non-organic soil that can be compacted.

Fig. 23-54 ■ Form for a foundation with all utility lines installed.

Fig. 23-55 ■ Methods of drawing walls and footings on foundation plans.

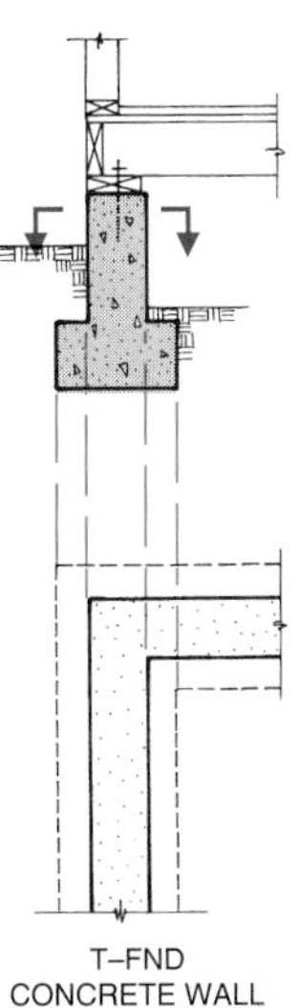

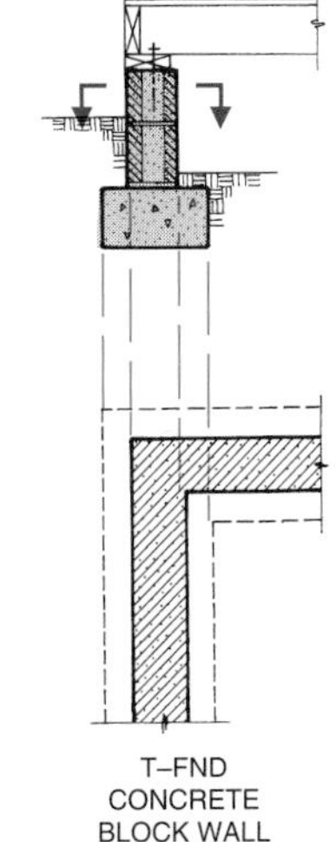

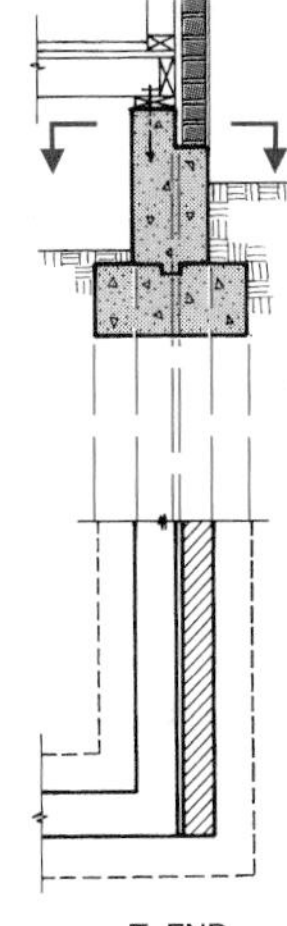

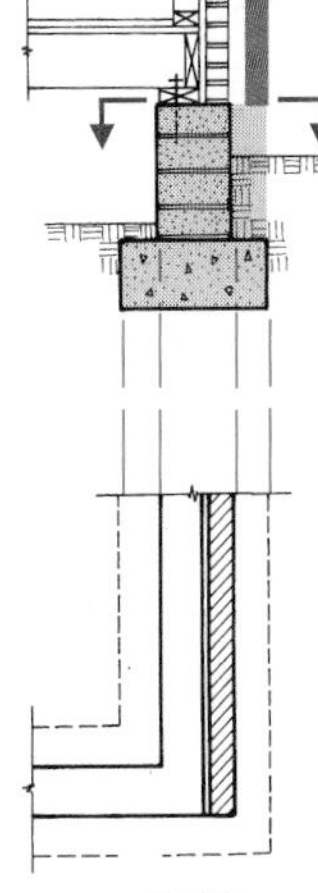

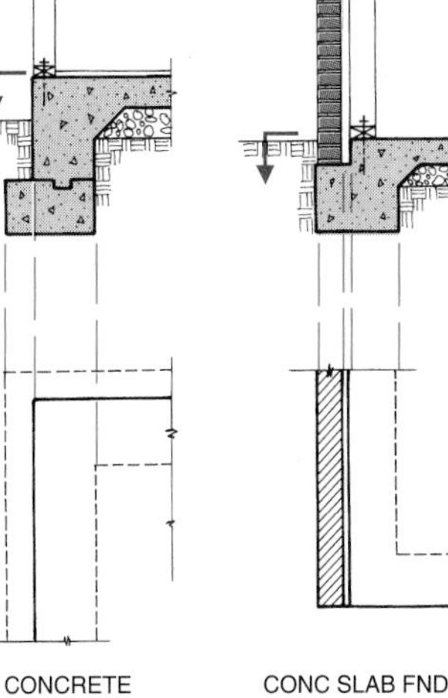

T-FOUNDATION DRAWINGS Figure 23-56 shows a T-foundation drawing. The sequence for drawing T-foundation or basement plans is similar to drawing floor plans. The outside perimeter line should be drawn first, then the interior partitions. Next, draw the parallel wall thickness lines, footing lines, and material symbols. Add the position of any piers, columns, footings, and beams. Include a floor joist directional label with dimensions.

SLAB FOUNDATION DRAWINGS Slab drawings show only the outer perimeter line with hidden lines to represent footings. Slab drawings must note the thickness of the slab, size and spacing of reinforcing bars and wire mesh, concrete PSI (pounds per square inch), and thickness of waterproof membrane. See Fig. 23-57.

Detail Drawings

Foundation plan views only show the location of all features on a horizontal plane. Most foundation details require elevation sections to convey vertical positions and dimensions. The most common types of foundation details are footing and sill details.

Fig. 23-56 ■ Plan for a T-foundation.

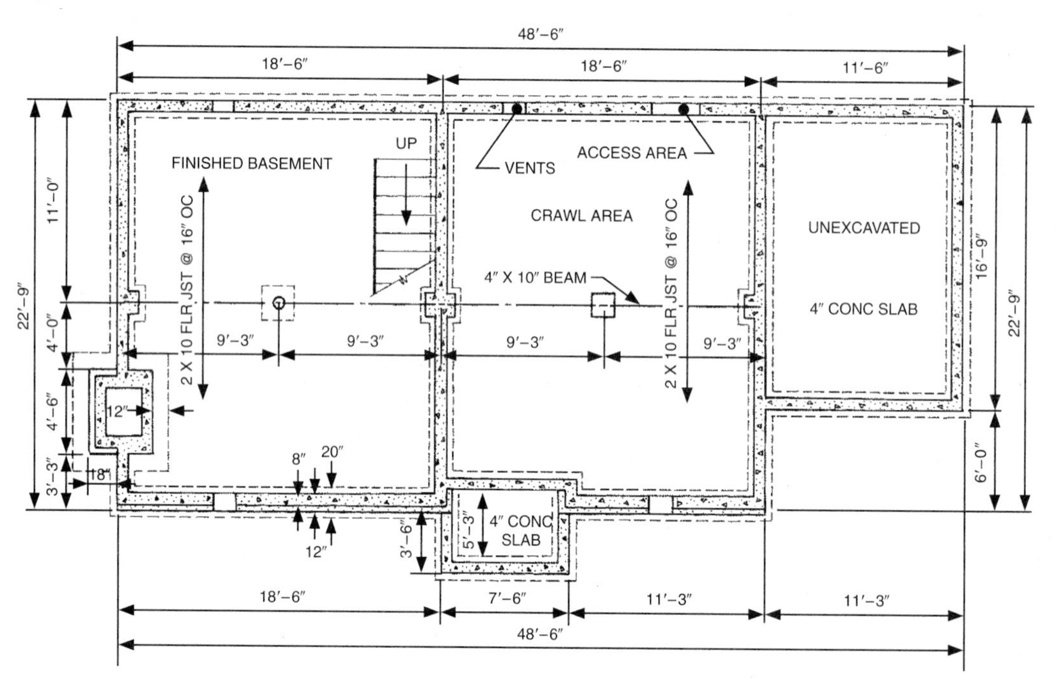

Fig. 23-57 ■ Slab foundation drawings. *Vardy Vincent*

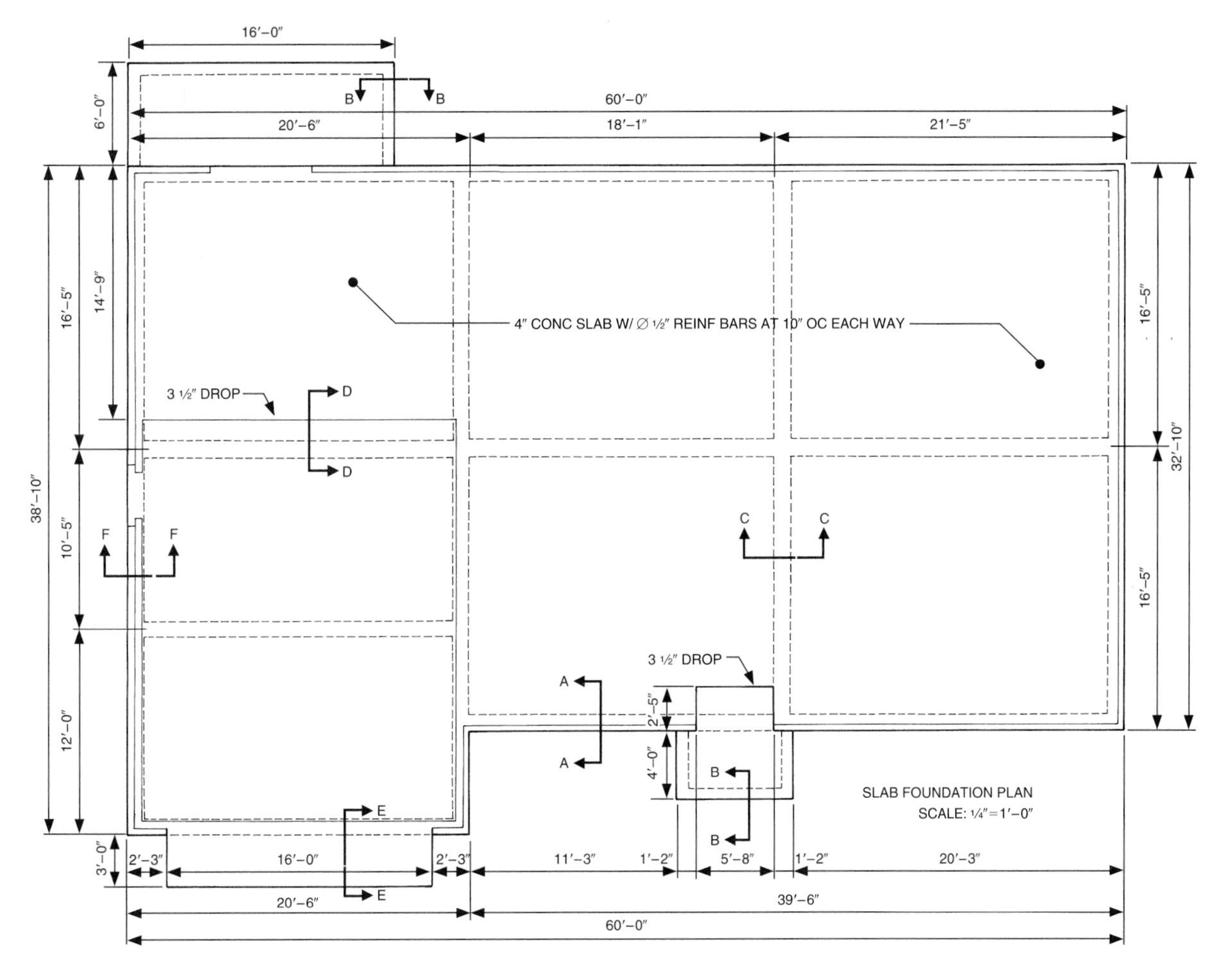

A. Plan with sections noted.

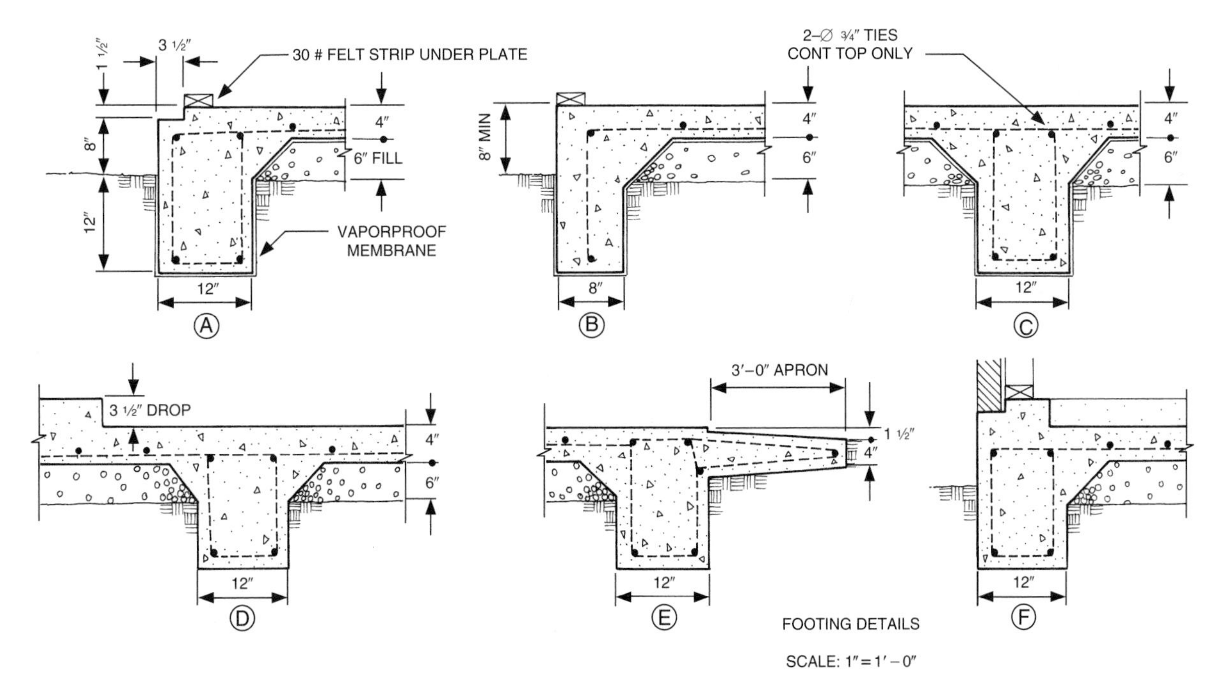

B. Footing details noted in part A.

Footing detail drawings show a vertical section through the footing, foundation wall, or slab. These drawings show the size, shape, and material used in footings. See Fig. 23-58. If many different footing types are used under partitions, an entire profile of the foundation may be drawn with the location of all footings, as shown in Fig. 23-59.

Fig. 23-58 ■ Footing detail drawings.

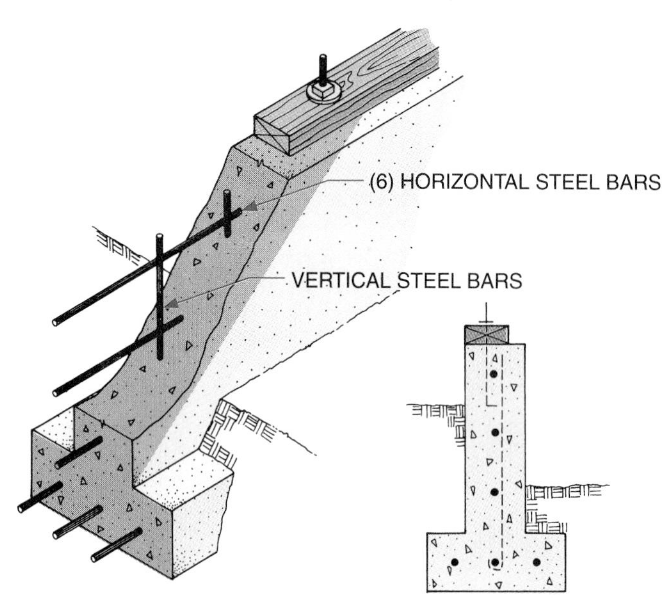

A. Solid concrete footing and foundation wall.

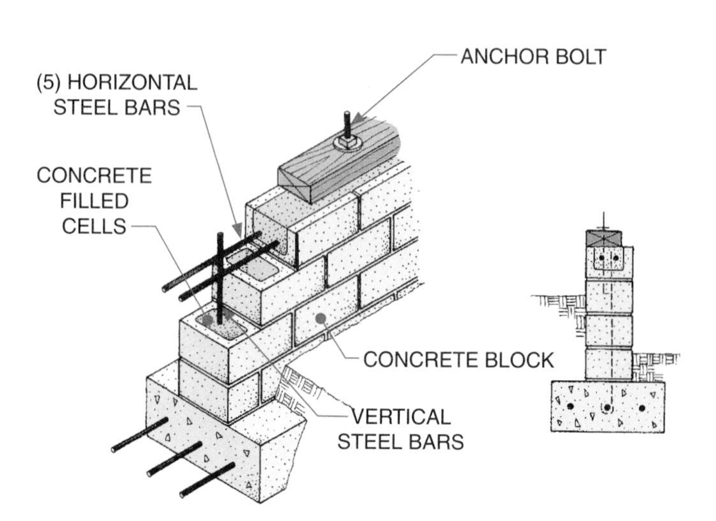

B. Concrete footing with a concrete block foundation wall.

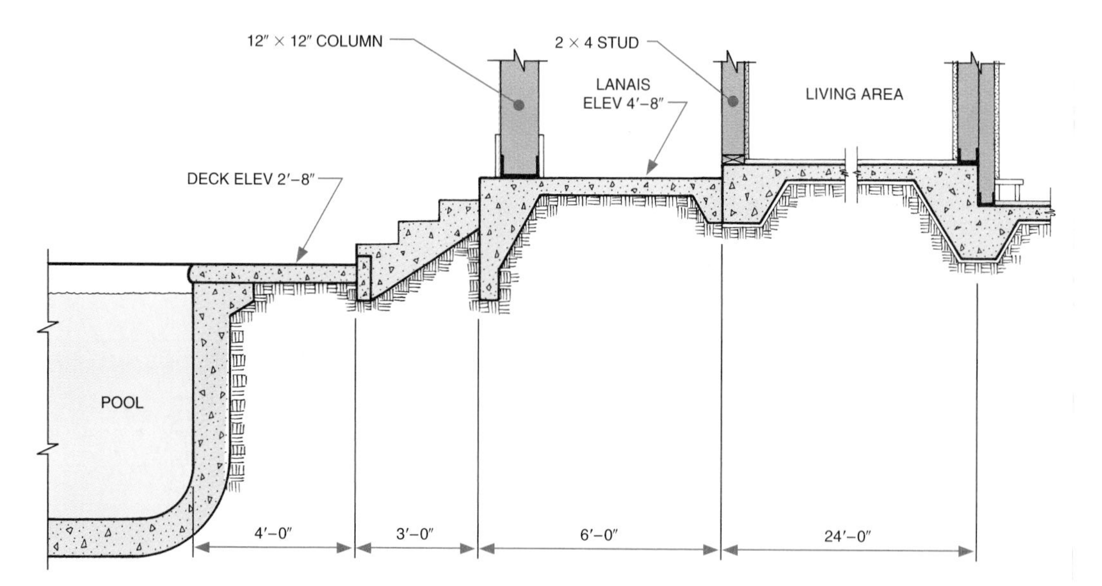

Fig. 23-59 ■ Footing profile of the floor plan area shown in Fig. 14-20.

Sill detail drawings show how the foundation, exterior wall, and floor system intersect. Sill details are viewed in elevation, pictorial, or plan sections. See Fig. 23-60.

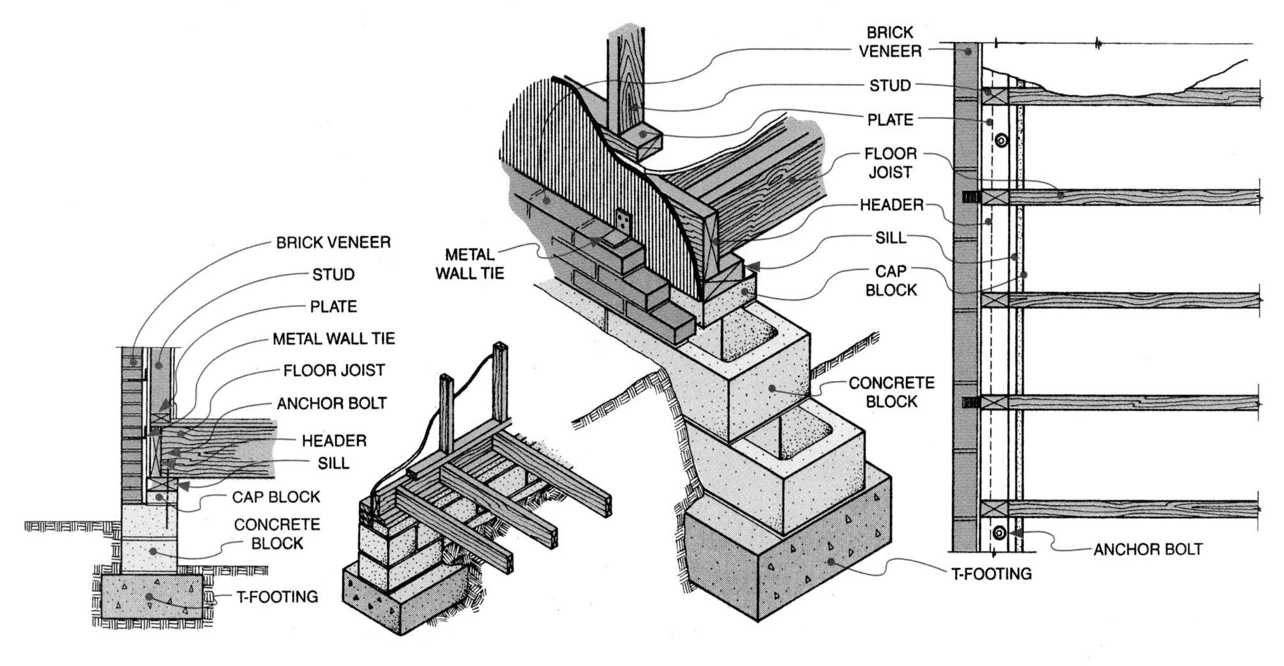

Fig. 23-60 ■ Brick veneer sill details.

CHAPTER 23 Exercises

1. Draw the foundation plan for the house you are designing.
2. Sketch a foundation plan for the design shown in Fig. 14-18, using the scale $\frac{1}{4}$″ = 1′-0″ for a T-foundation.
3. Sketch a foundation plan for the design shown in Fig. 14-18, using the scale $\frac{1}{4}$″ = 1′-0″ for a slab foundation.
4. Sketch a foundation plan for the design shown in Fig. 14-18, using the scale $\frac{1}{4}$″ = 1′-0″ for a pier foundation.

5. Complete Exercises 1 through 4 using a CAD system.
6. Draw a plan and elevation view of the fireplace in Fig. 23-38, using the scale $\frac{1}{2}$″ = 1′-0″. See Fig. 23-44 for typical dimensions.
7. Design a fireplace for the house you are designing.

8. Complete Exercises 6 and 7 using a CAD system.
9. Draw a sill and footing detail for the house you are designing.

CHAPTER 24

Wood-Frame Systems

Objectives

In this chapter, you will learn to:

- differentiate between skeleton-frame and post-and-beam construction.
- identify major characteristics of lumber, plywood, and structural timber.
- calculate the number of board feet in a piece of lumber.

Terms

balloon (eastern) framing
board foot
hardwoods
laminated timber
plank-and-beam construction
platform (western) framing
plywood
post-and-beam construction
skeleton-frame construction
softwoods

In wood-frame construction, wooden structural members are joined to make an open framework for the structure. This open framework is then covered with layers of other construction materials to form the solid surfaces of floors, walls, and roofs.

There are several varieties of wood-frame construction. This chapter covers skeleton-frame construction and post-and-beam construction.

■ **Most houses built in the United States and Canada use wood-frame construction.** *Westlight/Ron Watts*

Skeleton-Frame Construction

In **skeleton-frame construction**, small structural members are joined in such a way that they share the loads of the structure. When the structural members are covered, they form complete walls, floors, and roofs. Because of the limited size of wood materials, skeleton-frame is considered light construction. See Fig. 24-1.

Materials

Light construction materials include lumber, plywood, reconstituted wood, plastics, fasteners, and multimaterial components such as doors, windows, cabinets, plumbing, ductwork, and electrical fixtures.

LUMBER Construction lumber is classified according to grade, species, size, and whether it has been treated. Lumber

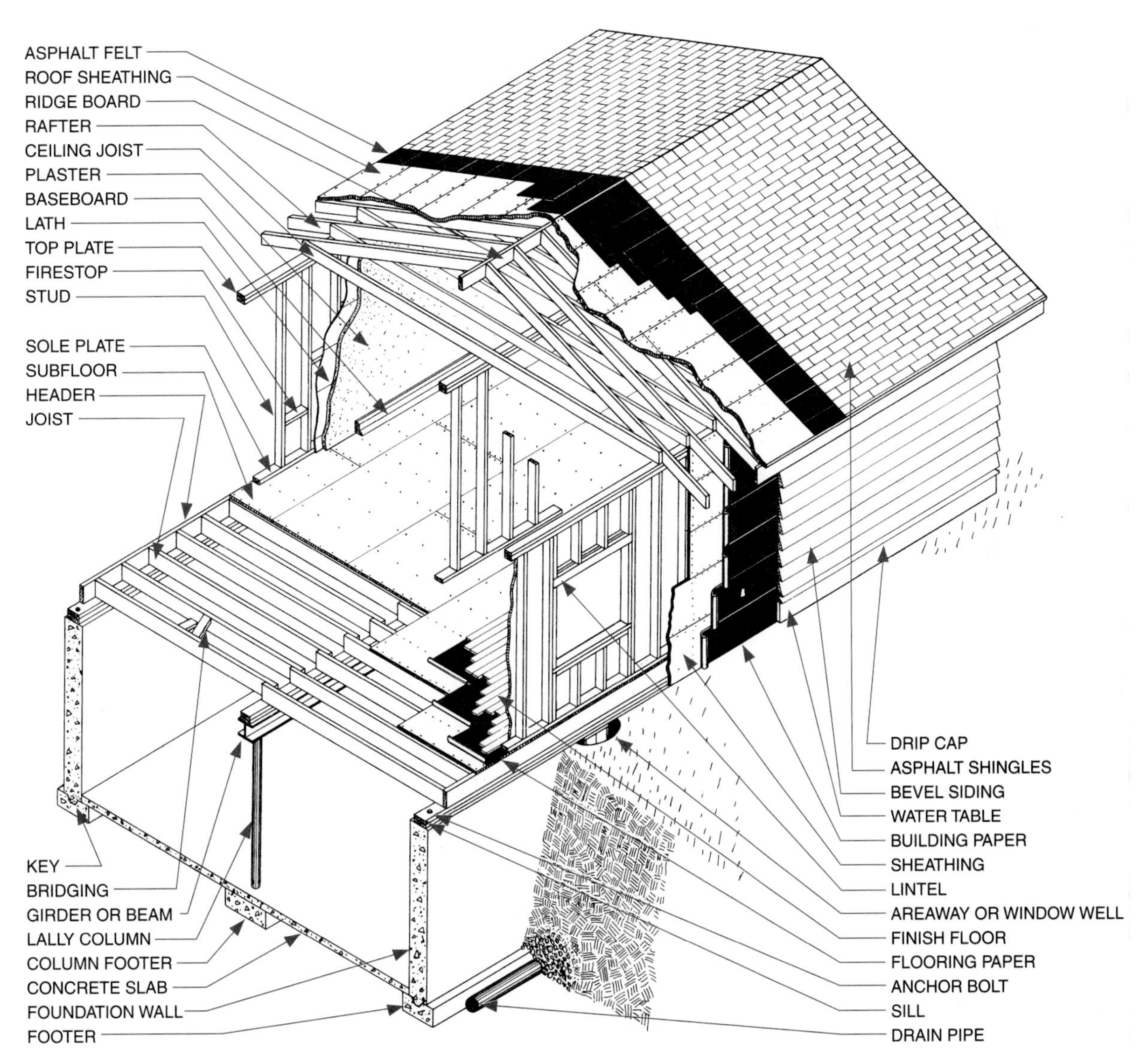

Fig. 24-1 ■ Typical skeleton-frame construction.

grades are determined by the number and location of defects, such as knots, checks, and splits, and by the degree of warp (deviation from a flat, even surface). For grading purposes, lumber is divided into two broad categories: hardwoods and softwoods.

Hardwoods are used for surfaces that must withstand much wear, such as flooring and railings. Hardwoods are also used to make items that require a fine natural finish, such as cabinets and furniture. Hardwoods come from broad-leaved trees. The species most commonly used in construction include oak, walnut, birch, cherry, mahogany, and maple. Hardwood is graded from highest quality to lowest quality depending on the amount of usable material in each piece. See Fig. 24-2. Hardwood lumber is available in lengths of 4′ to 16′, widths up to 12″, and thicknesses up to 2″.

Softwoods come from coniferous (needle-bearing) trees such as Douglas fir, pine, or cedar. They are used for structural members such as joists, rafters, studs, sheathing, and formwork. Most skeleton-frame lumber is softwood. Softwood lumber is divided into three grading classes: yard, structural, and factory (shop) lumber. *Yard lumber* is used for most light framing members, such as sheathing, bracing, subfloors, and casings. See Fig. 24-3. *Structural lumber,* as the name implies, is used for load-bearing

Fig. 24-2 Hardwood lumber grades.

Wood Grade	Quality Level
FAS firsts FAS seconds Select Common #1 Common #2 Common #3A Common #3B	High Quality ↓ Poor Quality

Fig. 24-3 Range of yard lumber grades.

Grade	Use
Selects and finish	Graded from the best side. Used for interior and exterior trim, molding, and woodwork where appearance is important.
B & BTR	Used where appearance is the major factor. Many pieces clear, but minor appearance defects allowed which do not detract from appearance.
C Select	Used for all types of interior woodwork. Appearance and usability slightly less than B & BTR.
D Select	Used where finishing requirements are less demanding. Many pieces have finish appearance on one side with larger defects on back.
Boards	Lumber with defects that detract from appearance but suitable for general construction.
No. 1 common (WWPA) Select merchantable (WCLIB)	All sound tight knots, with use determined by size and placement of knots. Used for exposed interior and exterior locations where knots are not objectionable.

continued

Fig. 24-3 Continued

Grade	Use
No. 2 common (WWPA) Construction (WCLIB)	All sound tight knots with some defects, such as stains, streaks, and patches of pitch, checks, and splits. Used as paneling and shelving, subfloors, and sheathing.
No. 3 common (WWPA) Standard (WCLIB)	Some unsound knots and other defects. Used for rough sheathing, shelving, fences, boxes, and crating.
No. 4 common (WWPA) Utility (WCLIB)	Loose knots and knotholes, up to 4″ wide. Used for general construction purposes, such as sheathing, bracing, low-cost fencing, and crating.
No. 5 common (WWPA) Economy (WCLIB)	Large knots or holes, unsound wood, massed pitch, splits, and other defects. Used for low-grade sheathing, bracing, and temporary construction. Pieces of higher-grade wood may be obtained by crosscutting or ripping boards without defects.

Fig. 24-4 Structural lumber grades.

Grade	Use
LF (Light Framing)	Used in thicknesses from 2″ to 4″ and widths from 3″ and 4″, for studs, joists, and rafters in light framing.
JP (Joints and Planks)	Used in thicknesses from 2″ to 4″ and widths over 2″, for joists and rafters to be loaded on either side, or for planking when laid flat.
B&S (Beams and Stringers)	Used in thicknesses from 2″ to 4″. Widths over 2″ must be loaded on narrow edge.
P&T (Posts and Timbers)	Used for posts or columns 5″ × 5″ and larger or where bending resistance is not critical.

members and is classified according to use and grades. Grades are based upon a lumber's stress-resistance. See Fig. 24-4. *Factory*, or *shop lumber*, consists of light members which are finished at a mill and used for trim, molding, and door and window sashes.

All lumber is graded by an authority at a lumber mill according to the American Lumber Standards. These standards assure that lumber is grade marked with a variety of appropriate, accurate information. Lumber is labeled with a mill identification number and by a certification association logo. Other identifying information includes the grade number, moisture content, and species classification, as shown in Fig. 24-5.

Structural lumber is defined as either rough or finished. Rough sizes represent the width and thickness of a piece of lumber as cut from a log. Rough lumber is also called *nominal* (name) *size* lumber. Finished lumber sizes represent the

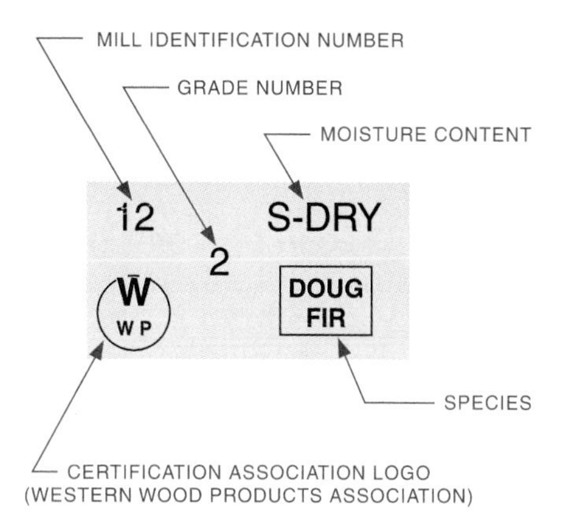

Fig. 24-5 ■ Typical lumber grade marks.

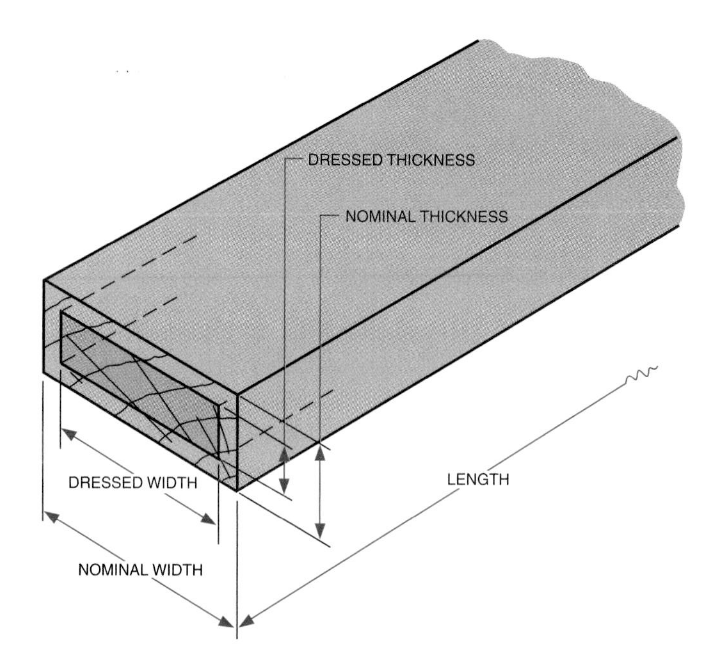

Fig. 24-6 ■ Nominal and finished (dressed) lumber sizes.

actual dimensions of a member after final surfacing. See Fig. 24-6. Finished lumber is also known as surfaced, *dressed,* dimensional, or actual size lumber.

When a drawing callout reads 2 × 4, builders know that the rough size is 2″ × 4″ but the actual size of the member is $1\frac{1}{2}'' \times 3\frac{1}{2}''$. Figure 24-7 shows the U.S. customary range of rough (nominal) and finished (actual) standard lumber sizes. Figure 24-8 shows the range of metric lumber sizes. Since lumber is not always surfaced on all sides, symbols (or codes) designate the number of sides that are surfaced. See Fig. 24-9. Figure 24-10 shows the nominal thickness of lumber sizes from 1″ to 16″ and the resulting (dressed) thickness of each.

Fig. 24-7 U.S. Customary lumber sizes.

Lumber Sizes in Inches

Nominal Size	2 × 4	2 × 6	2 × 8	2 × 10	2 × 12	4 × 6	4 × 8	4 × 10	6 × 6	6 × 8	6 × 10	8 × 8	8 × 10
Dressed Size	$1\frac{1}{2} \times 3\frac{1}{2}$	$1\frac{1}{2} \times 5\frac{1}{2}$	$1\frac{1}{2} \times 7\frac{1}{2}$	$1\frac{1}{2} \times 9\frac{1}{2}$	$1\frac{1}{2} \times 11\frac{1}{2}$	$3\frac{9}{16} \times 5\frac{1}{2}$	$3\frac{9}{16} \times 7\frac{1}{2}$	$3\frac{9}{16} \times 9\frac{1}{2}$	$5\frac{1}{2} \times 5\frac{1}{2}$	$5\frac{1}{2} \times 7\frac{1}{2}$	$5\frac{1}{2} \times 9\frac{1}{2}$	$7\frac{1}{2} \times 7\frac{1}{2}$	$7\frac{1}{2} \times 9\frac{1}{2}$

Board Sizes in Inches

Nominal Size	1 × 4	1 × 6	1 × 8	1 × 10	1 × 12
Actual Size—Common	$\frac{3}{4} \times 3\frac{9}{16}$	$\frac{3}{4} \times 5\frac{9}{16}$	$\frac{3}{4} \times 7\frac{1}{2}$	$\frac{3}{4} \times 9\frac{1}{2}$	$\frac{3}{4} \times 11\frac{1}{2}$
Actual Size—Shiplap	$\frac{3}{4} \times 3$	$\frac{3}{4} \times 4\frac{15}{16}$	$\frac{3}{4} \times 6\frac{7}{8}$	$\frac{3}{4} \times 8\frac{7}{8}$	$\frac{3}{4} \times 10\frac{7}{8}$
Actual Size—T&G	$\frac{3}{4} \times 3\frac{1}{4}$	$\frac{3}{4} \times 5\frac{3}{16}$	$\frac{3}{4} \times 7\frac{1}{8}$	$\frac{3}{4} \times 9\frac{1}{8}$	$\frac{3}{4} \times 11\frac{1}{8}$

2 × 4   2 × 6   2 × 8   2 × 10   2 × 12   4 × 6   4 × 8   4 × 10   6 × 6   6 × 8   6 × 10   8 × 8   8 × 10   1 × 4   1 × 6   1 × 8   1 × 10   1 × 12

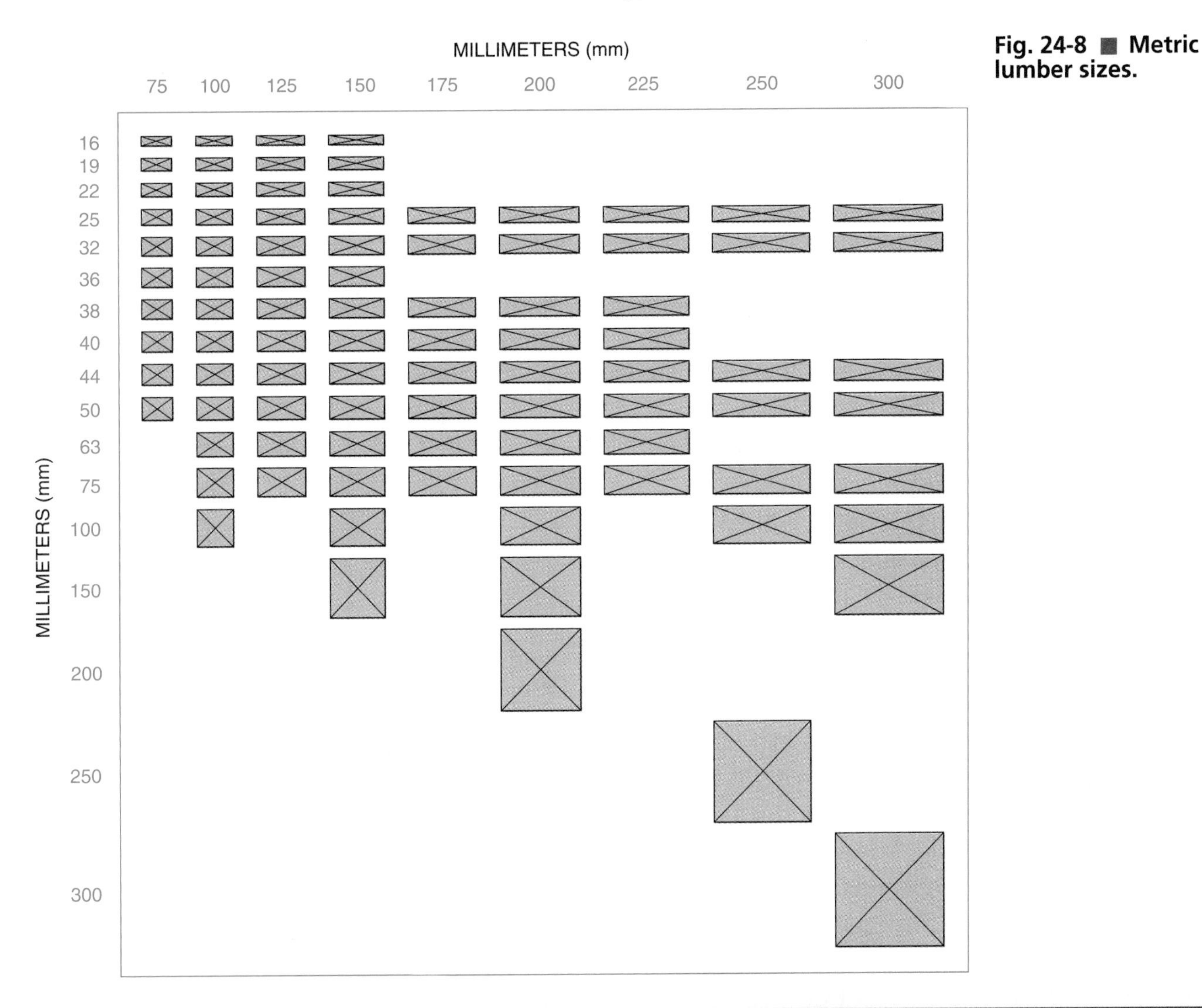

Fig. 24-8 ■ Metric lumber sizes.

Fig. 24-9 ■ Lumber surfacing codes.

Designation	Description
S1S	Surfaced on one side
S2S	Surfaced on two sides
S1E	Surfaced on one edge
S2E	Surfaced on two edges
S1S1E	Surfaced on one side, one edge
S2S1E	Surfaced on two sides, one edge
S4S	Surfaced on four sides

Fig. 24-10 ■ Lumber thickness 1″ through 16″

Nominal Thickness	Dressed Thickness
1″	¾″
2″	1½″
3″	2½″
4″	3½″
5″	4½″
6″	5½″
8″	7¼″
10″	9¼″
12″	11¼″
14″	13¼″
16″	15¼″

PLYWOOD Solid lumber is limited in width and has a tendency to warp, split, and check. **Plywood** can be made in wide sheets and is structurally stable. Plywood is manufactured from thin sheets (0.10″ to 1.25″) of wood laminated together with an adhesive, under high pressure. The number of layers (plies) varies from three to seven. The grain of each ply is laid perpendicular to the grain of each adjacent layer. The grain of both outside sheets always faces the same direction. This layering process greatly reduces the tendency of plywood to warp, check, split, splinter, and shrink. Plywood is available in individual 4′ × 8′ sheets or in continuous panels up to 50′ in length. It is made in thicknesses of 1/8″, 3/16″, 1/4″, 5/16″, 1/2″, 5/8″, 3/4″, 1″, 1-1/8″, and 1-1/4″.

All plywood is divided into two broad categories: exterior (waterproof) or interior. Plywood for structural use is made with surfaces of softwood. Plywood for cabinets and furniture has hardwood surfaces. Because of the different uses, construction-grade plywood (softwood) and veneer-grade plywood (hardwood) are identified by two different quality-rating systems.

Construction-grade plywood is unsanded. It is identified by grade levels based on structural strength, as shown in Fig. 24-11. Because of live-load differences, plywood panels used for structural purposes are marked with two numbers indicating the structural rating. See Fig. 24-12. The first number represents the maximum span (in inches) possible between supporting roof members. The second number represents the maximum allowable span when used for flooring.

Fig. 24-11 Plywood construction grades.

Grade	Description
Standard	For use as subflooring, roof sheathing, wall sheathing, and structural interior applications.
Structural Class I and II	For uses requiring resistance to tension, compression, and shear stress including box beams, stressed skin panels, and engineered diaphragms. High nail-holding quality and controlled grade and glue bonds.
CC Exterior	Meets all exterior plywood requirements.
BB Concrete-Form Panels, Class I and II	Edges sealed and oiled at the mill and used for concrete form panels.

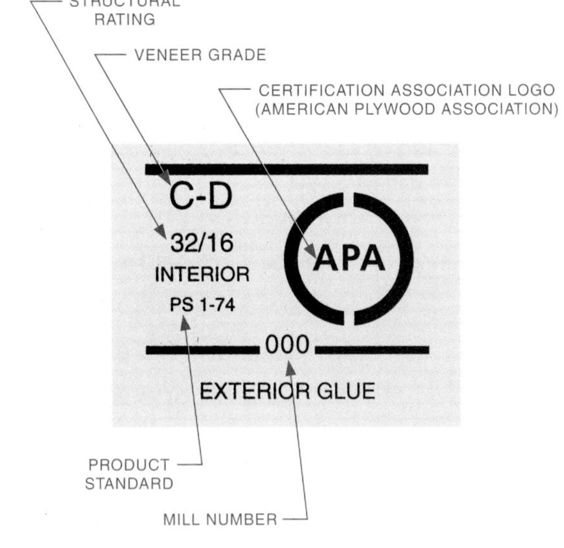

Fig. 24-12 ■ Construction plywood grade marks.

Since hardwood plywood is used for making cabinets and furniture, veneer plywood grades (groups) are classified by a letter indicating the number of knots, checks, stains, and open sections in each panel, as shown in Fig. 24-13. These letters are used to show only the group of the front and back plies (layers). On Fig. 24-14, notice the various kinds of information

Fig. 24-13 ■ Veneer plywood grades.

Grade	Description
Grade A	Paintable and smooth with no more than 18 neat boat, sled, or router type repairs made parallel with grain. Will accept natural finish.
Grade B	Solid surface with shims, circular repair plugs, or tight knots less than 1″ wide permitted.
Grade C	Tight knots of less than 1½″, knotholes less than 1″ wide, synthetic or wood repairs, limited splits, slices (gouges), discoloration, and sanding defects that do not impair strength permitted.
Grade C Plugged	Some broken grain, synthetic repairs, splits up to 1″ wide, knotholes and bareholds (other holes) up to ¼″ × ½″ permitted.
Grade D	Knots and knotholes up to 2½″ wide across grain or 3″ wide if within limits permitted but restricted to interior use.

Fig. 24-13 ■ Veneer plywood grades (groups). When front and back plies are different grades, letters are combined. For example, a grade of B-D means the front ply is B grade and the back ply is D grade.

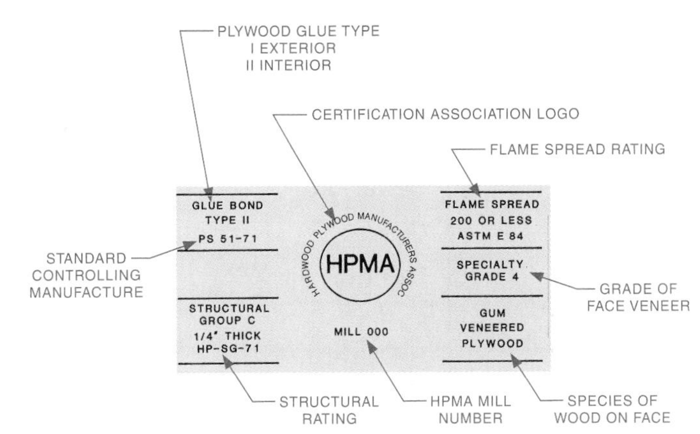

Fig. 24-14 ■ Typical hardwood plywood grade marks.

included in the product standard grade marks stamped on each sheet. Wood species are classified in groups. See Fig. 24-15.

When selecting hardwood plywood, several characteristics must be considered. Grain patterns, color and texture consistency, as well as specific species, smoothness, and finishability, must be matched for cabinets, paneling, and furniture.

Wood-Framing Methods

When multiple-level buildings are constructed using a wood skeleton frame, either the **platform (western) framing** method or the **balloon (eastern) framing** method is used. In platform framing, the second floor rests directly on first-floor exterior walls. See Fig. 24-16. In balloon-framed buildings, the first-floor joists rest directly on a sill plate, and the second floor joists bear on *ribbon (ledger) strips* set into the studs. The studs are continuous for the full height of the building, and floor joists butt against the sides of the studs. See Fig. 24-17.

Fig. 24-15 Wood species groups.

Group 1	Group 2	Group 3	Group 4	Group 5
Beech	Cedar, Port	Alder, Red	Aspen	Basswood
Birch	Cypress	Birch, Paper	Cedar	Poplar
Sweet	Douglas Fir 2[b]	Cedar, Alaska	Incense	Balsam
Yellow	Fir	Fir	Western Red	
Douglas Fir 1[a]	Balsam	Subalpine	Cottonwood	
Maple, Sugar	California Red	Hemlock	Pine	
Pine	White	Maple	Eastern	
Caribbean	Hemlock	Bigleaf	White	
Ocote	Lauan	Pine	Sugar	
Pine, South	Maple, Black	Jack		
Loblolly	Pine	Ponderosa		
Longleaf	Red	Spruce		
Shortleaf	Western White	Redwood		
Slash	Spruce	Spruce		
	Yellow Poplar			

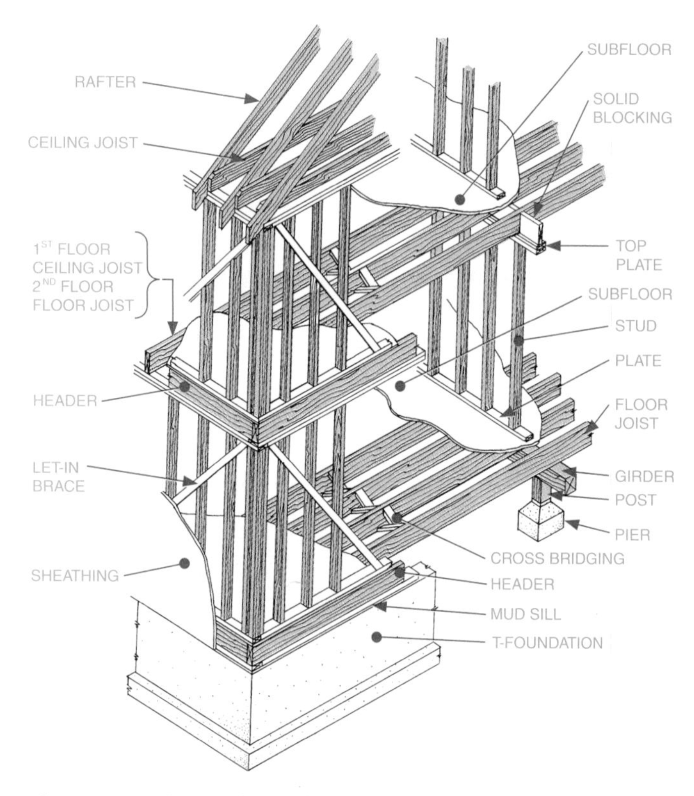

Fig. 24-16 ■ Platform framing.

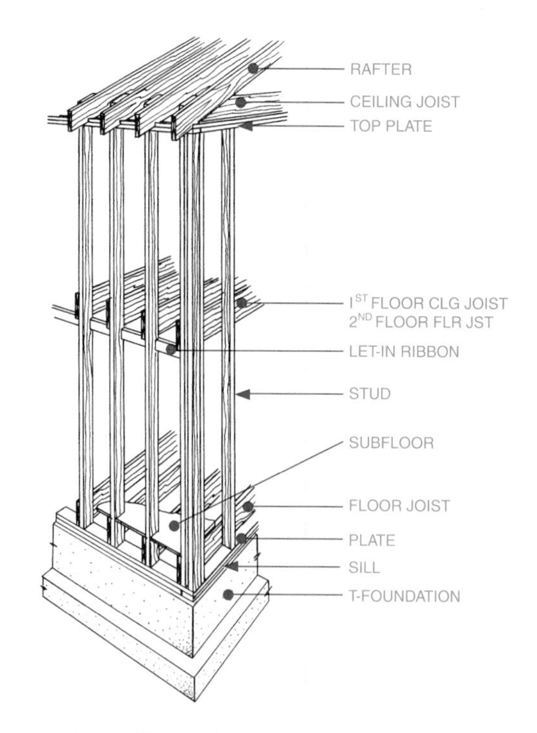

Fig. 24-17 ■ Balloon framing. This method is rarely used today.

Board-Foot Measure

Lumber is purchased in bulk by the board foot. One **board foot** is 1″ × 12″ × 12″. To determine the number of board feet in a given piece of lumber, multiply the thickness (in inches) × width (in inches) × length (in feet) and divide the result by 12:

Formula: $BF = \dfrac{T'' \times W'' \times L'}{12}$

Example: $BF = \dfrac{2'' \times 10'' \times 3'}{12} = 5\ BF$

2″
3′–0″
10″

Post-and-Beam Construction

Large timbers have been used in construction for centuries, usually for floor and roof systems in buildings of bearing wall design. The development of large glass sheets and sheathing materials as well as improvements in manufacturing and transporting large wood members have made new uses possible. Today heavy timbers may be used for walls as well as for floors and roofs. When heavy timbers are used in this manner, the construction method is called **post-and-beam construction.**

Post-and-beam construction uses larger structural members than skeleton-frame construction, and they are spaced farther apart. See Fig. 24-18. The larger spacing means that fewer members are needed.

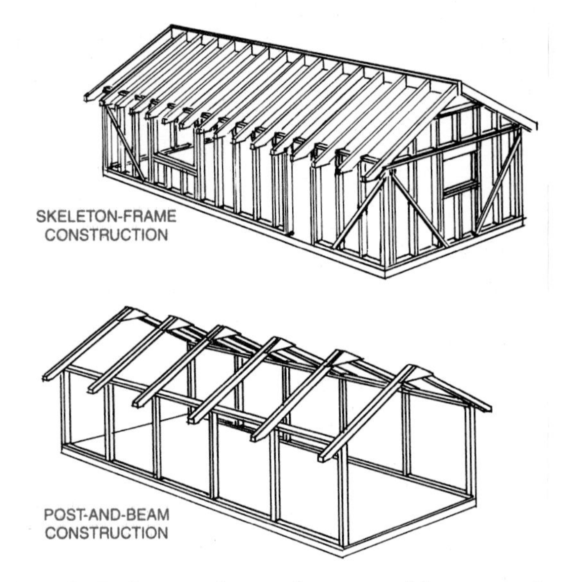

Fig. 24-18 ■ Comparison of post-and-beam and skeleton-frame construction.

There are also considerable savings in labor because fewer members are handled and fewer intersections connected.

Many post-and-beam members remain exposed in the finished building. To create a more pleasing visual effect, a better grade of lumber is specified. Rigid insulation must be used above, instead of under, the roof planks to expose the natural plank ceiling. Plumbing and electrical lines must be passed through cavities in columns and/or beams. Bearing partitions and other heavy dead loads, such as bathtubs, must be located over beams, or additional support framing must be used.

Post-and-beam construction relies upon the relationship of three basic components: *posts* or *columns, beams,* and *planks.* Vertical columns support horizontal beams. These beams support planks placed perpendicular to the beams, as in Fig. 24-19. Floor and roof systems are supported by the beams, which, as stated, are supported by posts or columns, which transfer loads to footings. Member sizes and spacing vary depending on load requirements.

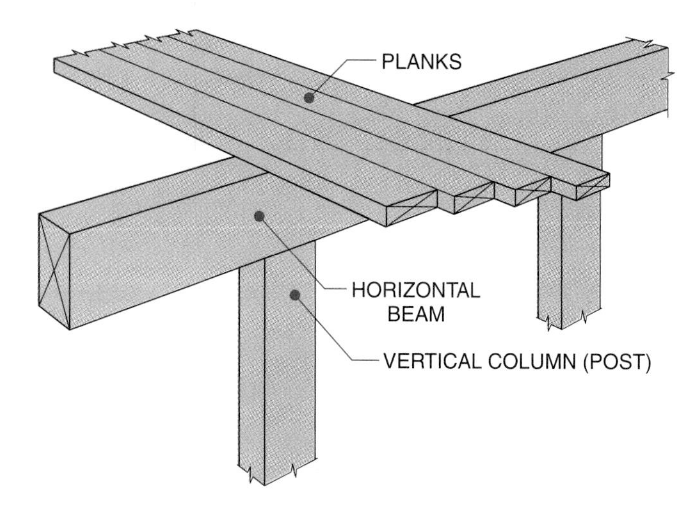

Fig. 24-19 ■ Basic components of heavy timber construction.

Floor Construction

Timber floor systems use heavy wood planks placed over widely spaced beams, as shown in Fig. 24-20. This type of floor system is called **plank-and-beam construction.** Each evenly spaced beam replaces several of the intermediate joists that would be used in a skeleton-frame floor system. For example, a 24′ distance may require 19 conventional floor joists spaced at 16″ OC (on center). See Fig. 24-21. However, only 7 beams placed 4′ OC may be needed to support the same loads. In this system, floor planks must be strong enough to avoid deflection at the middle or at midspan.

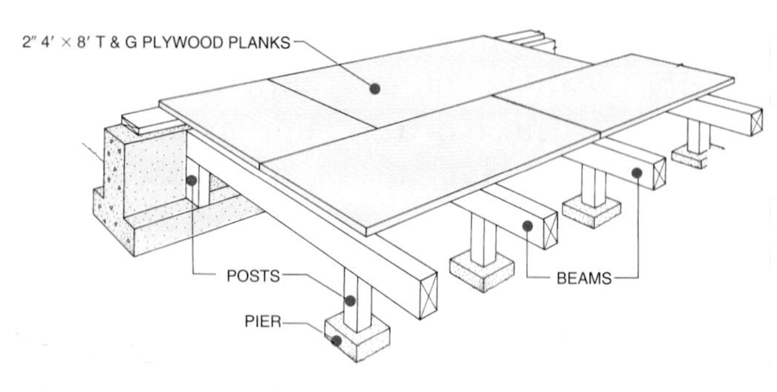

Fig. 24-20 ■ Plank-and-beam construction.

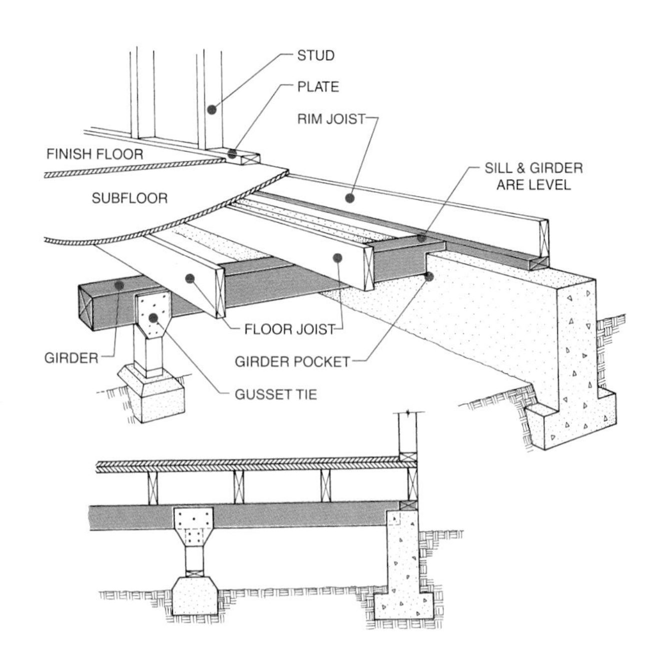

Fig. 24-21 ■ In this design, joists support the floor and girders support the joists. The girders, in turn, are supported by piers and posts in the center with girder pockets in the foundation for end support.

Wall Construction

Just as beams replace conventional joists in plank-and-beam floor systems, posts replace conventional studs in post-and-beam wall construction. The large open spans between the wall posts can be occupied by nonbearing material or components such as windows, doors, or insulating material. See Fig. 24-22. For this reason nonstructural elements in a post-and-beam outside wall are known as *curtain walls.* An example of the construction of wall column intersections is detailed in Fig. 24-23.

Roof Construction

In plank-and-beam roofs, beams replace conventional roof rafters and planks replace conventional roof sheathing. There are two types of plank-and-beam roof systems: longitudinal and transverse.

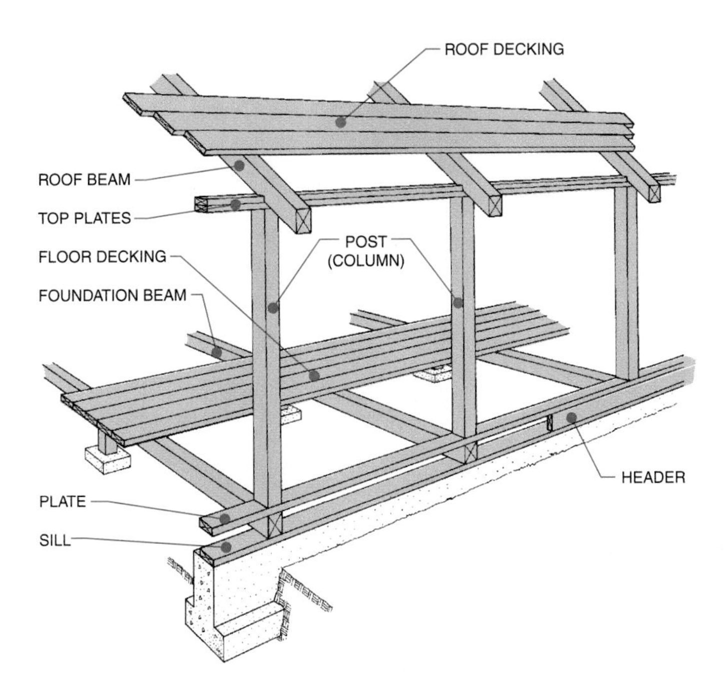

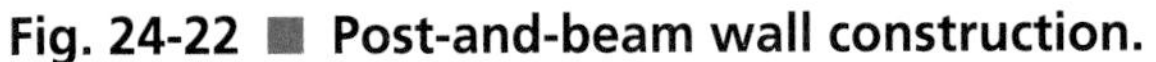
Fig. 24-22 ■ Post-and-beam wall construction.

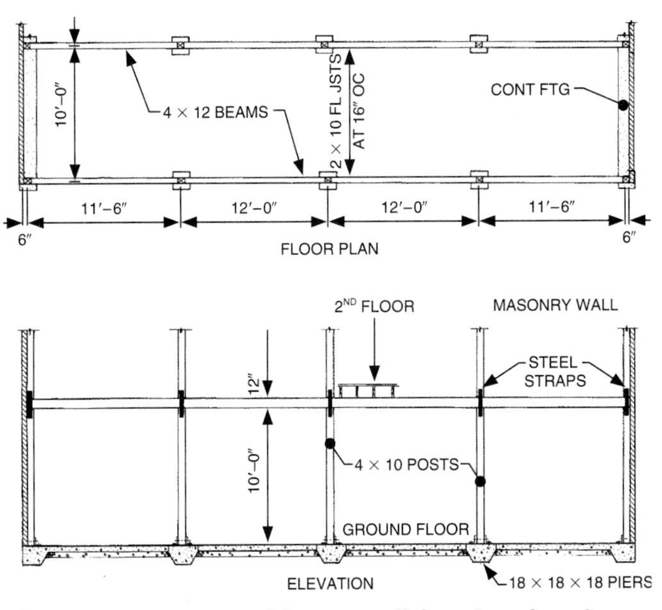

Fig. 24-23 ■ Post-and-beam wall framing drawings.

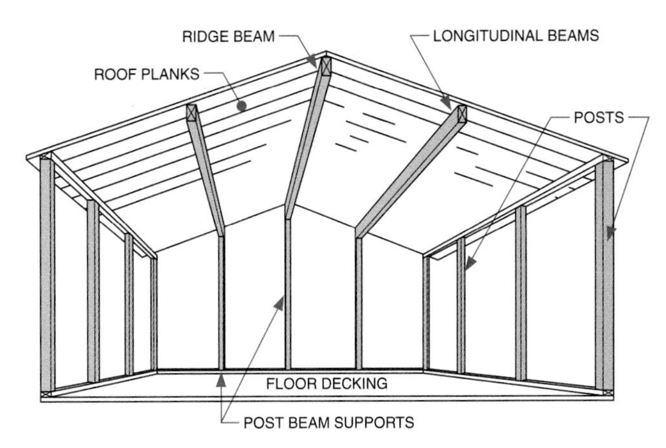

Fig. 24-24 ■ Longitudinal roof framing.

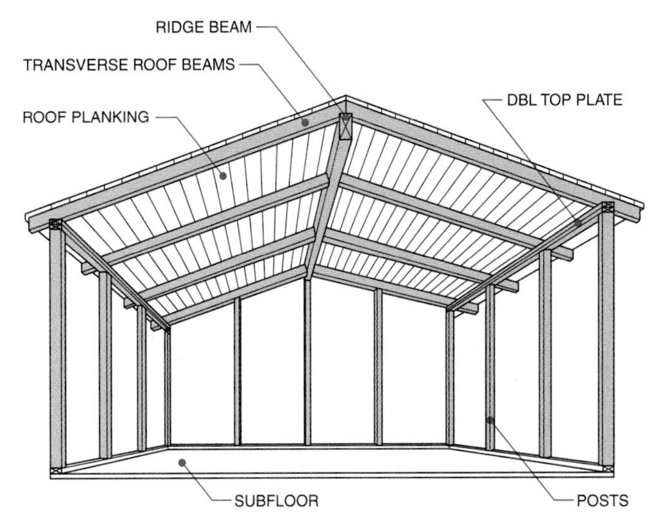

Fig. 24-25 ■ Transverse roof framing.

In longitudinal systems, roof beams are aligned parallel with the long axis of the building. See Fig. 24-24. In transverse systems, beams are aligned across the short width of the building as illustrated in Fig. 24-25. One end of the beam is supported by a post. The other end may rest on the top of a ridge beam, or it may be butted and fastened against the side of the ridge beam. Transverse beams either intersect a ridge beam on pitched roofs or lie flat across the span on flat roofs.

Structural Timber Members

There are three types of structural members used in contemporary post-and-beam construction: *solid*, *laminated*, and *fabricated* components.

SOLID MEMBERS Solid wood timbers are available in thicknesses which range from 3" to 12". However, the use of sizes over 8" is hampered by the tendency of large solid wood timbers to warp. One method of stabilizing larger solid wood members is to add steel plates to create *flitch beams* which are stronger and remain straighter. See Fig. 24-26.

Because planking is used in smaller widths (2″ to 4″), solid flooring is more commonly used than plywood panel flooring. Because of the impact of live load thrusts, solid planking is usually specified as tongue and groove (T & G). The T & G joint reduces deflection by tying the flooring planks together into one monolithic unit, as shown in Fig. 24-27.

LAMINATED MEMBERS When larger timbers are needed to support heavier weights or greater spans, glue-laminated timbers are often used. **Laminated timbers** are made from thin layers (less than 2″) of wood, glued together either vertically or in patterns. See Fig. 24-28. Laminated timbers are stronger than solid timbers because the grain direction is reversed (180°) in alternate layers. Because of its more consistent moisture content, there is less expansion in a laminated wood member than in solid wood. Glue-laminated members (or glulam for short) are manufactured in forms for columns, beams, and arches.

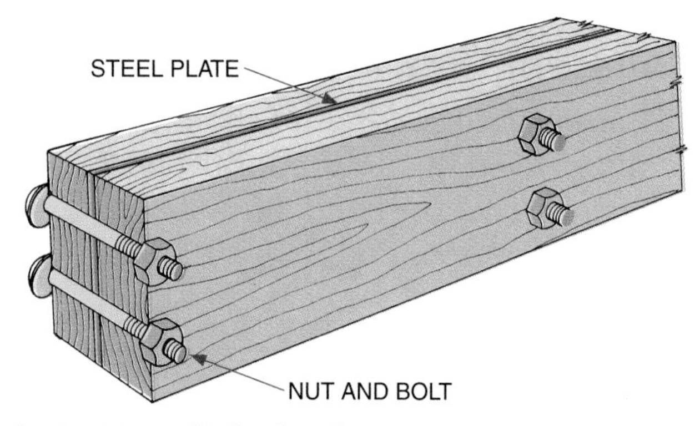

Fig. 24-26 ■ Flitch plate beam.

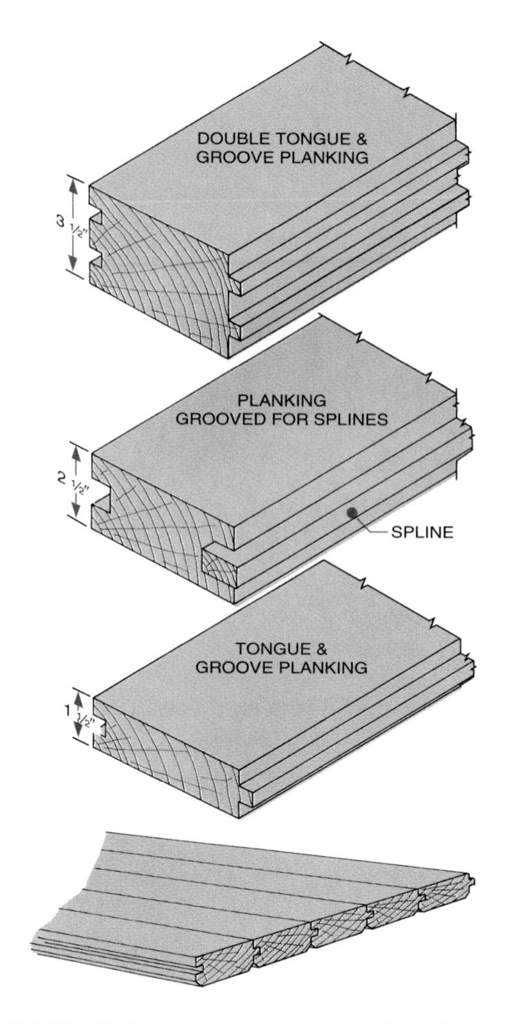

Fig. 24-27 ■ Tongue-and-groove planking.

In addition to the laminated members for posts and beams, laminated decking is also available in 2″ thicknesses and in 6″ to 12″ widths. Laminated decking is specified in nominal sizes on construction drawings. When decking material is laminated, the layers may be offset to fit together. Another method of lamination is to align the wood grain vertically and laminate the sides for greater resistance to loads. See Fig. 24-29.

Although lamination can create stronger, larger, and more structurally stable members, its most popular feature is its capability to be bent into a wide variety of structurally sound and aesthetically

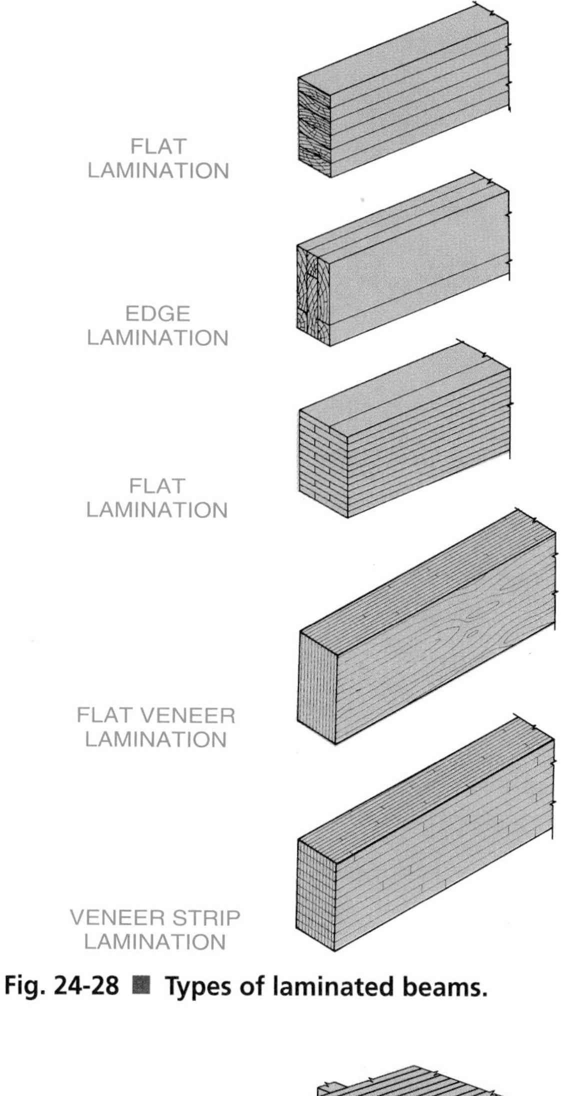

Fig. 24-28 ■ Types of laminated beams.

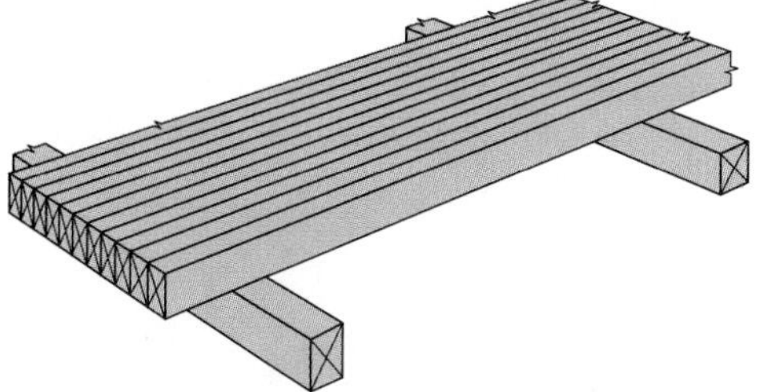

Fig. 24-29 ■ Side laminated decking.

pleasing shapes, as shown in Fig. 24-30. A variety of beam forms, including arches, are created by first bending thin, parallel layers of wood to a desired shape. Then the layers are glued and clamped together under pressure. When the glue dries, the member retains the new, bent form.

Fig. 24-30 ■ Laminated beam forms.

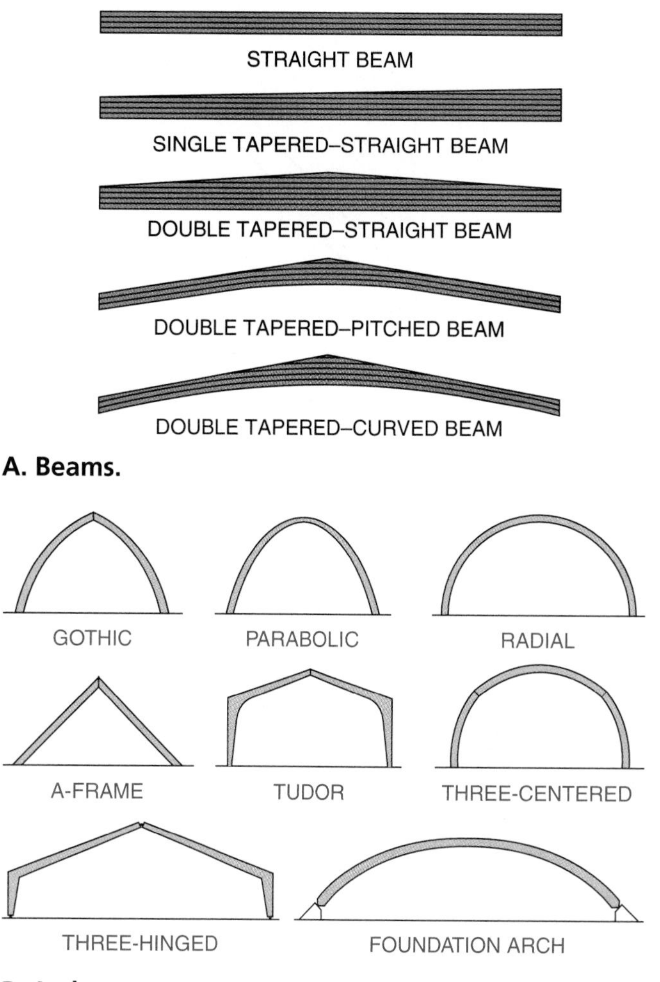

To indicate arches on architectural drawings, the base location of each arch is shown in the ground floor plan. The profile shape, including height and width dimensions, is drawn on elevation drawings and/or on elevation sectional drawings. See Fig. 24-31.

FABRICATED MEMBERS Many fabricated products and materials are used in place of solid or laminated wood members. These include a variety of wood I beams, truss joists, panels, box beams, strand lumber, and recycled plastic material.

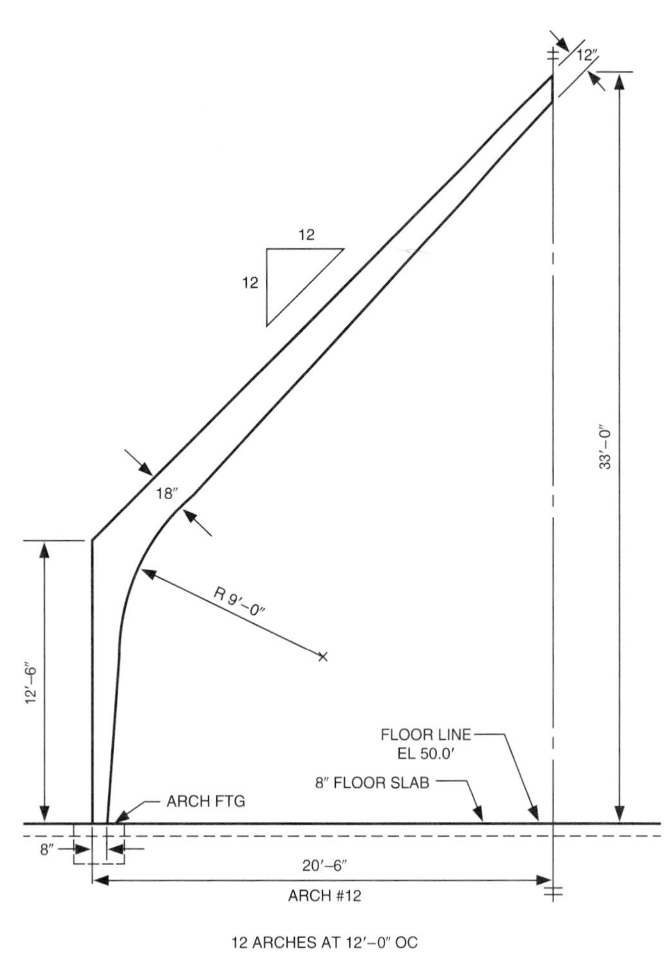

Fig. 24-31 ■ An arch shape is dimensioned in profile.

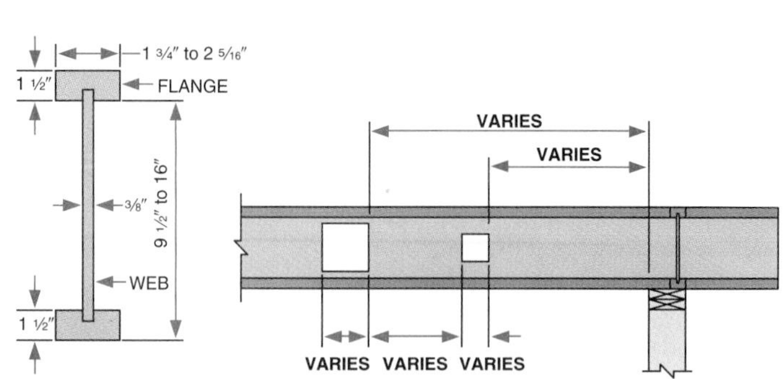

Fig. 24-32 ■ Special areas (knockouts) in an I beam, such as this TrusJoist®, can be removed without causing the member to lose strength. *TrusJoist MacMillan*

Fig. 24-33 ■ Truss joists.

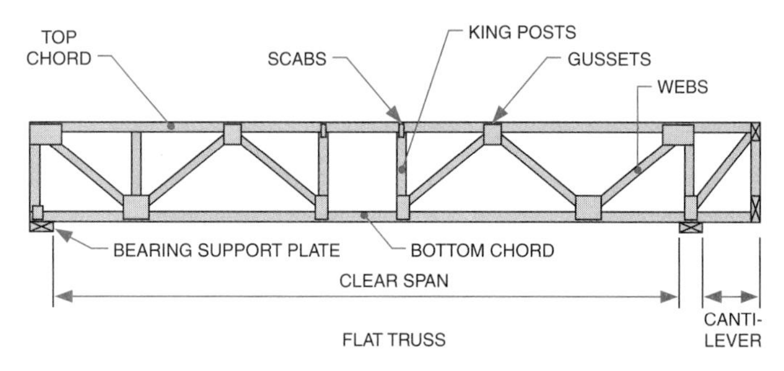

A. Parts of a truss joist.

B. Truss joists used as roof rafters.

Wood I beams, used for joists and rafters, are constructed with a plywood or strand board web which is inserted into a groove in a laminated or machine-stressed wood flange. This construction produces a very straight, lightweight, strong, and stable member. Knockout areas can be used to install HVAC, electrical, or plumbing lines without sacrificing strength. See Fig. 24-32. Wood I beams can span much longer distances than comparable size solid members.

Truss joists are constructed like a truss, but with parallel flanges usually made of 2 × 4s. See Fig. 24-33. These members can support greater loads than the same size solid lumber.

Stressed-skin or *sandwich panels* are often used in place of solid or laminated members. These lightweight prebuilt or site-built panels are made by gluing and/or nailing plywood sheets to structural member frames. Since panels can easily be constructed using standard plywood sizes, they are often used for floor, wall, and roof panels. Stressed-skin panels can also be used to make box beams for spans up to 120′ depending on load factors. Folded plate roofs and curved panels can also be either fabricated on the site or factory-built.

Foam sandwich panels are made of two plywood or strand board sheets adhered to a core of polyurethane or polystyrene. These panels function as framing, insulation, and sheathing and may sometimes be used as finished interior wallboard.

Strand lumber is made of cellulose fiber strips. The fibers from poor quality or small trees, rice, rye, wheat, or straw are crushed into long strands. After being combined with formaldehyde and adhesives, the strands are woven, compressed, and heat treated. The finished shapes are very strong and can span long distances. They are used primarily for columns, beams, and large headers.

Recycled thermoplastics are plastics that are ground, glued, and pressed into structural members. These thermoplastic members are capable of being sawed and glued, and they can accept and hold screws and nails. Contrary to wood, thermoplastic members will not rot, absorb moisture, expand, contract, or be infested by insects.

TIMBER CONNECTORS Because of heavy timber sizes, concentrated loads, and lateral thrust from winds and earthquakes, special joints and fasteners are required to attach post-and-beam members to the foundation and to other members. Nails are useful only as temporary holding devices, and lag screws can only be used in areas of limited stress. Figure 24-34 shows common joints used in post-and-beam construction.

Base anchors and plates are used to attach the base of heavy timber posts to the foundation and to prevent wood dete-

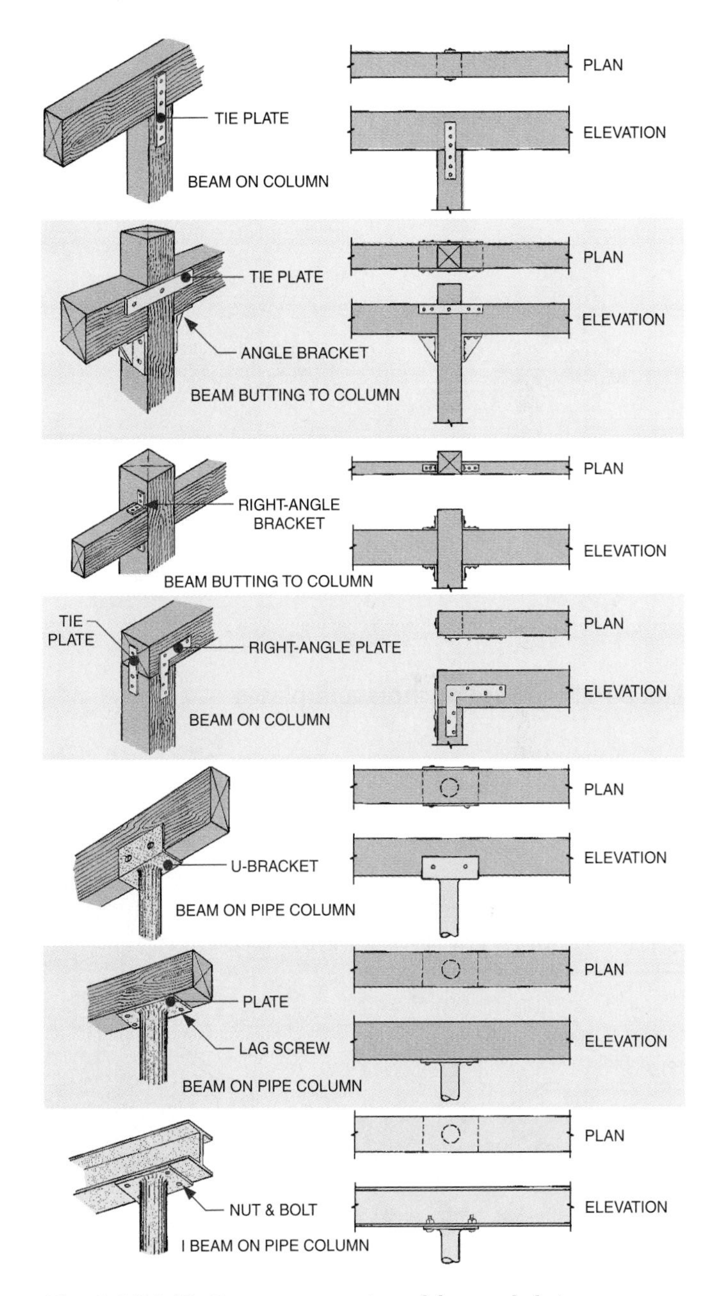

Fig. 24-34 ■ Common post-and-beam joints.

rioration. See Fig. 24-35. Timber brackets can be embedded into concrete as shown in Fig. 24-36, or timber can be placed into an impact-fastener base anchor. See Fig. 24-37.

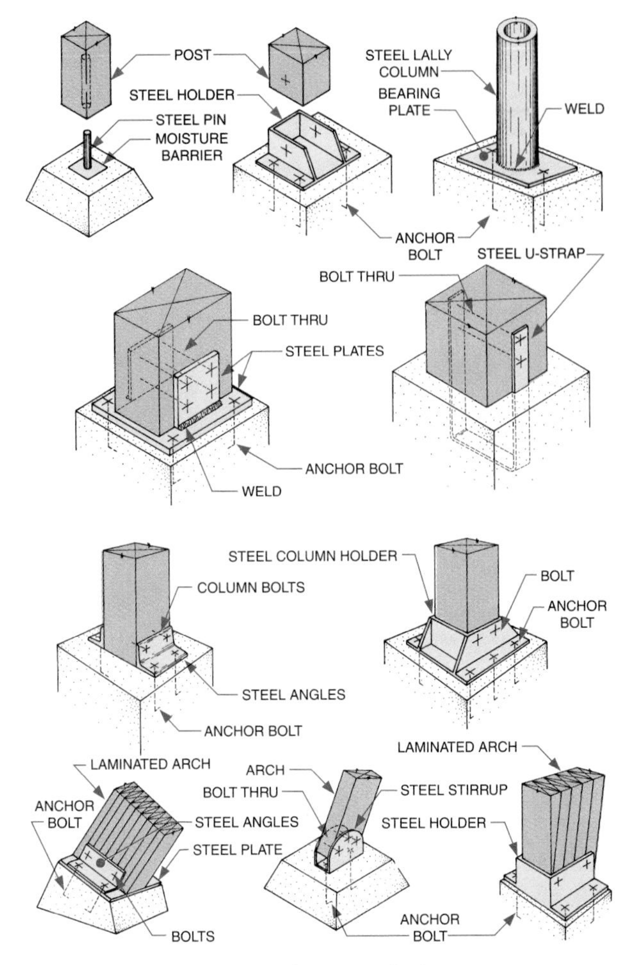

Fig. 24-35 ■ Base anchors and plates.

Fig. 24-37 ■ Base anchor for timber.

A. Before installation.

B. After installation.

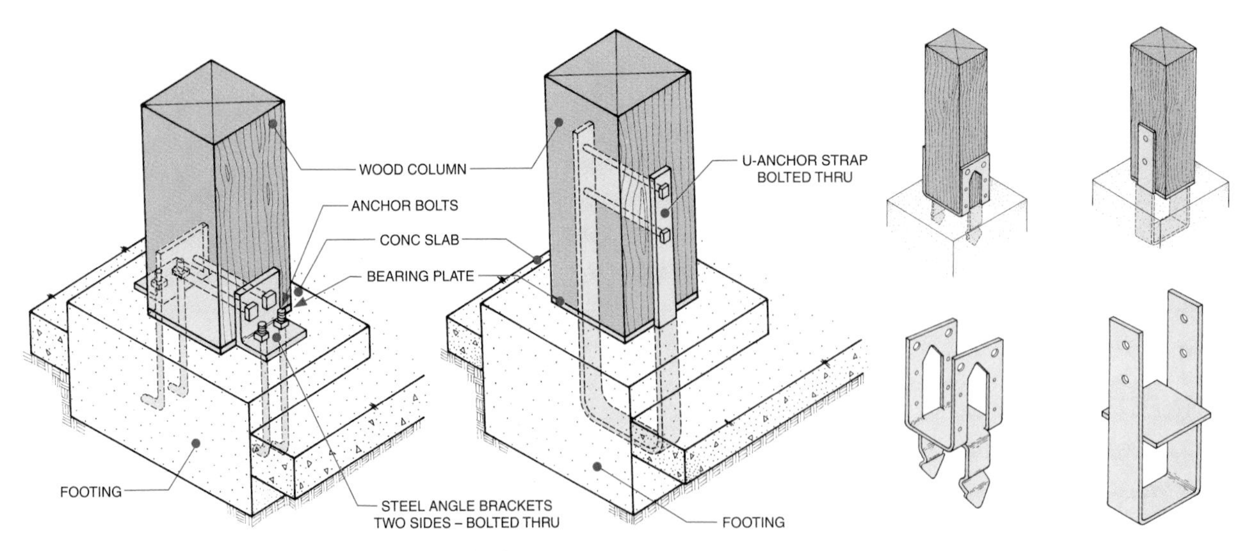

Fig. 24-36 ■ Timber brackets embedded in concrete.

Metal strap ties and gusset plates are used to attach posts to beams and to keep transverse roof beams aligned with ridge beams. Post-and-beam construction has excellent resistance to dead loads, which exert pressure directly downward. However, because of the large unsupported wall areas, lateral live loads can be a problem. Although diagonal ties or sheathing help control the lateral thrust, angle brackets should be used to fasten perpendicular intersections to prevent lateral movement between members and help provide rigidity to joints. Fig. 24-38.

Post caps are used extensively when posts intersect beams at a beam joint, as shown in Fig. 24-39. When the end of a member intersects the side of another member, without resting on it, metal hangers are usually specified. See Fig. 24-40. In some cases special truss clips may be used to prevent movement where two members intersect at angles other than 90°. See Fig. 24-41.

Because interior members in post-and-beam construction are often exposed, joints and fasteners may need to be

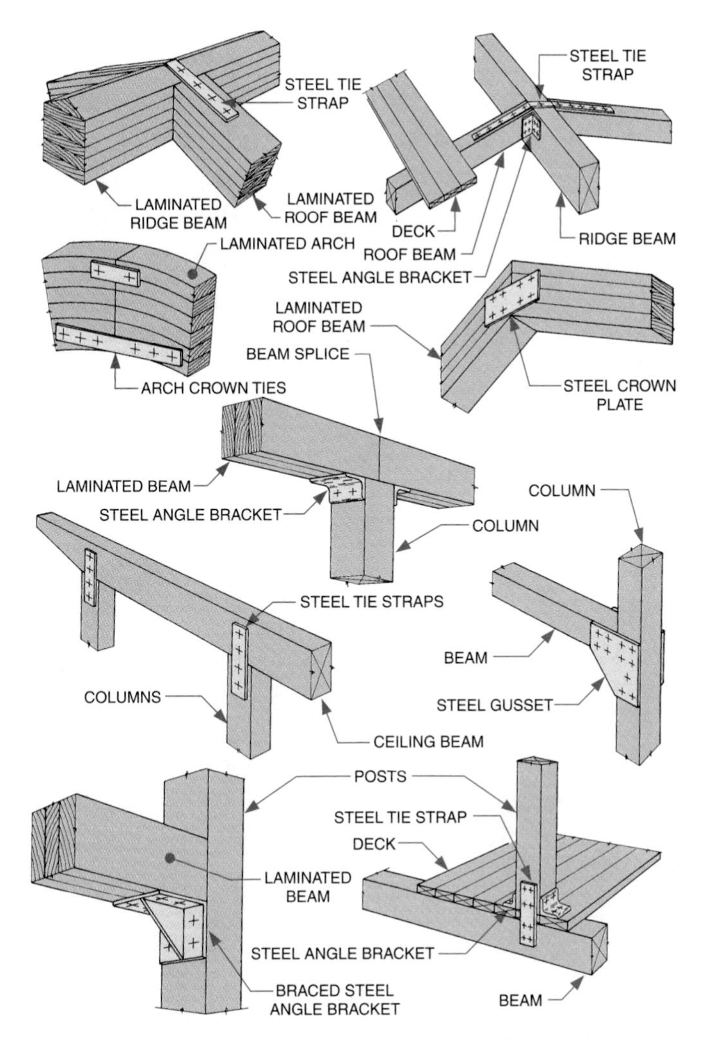

Fig. 24-38 ■ Straps, gussets, and angle brackets.

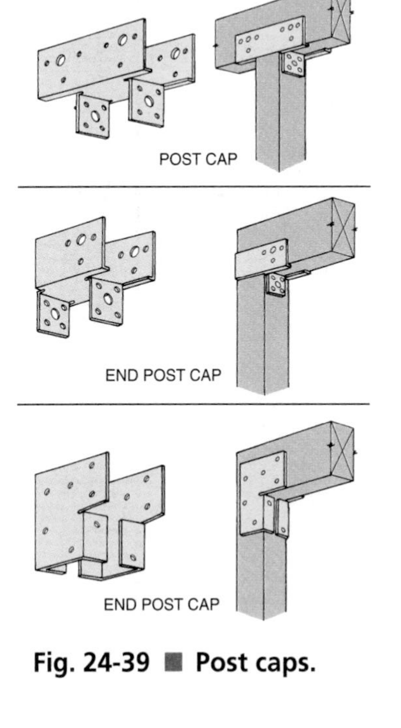

Fig. 24-39 ■ Post caps.

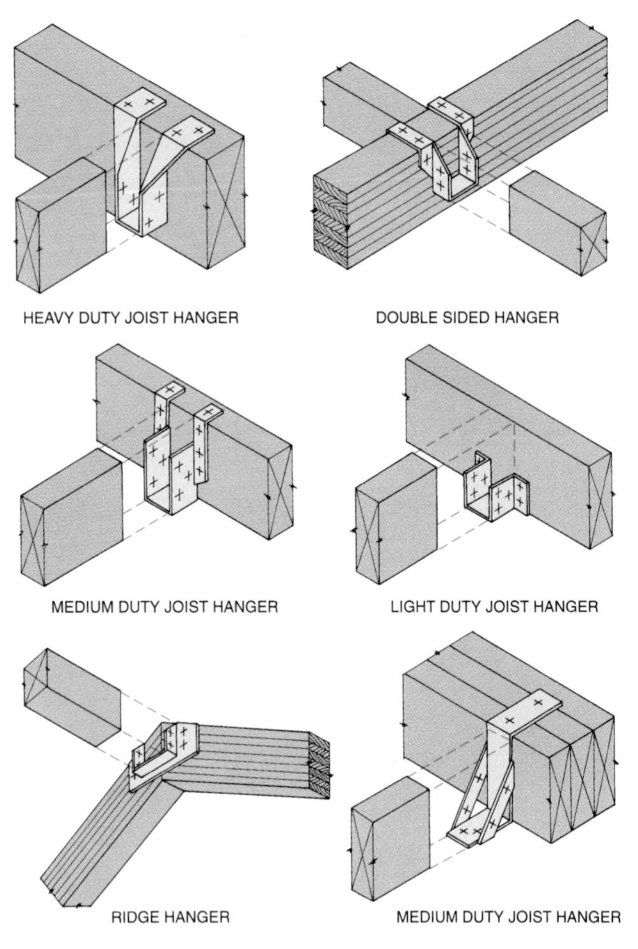

Fig. 24-40 ■ Hangers for joists and beams.

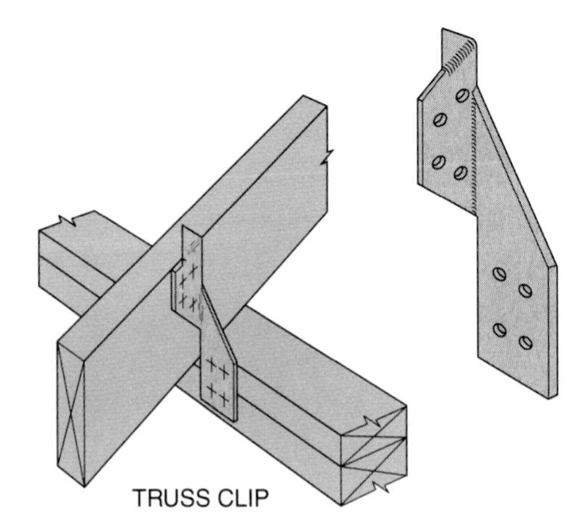

Fig. 24-41 ■ Truss clip.

hidden. Dowels, rods, and half-lap joints are often used for this purpose. However, these methods are costly to construct and install. An alternative is to use split-ring connectors which are extremely strong. They can be assembled with relative ease and the ring can be concealed with the use of dowels or bolts. When bolts are used, the ends can be counterbored and plugged once the final alignment is achieved. Figure 24-42 shows the types of mortise and tenon joints for exposed joints.

Fig. 24-42 ■ Types of mortise and tenon joists used in post-and-beam construction.

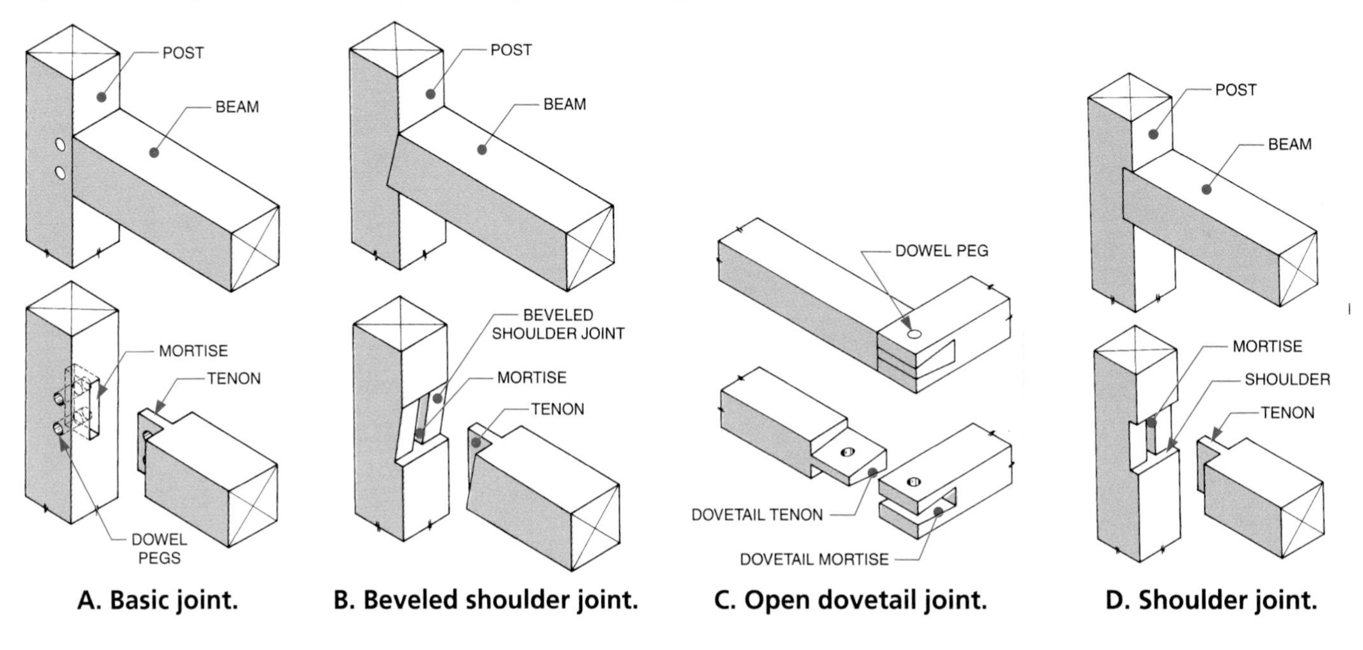

Fig. 24-42 ■ continued.

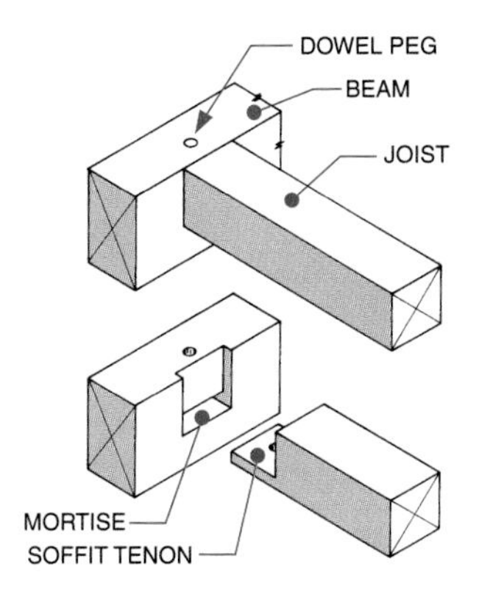

E. Soffit joint.

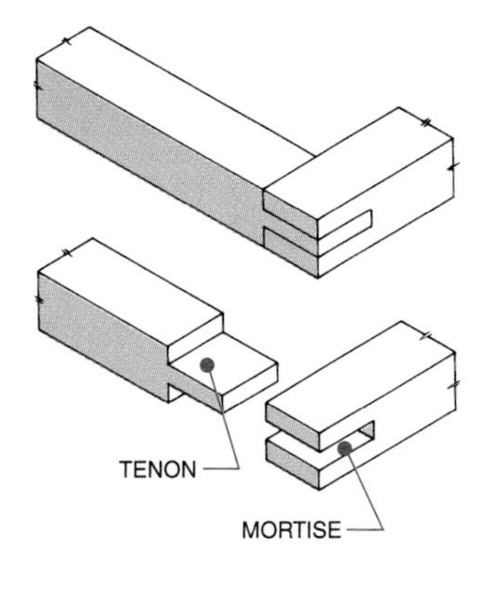

F. Open joint.

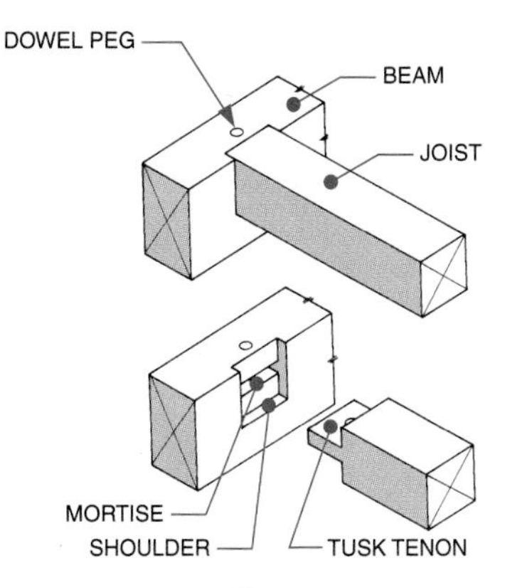

G. Tusk soffit joint.

CHAPTER 24 Exercises

1. Describe the major differences between skeleton-frame and post-and-beam construction.
2. Describe the differences between platform and balloon framing.
3. List six types of fabricated members used in light construction.

4. Select the lumber grade you will specify for the studs, rafters, sheathing, and joists for the house you are designing.
5. Draw an exterior wall section of the house you are designing.

6. Complete Exercise 5 using a CAD system.
7. Describe the difference between yard, structural, and factory lumber.
8. How many board feet are in 400 wood members 2″ × 6″ × 12′?
9. What is the dressed size of a 2 × 4 and a 4 × 6 wood member?
10. What is the range of plywood thicknesses?
11. Name three uses for hardwood and softwood in light construction.

CHAPTER

25

Masonry and Concrete Systems

Objectives

In this chapter, you will learn to:

- identify the types of masonry materials used in construction.
- describe four types of masonry walls.
- describe ways to strengthen concrete and prevent deflection.
- explain how concrete is used for slabs and other structural components.

Terms

aggregate
lally column
masonry bond
one-way slab system
post-tensioning
prestressing
pretensioning
two-way slab system

Masonry construction systems use brick, stone, concrete block, or clay tile products. Masonry units are arranged, usually row upon row, to form structures such as walls. *Concrete* construction systems use structural members made of poured or precast concrete.

■ **Effective use of masonry materials in construction can make a well-designed structure even more attractive.** *Alan Goldstein Photography*

Masonry Construction Systems

Construction systems that use masonry are usually combined with other systems, such as structural steel or skeleton-frame construction. Buildings usually are not constructed only with masonry materials because wood, steel, or reinforced concrete is needed for the large span floor and roof systems.

Masonry Materials

Masonry materials used for today's construction include a broad range of manufactured products. The many different types, sizes, shapes, and grades of masonry materials serve a variety of purposes.

BRICK Bricks are divided into two general categories: common brick and face brick. Color, texture, and dimensional tolerance are less consistent and critical for common brick than for face brick. Common brick is therefore less expensive and is generally used in unexposed construction areas. Common brick is graded according to structural characteristics. See Fig. 25-1.

Face brick is used in exposed areas that require dimensional accuracy and absorption control, as well as consistent color and texture. Face brick is therefore graded according to these characteristics. See Fig. 25-2. Many special types of face brick are available for specific construction needs; for example, glazed brick, fire brick, cored brick, and paving brick.

Bricks are also classified by their positioning in construction. See Fig. 25-3.

Most bricks are rectangular. Special shapes for sills, corners, and thresholds are also available or can be made to order. Bricks usually have holes, as shown in Fig. 25-4. Holes reduce the weight of

Grade	*Use*
SW	Used for maximum exposure to heavy snow, rain, and/or continuous freezing conditions.
MW	Used for average exposure to rain, snow, and moderate freezing conditions.
NW	Used for minimum exposure to rain, snow, and freezing conditions.

Fig. 25-1 ■ Grades of common brick.

Type	*Use*
FBX	Used where minimum size and color variations, and high mechanical standards are required.
FBS	Used where wide color variations and size variations are permissible or desired.
FBA	Used where wide variations in color, size, and texture are required or permissible.

Fig. 25-2 ■ Grades of face brick.

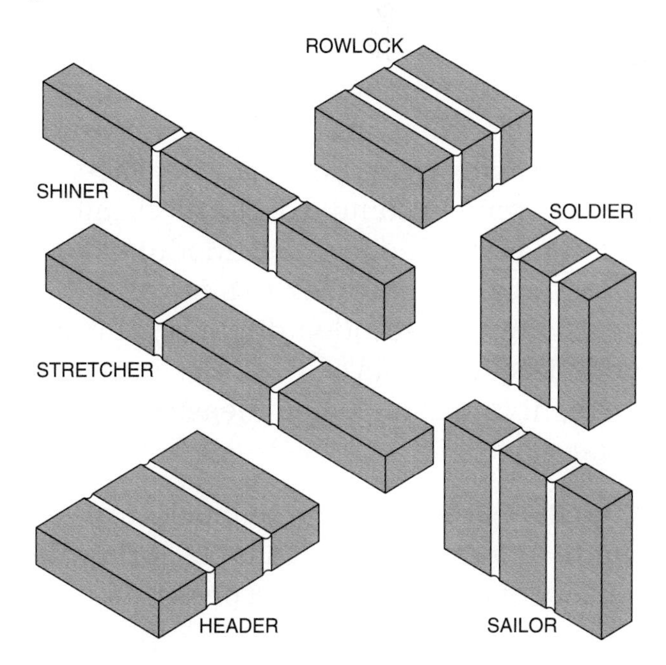

Fig. 25-3 ■ Brick classified by laid position.

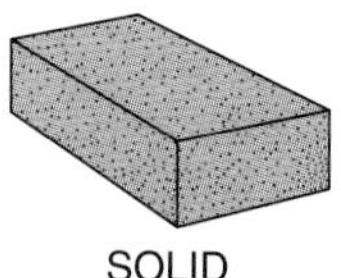

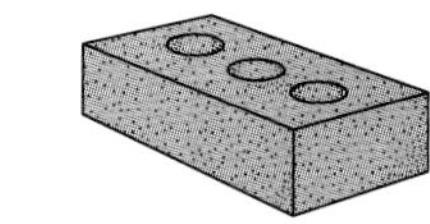

Fig. 25-4 ■ Bricks with holes are used most often in building construction because they weigh less than solid bricks.

Type	*Size*
Standard	$2\frac{1}{2}'' \times 3\frac{7}{8}'' \times 8\frac{1}{4}''$
	$2\frac{1}{4}'' \times 3\frac{3}{4}'' \times 8''$
Oversized	$3\frac{1}{4}'' \times 3\frac{1}{4}'' \times 10''$
Modular	$2\frac{1}{2}'' \times 3\frac{3}{4}'' \times 7\frac{3}{4}''$
($\frac{1}{4}''$ Joints)	$2\frac{5}{16}'' \times 3\frac{3}{4}'' \times 7\frac{3}{4}''$
	$2\frac{1}{2}'' \times 3\frac{3}{4}'' \times 11\frac{3}{4}''$
Modular	$2\frac{1}{4}'' \times 3\frac{1}{2}'' \times 7\frac{1}{2}''$
($\frac{1}{2}''$ Joints)	$2\frac{1}{4}'' \times 3\frac{1}{2}'' \times 11\frac{1}{2}''$
	$2\frac{1}{16}'' \times 3\frac{1}{2}'' \times 7\frac{1}{2}''$

Fig. 25-5 ■ Common brick sizes.

Type	*Size*
Standard	$2\frac{1}{2}'' \times 3\frac{1}{2}'' \times 11\frac{1}{2}''$
Norman	$2\frac{3}{16}'' \times 3\frac{1}{2}'' \times 11\frac{1}{2}''$ $2\frac{1}{4}'' \times 3'' \times 11\frac{11}{16}''$
Roman	$1\frac{1}{2}'' \times 3\frac{1}{2}'' \times 11\frac{1}{2}''$

Fig. 25-6 ■ Face brick sizes.

bricks and increase bonding. Masonry bonds are discussed later in this chapter.

Sizes differ among brick types. Common bricks come in standard, oversized, and modular sizes. See Fig. 25-5. Face bricks come in standard, Norman, and Roman sizes. See Fig. 25-6. Modular bricks are standardized to align on 4″ grids after mortar joint dimensions are added. Increments of 4″, 8″, 12″, and so forth, fit into established measured spaces.

CONCRETE BLOCK Concrete blocks are made in many different shapes for a wide variety of construction purposes, as shown in Fig. 25-7. Concrete blocks are either solid, hollow-core, or split-face for exposed surfaces. The weight, texture, and color of each block are determined by the types of aggregate used. **Aggregate** is a combination of sand and crushed rocks, slate, slag, or shale. When concrete masonry does not need to support heavy loads, lightweight masonry blocks are ideal. Lightweight blocks are molded by adding fly ash cinders to concrete.

Similar to modular bricks, concrete block is manufactured in modular sizes. That means the actual size of each block is ⅜; smaller than the space to be filled. The difference allows for the thickness of the mortar joint. For example, the dimensions of an 8″ × 8″ × 16″ concrete block are actually 7 ⅝″ × 7 ⅝″ × 15 ⅝″. Because their sizes are standardized, modular blocks can be used in conjunction with modular bricks. See Fig. 25-8.

STONE For centuries, natural stones were used as a major structural material. Today stone is primarily used decoratively except for landscape construction. Stone masonry is classified by the type of material, shape of cut, finish, and laying pattern. The most common types of stone used in construction are sandstone, limestone, granite, slate, and marble. These can be cut and arranged into a variety of patterns.

STRUCTURAL CLAY TILE Hollow-core structural tile units are larger than bricks and can be either load-bearing or non-load-bearing. Structural tiles are used for partitions, fireproofing, surfacing, or furring. See Fig. 25-9. Load-bearing tile is graded according to structural characteristics: LBX for tile exposed to weathering and LB for tile not exposed to weathering or frost.

Fig. 25-7 ■ Concrete block shapes and sizes.

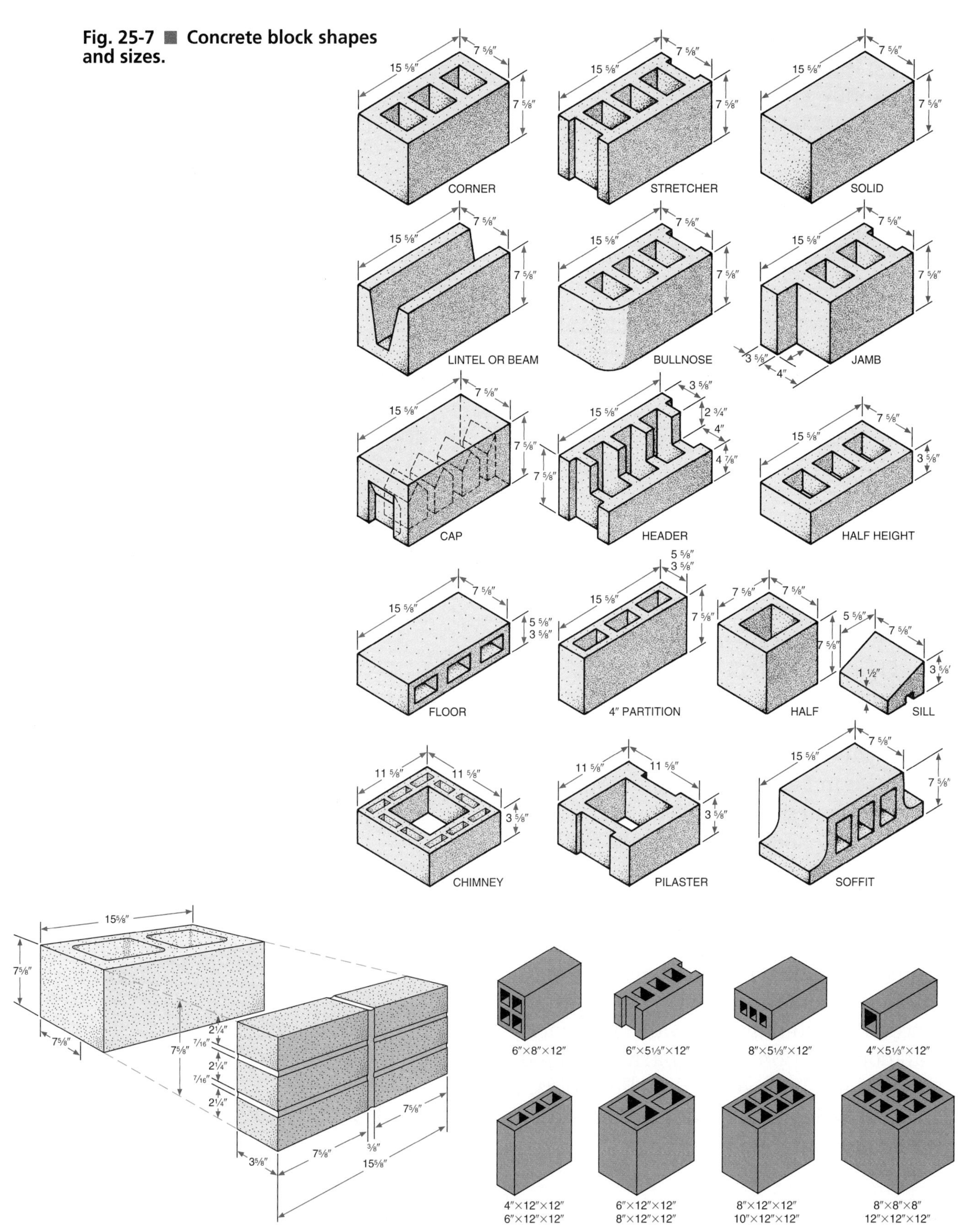

Fig. 25-8 ■ Modular concrete block and brick dimensions.

Fig. 25-9 ■ Structural tile is available in a variety of sizes and shapes.

Non-load-bearing facing tile is graded by surface texture, its stain resistance, color consistency, and dimensional accuracy: FTX for high quality, FTS for low quality.

Masonry Walls

Four basic types of masonry wall construction are solid, cavity, facing, and veneer.

SOLID MASONRY WALLS Most masonry bearing-wall construction is solid. Solid masonry construction can utilize almost any masonry material if it is laid flat to support loads. However, the material used must be able to withstand the loads involved.

Concrete block is commonly used for solid load-bearing walls. For heavy loads and/or high walls, steel reinforcing rods (rebars) are added and the block cells are filled with concrete for structural stability. See Fig. 25-10. When solid masonry walls are constructed with combinations of materials (such as concrete block and brick), steel reinforcement is mandatory, as shown in Fig. 25-11. Reinforcement between courses or between materials is particularly necessary for walls subject to earthquakes, heavy storms, wind, and lateral earth loads. See Fig. 25-12.

MASONRY-CAVITY WALLS To reduce dead loads and to improve temperature and humidity insulation, cavity walls are preferred to solid masonry walls. In cavity wall construction, two separate and parallel walls are built several inches apart. A structural tie, usually metal, bonds the walls together. See Fig. 25-13.

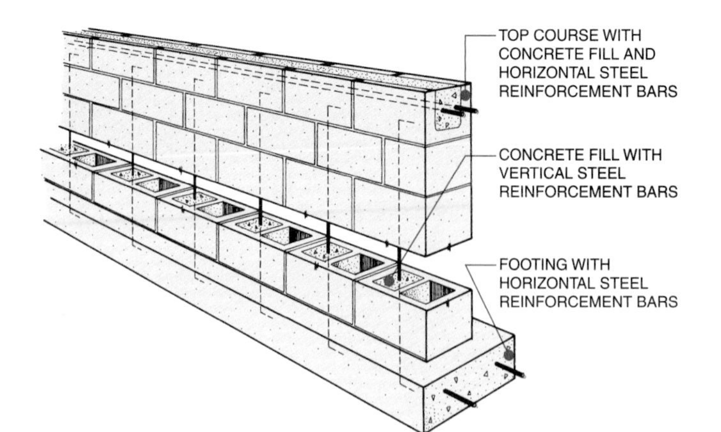

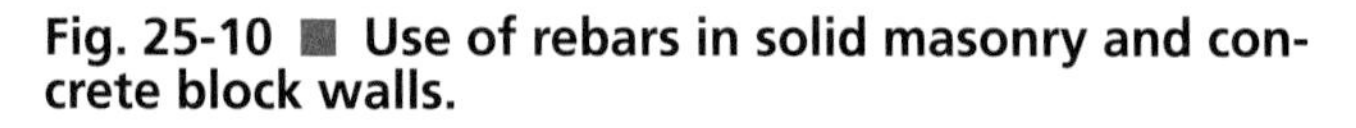

Fig. 25-10 ■ Use of rebars in solid masonry and concrete block walls.

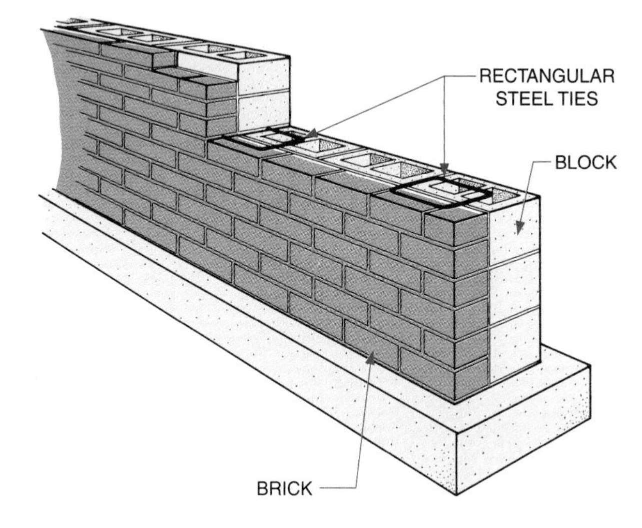

Fig. 25-11 ■ Steel ties used as reinforcement between different kinds of masonry walls.

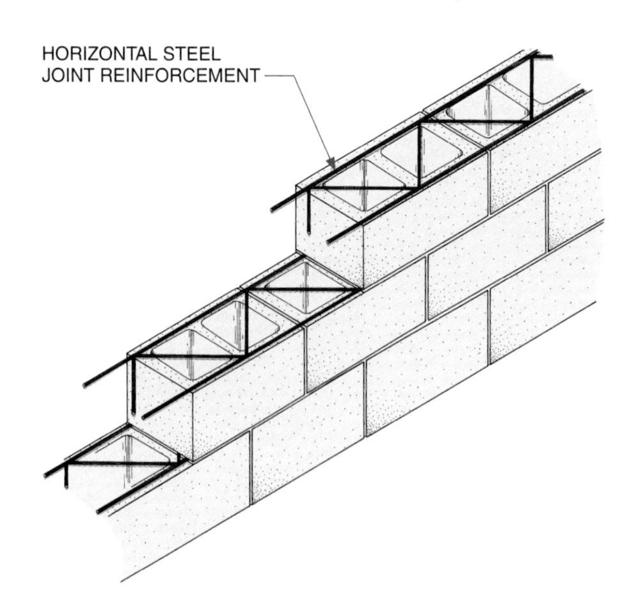

Fig. 25-12 ■ Reinforcement between courses of masonry.

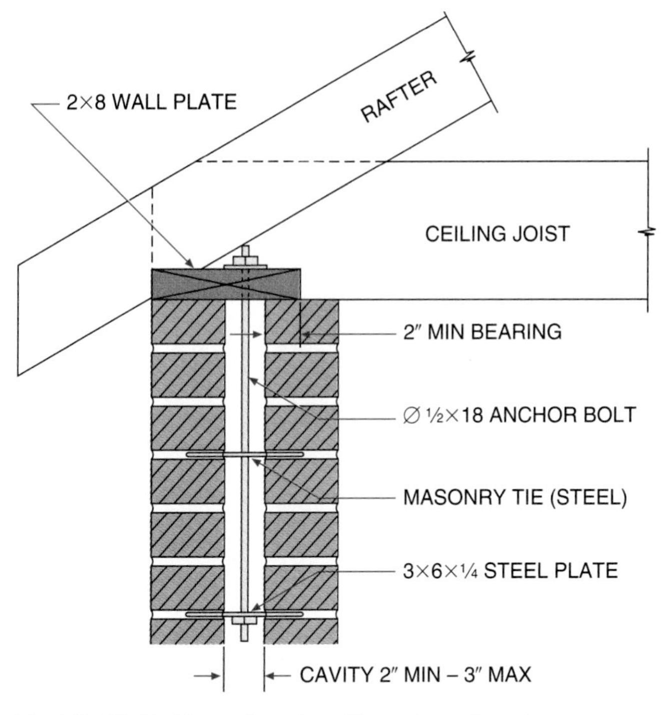

Fig. 25-13 ■ Elevation detail section showing a masonry-cavity wall.

MASONRY-FACED WALLS Walls are often faced with different masonry materials. Any type masonry wall can be faced with another facing material. For example, a faced wall may consist of common bricks faced with structural tile or concrete block faced with brick. Regardless of the material, the two walls are always bonded so that they become one wall structurally. The bonding material can be metal ties as shown in Fig. 25-14, steel reinforcing rods, or masonry units laid on end to intersect the opposite wall. Always remember that walls with different coefficients of expansion are never faced together. Their differing rates of expansion and contraction under extreme temperature-change conditions can cause cracks and damage the structural integrity of the wall.

MASONRY-VENEER WALLS Veneer walls, like faced masonry walls, include two separate walls constructed side by side. Unlike faced walls, the veneer wall is not tied to

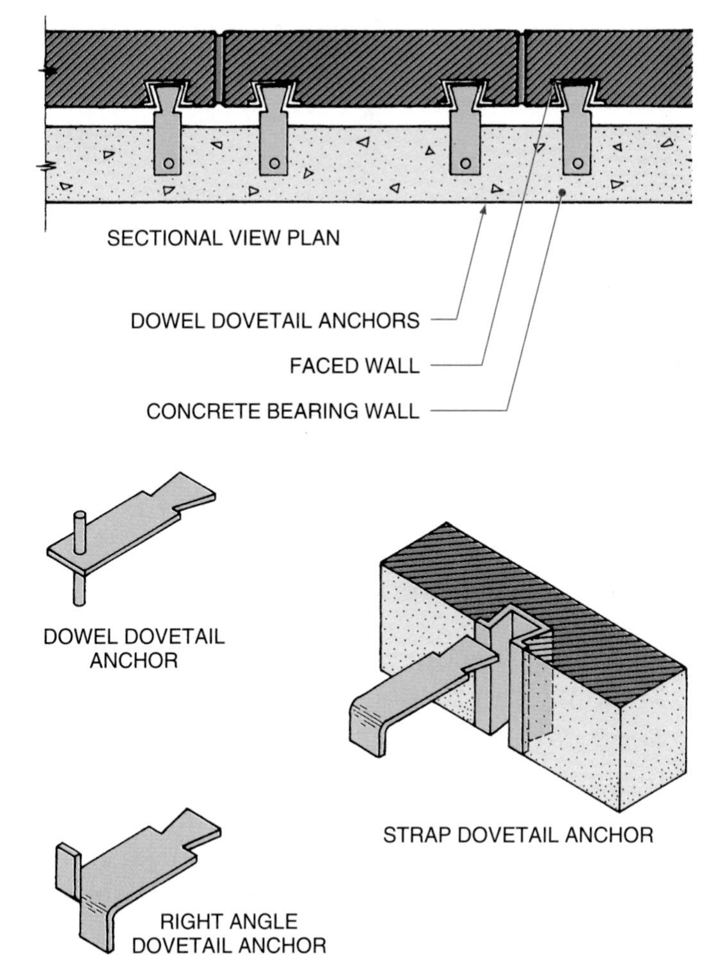

Fig. 25-14 ■ Use of metal ties to bond a masonry-faced wall.

the other wall to form a single structural unit. The veneer wall is simply a non-load-bearing decorative facade, although the two walls may be connected with masonry ties or adhesives. See Fig. 25-15.

A veneer wall may include two different masonry materials or include a skeleton-frame wall veneered with a masonry material. In the latter case, the space between the wood and masonry walls (usually 1″) may remain empty or may be filled with insulation, depending on climactic conditions. A wall detail or sectional drawing is usually prepared to provide this information.

MASONRY BONDS A **masonry bond** for walls is the pattern of arranging and

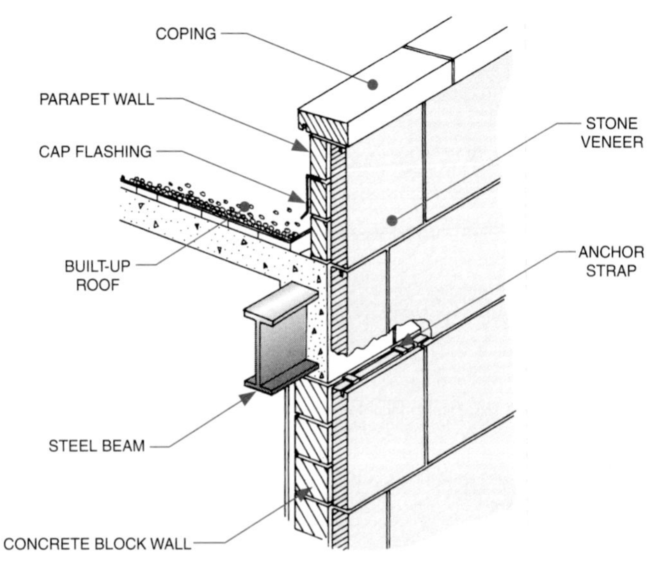

Fig. 25-15 ■ Masonry-veneer wall.

attaching masonry units in courses (rows). Masonry can be placed in a variety of bond patterns, as shown in Fig. 25-16. Different patterns can make the same size, shape, and material appear completely different. Various types of mortar joints are specified on construction drawings. See Fig. 25-17.

Concrete Construction Systems

Early Romans crushed and processed rocks to create cement for bonding their structures. Today, cement continues to be

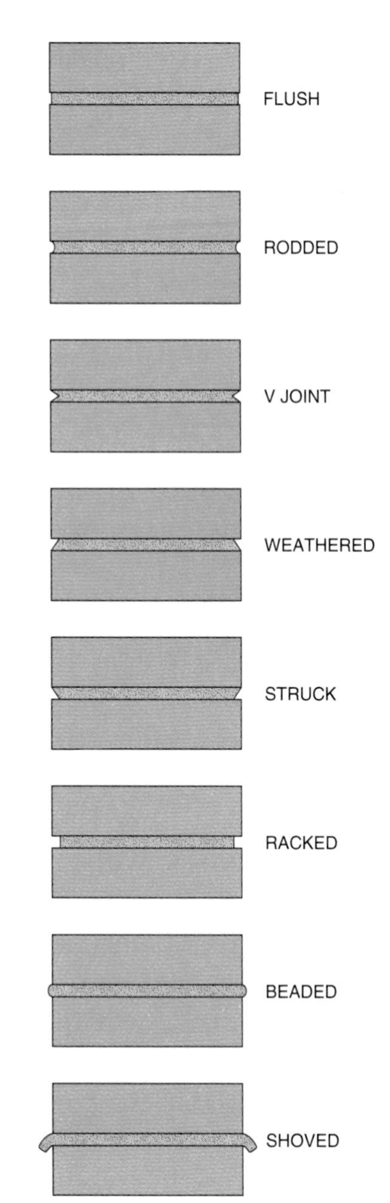

Fig. 25-17 ■ Types of mortar joints.

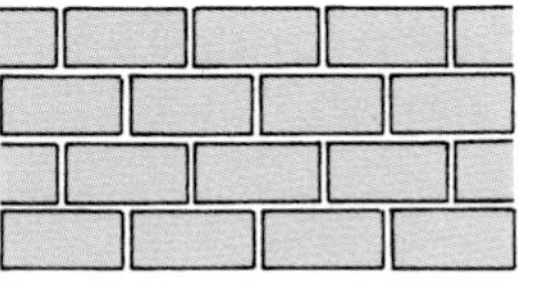

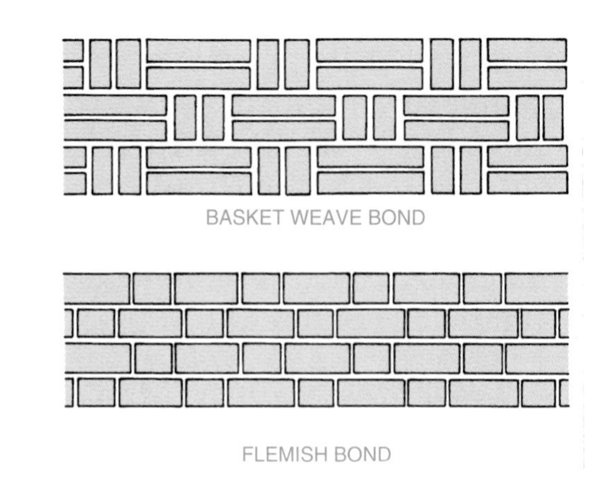

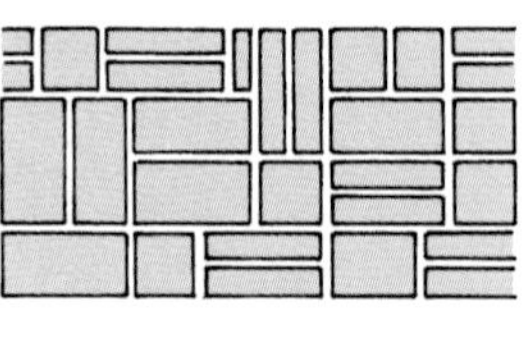

Fig. 25-16 ■ Common types of masonry bonds.

an important material in construction. Cement is manufactured primarily from clay and limestone. *Concrete*, so widely used in today's construction, is basically made from cement combined with water and aggregate (such as sand and gravel). Other ingredients may be added to improve the strength or workability of the concrete. See Fig. 25-18.

Different types of concrete are identified by their compressive strength, measured in pounds per square inch (PSI). Concrete strengths range from 2500 PSI to 4000 PSI for most residences and up to 8000 PSI for large industrial buildings. Concrete volume is measured in cubic yards.

Concrete can either be cast in place on-site or precast off-site and shipped to the site as finished girders, beams, slabs, columns, or other components. Either way, concrete may be reinforced, prestressed, or poured plain depending on the construction application and/or site conditions.

Reinforced Concrete

Because concrete is weak in tensile strength but has a strong resistance to compression stress, it was previously used only for nontension applications such as ground-level slabs, walks, or roadways. However, when materials with high tensile strength are added to the concrete, the tensile strength of the concrete is greatly increased, and its compression strength is doubled. Improvements in the reinforcing of concrete are mainly responsible for the increased use of concrete in all types of building construction.

REINFORCEMENT MATERIALS Reinforcement bars (rebars) and welded wire mesh provide the steel that converts concrete into reinforced concrete. The use of rebars and welded wire mesh was described in relation to foundations and slabs in Chapter 23.

Welded wire mesh (WWM) is designed to evenly distribute stress forces and prevent cracking. The spacing of the longitudinal and transverse wires and the size of each are noted in Figs. 25-19 and 25-20.

The use of rebars for reinforcing slabs was covered in Chapter 23. Rebars are also used in the formation of concrete beams, columns, walls, and suspended decks. See Fig. 25-21. A marking system

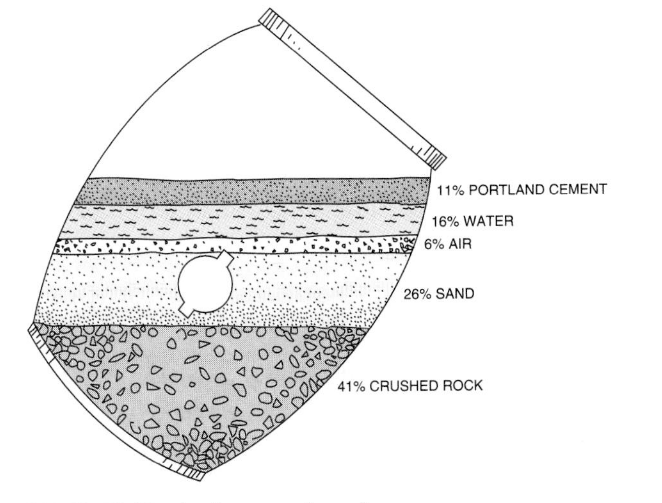

Fig. 25-18 ■ Typical concrete mix.

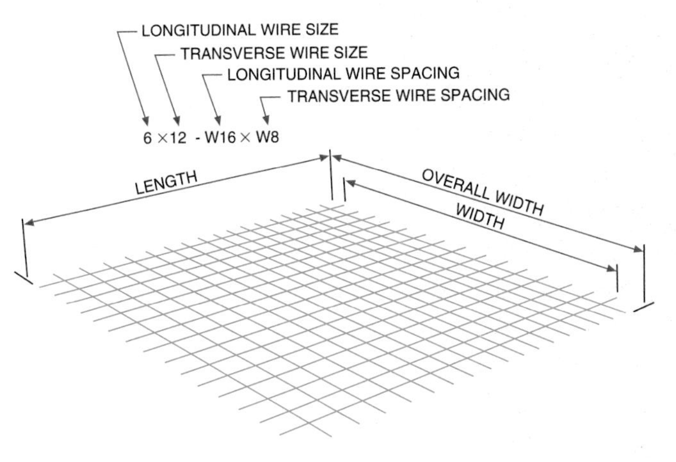

Fig. 25-19 ■ Method of designating welded wire mesh specifications.

Fig. 25-20 Welded wire mesh sizes and spacing.

∅ in inches		Gauge number	Actual size of wire	Woven wire mesh spacing
Decimal	Fraction			
.2437	1/4″	3	●	–
.2253	7/32″	4	●	–
.2070	13/64″	5	●	–
.1920	3/16″	6	●	2 1/2″
.1770	11/64″	7	●	2 1/4″
.1620	5/32″+	8	●	2″
.1483	5/32″–	9	●	1 3/4″
.1350	9/64″	10	●	1 1/2″
.1205	1/8″	11	●	1 1/4″
.1055	7/64″	12	●	1″
.0915	3/32″	13	●	–
.0800	5/64″	14	●	3/4″
.0625	1/16″	16	●	3/8″ or 1/2″

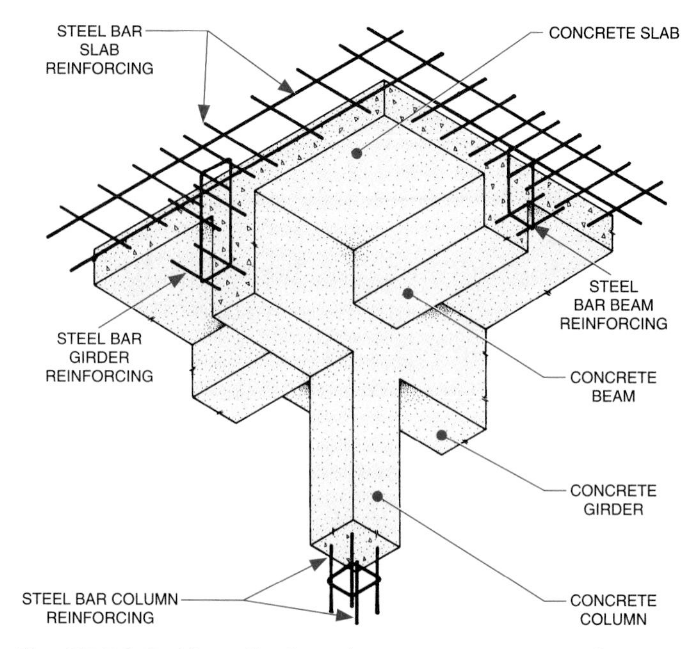

Fig. 25-21 ■ Use of rebars in concrete structural members.

identifies rebar manufacture, size, steel type, and PSI grade, as shown in Fig. 25-22.

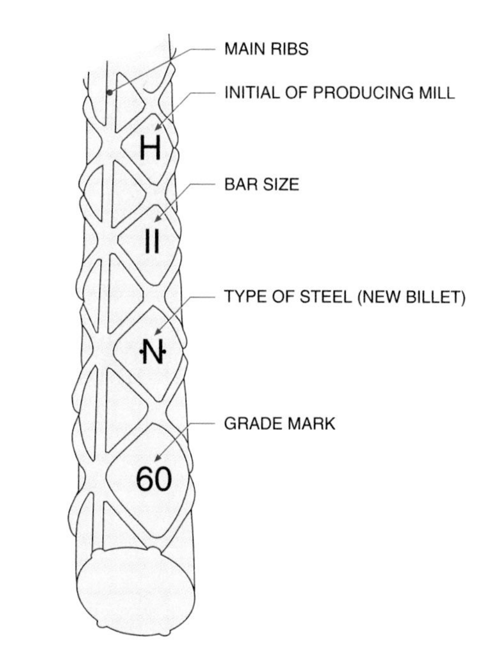

Fig. 25-22 ■ Rebar identification marking system.

Prestressed Concrete

When loads are added to concrete members, some deflection (sag) occurs in the center of the member. This happens to all materials under load. However, since concrete has very low tensile strength, excessive deflection can result in tension cracking or complete member failure. This is caused by compression of the upper side and tensioning (stretching) of the lower side. See Fig. 25-23. To counteract these unstable compression and tension stresses, concrete is often prestressed.

Prestressing is a method of compressing concrete so that both the upper and lower sides of a member remain in compression during loading. Prestressing can be accomplished either by pretensioning or post-tensioning.

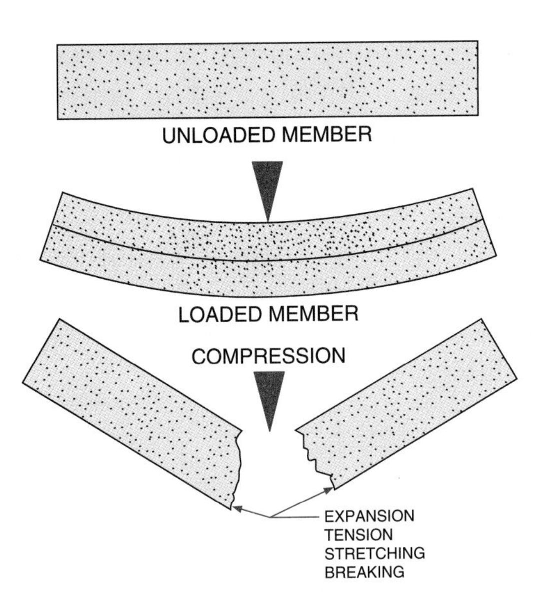

Fig. 25-23 ■ Effect of loading on concrete.

PRETENSIONING In **pretensioning,** deformed steel bars called tendons are stretched (tensioned) between anchors and the concrete is poured around the bars. Once the concrete has cured, the tension is released and the bars attempt to return to their original, shorter length. However, the concrete that has hardened around the bar grooves holds the deformed bars at nearly their stretched length. This creates a continual state of compressive stress that can be compared to pressing a row of blocks together. See Fig. 25-24.

As a further aid to prevent bending, concrete members are prestressed by draping tendons near the bottom of the member, as shown in Fig. 25-25. This bottom tension buckles the member upward so that when the anticipated loads are added, the beam straightens to a level position.

POST-TENSIONING **Post-tensioning** is done after concrete has cured. In post-tensioning, tendons are either placed inside tubes embedded in the concrete or the

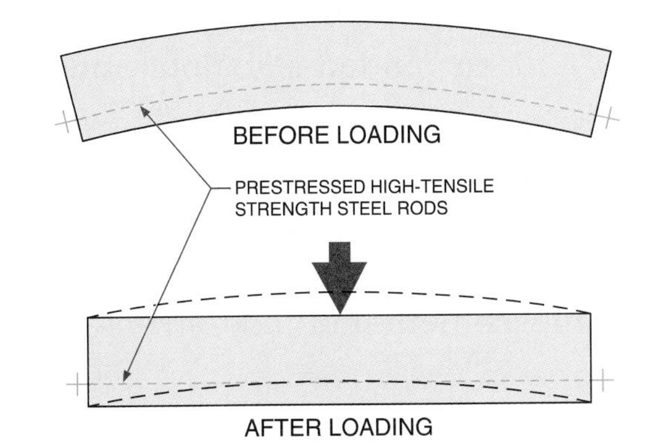

Fig. 25-25 ■ Prestressing with draped tendons.

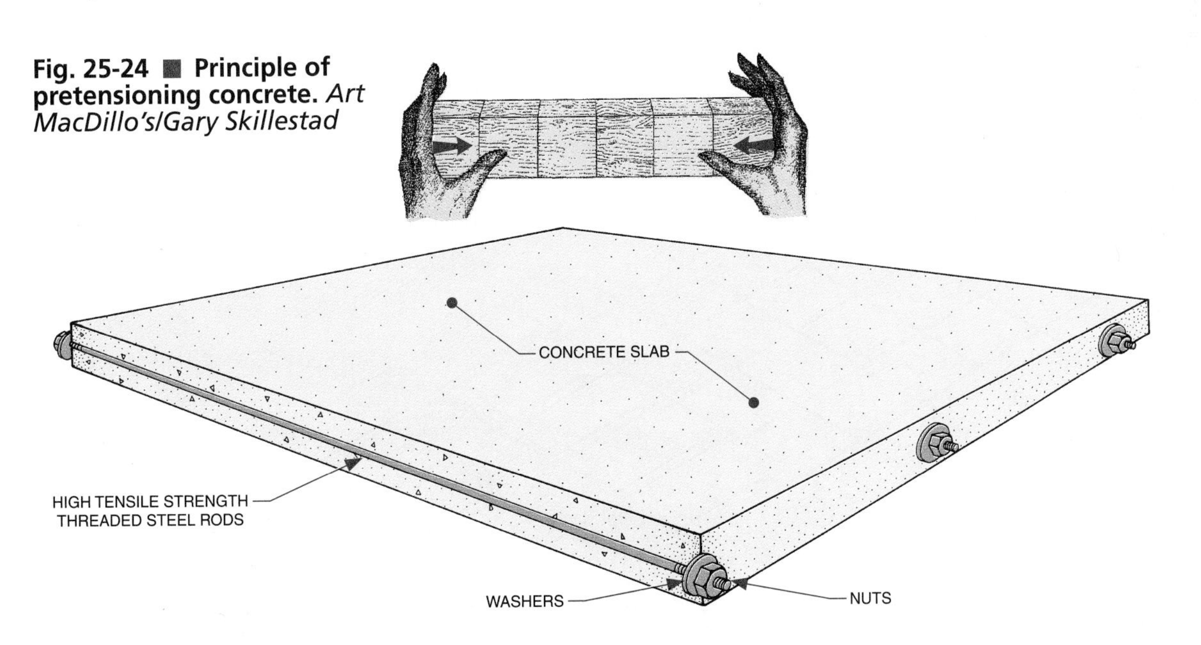

Fig. 25-24 ■ Principle of pretensioning concrete. *Art MacDillo's/Gary Skillestad*

tendons have been greased to allow slippage. The tendons are then stretched with hydraulic jacks and the ends anchored. This creates compressive stress because the ends of the tendons pull toward the center. Post-tensioning can be done at a factory or on-site to reduce shipping weight, especially for large members.

Concrete Structural Members

Concrete has been used for centuries for foundations, walls, and ground-supported slabs. Not until low-tensile-strength concrete was reinforced with high-tensile-strength steel, however, could concrete be used for structural components such as columns, beams, girders, and suspended-slab floor and roof systems.

CONCRETE SLABS Once slabs could only be poured in place at grade level, as shown in Fig. 25-26. However, with advances in steel reinforcement methods, structural slab members can now be manufactured off-site. Elevated slabs can also be poured using parallel rebars aligned with the beam direction. To prevent cracking due to temperature and moisture changes, rebars are also placed perpendicular to the load-supporting rebars. These rebars are known as temperature rebars (or bars). See Fig. 25-27.

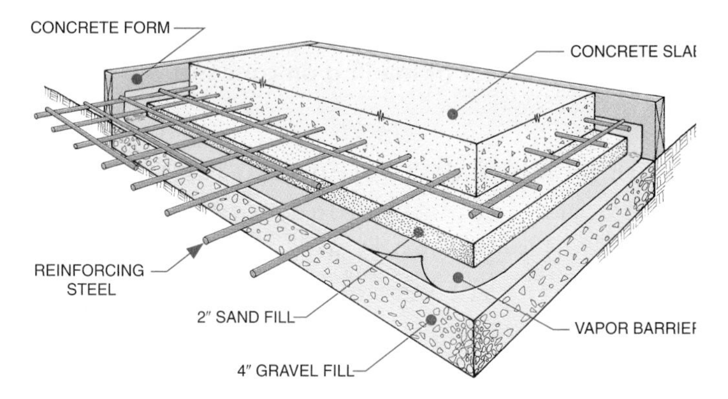

Fig. 25-26 ■ Typical grade level slab construction.

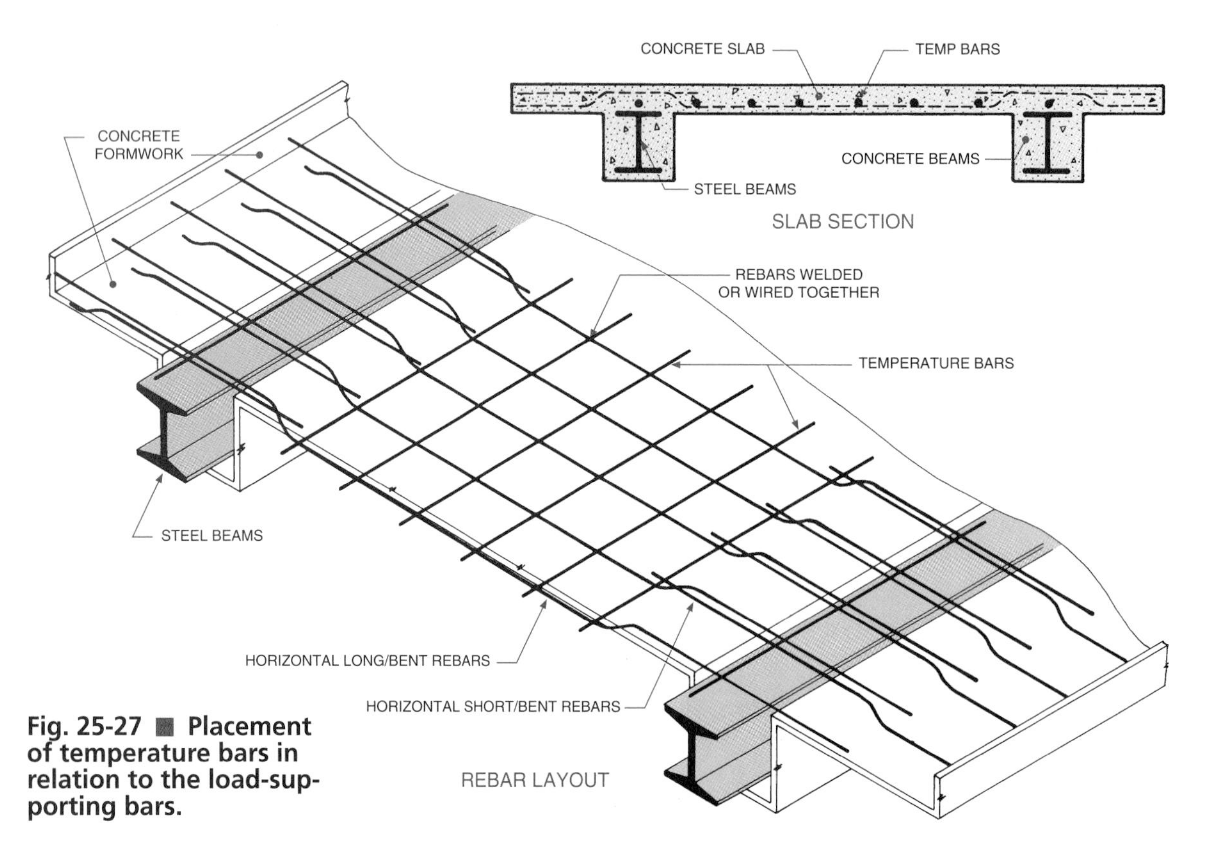

Fig. 25-27 ■ Placement of temperature bars in relation to the load-supporting bars.

COLUMNS Concrete columns are vertical members which support weights transferred from horizontal beams and girders. Concrete columns are made structurally sound by the addition of rebars. See Fig. 25-28. Another method is to fill a hollow steel column with concrete. Such a member is called a **lally column.**

Sectional drawings convey the exact relationship of column, beam, and reinforcement material, as shown in Fig. 25-29. If the exact position of each column is not dimensioned on a floor plan, column schedules are prepared. Column schedules include coding that is indexed to a column plan.

BEAMS AND GIRDERS Concrete girders are major horizontal members that rest on columns. Beams are horizontal members supported by girders or columns. Concrete beams and girders are reinforced with steel rebars to increase tensile strength. Some reinforced concrete beams are rectangular, but most are wider at the top.

Rebars or WWM (welded wire mesh) placed low in a beam will prevent cracking. Rebars or WWM placed too high in a beam may bend and result in cracking. For very heavy loads, a top and bottom row may be needed. This practice is called draped reinforcement. See Fig. 25-30.

The position of girders and beams is shown on floor plans with dotted lines or indexed to a beam schedule, similar to a column schedule.

LINTELS Short horizontal members that span the top of openings in a wall are known as lintels. Concrete lintels are either poured in a form, into lintel blocks, or precast as a small concrete beam.

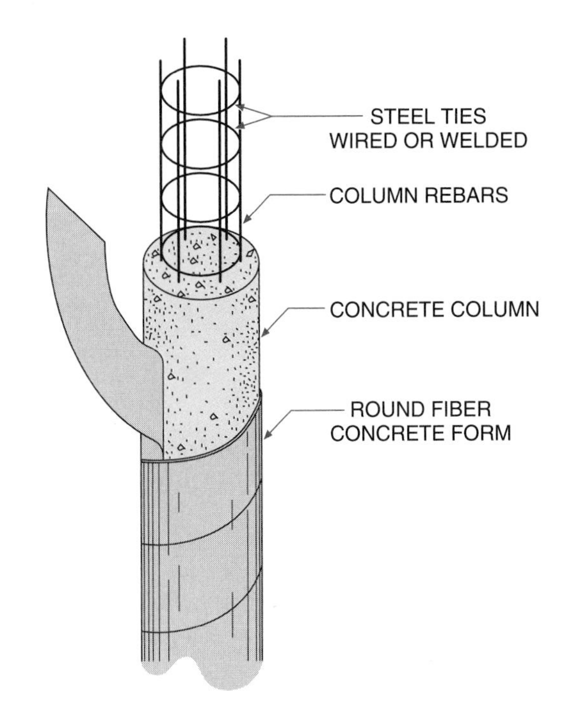

Fig. 25-28 ■ **Typical reinforced concrete column.**

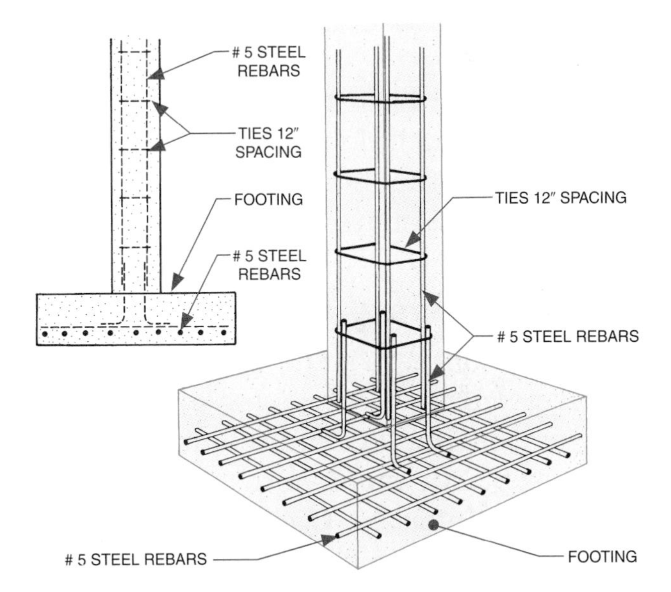

Fig. 25-29 ■ **Column section showing rebar placement.**

Cast-in-Place Concrete

Forms for pouring concrete for footings, foundations, slabs, and walls have been used for a long time. However, new developments in reinforced and prestressed

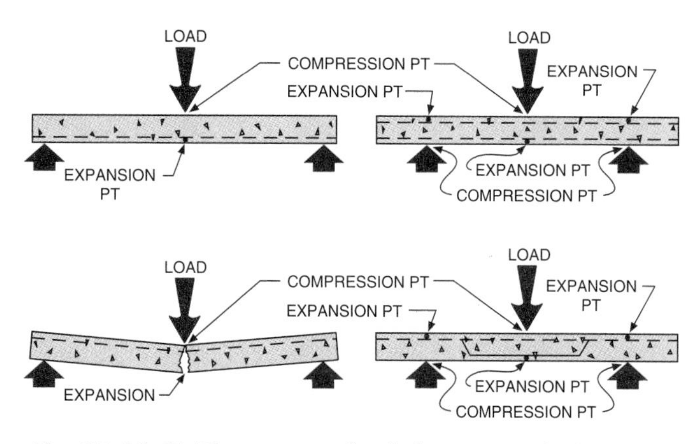

Fig. 25-30 ■ Placement of reinforcement in beams or suspended slabs to prevent cracking.

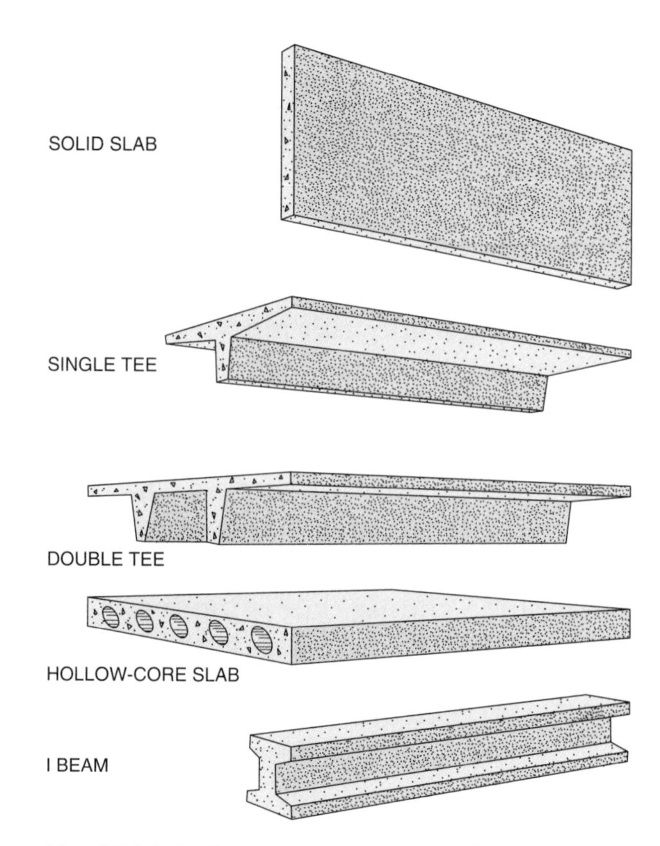

Fig. 25-31 ■ Precast concrete members.

concrete enable builders to erect structures with extremely complex contours. Concrete shells are a type of concrete system that uses poured reinforced concrete. A light steel structure is erected. Then concrete is poured or sprayed. The concrete holds the steel in place after hardening.

Drawings for cast-in-place concrete systems need to include the outline and dimensions of the finished job, including the position of rebars and joints.

Precast Concrete

Precast concrete is the opposite of cast-in-place concrete. Precasting of concrete involves pouring the concrete into wood, metal, or plastic molds. Precast concrete is usually reinforced. Once set, the molded concrete is placed in position in its hardened form.

Although concrete block is the most commonly used precast concrete material, it is considered a masonry material, like brick. Precast concrete structural members include wall panels, girders, beams, and a variety of slabs. See Fig. 25-31. Available in solid, hollow-core, and single and double tee shapes, precast slabs are used for walls, floors, and roof decks. Wall panels are solid precast units used either for bearing or non-bearing walls, depending on the amount of reinforcement. Since the exterior sides of concrete wall panels are usually exposed, special textured finishes are often applied during the casting process. These panels may be combined with layers of insulation to form a complete monolithic wall unit.

Concrete floor system drawings show only the dimensions of slabs that are poured in place. For precast systems, the locations of ribs and/or support beams are shown with dotted lines. See Fig. 25-32. All other information is found on detail and/or sectional drawings keyed to the general floor framing plan.

Slab Component Systems

Precast slab components or cast-in-place slabs are divided into two types: one-way systems and two-way systems.

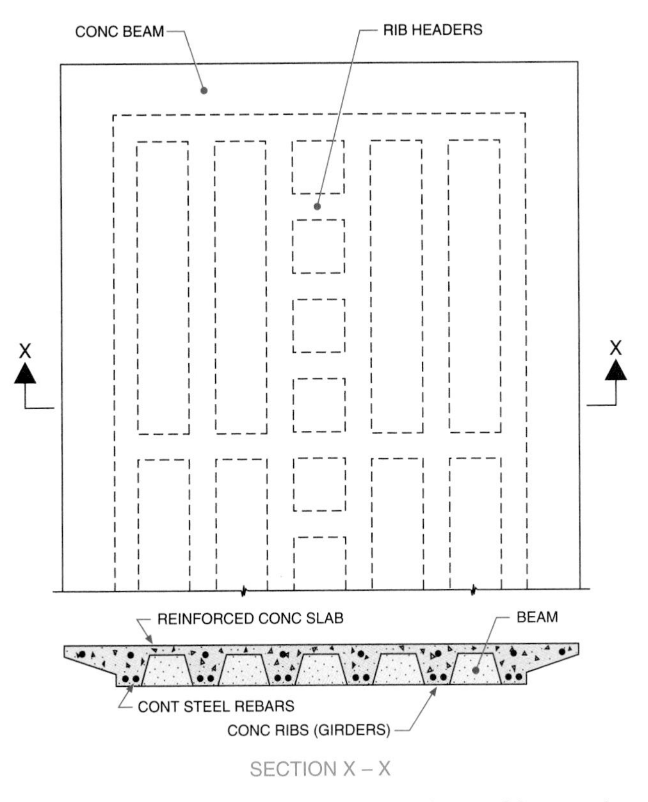

Fig. 25-32 ■ Method of showing ribs and beams in a precast slab drawing.

ONE-WAY SYSTEMS In **one-way slab systems** the rebars are all parallel. One-way system girders, which rest on columns, are parallel to the rebar alignment. One-way solid slabs are extremely heavy and are therefore impractical for most spans over 12 feet. To lighten the dead load, ribbed one-way slabs are often used. See Fig. 25-33. The ribbed slab is a thin slab (2″ to 3″ thick), supported by cast ribs. These units are constructed of precast slab tees or are cast in place. When ribbed slabs are to be poured in place, a ribbed slab plan shows the horizontal rib positions with dotted lines. Dotted lines are also used on detail drawings to show the position of rebars in the slab and in the ribs.

TWO-WAY SYSTEMS In **two-way slab systems** the rebars, girders, and beams are placed in perpendicular directions. See Fig. 25-34. When ribs extend in both directions, the system is known as a waffle slab.

Pan and *waffle* slabs are used to cast suspended floor and roof systems for spans up to 60 feet. A pan slab is created by pouring concrete into molded fiberglass forms on site. Waffle slabs may be poured on site or prefabricated off site. Temporary fiberglass or metal pans (domes) are placed, open side down, 4″ to 7″ apart on a temporary floor. No pans are

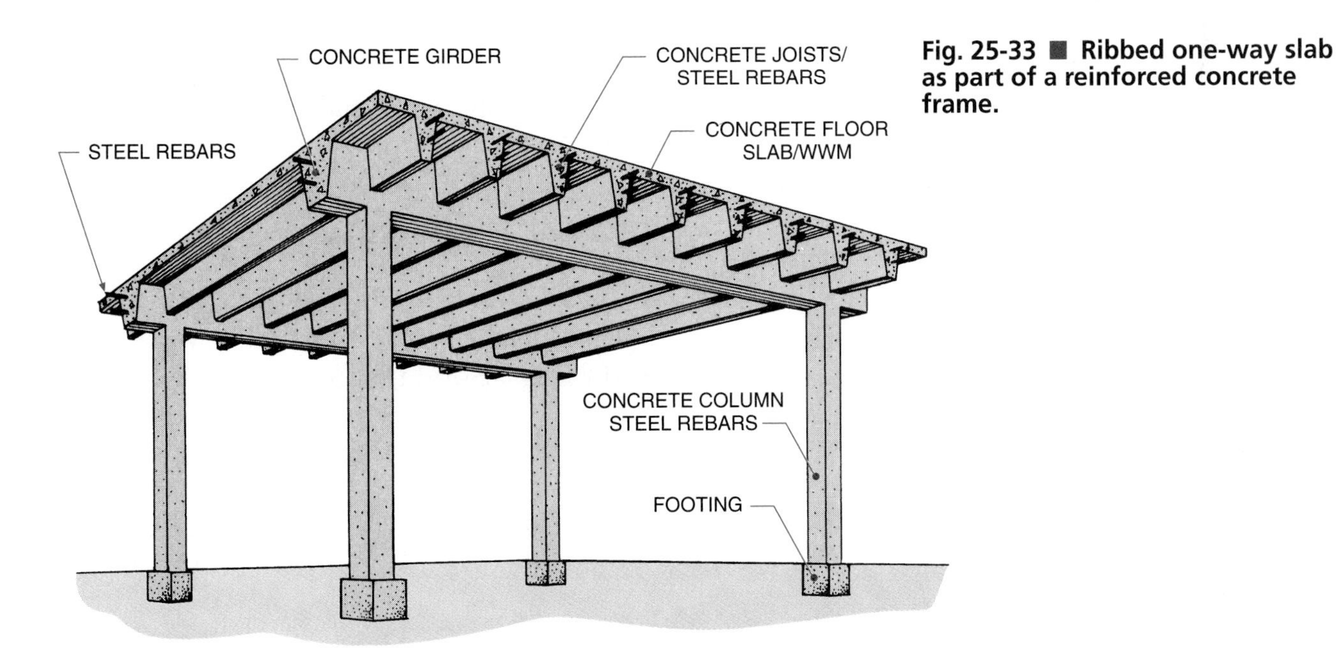

Fig. 25-33 ■ Ribbed one-way slab as part of a reinforced concrete frame.

Fig. 25-34 ■ Two-way slab system.

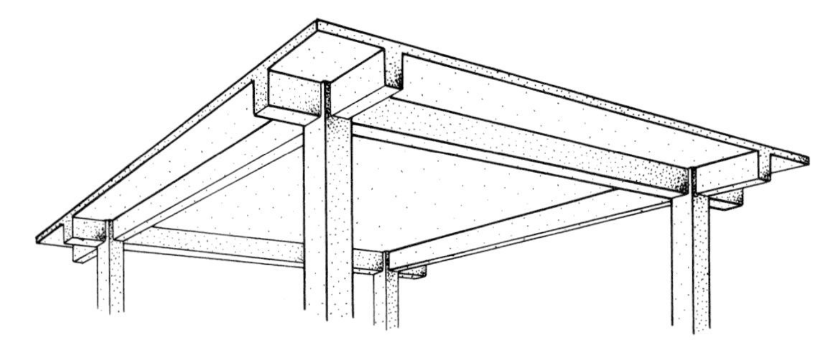

Fig. 25-35 ■ Waffle slab system.

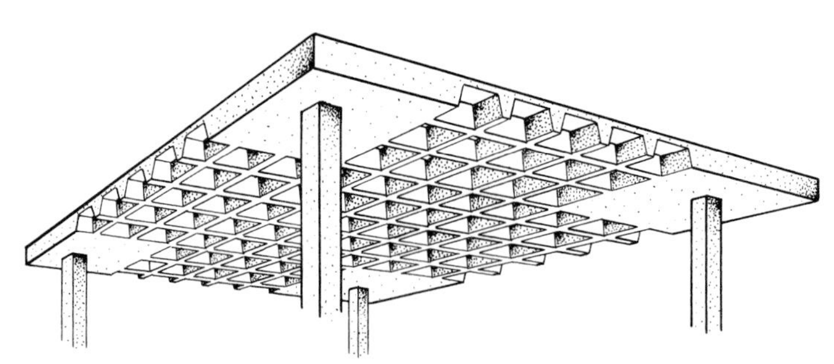

Fig. 25-36 ■ Flat slab system.

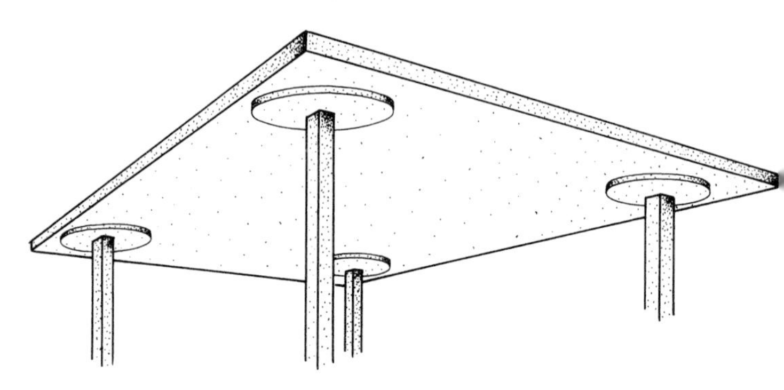

placed around columns. Rebars are added and concrete is then poured to a depth of several inches over the pans. After the concrete has cured, the pans and temporary flooring are removed, and a suspended waffled floor (or roof) results. See Fig. 25-35. This type of cast-in-place system is lightweight (for concrete), sound-resistant, fireproof, and economical.

A *flat slab* is a two-way slab unit that rests directly on columns without a girder or beam support. A flat slab floor (or roof) system is actually a series of individual slabs, with the center of each slab resting independently on a column. All of the slabs' weight is directed through the columns to footings. When the slabs are joined together, a unified floor is created. In flat slab systems the supporting columns are strengthened by the addition of a thicker slab area (*drop panel*) around columns. See Fig. 25-36. A column capitol, or flared head, also helps spread the slab loads onto a column in this type of construction.

Concrete Joints

Since concrete expands and contracts with changes in moisture levels and temperature, relief joints (expansion joints) are required to allow for these fluctuations. Some construction drawings and/or specifications indicate minimum dimensions for placement of expansion joints. Some drawings show specifically where joints are required or must be avoided.

Concrete Wall Systems

CONVENTIONAL CONCRETE WALLS Most concrete walls are poured between wood or metal forms. See Fig. 25-37. Most walls of

Fig. 25-37 ■ Concrete wall construction.

A. Concrete is poured into forms constructed on the site.

B. Finished walls.

this type are used with wood-frame construction. Reinforced concrete walls are also the main structural material for both walls and roofs in an earth-sheltered construction. See Fig. 25-38.

INSULATED CONCRETE WALLS Concrete walls can also be formed by pouring concrete between polystyrene panels, which become part of the finished wall. Other systems use interlocking hollow-core blocks that are filled with concrete.

Fig. 25-38 ■ Use of reinforced concrete walls in earth-sheltered building.

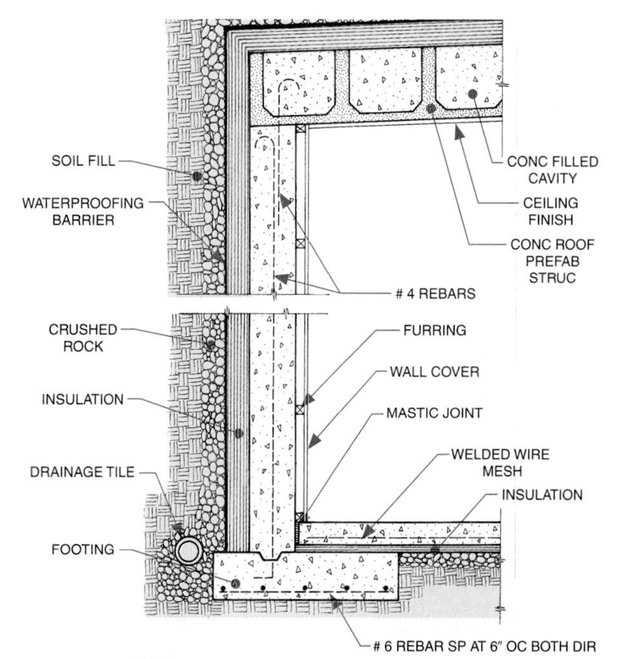

CHAPTER 25 Exercises

1. Name the types of brick and their uses.
2. Describe the types of masonry walls.
3. Describe the types of concrete construction systems.

4. Draw a wall section for the house you are designing using one of the masonry systems described in this chapter.

5. Complete Exercise 4 using a CAD system.

CHAPTER

26

Steel and Reinforced-Concrete Systems

Objectives

In this chapter, you will learn to:

- describe three types of steel construction and explain the basic purpose of each.
- identify manufactured steel forms and their function as structural members.
- read and interpret steel symbols, notations, identification, and measurements for working drawings.
- relate the types of fasteners and intersections of steel members to construction methods.

Terms

channel
large-span construction
L-shape (angle)
rolled steel
S-shape
steel cage construction
tee
W-shape

Steel can span greater distances and support greater loads than any other conventional building material. However, steel is extremely heavy and thus creates heavy dead loads. This chapter describes structural steel members and explains how they are fastened together to form building components.

■ **In steel construction, welding is just one method used to securely fasten structural members together.** *Gary Conner/Index Stock Photography Inc.*

Steel Building Construction

In steel construction, plates, bars, tubing, and rolled shapes are used for columns, girders, beams, and bases. There are three general types of steel construction systems: steel cage, large-span, and cable-supported.

When steel members are used in a manner similar to skeleton-frame wood members, the system is known as **steel cage construction.** The terms *steel skeleton-frame* and *steel cage* construction are often used interchangeably. The major structural members used in steel cage construction are columns, girders, and beams. The definitions of structural members remain the same for steel construction as for other types of construction. Columns are vertical members that rest on footings or piers. Girders are horizontal members that extend between columns. They are sometimes called *spandrel beams,* if they connect to columns erected on the perimeter. See Fig. 26-1. Beams are horizontal members placed on or between girders. *Purlins* are beams which connect roof trusses or rafters.

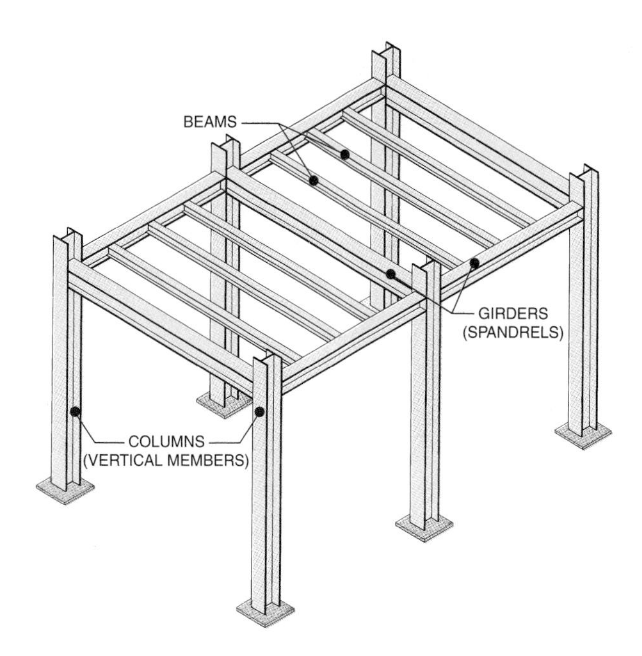

Fig. 26-1 ■ Steel girders which extend between columns are called spandrel beams (or spandrels).

Beams are supported by girders, which in turn are rigidly attached to columns, through which all loads are transmitted through bearing plates to footings. Since all live and dead loads are transmitted through the columns, there is no need for additional exterior or interior bearing walls in steel cage construction. This enables buildings to be built extremely high with a minimum of interior obstruction. It also allows for the use of large exterior walls that have no structural value. These walls, called curtain walls, are often glass.

Even steel cage construction cannot provide the enormous amount of unobstructed space needed in structures such as aircraft hangars, sports stadiums, and convention centers. For these structures, large trusses or arches are necessary to span long distances. Such construction is called **large-span construction**.

Structural steel is manufactured in many forms and types. Structural steel types are specified by their metallurgical characteristics and minimum stress yields, as designated by the American Society for Testing Materials (ASTM). See Fig. 26-2.

Preparing structural steel drawings requires a working knowledge and understanding of steel symbols, notations, identifications, drawing conventions, measurements, and fastening and intersection methods.

Fig. 26-2 Structural steel types.

ASTM* Type	Min Yield† Stress Point	Manufactured Forms	Description
A36	36,000 PSI	Sheets Plates Bars Shapes Rivets Nuts Bolts	A medium carbon steel that is the most commonly used structural steel. Suitable for buildings and general structures, and capable of welding and bolting.
A440	42,000 PSI	Plates Bars Shapes	A high strength, low alloy steel suitable for bolting and riveting, but not welding. Used for lightweight structures–high resistance to corrosion.
A441	40,000 PSI	Plates Bars Shapes	A high strength, low alloy steel modified to improve welding capabilities in lightweight buildings and bridges.
A572	41,000 PSI	Limited Types of Shapes Bars & Plates	A high strength, low alloy economical steel suitable for boltings, riveting, and welding with lightweight high toughness for buildings and bridges.
A242	42,000 PSI	Plates Bars Shapes	A durable, corrosion resistant, high strength, low alloy steel which is lightweight and used for buildings and bridges exposed to weather. Can be welded with special electrodes.
A588	42,000 PSI	Plates Bars Shapes	A lightweight corrosion, high strength, low alloy steel with high durability in high thicknesses used for exposed steel.
A514	90,000 PSI	Limited Shapes & Plates	A quenched and tempered alloy steel with varying strength, width, thickness, and type.
A570	25,000 PSI	Plates Light Shapes	A light gauge steel used primarily for decking, siding, and light structural members.
A606	45,000 PSI	Plates	A high strength, low alloy sheet and strip steel with high atmospheric corrosion resistance.

*American Society for Testing Materials

†The amount of stress that causes the steel to begin to deform.

Steel Structural Members

Steel used for structural purposes is manufactured in plates, bars, pipes, tubing, and a variety of other shapes. Many of these items are formed by passing steel between a series of rollers. Steel that has been shaped by this method is called **rolled steel.**

Plates

Structural plates are flat sheets of rolled steel. These sheets range in thickness from 1/8″ to 3″ and in width from 8″ to 60″. Plates are specified by thickness, width, and length in that order. See Fig. 26-3.

Plates are used as webs in built-up girders and columns, as shown in Fig. 26-4,

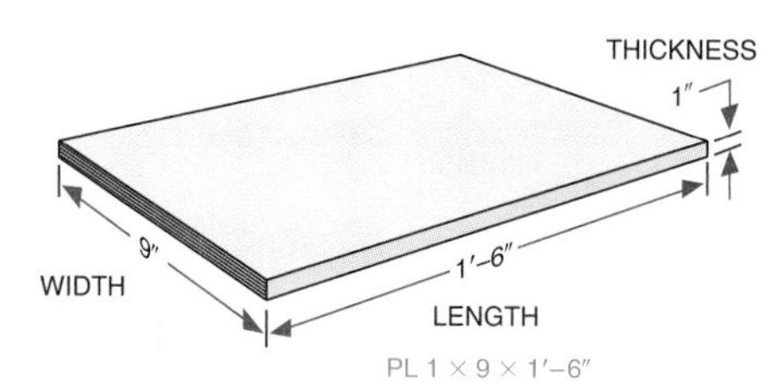

Fig. 26-3 ■ Plate specifications.

and to reinforce other webs or flanges of structural steel shapes. Bearing plates provide bearing surfaces between columns and concrete footings. See Fig. 26-5.

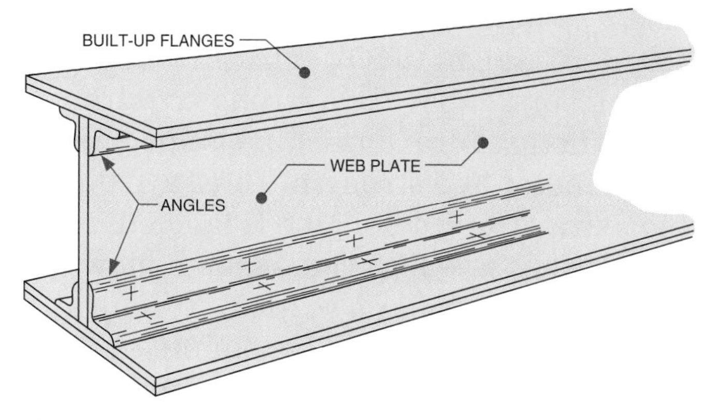

Fig. 26-4 ■ Built-up steel plate girder.

Bars

Steel bars used for structural purposes are available in round, square, hexagonal, and flat (rectangular) cross-section shapes. See Fig. 26-6. Square, hexagonal, and round bars are manufactured in 1/16" increments from 1/16" to 12". Flat bars are manufactured in 1/4" increments up to 8". Round bars are specified by diameter; square bars by width or gauge number; flat bars by width and thickness; and hexagonal bars by the distance across the flats (AF). Steel bars are used primarily for bracing other structural components and for concrete reinforcement.

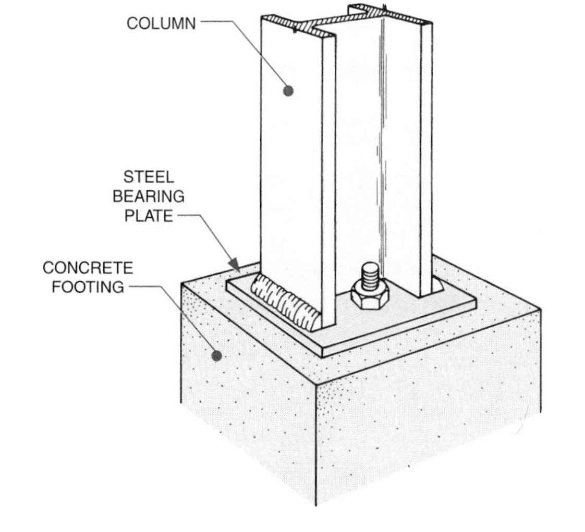

Fig. 26-5 ■ Bearing plate.

Steel Pipe and Structural Tubing

Steel pipe and tubing are used extensively in exposed areas because of their clean, pleasing lines. Structural pipe and tubing is available in round, square, and rectangular cross-section shapes, as shown in Fig. 26-7.

Hollow steel pipe is manufactured in sizes from 1/2" to 12" (inside diameter) and in three strength classes. Strength classes

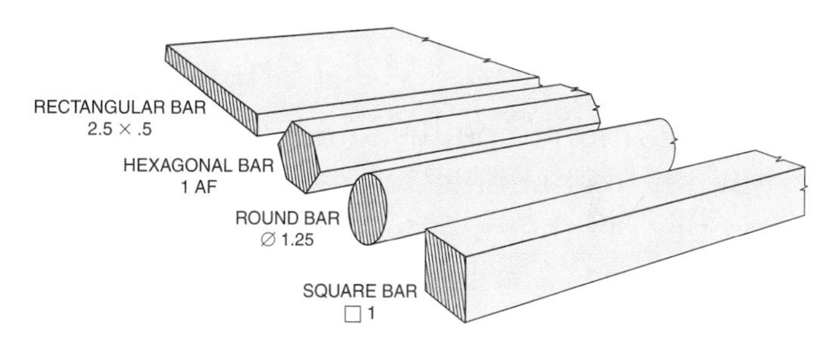

Fig. 26-6 ■ Types of steel bars.

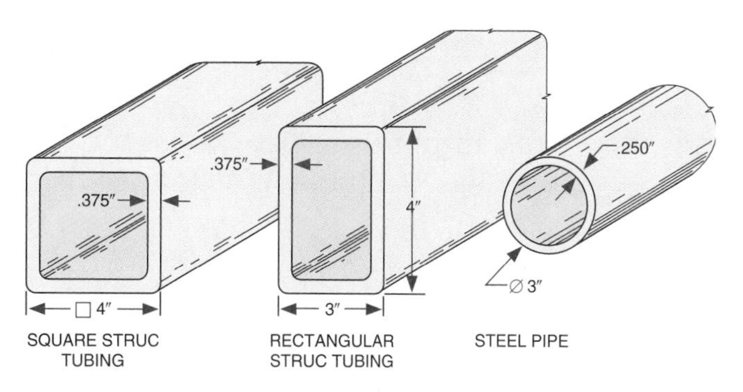

Fig. 26-7 ■ Steel pipe and tubing.

relate to wall thicknesses and are either standard weight (STO), extra strong (x-strong), or double extra strong (xx-strong). On structural drawings, pipe is specified by diameter and strength. Thus, a 4″ double extra-strong steel pipe is labeled: pipe 4 xx-strong. See Fig. 26-8. When hollow steel pipe is used as a vertical structural support, it is called a lally column.

Square structural tubing is specified by cross-section width and thickness and is available in sizes from 2″ × 2″ to 10″ × 10″ outside dimension (OD). Rectangular tubing sizes range from 3″ × 2″ to 12″ × 8″ (OD). Structural tubing is specified by the symbol TS followed by the width, thickness, and wall thickness. A rectangular structural tube 4″ wide and 3″ thick with a wall thickness of ¼″ is therefore labeled: TS 4 × 3 × .25. Round structural tubing is specified by outside diameter and wall thickness. For example, a 4″-diameter tube with a ¼″ wall thickness is labeled: 4 OD × .25.

Other Structural Steel Shapes

In addition to plates and bars, steel is rolled into channels, tee sections, and a number of other shapes. See Fig. 26-9. Steel shapes are designated on construction drawings by shape symbol, depth in inches, and weight in pounds per foot. Figure 26-10 shows the most common

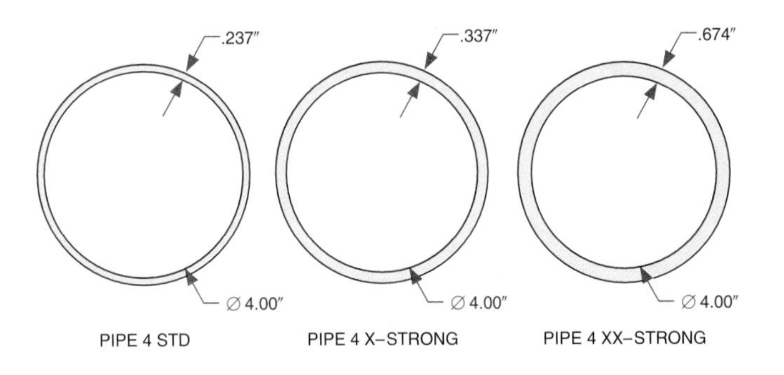

Fig. 26-8 ■ Steel pipe thickness.

Fig. 26-9 ■ Structural steel designations.

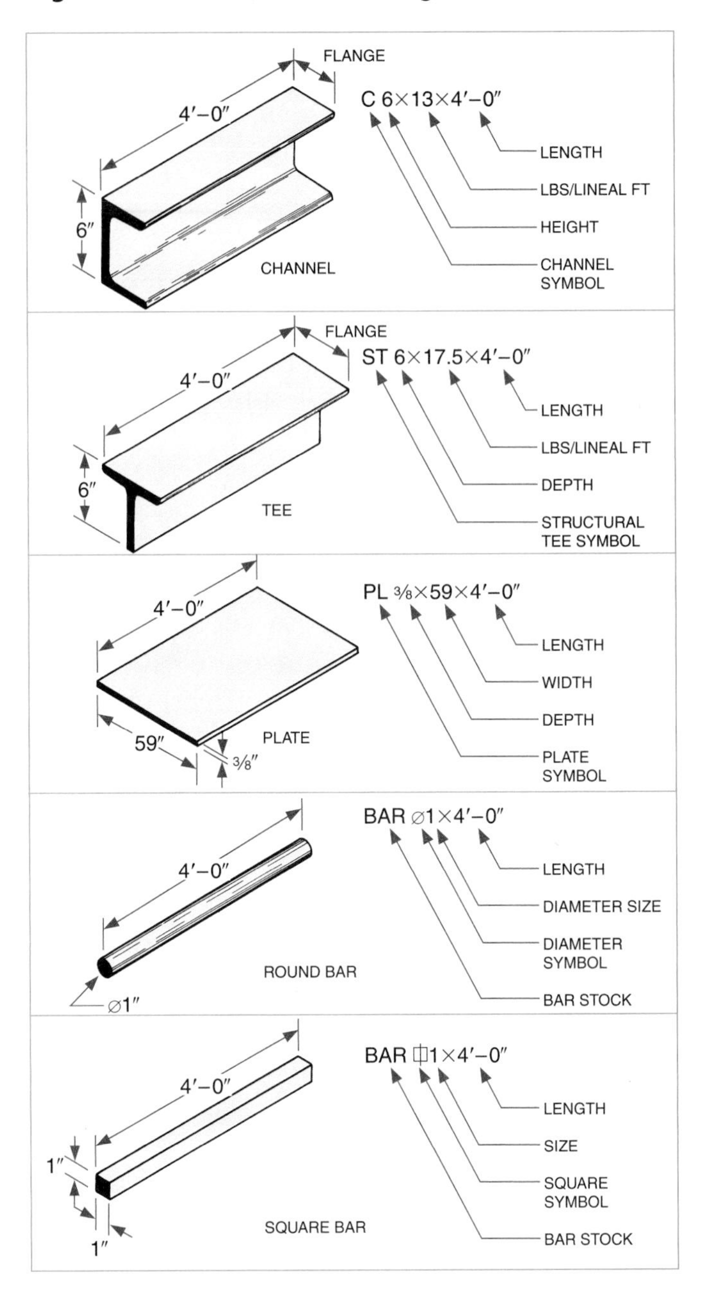

structural steel shapes and the related drawing symbols.

L-shapes (angles) are structural steel members rolled in the (cross-section) shape of the letter L with legs of equal or unequal length. (Equal leg lengths are available in sizes of 1″ to 8″. Unequal leg lengths range from 1¾″ to 9″.) Whether equal or unequal, the thickness of each leg (called wall thickness) is always the

NAME	SECTIONAL FORM	SYMBOL	PICTORIAL
WIDE FLANGE		W	
AMERICAN STANDARD BEAM		S	
TEE		T	
ANGLE		L	
ZEE		Z	
AMERICAN STANDARD CHANNEL		C	
BULB ANGLE		BL	
LALLY COLUMN		◎	
SQUARE BAR		⌽	
ROUND BAR		ϕ	
PLATE		℄	

Fig. 26-10 ■ Structural steel shapes.

same. L-shapes (angles) are specified on drawings by the symbol L followed by the length of each leg, followed by the wall thickness.

All inch marks are omitted on shape notations used on structural drawings since all sizes are assumed to be in inches unless otherwise specified. For example, an L-shape member with one 2″ and one 3″ leg and a wall thickness of ½″ is specified: L2 × 3 × .5. L-shape members are used as components in built-up beams, columns, and trusses. They are also used for connectors and as lintels in light- or short-span construction.

Channels are rolled into a cross-section shape resembling the letter U, with the inner faces of flanges shaped with a 2/12 pitch. Channels are classified by depth, from 3″ to 15″. There are two types of channels specified for structural use: American Standard channels (C) and Miscellaneous channels (MC). Channels are specified by symbol (C or MC), followed by the depth times the weight per foot. An 8″-deep Standard channel that weighs 11.2 lb/ft is labeled: C 8 × 11.2. Channels are used for roof purlins, lintels, truss chords, and to frame-in floor and roof openings.

S-shapes (formerly I beams) are rolled in the shape of a capital letter I. American Standard shapes have narrow flanges with a 2/12 inside pitch. S-shapes are classified by the depth of the web and the weight per foot. The web is the portion between the flanges. Web depths range from 3″ to 24″. S-shapes are designated by their symbol (S) followed by the web depth and the weight per lineal foot. See Fig. 26-11. For example, an S-shape member with a 14″-deep web that weighs 56 lb/ft is labeled: S 14 × 56. On some drawings, the length may be added to the designation

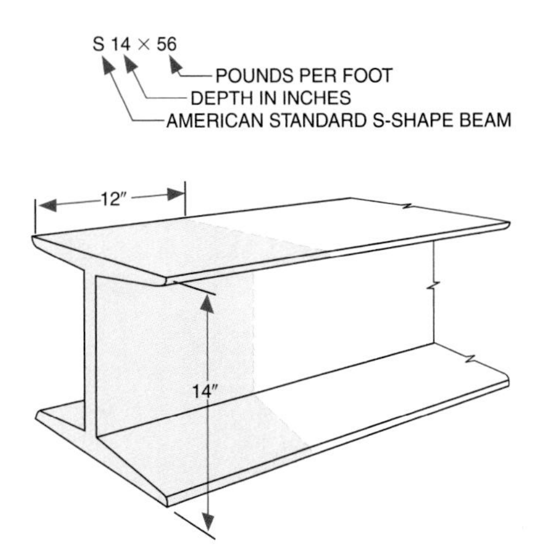

Fig. 26-11 ■ S-shape designations.

rather than as a dimension on the drawing. S-shapes are used extensively as columns because of their symmetry. Their narrow flanges are applicable to many designs where size restrictions are a problem.

W-shapes (formerly wide-flange or H beams) are similar to S-shapes but with wider flanges and comparatively thinner webs. Their capacity to resist bending is greater than that of S-shapes. W-shapes are designated in the same manner as S-shapes. See Fig. 26-12. For example, W 18 × 62 describes a W-shape member with an 18″-deep web weighing 62 lb/ft. W-shapes are available in depths from 4″ to 36″. Lighter-weight versions of W-shape members are known as M-shapes.

Structural **tees** are made by cutting through the web of an S, W, or M-shape, although some tees are rolled to order. If the web is cut exactly through the center, two identical tees result. The symbol for a tee is the capital letter T. On structural drawings the tee symbol includes the shape from which the tee was cut (S, W, or M) followed by the letter T, the depth of cut (from web to flange), and the weight per foot. Therefore a tee-shape member cut in half from a W 12 × 50 would be specified WT 6 × 25. (6 is half the depth and 25 is half the weight per foot.) Tees are most commonly used for truss chords and to support concrete reinforcement rods.

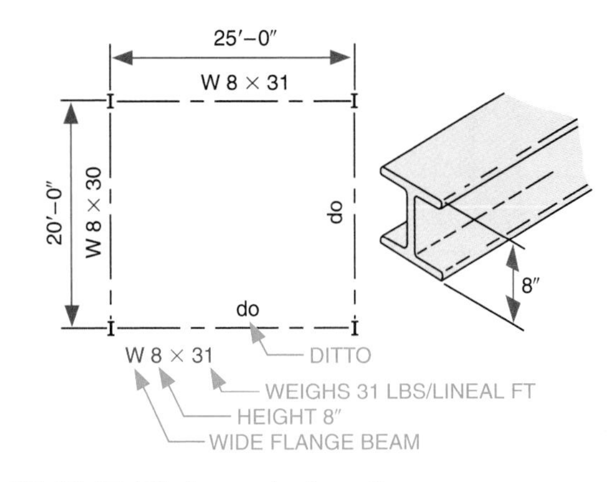

Fig. 26-12 ■ W-shape designations.

Metal studs, as shown in Fig. 26-13, are manufactured in different forms. Stud thicknesses vary depending on the loads to be supported. Widths are manufactured to be identical with conventional wood stud sizes. Fireproof steel framing can be substituted for conventional wood framing. It is insect-proof and will not decay, expand, contract, warp, or split.

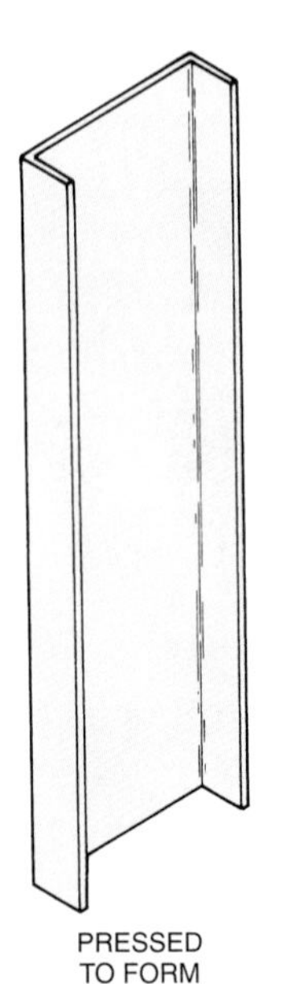

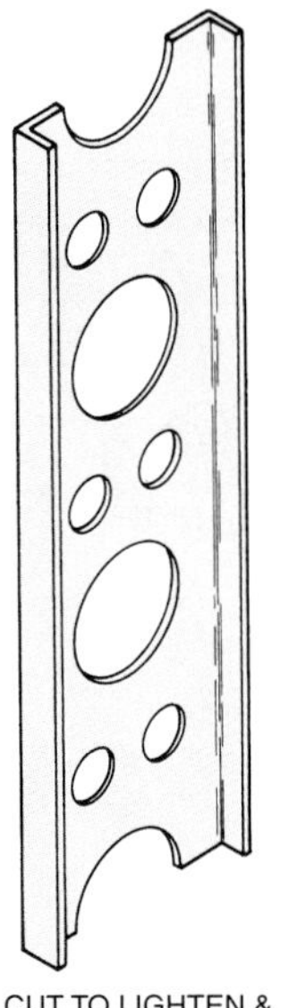

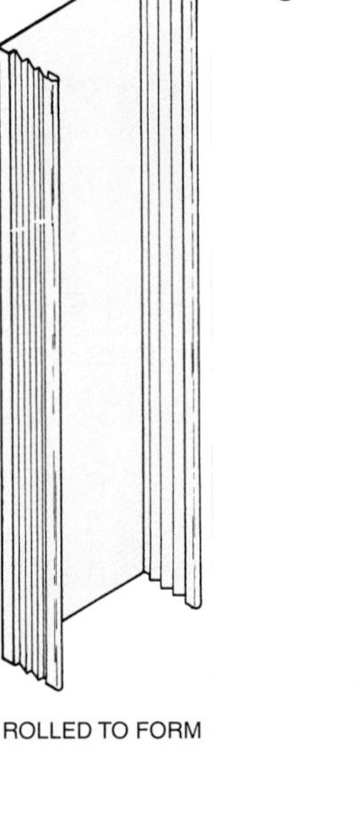

Fig. 26-13 ■ Types of metal studs.

Steel Fasteners and Intersections

Major structural steel members depend on a wide variety of joining methods and devices to function as a structurally stable frame. This includes the use of brackets, rivets, bolts, and welds to attach members to each other and to foundation piers and footings. See Fig. 26-14. Some steel components can be assembled before shipping to a site. These are usually assembled and welded at a fabrication shop. All other members are assembled and permanently fastened at the building site. For example, some brackets may be welded to a girder at a shop, then bolted or welded to a column at the site during construction.

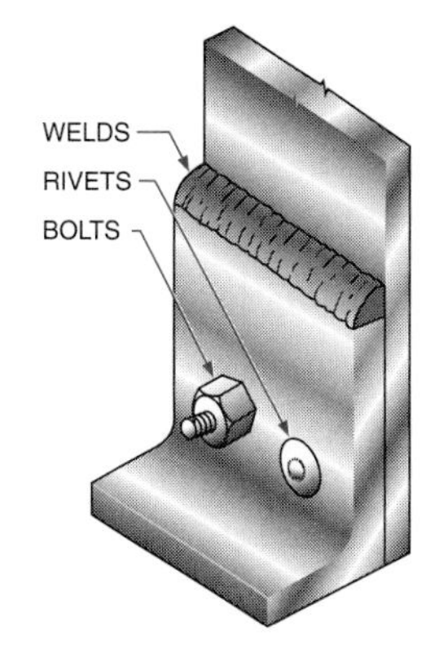

Fig. 26-14 ■ Methods of joining steel members.

Brackets

Most structural steel members intersect at right angles. Many different types of brackets are used to provide a perpendicular surface for bolting, riveting, or welding. Angles, L-shapes, and bent or welded plates are used for this purpose. Angles, nuts, and bolts help join a girder to the top of a column. Bracket information on a structural drawing shows the size of the bracket legs followed by the thickness, width, shape symbol, and fastening device information. See Fig. 26-15. If brackets are to be welded to a member at a fabrication shop, a detail drawing is not provided in the field. Only the assembled intersection of the joint is drawn for field reference. Only shop fabricators are provided with a complete set of details.

Rivets

Rivets are used to connect steel members. See Fig. 26-16. The type of rivet

Fig. 26-15 ■ Angle bracket and bolt notations.

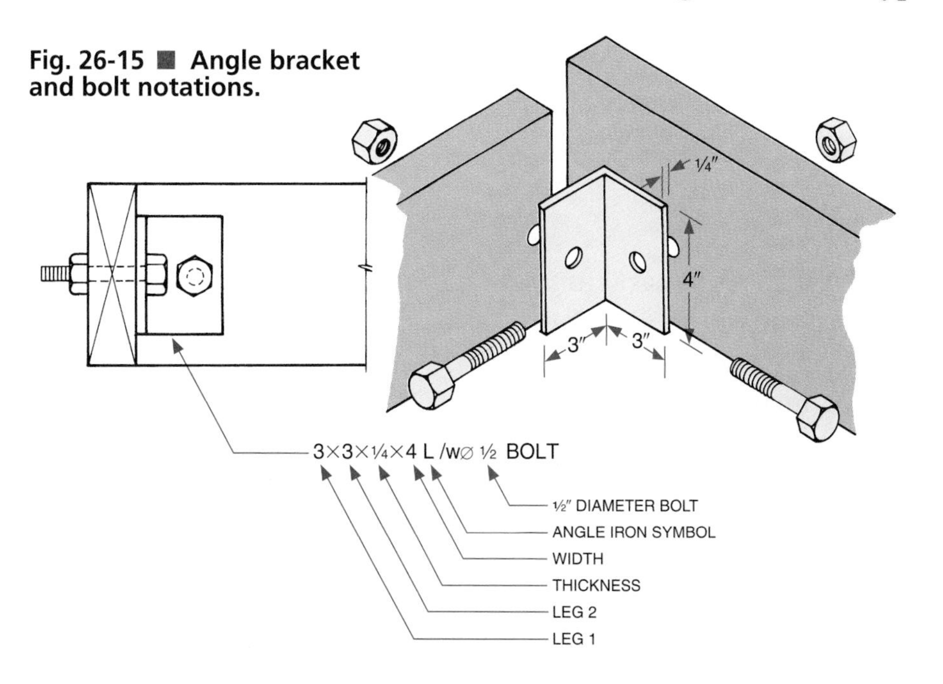

Fig. 26-16 ■ Types of rivets.

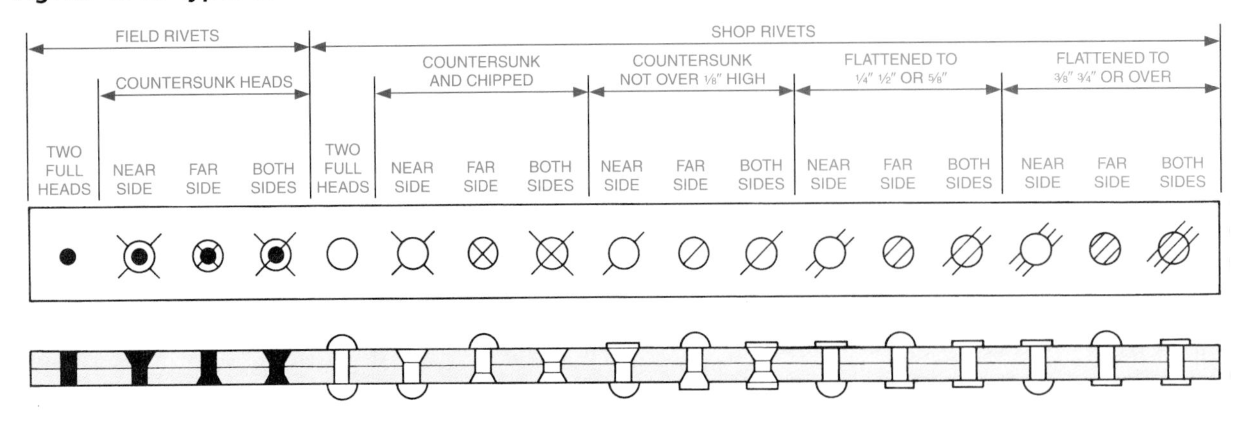

A. Field rivets and shop rivets.

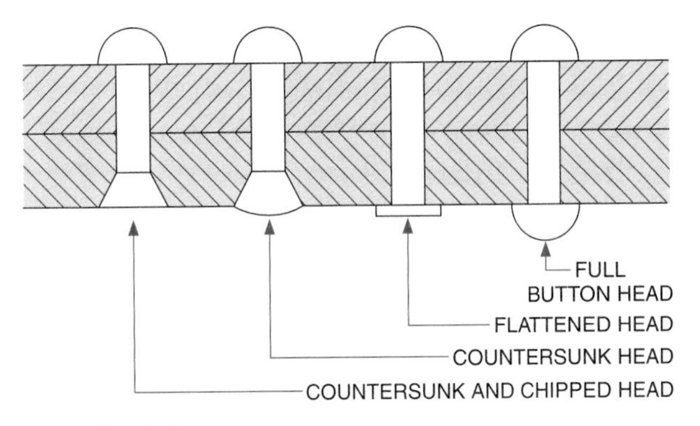

B. Bucked rivets.

specified is shown at the end of the drawing notation. Rivets are made of soft steel and, when heated rivets are cooled, they tend to shrink. The shrinking decreases the tightness of the joint. Consequently, bolts are now used more extensively than rivets in the erection of structural steel.

Bolts

High-strength bolts and nuts can carry loads equal to rivets of the same size, but they can be turned tighter because of their high tensile strength. Bolts used in steel construction are either high-strength or unfinished bolts. High-strength bolts are used to connect extremely heavy load-bearing members such as girders, beams, and columns. See Fig. 26-17. They are also used to attach members where shear loads are transmitted through the bolts.

Unfinished bolts are used for lighter connections, where loads are transmitted directly from member to member. For example, unfinished bolts could be used when a beam rests directly on a girder because there is no vertical shear load on the bolts holding these two members together. Unfinished bolts are also used to anchor column base plates to footings, as shown in Fig. 26-18, since there is also no shear stress at this location.

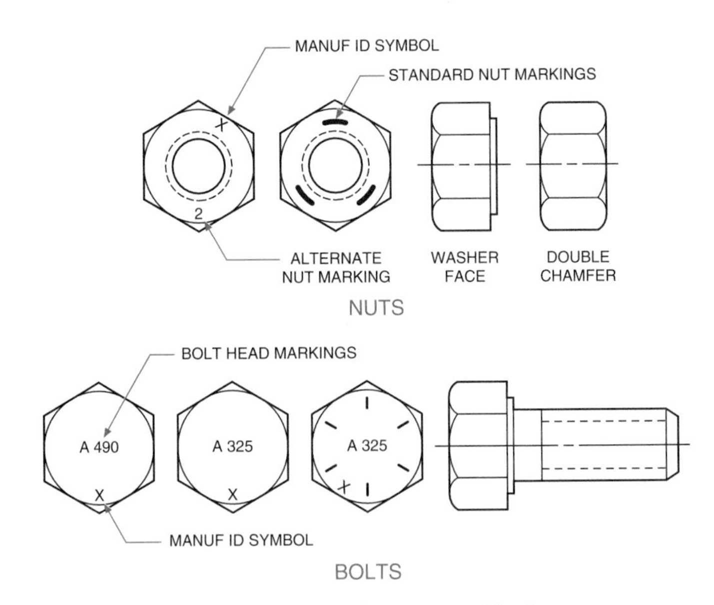

Fig. 26-17 ■ High-strength nuts and bolts.

Fig. 26-18 ■ Unfinished bolts used on column bases.

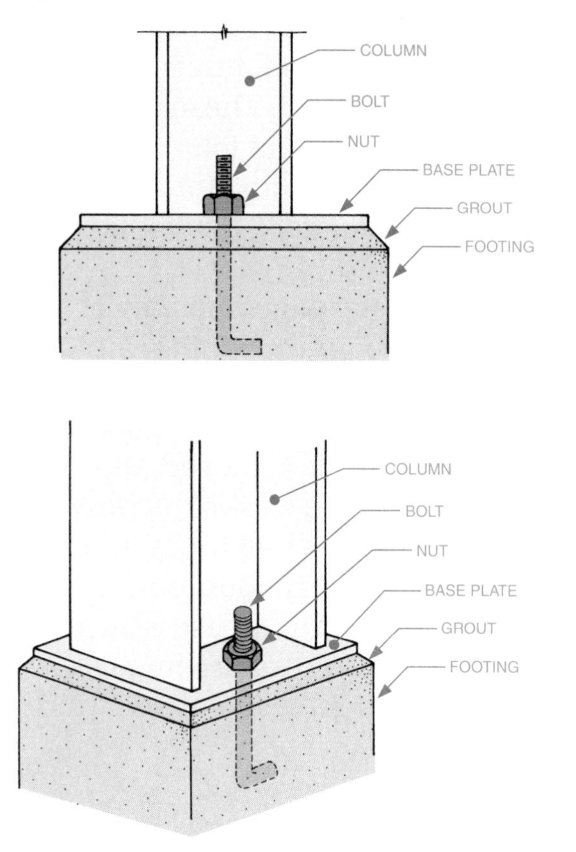

Welds

Welding is a popular method of connecting structural steel members. It has some advantage over bolting. For example, fabrication is simplified by reducing the number of individual parts to be cut, punched with holes, handled, and installed. The major types of welds and their symbols are illustrated in Fig. 26-19.

The convention used to locate welding information on drawings is a horizontal reference line with a sloping arrow directed to the joint. See Fig. 26-20. The arrow may be directed right or left, upward or downward, but always at an angle to the reference line. If no extra marking is shown, a shop weld is assumed. Other drawing symbols give additional instructions. A triangular flag indicates a field weld. An open circle means weld all around the member. If the

Fig. 26-19 ■ Types of welds.

WELD SYM	WELD NAME
◺	FILLET
▭	PLUG/SLOT
○	SPOT/PROJECTION
⊖	SEAM
◡	BACK/BACKING
◡◡	SURFACING
//	SCARF (BRAZING)
J L	FLANGE – EDGE
I L	FLANGE – CORNER
II	GROOVE – SQUARE
V	GROOVE – V
I/	GROOVE – BEVEL
Y	GROOVE – U
I/	GROOVE – J
) (	GROOVE – FLARE V
I (	GROOVE – FLARE BEVEL

Fig. 26-20 ■ A welding symbol showing the information contained in a detail drawing.

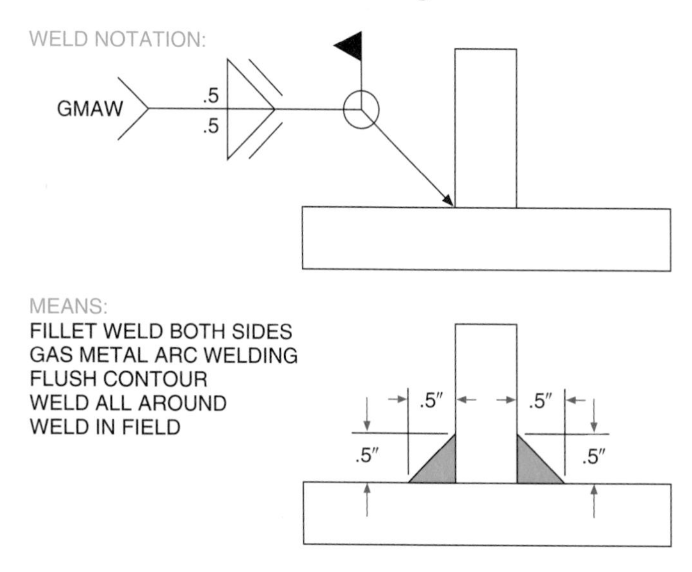

open circle is shown around the base of the flag, it means weld all around in the field.

The basic weld symbols or supplementary weld symbols are located midway on the horizontal reference line. See Fig. 26-21. The symbol is located below the line if the weld is to be placed on the near side of the workpiece where the arrow points. The symbol is placed above the line if the weld is to be placed on the far side. The symbol is placed above and below if both sides are to be welded. The side of the weld (or its depth) is indicated to the left of the basic symbol. The length of the weld is shown to the right of the symbol. When long joints are used, intermittent welds are often specified. These are indicated by the length of weld followed by the center-to-center spacing (pitch). Such welds are usually staggered on either side of a joint.

The tail of the reference line may contain information about the kind of material or process required. This feature is not often used on structural steel details. When no information is required, the tail is omitted. Figure 26-22 shows the position of information on a welding symbol.

Although steel construction is made extremely rigid through the use of regular fasteners, special cross-bracing is often required to counteract lateral wind loads.

Fig. 26-21 ■ Weld symbols.

F – Finish symbol
⌒ – Contour symbol
A – Groove angle: included angle of countersink for plug welds
R – Root opening: depth of filling for plug and slot welds
S – Depth of preparation
– Size or strength for specific welds
– Height of weld reinforcement
– Radii of flare-bevel grooves
– Radii of flare-V grooves
– Angle of joint (brazed welds)
(E) – Effective throat
T – Specific process or reference
L – Length of weld
– Length of overlap (brazed joints)
P – Pitch of welds (center-to-center spacing)
1 – Weld located on opposite side of arrow
2 – Weld located on same side of arrow
(N) – Number of spot or projection welds
⚑ – Weld made in field
o – Weld all around

A. Legend of weld symbols.

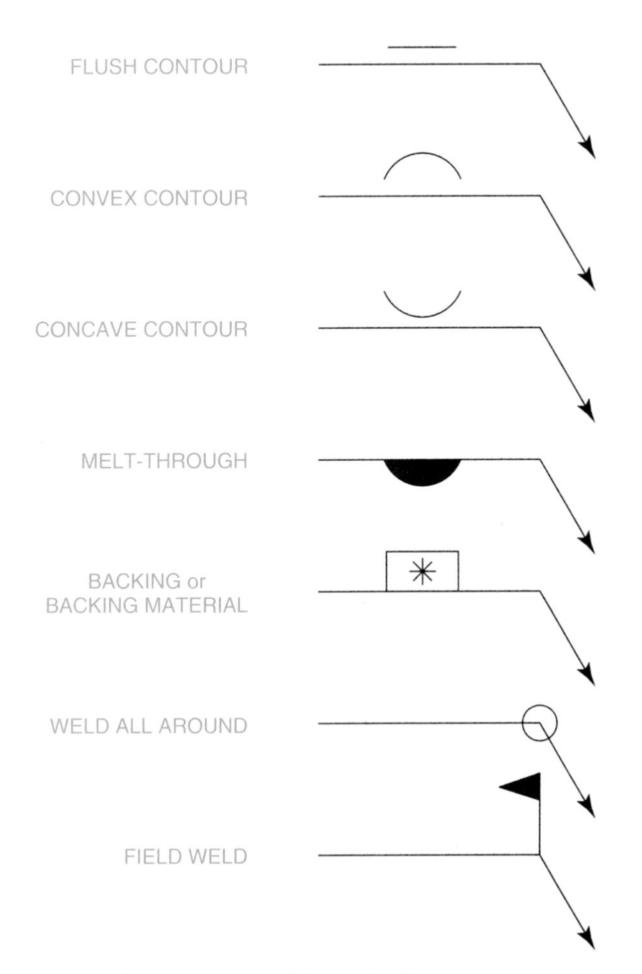

B. Supplementary weld symbols.

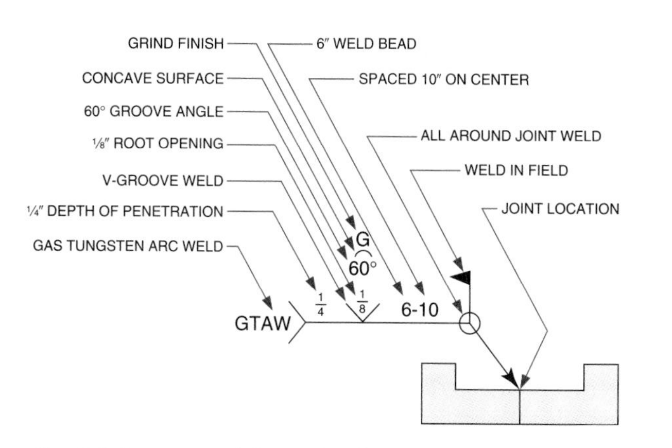

Fig. 26-22 ■ Position of information on a weld symbol.

Structural Steel Drawing Conventions

Structural steel drawings are of several types: design drawings, working (shop) drawings, and erection drawings. *Design (schematic) drawings* are very symbolic and show only the position of each structural member with a single line. Notations describing each member's shape, size, and weight are included on each line. When several members with identical characteristics are aligned, the successive lines are labeled with a ditto symbol (DO) indicating that the shape, size, and weight of the member are identical to the previous member's. See Fig. 26-23.

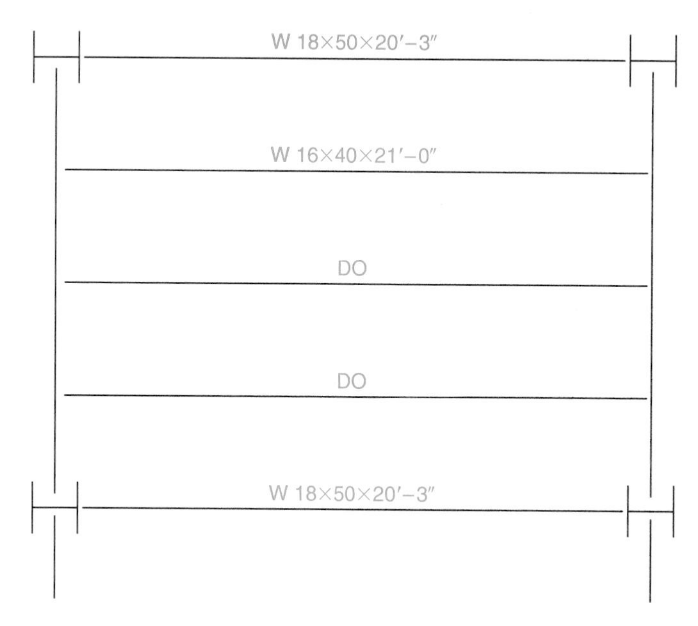

Fig. 26-23 ■ Schematic drawing of structural steel floor framing members.

Working (shop) drawings are complete orthographic engineering drawings showing the exact size and shape of each member, including every cut, hole, and method of fastening. *Erection drawings* show the method and order of assembling each member, which is coded for easy field identification. The specific methods used to prepare structural steel floor, wall, and roof framing drawings are covered in Chapters 28, 29, and 30.

CHAPTER 26 Exercises

1. Describe the differences between a design drawing, shop drawing, and erection drawing.
2. List the advantages and disadvantages of steel construction.
3. Describe the difference between beams, girders, columns, and purlins.
4. Name the types of structural steel shapes.
5. Draw the symbols for these steel shapes: wide flange beam, tee, angle, channel, and square bar.

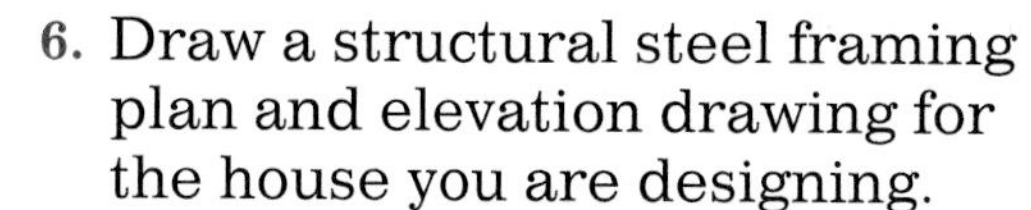

6. Draw a structural steel framing plan and elevation drawing for the house you are designing.

7. Complete Exercise 6 using a CAD system.
8. Name the types of fastening methods used in steel construction.
9. Draw the symbols for these welds: fillet, spot, seam, edge flange, V groove, and backing.

Structural Engineer

Structural engineers are civil engineers who specialize in structures. They use calculations to design the frames of buildings and other structures, such as bridges and tunnels. This professional area is considered one of the most complex, because of the high competence required in physics and math.

> "Structural engineering explains why different materials combine with each other to make structures work."

Nelson Baker

"Through my growing-up years, I was always interested in how things fit together," says Nelson Baker. So interested was he that he made it his life's work—he became a structural engineer.

Nelson first worked as a structural engineer in Seattle, Washington, where he designed bridges and tunnels. There his most challenging project was working on a three-layer tunnel. "It was an interesting project," he says. "The top layer of the tunnel was designed for pedestrians and bicycles, the middle layer for three lanes of auto traffic, and the lower level for two lanes of higher occupancy vehicle traffic, such as buses." Adding to the complexity of the design, the tunnel had to be connected on one end to a floating bridge and on the opposite end to Interstate 90.

Another challenge was created because the soil around the tunnel was consolidated clay. When part of the soil was removed, the surrounding soil expanded. "A six-inch diameter hole one day would be only a four-inch hole the next," says Nelson. "That made us wonder what the soil would do and what kind of forces would act on our tunnel because of the soil. There were many unknowns." The structural engineers built an expensive test tunnel to find out how the soils would behave. Such testing at the beginning saved time and money when the project was constructed.

The tunnel project was particularly interesting to Nelson because of the level of geometry involved. "If you like geometry, it was an absolute joy," says Nelson. "Everything was on a super elevation, a vertical curve, a horizontal curve, the tunnel was cylindrical—there wasn't any easy geometry at all. You had to use math and geometry to find out where you were in the tunnel." Perhaps that's one reason why the publication *Engineering News Record* gave it the "Outstanding Engineering Award of the Year."

Ted Mishima/Georgia Institute of Technology

CHAPTER 27

Disaster Prevention Design

Hurricanes, tornadoes, high winds, floods, blizzards, gas leakage, wildfires: these disasters are not preventable. However, precautions can be designed into a structure to prevent or minimize the damage they cause.

■ **When a disaster such as a hurricane strikes, structural damage can be severe. The use of appropriate design and construction techniques can often limit damage and save millions of dollars.** © *Dennie Cody/FPG International Corp.*

Objectives

In this chapter, you will learn to:

- describe the measures that can be taken during construction to minimize potential damage from natural disasters.
- describe how to prevent gas leakage.
- name ways to provide fire protection for a structure and its residents.
- discuss methods for ensuring clean air and water in a building.

Terms

carbon monoxide
Environmental Protection Agency (EPA)
radon
Underwriters Laboratories (UL)

Preventing Wind Damage

Winds do not need to reach hurricane or tornado velocity to seriously damage or demolish a structure. To design a structure with maximum wind resistance, a continuous and strong structural link must be made from the foundation to the roof. See Fig. 27-1. Many methods are employed to achieve a strong structure—depending upon the type of construction.

To maximize structural strength requires vertical rebars and continual pours during concrete construction. In wood construction, the effective placement of timber connectors is vital.

Sills must be fastened firmly to the foundation using foundation anchors. See Fig. 27-2. Long anchors provide additional control by connecting the foundation and sill to the wall framing. Tie straps help hold the wall framing to the sill, as shown in Fig. 27-3. Additional blocking on the sill and between vertical members also helps anchorage and provides added rigidity to the structural frame. Steel ties between joists and the foundation provide protection against uplift. See Fig. 27-4.

Fig. 27-2 ■ Use of foundation anchors.

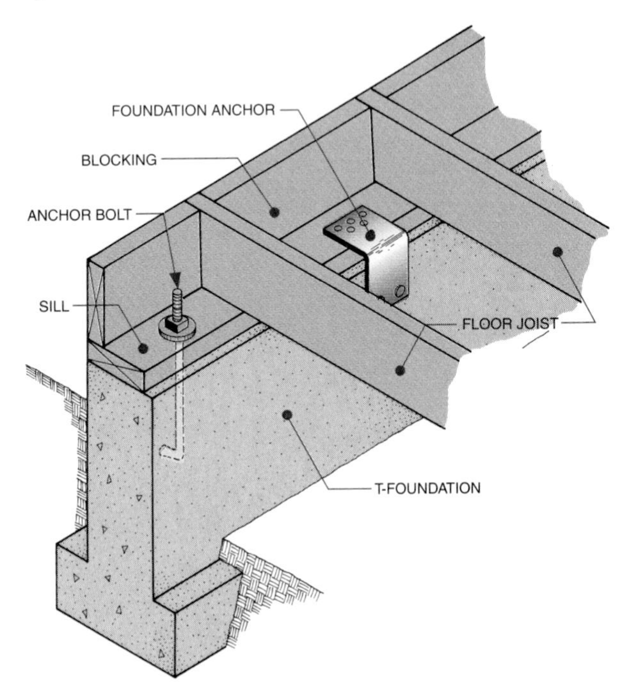

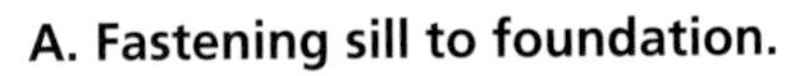

A. Fastening sill to foundation.

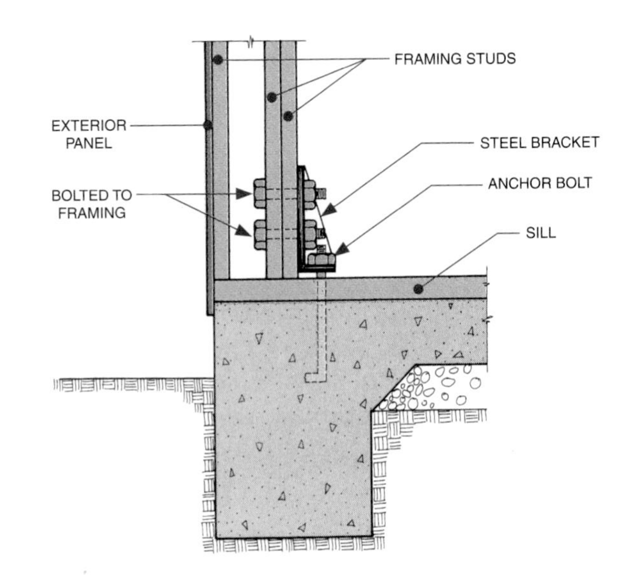

B. Fastening sill and foundation to framing.

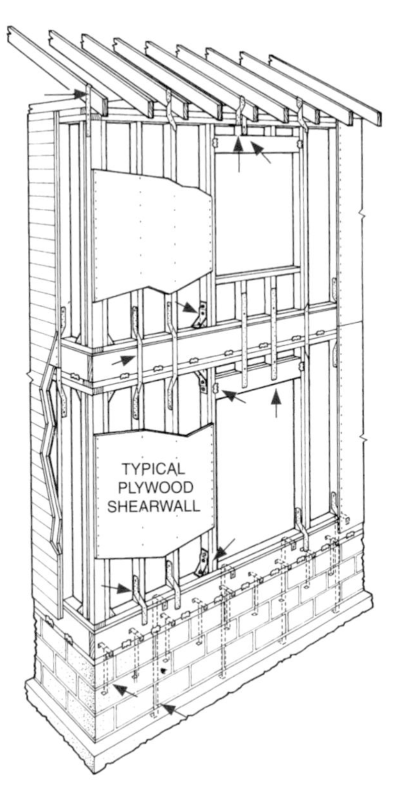

Fig. 27-1 ■ Continuous load transfer design using structural link connectors. *Simpson Strong-Tie Co. Inc.*

Fig. 27-3 ■ Sill tie-down straps.

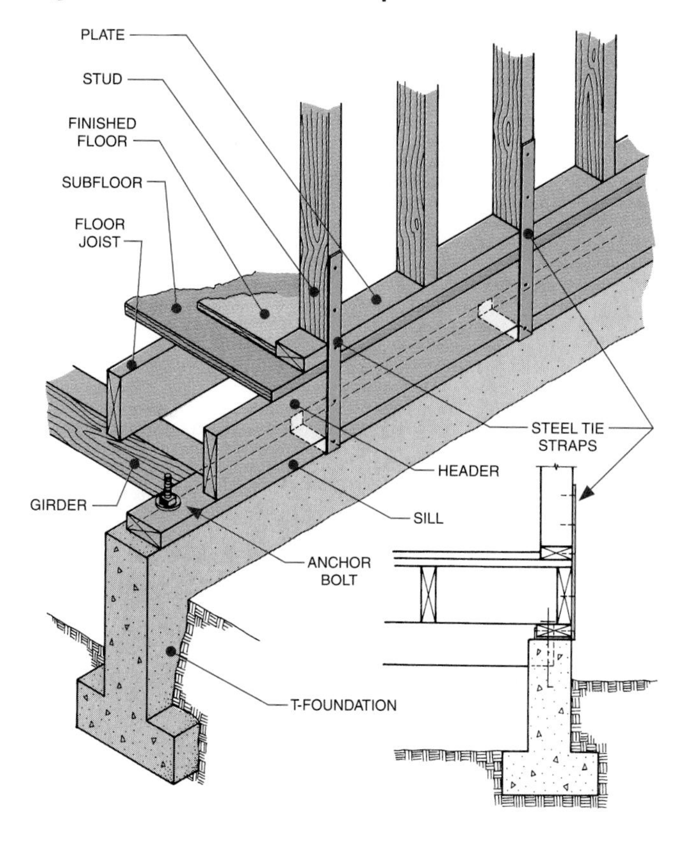

Fig. 27-4 ■ Use of ties to prevent uplift.

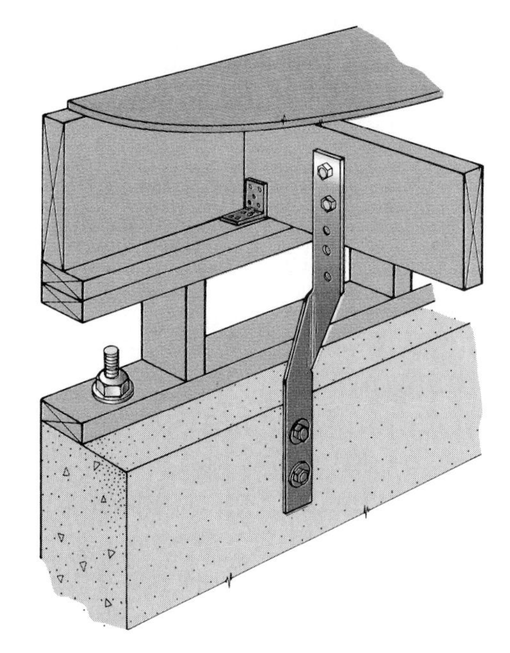

Non-masonry wall surfaces should be made of shear stress plywood or a material with equal wind velocity rating. See Fig. 27-5. Windows must also be rated to withstand the same wind velocity as the structural walls.

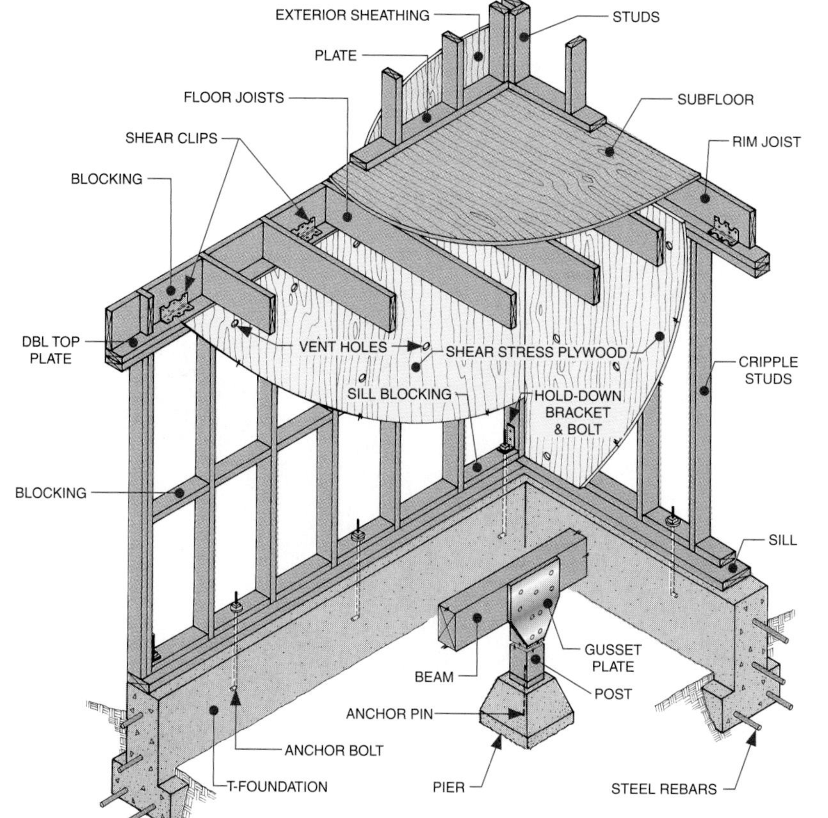

Fig. 27-5 ■ Methods used to protect against wind and seismic (earthquake) damage.

On bi-level platform frames, the two levels must be connected in order to transfer the uplift forces from the upper studs to the lower studs. Figure 27-6 shows several connecting methods. Bolt hold downs between joists and studs on both levels will also provide this connection. When bolts are used, they must be firmly fastened with nuts.

Connections between wall framing and roof framing are also critical. Winds trapped under overhangs can damage or break away soffits and fascia boards. Any area where wind can be trapped should be ventilated to allow the air to escape. Roof sheathing should be connected with hurricane clips. Roofing screws or nails, not staples, must be used to attach sheathing, soffit, and fascia material. In high wind areas, wind shear shingles should be specified and connectors used to attach top plates and rafters to studs. See Fig. 27-7. Structural straps tie studs and joists together, as shown in Fig. 27-8.

Fig. 27-6 ■ Platform framing connectors.

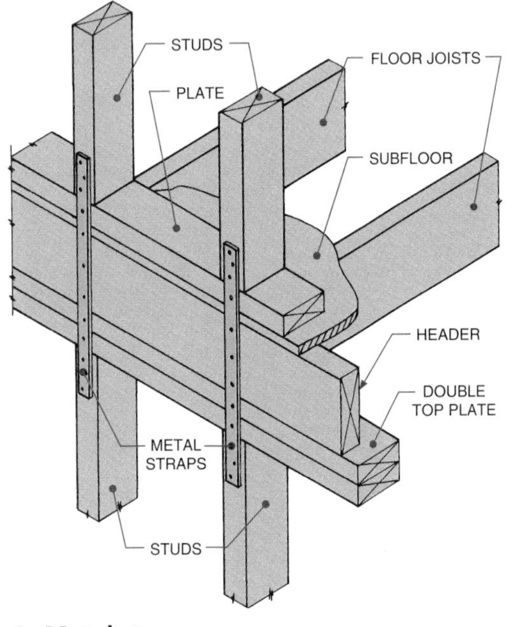

A. Metal straps.

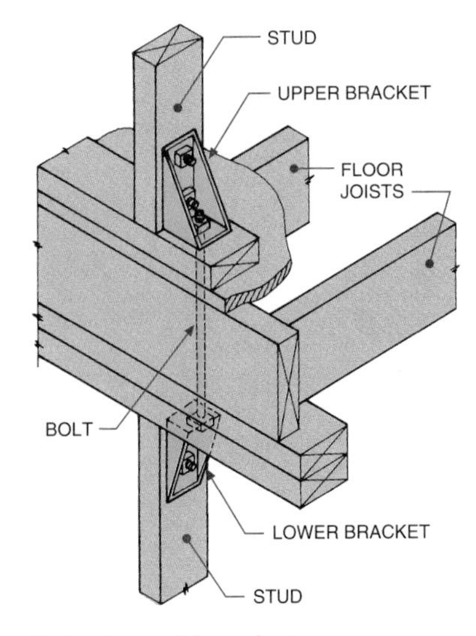

B. Bolt and brackets.

Fig. 27-7 ■ Rafter-plate connectors.

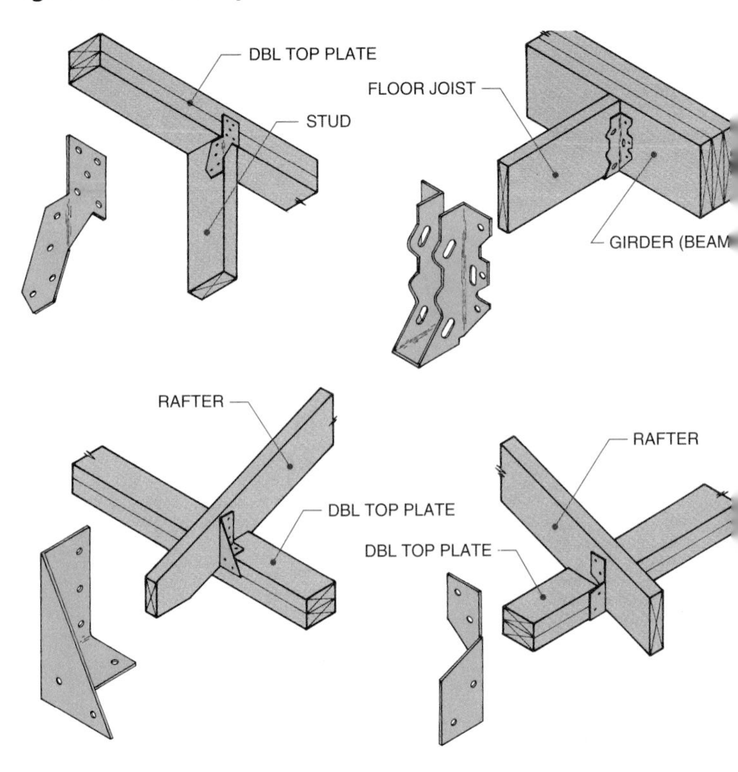

Preventing Earthquake Damage

The structural features designed to reduce wind damage also apply to earthquake damage reduction. Figure 27-9 illustrates many of these features. How-

Fig. 27-8 ■ Joist-stud structural tie.

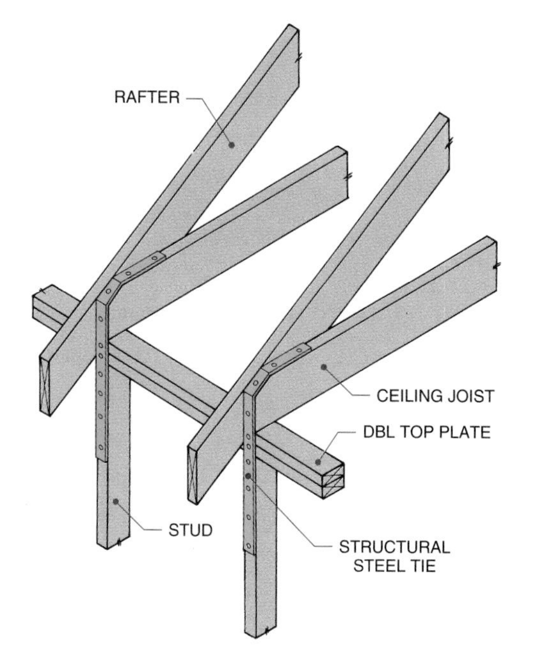

Fig. 27-9 ■ Hurricane- and earthquake-resistant wall design.

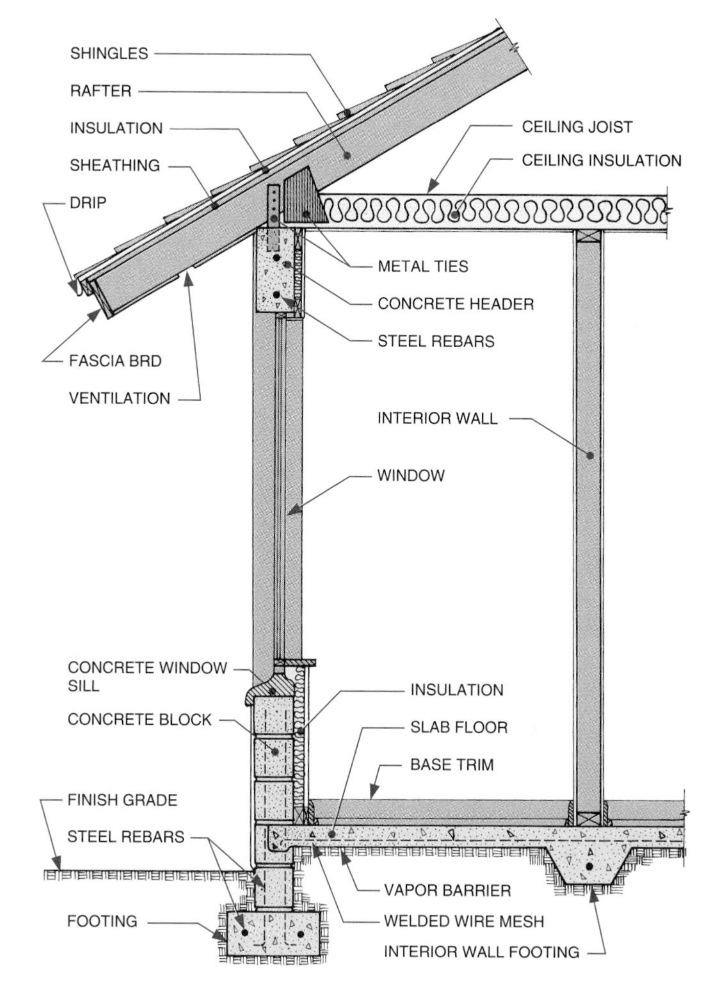

ever, special attention is needed for earthquake protection as follows:

1. Design a continuous structural link from foundation to roof according to the same principles as those for preventing wind damage.
2. Specify that all gas appliances are to be connected with flexible lines.
3. Fasten built-in appliances to structural members, not to wallboard or trim.
4. Don't design masonry chimney materials above the roof line. Preferably avoid masonry chimneys at all levels.
5. Specify push-type cabinet latches (touch latch).
6. Specify hook and eye latches on workshop cabinets.
7. Specify security film for windows—to prevent shattering.
8. Use ball-bearing supports to prevent columns from flexing or bending. See Fig. 27-10.

Fig. 27-10 ■ Ball-bearing supports below columns help the structure absorb the movement of the ground during an earthquake.

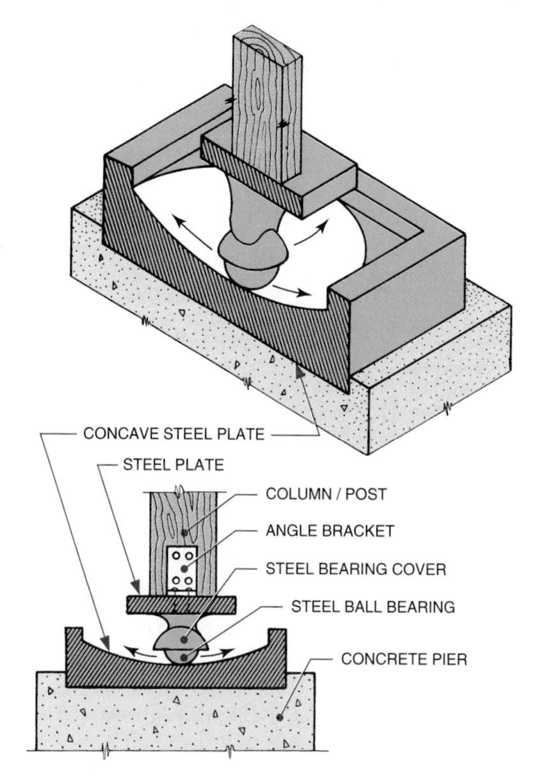

9. Footings should cover the largest horizontal area possible.
10. Avoid using suspended floor systems which could collapse as one unit on the floors below.
11. Ensure good site drainage.
12. Strap all gas, water, and vent pipes to framing members.
13. Bolt all tall, heavy wall furniture into wall studs.

Preventing Gas Leakage

The nature of gas is to expand and fill all available space. When gas escapes from containers, pipes, or the soil, it may be trapped in a sealed building. These trapped gases can seriously injure or kill people who inhale them. If ignited with a spark or flame, some gases can explode and cause great damage.

Natural gas, propane, carbon monoxide, and radon are potentially the most dangerous gases. Gas fumes created by many synthetic building materials and interior furnishings are suspected of causing long-range health problems.

Natural and Propane Gas

To minimize the potential for gas leakage, specify strapping pipes to structural members and use flexible lines to all appliances. Locate propane containers as far from structures as possible. Specify electronic pilots for all gas appliances.

Carbon Monoxide (CO)

Carbon monoxide is a colorless, odorless gas produced by combustion. Poisoning from CO gas causes brain damage at low levels and death at high levels. Carbon monoxide is produced by all combustion appliances. It is not a problem unless the gas leaks into the structure instead of venting to the outdoors. An inspection should be specified and CO detectors installed during construction. Detectors should be mounted on ceilings on each story.

Radon

Radon is a colorless gas produced by the natural decay of the element radium. This radioactive gas enters a structure from the soil. To minimize risk, specify polyethylene membranes under floors and slabs. Specify total parging (protective plasterwork) on all walls below the grade line. Specify a separate outside air source for fireplaces to avoid drawing radon in through small cracks in the floor system. Ventilate crawlspaces to the outside. Provide a fan-driven air escape vent to avoid gas buildup. Test for radon before construction begins. In areas with high levels of radon, a central vent from under the foundation through the roof can divert gas from entering the house. See Fig. 27-11.

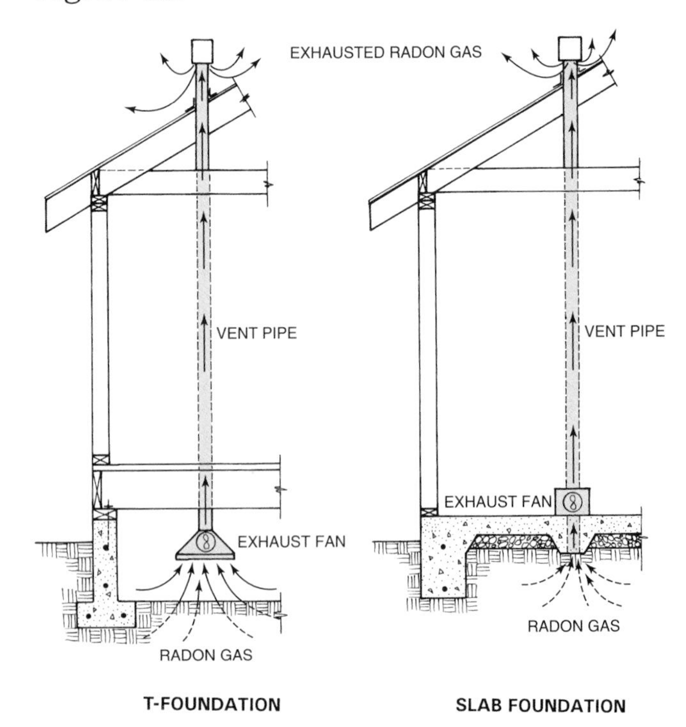

Fig. 27-11 ■ Radon prevention techniques.

Fire Prevention and Control

The best fire protection is to observe building codes and to use electrical and gas devices approved by **Underwriters Laboratories (UL)**, a nonprofit agency that tests products for safety. All electrical wiring and devices must also be grounded according to UL standards. Fire control measures include specifying smoke detectors on each level of the house. Fire extinguishers and water hoses should be specified for the kitchen, laundry, garage, and workshops.

Water and Air Purification

To provide the maximum pure air, specify electronic air filters for all furnaces, air conditioners, and the air exchangers inside heat pumps. Air cleaners may be installed to quickly purify contaminated air.

Water quality varies greatly from one community to another. To maximize water purity, specify a water system which may include a water softener, carbon filter, ultraviolet filter, and/or reverse osmosis unit. You may specify a dedicated faucet for pure drinking water and for the refrigerator ice-maker.

Other Safety Considerations

Although the effects of synthetic material vapors may not be immediate, the long range health effects may be serious. Avoid the interior use of exposed and unsealed building materials that contain toxic materials as defined by the **Environmental Protection Agency (EPA)**, a federal regulatory agency. Use solid wood products where possible.

Household pests can be controlled in the design process by specifying the installation of wall cavity tubes. These tubes safely carry pesticides into the walls without detectable levels entering living spaces.

CHAPTER 27 Exercises

1. List the design features you will use in the house you are designing to prevent wind, earthquake, gas, and fire damage.
2. Sketch structural methods of preventing wind and earthquake damage.
3. Complete Exercise 2 using a CAD system.

4. Explain how you will provide for water and air purification in the house you are designing.
5. Sketch a ventilating system you will use to provide adequate ventilation in the house you are designing.

PART 7

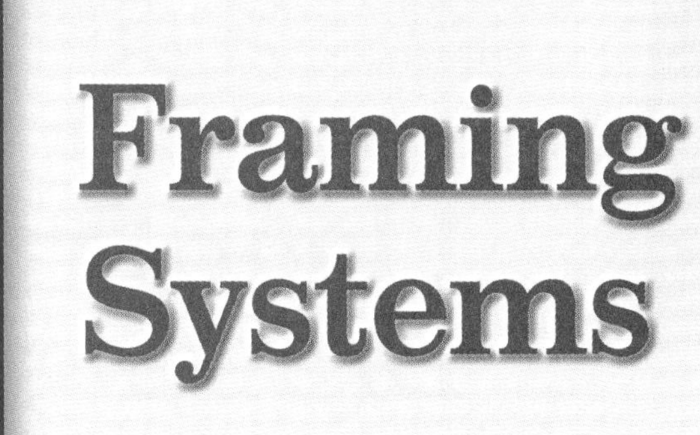

CHAPTER
28 Floor Framing Drawings
29 Wall Framing Drawings
30 Roof Framing Drawings

CHAPTER

28

Floor Framing Drawings

Objectives

In this chapter, you will learn to:

- identify the components of floor systems.
- draw a floor framing plan that shows all structural parts.
- draw details of sills, supports, and stairwells.

Terms

bays
blocking
decking
I-joists
lookout joists
truss joists

The design of floor framing systems demands careful calculation of the live and dead loads acting on a floor. The exact size and spacing of floor framing members, plus the most appropriate materials, must be selected.

Some floor framing drawings are prepared to show only the structural support for the floor platform. Others may illustrate details of construction, such as the attachment of the floor frame to the foundation.

■ **Floors must be designed to support heavy loads.**
© E. Alan McGee FPG International Corp.

Types of Platform Floor Systems

Systems that are supported by foundation walls and/or beams or girders are called platform floor systems. These differ from ground level slab floors that are structurally part of the foundation. (Refer back to Chapter 23.) Platform floor systems are divided into three types: conventional, heavy timber (plank-and-beam), and panelized floor systems.

- *Conventional systems.* Conventionally framed platform systems provide a flexible method of floor framing for a wide variety of designs. Floor joists are usually spaced 16″ (406 mm) apart and are supported by side walls of the foundation and/or by girders. See Fig. 28-1.
- *Plank-and-beam systems.* The plank-and-beam method of floor framing uses fewer and larger members than conventional framing. Because of the increased size and the rigidity of the larger members, longer distances can be spanned. Unlike conventional framing, no cross bridging is needed between joists. Extended blocking provides the needed rigidity. See Fig. 28-2.
- *Panelized systems.* Panelized floor systems are composed of preassembled sandwich panels. The panels are made from a variety of skin and core materials. Panelized systems are used for long, clear spans. The main advantage of panelized systems is the reduction of on-site construction costs. See Fig. 28-3.

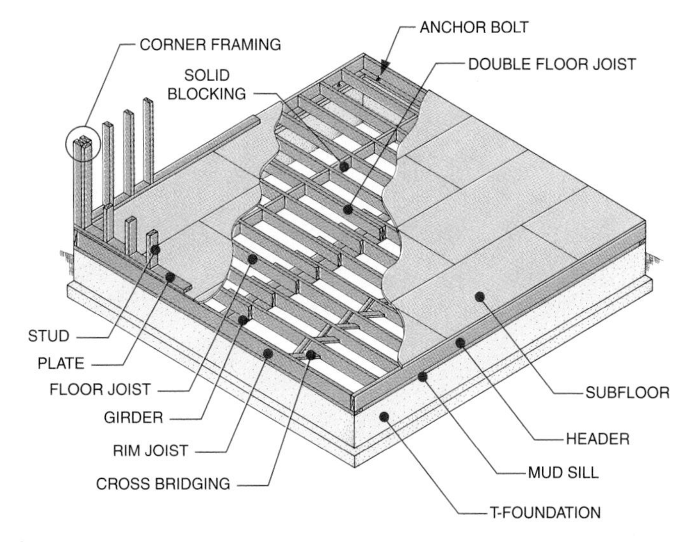

Fig. 28-1 ■ Conventional platform floor system.

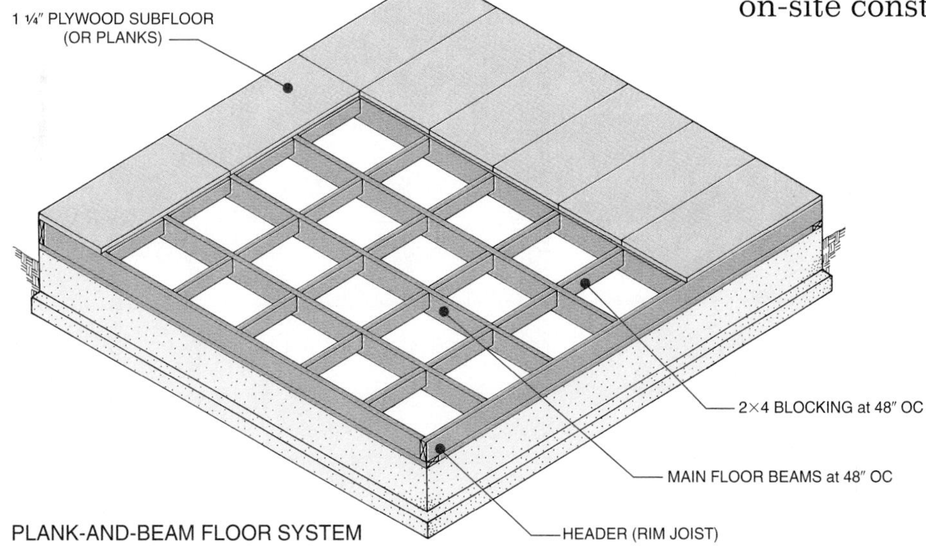

Fig. 28-2 ■ Plank-and-beam platform floor system.

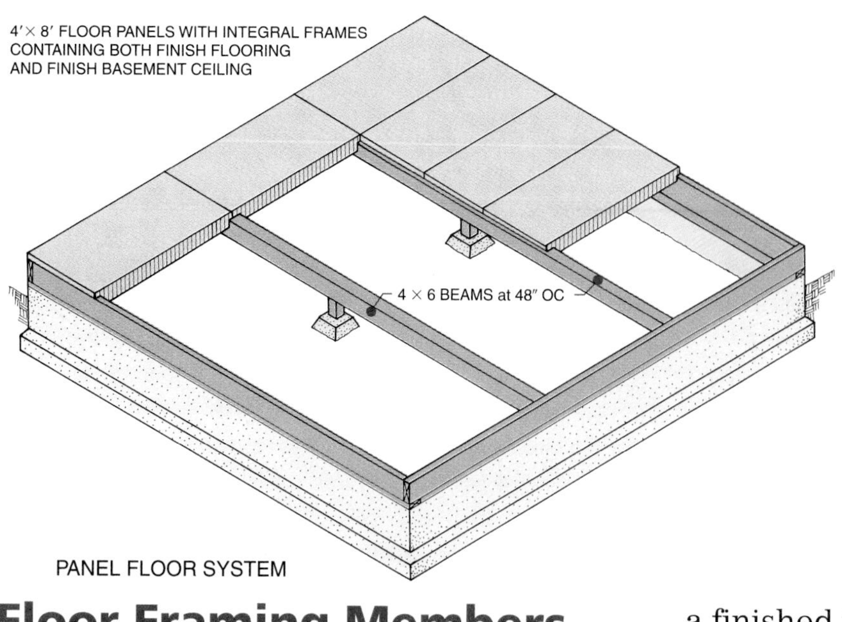

Fig. 28-3 ■ Panelized platform floor system.

Floor Framing Members

Floor framing members for platform floors consist of decking, joists, and beams or girders. Supporting walls and columns also are part of the floor framing system.

Decking

Decking is the surface of a floor system. Decking usually consists of a subfloor and a finished floor, although in some plank systems these are combined. Subfloor decking materials range from wood boards and plywood sheets to concrete slabs or corrugated steel sheets. The finished floor may be wood, ceramic, vinyl, concrete, or carpeting.

WOOD DECKING Boards or plywood sheets used as subflooring are placed directly over the joists. See Fig. 28-4. Unless the

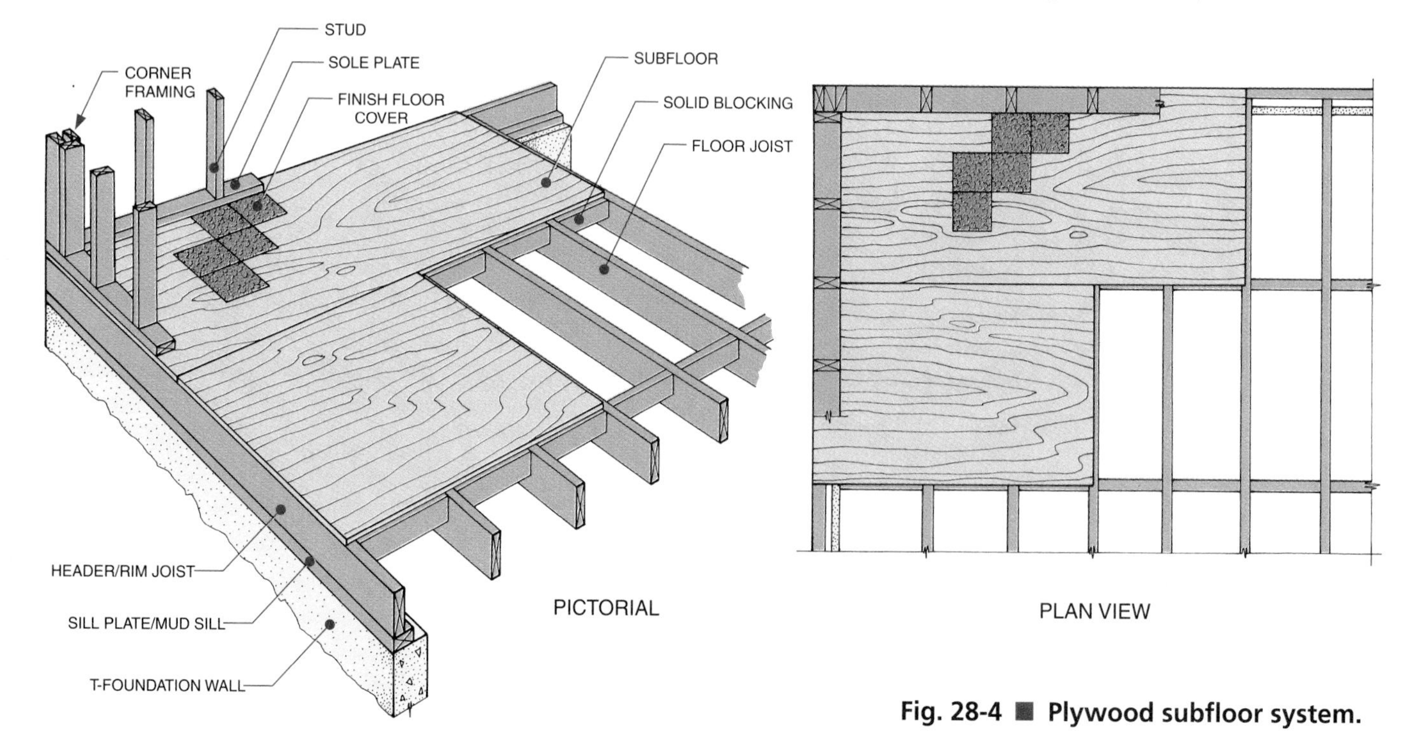

Fig. 28-4 ■ Plywood subfloor system.

edges of plywood sheets are tongue-and-grooved, blocking may be needed beneath the joints. **Blocking** consists of short pieces of lumber nailed between the joists. The blocking provides additional support for the subfloor joints. Sole plates for exterior and interior walls are laid directly on the subflooring.

The functions of the subfloor are to:

- Increase the strength of the floor and provide a surface for laying a finished floor.
- Help to stiffen the position of floor joists.
- Serve as a working surface during construction.
- Help to deaden sound.
- Prevent dust from rising through the floor.
- Help insulate.
- Act as a buffer to soften and reduce the hard impact of slab floor construction. See Fig. 28-5.

Finished flooring is installed over the subfloor. The finished floor provides a wearing surface over the subfloor. If there is no subfloor, the finished floor must be tongue-and-groove boards 1½″ to 2″ thick. Hardwood such as oak, maple, beech, and birch, is used for finished floors over wood subfloors. Vinyl, ceramic tile, marble, and carpeting are also used.

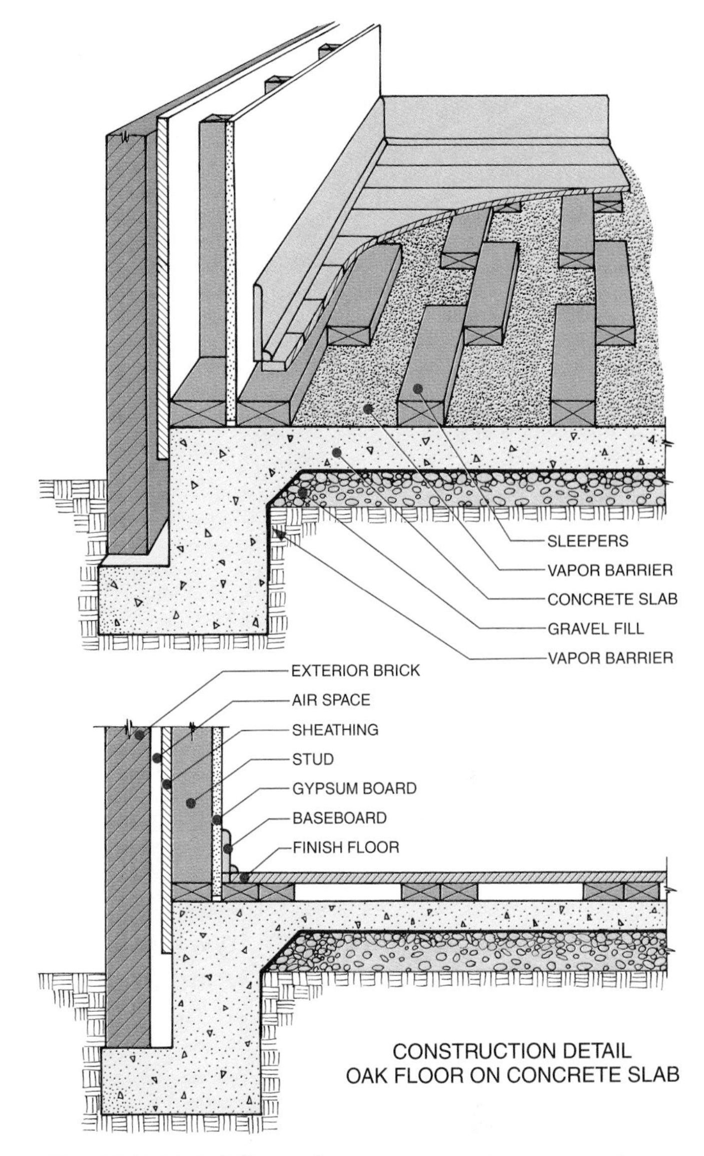

Fig. 28-5 ■ Subfloor sleepers separate a concrete slab from the finished floor.

STEEL AND CONCRETE DECKING Steel decks for floors (and roofs) use corrugated sheets, interlocking galvanized steel panels, or cellular units over steel beams. See Fig. 28-6. Steel deck details or sectional drawings are usually prepared to show the relationship of the decking to the structural support members.

When steel subfloors are used, they are usually constructed of corrugated sheet steel. These subfloors act as platform surfaces during construction and also provide the necessary subfloor surface for a concrete slab floor. Steel subfloors are not always needed with concrete slabs. The concrete slabs can function as subfloors if the concrete floor is precast with reinforcement bars.

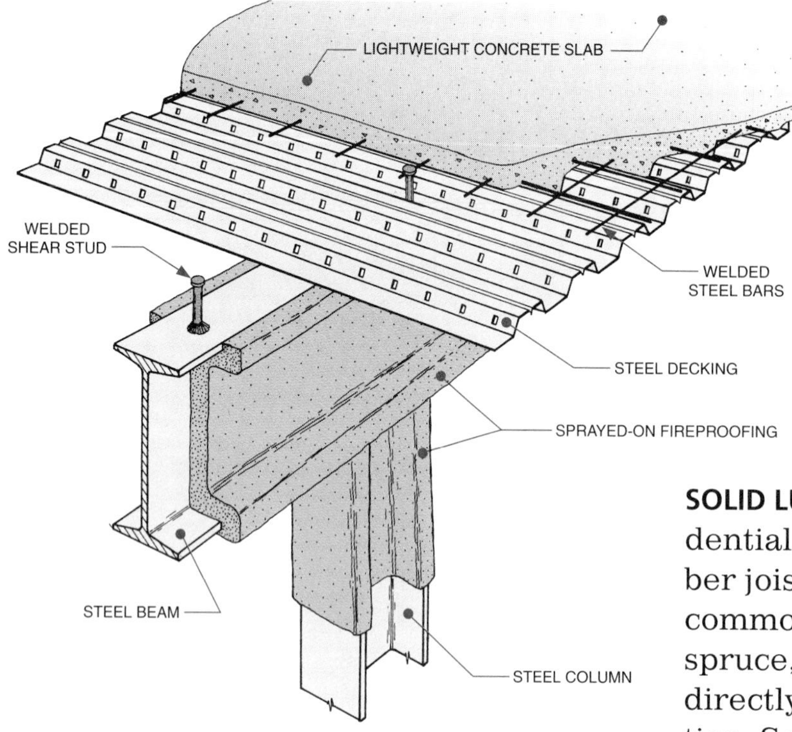

Fig. 28-6 ■ Steel deck details.

Joists

Floor joists are horizontal members that rest on a wall and/or girder (beam) and support the floor decking. Floor joists must support the maximum live load of the floor. Many variations of wood and steel joists are manufactured.

SOLID LUMBER JOISTS Conventional residential framing normally uses solid lumber joists. Solid lumber joists are most commonly made from Douglas fir, pine, spruce, or hemlock. All joists must rest directly on a girder, beam, wall, or foundation. See Fig. 28-7.

Double joists, known as *trimmers,* are used under partitions (interior walls) to provide added support. Sometimes a small space is left between these joists to provide a channel for electrical or piping access. See Fig. 28-8. Joists are often intersected at right angles to reduce spans. At

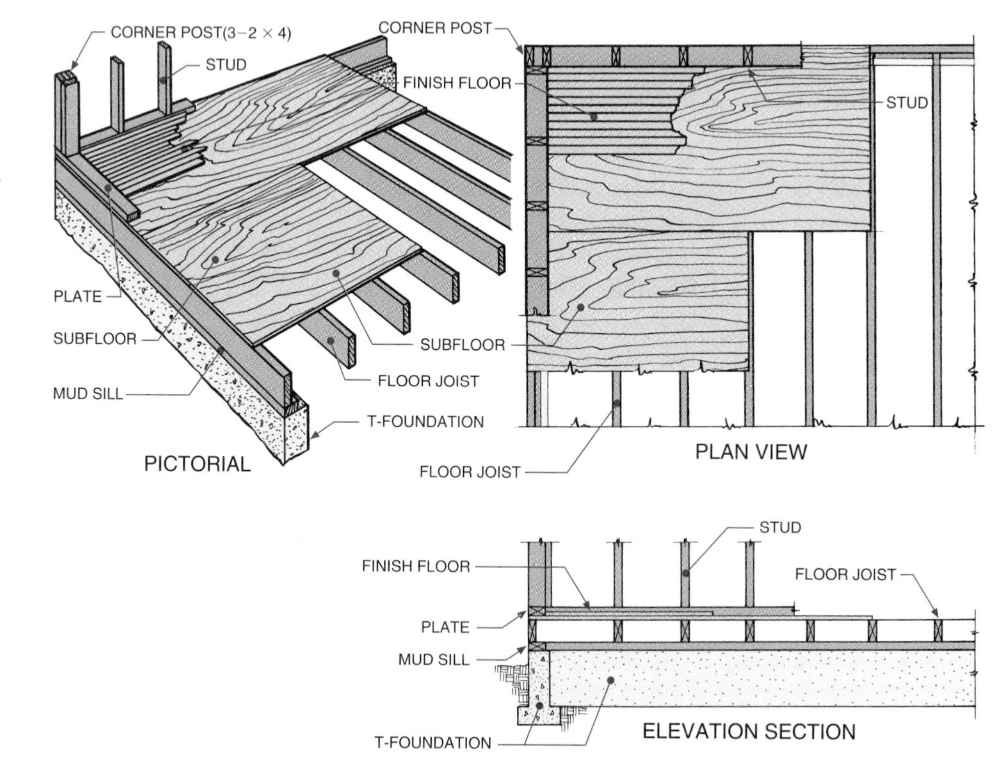

Fig. 28-7 ■ Use of solid lumber joists in a conventionally framed floor system.

these intersections, double joists and joist hangers are used to attach joists to trimmers or beams. See Fig. 28-9.

To prevent drift and warp and to distribute loads more evenly, bridging is used between joists. See Fig. 28-10.

Wherever joists need to be cut for an opening, such as a stairwell, chimney, or hearth, it is necessary to provide trimmers and auxiliary joists called *headers*. Headers are placed at right angles to trimmers to support the ends of joists that are cut. A header cannot be of greater depth than joists. Therefore headers are usually doubled (placed side by side) to compensate for additional loads. See Fig. 28-11. Double headers and trimmers are also used for floor, ceiling, and roof openings around fireplaces, chimneys, stairwells, and skylights.

The size, spacing, and strength needed for joists depend on the loads acting on a floor. (Review Chapter 22 and see Appendix A for the physical effects of loads, material strength, member size, and spacing.) Standard joist sizes for most residential construction range from 2 × 6 to 2 × 14. Normal residential spacing of wood

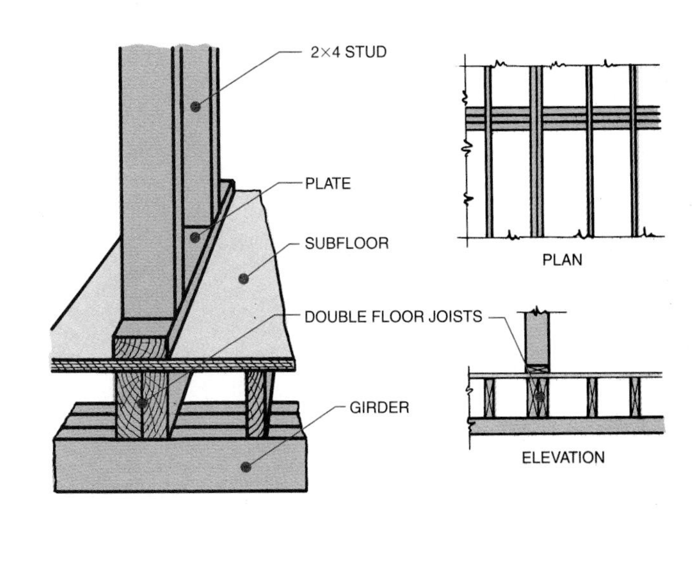

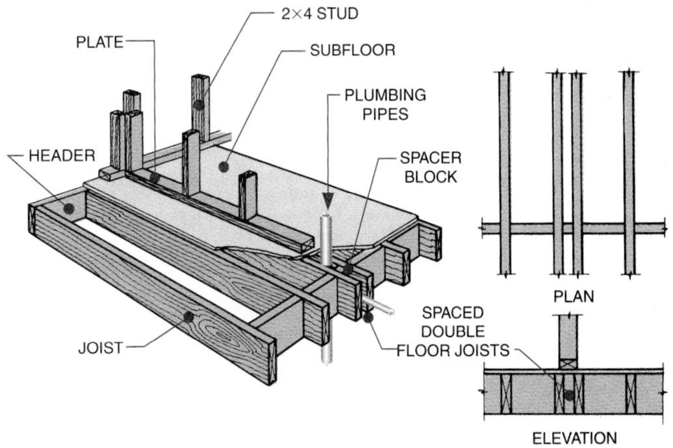

Fig. 28-8 ■ Trimmers (double joists) are used under partitions.

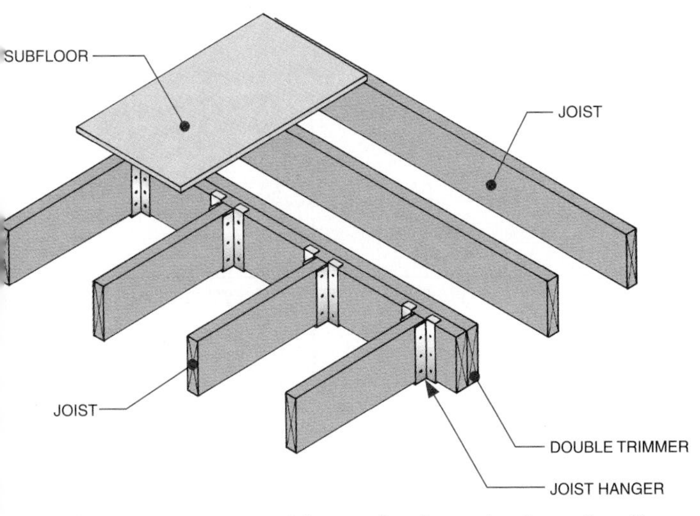

Fig. 28-9 ■ Assembly method used when the direction of floor joists is changed.

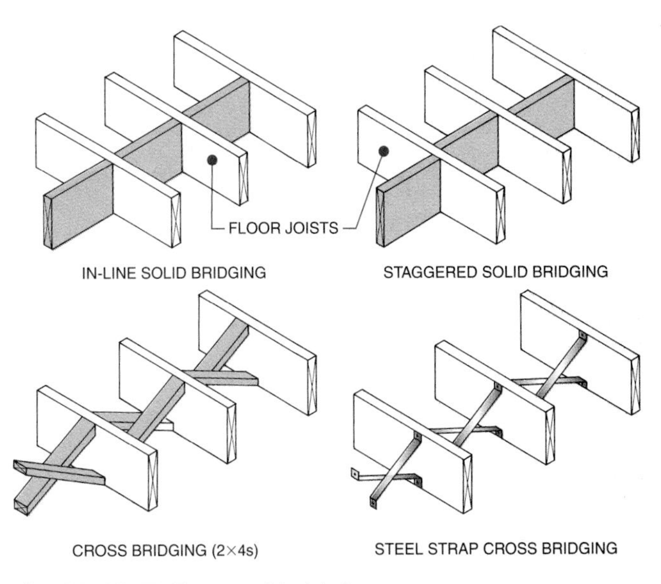

Fig. 28-10 ■ Types of bridging.

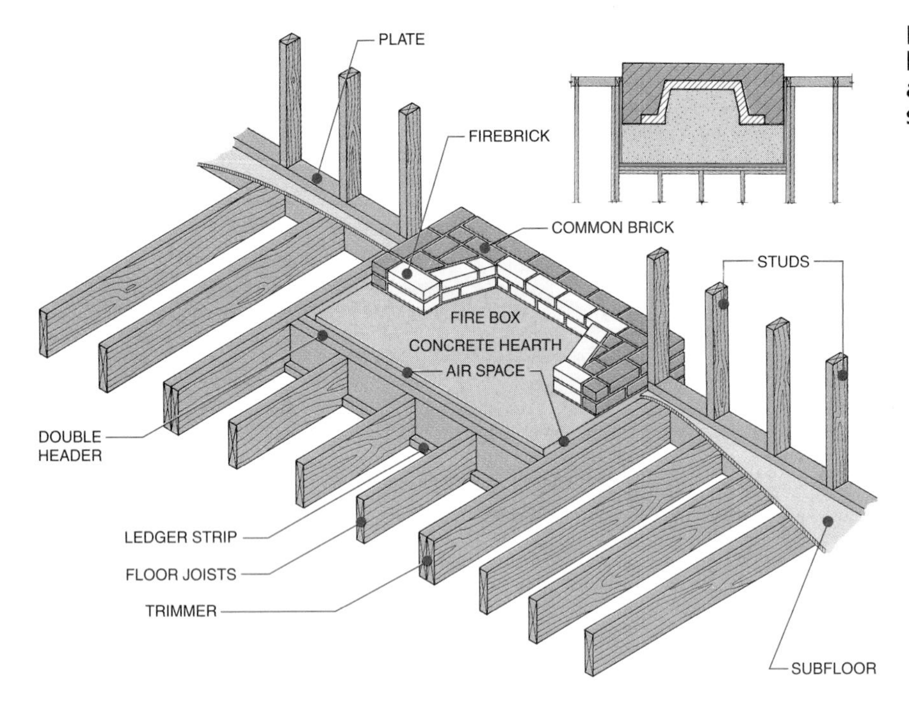

Fig. 28-11 ■ Trimmers and double headers are used around a fireplace and hearth to provide additional support.

joists is from 12″ to 24″ OC (on center). The most common spacing is 16" (OC).

Live loads bear directly on the decking and joists. Therefore, the total live load for the room containing the heaviest furniture and the heaviest traffic should be used to compute the total load for the entire floor.

To find the live load in pounds per square foot, divide the total room load in pounds by the number of square feet supporting the load. The average live load for most residences is between 40 and 50 pounds per square foot. See Appendix A, "Mathematical Calculations," for calculating loads.

Engineering tables are used to select sizes and spacing for structural members of standard grade wood. Figure 28-12 is a table used to select joist sizes and spac-

Fig. 28-12 ■ Allowable spans for spacing of floor joists.

		Maximum span (feet and inches)			
Standard sizes	**Spacing (OC)**	**Group 1** top quality	**Group 2**	**Group 3**	**Group 4** lowest quality
2 × 6	12″ 16″ 24″	10′–6″ 9′–6″ 7′–6″	9′–0″ 8′–0″ 6′–6″	7′–6″ 6′–6″ 5′–6″	5′–6″ 5′–0″ 4′–0″
2 × 8	12″ 16″ 24″	14′–0″ 12′–6″ 10′–0″	12′–6″ 11′–0″ 9′–0″	10′–6″ 9′–0″ 7′–6″	8′–0″ 7′–0″ 6′–0″
2 × 10	12″ 16″ 24″	17′–6″ 15′–6″ 13′–0″	16′–6″ 14′–6″ 12′–0″	13′–6″ 12′–0″ 10′–0″	10′–6″ 9′–6″ 7′–6″
2 × 12	12″ 16″ 24″	21′–0″ 18′–0″ 15′–0″	21′–0″ 18′–0″ 15′–0″	17′–6″ 15′–6″ 12′–6″	13′–6″ 12′–0″ 10′–0″

ing. To use this table, first select the wood group. Numbers indicate the quality of lumber. Number 1 is top quality, and number 4 is lowest quality. Next select the shortest span the joists must cross. The joist size and spacing (distance between members) is shown on the left. For example, for group 1, if the shortest span is 12′-0″, then the smallest joist that can be used is a 2 × 8 at 16″ OC, which has a maximum span of 12′-6″. Figure 28-13 shows the normal girder spans for standard grade wood, with and without supporting walls. Always remember: As the size, spacing, and load vary, the spans must vary accordingly. If the span is changed, the joist and girder spacing must also change. You should not overdesign but you MUST NOT underdesign.

FLOOR TRUSS JOISTS **Truss joists** have long, usually parallel, top and bottom chords connected by shorter pieces called webs. Triangular web patterns give truss joists the ability to span long distances, and they weigh less than solid lumber. Truss joists have open spaces through which plumbing and electrical lines can be run. They can also be designed to accommodate heating and air-conditioning ducts. Truss joists are manufactured using stress grade lumber or a combination of wood, metal, and/or composite materials. See Figs. 28-14 and 28-15.

Fig. 28-13 ■ Allowable spans for girders.

Girder size	Supporting walls	No wall support
4 × 4	3′–6″ 3′–0″	4′–0″ 3′–6″
4 × 6	5′–6″ 4′–6″	6′–6″ 5′–6″
4 × 8	7′–0″ 6′–0″	8′–6″ 7′–6″

Fig. 28-14 ■ Truss joists fabricated with lumber flanges (chords) and steel webs.

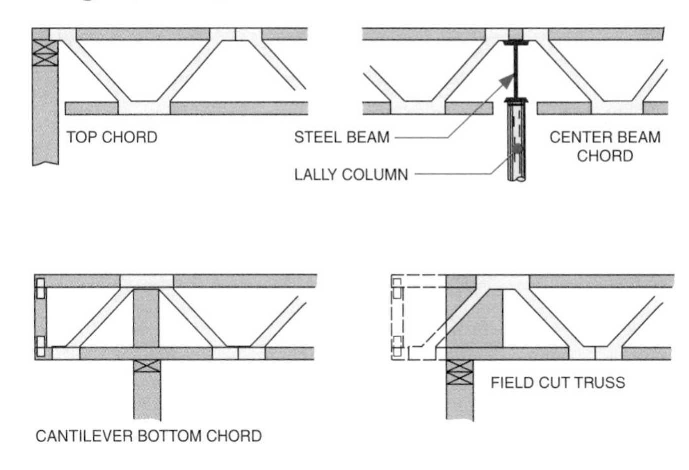

Fig. 28-15 ■ Truss joists.

A. For second floor framing.

B. For larger cantilevered extensions.

FLOOR I-JOISTS Structurally, **I-joists** are similar to steel I beams (S beams). Joist weight is reduced through the use of thin webs without sacrificing the strength and stability provided by the flanges. See Fig. 28-16. I-joist webs are usually made of stranded wood. The flanges are made of laminated or solid lumber. The web area can be cut, within limits, to receive HVAC, plumbing, or electrical lines without sacrificing strength. Figure 28-17 shows the use of I-joists with laminated headers and beams in a platform floor system.

LAMINATED JOISTS Laminated members are made by bonding layers of material together. Solid and parallel strand lumber are used to manufacture laminated joists, headers, and beams that can span up to 60′. Laminated joists are extremely straight, dimensionally stable, and without checks, cracks, or twists.

STEEL JOISTS Joists made of steel are either *bent sheet steel* or *open web* joists. See Fig. 28-18. Open web steel joists are more common and consist of angles and bars welded into truss shapes.

There are three types of steel open web joists: short-span, long-span, and deep long-span. Since joists are closely spaced, one note is used on drawings to give the number, classification, spacing, and length of all joists in a series. Only the first few joists in a series are usually noted. For example, if there are 8 short-span joists spaced at 3′ intervals over a 24′ distance, the note should read: 8SP @ 3′-0″ = 24′-0″. In addition, a notation is placed on a line representing the joists' direction and includes the length, class of the joist, and load range.

Girders and Beams

As explained previously, girders are the largest horizontal support members and rest on columns, posts, or exterior walls. In heavy construction, beams are the

Fig. 28-16 ■ I-joists compared with regular joists. *TrusJoist Macmillan*

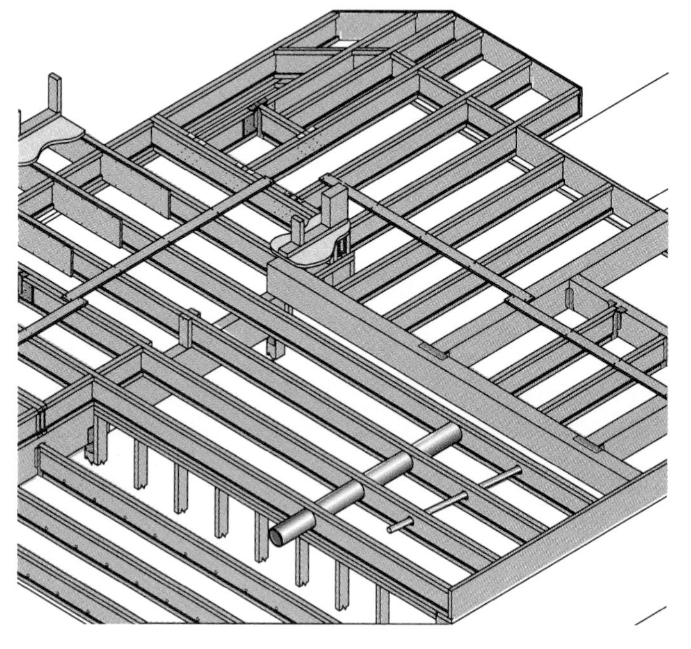

Fig. 28-17 ■ I-joists in a platform floor system. Note how pipes can be run through holes cut in the web. *TrusJoist Macmillan*

Fig. 28-18 ■ Types of steel joists.

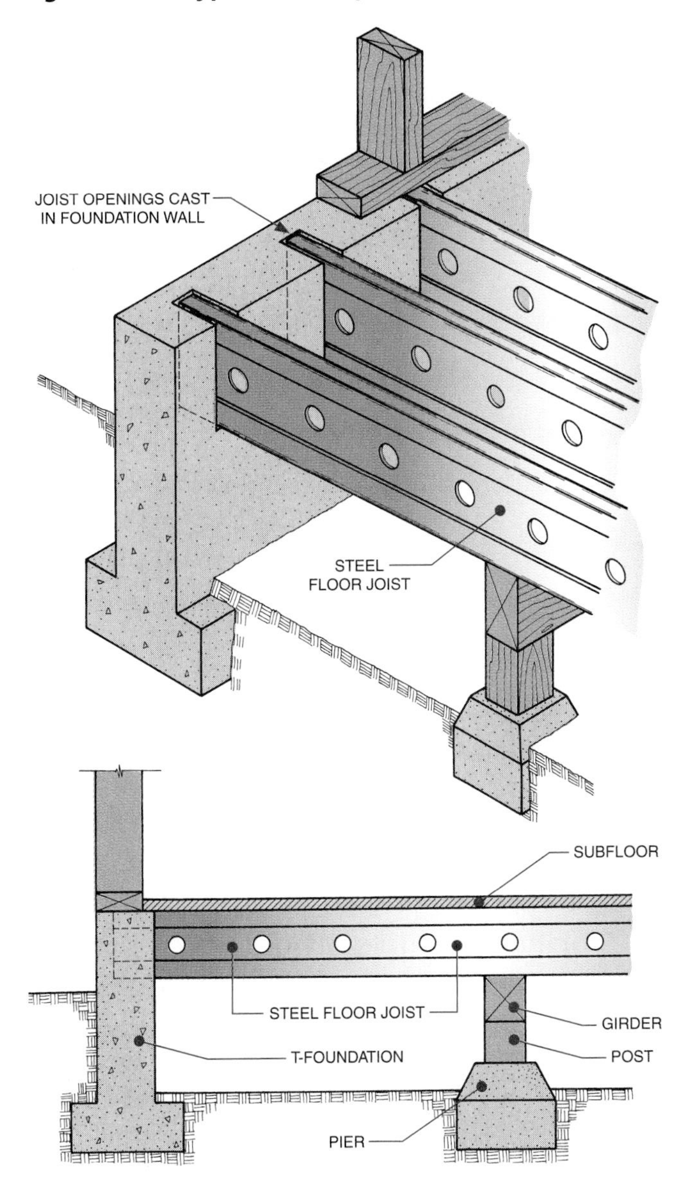

A. Bent sheet.

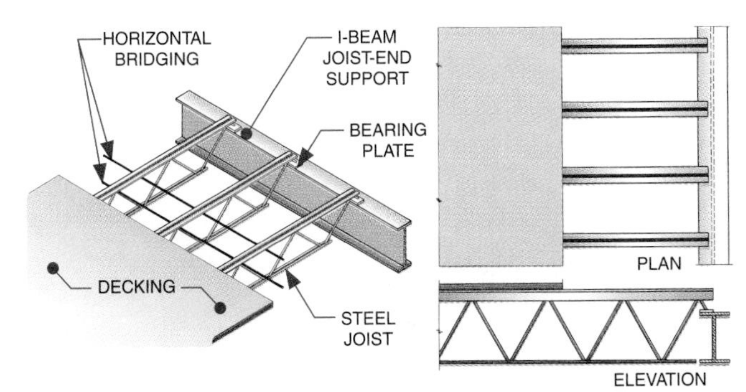

B. Open web.

members that span the distances between girders. In residential timber construction, the terms beam and girder are often used interchangeably.

WOOD GIRDERS AND BEAMS Girders (or beams) used in wood construction are either built up from solid lumber members, made of solid heavy timber, or laminated or fabricated, often with additional support or construction features. See Fig. 28-19. Figure 28-20 shows how wood girders (beams) are used to support joists.

When two or more girders are placed end-to-end to span the distance between outside supports, the joints between the girders must be placed directly over supporting columns or posts. Built-up girder members must be overlapped. Heavy timber girders should be half-lapped over

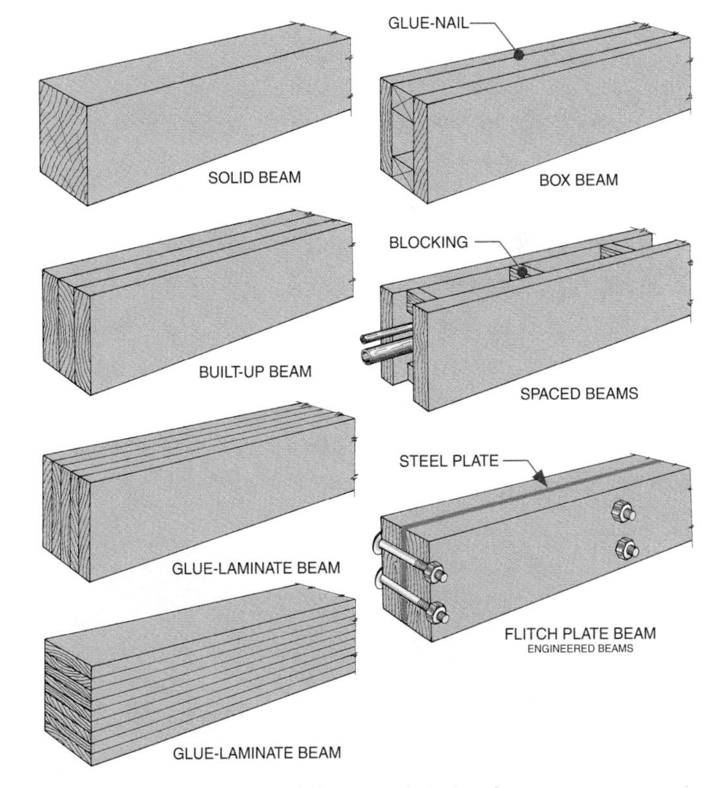

Fig. 28-19 ■ Types of beams (girders).

Fig. 28-20 ■ Methods of using wood girders to support joists.

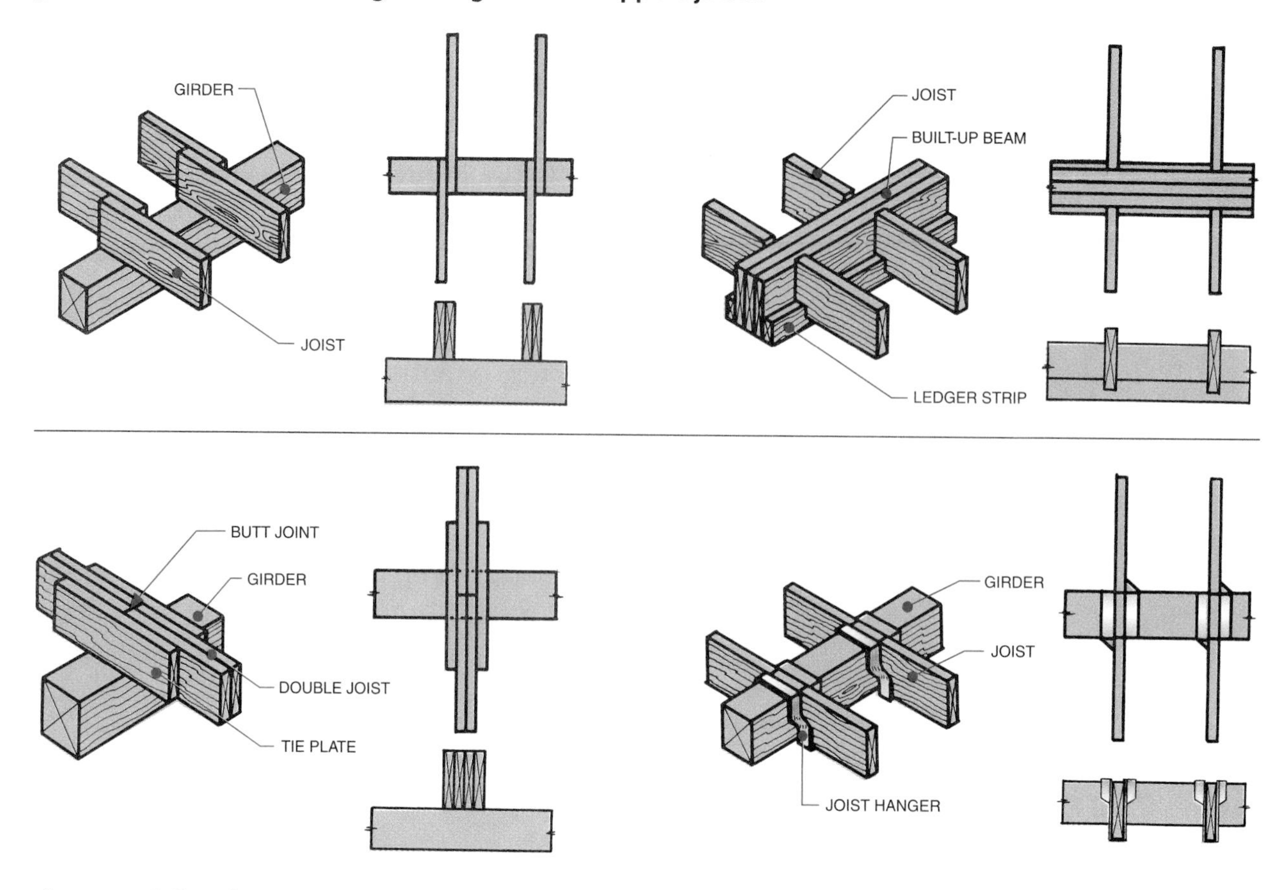

columns. Members can be spliced to reduce compression, tension, bending, and torque forces. See Fig. 28-21. Second-floor and higher-level girders are supported by bearing partitions or by columns aligned with lower-level columns. See Fig. 28-22.

Post-and-beam floor systems have no joists. In this system, the girders and blocking perform the function of joists. Girders rest directly on posts, and the subflooring rests directly on girders. In this type of construction, girders need to be spaced more closely together than girders that support joists. See Fig. 28-23.

STEEL GIRDERS AND BEAMS Steel girders and beams may be solid steel (S-shape or W-shape) or built-up steel assemblies. They are bolted or welded together to

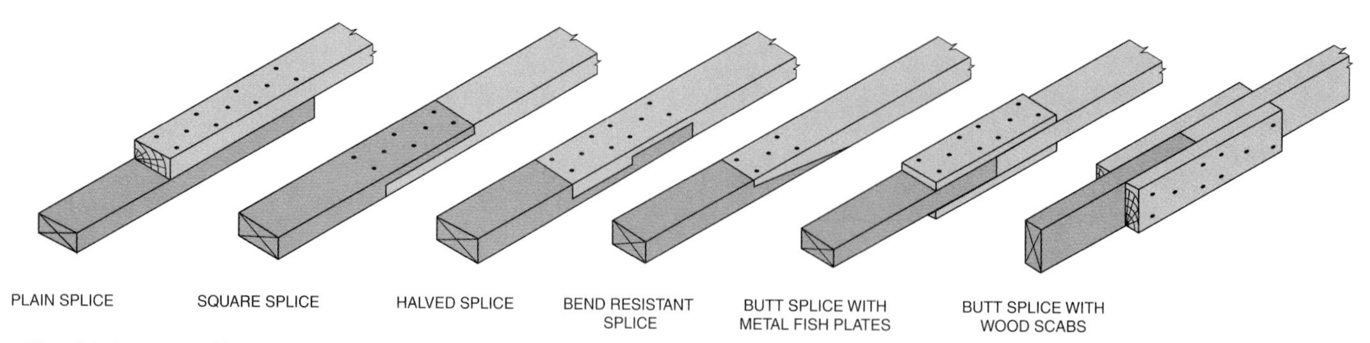

Fig. 28-21 ■ Splices that resist structural forces.

Fig. 28-22 ■ Girder and beam support methods at two different levels.

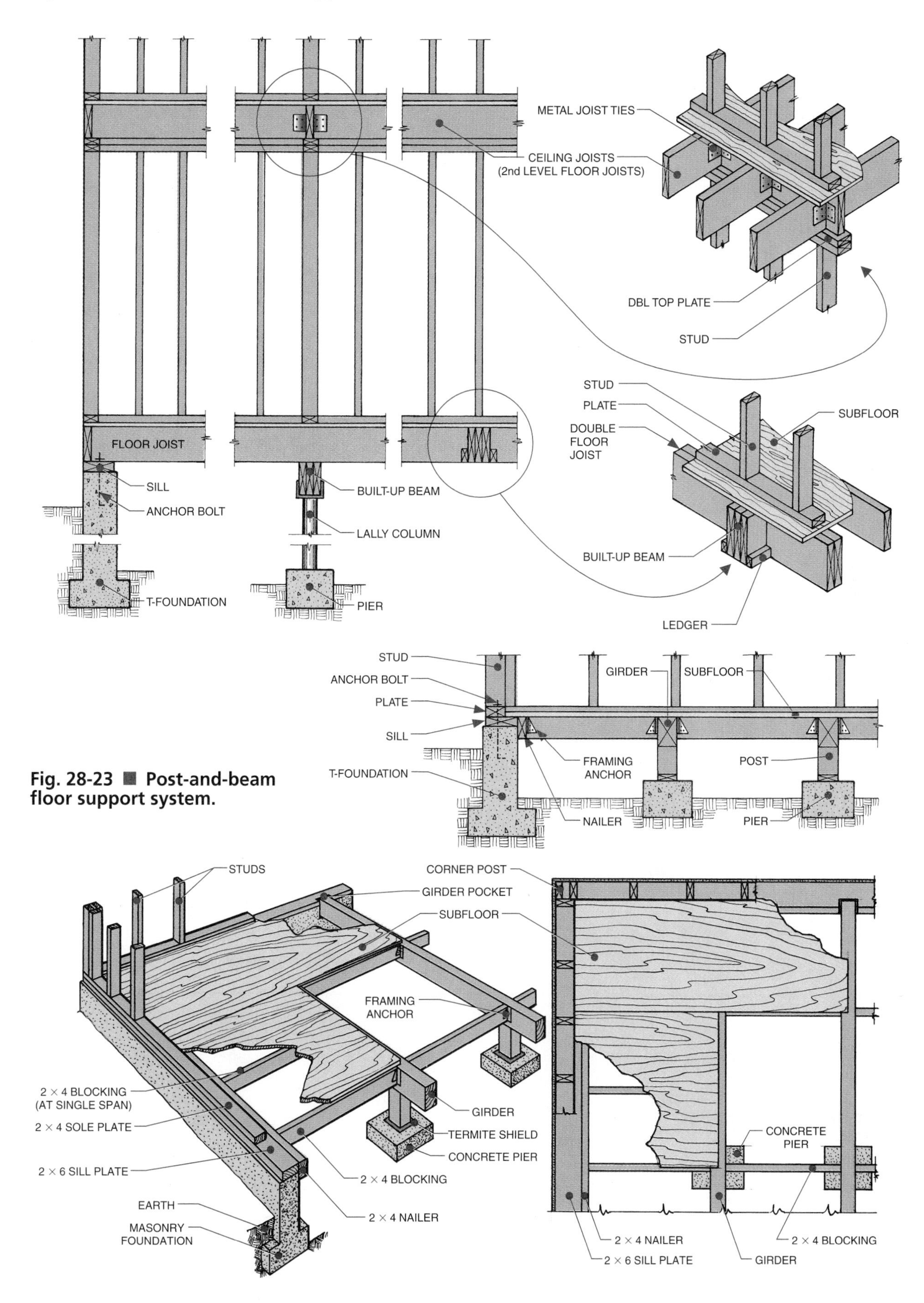

Fig. 28-23 ■ Post-and-beam floor support system.

form the major structural element of a steel cage construction. See Fig. 28-24. In wood construction, steel girders may be used in addition to or in place of wood girders.

Floor Framing Plans

If a set of architectural drawings does not include a separate floor framing plan, the builder, not the designer, determines the framing design. Floor framing plans for wood framing and steel framing use different conventions and symbols.

Floor Framing Plans for Wood

Floor framing plans for wood structures range from simple to very detailed, as shown in Fig. 28-25. In some cases, the direction of joists and girders may simply be shown on the floor plan (Fig. 28-25A). In the most detailed floor framing plans, each structural member is represented by a double line to show exact thicknesses (Fig. 28-25B).

The more simplified plan (Fig. 28-25C) is a short-cut method of drawing floor framing plans. A single line is used to designate each member. Chimney and stair openings are shown by diagonals. Only the outline of the foundation and post locations is shown.

The abbreviated floor framing plan (Fig. 28-25D) simply shows the entire area where the uniformly distributed joists are placed. The direction of joists is shown with arrows, and notes indicate the size and spacing of the joists. This type of framing plan is usually accompanied by numerous detail drawings.

While floor framing may be shown on drawings, the method of cutting and fitting the subfloor and finished floor panels is usually left up to the builder. When off-site or mass-produced floor systems are to be installed, a floor panel layout may be prepared. Its purpose is to ensure maximum use of materials and minimum waste.

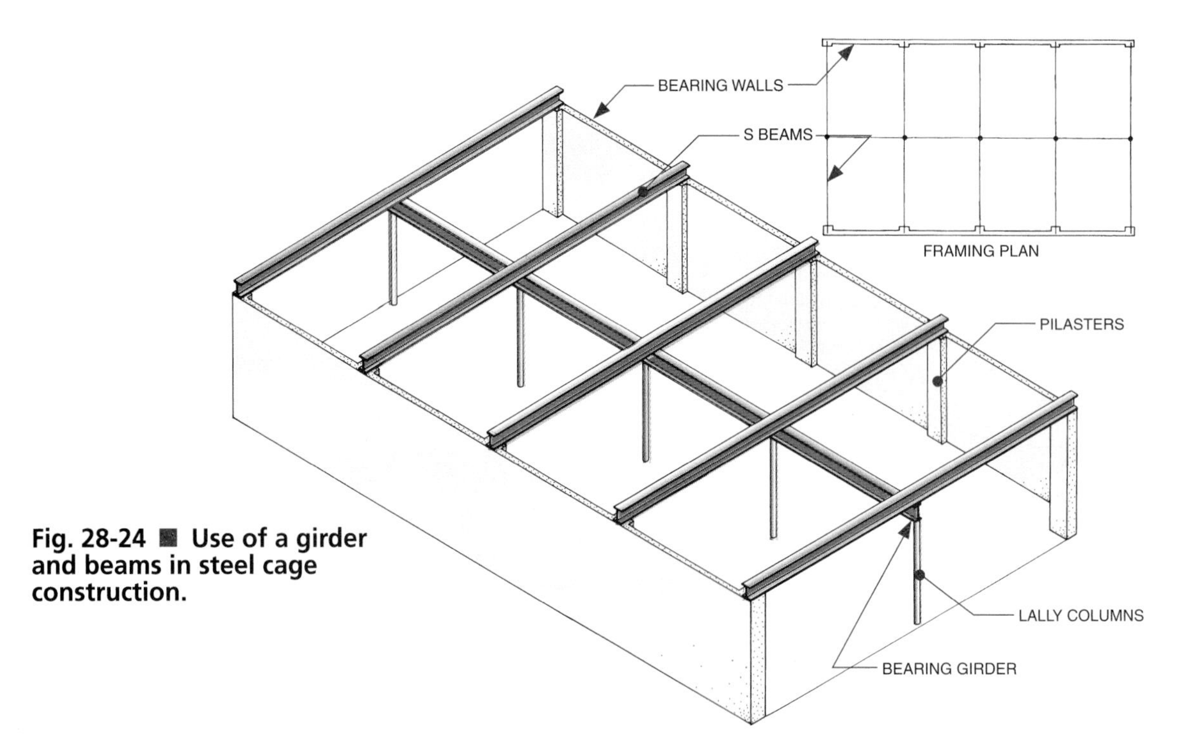

Fig. 28-24 ■ Use of a girder and beams in steel cage construction.

Fig. 28-25 ■ Floor framing plans.

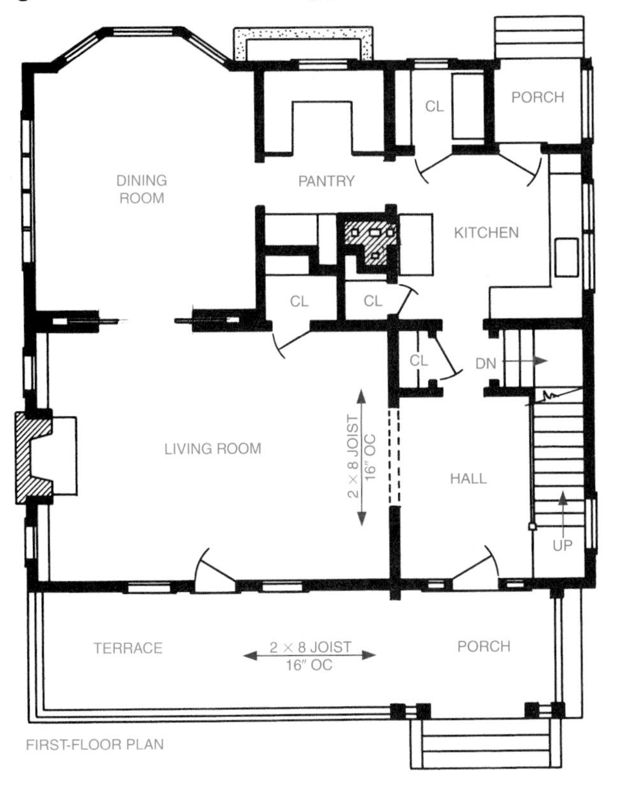

A. A method of showing joist direction on a floor plan.

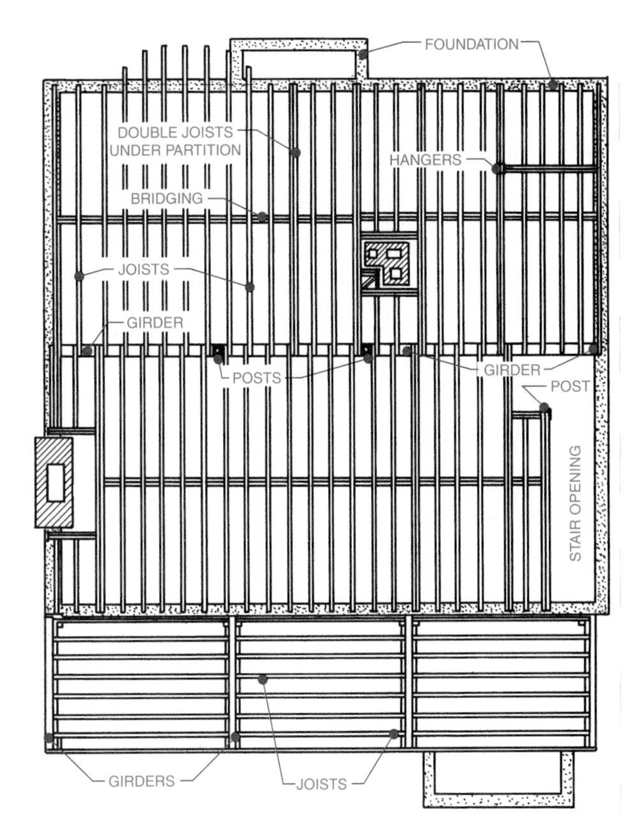

B. A floor framing plan showing material thickness with double lines.

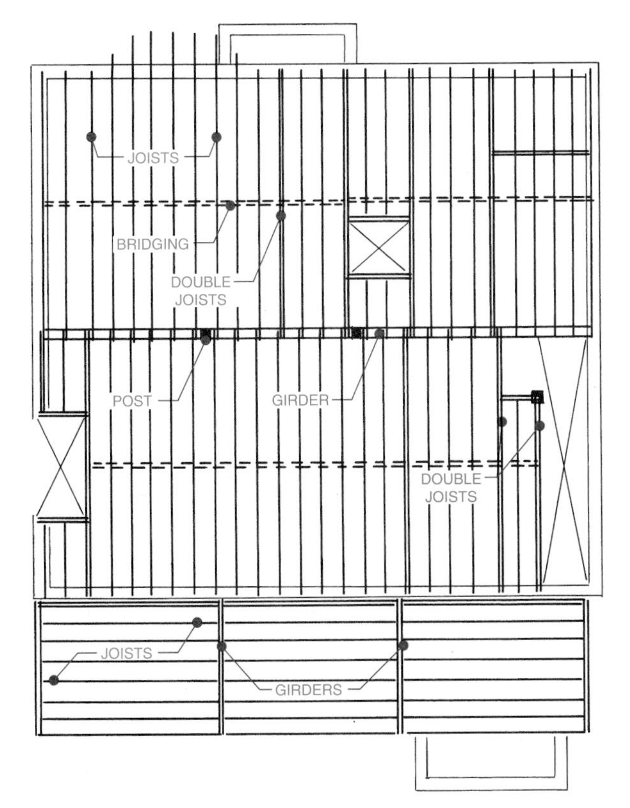

C. A simplified method of drawing floor framing plans with single lines.

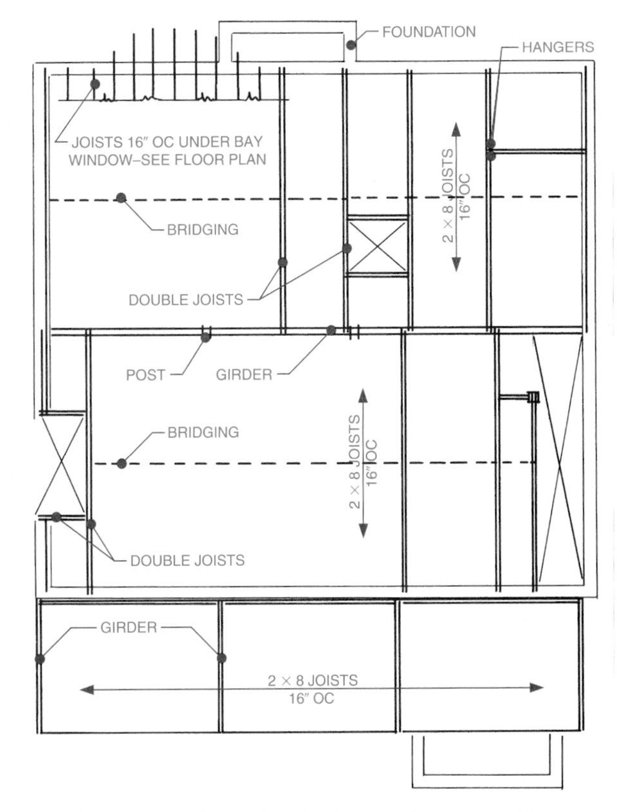

D. An abbreviated method of drawing floor framing plans.

SECOND-FLOOR PLANS Floor framing details for second floors or above are usually shown with a full section through an exterior wall. In balloon (eastern) framing the studs are continuous from the foundation to the eave. See Fig. 28-26. The second-level joists are supported by a ribbon board that is recessed and nailed or screwed directly to the studs. In construction for western (platform) framing, second-floor joists are placed on a top plate that rests on the platform (subfloor), as shown in Fig. 28-27.

When a combination of exterior covering materials is used, the relationship between the floor system and the exterior wall is shown on an elevation section. See Fig. 28-28.

If an upper-level floor is cantilevered over the first floor, the second-floor joists are either parallel or perpendicular to

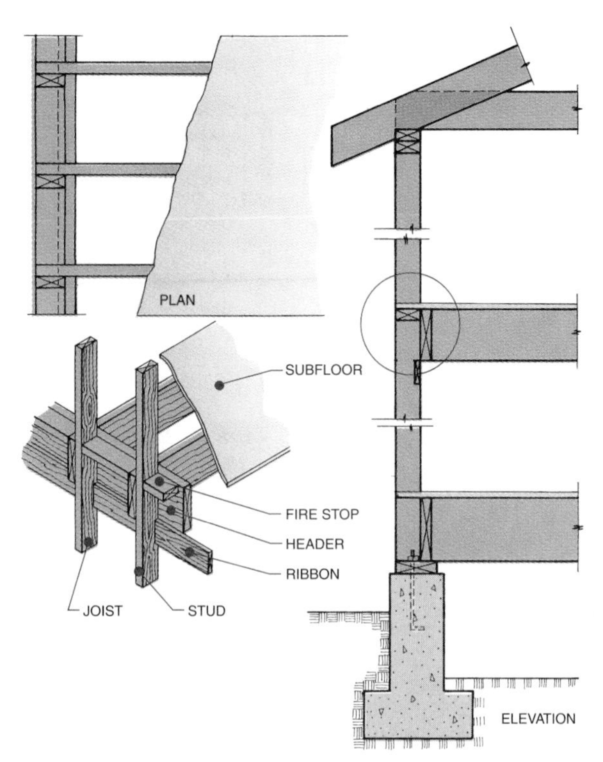

Fig. 28-26 ■ Second-floor framing details of balloon framing.

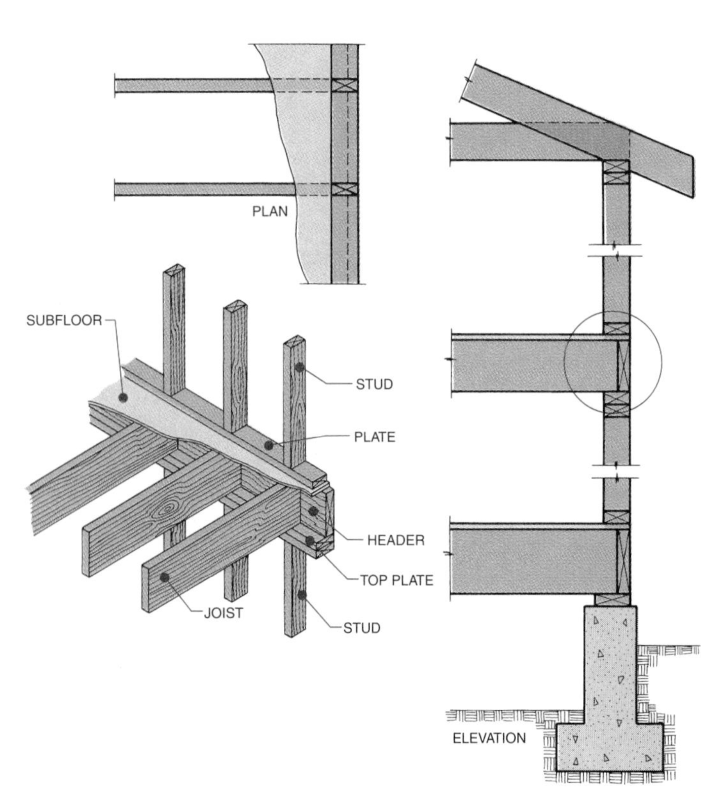

Fig. 28-27 ■ Second-floor framing details of platform framing.

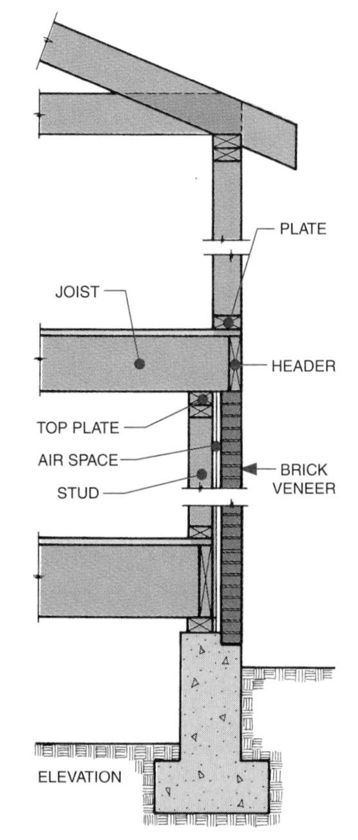

Fig. 28-28 ■ In this drawing, brick veneer covers the first level. The second-floor header and joists rest directly on top of the brick veneer.

the first-floor top plate that supports the second floor. See Fig. 28-29. When the joists are perpendicular to the wall, the construction is simple. When the joists are parallel to the wall, **lookout joists** must be used to support the cantilevered second-floor extension.

DETAILS Most floor framing plans are easily interpreted by experienced builders. Some plans may require additional details to explain a construction method. Details are drawn to eliminate the possibility of error in interpretation or to explain a unique condition. Details may be enlargements of what is already on the floor framing plan. They may be prepared for dimensioning purposes, or they may show a view from a different angle for better interpretation. On the floor framing plan in Fig. 28-30, page 544, the circles indicate areas for which detail drawings have been made.

- *Sill support.*

The sill is the transition between the foundation and the exterior walls of a structure. Drawings that show sill construction details reveal not only the construction of the sill, but also the method of attaching the sill to the foundation. A sill detail is included in most sets of architectural plans. Some sill details are drawn in pictorial form. Pictorial drawings are easy to interpret but are difficult and time-consuming to draw and dimension, and they are not orthographically accurate. Therefore, most sill details are prepared in two-dimensional sectional form. Figure 28-31 shows a detail drawing in plan, elevation, and pictorial form.

The floor area in a sill detail sectional drawing usually shows at least one joist. This is done to show the direction of the joist and its size and placement in relationship to the placement of the subfloor and finished floor. For example, the floor framing-plan in Fig. 28-31, page 545, is needed to indicate the spacing of joists and studs. The elevation section shows the intersections between the foundation sill and exterior wall. For this reason, both a plan and an elevation are needed to more fully describe this type of construction.

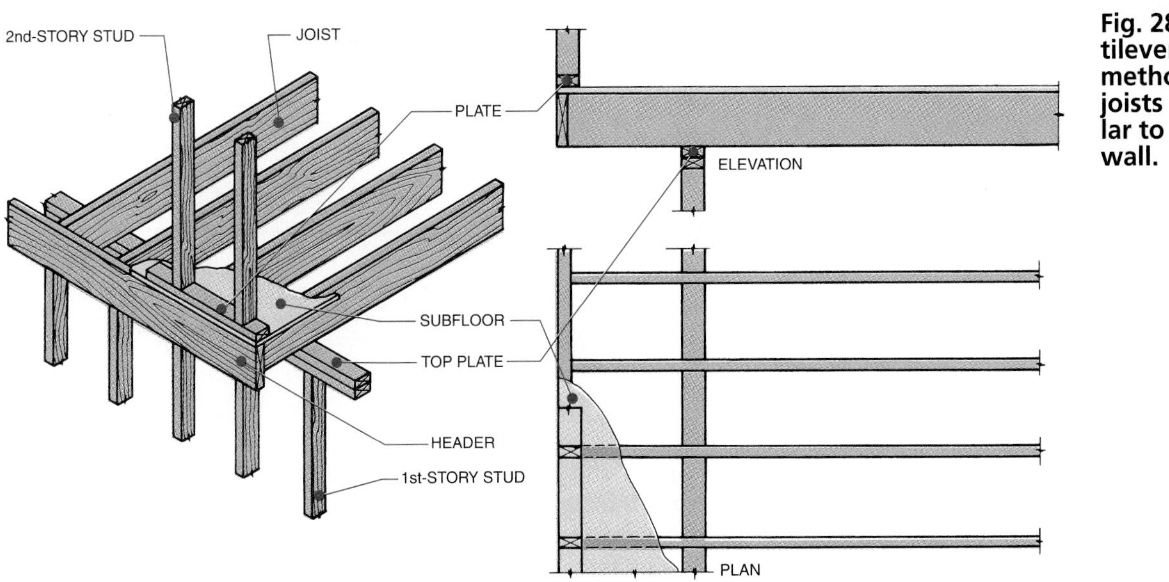

Fig. 28-29 ■ Cantilever-framing methods with joists perpendicular to the exterior wall.

Fig. 28-30 ■ Floor framing plan and details.

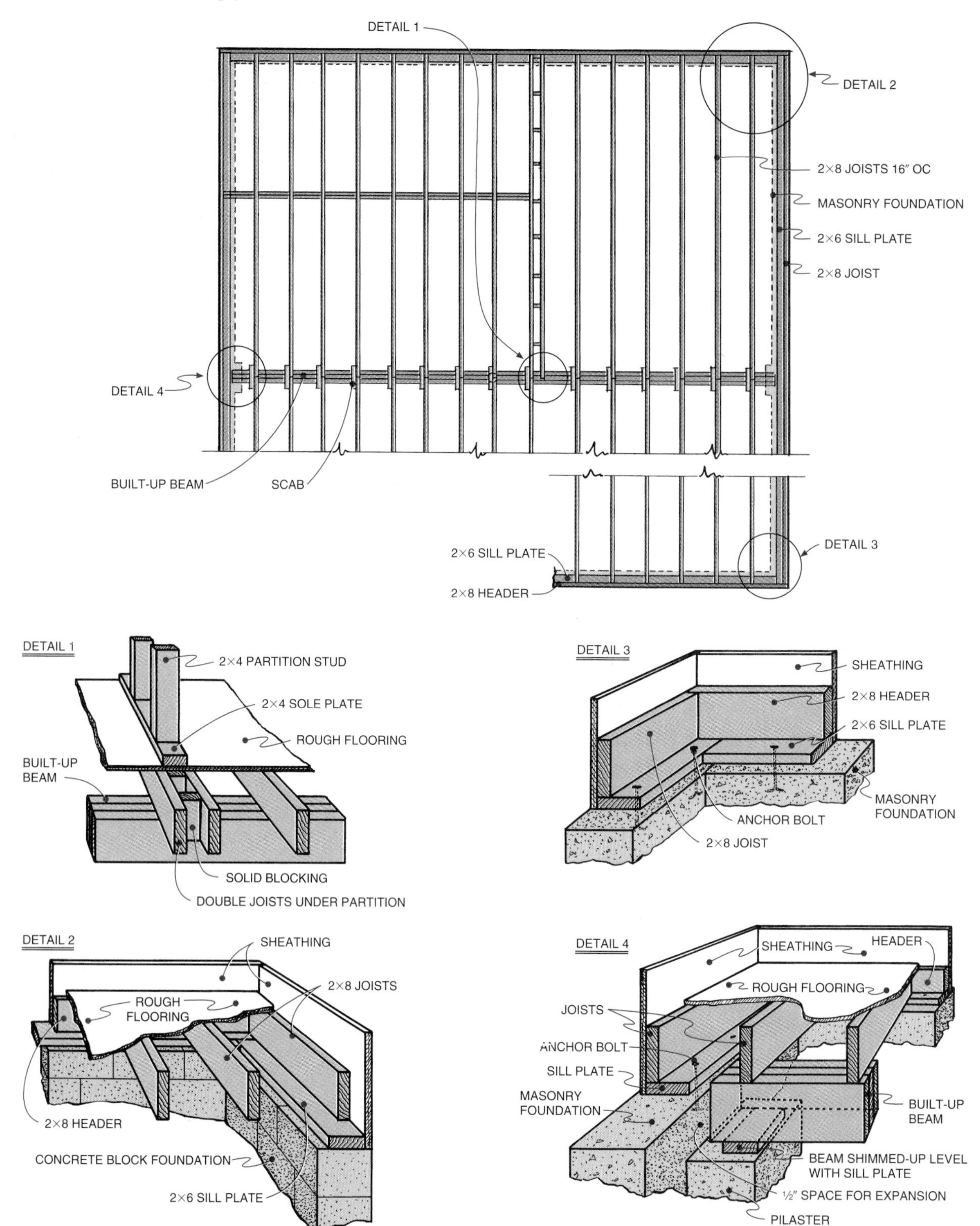

Detail 1: Relationship of the built-up girder, the double joist under the partition, and the solid blocking.
Detail 2: Sill construction in relation to the floor joist, rough flooring, and foundation.
Detail 3: Attachment of a typical box sill to a masonry foundation.
Detail 4: Method of supporting the built-up beam with a pilaster and the tie-in with the box sill and joist.

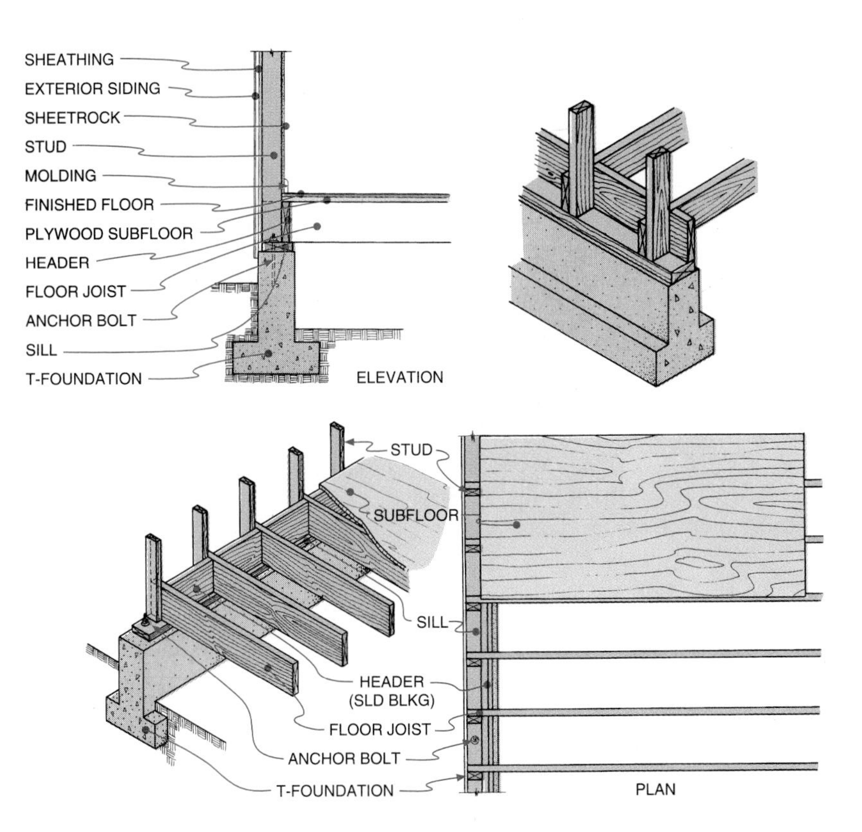

Fig. 28-31 ■ Sill detail of a balloon framed floor system.

Sill details are also required to show how materials are joined, such as masonry, wood, precast concrete, and structural steel. Special design features are shown on details as well. For example, firecuts are necessary in masonry walls. See Figs. 28-32 and 28-33. Sill details are also needed to show the intersection between girders and foundation walls. Figure 28-34 shows a box sill used to

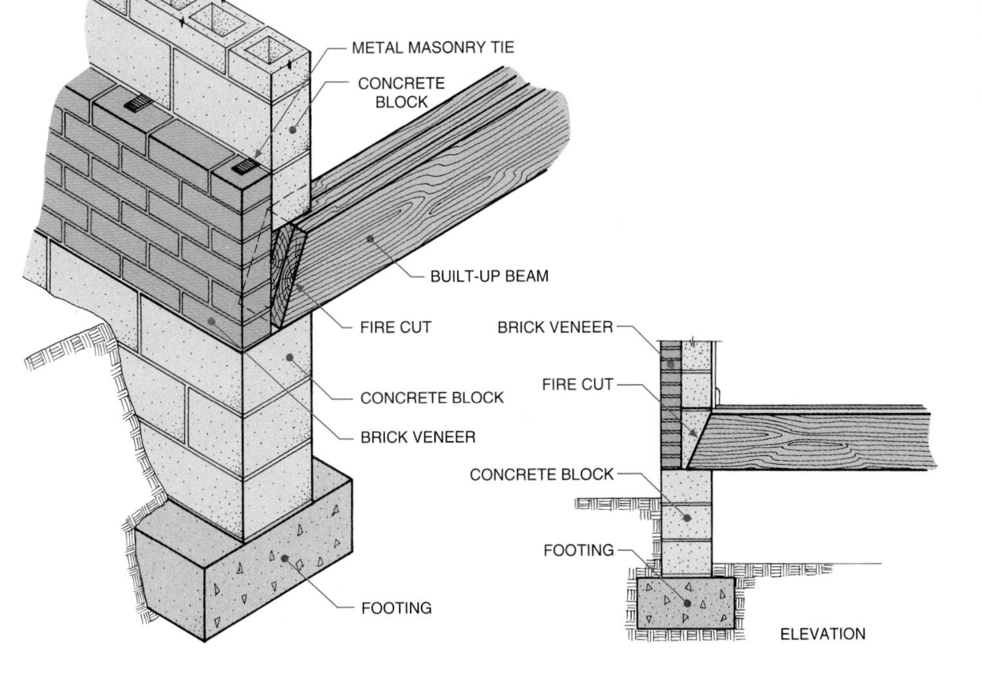

Fig. 28-32 ■ Plan and elevation sill details of brick veneer construction. Note that joists joined with masonry walls must be firecut.

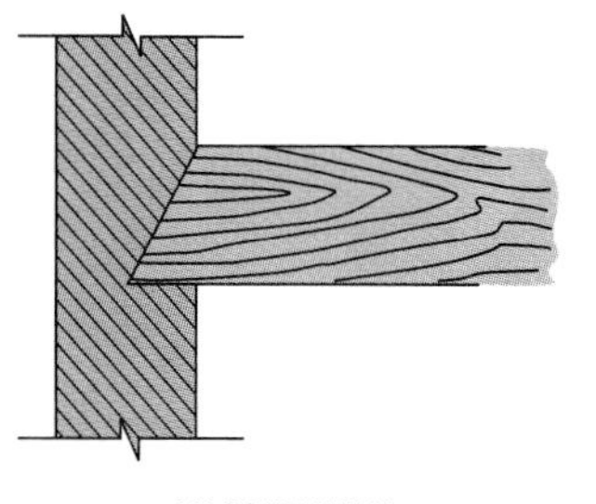

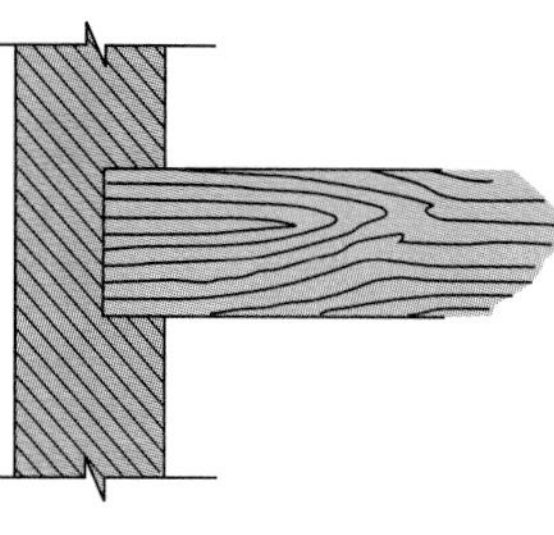

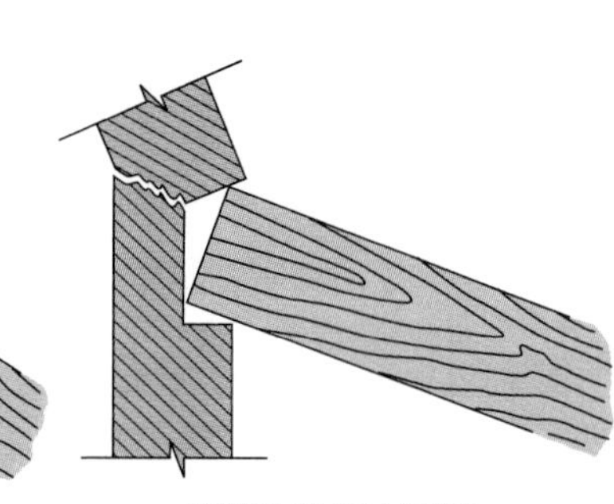

Fig. 28-33 ■ Joists must be fire cut to prevent the destruction of the masonry wall if the beam collapses.

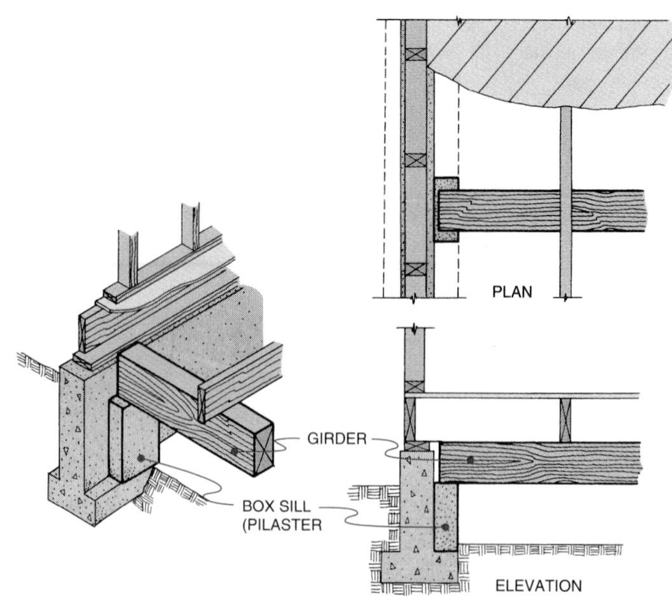

Fig. 28-34 ■ Girder supported with a box sill on an exterior T-foundation wall.

support a girder. Some details may be expanded to show additional details. See Fig. 28-35.

- *Intermediate support.*

If girders cannot safely span the distance between exterior supports, intermediate vertical supports (such as wood or steel columns, piers, or bearing walls) must be used to reduce the span. Detail drawings for intermediate supports consist of sections, elevations, or plan views to show the intersections between footings, vertical members, and horizontal girders. Intermediate framing details are

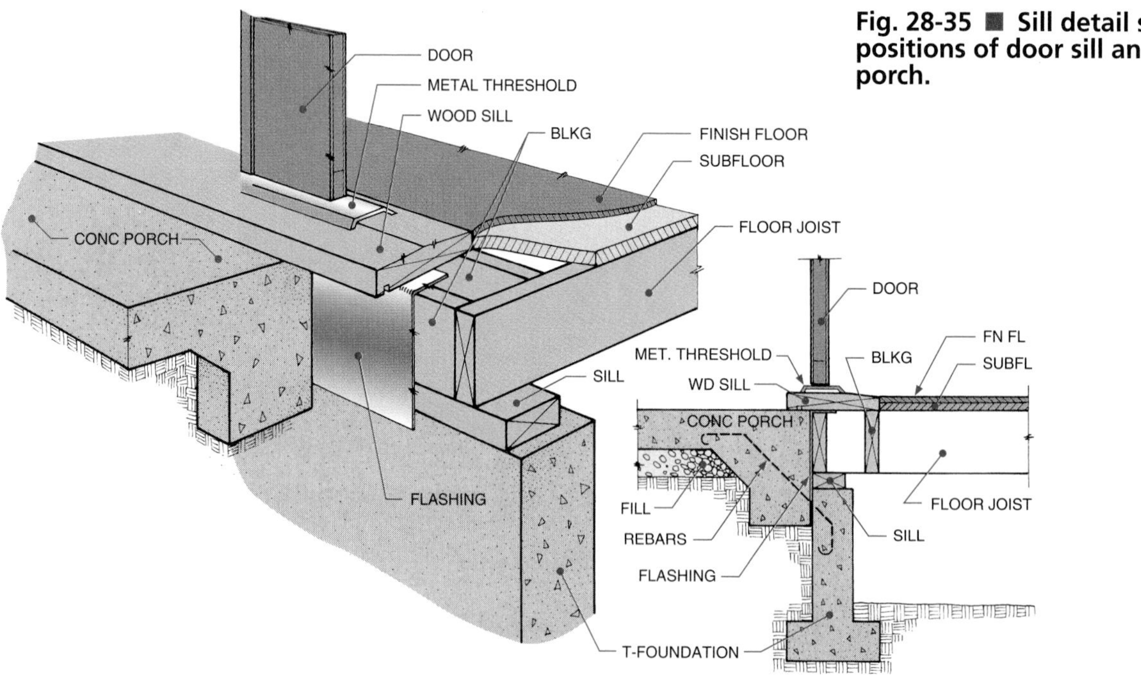

Fig. 28-35 ■ Sill detail showing the positions of door sill and concrete porch.

needed, for example, where level changes in floor level require special support. See Fig. 28-36.

- *Headers and trimmers.*

Building codes usually require headers and trimmers around stairwells, chimneys, and hearths and under heavy dead loads such as bathtubs, waterbeds, and masonry furniture. To ensure headers and trimmers are used correctly, detail drawings are prepared. For example, stairwell floor framing details show the position of joists, headers, and trimmers. See Fig. 28-37.

Since stairwell openings must be precisely shown on the floor framing plan, the stair system must be designed before the floor framing plan is drawn. See Fig. 28-38.

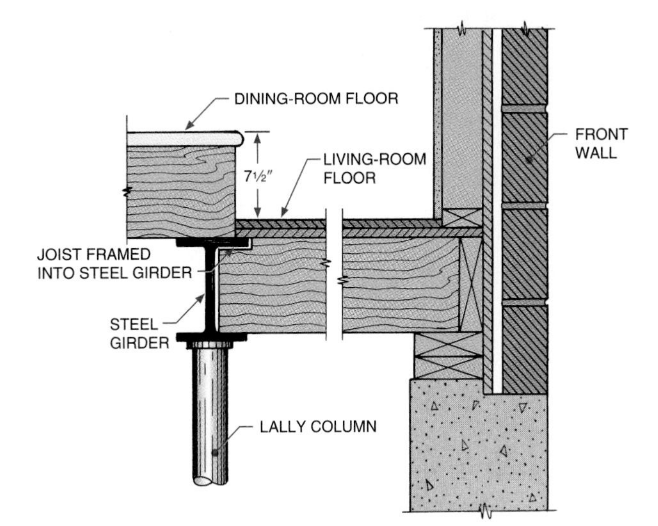

Fig. 28-36 ■ An elevation section of intermediate support details for a floor-level change.

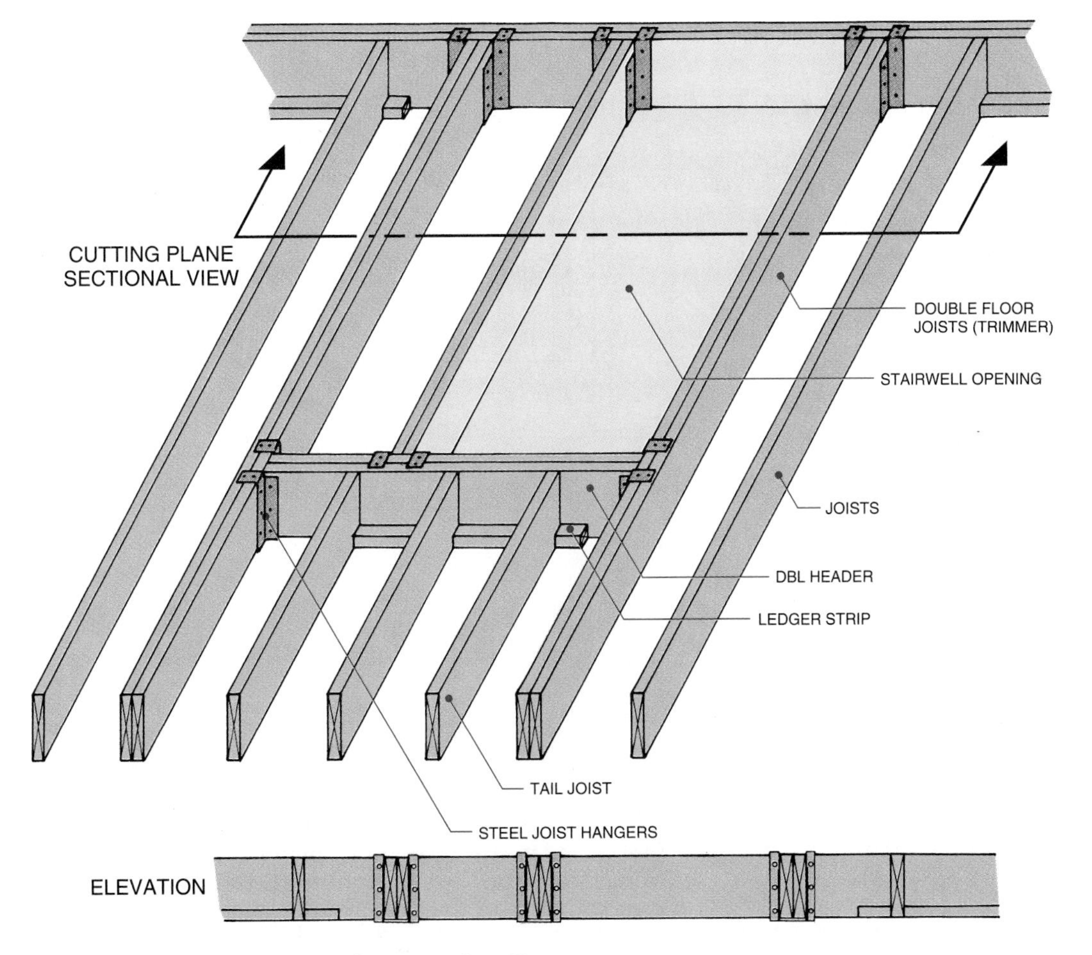

Fig. 28-37 ■ Stairwell floor framing detail.

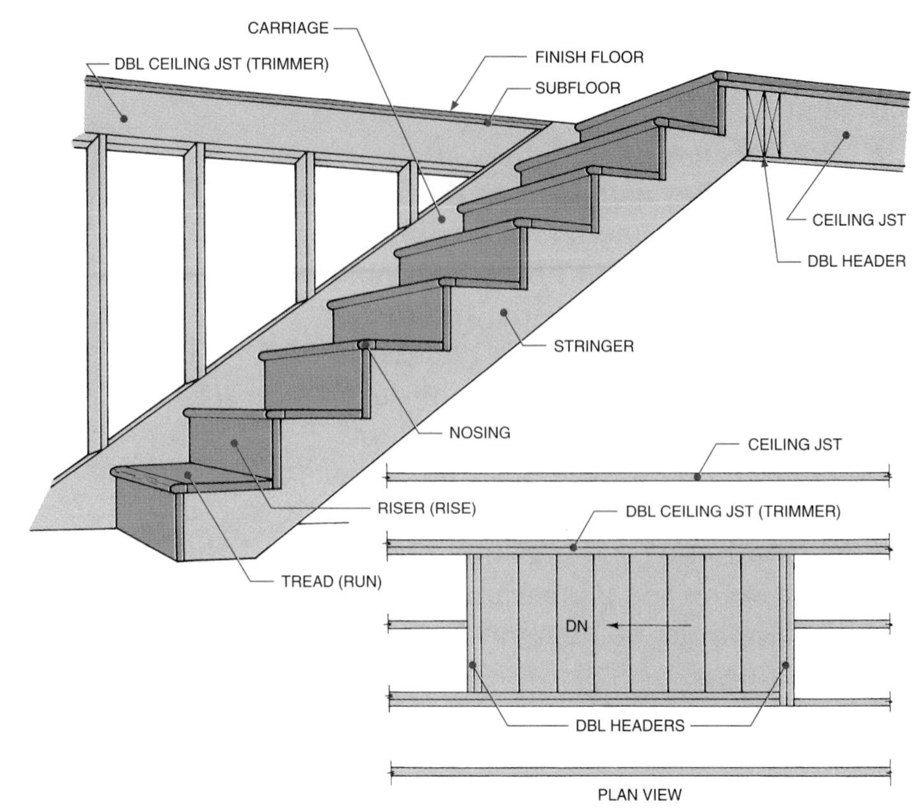

Fig. 28-38 ■ Plan and pictorial views of a typical stairwell opening.

During the floor plan design stage, the exact dimensions of an entire stair structure are determined. (Refer back to Chapter 14 for calculating stair dimensions.)

Floor Framing Plans for Steel

Steel floor framing plans are similar to other plans, except the exact position of every column, girder, and beam is classified and dimensioned. Grid systems are used to identify the position of each member. See Fig. 28-39.

Floor framing systems for steel construction include girders, beams, joists, and decking materials. A separate floor framing plan is prepared for each floor of a multilevel building. Although the framing for many floors may be nearly identical, this cannot be assumed unless specified. Usually there are slight differences on each floor plan. For this reason CAD layering is ideal for preparing high-rise structural steel floor framing plans.

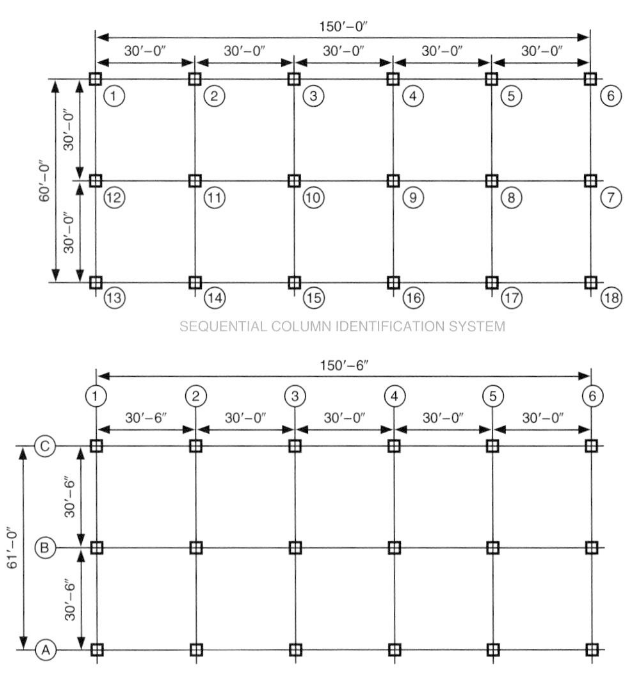

Fig. 28-39 ■ Systems used to identify steel framing members. In the sequential system, columns are numbered from left to right in rows from building rear to front. In the grid system, members are identified by letters that read from building front to rear, while numbers that read from left to right identify perpendicular rows.

Layering (or pin graphics) as used in conventional drafting allows the drafter to draw a base floor plan and make specific floor changes without redrawing each floor separately. This is done by drawing each floor plan on clear acetate. Drawings are aligned vertically in layers with registration pins. The same spacing must be used between grid lines to ensure alignment of columns and other vertically oriented features, such as stairwells, plumbing lines, HVAC ducts, and electrical conduits. Each floor framing plan shows the position of each column that passes through the floor.

In drawing steel floor framing plans, major members, such as girders and beams, are shown with a solid heavy line with the identifying notation placed directly on or under the line. The length of each line represents the length of each member. If a continuous beam passes over a girder, a solid unbroken line is drawn through the girder line. However, if the beam stops and is connected to the girder, the beam line is broken. See Fig. 28-40. Remember that solid lines represent continuous members. Broken lines indicate that the member intersects or is under a continuous member.

Spaces created between rows of members in two directions are known as **bays.** See Fig. 28-41.

Three methods of dimensioning are used on structural steel drawings. In the first method, a description of each member includes the length placed directly on each schematic line. The second method uses notations to show only the shop size (width) and weight. Dimension lines are used to show the position and length of each member. The third method uses a coding system that relates each member to a schedule containing all pertinent information. See Fig. 28-42.

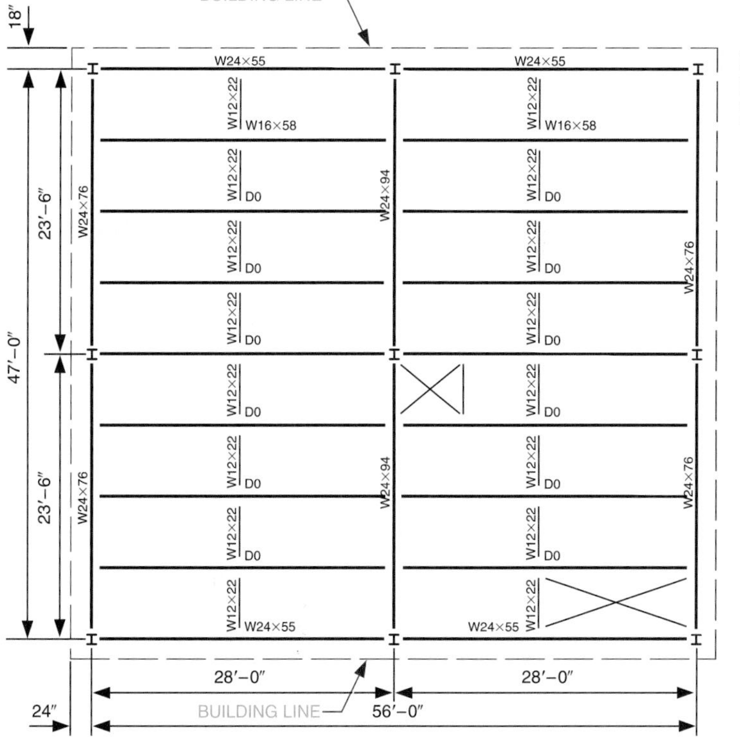

Fig. 28-40 ■ **On steel floor framing plans, solid lines represent continuous members, and broken lines represent intersecting members.**

Fig. 28-41 ■ Construction bays.

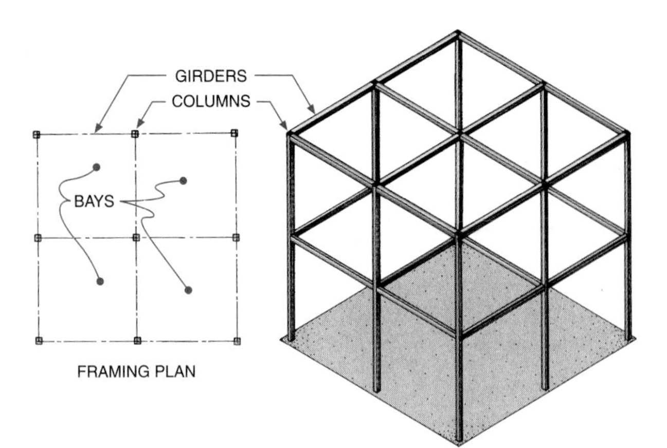

A. Bays with framing plan.

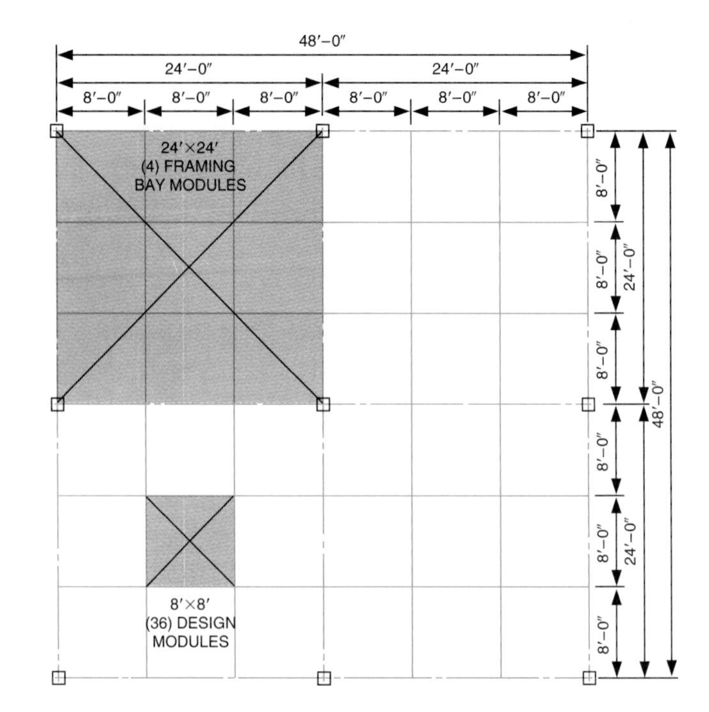

B. Dimensioning of bay modules.

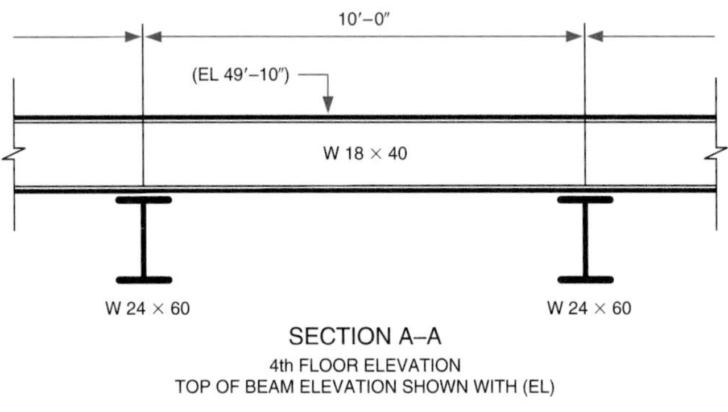

Fig. 28-42 ■ Method of identifying steel beams and girders.

CHAPTER 28 Exercises

1. Draw a simplified plan view of a floor system shown in this chapter.
2. Redesign a floor framing plan shown in this chapter with the joists aligned in the opposite direction.
3. Add a 12′ × 14′ room to the upper left corner of a 16′ × 20′ room. Draw the outlines and show the position of joists and girders.

4. Develop a complete floor framing plan for one floor or as many floors as you are designing in your house plans.

5. Using a CAD system, complete an abbreviated floor framing plan.
6. Draw the detail of a sill support in this chapter. Include callouts for all the parts shown.

CHAPTER 29

Wall Framing Drawings

Wall framing provides the base to which coverings, such as siding and drywall, are attached. Typically, exterior wall framing supports the roof and ceiling loads. In some designs, the interior walls also help support these loads.

Wall framing drawings may consist of exterior and interior elevations, column and stud layouts, and details.

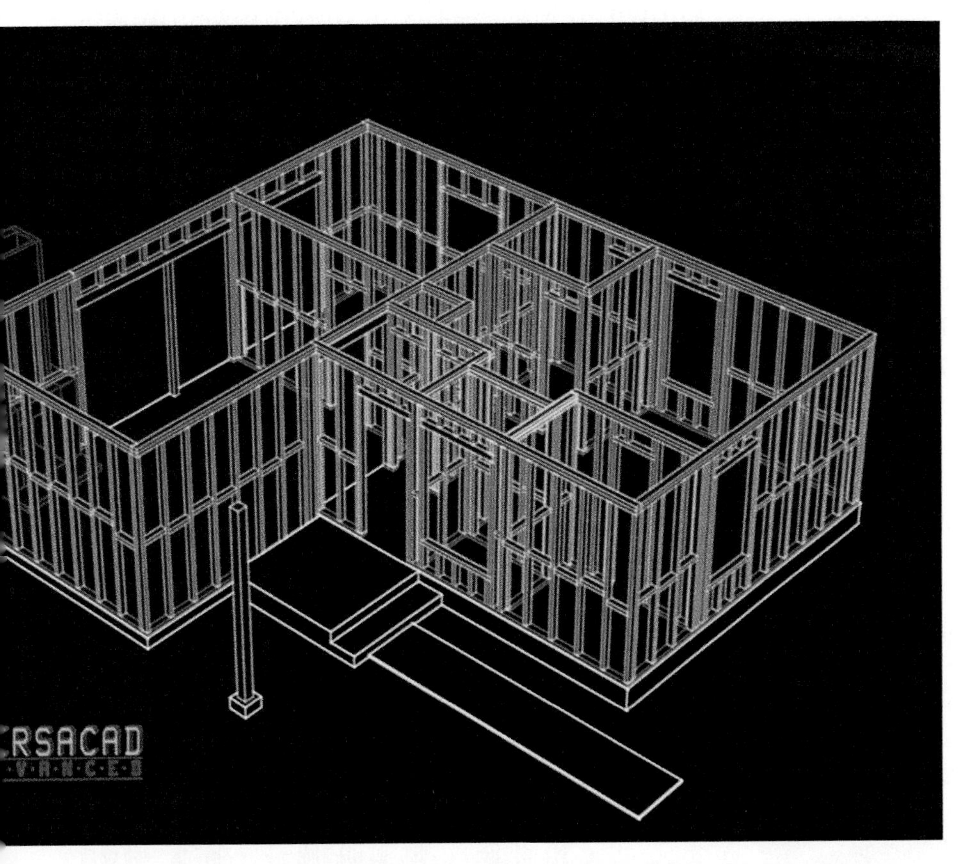

■ **Shown here is a three-dimensional wall framing drawing generated by a computer.** *VERSACAD*

Objectives

In this chapter, you will learn to:

- **draw an exterior wall framing elevation and plan.**
- **draw an interior wall framing elevation and plan.**
- **draw details and sections of walls.**
- **draw wall intersections.**

Terms

board and batten siding
bracing
column schedule
drywall construction
framing elevation drawings
lap siding
rough opening
siding
stud layout

Exterior Walls

Exterior walls for most wood-frame residential buildings use either skeleton or post-and-beam construction. The typical method of erecting walls for most buildings follows a conventional braced-frame system. See Fig. 29-1. (Bracing is discussed on page 555.) Prefabricated components have led to variations in the methods of erecting walls. Manufactured panels range from a basic wall-frame panel to a completed wall that includes plumbing, electrical work, doors, and windows. See Fig. 29-2. Whether a structure is prefabricated or field constructed, the preparation of exterior framing drawings is the same.

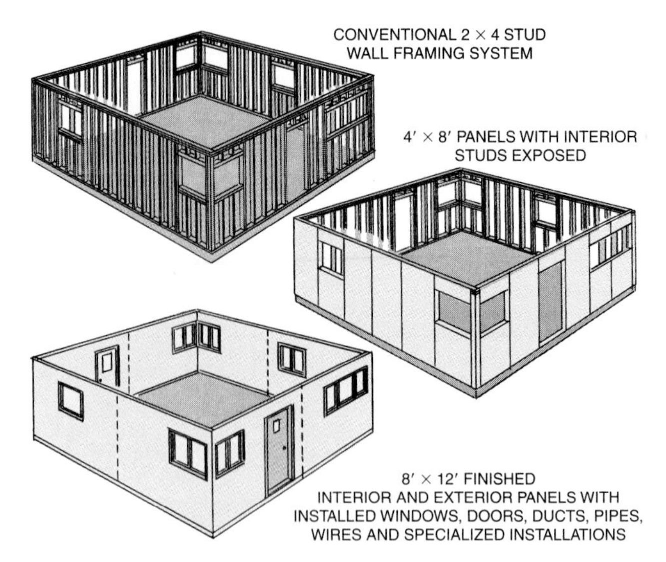

Fig. 29-2 ■ Methods of wall construction.

Framing Elevations

Exterior walls are best constructed when a framing elevation drawing is used as a guide. A wall framing elevation drawing is the same as an elevation of a building with the siding materials removed. Figure 29-3 shows the basic wood framing members included in framing elevations.

To draw framing elevations, project lines from floor plans and elevations, as shown in Fig. 29-4. The elevation supplies all the projection points for the horizontal framing members. The floor plan provides all the points of projection for locating the vertical members. Since floor-plan wall thicknesses normally include the thickness of siding materials, care should be taken to project the outside of the *framing* line to the framing elevation drawing and not the outside of the siding line.

When drawing door and window **rough openings** (framing openings), first check the sizes listed in manufacturing specifications or door and window schedules. Rough openings are slightly larger than

Fig. 29-1 ■ Skeleton frame construction.

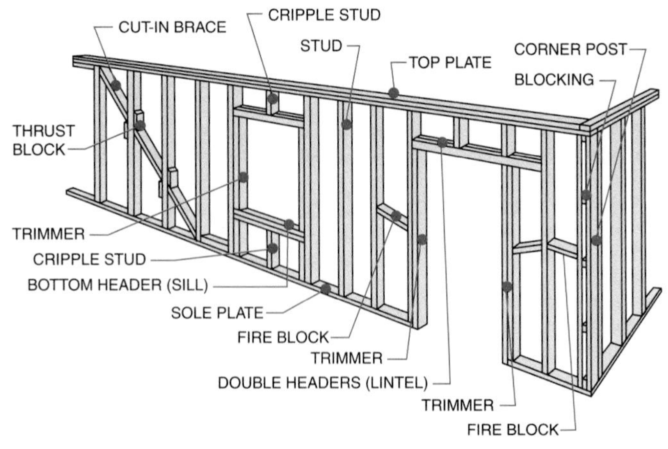

Fig. 29-3 ■ Wall framing members.

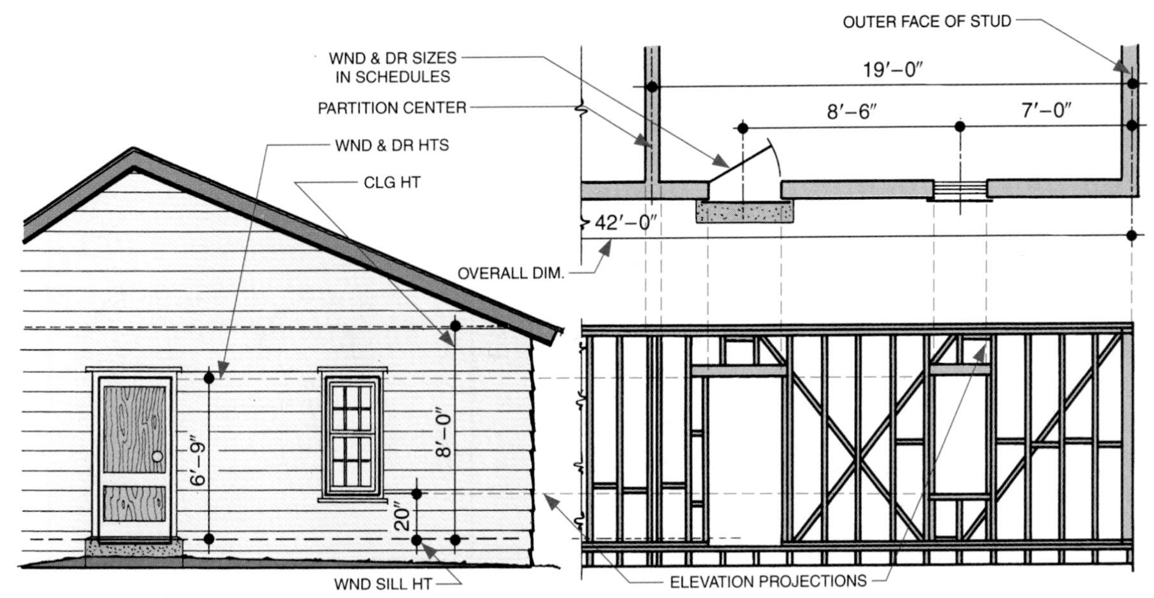

Fig. 29-4 ■ **Projection of an exterior framing elevation from a floor plan and elevation drawing.**

the size of doors or windows. Then project the position of the top, bottom, and sides of the door and window openings from the floor plan and elevation. Figure 29-5 shows examples of exterior wall framing elevations.

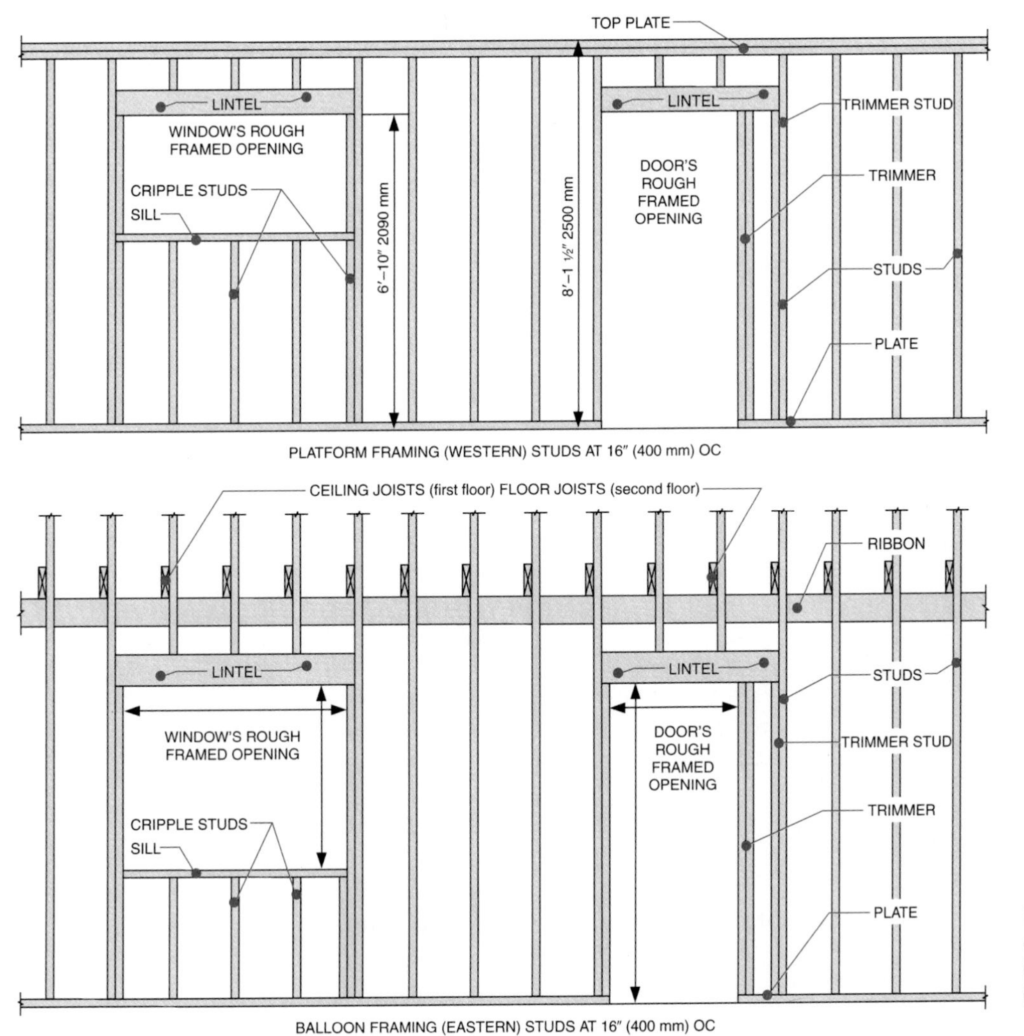

Fig. 29-5 ■ **Exterior elevations for balloon (eastern) and platform (western) framing.**

In drawing steel-framed elevations, only the locations of structural members are included. See Fig. 29-6. Information about coverings for curtain walls is shown on wall elevations and wall details.

In Fig. 29-7, the framing elevation is incorporated into a sectional drawing of the structure. This drawing shows an elevation section of the framing from the foundation-floor system to the roof construction. In a sectional drawing, any members intersected by a cutting-plane line, such as the joists, are indicated by crossed diagonals. (Note the double blocking in the lower left portion of Fig. 29-7.)

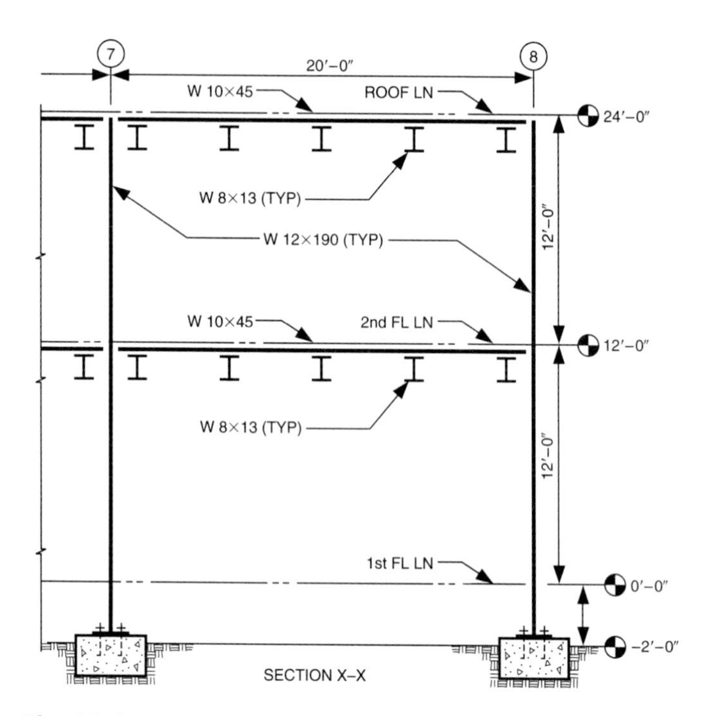

Fig. 29-6 ■ Structural members shown on a steel elevation drawing.

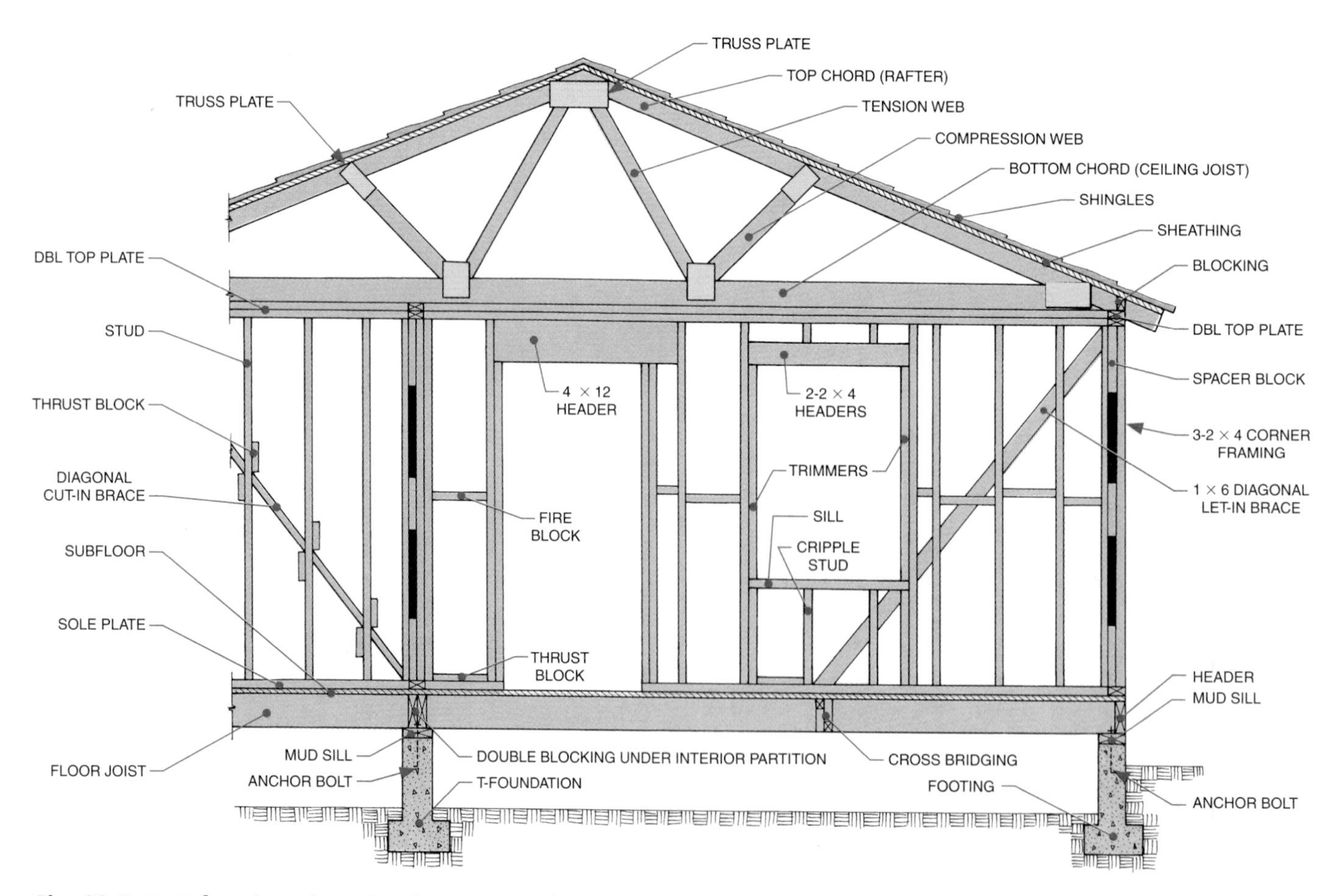

Fig. 29-7 ■ A framing elevation incorporated into a complete section of a building.

BRACING To make wall frames rigid, lumber members are attached at an inclined angle. Members used for this purpose are called **bracing.** The bracing may be placed on the inside or outside of the wall, or between the studs. See Fig. 29-8. Steel straps are often used for corner bracing as shown in Fig. 29-9.

Difficulties often occur in interpreting the true position of headers, cripple studs, plates, and trimmers. Figure 29-10 shows how to illustrate the position of these members on framing elevation drawings to ensure proper interpretation.

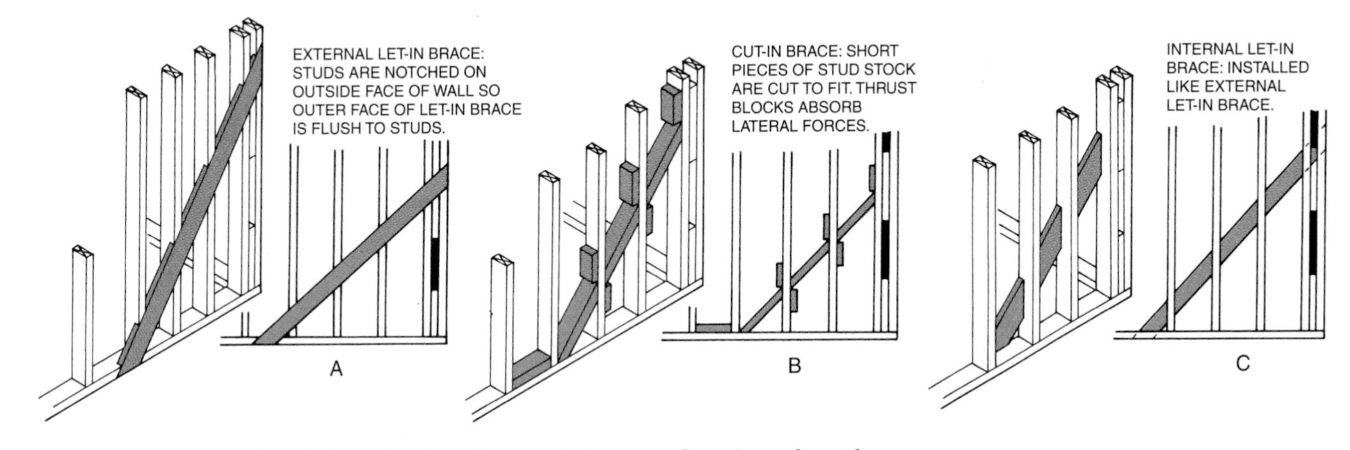

Fig. 29-8 ■ Methods of showing bracing positions on framing elevations.

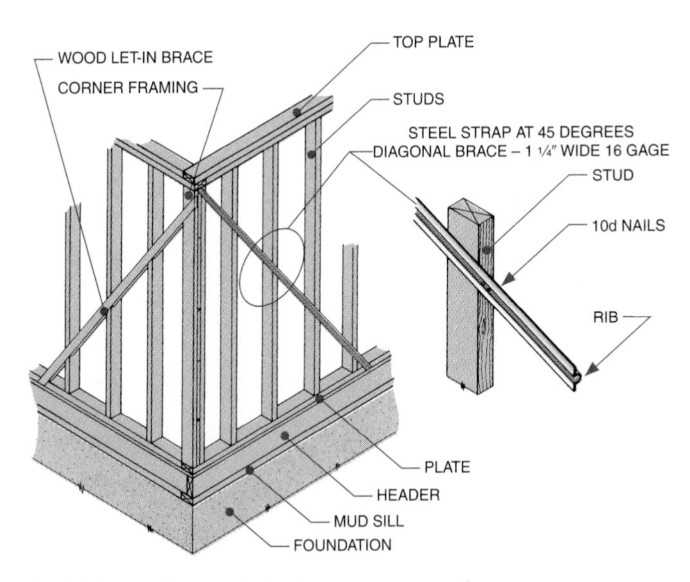

Fig. 29-9 ■ Use of steel strap corner brace.

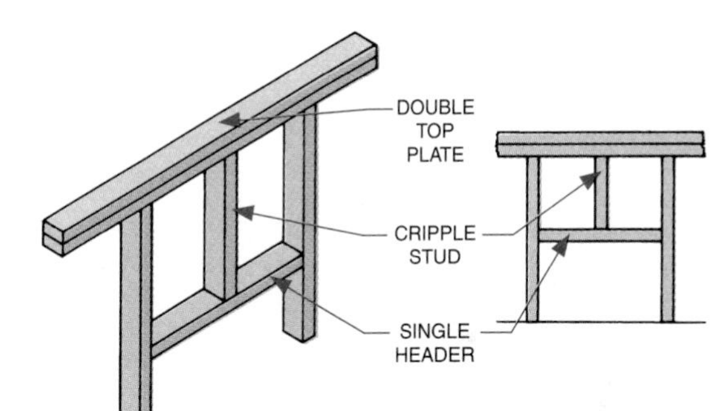

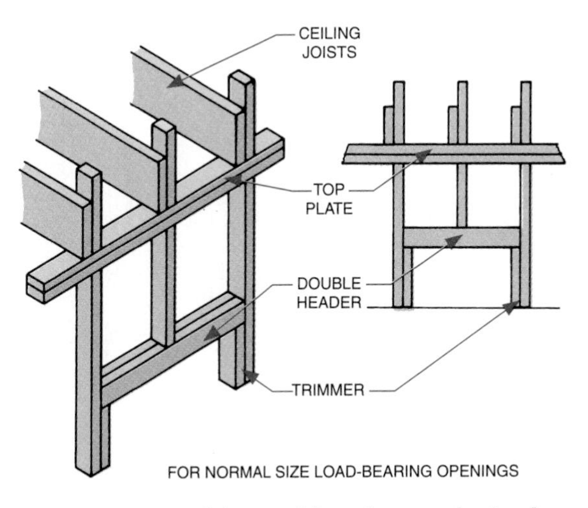

Fig. 29-10 ■ Positions of headers and cripples.

PANELS Panel elevations show the attachment of the sheathing panels to the framing. To show the relationship between the panel layout and the framing, panel drawings and framing drawings are usually combined in one drawing. Diagonal dotted lines indicate the position of the panels. When only the panel layouts are shown, the outline of the panels and the diagonals are drawn solid. See Fig. 29-11.

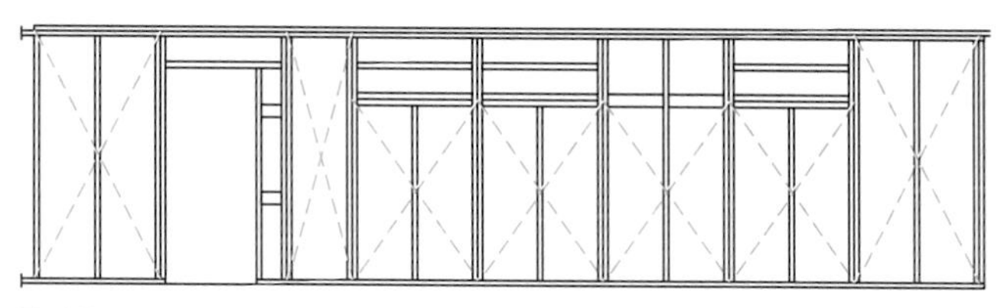

THIS DRAWING PROCEDURE COMBINES FRAMING AND PANEL LAYOUT. DIMENSIONS AND SPECIFICATIONS USUALLY SHOWN ARE OMITTED FOR CLARITY.

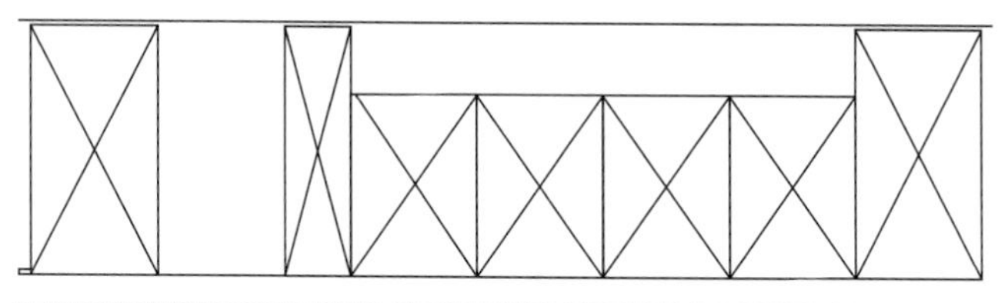

THIS DRAWING PROCEDURE IS ACCOMPANIED BY SEPARATE FRAMING LAYOUT. NOTES AND DIMENSIONS (NOT SHOWN HERE) REFER ONLY TO PANEL SIZES AND SPECIFICATIONS.

Fig. 29-11 ■ Diagonal lines are used to indicate panel positions.

DIMENSIONS Dimensions on framing elevations include overall widths, heights, and spacing of all studs. See Figs. 29-12 and 29-13. Control dimensions for the size of all rough openings (RO) for windows are also indicated. If the spacing of studs does not automatically provide the rough opening necessary for windows, the rough-opening width of windows must also be dimensioned. Framing dimensions for a standard 8′ ceiling height using standard 93″ studs are shown in Fig. 29-14, page 558.

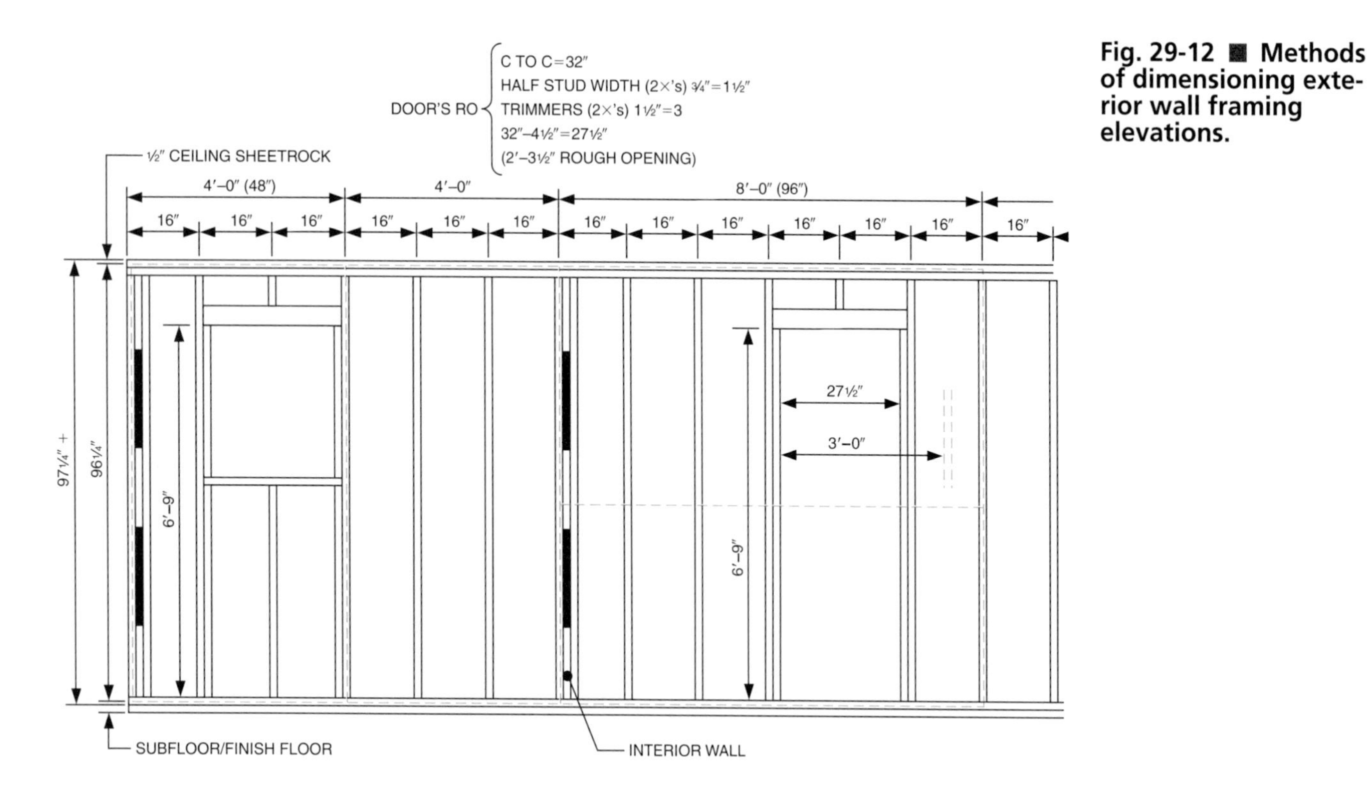

Fig. 29-12 ■ Methods of dimensioning exterior wall framing elevations.

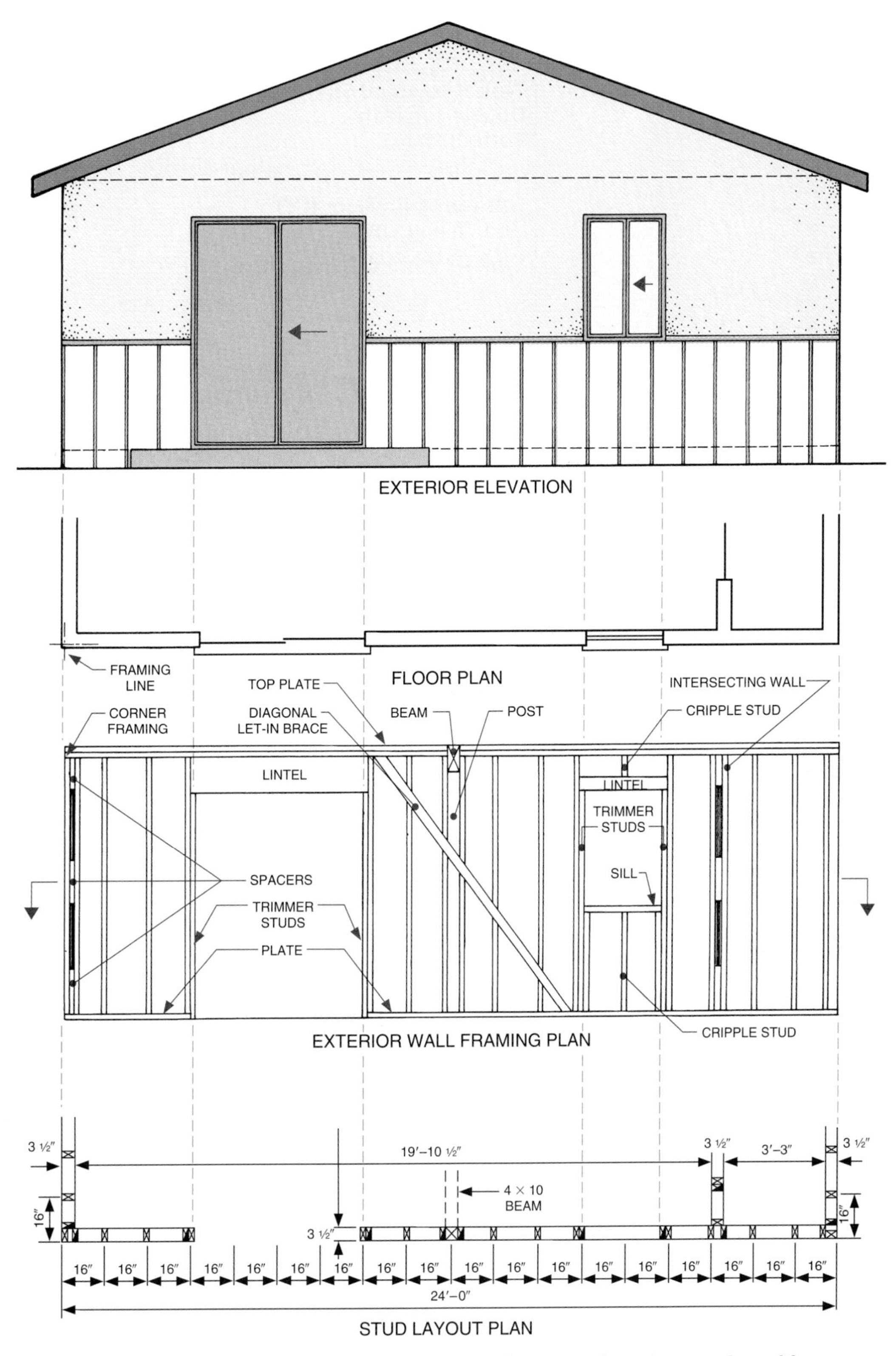

Fig. 29-13 ■ The relationship of the elevation, framing elevation, and stud layout.

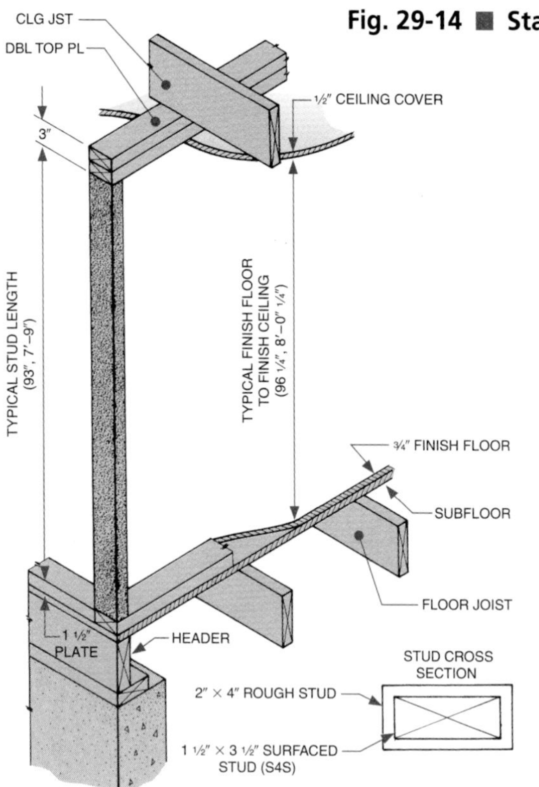

Fig. 29-14 ■ Standard framing height dimensions.

Typical Standard Stud Height ..	93"
Double Top Plate	3"
Bottom Plate	1 ½"
Total	97 ½"
Less the Finish Floor (¾") And Ceiling Cover (½")	−1 ¼"
Typical Finished Ceiling Ht	96 ¼" (8'–0 ¼")

Fig. 29-15 ■ Sectional wall framing drawings.

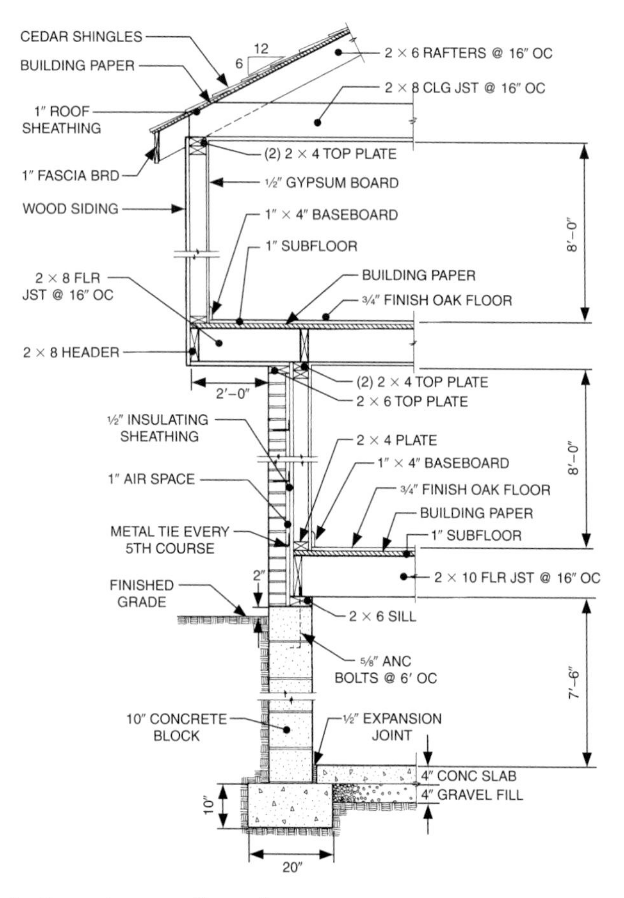

A. Complete wall section.

Detail Drawings

Not all the information needed to frame an exterior wall can be shown on a framing elevation. Many details must be shown through the use of sectional drawings, exploded views, or pictorials.

SECTIONS Sectional wall framing drawings are either complete sections, partial sections, or removed sections indexed from a plan or elevation drawing. See Fig. 29-15.

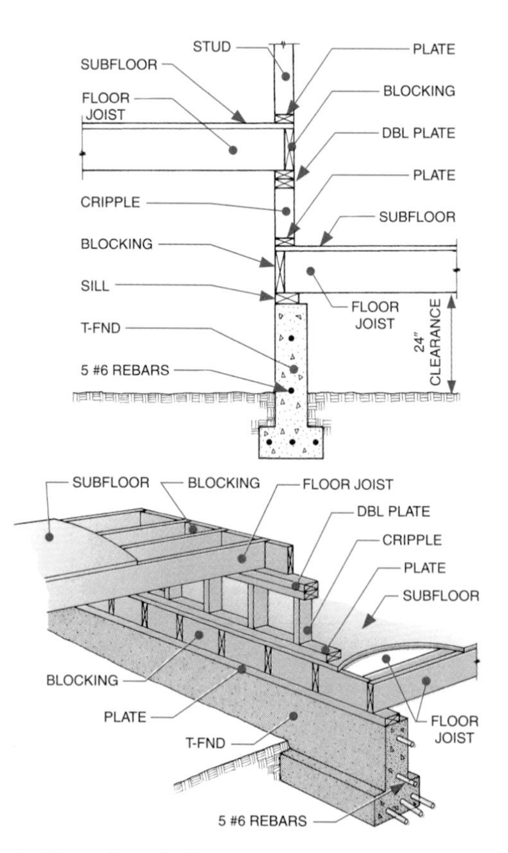

B. Floor level change in elevation and pictorial sections.

With sectional breaks, such as on a full wall section, a larger scale can be used to allow more detail. Elevation drawings may also be partially sectioned to reveal construction details, as shown in Fig. 29-16.

EXPLODED VIEWS Occasionally, exploded views are drawn to show internal wall framing construction if it is hidden when the total assembly is drawn. See Fig. 29-17. This method of detailing, however, is generally used in cabinet work.

Fig. 29-16 ■ A removed section.

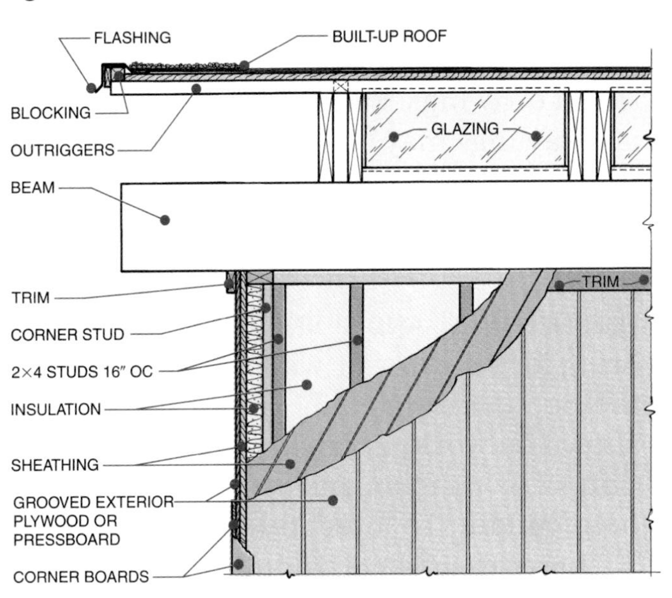

Fig. 29-17 ■ Exploded views.

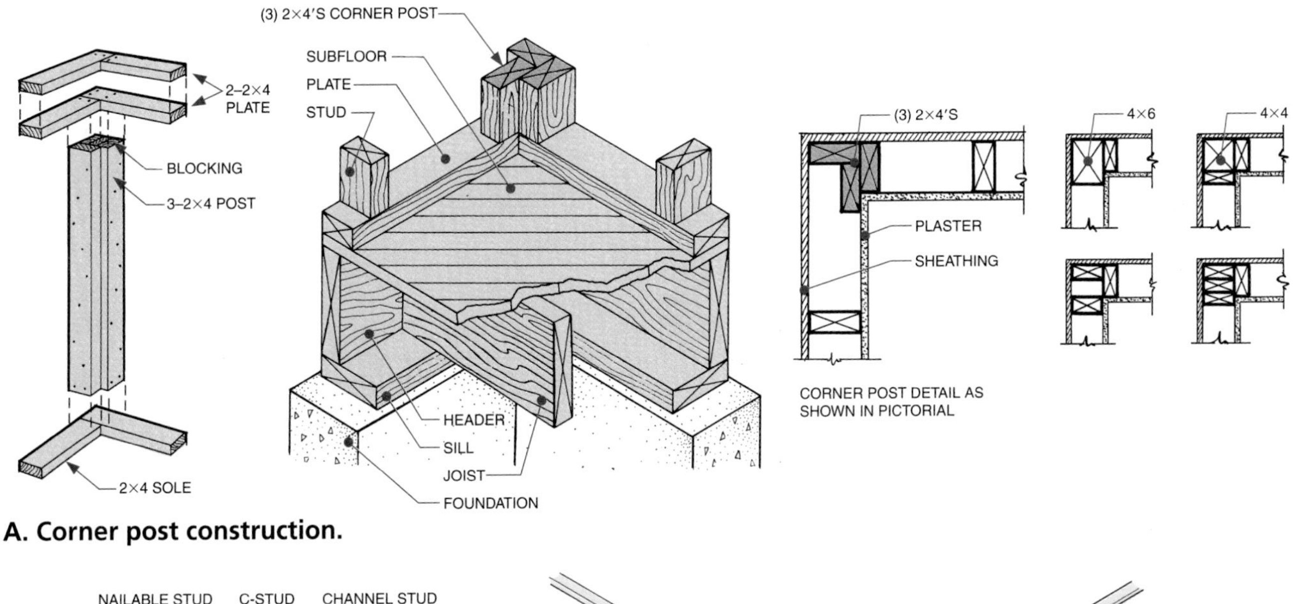

A. Corner post construction.

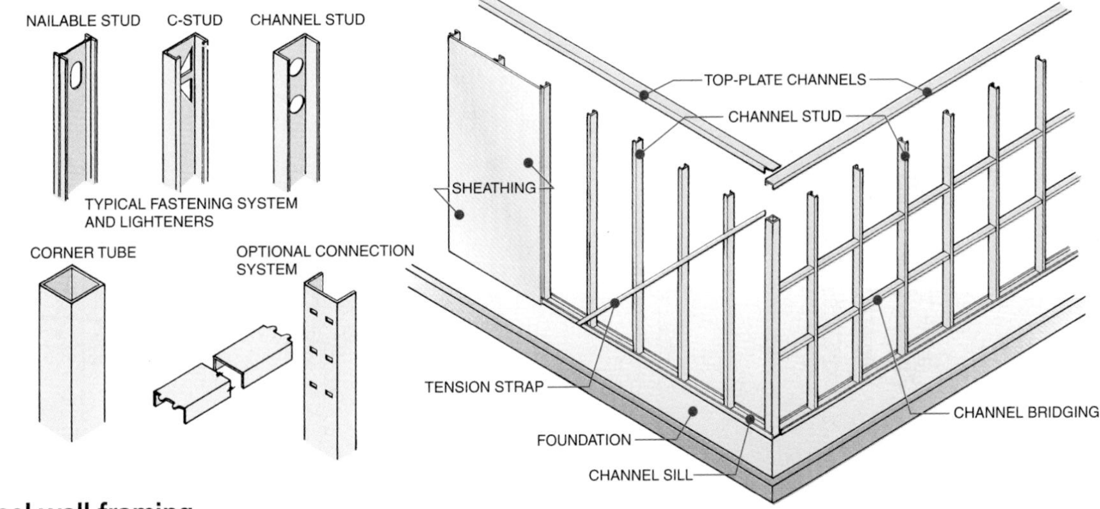

B. Steel wall framing.

PICTORIALS Pictorial framing drawings help eliminate construction errors due to misreading of plans, elevations, and sectional drawings. Full wall framing pictorial drawings may be used for this purpose. However, most pictorial details are limited to a single intersection detail or easily misunderstood construction feature.

FINISHED WALL AND SIDING DETAILS The covering for an exterior wall is called **siding.** Siding material details are most often shown on working drawings or on elevation sections. Plan and pictorial sections help builders to interpret drawings.

- **Lap siding** is horizontal siding applied over sheathing. Each piece covers (overlaps) part of the piece below it. Lap siding is available in solid redwood, cedar, and pine. Other materials are also used extensively, such as aluminum, steel, and fabricated boards. See Fig. 29-18.
- **Board and batten siding** is vertical siding that originally consisted of a vertical board placed over studs, with the joints covered with vertical batten boards. Today, many board and batten sidings are made of plywood or strandboard sheets with battens over the 48″ joints. Additional battens at 16″ intervals add consistency and help stabilize the vertical sheets. See Fig. 29-19.

A variation of vertical sheet siding is *grooved panel boards.* These are available with a wide variety of surface textures, grades, and groove sizes and shapes. Panels are also available which simulate the appearance of bevel or lap siding or shingles.

Shingle siding is applied as individual overlapped shingle boards or as prefabricated shingle siding sheets. In either case they are applied over insulated sheathing.

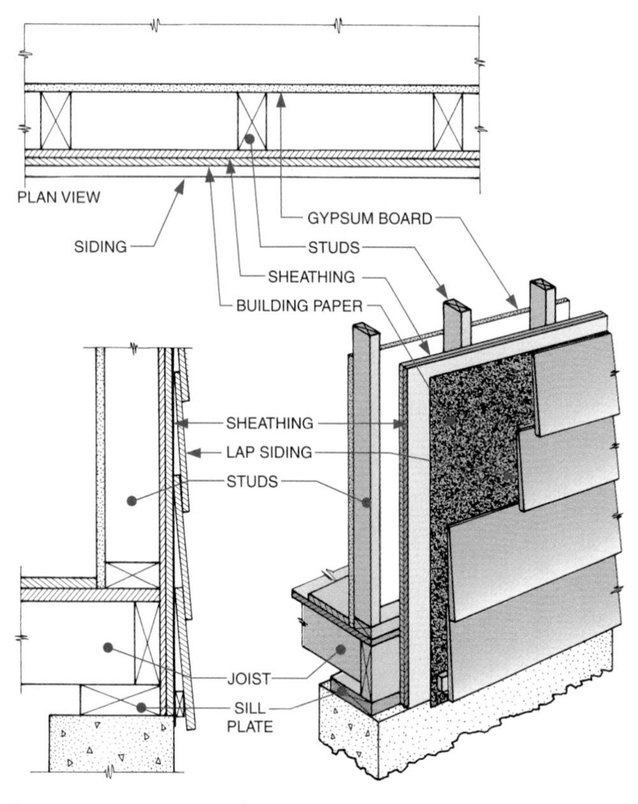

Fig. 29-18 ■ Lap siding.

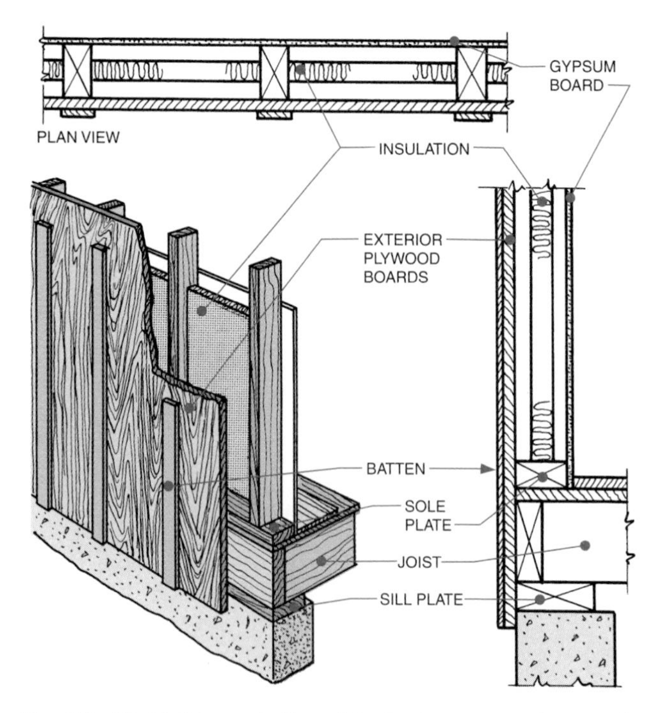

Fig. 29-19 ■ The relationship between a plan and an elevation section of a board-and-batten wall.

- *Stucco siding*, Fig. 29-20, is applied with a trowel. In wood construction, steel mesh must be applied over insulated sheathing to provide a base for the stucco. Stucco can be applied directly to concrete block without the use of mesh. Stucco is available in traditional three-coat portland cement formula or in synthetic or insulated-finish system (EIFS) form.
- *Solid masonry wall facing* materials include stucco, brick, stone, aluminum, steel, vinyl, or polyurethane siding. These are applied over solid concrete or concrete block. Brick is a material that can be used both structurally and as a finished wall face. Figure 29-21 shows a solid brick wall that performs both functions. Some brick walls include cavities (spaces between) to provide insulation, reduce dead loads, and lower material costs.

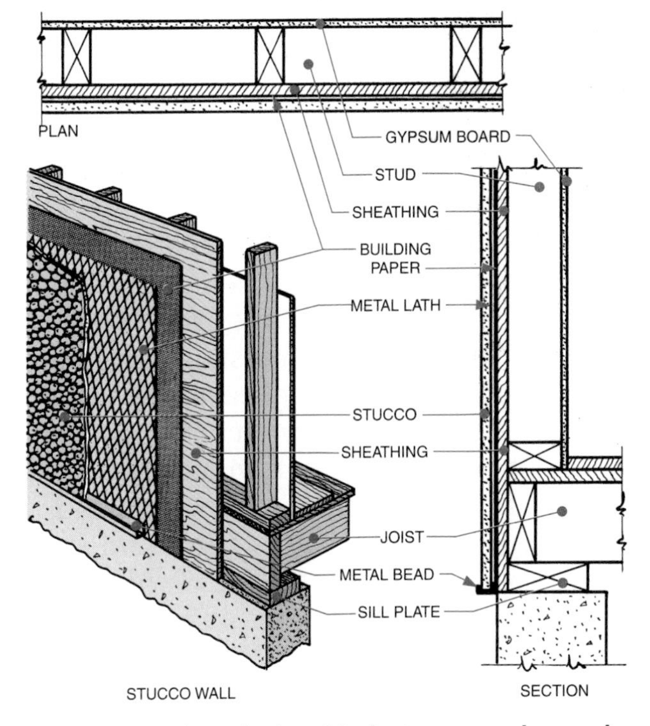

Fig. 29-20 ■ The relationship between a plan and an elevation section of a stucco wall.

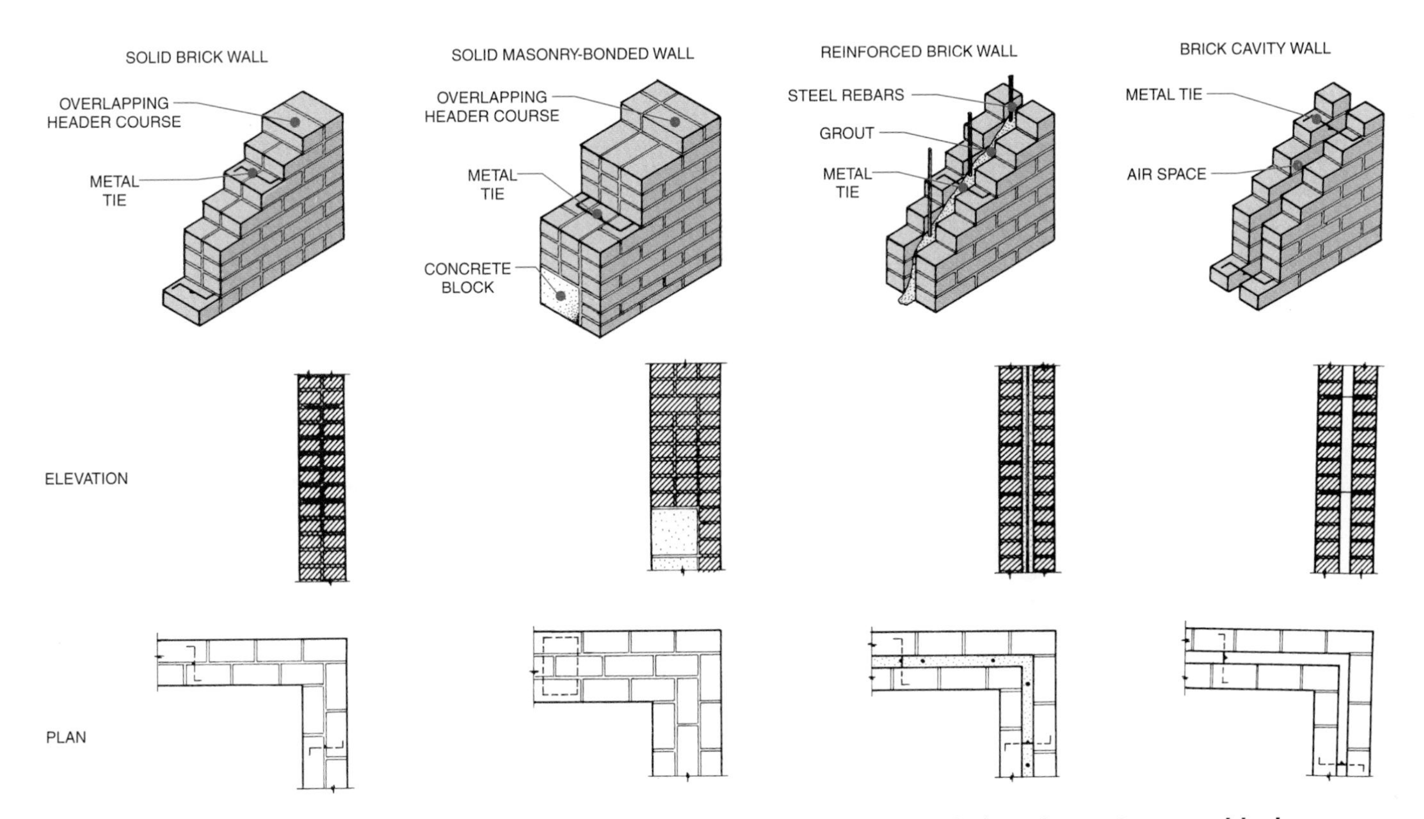

Fig. 29-21 ■ Solid brick wall construction shown in pictorial, elevation, and plan views. Concrete blocks may be used in place of one wall of bricks in cavity wall construction.

- *Masonry veneer walls* provide brick or stone facing to wood-framed walls. See Fig. 29-22. An entire wall may be a brick veneer wall. Where added support is required, a concrete block veneer wall may be used.
- *Curtain walls* are used where structural steel or post-and-beam construction provides large wall spaces that are not part of the load-bearing structure. See Fig. 29-23. One of the greatest advantages of steel-cage construction is the unobstructed space provided by curtain walls. Building loads are transmitted through columns, so the remaining open wall space can be filled with any type of nonbearing (curtain) panels. These panels are usually prefabricated in modular units. Therefore wall framing plans show only the position of modular units and not the construction details.

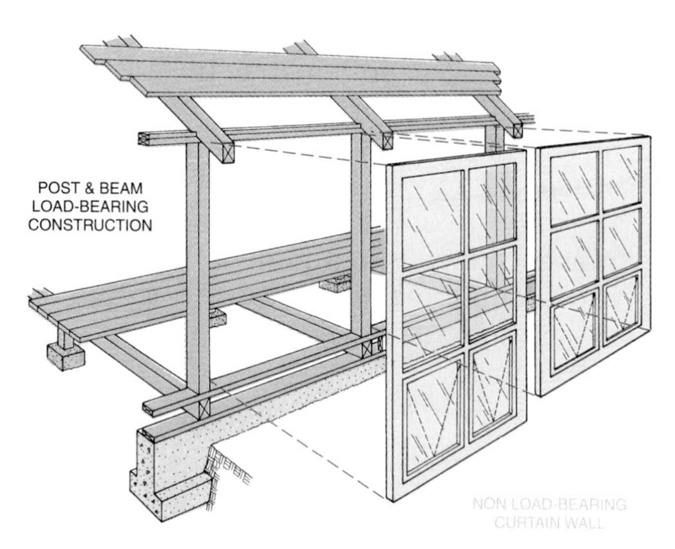

Fig. 29-23 ■ Curtain wall construction.

- *Other* types of wall construction include log wall and rammed earth designs. *Log wall* construction is shown in Fig. 29-24. To further insulate, sheathing and drywall panels can be attached to plumbed furring strips.

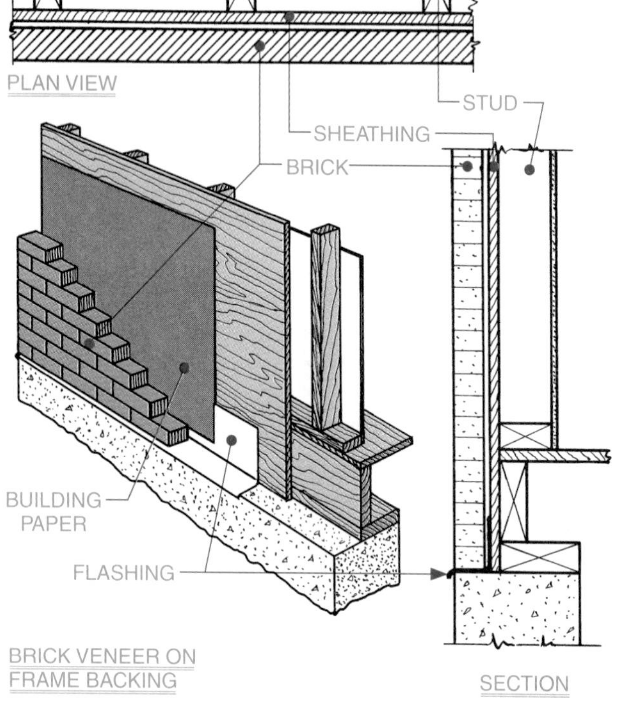

Fig. 29-22 ■ Brick veneer facing on wood frame wall construction.

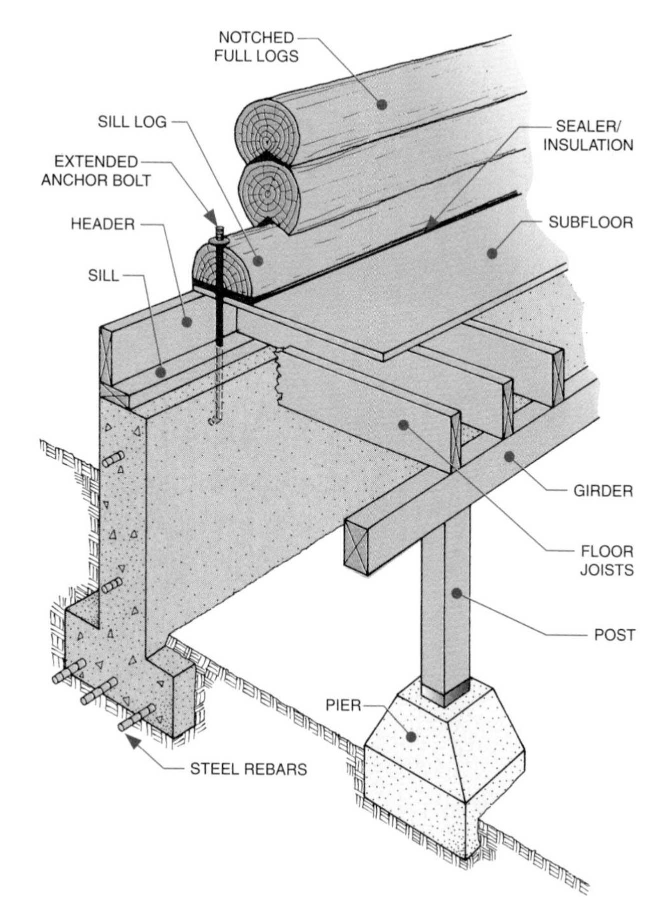

Fig. 29-24 ■ Log wall construction.

Rammed earth designs are gaining popularity in mild climates. Rammed earth walls are created by tamping a soil-cement mixture into two-foot thick forms. Some walls are formed between load bearing post-and-beam members. Others use rebar grids embedded into foundation footings. Blocks manufactured from heat-compressed straw fibers may be used in the same manner.

Window Framing Drawings

Framing members around a window must not transfer structural movement or thermal stress to the glass. To design the framing needed to support each window, an understanding of the major components of a window is necessary. See Figs. 29-25 and 29-26. Window construction details are usually shown on a head, jamb, rail, or sill section. See Fig. 29-27. The

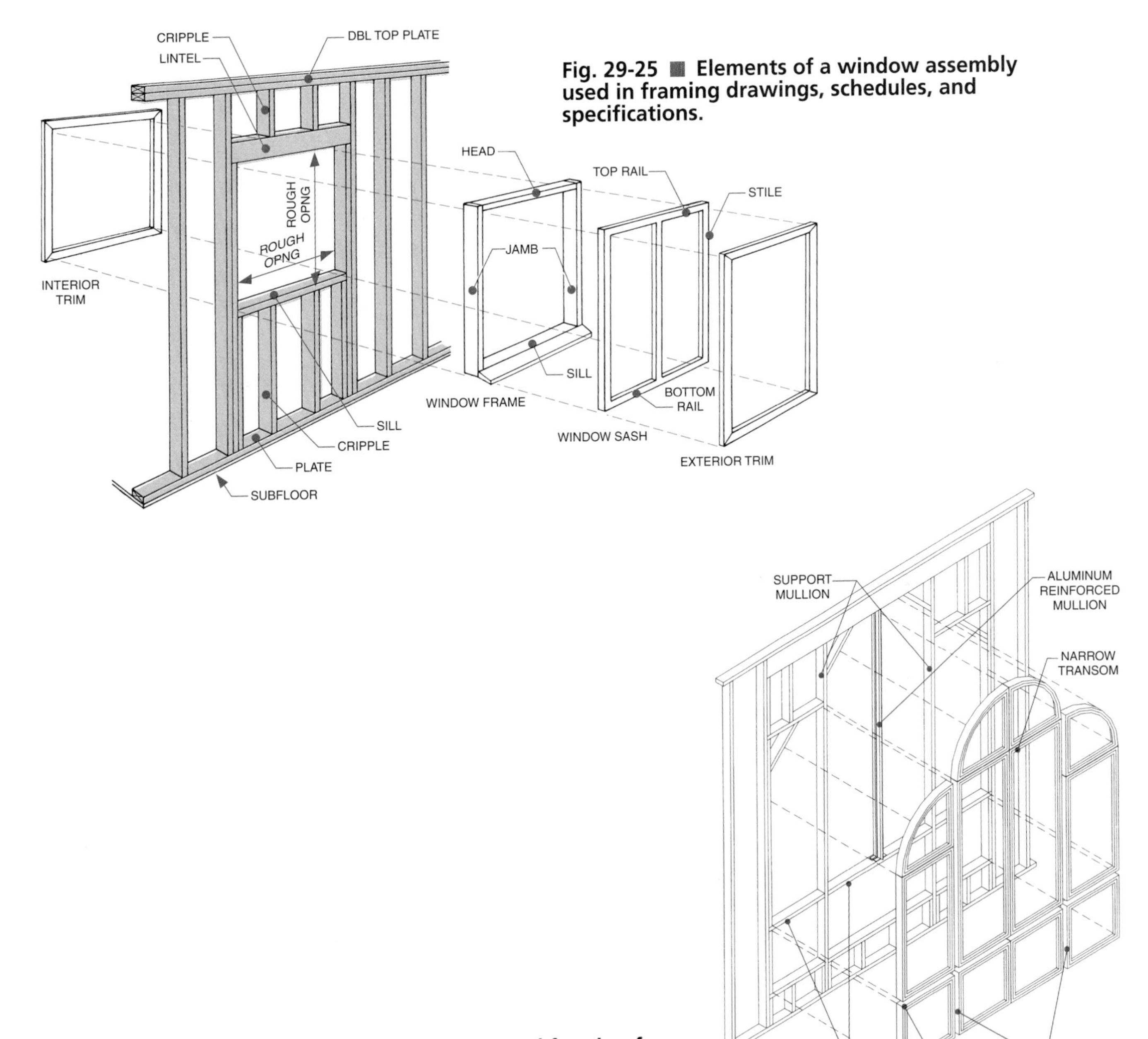

Fig. 29-25 ■ Elements of a window assembly used in framing drawings, schedules, and specifications.

Fig. 29-26 ■ Components and framing for a complex manufactured window. ***Andersen Windows, Inc.***

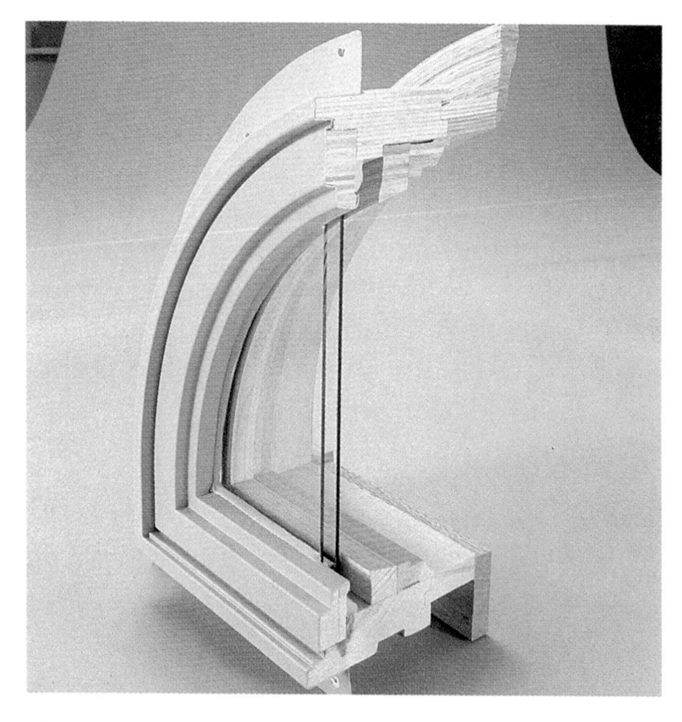

Fig. 29-27 ■ Sill and head sections of a manufactured window. *Andersen Windows, Inc.*

positions of trimmers, headers, and sill-support members are also shown on these details. The method of weatherproofing between window components and panel framing is illustrated in Fig. 29-28.

There are hundreds of window manufacturers, and each has hundreds of window styles and sizes. It is therefore necessary to refer to the manufacturer to determine the exact dimensions and rough opening for each window. This information is usually included in schedules and specifications. (See Chapter 35.) Refer to Figs. 29-29 and 29-30 for rough opening dimensions for common window sizes.

When fixed windows or unusual window shapes or sizes are to be constructed in the field—or even at a factory—complete framing details must be drawn. Unusual and nonstandard sizes and components require more complete detail drawings. *Fixed sheet glass* thicknesses range from $\frac{3}{32}$" to $\frac{7}{16}$". Widths range from 40" to 60" and

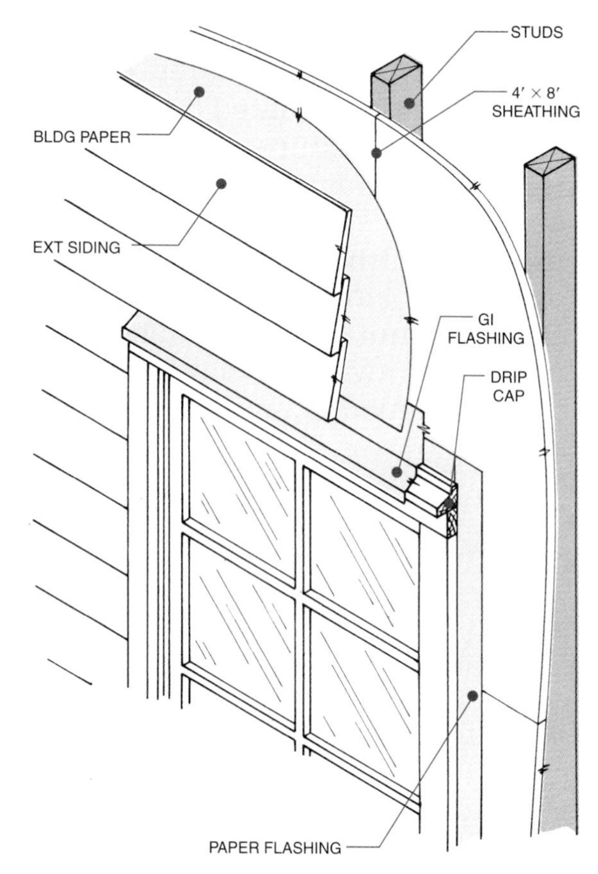

Fig. 29-28 ■ Typical window waterproofing installation.

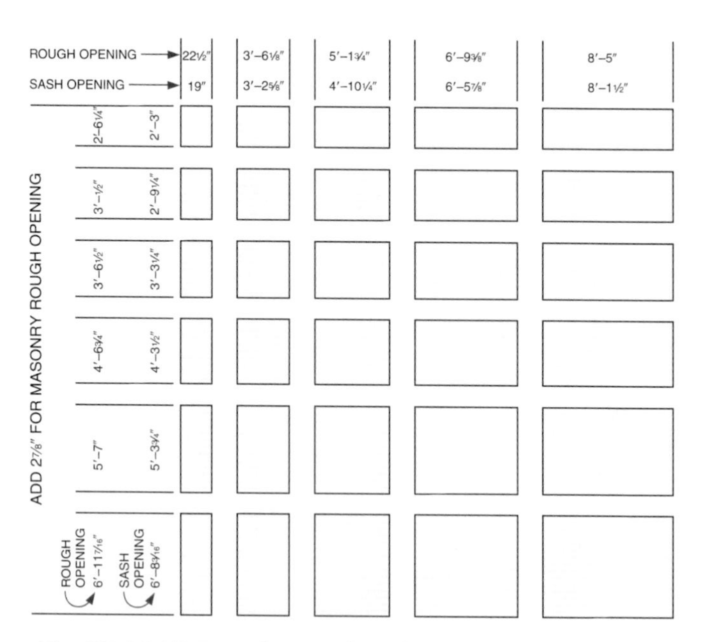

Fig. 29-29 ■ Rough opening dimensions for common window sizes.

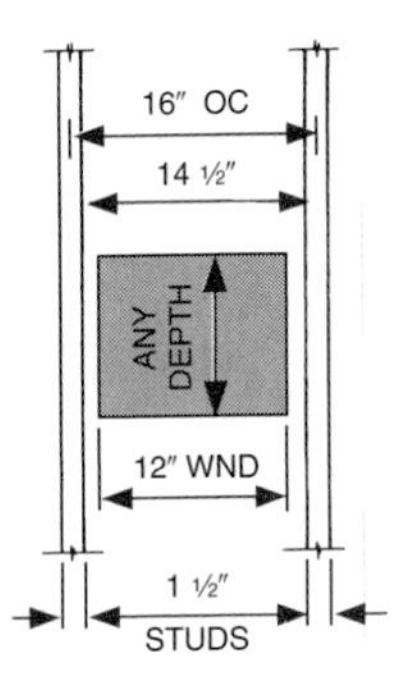

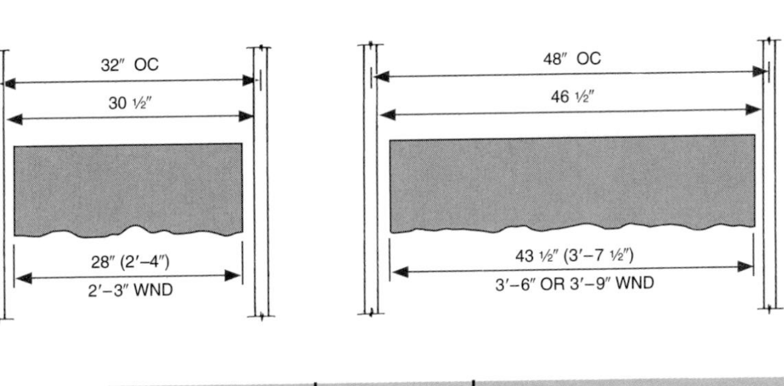

Fig. 29-30 ■ Rough openings and modular sizes for doors and windows.

Modular stud spacing @ 16″ OC	Rough opening	Window/door width dimensions
16″	14 ½″	12″
32″	30 ½″	28″ 2′–3″ or 2′–4″
48″	46 ½″	44″ 3′–8″ or 3′–9″
64″	62 ½″	60″ 5′–0″
80″	78 ½″	76″ 6′–3″ or 6′–4″
96″	94 ½″	92″ 7′–8″ or 7′–9″
112″	110 ½″	108″ 9′–0″
128″	126 ½″	124″ 10′–3″ or 10′–4″

NOTE: NON-MODULAR STUDS MAY ACCOMMODATE ANY SIZE ROUGH OPENING

PLAN 1″ TO 1 ½″ CLEARANCE (4S) TO ENSURE FIT. SECURE WITH SHIMS AND FASTENERS.

lengths range from 50″ to 120″. *Plate glass* 1/8″ to ½″ thick) ranges from 80″ × 130″ to 125″ × 280″. *Glass block* windows or walls also require a head, jamb, and sill detail. See Fig. 29-31. Glass block rough openings must allow space for the block size, plus mortar or channel space as prescribed by the manufacturer.

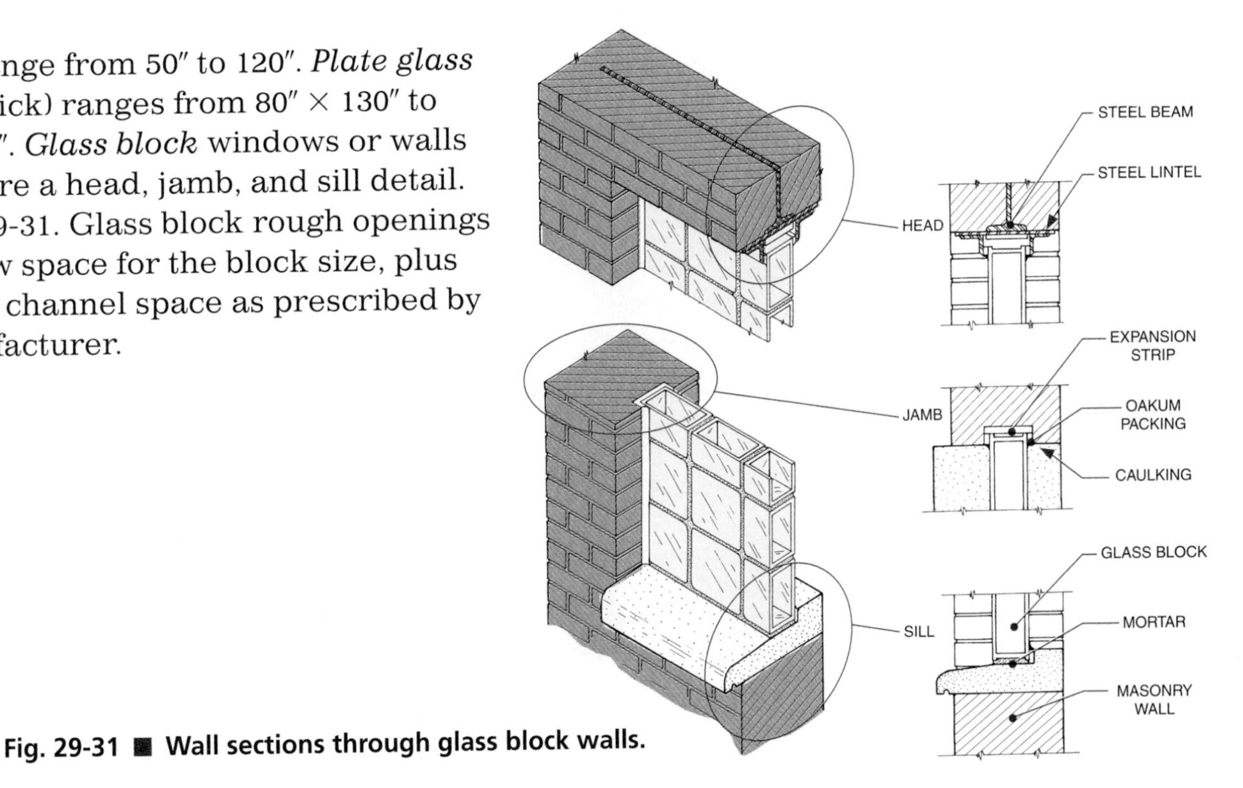

Fig. 29-31 ■ Wall sections through glass block walls.

Door Framing Drawings

Components of a door assembly include the wall framing with the rough opening, the door frame (head and side jambs), the door, and sometimes the sill (threshold). Doors are usually prehung (attached with hinges) to the jamb, and the complete assembly is fit into a rough opening (RO) in a framed wall. See Fig. 29-32. A lintel must be placed above the head jamb to prevent the jamb from sagging. See Fig. 29-33. Lintels distribute building loads to vertical support members such as studs or posts.

Modular door units may include multiple doors and windows that must also fit into a rough opening. See Fig. 29-34. Specifying, dimensioning, and constructing accurate rough openings for these units is critical to successful door functioning. See Fig. 29-35. Like windows, rough openings for doors are found in manufacturer's specifications which are included in a door schedule. (See Chapter 35.) Add 3½″ to the width and 1½″ to the height of a door, if the rough opening is not specified by a manufacturer.

Head, sill (threshold), and jamb sections are just as effective in describing door framing construction as they are in show-

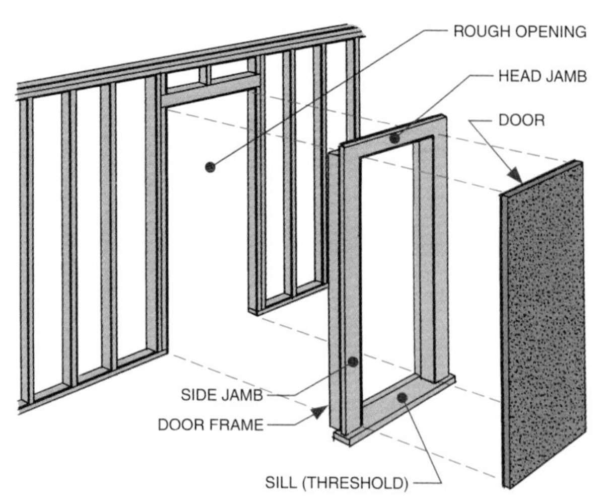

Fig. 29-32 ■ The relationship of the door jamb to door and rough opening.

Fig. 29-33 ■ Door lintels.

A. A heavy door lintel.

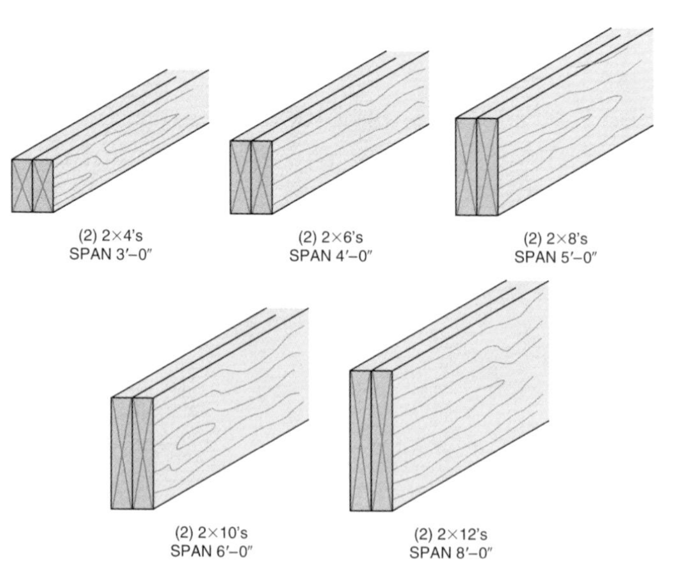

B. Typical wood lintel sizes.

Fig. 29-34 ■ Prehung swinging double French doors. *Marvin Windows & Doors*

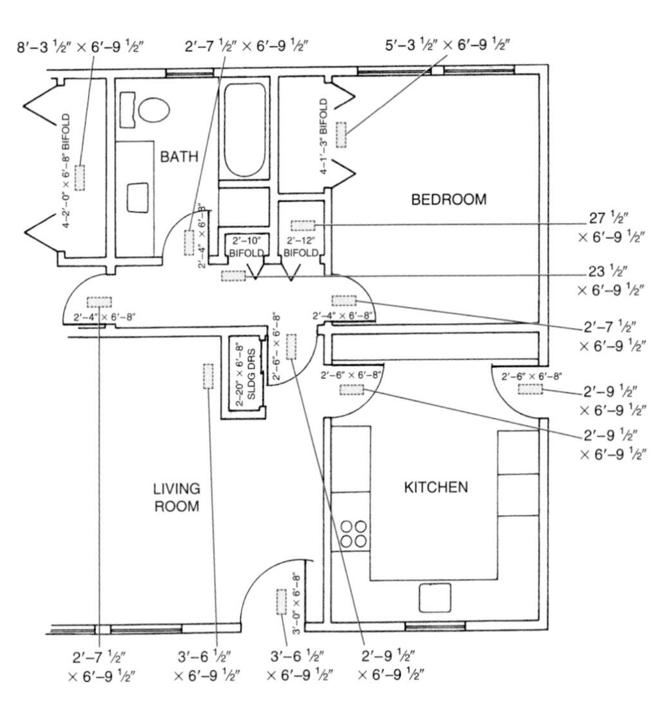

Fig. 29-35 ■ Rough opening dimensions for standard size doors.

Fig. 29-36 ■ Framing for sliding doors.

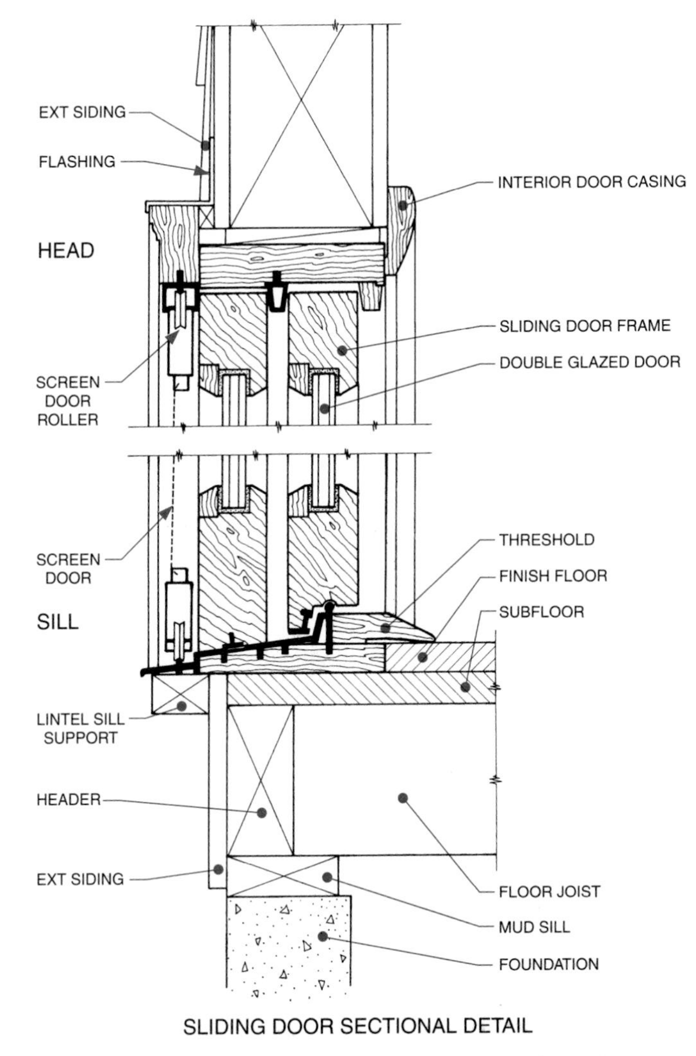

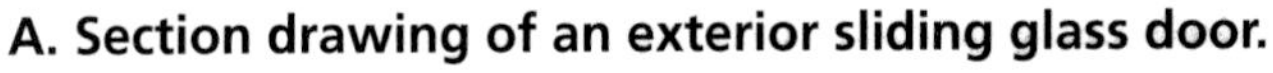

A. Section drawing of an exterior sliding glass door.

ing window framing details. Since doors extend to the floor, the relationship of the floor framing system to the position of the door is critical. The method of intersecting the door and hinge with the wall framing is also important.

Framing for sliding doors also requires careful detailing of intersections. See Fig. 29-36. If thresholds are not part of the exterior door assembly, a separate detail

B. Typical exterior sliding glass door components. *Marvin Windows & Doors*

is necessary to show the exact type and alignment of the door and the threshold. See Fig. 29-37.

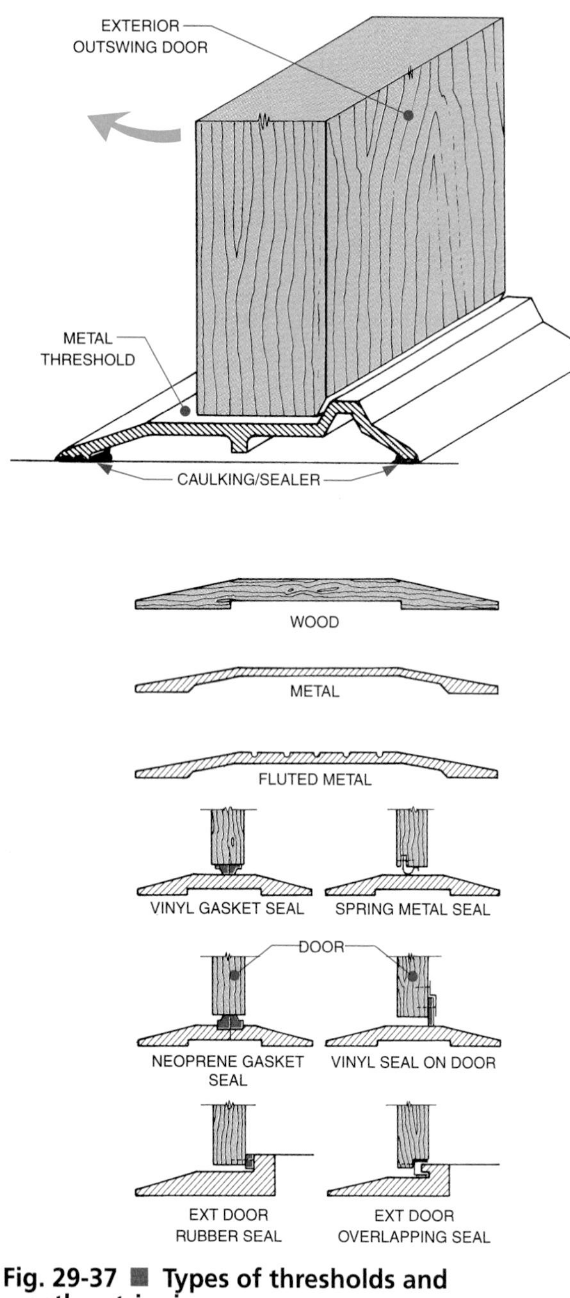

Fig. 29-37 ■ Types of thresholds and weatherstripping.

Interior Walls

Interior framing drawings include plan, elevation, and pictorial drawings of partitions and wall coverings. Detail drawings of interior partitions may also show intersections between walls and ceilings, floors, windows, and doors. See Fig. 29-38.

Framing Elevations

Framing elevation drawings show direct views of the framing. They are most effective for showing the construction of interior partitions. Interior partitions are projected from a floor plan, and are always projected from the center of the room. To ensure the correct interpretation of the partition, each interior

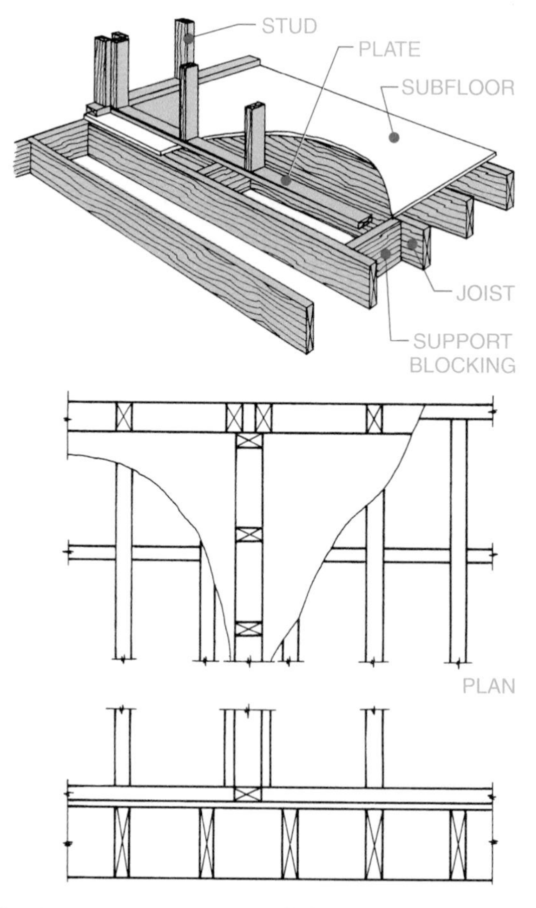

Fig. 29-38 ■ Interior wall framing detail.

elevation drawing should include a label that indicates the room, a reference number, and/or the compass orientation of the wall. See Figs. 29-39. If either the room name, compass direction, or reference is omitted, the elevation may easily be misinterpreted and confused with a similar wall in another room.

A complete study of the floor plan, elevation, plumbing diagrams, and electrical plans should be made prior to the preparation of interior wall framing drawings. Provisions must be made in framing drawings for special needs and to allow openings for electrical, plumbing, and HVAC installations. See Figs. 29-40 and 29-41. When a stud must be broken to accommodate various items, the framing drawing must show the recommended construction. A structural stud should never have more than half its thickness removed.

Fig. 29-39 ■ An elevation section is often the best method for showing special framing needs.

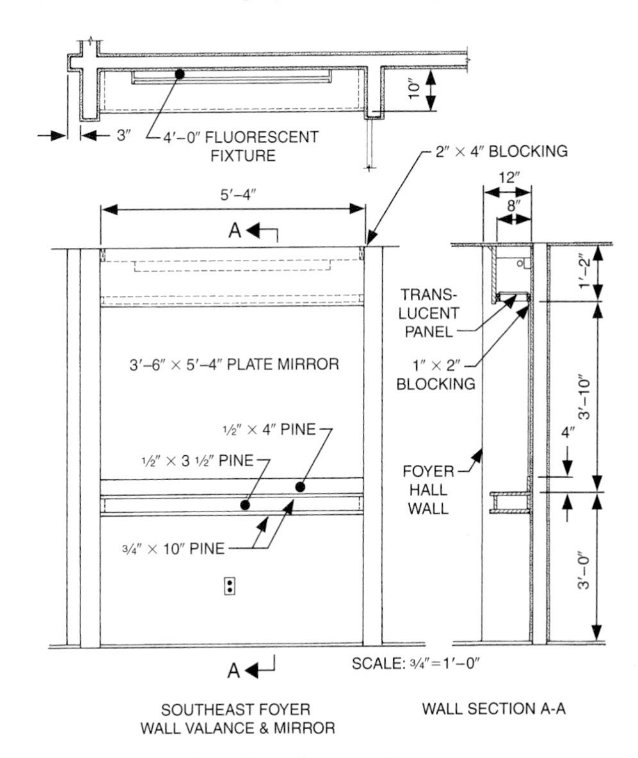

A. Special partition framing sections.

B. Same area under construction. *James Eismont Photo*

C. Completed area. *James Eismont Photo*

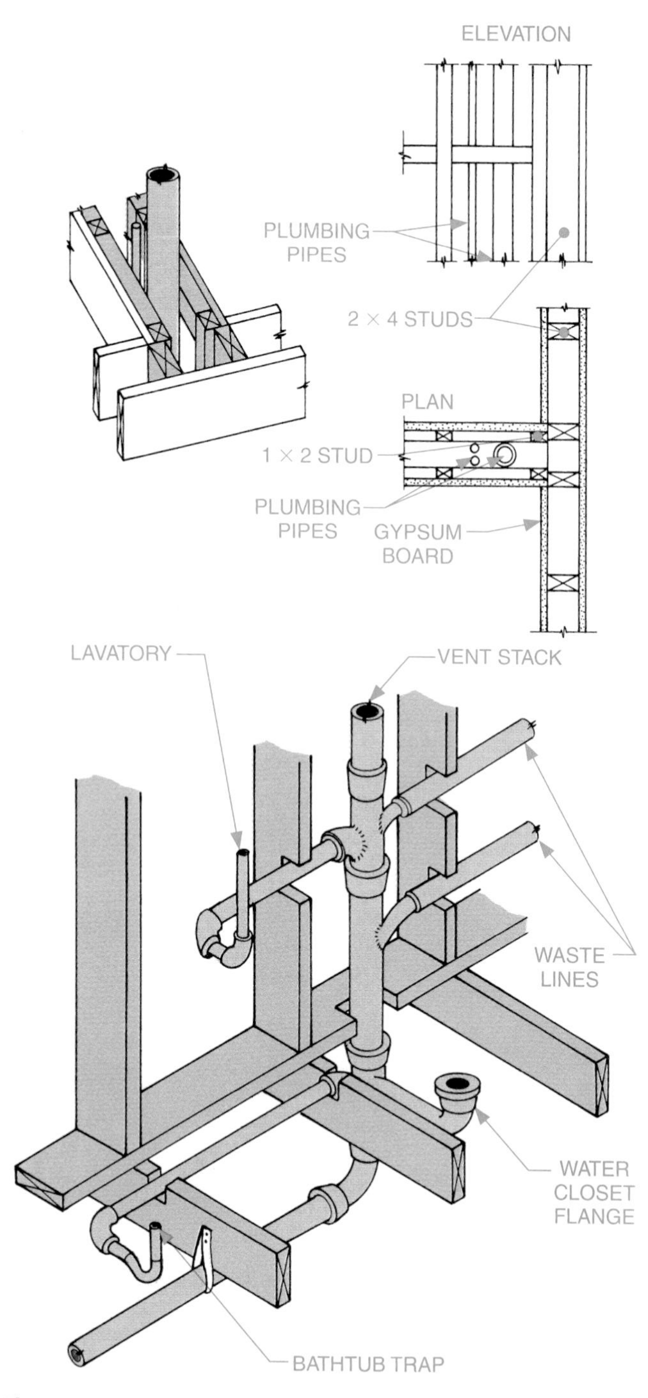

Fig. 29-40 ■ **Framing allowances for plumbing lines.**

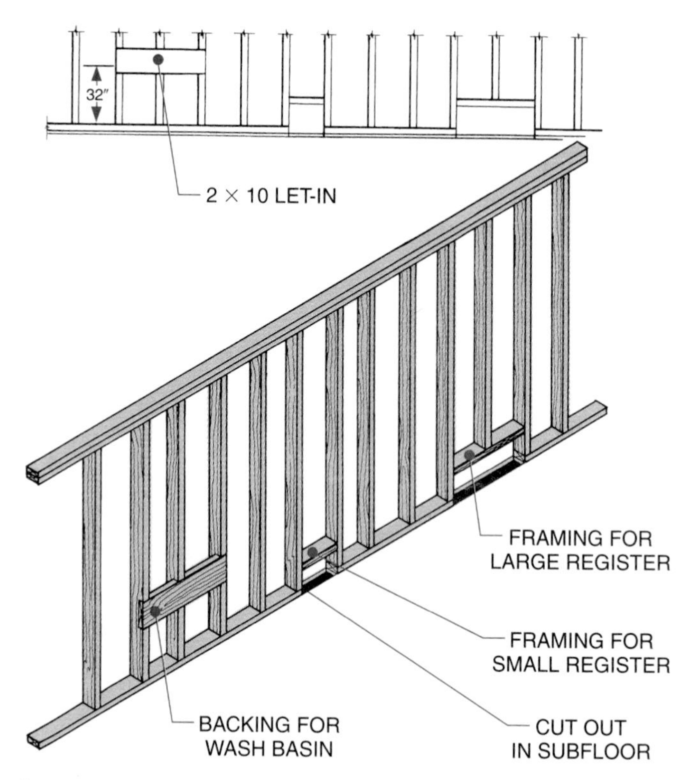

Fig. 29-41 ■ Wall framing plan showing special framing needs.

Built-in wall items may only require a partial framing elevation if location dimensions are included in an interior wall elevation or the heights are noted on a floor plan. See Fig. 29-42.

COLUMNS Steel, concrete, masonry, and wood columns or posts perform the function of load-bearing partitions. They support girders and beams upon which floor decking rests. Horizontal members can also be supported at the same height as the post through the use of hangers or blocking. See Fig. 29-43.

Steel column positions may be shown on floor plans or on elevation drawings. On a floor plan, the style, size, and weight may be noted on each outline of the column, such as A1, B2. This information may also be shown on a column schedule. **Column schedules** are schematic elevation drawings showing the entire height of a building and the elevation of each floor, base plate, and column splice. The type, depth, weight, and length of columns with common characteristics are shown under the column mark for each column. Figure 29-44 shows a pictorial floor plan with column locations related to the column schedule shown in Fig. 29-45 on page 572. There are 13 columns in this plan.

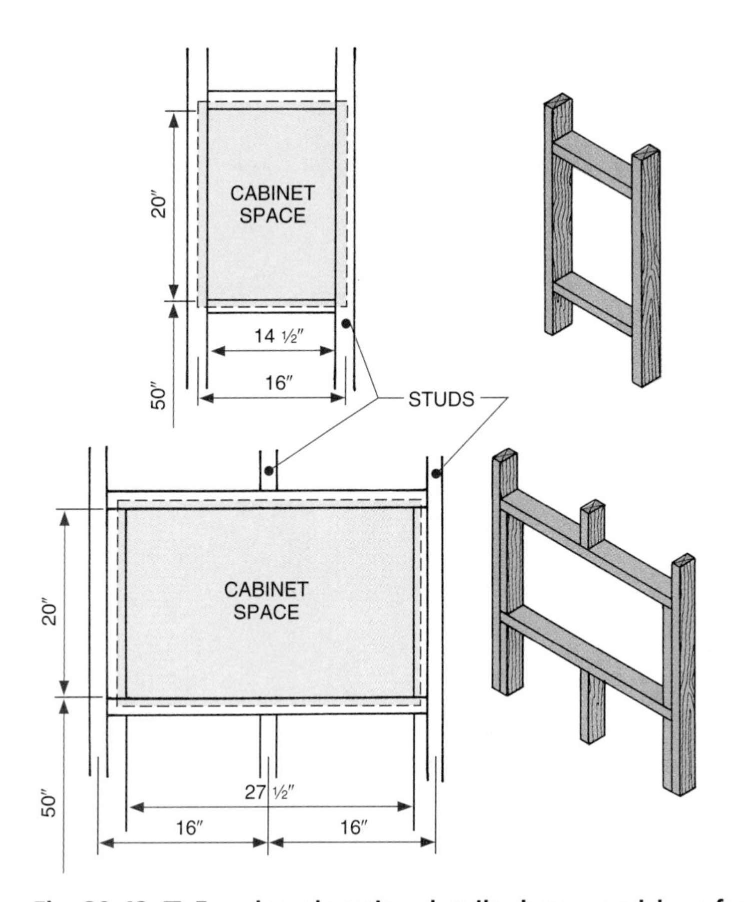

Fig. 29-42 ■ Framing elevation details show provisions for built-in items.

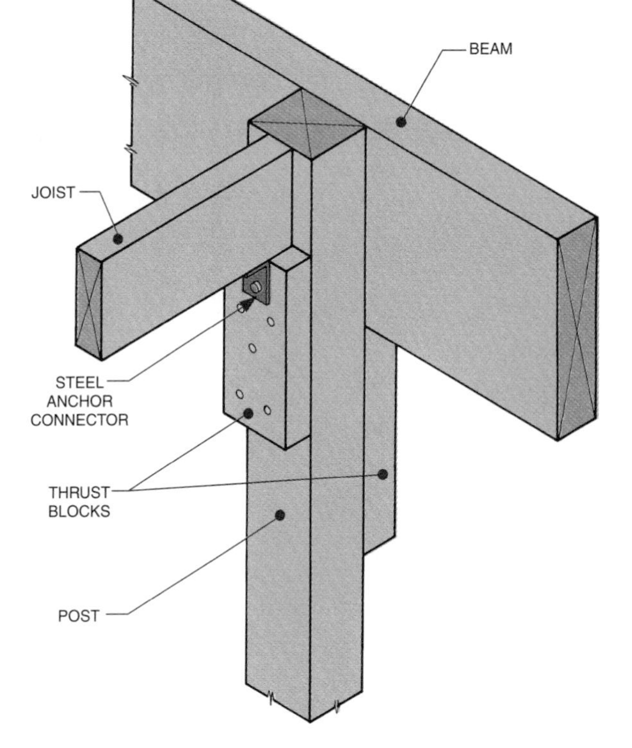

Fig. 29-43 ■ Post support of joists and beams at the same level.

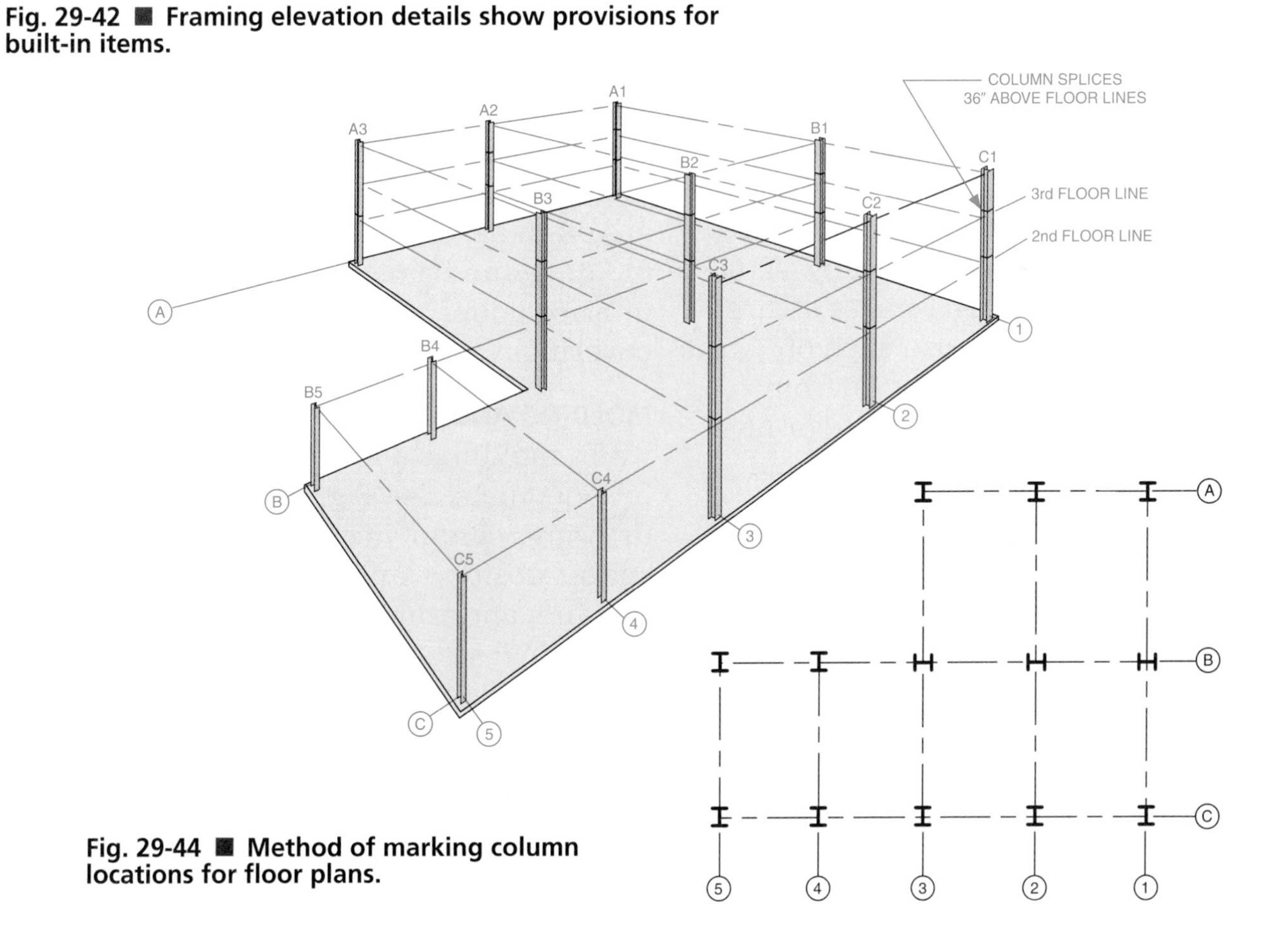

Fig. 29-44 ■ Method of marking column locations for floor plans.

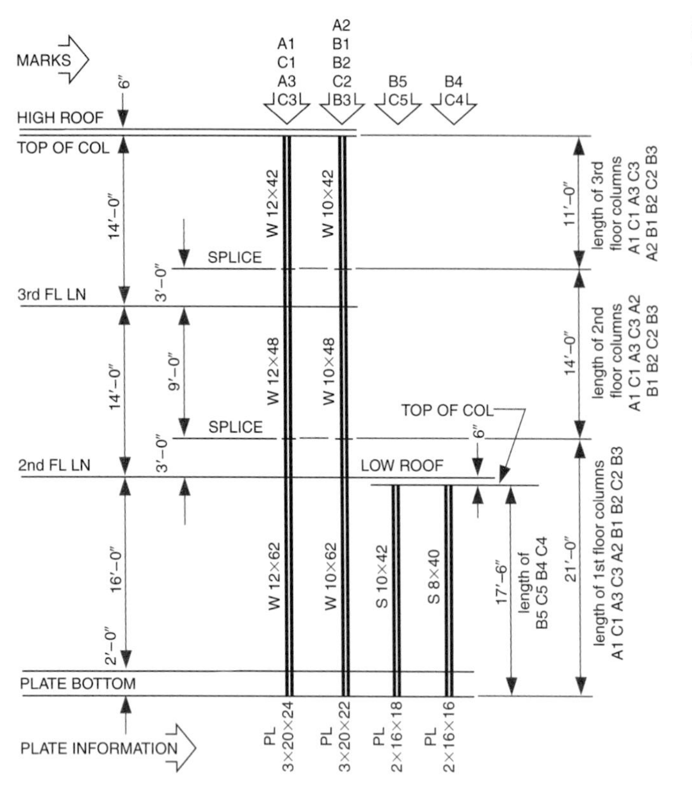

Fig. 29-45 ■ Column schedule showing height, size, and type of columns for each column mark.

Columns with common specifications are grouped together at the top of the schedule. Under each grouping a heavy vertical line represents the height of each column, with the type, size, and weight noted on each. For example, columns with marks A1, C1, A3, and C3 are all 12″-wide flange shapes and extend vertically 46′-0″ from base to top (2′ + 16′ + 14′ + 14′). Three individual columns comprise each of these. The bottom length is 21′-0″ from plate to first splice (3′ above the floor line), 14′-0″ from the first splice to the second splice, and 11′-0″ from the second splice to the top. The second row of column marks (A2, B1, B2, C2, and B3) have the same lengths as the first row but are 10″-wide flange shapes. The third-row and fourth-row columns are continuous 17′-6″, without splices; row three columns (B5, C5) are 10″ S-shapes and row four columns (B4, C4) are 8″ S-shapes.

Details

Math Connection

Additional drawings of wall construction, besides the basic structural framing, are often needed. Such drawings include molding and trim, detail intersections, interior doors, stair elevations, and wall coverings.

MOLDING AND TRIM For finished interior walls, moldings and trim are used at intersections. See Fig. 29-46. Small-scaled drawings cannot accurately show the size, shape, position, and material used for molding and trim. Thus, detail sections are prepared, such as for crown (ceiling) and base (floor) moldings. See Fig. 29-47. Where more intricate designs are used, a detail drawing that shows the different molding segments is prepared. See Fig. 29-48.

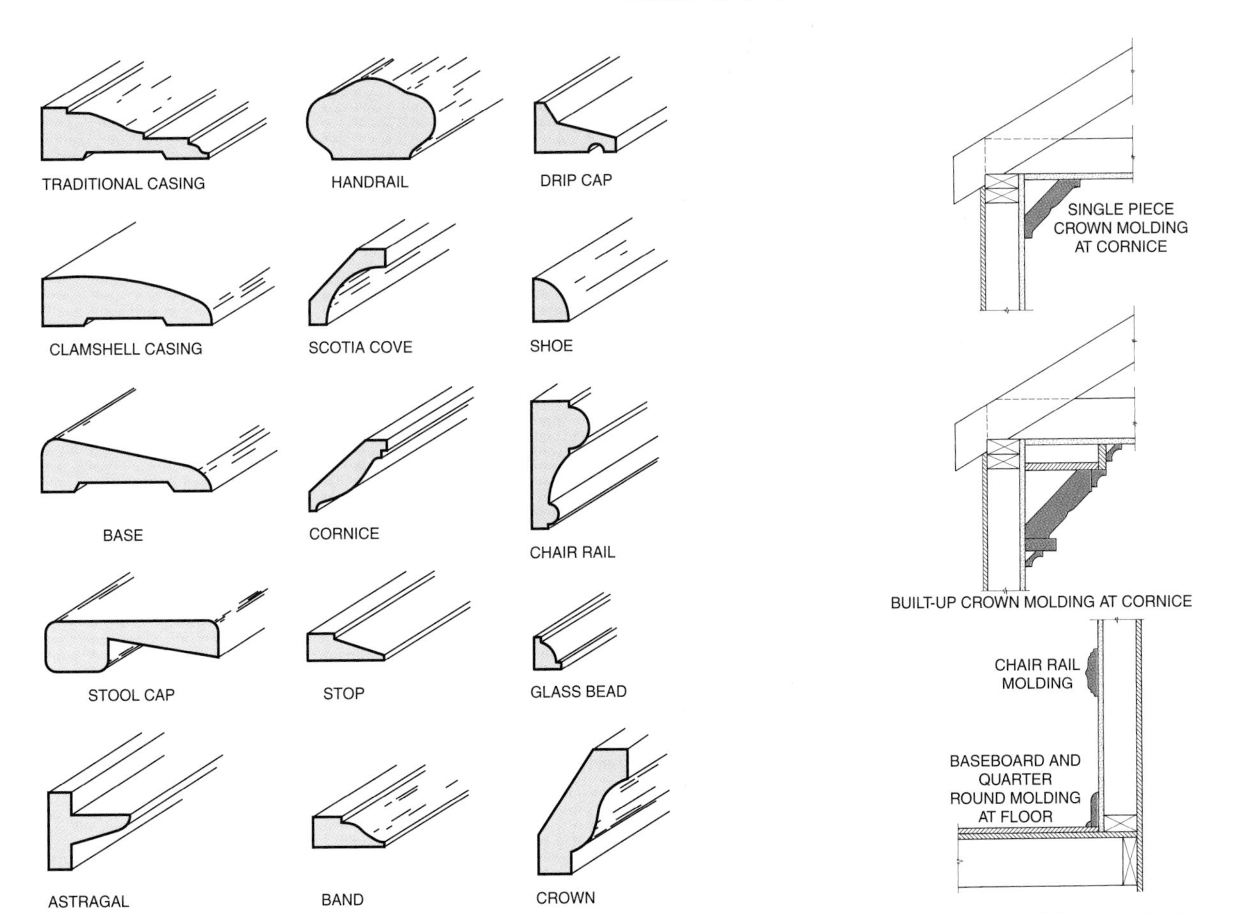

Fig. 29-46 ■ Standard molding shapes.

Fig. 29-47 ■ Common molding and trim sections.

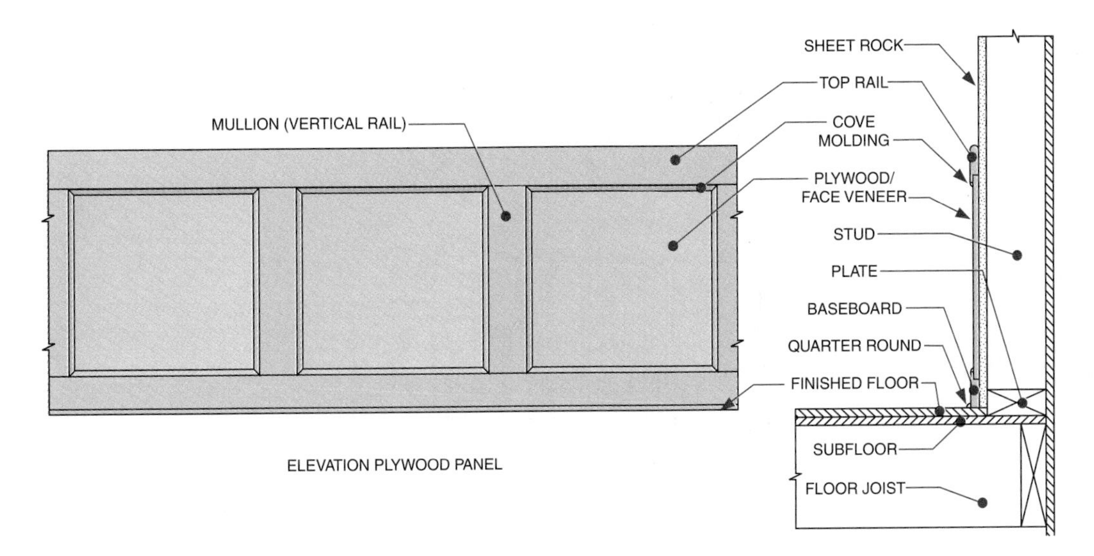

Fig. 29-48 ■ Combination molding plan details.

INTERIOR DOORS Pictorial or orthographic section drawings of the jamb, sill, and head should be prepared to illustrate the framing around interior doors. See Fig. 29-49. A detailed drawing need not be prepared for each door, but one should be prepared for each *type* of door used. The position of headers, particularly to support closet sliding doors and pocket doors, is critical to the vertical fit and horizontal matching of doors. Drawings should also be keyed to the door schedule for identification. A pictorial section of an interior-door jamb may be needed to show variations because of different wall coverings, such as plaster, gypsum-board, or paneling. See Fig. 29-50.

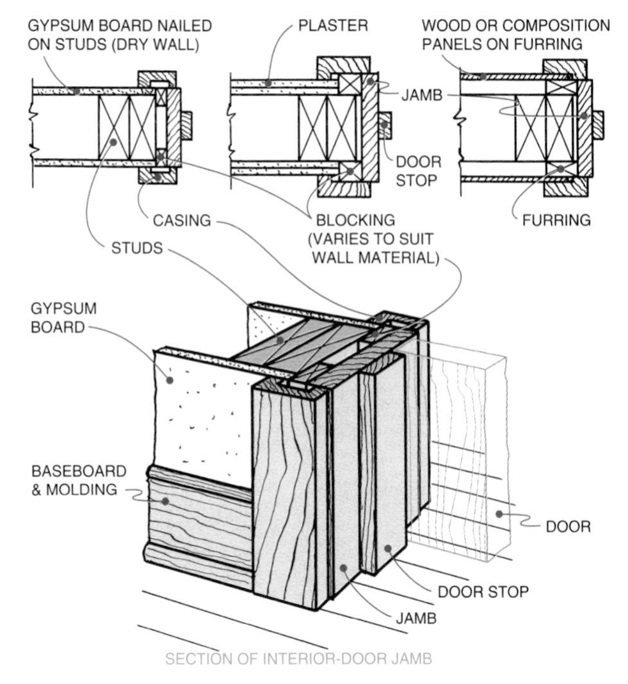

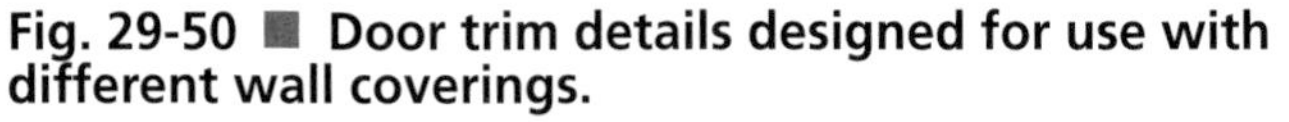
Fig. 29-50 ■ Door trim details designed for use with different wall coverings.

STAIRS Just as rough openings must be allowed for windows and doors, floor openings and wall space must be planned for stairwell framing. A sectional wall elevation drawing is often used to describe the vertical distances that are not found on floor framing plans. See Fig. 29-51. A pictorial view further illustrates a stair elevation. See Fig. 29-52, page 576.

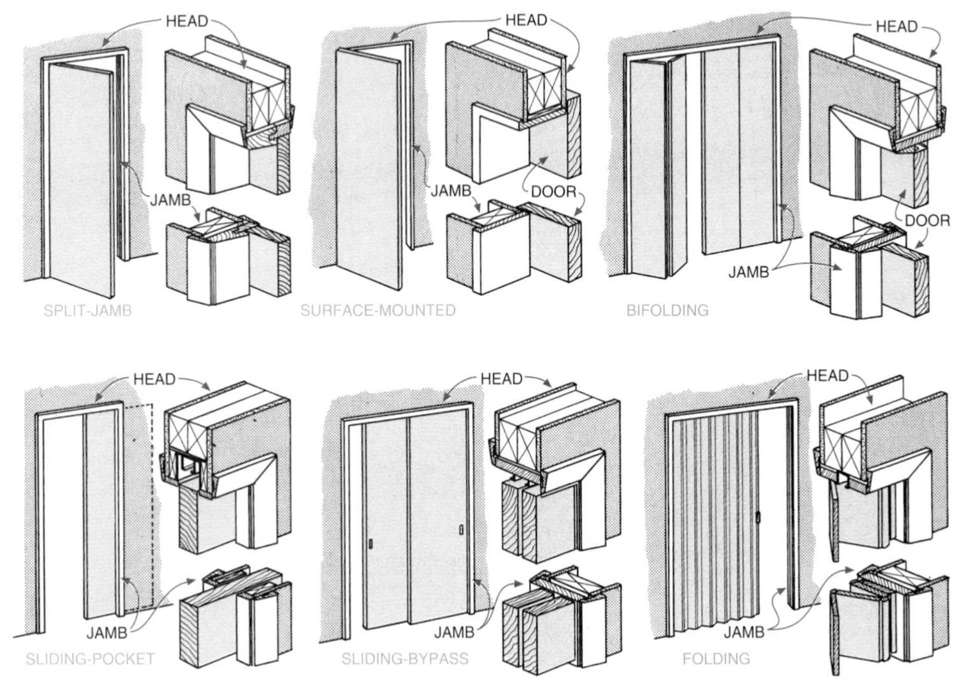

Fig. 29-49 ■ Framing and trim details for interior doors.

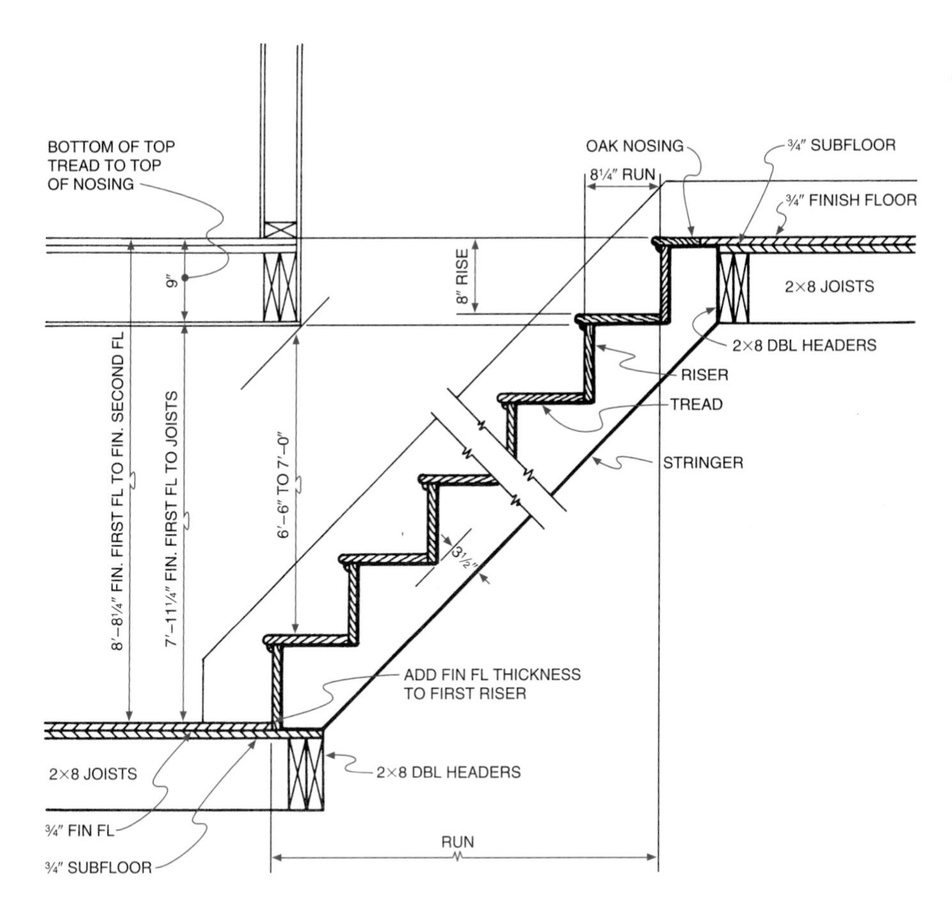

Fig. 29-51 ■ Elevation section drawing of a stair assembly.

Tread width is shown on floor plans and floor framing plans. Riser height is shown on interior elevations and on interior wall framing drawings. To calculate the number of risers use the following formula:

FORMULA:

$$\frac{\text{height of stairs}}{\text{height of each riser}} = \text{riser number}$$

EXAMPLE: 9′-4″ = 108″ + 4″ = 112″

$$\frac{112''}{7''} = 16 \text{ risers}$$

OR $\frac{115''}{7''} = 16.4$ risers

(adjust 16 risers at 7.15″ each)

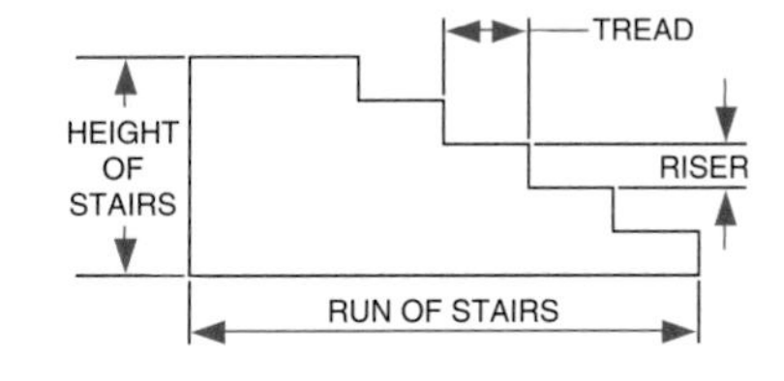

See Chapter 14 for more information about calculating stair dimensions.

WALL COVERINGS Basic types of wall-covering materials used for finished interior walls include plaster, drywall, paneling, tile, and masonry. Each type requires a different method of attachment to the wall.

- *Plaster* is applied to interior walls over wire lath or gypsum sheet lath. Plaster walls are very strong and sound-absorbing. Plaster is also decay-proof and termite-proof. However, plaster walls crack easily, they take months to dry, and the installation costs are high.
- **Drywall construction** is a system of interior wall finishing using prefabricated sheets of materials. A variety of manufactured materials may be used, such as fiberboard, gypsum wallboard, stranded lumber sheets, sheet-rock, and ply-

Fig. 29-52 ■ Pictorial interpretations of stair elevation drawings.

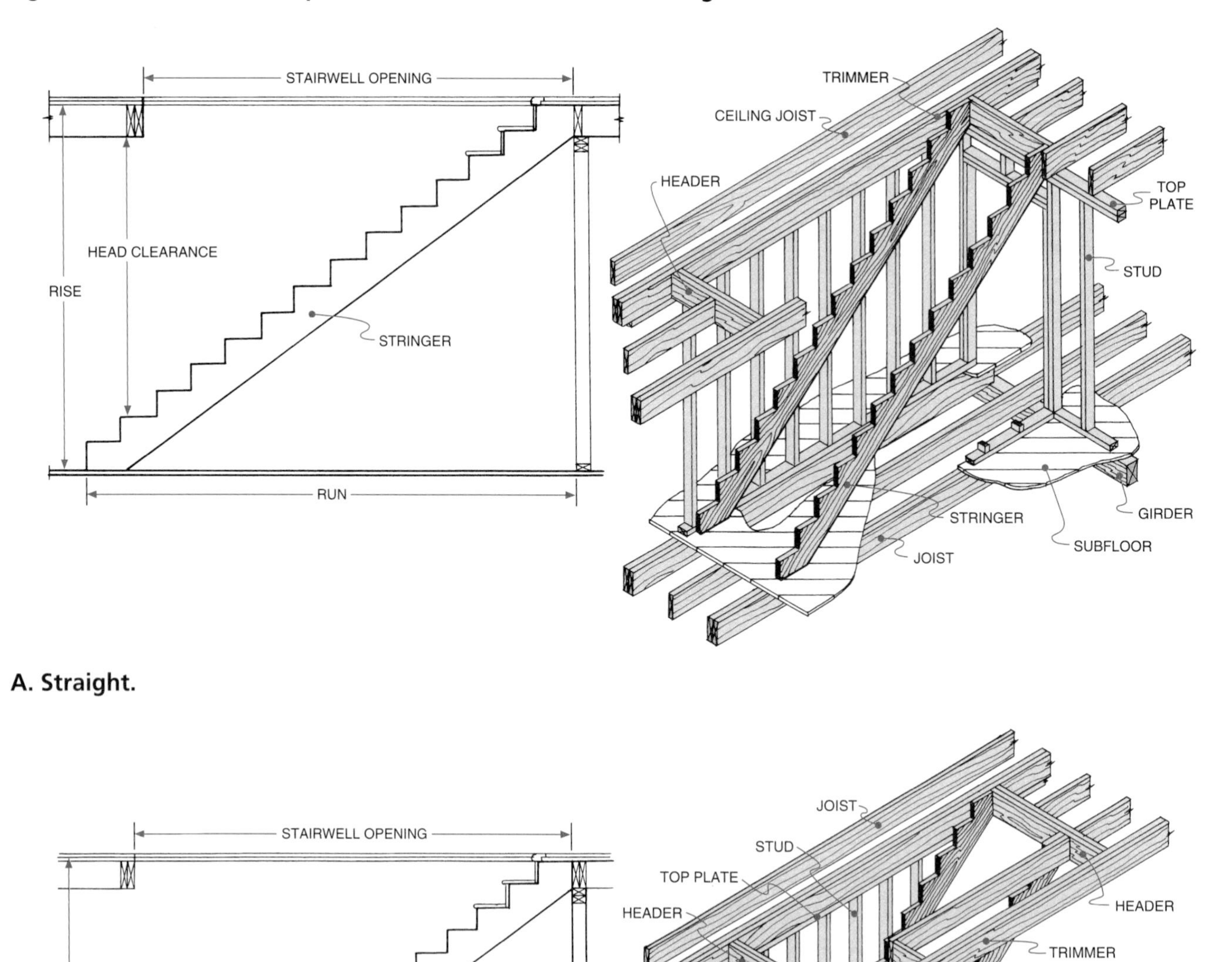

A. Straight.

B. L-shaped.

wood. Drywall is nailed directly to studs. Then the drywall joints are finished. See Fig. 29-53. Drywall thicknesses range from ¼″ to 1″. Width and length range from 2′ × 8′ to 4′ × 14′.

- When *paneling* is used as an interior finish, horizontal furring strips should be placed on the studs to provide a nailing or gluing surface for the paneling. See Fig. 29-54. The type of joint used between panels should be determined by developing a separate detail. The method of intersecting the outside corners of paneling should also be

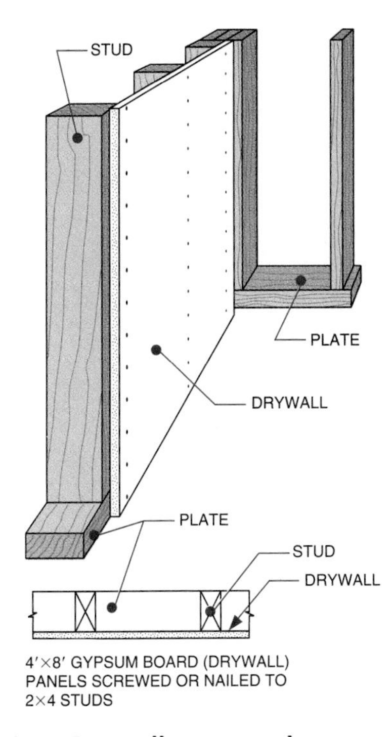

Fig. 29-53 ■ Drywall construction.

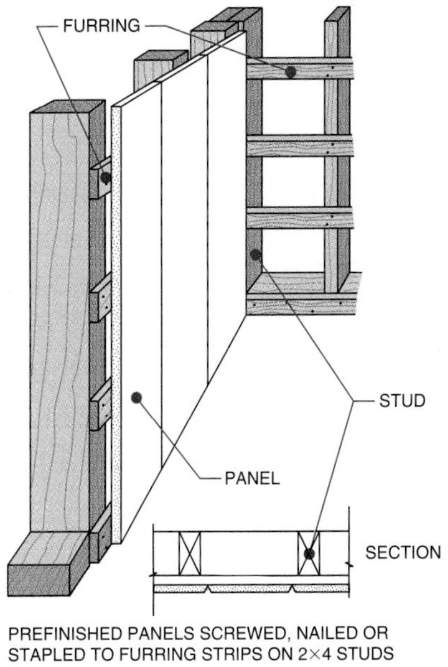

PREFINISHED PANELS SCREWED, NAILED OR STAPLED TO FURRING STRIPS ON 2×4 STUDS

Fig. 29-54 ■ Furring strips provide a horizontal surface for attaching paneling.

detailed. Outside corners can be intersected by mitering, overlapping, or exposing the paneling. Corner boards, metal strips, or molding may be used on the intersections. Inside corner intersections can be constructed by butting the wall coverings or by using corner moldings.

Stud Layouts

A horizontal framing section called a **stud layout** is a plan similar to a floor plan, except it shows the position of each wall framing member, exterior and interior. Stud layouts are used to show how studs are spaced on the plan and how interior partitions fit together.

A stud layout is a horizontal section through walls and partitions. The cutting-plane line is placed approximately at the midpoint of the panel elevation. See Fig. 29-55. The exact position of each stud that

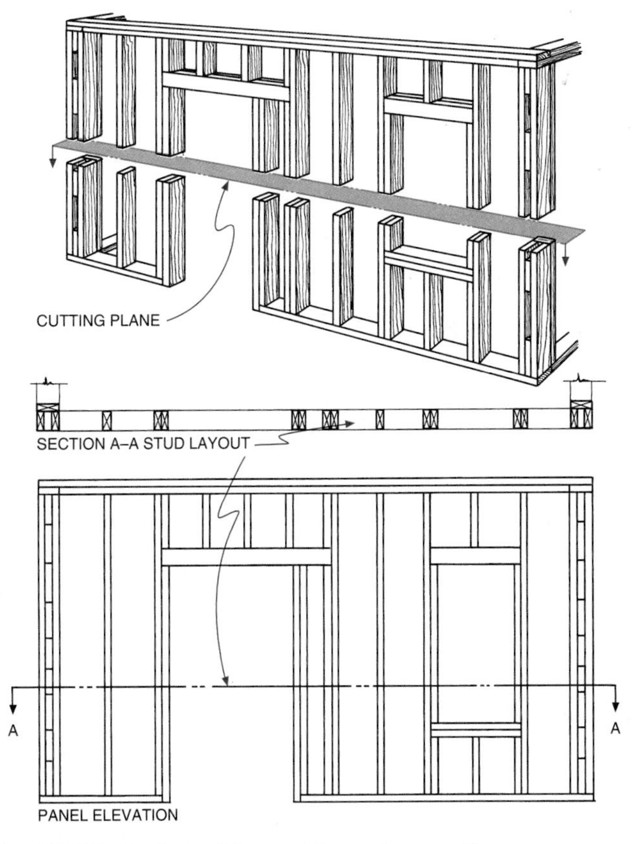

Fig. 29-55 ■ A stud layout is a plan section as viewed through a wall framing elevation.

falls on an established center (16″, 32″, or 48″) is usually shown by diagonal crossed lines. Studs other than those on 16″ centers are shown with different symbols. Different symbols also identify studs that are different in size, such as blocking and short pieces of stud stock. See Fig. 29-56. A coding system of this type eliminates the need for dimensioning the position of each stud. The practice of coding studs and other members in the stud plan also eliminates the need for repeating dimensions of each stud.

To conserve space, non-bearing studs are sometimes turned so they are flat. This rotation should be reflected in the stud layout. See Fig. 29-57.

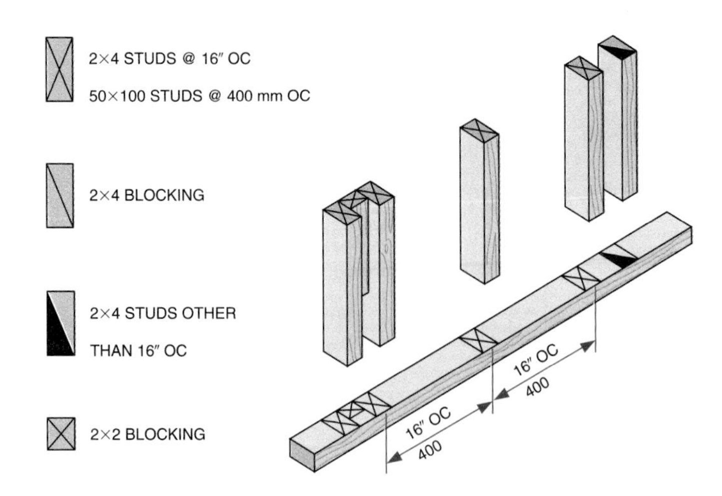

Fig. 29-56 ■ Stud layout symbols.

Dimensions

Detailed dimensions are normally shown on a stud symbol key or on a separate enlarged detail. Distances that are typically dimensioned on a stud layout include the following:

- inside framing dimensions of each room (stud to stud)
- framing width of the halls
- rough opening for doors and arches
- length of each partition
- width of partition where dimension lines pass through from room to room. This provides a double check to ensure that the room dimensions plus the partitioned dimensions add up to the overall dimension. When a stud layout is available, it may be used on the job to establish partition positions.

Stud Details

Stud layouts are of two types: *complete plans,* which show the position of all framing members on the floor plan, and *stud details,* which show only the position and relationship of some studs or framing intersections. For example, the position of each stud in a corner-post layout is frequently detailed in a plan view. See Fig. 29-58.

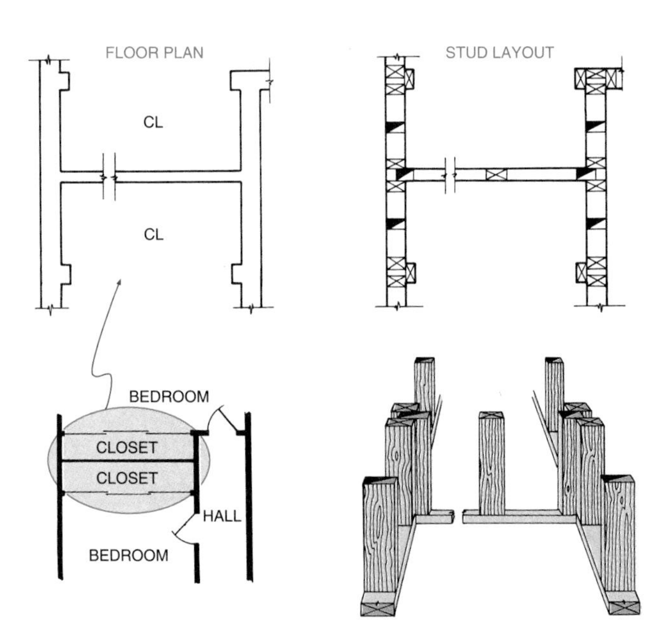

Fig. 29-57 ■ Non-bearing studs may be placed flat to conserve space.

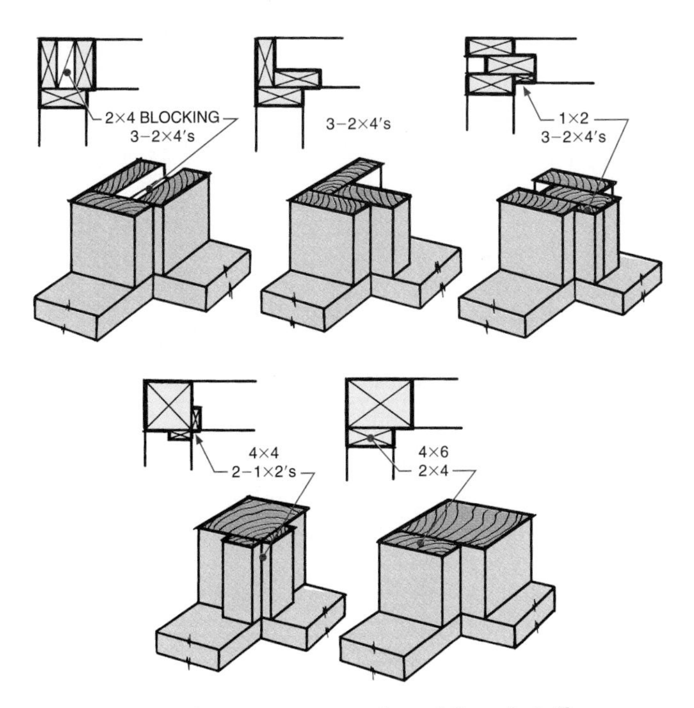

Fig. 29-58 ■ Corner post stud position details.

Occasionally, siding and inside-wall covering materials are included on this plan. Preparing this type of drawing without showing the covering materials, however, is the quickest way to show corner-post construction. If wall coverings are included on the detail, the complete wall thickness can be drawn. In a plan section, care should be taken to show the exact position of blocking because it may not pass through the cutting-plane line. Blocking should be labeled or a symbol used to prevent the possibility of mistaking it for a full-length stud. See Fig. 29-59.

When laying out the position of all studs, remember that the finished dimensions of a 2 × 4 stud are actually 1½″ × 3½″. For the exact dressed sizes of other rough stock, refer back to Fig. 24-6.

Modular Plans

In drawing modular framing plans, partitions must fit in relation to modular grid lines. Allowances for exterior wall thicknesses need to be indicated. See Fig. 29-60. Space must also be provided for such items as door and window placement, medicine cabinets, closets, fireplaces, and plumbing runs.

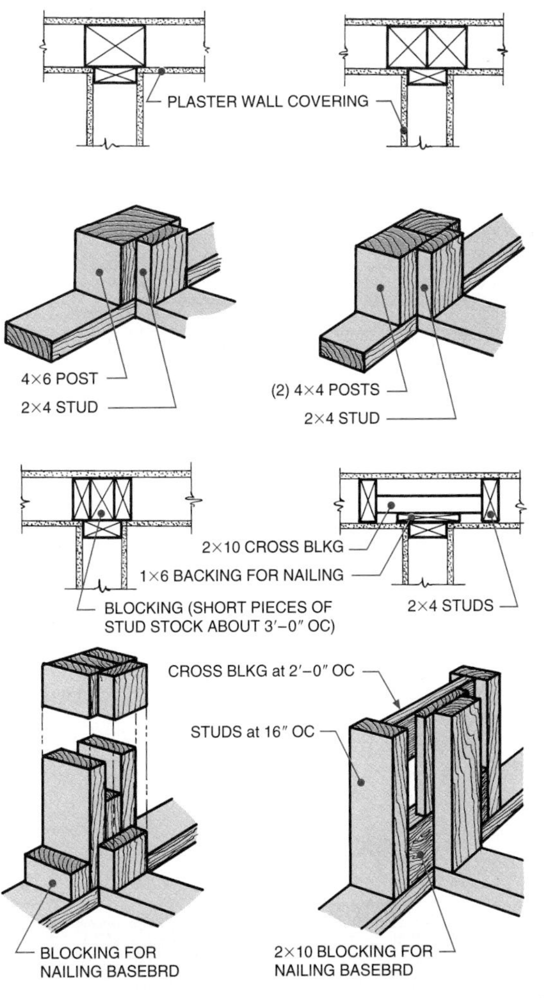

Fig. 29-59 ■ Wall covering and blocking can be shown on intersection details.

Steel Studs

Steel studs are one-half the weight of wood studs. They are stronger, won't warp or split, and are moisture and insect resistant. Some non-bearing steel studs are pre-punched for electrical or plumbing lines or for attachment to wood sills, plates, or other studs. See Fig. 29-61. A steel symbol should be added to stud layouts where steel studs are specified.

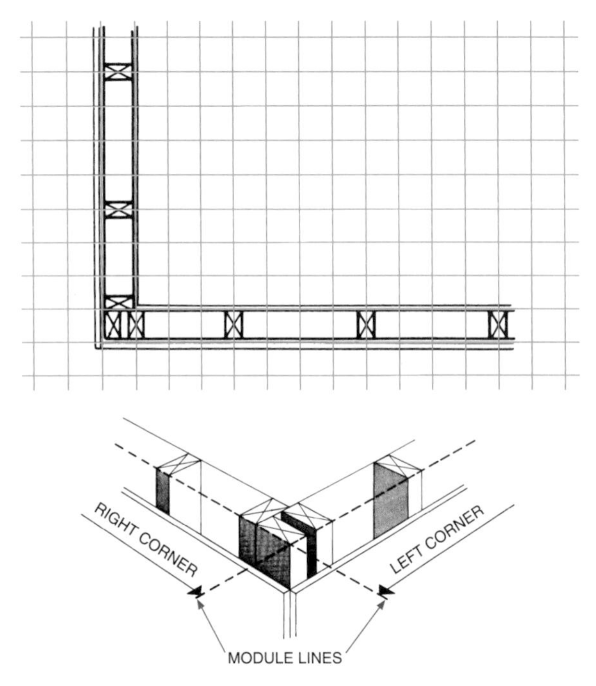

Fig. 29-60 ■ Modular grid applied to stud layout.

Fig. 29-61 ■ Non-bearing steel stud wall.

CHAPTER 29 Exercises

1. Prepare an exterior wall framing plan for a home of your own design.
2. Draw a stud layout (16″ OC) for a floor plan. Indicate the rough openings for doors and windows. Show a corner post and intersection detail.
3. Complete Exercise 2 using a CAD system.
4. Using a CAD system, draw a window detail of a window component: a head, jamb, or sill.
5. Draw a framing elevation for a double swinging door. Include the dimensions.
6. Project a framing elevation drawing of a kitchen wall in Fig. 16-24.

7. Draw an interior wall framing plan for a kitchen, bath, and living area wall of the house you are designing.
8. Prepare a stud layout ($\frac{1}{4}$″ = 1′-0″). for a home of your own design.
9. Using a CAD system, complete a stud detail of a corner post and another intersection.

CHAPTER 30

Roof Framing Drawings

Roof styles developed through the centuries. Pitches (slopes) of roofs were changed, gutters and downspouts were added for better drainage, and overhangs were extended to provide more protection from the sun. The size of roofs and the types of materials changed accordingly. The structure of a roof affects the choice of roof framing and roof covering materials, as well as the interior ceiling systems.

■ **The shape of the roof is an important factor in the overall design of a structure.** *Stock Market/© Frank Rossotto*

Objectives

In this chapter, you will learn to:

- **describe roof framing members, components, and methods.**
- **calculate roof pitch.**
- **draw a roof framing plan showing structural members, sizes, pitch, and spacing.**
- **draw framing details and elevations.**

Terms

cornice
fascia
flashing
pitch
rafter
ridge board
rise
run
soffit
span

Roof Function

The main function of a roof is to provide protection from rain, snow, sun, and hot or cold temperatures. In a cold climate, a roof is designed to withstand heavy snow loads. In a tropical climate, a roof provides protection mainly from sun and rain.

The walls of a structure are given stability by their attachment to the ground and to the roof. Buildings are not structurally sound without roofs. As explained in Chapter 22, live loads that act on the roof include wind loads and snow loads, which vary greatly from one geographical area to another. Dead loads that bear on roofs include the weight of the structural members and coverings. All loads are computed by pounds per square foot, or kilograms per square meter if metric measurements are used. (For computation of loads for an entire structure, see Appendix A, "Mathematical Calculations.")

Roof Framing Members

The structural members of a roof must be strong enough to withstand many types of loads.

Wood Roof Members

Beams used for roof construction may be made from the same materials used for floor framing. A ridge beam, also called ridge or **ridge board,** is the top member in the roof assembly. See Fig. 30-1. In post-and-beam construction, the ridge beam or board may be exposed on the inside of a building.

Roof **rafters** intersect ridge boards, as in Fig. 30-2, and rest on the tops of outside walls. Rafters may be selected from the same materials as floor joists. They may be solid lumber, truss joists, I-joists, laminated, and stranded lumber.

Fig. 30-1 ■ Members of a roof assembly. Note that the top member is a ridge board (beam).

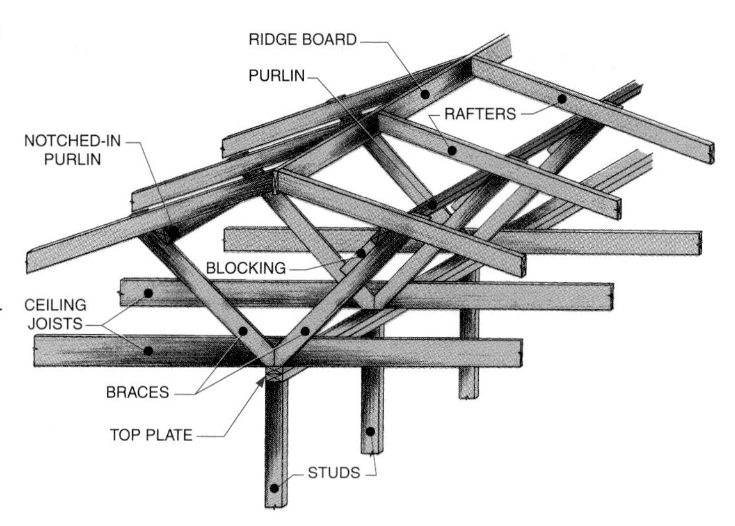

Steel Roof Members

Roofs framed with structural steel use steel girders, beams, and joists in the same manner as steel-framed floors. Most steel-framed roofs are flat, but steel ridge beams and rafters are also used on large pitched (angled) roofs.

Concrete Roof Members

At one time concrete was never used in roof construction because normal spans could not be achieved. Technological developments have dramatically increased the use of concrete for roofs.

REINFORCED CONCRETE Reinforced concrete is either precast or poured-on-site concrete with steel reinforcing rods inserted for stability and rigidity. Reinforced concrete slabs are used extensively for roof systems where short spans make prestressing unnecessary. A wood-framed flat

Fig. 30-2 ■ Typical ridge board intersections.

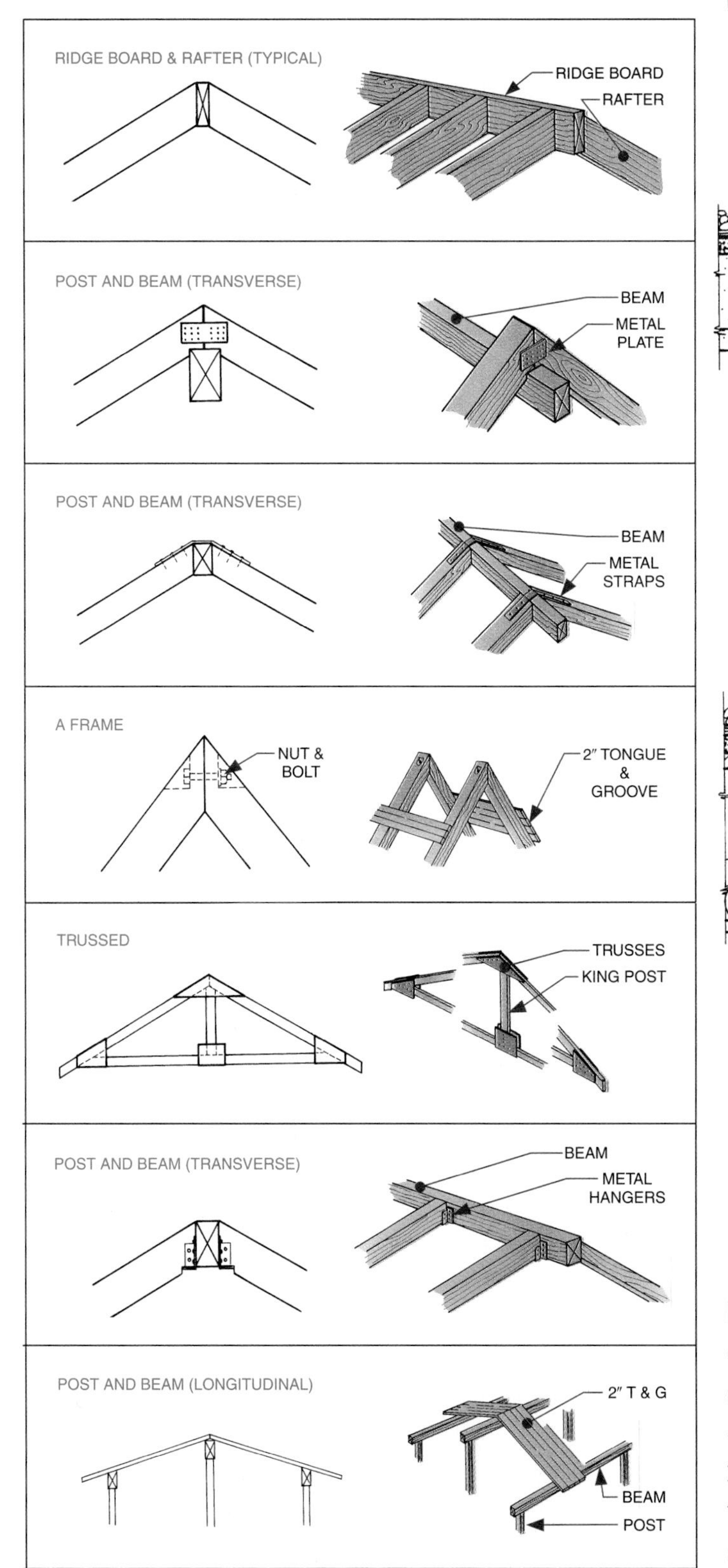

Fig. 30-3 ■ Comparison of a reinforced concrete roof and a wood-framed roof.

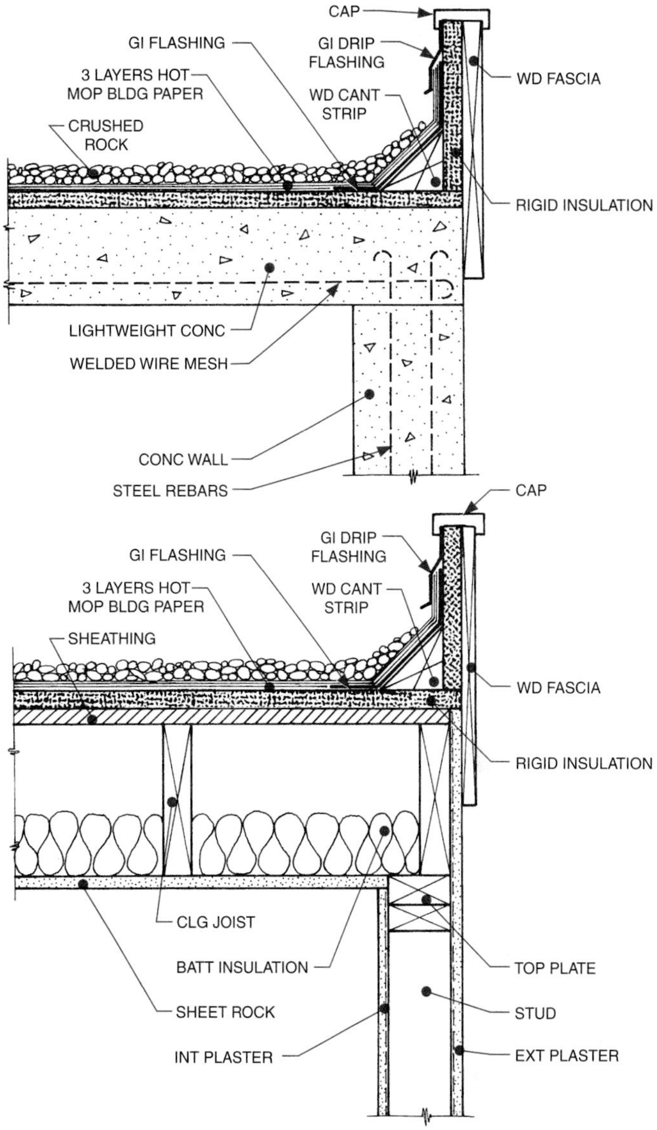

roof and a reinforced concrete roof may have the same covering. See Fig. 30-3.

PRESTRESSED CONCRETE A prestressed concrete roof, beam, or slab is made by stretching steel rods and pouring concrete around them. When the stretched rods are released, the rods try to return to their original shape, and this creates tension on the concrete. Prestressing strengthens the concrete and allows beams to have a lower ratio of depth to span.

PRECAST CONCRETE Precast concrete roof members may be poured off-site into molds, with steel reinforcement. When high stresses will not be incurred, precasting without prestressing is acceptable for most construction.

CONCRETE SHELLS Concrete shells are curved sheets of lightweight concrete poured or sprayed onto steel mesh and bars that provide a temporary form until the concrete hardens. Once the concrete hardens, the steel is locked in place and the structure is rigid and stable.

Roof Framing Components

Trusses

Trusses for roofs are prefabricated components that perform all the functions of rafters, collar beams, and ceiling joists. Trusses consist of a top chord and bottom chord, joined with diagonal and vertical members called *webs*. See Fig. 30-4. These truss members are held rigidly in place with bolts, ring connectors, and/or gussets. Plywood gussets or sheet-metal gussets are used in light construction. See Fig. 30-5. Heavy timber trusses require heavy-duty steel gusset plates that are welded together and bolted through the top chord.

Widely spaced trusses are tied together horizontally to make the roof system structurally stable. Horizontal members known as purlins are used for this purpose. See Fig. 30-6. On a trussed roof, purlins perform the same function as joist bridging in conventional roof construction.

Trusses prevent sags and cracks in the roof system because they are structurally independent and resist both compression and tension forces. See Fig. 30-7. (Figures 30-8 through 30-10 are on pages 586-587.) Since trusses can span larger distances than conventionally framed roofs, the spacing of interior partitions is more flexible. Fewer or no bearing partitions may be needed. Trusses may rest on steel beams

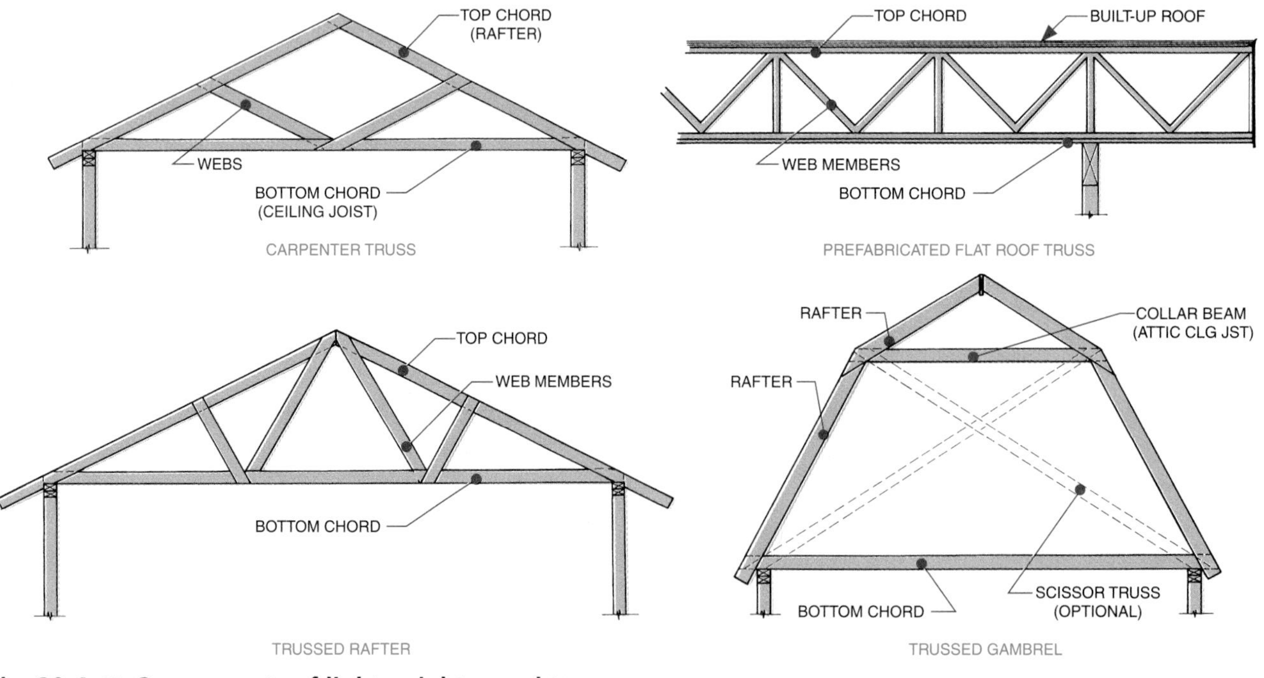

Fig. 30-4 ■ Components of lightweight wood trusses.

Fig. 30-5 ■ Truss members are most often held in place by gussets.

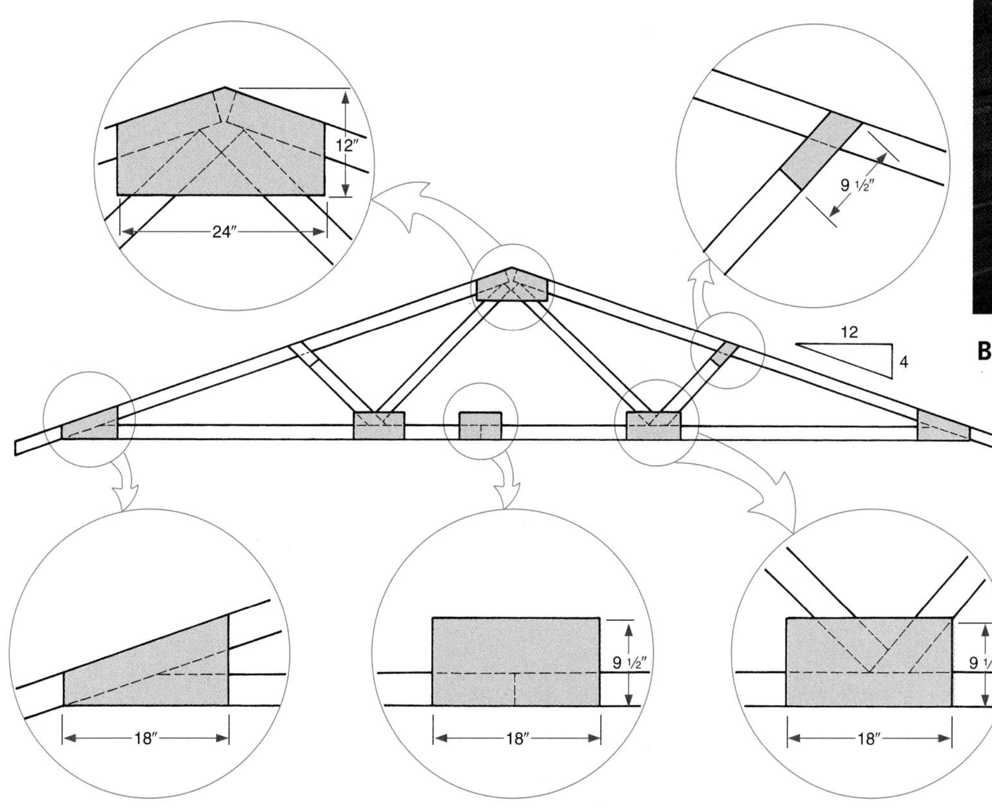

A. Use of plywood gusset plate connectors.

B. Use of sheet-metal gussets.

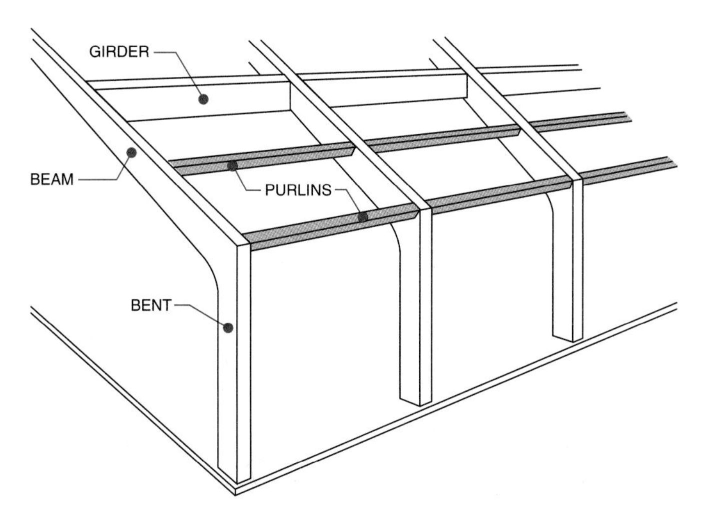

Fig. 30-6 ■ Horizontal members called purlins may be used to provide stability between rafters, trusses, or arches.

or columns, on masonry walls, or on exterior wood-framed walls. See Fig. 30-8.

Truss type and design depend on the length of span, room height requirements, spacing, roof pitch, live and dead loads, and cost factors. Several types of trusses are manufactured for structural steel construction. See Fig. 30-9. A variety of truss designs are used in light residential construction. See Fig. 30-10A. Specifications for a Fink truss are found in Fig. 30-10B. More complete truss specifications are found in Appendix A, "Mathematical Calculations."

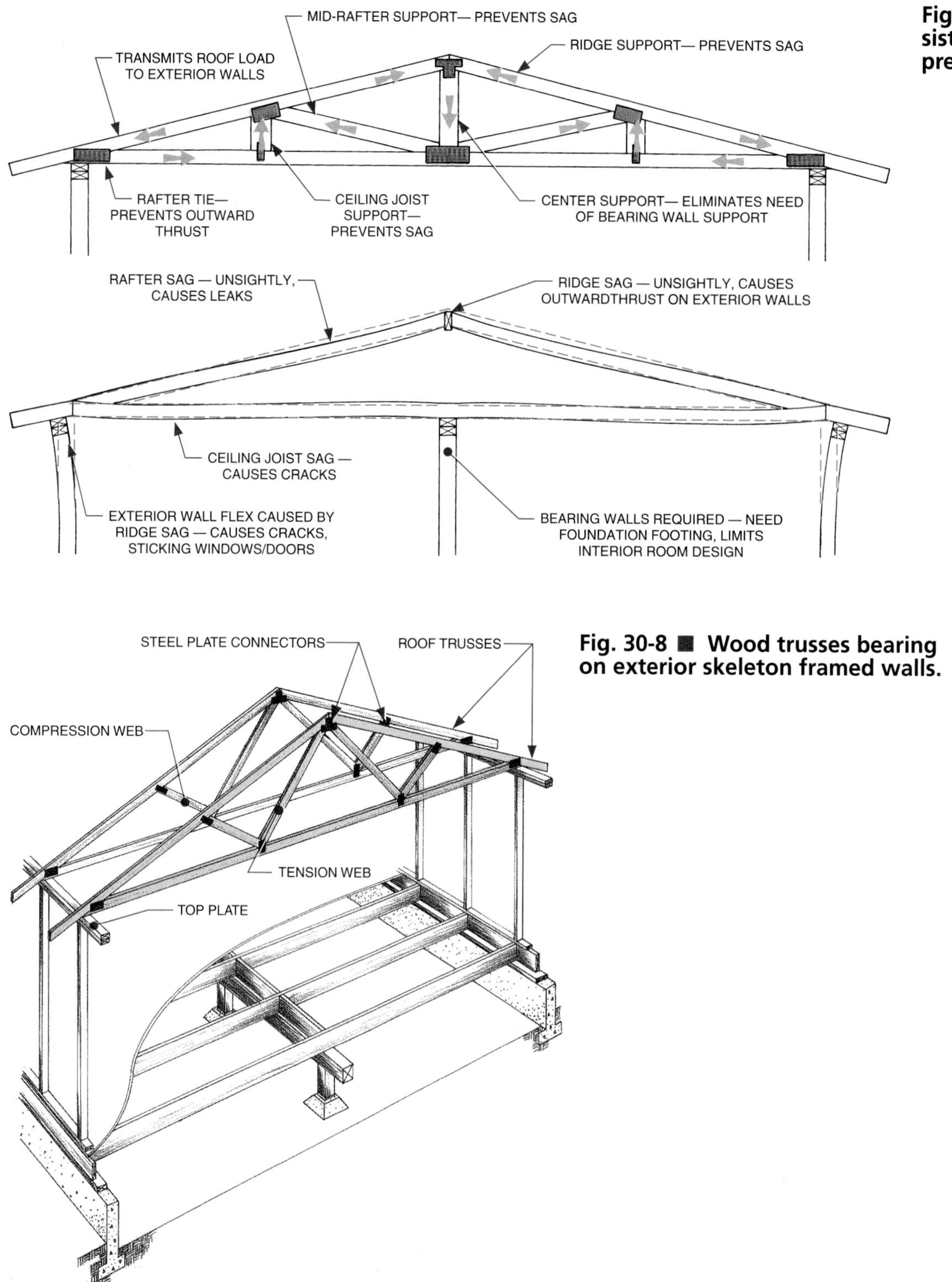

Fig. 30-7 ■ Trusses resist tension and compression forces.

Fig. 30-8 ■ Wood trusses bearing on exterior skeleton framed walls.

Fig. 30-9 ■ Standard types of steel trusses.

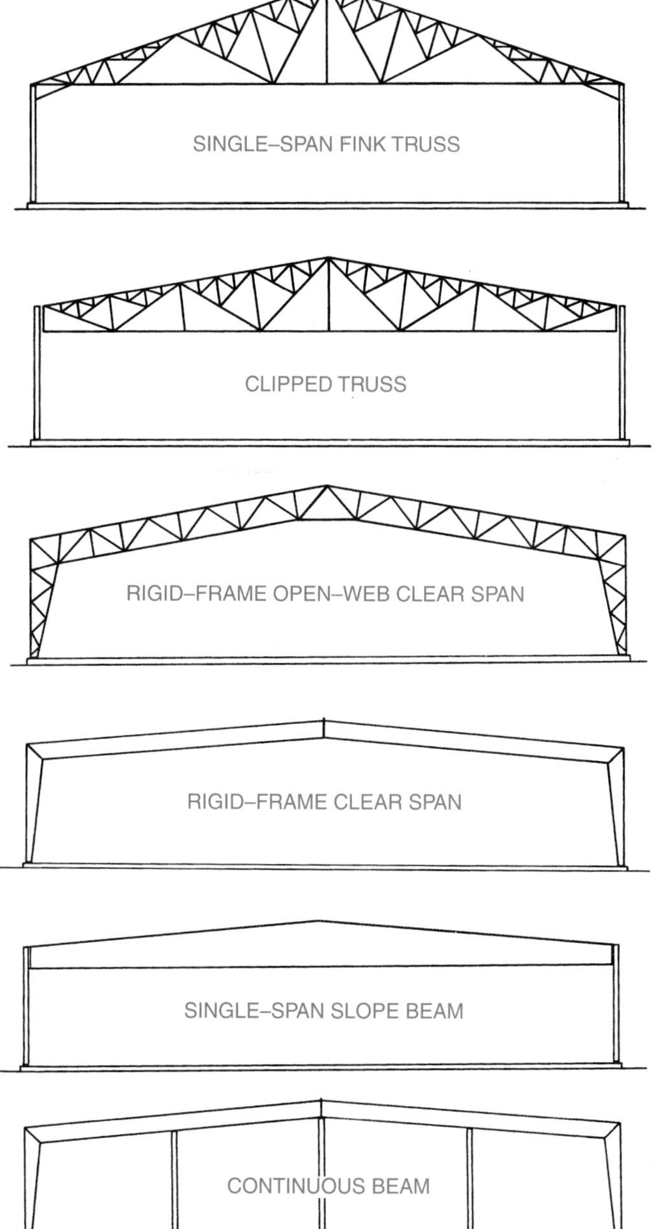

Fig. 30-10 ■ Typical trusses and specifications used in light construction.

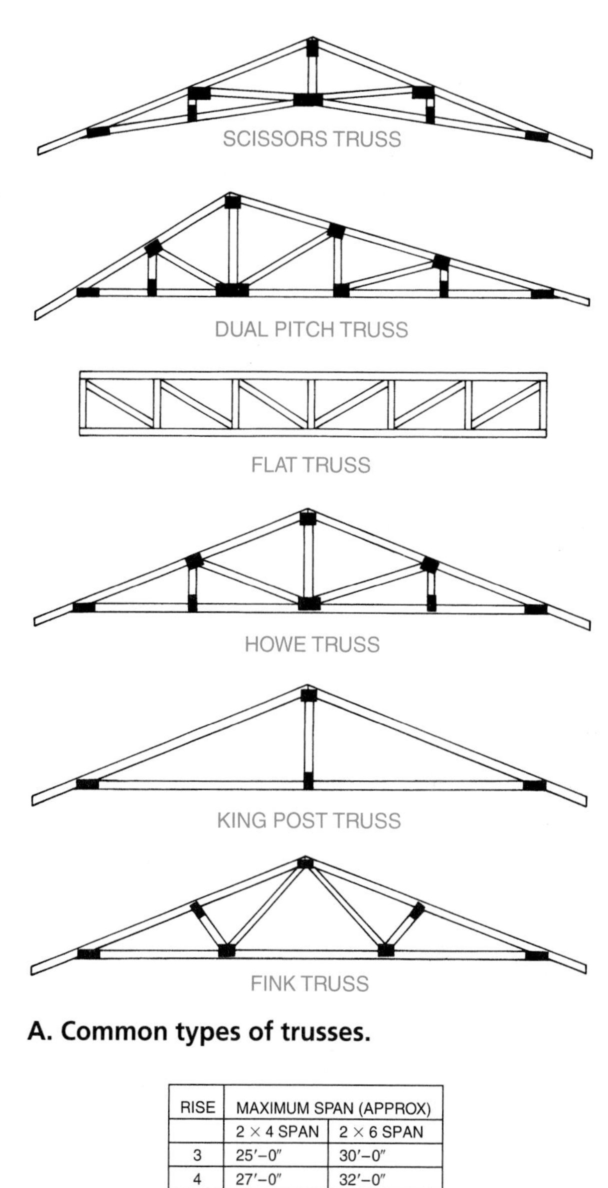

A. Common types of trusses.

RISE	MAXIMUM SPAN (APPROX)	
	2 × 4 SPAN	2 × 6 SPAN
3	25′–0″	30′–0″
4	27′–0″	32′–0″
5	30′–0″	35′–0″
6	32′–0″	38′–0″

B. Specifications for a Fink truss.

Roof Panels

There are many forms of lightweight, prefabricated roof panels. See Fig. 30-11. These units can be designed to resist loads, span great distances, or eliminate the need for trusses.

STRESSED SKIN ROOF PANELS These panels are constructed of plywood or stranded panels and seasoned lumber. The framing plywood skin acts as a unit to resist loads. Glued joints transmit the shear stresses, making it possible for the structure to act as one piece.

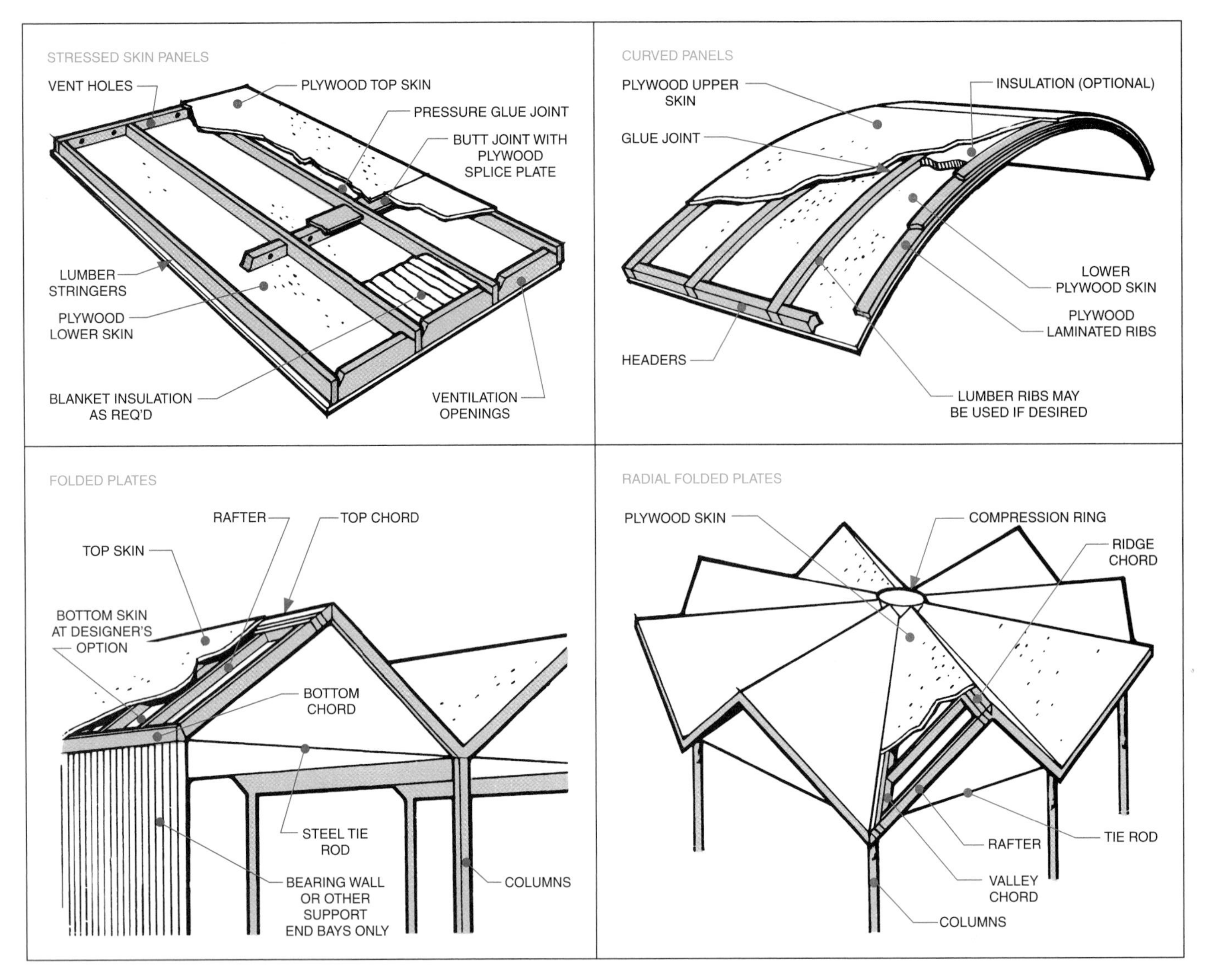

Fig. 30-11 ■ Commonly used prefabricated roof units.

CURVED PANELS The three types of curved panels are the sandwich (or honeycomb paper-core) panel, the hollow-stressed end panel, and the solid-core panel. The arching action of these panels permits spanning great distances with a relatively thin cross section.

FOLDED PLATE ROOFS These roofs are thin skins of plywood reinforced by rafters to form shell structures that can utilize the strength of plywood. The use of folded plate roofs eliminates trusses and other roof members. The tilted plates lean against one another, acting as giant V-shaped beams which are supported by walls or columns.

Cornices

The area of the roof that intersects with the outside walls and extends to the end of the roof overhang is the **cornice**. Detail drawings are necessary to show cornice areas. These detail drawings include part of the wall framing, roof framing, and methods of attaching the roof structure to the wall. See Fig. 30-12.

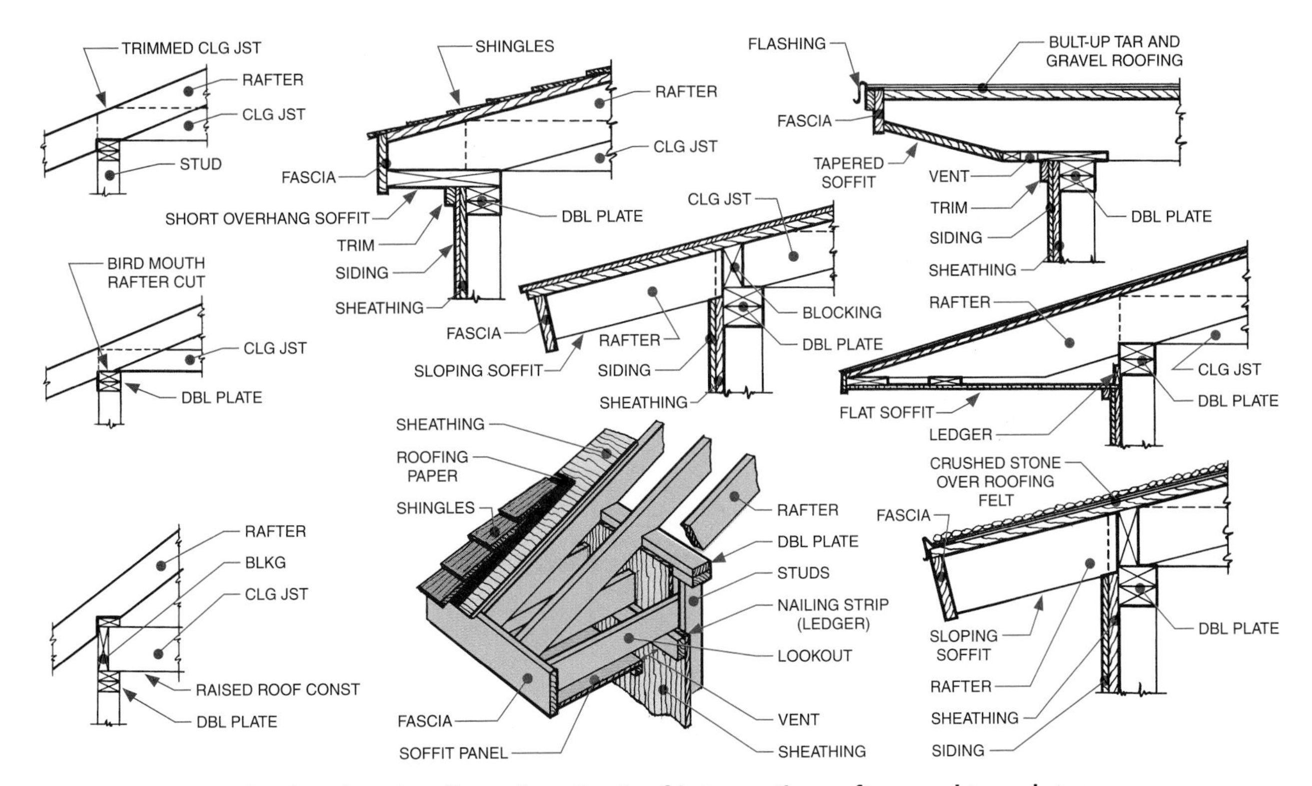

Fig. 30-12 ■ Cornice framing details and methods of intersecting rafters and top plates.

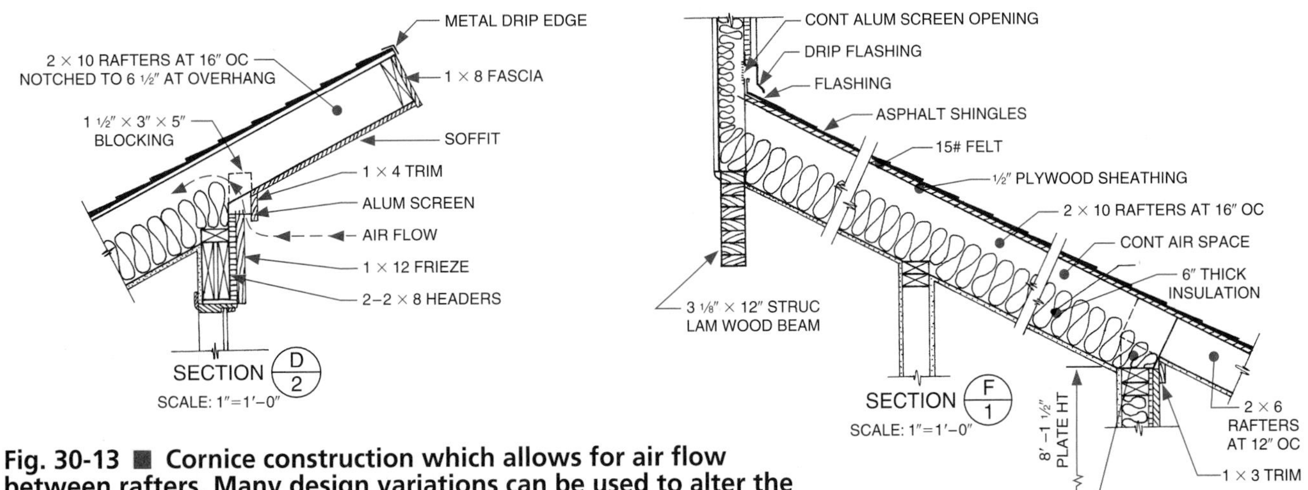

Fig. 30-13 ■ Cornice construction which allows for air flow between rafters. Many design variations can be used to alter the size and shape of the cornice.

Several types of rafter cuts are used at intersections to help hold rafters onto top wall plates. For example, the *bird mouth cut* shown in Fig. 30-12 provides a level surface for the intersection of the rafters and top plates. The area bearing on the plate should not be less than 3″ (75 mm).

The outer vertical edge of an overhang is the **fascia**, and the horizontal bottom of an overhang structure is the **soffit**. Figure 30-12 shows four types of soffit design. Soffits are made from plywood or sheet-metal panels. Plywood soffit panels should contain screened openings to allow air to pass between rafters or to circulate through crawl spaces if rafters are exposed. Sheet-metal soffits are available in solid or ventilated designs to allow air flow. See Fig. 30-13.

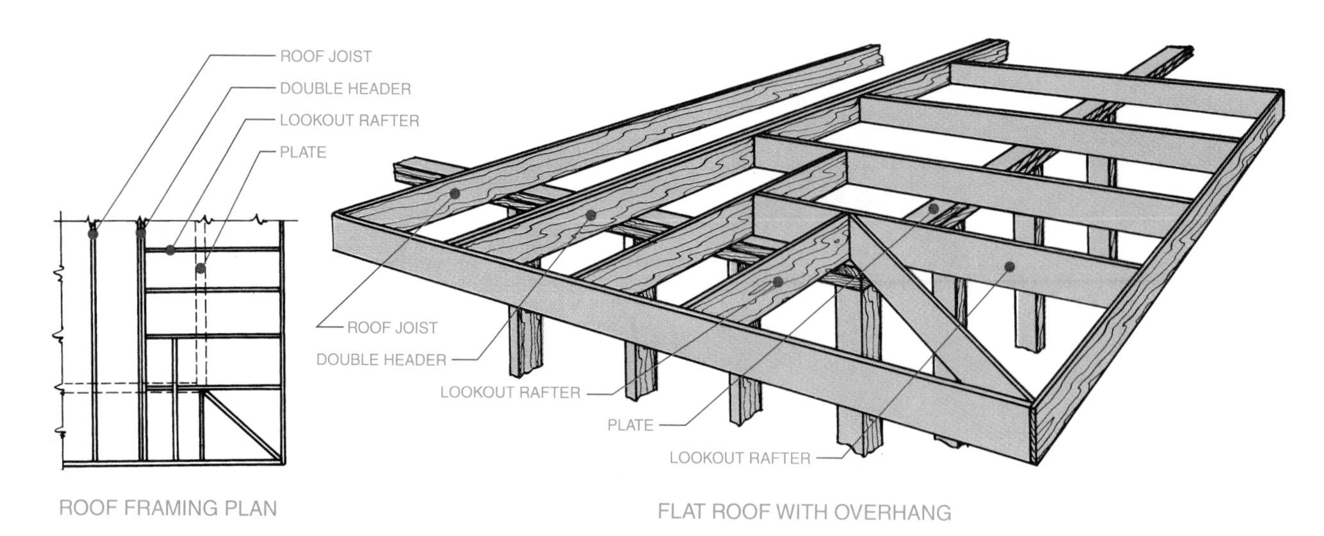

Fig. 30-14 ■ Lookout rafters are used to extend the overhang perpendicular to the common rafters.

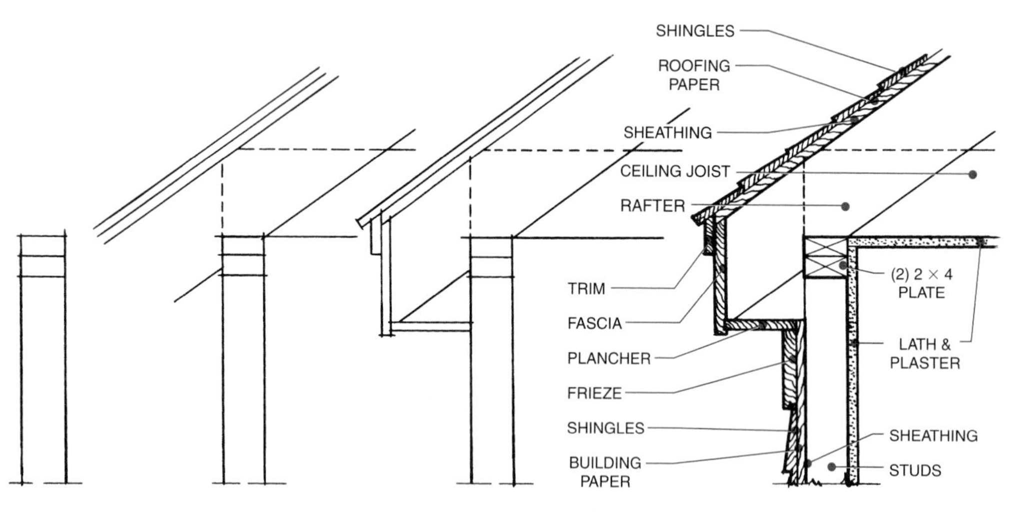

Fig. 30-15 ■ Steps in drawing a cornice detail section.

Cantilevered *lookout rafters* are used where common rafter extensions cannot create an overhang. See Fig. 30-14. Lookout rafters are placed perpendicular to the common rafter direction.

Figure 30-15 shows the steps in laying out and drawing a cornice detail.

Collar Beams and Knee Walls

Collar beams provide a tie between opposing rafters. They may be placed on every rafter or every second or third rafter, or more. Collar beams are used to reduce the rafter stress that occurs between the top plate and the top of the rafter. They lock rafters into position. They may also act as ceiling joists for finished attics. On low-pitched roofs, collar beams may be required to counteract the lateral (outward) thrust of joists. See Fig. 30-16.

Knee walls are vertical studs that project from ceiling joists or attic floors to roof rafters. See Fig. 30-17. Knee walls add rigidity to the rafters and may also provide half-wall framing for finished attics.

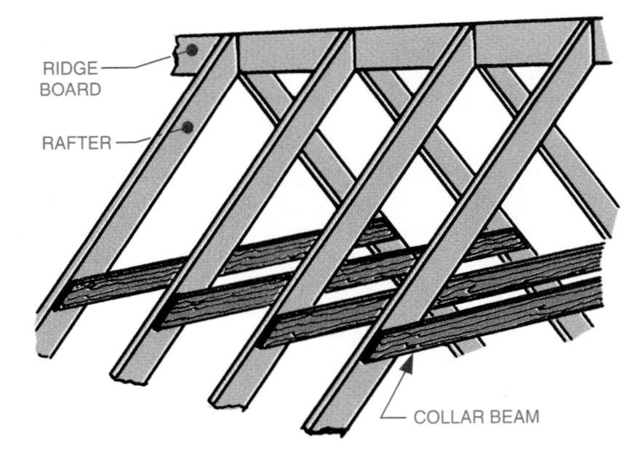

Fig. 30-16 ■ Collar beams reduce the amount of stress on rafters.

Dormers

Parts of the roof framing often extend above the roof line. Dormer rafters are one example. These are sometimes drawn with dotted lines. The details of intersecting dormer roof framing with dormer walls is shown in the framing plan. See Fig. 30-18.

Dormer rafters and walls do not lie in the same plane as the remainder of the roof rafters. A framing elevation drawing is therefore needed to show the exact position of the dormer members, their

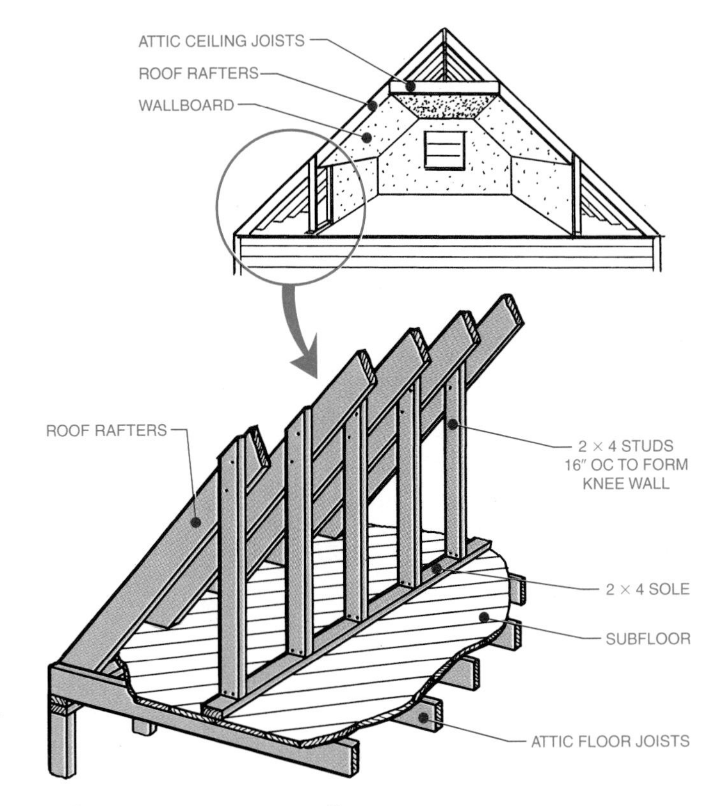

Fig. 30-17 ■ Knee walls.

Fig. 30-18 ■ Dormer framing plan.

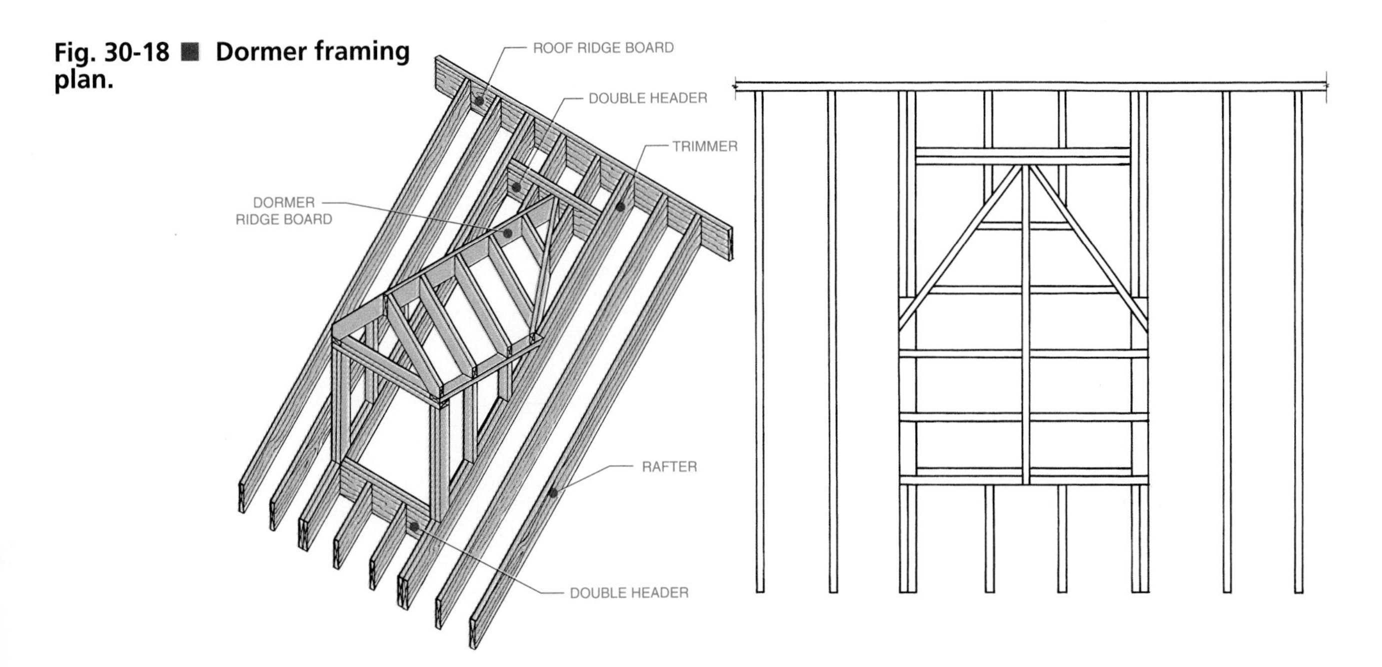

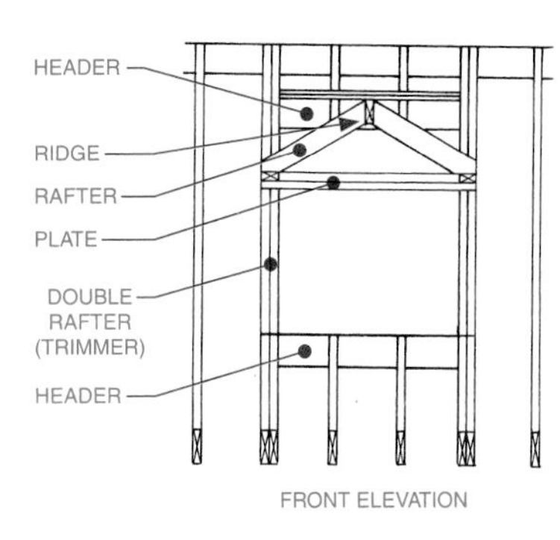

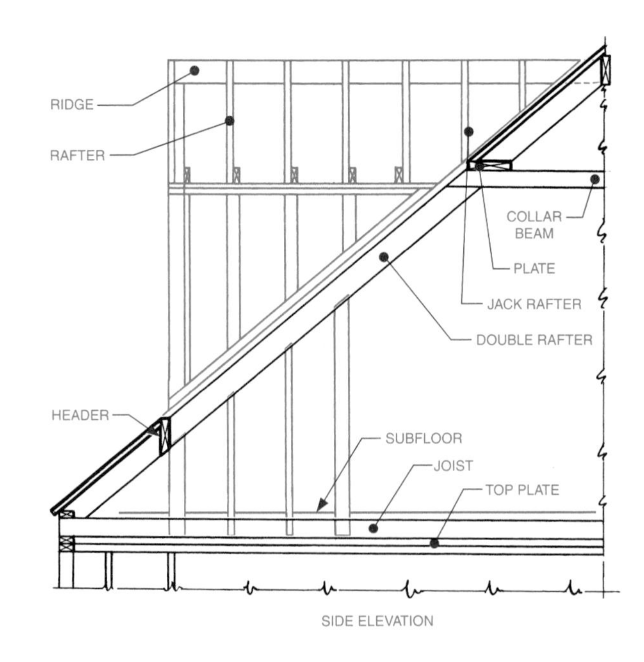

Fig. 30-19 ■ Front and side framing elevations of a gable dormer.

tie-in with the common roof rafters, and with other roof framing members. See Fig. 30-19.

Chimney Details

If a detailed roof framing plan is not prepared, chimney framing roof details are often on separate detail drawings. Chimney details include the position of ceiling joists, roof rafters, and type of construction. See Fig. 30-20.

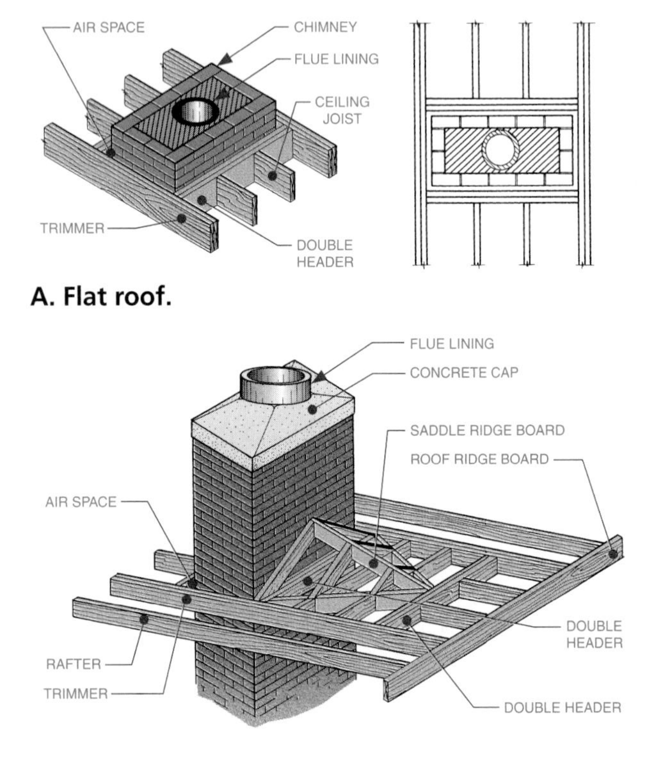

A. Flat roof.

Fig. 30-20 ■ Chimney intersection details.

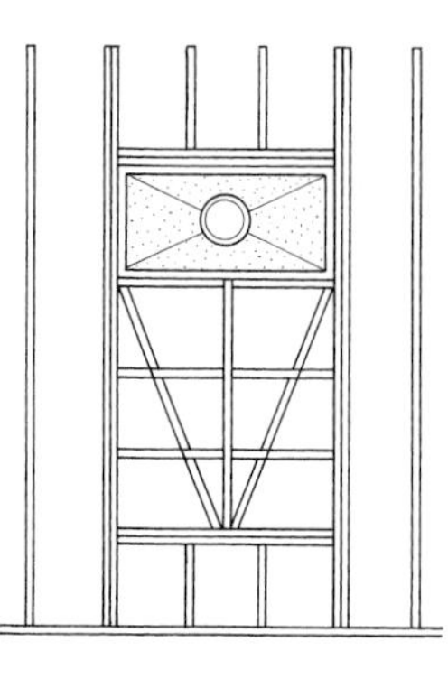

B. Roof saddle framing.

Roof Framing Drawings

Plans

To convey the structural design of a roof to a builder, plans must be accurate and complete. A *roof plan* shows only the outline and the major object lines of the roof. A roof plan is not a framing plan but a plan view of the roof. A roof plan can be used, however, as the basic outline to develop a roof framing plan. A *roof framing plan* exposes the exact position and spacing of each structural member. See Fig. 30-21.

Roof framing plans may be either simplified, single-line plans or detailed, double-line plans. A single-line plan is acceptable only to show the general relationship and spacing of the structural members. Again, see Fig. 30-21. When more details concerning the exact construction of intersections and joints are needed, a double-line plan showing the thickness of each member should be prepared. See Fig. 30-22. This type of plan is necessary to indicate the relative

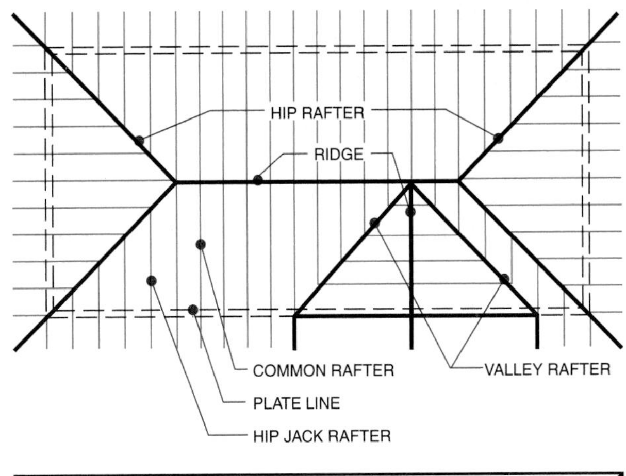

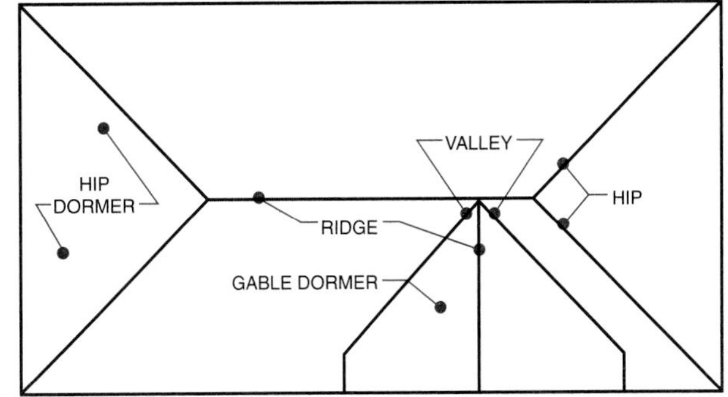

Fig. 30-21 ■ **Roof framing plan compared to a roof plan.**

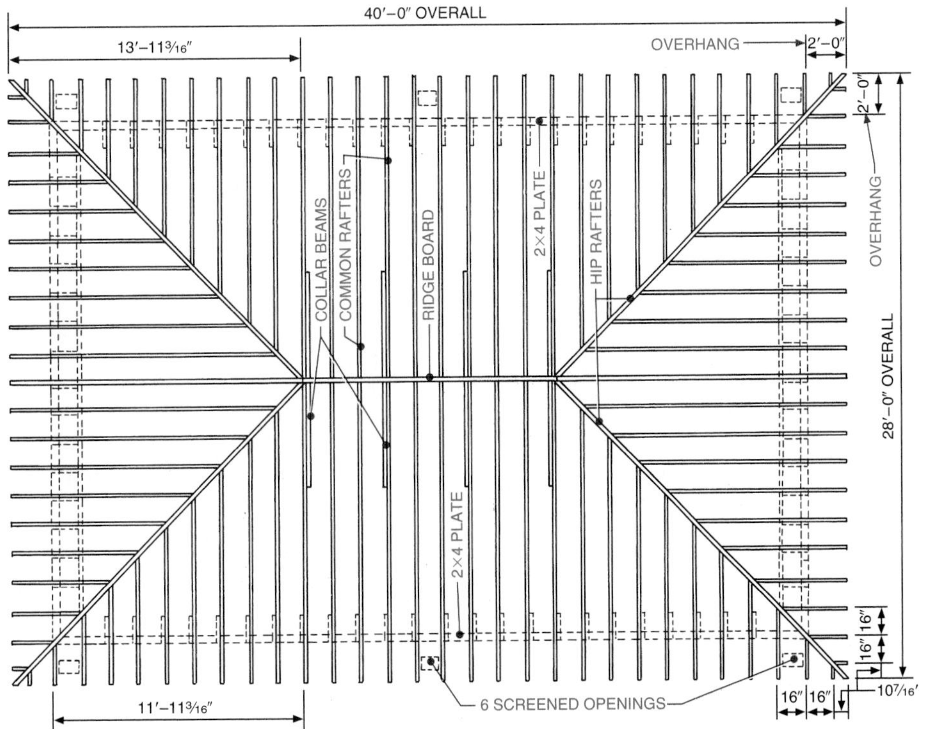

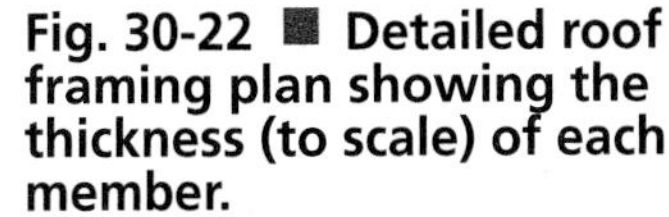

Fig. 30-22 ■ **Detailed roof framing plan showing the thickness (to scale) of each member.**

placement of one member compared with another; that is, to indicate whether one member passes over or under another. In a double-line plan, the width of ridge boards, rafters, headers, and plates should be drawn to the exact scale. If a complete roof framing plan of this type cannot fully describe construction framing details, then additional removed pictorial or elevation drawings should be prepared. See Fig. 30-23. These separate drawings can also be enlarged, for example, to detail dimensions.

In contrast to a true orthographic projection, only the outline of the top of the rafters is shown on roof framing plans. Areas underneath the rafters are shown by dotted lines. Roof framing plans show only horizontal relationships of members, such as thickness, length, and horizontal spacing. See Fig. 30-24. In a top view (plan view), you cannot show vertical dimensions such as structural heights and pitches.

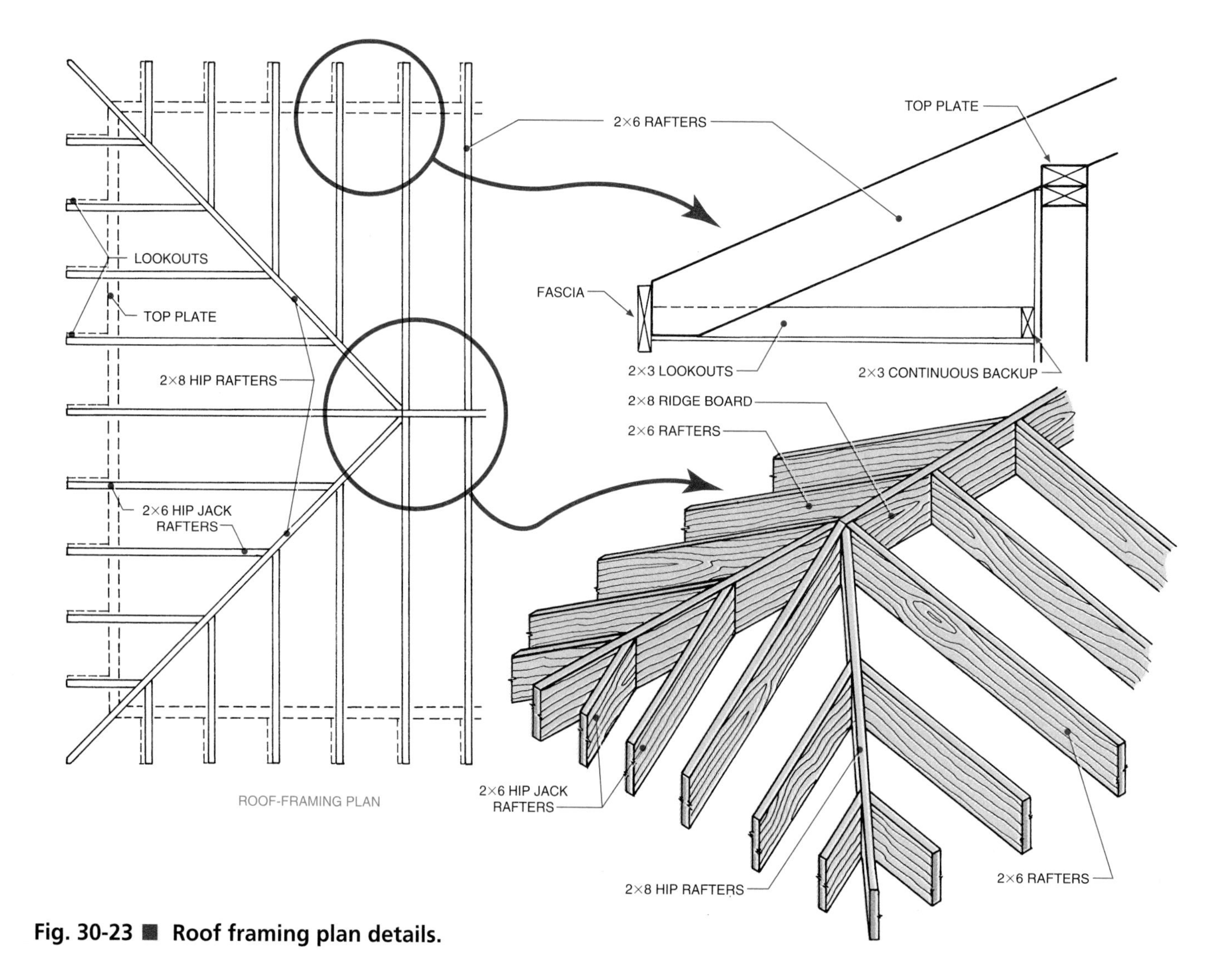

Fig. 30-23 ■ Roof framing plan details.

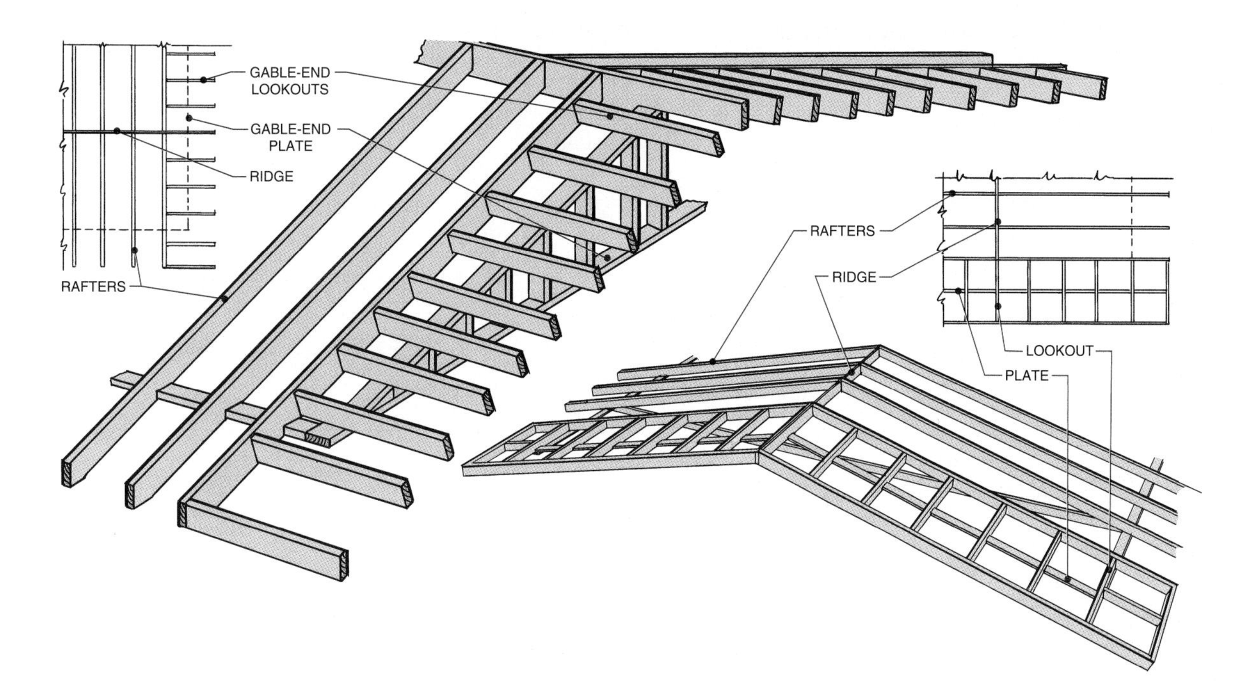

Fig. 30-24 ■ Roof framing plans show only the top of each member, regardless of angle.

Elevations

The angle or vertical position of any roof framing member should be shown on a roof framing elevation. If a comparison of different heights and pitches is desired, a *composite framing-elevation* drawing should be prepared. This elevation is one that can be projected from the roof framing plan or from corresponding lines on the elevation drawings. See Fig. 30-25.

Dimensions

Dimensions on roof framing plans usually include the size and spacing of framing members and the major distances (spans) between framing components. See Fig. 30-26A. Regular spacing of structural members, such as roof rafters, floor joists, and wall studs, is not dimensioned if these fall on modular increments. Notes may be used on framing drawings to show the size and spacing of members, as shown in Fig. 30-26B. On detail drawings, overall dimensions are not given. Instead, the key dis-

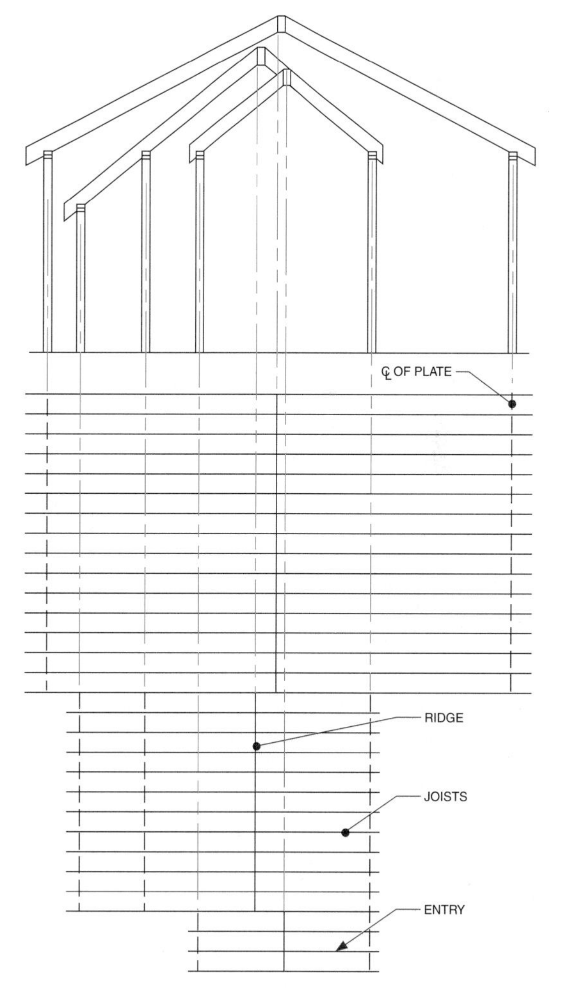

Fig. 30-25 ■ Method of projecting a roof framing elevation.

Fig. 30-26 ■ Methods of indicating sizes and spacing of roof framing members.

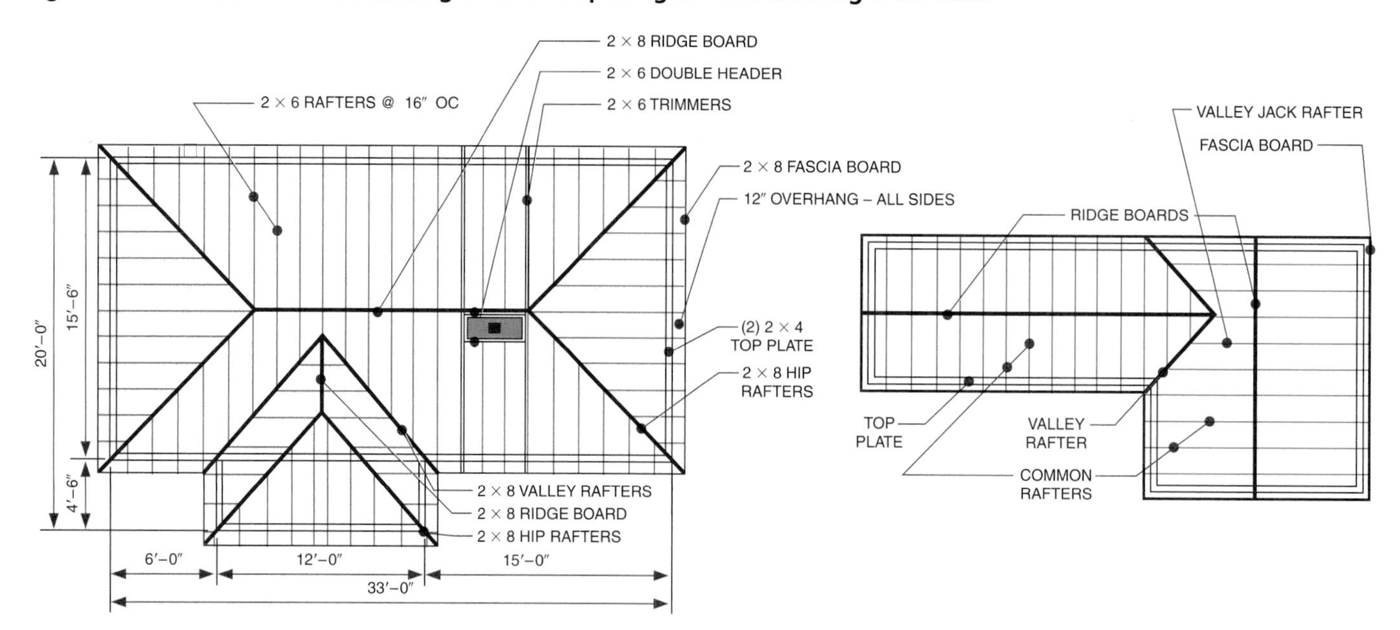

A. Framing plan dimensioned with overall dimensions, subdimensions, and sizes of all framing materials.

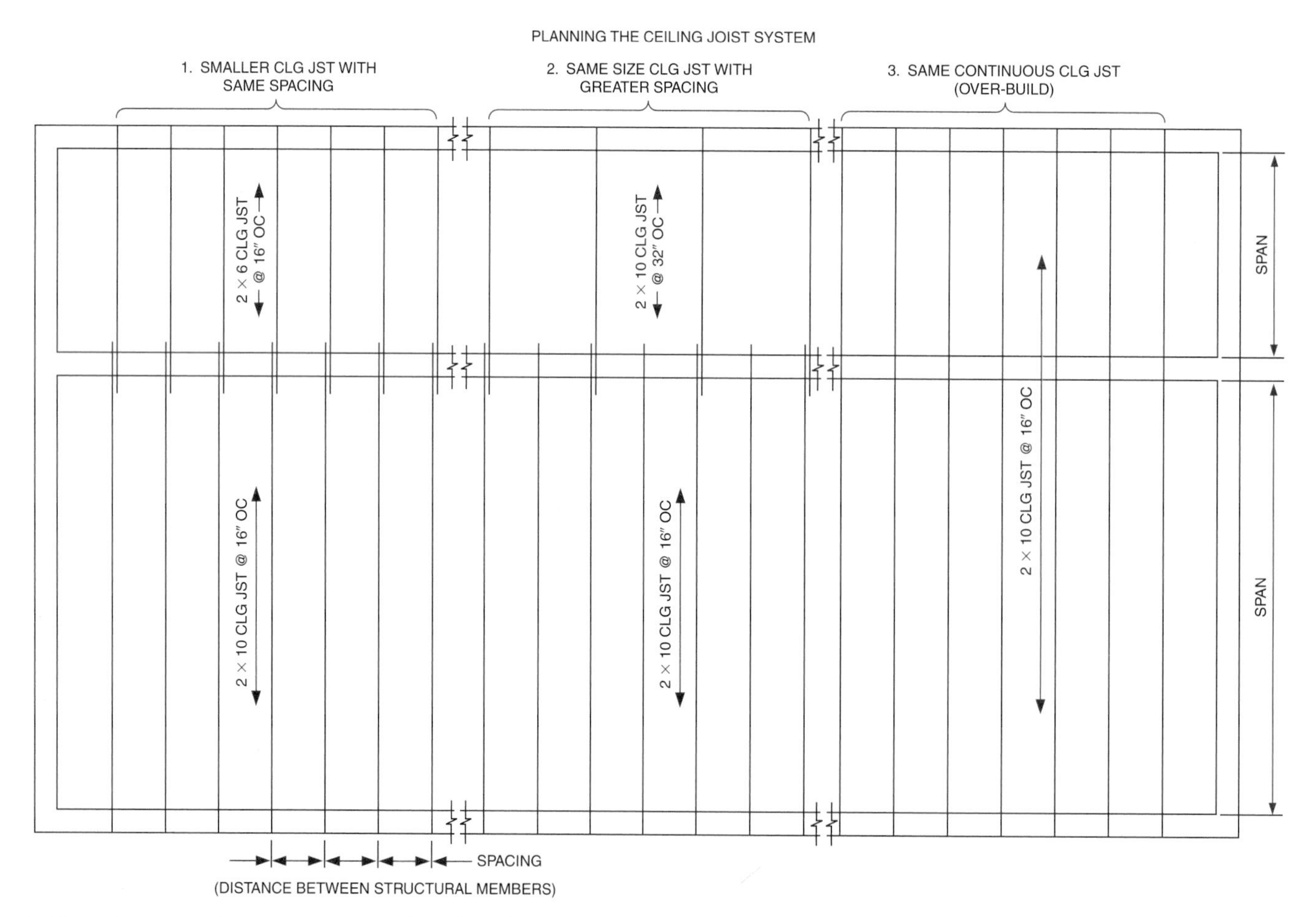

B. Notations used on the roof and ceiling framing plans.

tances between structural levels and horizontal distances are shown.

If material sizes are not given on the framing drawing, they should be included in the specifications. Figure 30-27 shows typical light construction roof rafter sizes, spans, and spacing for a pitch of 4/12 or more. Similar information for ceiling joists is in Fig. 30-28. All roof member sizes and spacing are based on spans and loads. Refer to Appendix A, "Mathematical Calculations."

Fig. 30-27 Rafter spans for roofs with a pitch greater than 4/12.

Size rafter	Spacing rafter	Maximum span			
		Group 1	Group 2	Group 3	Group 4
2 × 4	12″	10′–0″	9′–0″	7′–0″	4′–0″
	16″	9′–0″	7′–6″	6′–0″	3′–6″
	24″	7′–6″	6′–6″	5′–0″	3′–0″
	32″	6′–6″	5′–6″	4′–6″	2′–6″
2 × 6	12″	17′–6″	15′–0″	12′–6″	9′–0″
	16″	15′–6″	13′–0″	11′–0″	8′–0″
	24″	12′–6″	11′–0″	9′–0″	6′–6″
	32″	11′–0″	9′–6″	8′–0″	5′–6″
2 × 8	12″	23′–0″	20′–0″	17′–0″	13′–0″
	16″	20′–0″	18′–0″	15′–0″	11′–6″
	24″	17′–0″	15′–0″	12′–6″	9′–6″
	32″	14′–6″	13′–0″	11′–0″	8′–6″
2 × 10	12″	28′–6″	26′–6″	22′–0″	17′–6″
	16″	25′–6″	23′–6″	19′–6″	15′–6″
	24″	21′–0″	19′–6″	16′–0″	12′–6″
	32″	18′–6″	17′–0″	14′–0″	11′–0″

Fig. 30-28 Wood ceiling joist spans.

Size of ceiling joists	Spacing of ceiling joists	Maximum span			
		Wood Group 1	Wood Group 2	Wood Group 3	Wood Group 4
2 × 4	12″	11′–6″	11′–0″	9′–6″	5′–6″
	16″	10′–6″	10′–0″	8′–6″	5′–0″
2 × 6	12″	18′–0″	16′–6″	15′–6″	12′–6″
	16″	16′–0″	15′–0″	14′–6″	11′–0″
2 × 8	12″	24′–0″	22′–6″	21′–0″	19′–0″
	16″	21′–6″	20′–6″	19′–0″	16′–6″

Roof Pitch

Pitch is the angle between the roof's surface (top plate to ridge board) and the horizontal plane. The **rise** is the vertical distance from the top plate to the roof's ridge. The **run** is the horizontal distance from the top plate to the ridge. It is expressed in units of 12. The **span** is the full horizontal distance between outside supports. It is double the run.

Figure 30-29 shows how to determine units of rise (vertical distance) per units run (horizontal distance). On drawings, this figure is shown in a slope diagram (triangle) near the line of the roof along with the run unit number. For example, in Fig. 30-29, the roof shown on the left rises 6 units vertically for every 12 units of horizontal distance (run). When units are inches as shown in the equation, the roof rises 6″ per foot (12″) of run.

Pitch is the ratio of the *actual* rise to the *actual* span. It is also the ratio of the *units* of rise to *units* of span (double the units of run). Refer again to Fig. 30-29. In the drawing on the right, the pitch is 10/30 (ratio of actual rise to actual span) which reduces to 1/3. It is also 8/24 (unit ratio of rise to span) which also reduces to 1/3.

Roof pitches vary greatly. For example, a roof with a slope of 12/12 is steep. The pitch would be 12/24 or 1/2. A roof with a slope of 8/12 is moderately sloped. The pitch would be 8/24 or 1/3. A roof with a slope of 2/12 is nearly flat. The pitch would be 2/24 or 1/12. See Fig. 30-30.

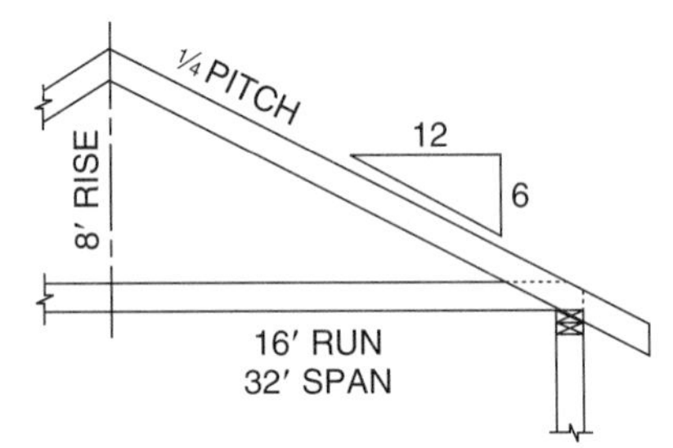

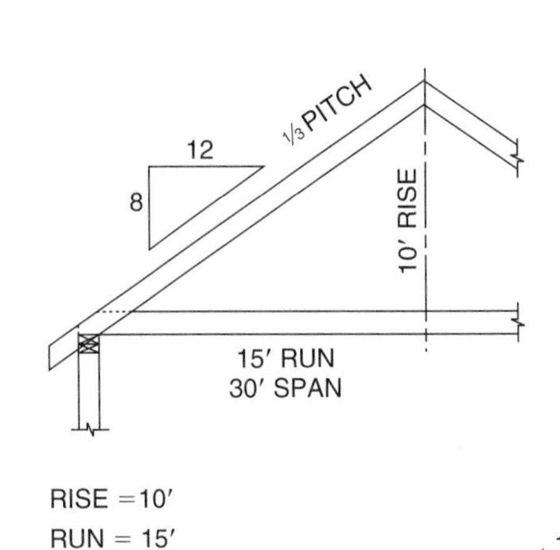

RISE = 8′
RUN = 16′
PITCH = 8 ÷ 32 = ¼
RISE/FT RUN = $\frac{8 \times 12}{16}$ = 6″
6/24 = ¼ PITCH

RISE = 10′
RUN = 15′
PITCH = 10 ÷ 30 = ⅓
RISE/FT RUN = $\frac{10 \times 12}{15}$ = 8″
8/24 = ⅓ PITCH

TYPICAL PITCHES	½	⅓	¼	⅙
RISE/FT RUN	12	8	6	4

Fig. 30-29 ■ Determining roof pitch.

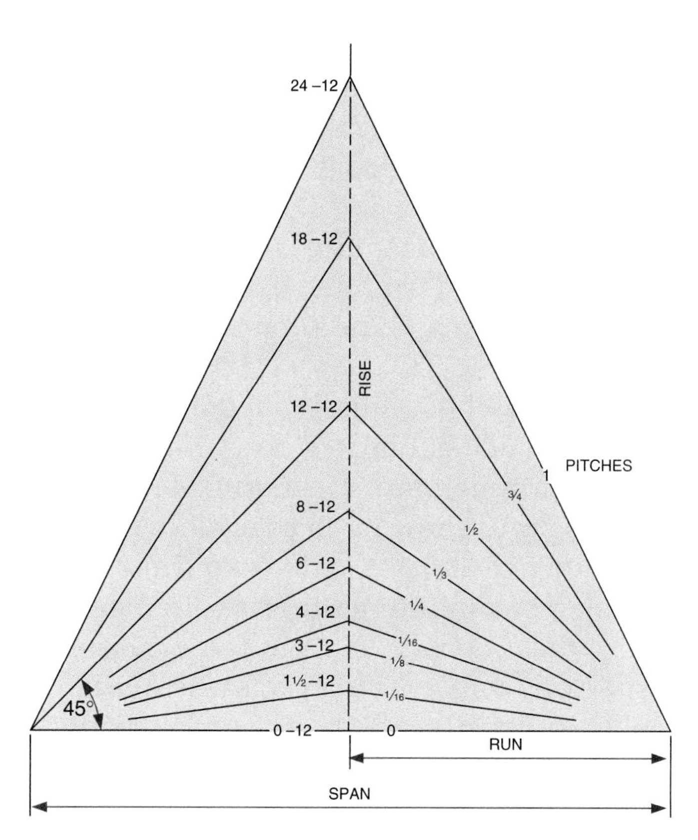

Fig. 30-30 ■ Commonly used roof pitches.

Roof Framing Methods for Wood

Wood framing can be divided into conventional and heavy timber methods, regardless of roof shape. Both methods are used extensively in all types of wood-framed roofs.

The *conventional* method of roof framing consists of roof rafters or trusses spaced at small intervals, such as 16″ on center. See Fig. 30-31. These roof rafters are perpendicular to the ridge board and align with the partition studs placed on the same centers.

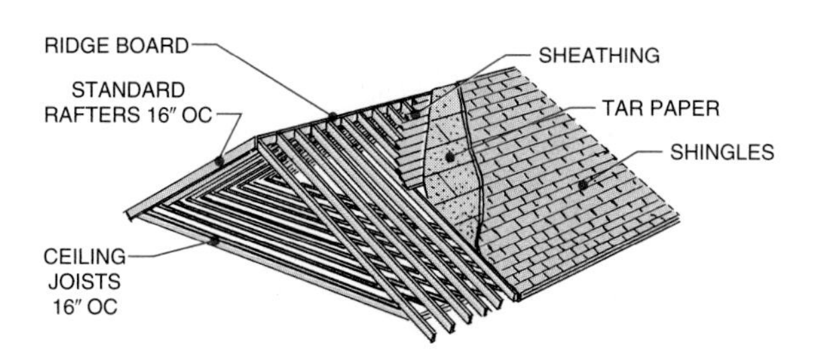

Fig. 30-31 ■ Conventionally framed wood roof.

An adaptation of this conventional method of constructing roofs is to substitute roof trusses for conventional rafters and ceiling joists. Trusses create a much more rigid roof, but they make it impossible to use space between ceiling joists and rafters for an attic or crawl-space storage. See Fig. 30-32.

Heavy-timber construction, another method of roof framing, consists of posts that support beams. Longitudinal beam sizes vary with the span and spacing of beams. The beams are installed parallel to the ridge board, or ridge beam. This is called longitudinal roof beam construction. See Fig. 30-33. Beams may also be placed perpendicular (transverse) to the

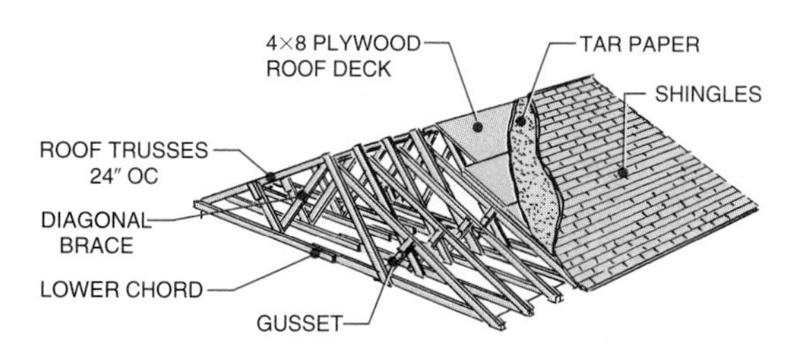

Fig. 30-32 ■ Wood trussed roof.

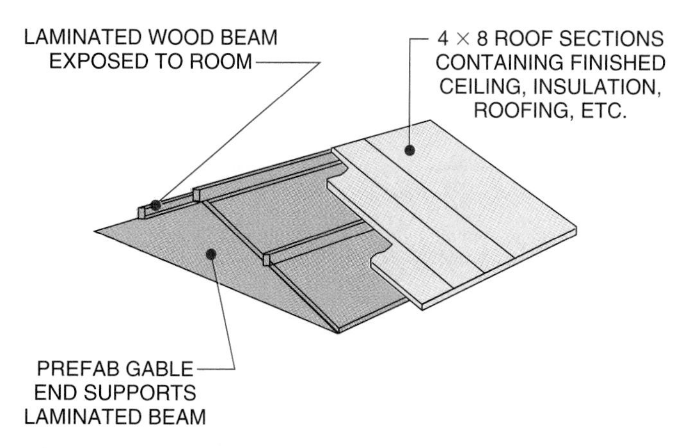

Fig. 30-33 ■ Longitudinal post-and-beam roof construction.

ridge beam. Planks are then placed across the beams. The planks can serve as a ceiling, as well as a base for roofing that will shed water. When planks are selected for appearance, the only ceiling treatment needed is a finish on the planks and beams.

Roof Framing Types

The most common roof framing types are *gable, hip, shed,* and *flat*. Other types, such as mansard and A-frame, are variations of the four common types and are framed in a similar manner.

Gable Roof Framing

Gable roof framing consists of rafters that form two inclined planes extending from the outer walls to a ridge. The drawing shown in Fig. 30-34 shows gable roof framing on the right end and hip roof framing on the left end. When trusses or truss joists are substituted for common rafters, a ridge board is not used. See Fig. 30-35.

A-frame roof framing is similar to conventional gable framing, except the roof rafters rest on a foundation rather than on top plates.

Gable pitches range from 2/12 to 12/12. The pitch of a gable roof does not show on a roof framing plan. A framing elevation is necessary to show pitch and gable framing. See Fig. 30-36. A *gable end* (rake) is the side of a building that rises to meet the ridge. See Fig. 30-37. In some cases, especially on low-pitch roofs, the entire gable-end wall from the floor to the ridge can be panelized with varying lengths of studs. However, it is more common to prepare a rectangular wall panel and erect separate studs.

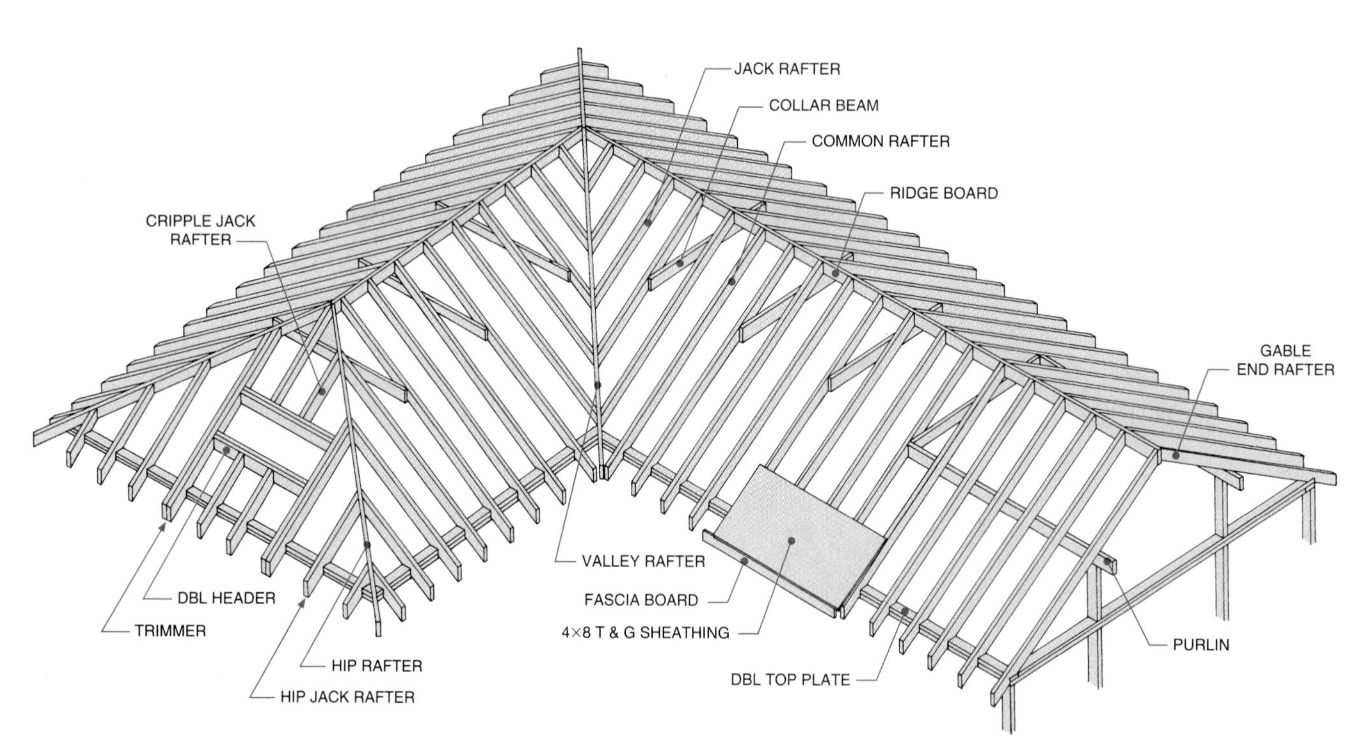

Fig. 30-34 ■ Structural members of a gable roof (right end) and a hip roof (left end).

Fig. 30-35 ■ Use of truss-joist rafters.

2×6 BLOCKING
2×6 LOOKOUTS
2×6 RAFTERS @ 16″ OC
EXTERIOR BEARING WALLS
TYP 12″ OH
2×8 RIDGE BOARD
TYP 12″ GABLE END OH

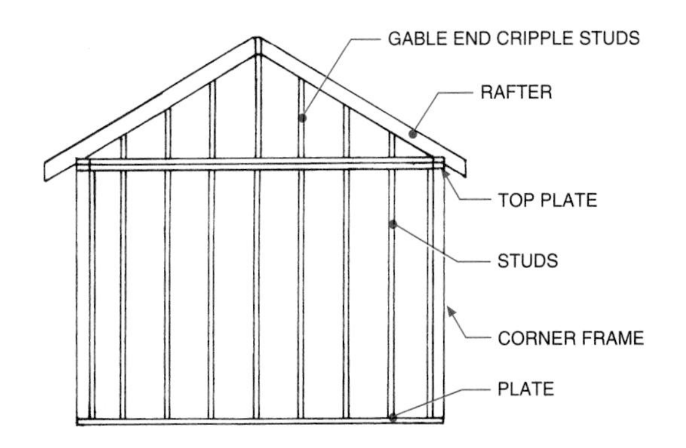

Fig. 30-36 ■ Gable-end framing drawing.

Fig. 30-37 ■ Gable-end roof construction.

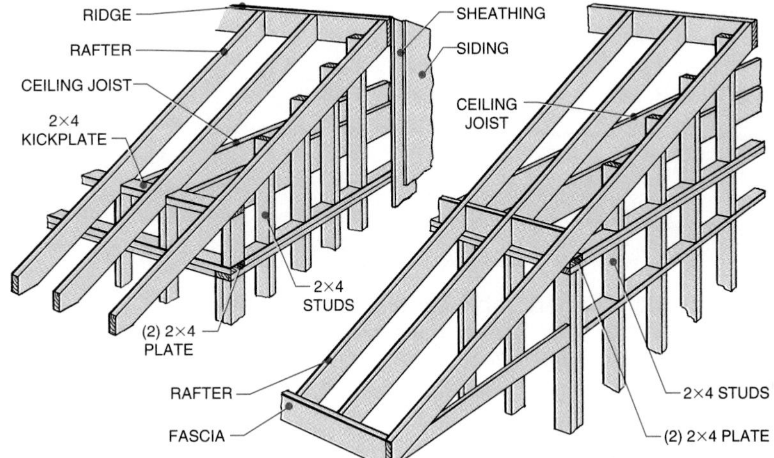

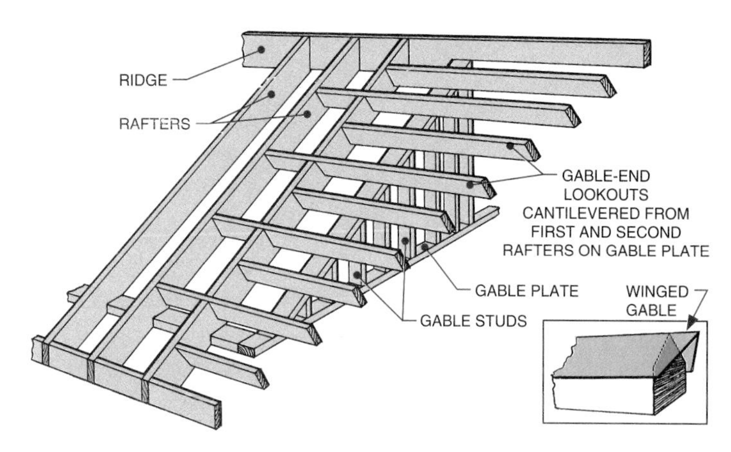

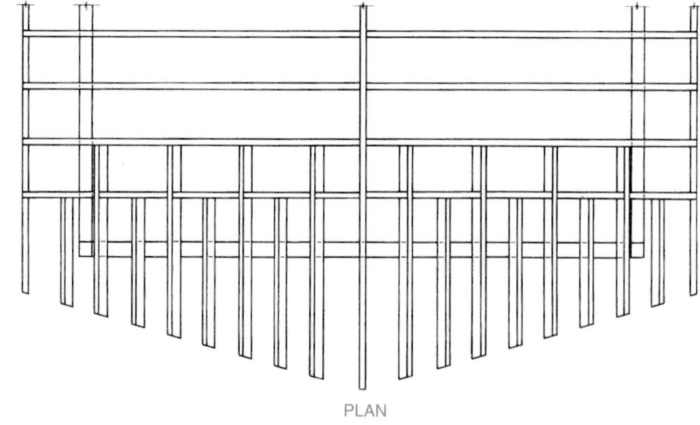

Fig. 30-38 ■ Winged gable lookout construction.

To design an overhang on a gable end, *lookouts* are used. Lookouts are short rafters placed perpendicular to the first or second common rafters. *Winged gables,* as shown in Fig. 30-38, use lookouts of varying lengths to form a triangular overhang. Large winged gable ends may require additional columns and beam support. See Fig. 30-39.

Gambrel Roof Framing

Gambrel (barn) roofs are a variation of gable roofs but use double-pitched rafters. See Fig. 30-40. The pitch of the lower part is always steeper than the pitch of the top part. Purlins may be used to stabilize pitch intersections. Gussets or prefabricated truss joists may also be used for gambrel roofs.

Fig. 30-39 ■ Winged gable with column and beam supports. *Two Creek Ranch, Fayetteville, TX. Lindi Shrovik, Coldwell Banker*

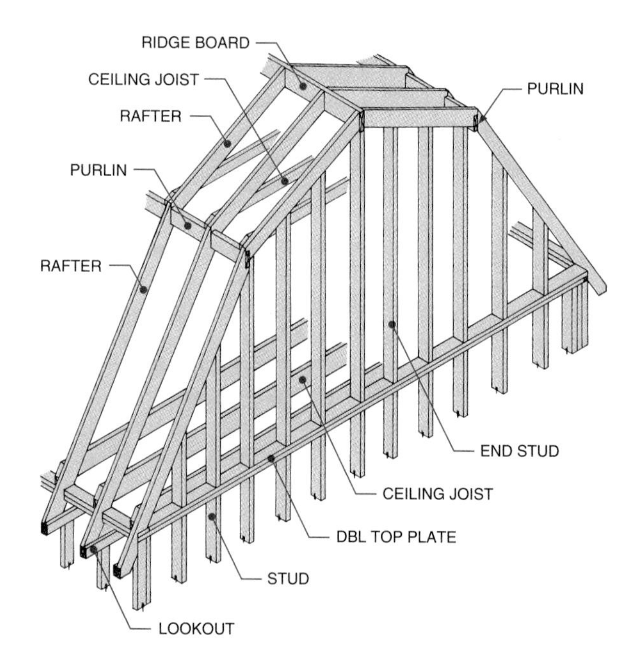

Fig. 30-40 ■ Gambrel roof framing.

Hip Roof Framing

Hip roof framing is similar to gable roof framing except that the roof slopes in four directions instead of intersecting gable-end walls. See Fig. 30-41. Also refer back to Fig. 30-34. Where two adjacent slopes meet, a hip is formed on the external angle. A *hip rafter* extends from the ridge board, over the top plate, and to the edge of the overhang. A hip rafter supports the ends of the shorter hip-jack rafters.

The internal angle formed by the intersection of two slopes of the roof is known as the *valley*. A *valley rafter* is used to form this angle. See Fig. 30-42.

Hip rafters and valley rafters are normally 2″ (50 mm) deeper or 1″ (25 mm) wider than the common rafters, for spans up to 12′ (3.7 m). For spans over 12′, the hip and valley rafters should be double the width of common rafters.

Jack rafters are rafters that extend from the wall plate to the hip or valley rafter. They are always shorter than common rafters, which extend from the top plate to the ridge.

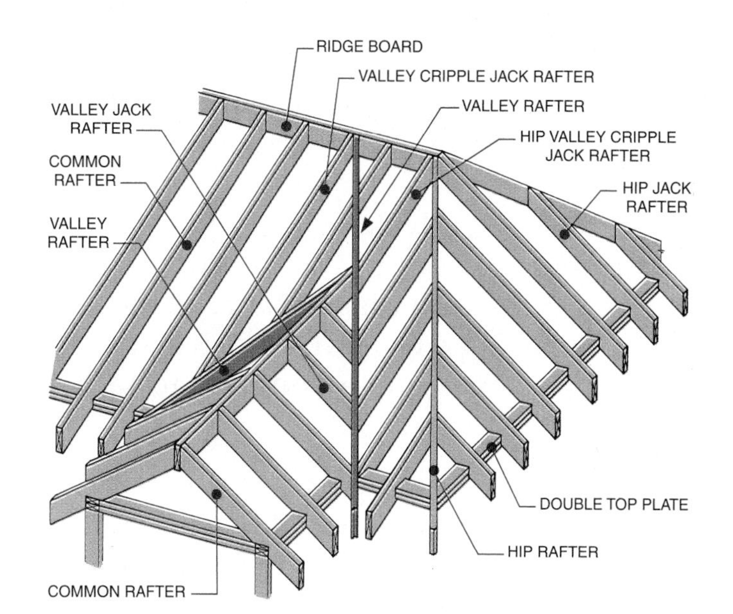

Fig. 30-42 ■ Valley rafters frame the internal angles of a hip roof. Hip rafters are used on the external angle.

Dutch Hip Roof Framing

A Dutch hip roof is framed the same as a gable roof in the center and as a partial hip roof on the ends. See Fig. 30-43. One of the many variations of this basic design is a mansard roof, a double-pitched hip roof. Refer back to the many roof styles shown in Fig. 15-4.

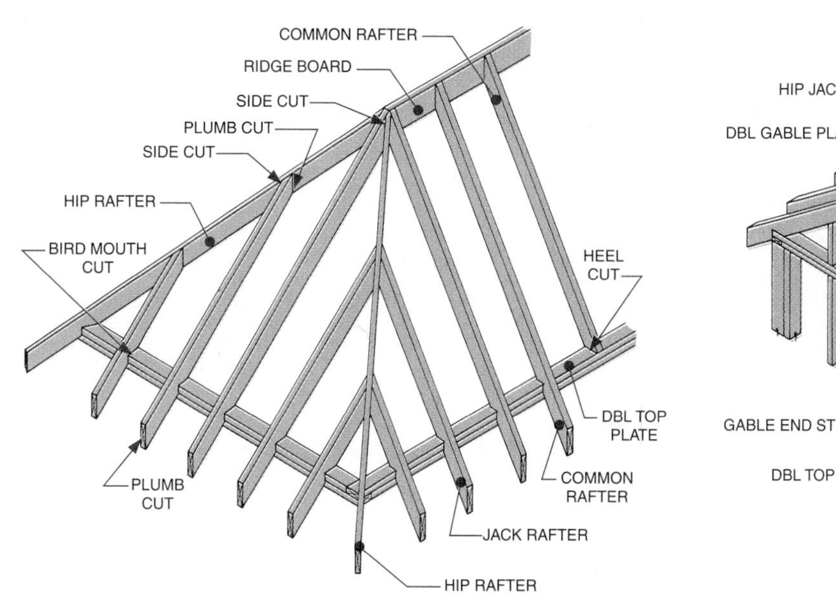

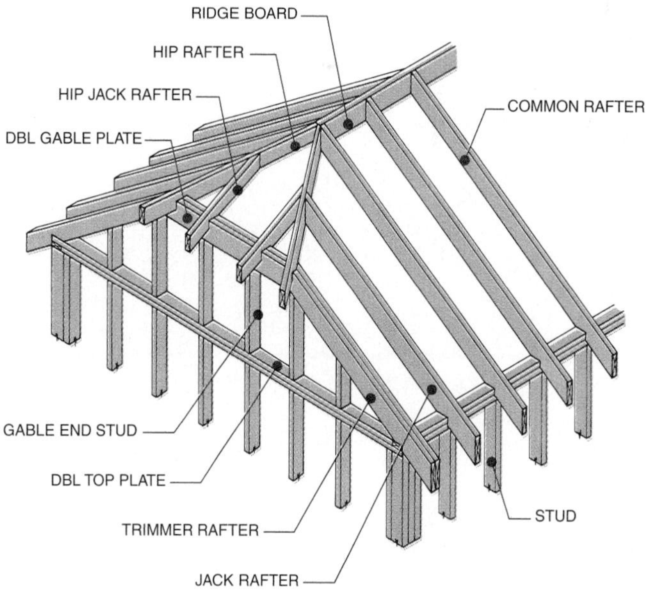

Fig. 30-41 ■ Hip roof framing members.

RIDGE BOARD
HIP RAFTER
HIP JACK RAFTER
DBL GABLE PLATE
COMMON RAFTER
GABLE END STUD
DBL TOP PLATE
TRIMMER RAFTER
JACK RAFTER
STUD

Fig. 30-43 ■ Dutch hip roof framing.

Shed Roof Framing

A shed roof is a roof that slants in only one direction. A gable roof is actually two shed roofs, sloping in opposite directions. Shed roof rafter design is the same as rafter design for gable roofs except that the run of the rafter is the same as the span. Some shed roofs are nearly flat. Shed roof slopes range from 2/12 to 12/12.

Flat Roof Framing

In conventional flat roof framing, rafters are similar to floor joists. Rafters span from wall to wall or from exterior wall to bearing partitions, columns, or beams. In heavy-timber construction, beams may be used to span these distances. Roof decking is laid directly on the beams. See Fig. 30-44. Flat roofs must be designed for maximum snow loads since the snow will not slide off. Instead, it must melt and drain away.

Flat roofs are usually not absolutely flat. Most flat roofs have a slight slope $\frac{1}{8}$" per foot to $\frac{1}{2}$" per foot) to allow for drainage. Some roofs are flat to allow a specific level of water to remain for insulation. Flat roof drainage must be provided through internal or external downspouts. A *cant strip* should be located on flat roof perimeters to stop water from flowing over the sides rather than through downspouts. Cant strips also help waterproof joints at the wall. See Fig. 30-45.

Fig. 30-44 ■ Heavy timber flat roof construction.

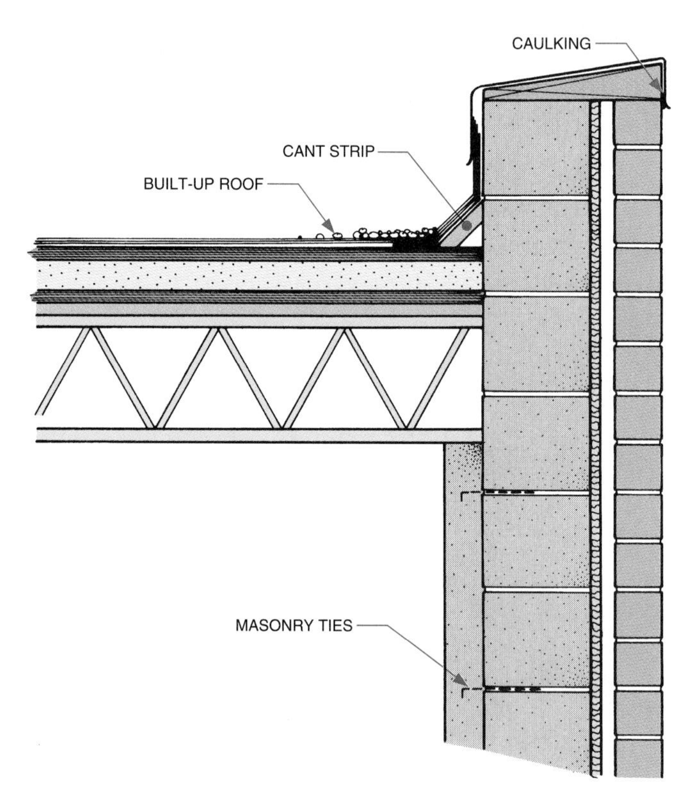

Fig. 30-45 ■ Flat steel roof intersecting a masonry wall parapet.

A flat roof may stop at a wall intersection or continue to form an overhang. Overhangs may include lookouts on the cantilevered ends, perpendicular to the rafter direction. A flat roof may also intersect a *parapet* (short wall), as shown in Fig. 30-45. Parapets are used to hide the roofing surface vents and any roof-mounted mechanical equipment. Parapets are also used to simulate mansard roofs, although true mansard roofs are double-pitched hips.

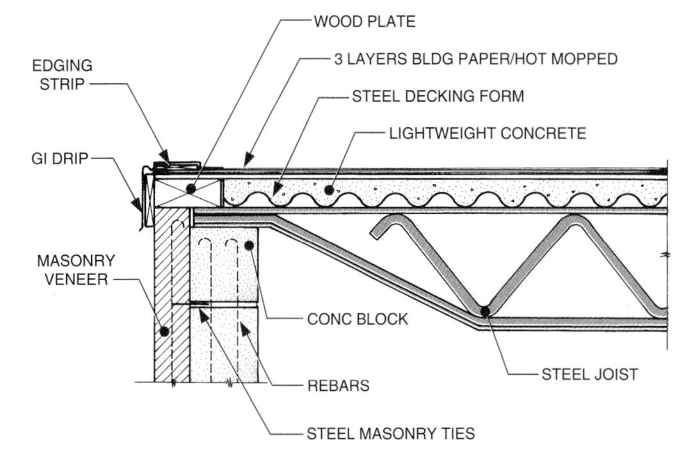

Fig. 30-46 ■ Open web steel joist anchored on a masonry exterior wall.

Steel and Concrete Roof Framing Methods

Structural steel framing methods are especially appropriate for flat roofs. Steel roof framing systems are very similar to floor framing systems. Girders, beams, joists, and decking are used in the same manner. Only the covering and cornice intersecting details are different.

Steel joists, in light construction, rest directly on concrete or on masonry bearing walls. See Fig. 30-46. In heavy construction, steel joists rest on steel girders or columns. Similar to other flat roofs, steel flat roof construction either meets an outside wall to form a right angle, intersects an outside wall to form a parapet, or extends to form an overhang. See Fig. 30-47.

Steel joists can span up to 40 feet depending on load factors. For larger spans, steel bents, arches, or space frames are used. Rigid steel bent frames are either straight single-span, shaped, or multiple-span, as shown in Fig. 30-48.

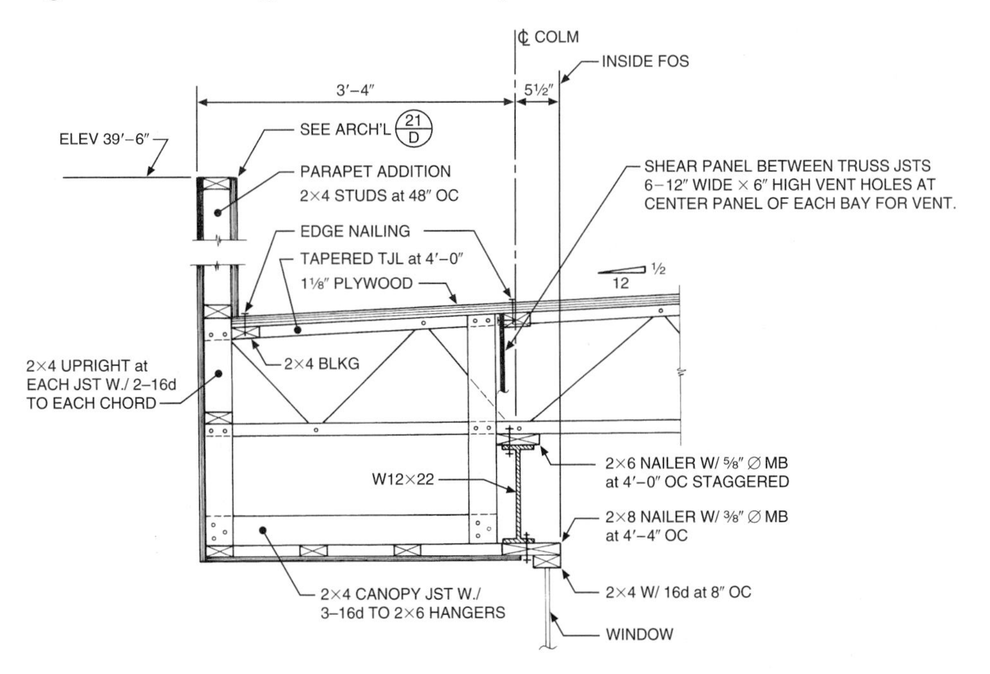

Fig. 30-47 ■ Steel joist overhang which is also framed to create a wood parapet.

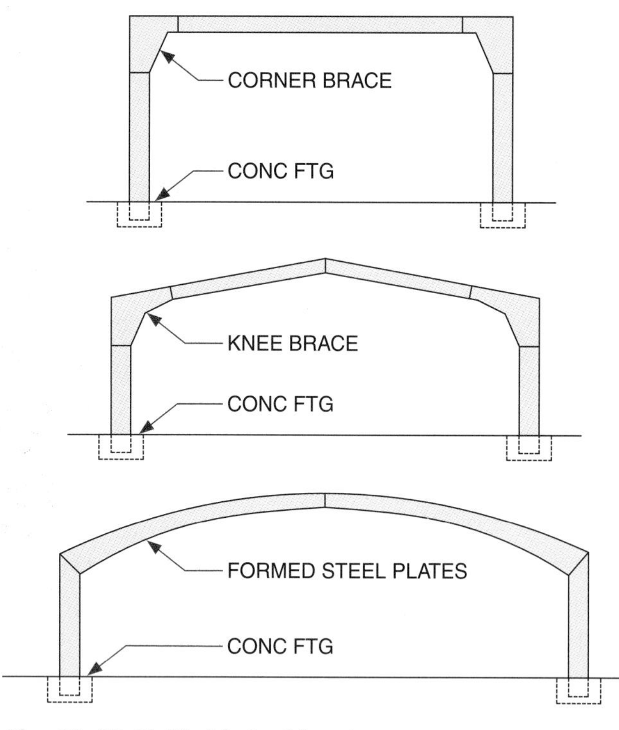

Fig. 30-48 ■ Rigid steel bents.

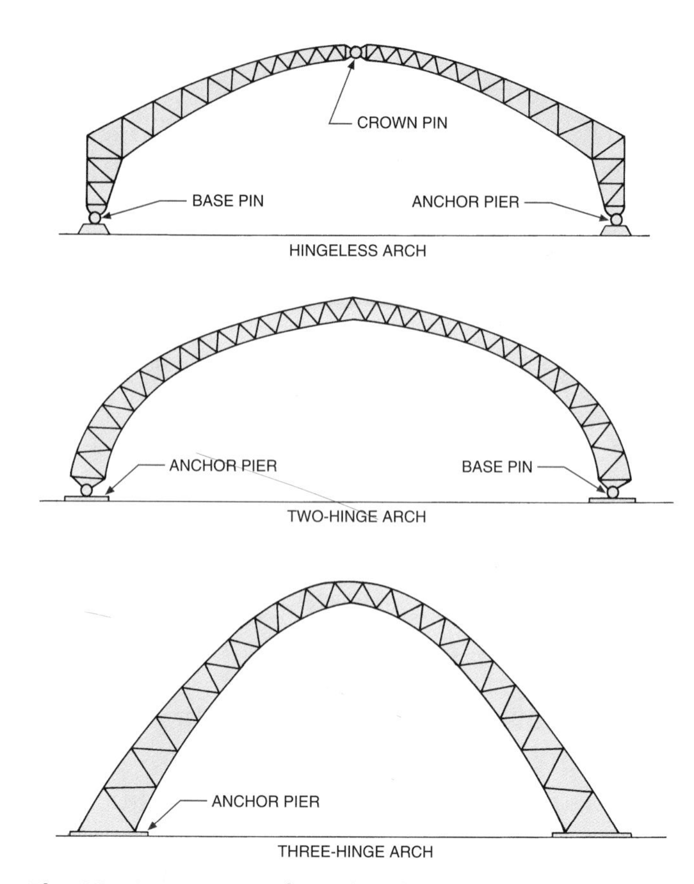

Fig. 30-49 ■ Types of steel arches.

Arches are bent trusses. See Fig. 30-49. Arch details are shown on structural drawings the same way truss details are shown. Space frames are three-dimensional trusses formed by connecting series of triangular polyhedrons. See Fig. 30-50. Space frames, because of their light weight and ability to resist bending, can span extremely large distances. Since their load-bearing capacity is limited, they are used primarily for roof and not floor systems.

Precast concrete joists are used where heavy loads and short spans exist. Filler blocks and a poured slab create a monolithic roof or floor.

Steel roof framing plans are prepared similar to single line wood framing plans. The lines, symbols, and notations are the same as those used on steel floor framing plans.

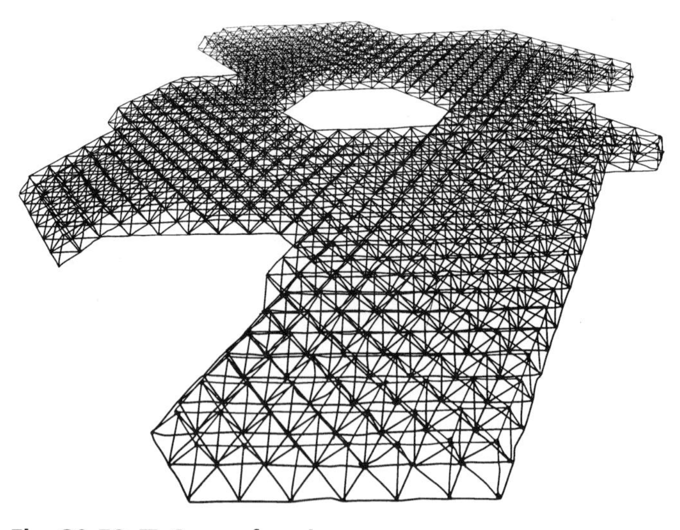

Fig. 30-50 ■ Space framing.

Roof-covering Materials

Roof coverings protect buildings from rain, snow, wind, heat, and cold. Materials used to cover pitched roofs include wood, fiberglass, fiber, cement, asphalt, and composition shingles. On heavier roofs, ceramic tile or slate may also be used. Roll roofing or other sheet material, such as galvanized iron, aluminum, copper, and tin, may also be used for flat or low-pitched roofs. Built-up roofing of felt and gravel are used extensively on flat or low-pitched roofs. Figure 30-51 lists the characteristics of common roof covering materials.

The weight of roofing materials is important in computing dead loads. A heavier roofing surface makes the roof more permanent than does a lighter surface. Generally, heavier roofing materials last longer than lighter materials. Therefore, heavy roofing, such as strip or individual shingles, is superior. Roof covering materials are classified by their weight per 100 square feet (100 sq. ft. equals 1 *square*). Thus 30-lb. roofing felt weighs 30 lbs. per 100 square feet.

Sheathing

Roof sheathing consists of lumber boards, gypsum, fiberboard, or plywood sheets nailed directly to roof rafters. Sheathing adds rigidity to the roof and provides a surface for the attachment of waterproofing materials. In humid areas, sheathing boards or panels are sometimes spaced slightly apart to provide ventilation and to prevent shingle rot.

The joints of sheathing panels are always staggered, and only exterior grade material is used. Most codes in high wind areas require sheathing to be applied with nails or screws not staples. Hurricane clips that lock the panels together may

TYPE	TYPICAL FORM DESCRIPTION	TYPICAL WIDTHS	AVERAGE SPANS	REMARKS
TONGUE & GROOVE (T & G) WOOD-LAMINATE	2″, 3″ & 4″	6″	8′ 14′ 20′	MAXIMUM 20′ LENGTHS.
T & G PRECAST GYPSUM	2″	15″	7′	MAXIMUM 10′ LENGTHS. AVAILABLE WITH METAL EDGES.
CORRUGATED STEEL	3″ 8″	24″	15′	MAXIMUM 10′ LENGTHS. ECONOMICAL FOR MEDIUM SPANS.
PRECAST CONCRETE	WWM 1″ 3½″ 2″	24″	8′	MAY BE CAST TO ANY LENGTH AND THICKNESS.
TONGUE & GROOVE PRECAST WOOD FIBER & CEMENT	2″, 3½″ & 3″	30″	3′ 4′ 5′	MATERIAL IS NAILABLE.

Fig. 30-51 ■ Characteristics of roof deck materials used in light construction.

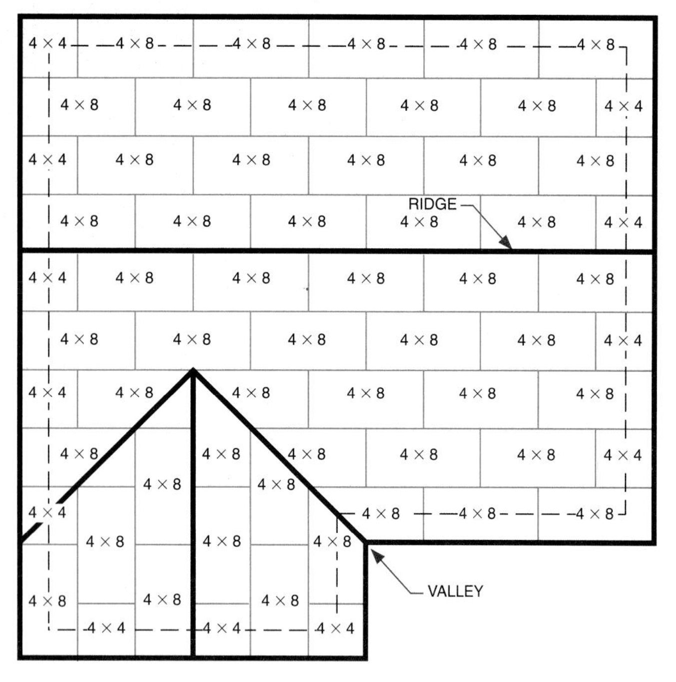

Fig. 30-52 ■ Roof sheathing plans are prepared to ensure that the sheets are laid out as planned.

also be required. In large housing developments, a *roof sheathing plan* is often prepared to plan the best possible arrangement with the minimum amount of waste. See Fig. 30-52.

Roll Roofing

Roll (continuous membrane) roofing may be used as an underlayment for shingles or as a finished roofing material for slopes up to 3/12. Roll roofing material used as an underlayment includes asphalt and saturated felt. The underlayment serves as a barrier against moisture and wind. As a finished roofing material, copper, aluminum, or galvanized steel is used in rolls or sheets. Seams are sealed to provide a watertight surface.

Roof Shingles, Shakes, and Tiles

Roof shingles and shakes are made from asphalt, cement, fiberglass, cedar, or bonded wood fibers. Shingles are laid over building felt that covers the sheathing. See Fig. 30-53. Shingles are available in a variety of patterns. See Fig. 30-54. Most shingles are not recommended for pitches less than 4/12.

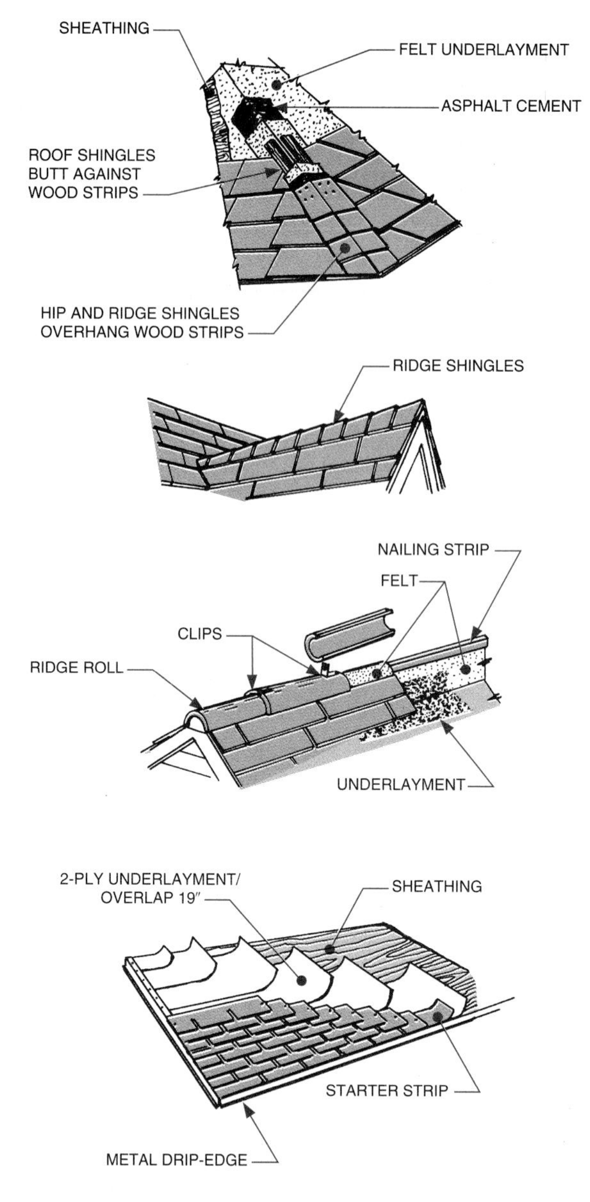

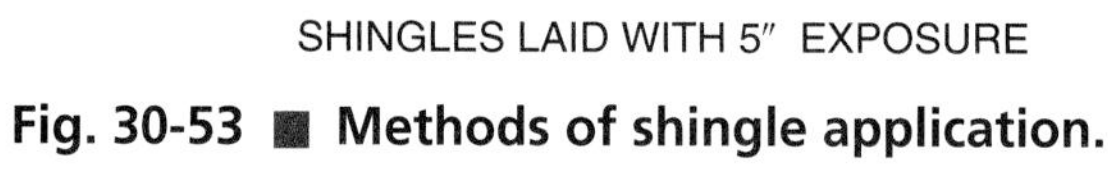

Fig. 30-53 ■ Methods of shingle application.

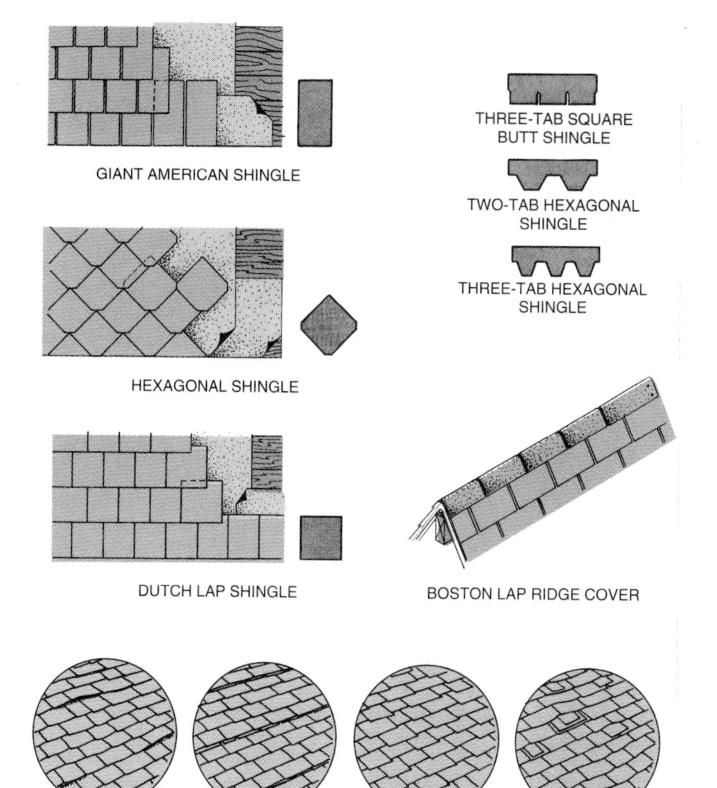

Fig. 30-54 ■ Common shingle patterns.

Shingles are classified by weight per 100 square feet. Shingles range in weight from 180 lbs. to 390 lbs. per square (*i.e.*, 100 sq. ft.) for residential roofs. The average residential asphalt shingle is 245 lbs. per square. Shingles are also classified by special features such as their resistance to mildew, wind, hail, and water and by their life expectancy, usually 20 or 25 years.

Tiles are manufactured from clay-ceramic (shown in Fig. 30-55), cement, and polystyrene. Copper, aluminum, and galvanized steel panels are also available in patterns that simulate shingles and tiles.

Built-up Roofs

Built-up roof coverings are used on flat or extremely low-pitched roofs. Built-up roofing cannot be used on high-pitched roofs because the gravel will wear off

Fig. 30-55 ■ Tile roofing used effectively on a well-designed residence. *John B. Scholz, Architect*

during rain and high winds. Because rain may be driven into gravel crevices and snow may not quickly melt from these roofs, complete waterproofing is essential.

Built-up roofs may have three, four, or five layers of roofing felt, sealed with hot-mopped tar or asphalt, between coatings. The final layer of tar or asphalt is then covered with roofing gravel or a top sheet of roll roofing. See Fig. 30-56.

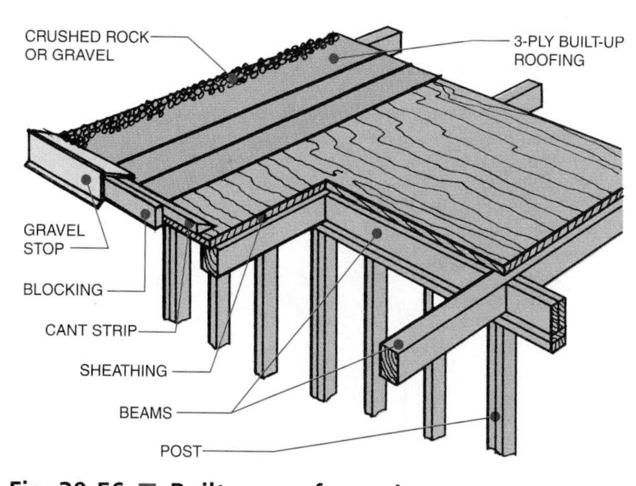

Fig. 30-56 ■ Built-up roof covering.

Roof Appendages

Gutters, flashing, vents, fascia covers, skylights, and downspouts are additions to many roofs. Appendages perform several functions, such as to control water flow, ventilate building fumes, admit light, and cover rough lumber.

Gutters and Downspouts

Gutters are troughs designed to carry water to downspouts, where it can be emptied into a sewer system or away from the building. Materials used most commonly for gutters are sheet metal, cedar, redwood, and plastic. Gutters may be built into the roof structure, as shown in Fig. 30-57. Sheet metal or wood gutters may also be attached or hung from the fascia board. All gutters should be pitched at least 1:20 to provide for drainage to downspouts. See Fig. 30-58. Gutters must be kept below the roof slope line to prevent snow and ice from accumulating. In selecting gutters and downspouts, care must be taken to ensure that their size is adequate for the local rainfall.

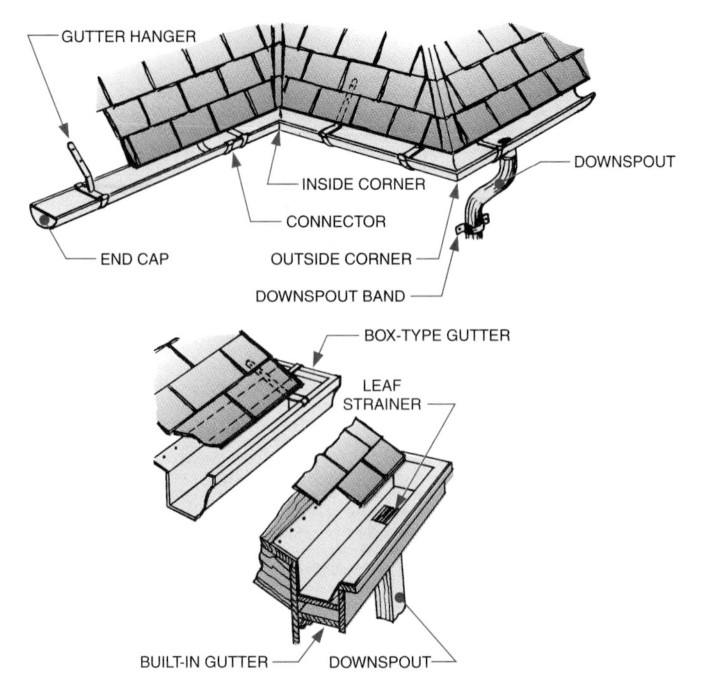

Fig. 30-58 ■ Hanging gutter assembly with downspout.

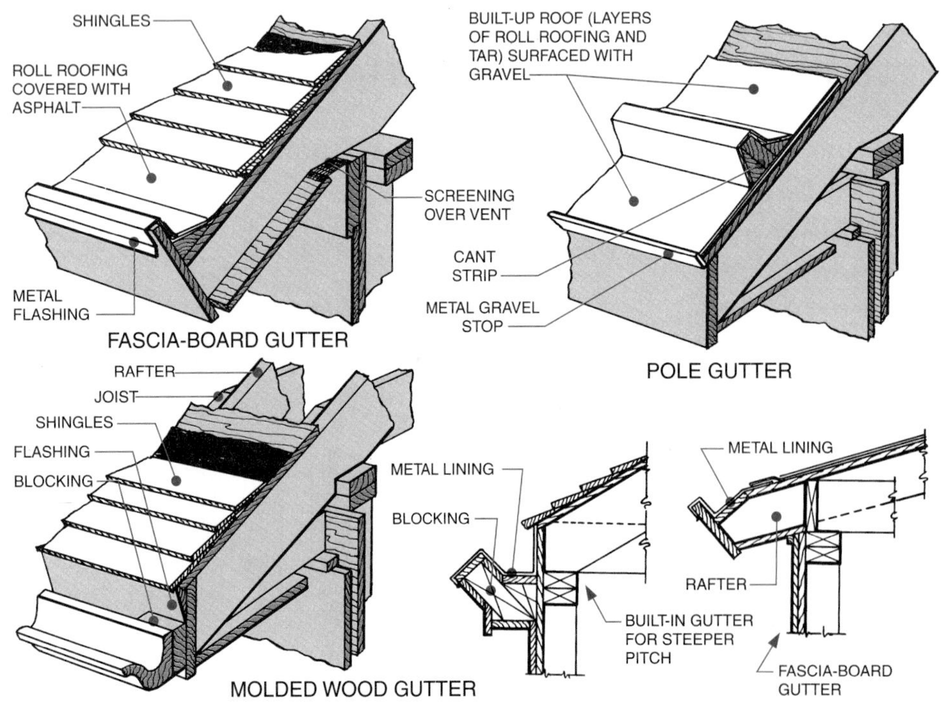

Fig. 30-57 ■ Built-in gutters.

A *roof drainage plan* should show the runoff direction of water for all roof segments. The downslope is shown by arrow direction. See Fig. 30-59. Gutter positions can then be planned accordingly. Some pitched roofs with large overhangs may not need gutters, if drainage is adequate and runoff doesn't fall on outdoor traffic or living areas.

Flashing

Joints where roof covering materials intersect a ridge, hip, valley, chimney, wall, vent, skylight, or parapet must be flashed. **Flashing** is additional covering used under a joint to provide complete waterproofing. Roll-roofing, shingles, or sheet metal is used as flashing material. For sheet-metal flashing, watertight sheet-metal joints should be made. See Fig. 30-60. Chimney flashing is frequently bonded into the mortar joint and caulked under shingles to provide a waterproof joint.

Roof Ventilation

Areas where excessive heat and moisture are trapped must be ventilated. Attics and crawl spaces can be ventilated by placing individual roof vents and/or a

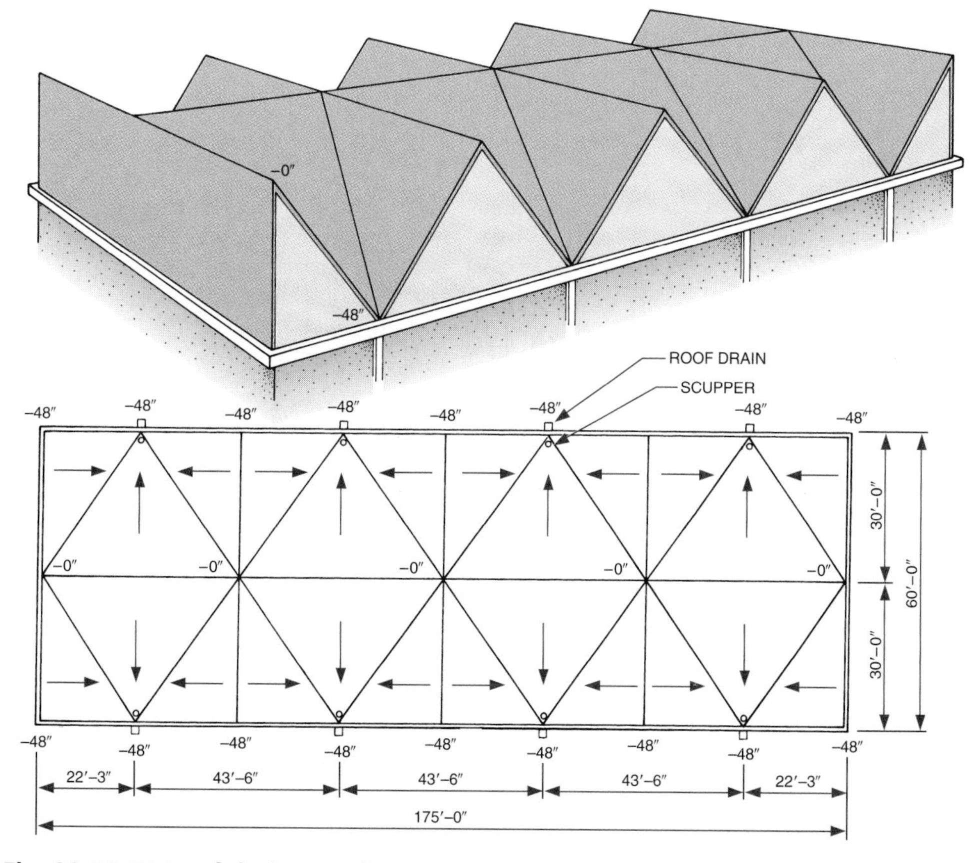

Fig. 30-59 ■ Roof drainage plan.

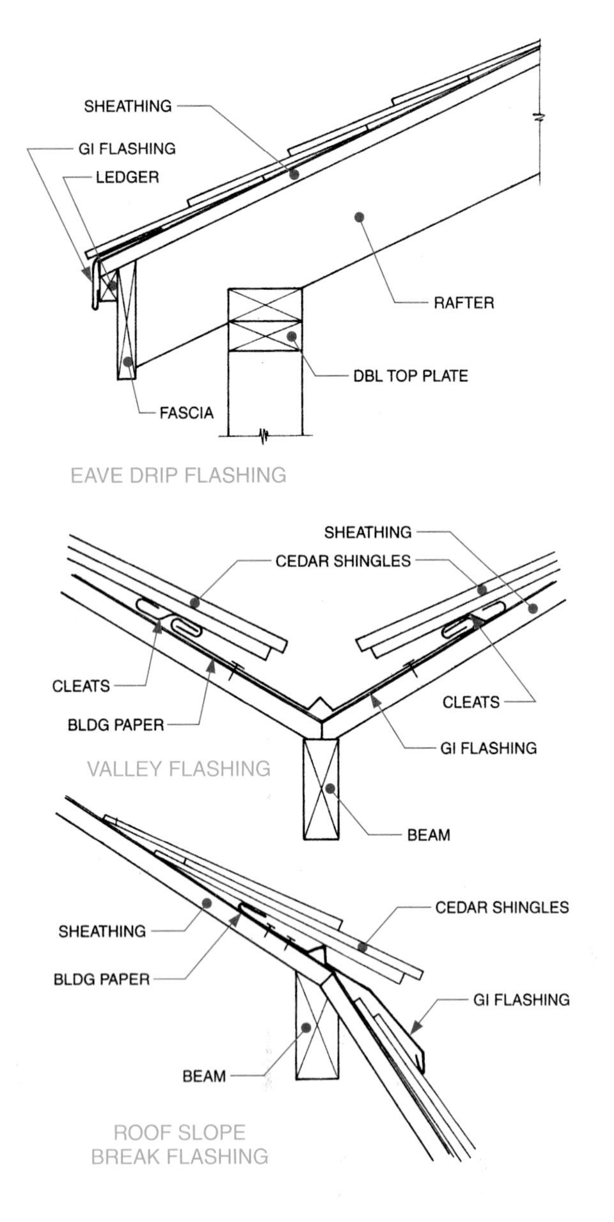

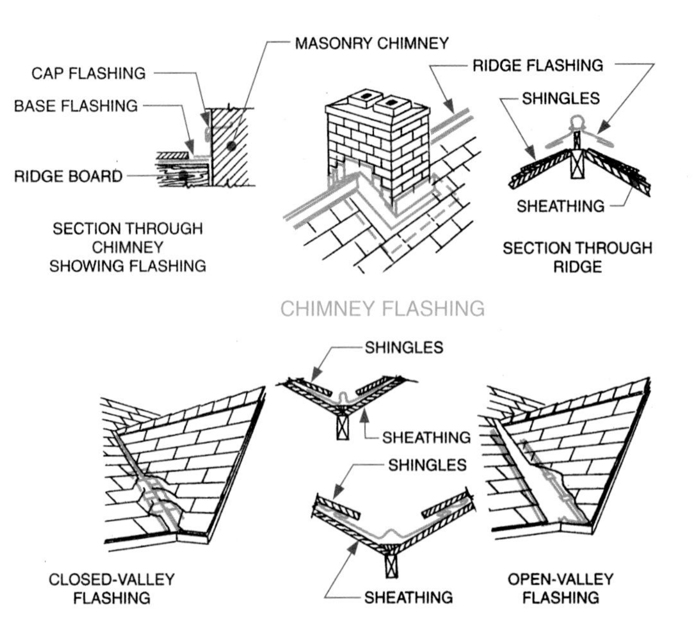

Fig. 30-60 ■ Sheet-metal flashing applications.

cupola connecting the inside area with the outside. Vent areas should be 1/150th of the enclosed area or 1/300th if the area has a vapor barrier. The space between rafters also needs to be ventilated. This is best done by creating an air flow between ventilated soffits and a continuous ridge vent. See Fig. 30-61.

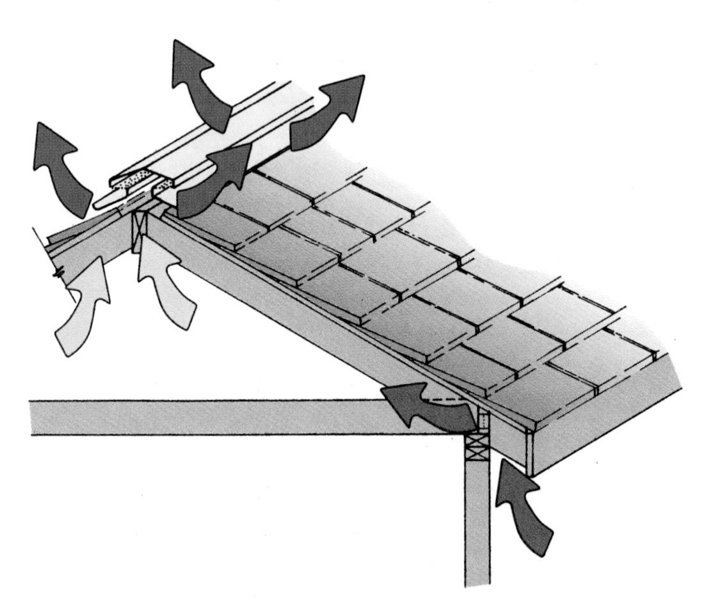

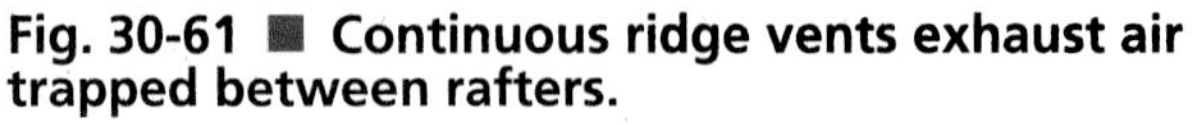
Fig. 30-61 ■ Continuous ridge vents exhaust air trapped between rafters.

Skylights

Roof framing for skylights, as shown in Fig. 30-62, is the same as for chimneys or any mechanical equipment requiring a break in the rafter and roof covering pattern. All openings must be flashed. Refer back to Fig. 30-60.

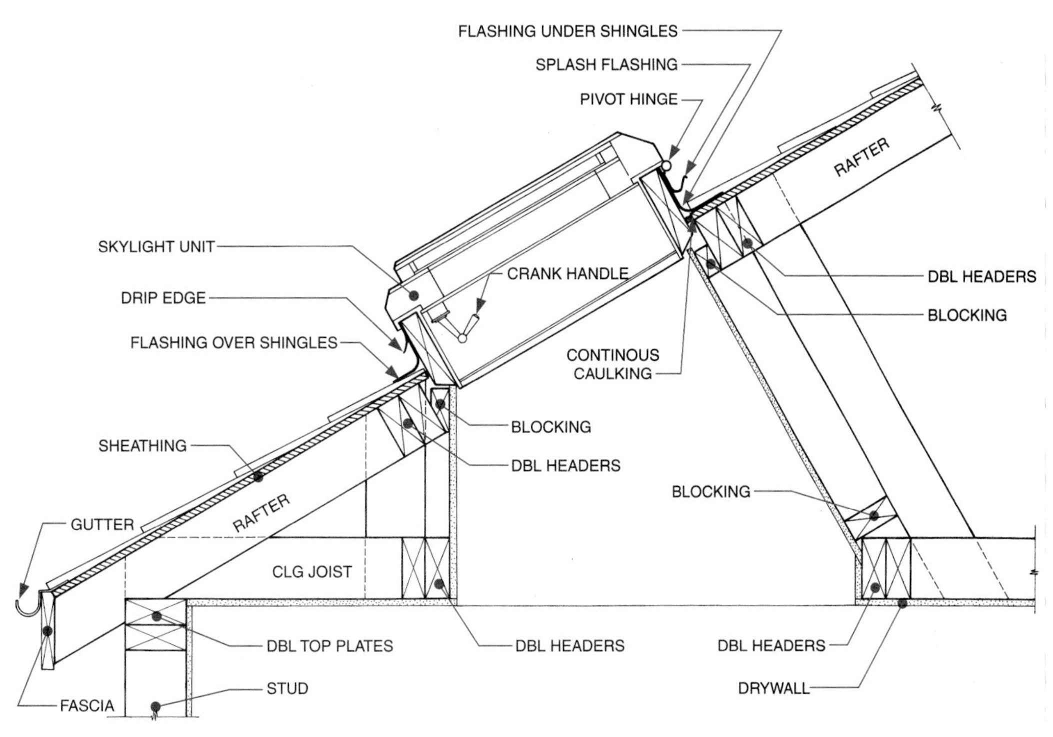

Fig. 30-62 ■ Roof framing and flashing for a skylight.

CHAPTER 30 Exercises

1. What are the two types of roof framing plans? What is the function of each?
2. Prepare a roof framing plan for a house of your own design. Label the members.
3. What is the roof pitch of a roof with a rise of 8′ and a run of 24′?
4. Name the main parts of a roof truss. Tell the advantages and disadvantages of using trusses for roofs.
5. Draw a cornice section for the house of your design. Include the roof covering.
6. Sketch a roof plan for a gable roof. Include the dimensions of the run.

7. Specify the type of roofing to be used in the house of your design. Explain why you chose the type of material: wood, steel, and/or concrete.
8. How many square feet of shingles are in four squares?

9. Using a CAD system, prepare a roof framing plan for a house of your own design, a cornice section, and/or a roof detail in this chapter.

PART 8

Electrical and Mechanical Design and Drawings

CHAPTER

31 **Electrical Design and Drawings**
32 **Comfort-Control Systems (HVAC)**
33 **Plumbing Drawings**

CHAPTER

31

Electrical Design and Drawings

Objectives

In this chapter, you will learn to:

- plan and draw electrical circuits for a house on a floor plan.
- plan and draw lighting for each room in a house.
- calculate electrical measurements for each circuit.
- draw electrical symbols.
- design and draw an electronic building control system.

Terms

ampere
circuit
circuit breaker
conductors
fluorescent
footcandle
incandescent
insulators
kilowatt-hour
lux
ohms
volt
watt

Buildings cannot function as they should without electricity. The design and drawing of electrical systems requires a knowledge of electrical power distribution, wiring circuits, lighting methods, and electrical drawing symbols and conventions.

■ **The design of a structure is made more dramatic through creative use of lighting.** *IMHOTEP-Alfred Karram*

Electrical Principles

Understanding electrical principles is vital to designing safe and efficient architectural electrical systems.

Power Distribution

Electric power is generated from several sources of energy: wind, water, nuclear, fossil fuel, solar (photovoltaic), and geothermal. Photovoltaic cells convert solar energy directly into an electric current. All other energy sources are harnessed to produce a rotary mechanical motion that drives electric generators. The generators convert movement into electricity. Transformers are used to "step up" (increase) the electrical power to very high voltages (hundreds of thousands of volts) for transmission by wires over long distances. Wherever the transmission lines enter an industrial or residential community for local power distribution, large transformers are used to "step down" the voltage to a few thousand volts. Smaller transformers set on poles or in underground vaults are used for final distribution to small groups of houses or individual factories. Usually 110 and 220 volts are delivered to residences.

Electrical Measurements

Math Connection

Electrical properties can be measured with instruments. The terms used to describe units of electricity—volt, ampere, and watt—are used in both metric and customary systems.

A **volt** is the unit of electrical *pressure* or potential. This pressure makes electricity flow through a wire. For a particular electrical load, the higher the voltage, the greater will be the amount of electricity that will flow.

The term for flow of electricity is *current*. An **ampere,** or amp, is the unit used to measure the magnitude of an electric current. An ampere is defined as the specific quantity of electrons passing a point in one second. The amount of current, in amperes, that will flow through a circuit must be known in order to determine proper wire sizes and the current rating of circuit breakers and fuses.

The amount of *power* required to light lamps, heat water, turn motors, and do all types of work is measured in **watts.** Wattage depends on both potential and current. Current (in amperes) multiplied by potential (in volts) equals power (in watts).

$$\text{amperes} \times \text{volts} = \text{watts}$$

The actual energy used (the watts utilized) for work performed is the basis for figuring the cost of electricity. The unit used to measure the consumption of electrical energy is the **kilowatt-hour.** A kilowatt is 1000 watts. An hour, of course, is a unit of time. A 1000-watt hand iron operating for one hour consumes one kilowatt-hour (1 kWh). The device used to measure the kilowatt-hours consumed is the watt-hour meter.

Electricity flowing through a material always meets with some resistance. Materials such as wood, glass, and plastic have a high resistance. They are good **insulators.** Copper, aluminum, and silver have low resistance and are therefore good **conductors** of electricity. Most electrical wiring consists of copper or aluminum surrounded by plastic. The plastic keeps the electricity from flowing where it isn't wanted.

The amount of electrical resistance is measured in **ohms**. The electron flow (or current in amperes) through a circuit is equal to the voltage (number of volts) divided by the resistance (ohms). This can be expressed in the formula:

$I = \frac{E}{R}$ or $E = IR$ or $R = \frac{E}{I}$

I = current (amperes)
E = electromotive force (volts)
R = resistance (ohms)

Example:
If the current is 10 amperes and the electromotive force is 120 volts, what is the resistance?

$R = \frac{E}{I}$

$R = \frac{120}{10}$

R = 12 ohms

Service Entrance

Math Connection

Power is supplied to a building through a service entrance. Three heavy wires, together called the drop, extend from a utility pole or an underground source to the structure. These wires are twisted into a cable. At the building, overhead wires are fastened to the structure and spliced to service entrance wires that enter a conduit through a service head, as shown in Fig. 31-1.

In planning overhead service drop paths, minimum height requirements for connector lines must be carefully followed. See Fig. 31-2. If these distances cannot be maintained, rigid conduit, electrical metallic tubing, or busways (channels, ducts) must be used.

If the service is supplied underground, three wires are placed in a rigid conduit.

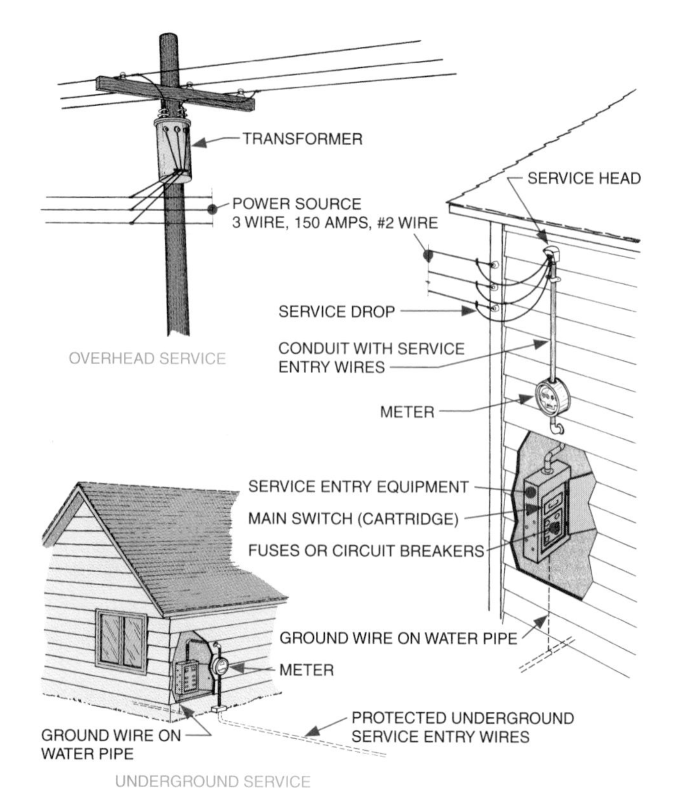

Fig. 31-1 ■ Electrical distribution to buildings.

An underground service conduit is brought to the meter socket. An underground service entrance includes a watt-hour meter, main breaker, and lightning protection. Automatic brownout equipment is also required by many codes for new construction. All electrical systems must be grounded through the service entrance.

Service Distribution

Math Connection

Electrical current is delivered throughout a building through a *distribution panel,* or service panel. See Fig. 31-3. The size of a distribution panel (in amperes) is determined by the total load requirements (watts) of the entire building. Watts can be converted to amperes by dividing

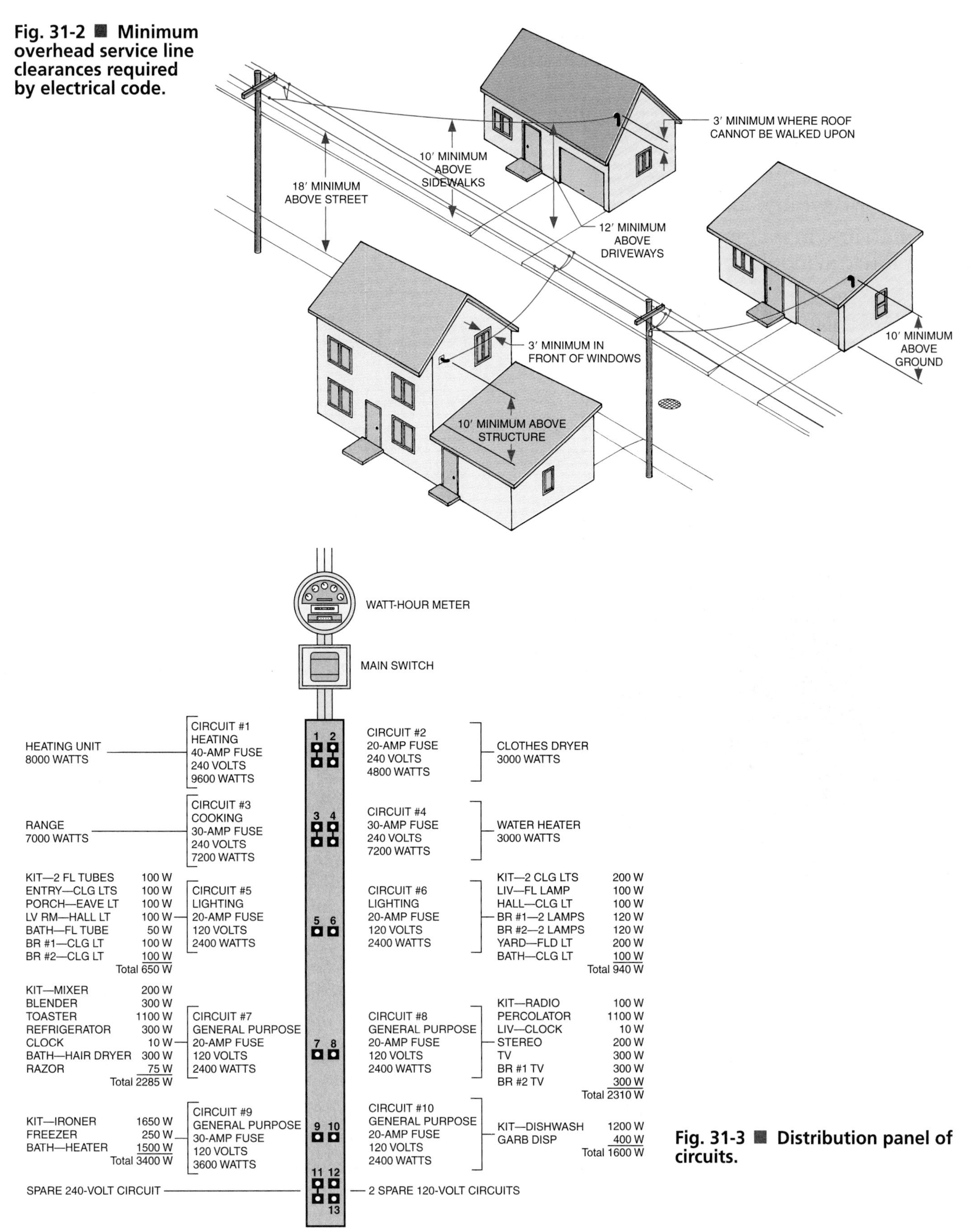

Fig. 31-2 ■ Minimum overhead service line clearances required by electrical code.

Fig. 31-3 ■ Distribution panel of circuits.

the total (and future) watts needed by the amount of voltage delivered to the distribution box:

Formula: $A = \frac{W}{V}$ or $W = AV$ or $V = \frac{W}{A}$

W = metric symbol for watts
V = metric symbol for volts
A = metric symbol for amperes

Example: $\frac{35000\ W}{240\ V} = 145\ A$

Most residences require a distribution panel with a capacity of 100 to 200 amps. The National Electrical Code (NEC) minimum for new residential construction is 60 amps. To compute the total load requirements, the watts needed for each circuit must first be determined.

Branch Circuits

From the distribution panel, electricity is routed to the rest of the building through *branch circuits.* A **circuit** is a circular path that electricity follows from the power supply source to a light, appliance, or other electrical device and back again to the power supply source. See Fig. 31-4. If the electrical load for an entire building were placed on one circuit, overloading would leave the entire building without power. Thus branch circuits are used. Each circuit delivers electricity to a limited number of outlets or devices.

Each circuit is protected with a **circuit breaker**. A circuit breaker is a device that opens (disconnects) a circuit when the current exceeds a certain amount. Without a circuit breaker, excessive electrical loads could cause the wiring to overheat and start a fire. When a breaker opens, or "trips," the power to the branch circuit is

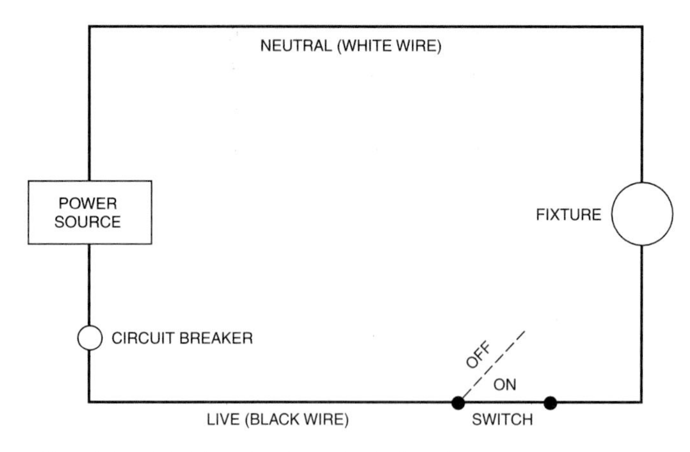

Fig. 31-4 ■ Simple electrical circuit.

disconnected. Similarly, if the sum of the current drawn by the branch circuits exceeds the rating of the main circuit breaker, the main breaker will trip. This protects the service-entrance wires and equipment from overheating and damage. Older homes often have fuses instead of circuit breakers. They serve the same purpose, but overloaded fuses must be replaced. Circuit breakers that trip can be reset.

Branch circuits are divided into three types by the National Electrical Code: lighting circuits, small-appliance circuits, and individual circuits.

LIGHTING CIRCUITS Lighting circuits are connected to lighting outlets for the entire building. Different lights in each room are usually on different circuits so that if one circuit breaker trips, the room will not be in total darkness.

In all dwellings other than hotels, the NEC requires a minimum general lighting load of 3 watts per square foot of floor space. However, the amount of wattage demanded at one time (demand factor) is calculated at 100 percent only for the first 3000 watts; 35 percent is used for the second 17,000 watts; and 25 percent is used for commercial demands over 120,000 watts. Thus, the general lighting load planned for a 1500 sq. ft. house would be 3525 watts, not the full 4500 watts. It is calculated as follows:

1500 sq. ft. × 3 W = 4500 W

First 3000 W × 100% = 3000 W

Next 1500 W × 35% = 525 W

Total 4500 W 3525 W

If each branch circuit supplies 1800 watts (120 V × 15 A = 1800 W), a 1500 sq. ft. house should have two 1800-watt general lighting circuits. See Fig. 31-5. Lighting circuits are also used for small devices such as clocks and radios. However, since all lights and other items on the circuit are probably not going to be used at the same time, it is not necessary to provide a service capable of supplying the full load.

SMALL-APPLIANCE CIRCUITS These circuits provide power to outlets wherever small appliances are likely to be connected. Small appliances include items such as toasters, electric skillets, irons, electric shavers, portable tools, and computers. Appliance circuits are not designed to also support lighting needs. See Fig. 31-6. The NEC requires a minimum of two small-appliance circuits in a residence. Each circuit is usually computed as a 3600-watt load.

INDIVIDUAL CIRCUITS Individual dedicated circuits are designed to serve a single large electrical appliance or device, such as electric ranges, automatic heating units, built-in electric heaters, and workshop outlets. Large motor-driven appliances, such as washers, garbage disposals, and dishwashers, also use individual circuits. These circuits are designed to provide sufficient power for starting loads. When a motor starts, it needs an extra surge of power to bring it to full speed. This is called a *starting load.*

A separate circuit (20 amps) is required in a laundry area to provide power for the washing machine and the dryer. Because of the danger of water leakage, a ground-fault circuit interrupter receptacle is recommended.

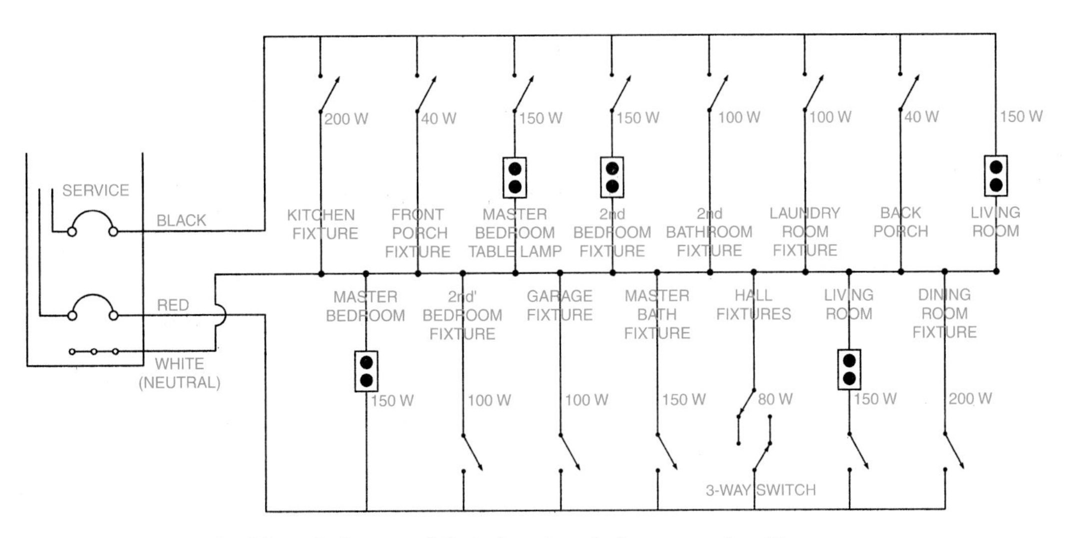

Fig. 31-5 ■ Typical breakdown of lighting loads for one circuit.

Fig. 31-6 Load requirements of common electrical appliances.

	Typical connected watts	Volts	Wires	Circuit breaker or fuse	Outlets on circuit	Outlet type	Notes
KITCHEN							
Range	12,500	120/240	3 #6 + GND	50A	1	14-50R	
Oven (built-in)	4,500	120/240	3 #10 + GND	30A	1	14-30R	#1
Range top	6,000	120/240	3 #10 + GND	30A	1	14-30R	#1
Dishwasher	1,500	120	2 #12 + GND	20A	1	5-15R	#2
Waste disposer	800	120	2 #12 + GND	20A	1	5-15R	#2
Trash compactor	1,200	120	2 #12 + GND	20A	1	5-15R	#2
Microwave oven	1,450	120	2 #12 + GND	20A	1 or more	5-15R	
Broiler	1,500	120	2 #12 + GND	20A	1 or more	5-15R	#3
Fryer	1,300	120	2 #12 + GND	20A	1 or more	5-15R	#3
Coffeemaker	1,000	120	2 #12 + GND	20A	1 or more	5-15R	#3
Refrigerator/freezer 16–25 cubic feet	800	120	2 #12 + GND	20A	1 or more	5-15R	#4
Freezer chest or upright 14–25 cubic feet	600	120	2 #12 + GND	20A	1 or more	5-15R	#4
Roaster-broiler	1,500	120	2 #12 + GND	20A	1 or more	5-15R	#3
Waffle iron	1,000	120	2 #12 + GND	20A	1 or more	5-15R	#3
FIXED UTILITIES							
Fixed lighting	1,200	120	2 #12	20A	1 or more		#10
Window air conditioner							
14 000 Btu	1,400	120	2 #12 + GND	20A	1	5-15R	#11
25 000	3,600	240	2 #12 + GND	20A	1	6-20R	#11
29 000	4,300	240	2 #10 + GND	30A	1	6-30R	#11
Central air conditioner							
23 000 Btu	2,200	240					#6
57 000 Btu	5,800	240					#6
Heat pump	14,000	240					#6
Sump pump	300	120	2 #12	20A	1 or more	5-15R	#1
Heating plant oil or gas	600	120	2 #12	20A	—	—	#6
Fixed bathroom heater	1,500	120	2 #12	20A	—	—	#6
Attic fan	300	120	2 #12	20A	1	5-15R	
Dehumidifier	350	120	2 #12	20A	1 or more	5-15R	#1

continued

Fig. 31-6 Continued

	Typical connected watts	Volts	Wires	Circuit breaker or fuse	Outlets on circuit	Outlet type	Notes
LAUNDRY							
Washing machine	1,200	120	2 #12 + GND	20A	1 or more	5-15R	#5
Dryer all-electric	5,200	120/240	3 #10 + GND	30A	1	14-30R	#1
Dryer gas/electric	500	120	2 #12 + GND	20A	1 or more	5-15R	#5
Ironer	1,650	120	2 #12 + GND	20A	1 or more	5-15R	
Hand iron	1,000	120	2 #12 + GND	20A	1 or more	5-15R	
Water heater	3,000–6,000					DIRECT	#6
LIVING AREAS							
Workshop	1,500	120	2 #12 + GND	20A	1 or more	5-15R	#7
Portable heater	1,300	120	2 #12 + GND	20A	1	5-15R	#3
Television	300	120	2 #12 + GND	20A	1 or more	5-15R	#8
Portable lighting	1,200	120	2 #12 + GND	20A	1 or more	5-15R	#9
Band saw	300	120	2 #12 + GND	20A	1 or more	5-15R	#6
Table saw	1,000	120/240	2 #12 + GND	20A	1	5-15R	#6

NOTES
#1 May be direct-connected.
#2 May be direct-connected on a single circuit; otherwise, grounded receptacles required.
#3 Heavy-duty appliances regularly used at one location should have a separated circuit. Only one such unit should be attached to a single circuit at a time.
#4 Separate circuit serving only refrigerator and freezer is recommended.
#5 Grounding-type receptacle required. Separate circuit is recommended.
#6 Consult manufacturer for recommended connections.
#7 Separate circuit recommended.
#8 Should not be connected to appliance circuits.
#9 Provide one circuit for each 500 sq. ft (46 m^2). Divided receptacle may be switched.
#10 Provide at least one circuit for each 1200 watts of fixed lighting.
#11 Consider 20-amp, 3-wire circuits to all window-type air conditioners. Outlets may then be adapted to individual 120- or 240-volt units. This scheme will work for all but the very largest units.

Ground-Fault Circuit Interrupter (GFCI)

A GFCI receptacle must be located wherever there is a possibility for people to ground themselves and be shocked by the electrical current flowing through their body to the ground. The purpose of a GFCI receptacle is to cut off the current at the outlet. When the GFCI receptacle senses any change of current, it immediately trips a switch to interrupt the current. It operates faster and is safer than the circuit breaker switch or fuse at the power entry panel. A GFCI valve will trip in 1/40 second when an extremely small current variation (ground fault) of 0.005 amps is reached.

In new construction, GFCI receptacles must be located with each convenience outlet near water sources and/or pipes in the bathroom, kitchen, garage, laundry, and outdoors. Any receptacle located within 10′ or within 15′ of the inside of a permanently installed swimming pool must also be wired through a GFCI. GFCIs are also required if outlets are placed in unfinished crawl spaces below grade level.

Electrical Conductors

Wires used to conduct electricity are classified by the type of wire material, the insulation material, and the wire size. The size of the wire used in a circuit depends on the current to be carried by the circuit. See Fig. 31-7. Although the meter voltage is 120V and 240V, wiring resistance reduces the voltage at the receptacles to approximately 110V and 220V. Sizes 6 through 2/0 are used for 240-volt (240V) service entrance and circuits. The exact size depends on the capacity of the service panel. Sizes 10 through 14 are used for 120V and 240V lighting and small-appliance circuits. Sizes 16 and 18 are used for low voltage items such as thermostats and doorbells.

A low-voltage switching system may be used to turn on or off any fixture, appliance, or light. Because of the low voltage, extremely small wires are used to hook up the switch to the fixture.

Wire size is critical. If a wire is too small for the current applied, excessive resistance (overload) can result. This may cause the insulation to overheat and break down, causing a potential fire hazard. When selecting or preparing to use appliances, it is important to check the UL (Underwriters' Laboratories) ratings to learn the proper wiring requirements.

Aluminum wire is lighter and less expensive than copper, but many codes

Fig. 31-7 ■ Selecting electrical conductor.

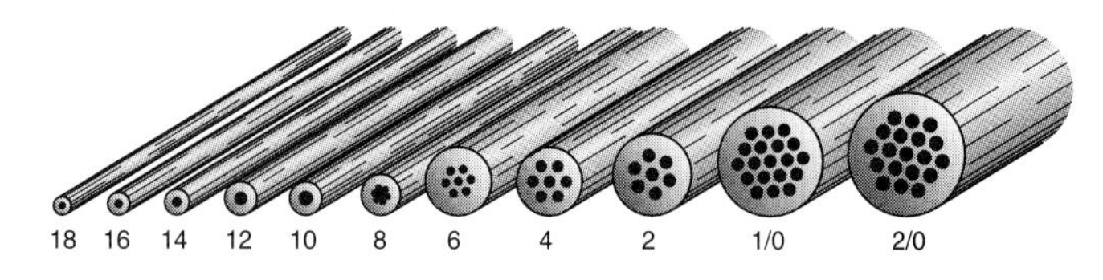

A. Sizes of copper wire conductors.

Fig. 31-7B Relationship of amperes and wire size.

BRANCH CIRCUITS			
		Current rating (amperes)	
Size	Diameter (inches)	Copper	Aluminum
14	0.064	15	—
12	0.081	20	15
10	0.102	30	25
8	0.129	40	30
6	0.162	55	40
4	0.204	70	55
3	0.229	80	65
2	0.258	95	75
1	0.289	110	85
0	0.325	125	100
00 (2/0)	0.365	145	115
000 (3/0)	0.410	165	130
0000 (4/0)	0.460	195	155

SERVICE ENTRANCE		
	Service rating current (amperes)	
3-Wire Service Size (each wire)	Copper	Aluminum
4	100	—
3	110	—
2	125	100
1	150	110
0	175	125
00 (2/0)	200	150
000 (3/0)	—	175
0000 (4/0)	—	200

apply stricter rules to the use of aluminum for residential work. Insulation is available in flexible metal armored or nonmetal sheathed form. For underground or exterior exposed wiring, wires must be encased in rigid or flexible metal or PVC (plastic) conduits.

Calculating Total System Requirements

The installation of the proper size of service entrance equipment and branch circuits are dependent upon the square footage of the residence, number of appliances, lighting, and future expansion allowances. To find the total amp service needed for an entire building, first determine the total number of watts needed for each circuit. Add these to find the total watts needed for the building. For example, to calculate the size of the service entrance for a 2000 sq. ft. residence, list the amount of wattage to be used as follows:

- Lighting circuits (typical)
  2000 sq. ft. uses 3 watts per sq. ft. = 4,050 watts
- Convenience outlets
  2 circuits in service area (120V × 20A) = 4,800 watts
  2 circuits in sleeping area (120V × 20A) = 4,800 watts
  2 circuits in living area (120V × 20A) = 4,800 watts
- Dedicated circuits

electric range	=	10,000 watts
electric dryer	=	5,000 watts
washing machine	=	1,000 watts
dishwasher	=	1,000 watts
forced air unit	=	1,000 watts
electric water heater	=	2,000 watts
	total	38,450 watts

To find the required service panel amps needed, divide the total watts by the available voltage (240V):

38,450 watts ÷ 240 = 160 amps

Service panels are available with capacities of 30, 40, 50, 60, 70, 100, 125, 150, 175, and 200 amps. The next highest panel above the required amps should be chosen to allow for future expansion. In this case, the next highest is 175-amp service.

Lighting Design

Functional lighting design must consider the interaction among eyesight, objects, and light sources. Good lighting design provides sufficient but not excessive light. Glare from unshielded bulbs or improperly placed lighting should be avoided. Excessive contrast between light and shadows within the same room should also be avoided, especially in work areas.

For centuries, candles and oil lamps were the major source of artificial light. Although candles continue to function for special effects, the major sources of light today are incandescent and fluorescent lamps. **Incandescent** lamps have a filament (a very thin wire) that gives off light when heated. **Fluorescent** lamps have an inner coating that gives off visible light when exposed to ultraviolet light. The ultraviolet light is released by a gas inside the fluorescent tube. Incandescent lamps concentrate the light source, while fluorescent lamps provide linear patterns of light. Fluorescent lamps give a uniform glareless light that is ideal for large working areas. Fluorescent lamps give more light per watt, last seven times longer, and generate less heat than incandescent lamps.

Light Measurements

Human eyes adapt to varying intensities of light. However, they must be given enough time to adjust slowly to different light levels. Sudden extreme changes of light may cause discomfort.

Light intensity is measured in units called footcandles. A **footcandle** is equal to the amount of light a candle casts on an object one foot away. See Fig. 31-8. Ten footcandles (10 fc) equals the amount of light that ten candles throw on a surface one foot away. In the metric system, the standard unit of illumination is the **lux** (lx). One lux is equal to 0.093 fc. To convert footcandles to lux, multiply by 10.764. See Fig. 31-9.

Types of Lighting

The three basic types of lighting are general lighting, specific lighting, and decorative lighting. Good examples of all three types of lighting can be found in Part Three of this text.

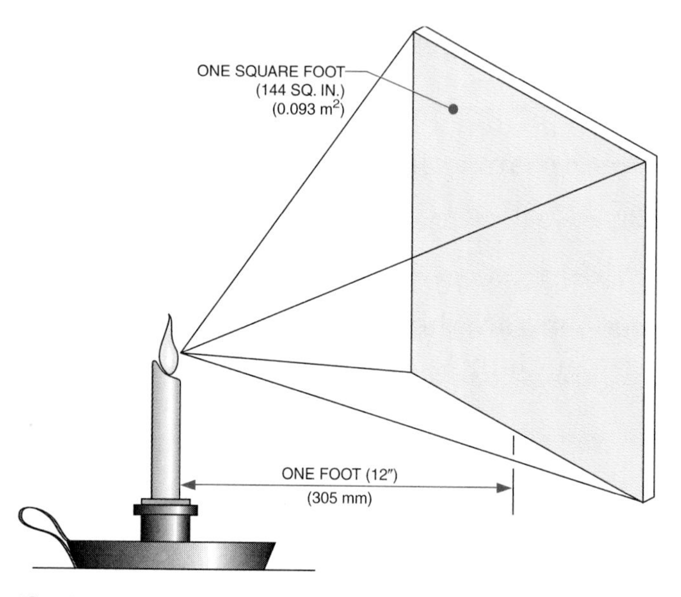

Fig. 31-8 ■ One footcandle of light.

SUNLIGHT	
Beaches, open fields	10,000 FC (107 640 LX)
Tree shade	1,000 FC (10 764 LX)
Open park	500 FC (5382 LX)
Inside 3′ from window	200 FC (2153 LX)
Inside center of room	10 FC (108 LX)
ACCEPTED ARTIFICIAL LIGHT LEVELS	
Casual visual tasks, conversation, watching TV, listening to music	10–20 FC (108–215 LX)
Easy reading, sewing, knitting, house cleaning	20–30 FC (215–323 LX)
Reading newspapers, kitchen & laundry work, keyboarding	30–50 FC (323–538 LX)
Prolonged reading, machine sewing, hobbies, homework	50–70 FC (538–753 LX)
Prolonged detailed tasks such as fine sewing, reading fine print, drafting	70–200 FC (753–2153 LX)

Fig. 31-9 ■ Comparison of sunlight and artificial light levels.

GENERAL LIGHTING General lighting provides overall illumination and radiates a comfortable level of brightness for an entire room. See Fig. 31-10. General lighting replaces sunlight and is provided primarily with chandeliers, ceiling or wall-mounted fixtures, and track lights. To avoid contrast and glare, general lighting should be diffused through the use of fixtures that totally hide the light source or that spread light through panels. Close spacing of hanging fixtures also creates diffuse lighting. Another solution is to use adjustable fixtures so that the light can be directed away from eye contact.

Where possible, daylight should be included as a part of the general lighting plan during daylight hours. If adequate window light is not available, the use of skylights should be considered.

Fig. 31-10 ■ Well diffused indoor general lighting combined with decorative lighting. *Carl's Furniture Showrooms Inc. Design by Teri Kennedy ASID and Linda Dragin ASID. Photo by Lee Gordon*

The intensity of general lighting should be between 5 and 10 fc (54 to 108 lx). A higher level of general lighting should be used in the service area and bathrooms. Many general lighting fixtures can also be used for decorative lighting by a connection to dimmer switches.

SPECIFIC LIGHTING Light directed to a specific area or located to support a particular task is known as specific, local, or task lighting. See Fig. 31-11. Specific lighting helps in performing such tasks as reading, sewing, shaving, computer work, and home theater viewing. It also adds to the general lighting level. Track lighting and portable lamps provide sources of specific indoor lighting.

DECORATIVE LIGHTING Bright lights are stimulating, while low levels of light are quieting. Decorative lighting is used to create atmosphere and interest. Indoor decorative lights are often directed on plants, bookshelves, pictures, wall textures, fireplaces, or any architectural feature worthy of emphasis. Some decorative lighting can be used as general lighting through the use of dimmer switches.

Fig. 31-11 ■ Specific lighting on each step makes traveling the stairs easier and safer. *Intermatic Inc.*

Outdoor decorative lighting can be most dramatic. Exterior structural and landscape features can be accented by well-placed lights. Outdoor lighting is used to light and accent wall textures, trees, shrubs, architectural features, pools, fountains, and sculptures. See Fig. 31-12. Outdoor lighting is especially

Fig. 31-12 ■ Dramatic example of outdoor decorative lighting. *Intermatic Inc.*

needed to provide a safe view of stairs, walks, and driveways.

Remember to conceal light sources and don't overlight. Use waterproof devices and an automatic timing device to turn lights on and off.

Light Distribution

Light from any artificial source can be distributed (dispersed or directed) in five different ways: direct, indirect, semidirect, semi-indirect, and diffused. See Fig. 31-13. *Direct* light shines directly on an object from a light source. Indirect light is reflected from surfaces. *Semidirect* light shines mainly down as direct light, but a small portion of it is directed upward as indirect light. *Semi-indirect* light is mostly reflected, but some light shines directly. *Diffused* light is spread evenly in all directions with the light source (bulb) not visible.

Reflection

All objects absorb and reflect light. Some white surfaces reflect 94 percent of the light that strikes them. Some black surfaces reflect only 2 percent. The remainder of the light is absorbed. All surfaces in a room act as a secondary source of light when light is reflected. Refer again to Fig. 31-10. Excessive reflection causes glare. Glare can be eliminated from this secondary source by using matte (dull) finish surfaces and by avoiding exposed light bulbs. Eliminating excessive glare is essential in designing adequate lighting.

Structural Light Fixtures

Light fixtures are either portable plug-in lamps or structural fixtures. Structural fixtures are wired and built into a building hard-wired. These must therefore be shown on electrical plans and specifications. Structural fixtures may be located on ceilings, on interior and exterior walls, and on the grounds around the building.

Different light patterns are produced, depending upon the type of light fixture. Figure 31-14 illustrates the types of structural light fixtures described in the following paragraphs.

- Soffit lighting is used to direct more light to wall surfaces and to horizontal surfaces, such as kitchen and bath countertops, wall desks, music centers, and computer centers.

Fig. 31-13 ■ Methods of light dispersement.

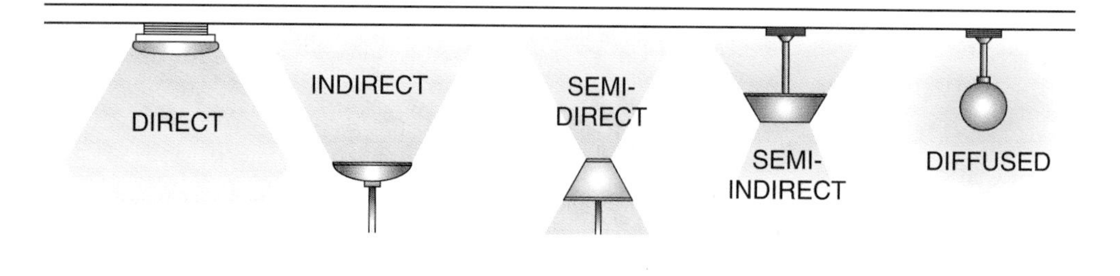

Fig. 31-14 ■ Structural light fixture details.

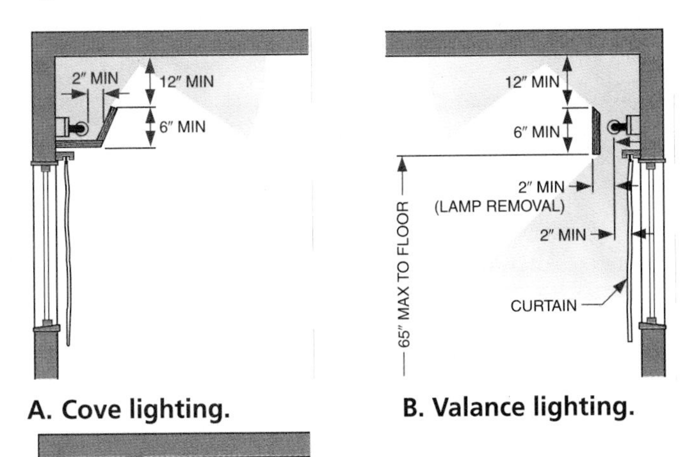

A. Cove lighting.

B. Valance lighting.

LOCATION	USE	CAVITY DIMENSIONS DEPTH	WIDTH	LENGTH
KITCHEN	Over sink or work center	8″ to 12″	12″	38″ Min.
BATH OR DRESSING ROOM	Over large mirror	8″	18″ to 24″	Length of mirror
LIVING AREA	Over piano, desk, sofa, or other seeing area	10″	Fit space available 12″ Min.	Fit space available 50″ Min.

C. Soffit lighting.

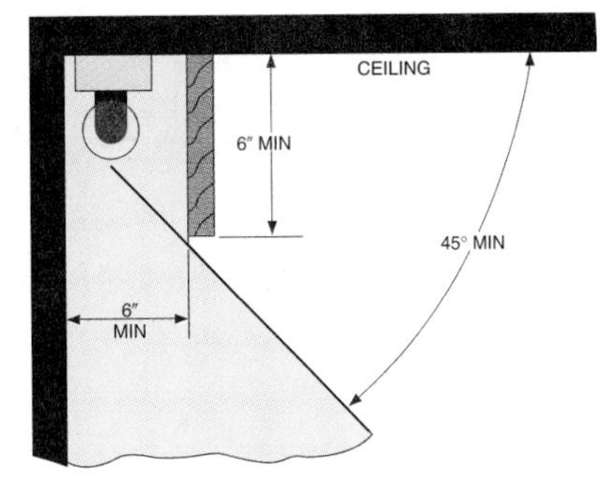

D. Cornice lighting.

- Cove lighting directs light (usually fluorescent) onto ceiling surfaces and indirectly reflects light into the center of a room. The soffit should hide the fixture from view from any position in the room.
- Valance lighting directs light upward to the ceiling and down over the wall or window treatment. Valance faceboards can be flat, scalloped, notched, perforated, papered, upholstered, painted, or trimmed with molding.
- Cornice lighting directs all light downward. It is similar to soffit lighting, except cornice lights are totally exposed at the bottom.

WALL FIXTURES Wall fixtures are used as a source of general lighting, as well as decorative lighting when attached to a dimmer switch. Wall spotlights or fluorescent fixtures may also be used as task lighting. Wall spotlights for accents, diffusing fixtures for general lighting, and sconces are used extensively on walls. See Fig. 31-15. Vanity lights and fluorescent tube lights are also used on walls as task lighting.

CEILING FIXTURES A wide variety of lighting fixtures are designed for ceiling installation. Many optional designs are possible within each type. See Fig. 31-16. Likewise, track-lighting units are available in a variety of shapes, materials, and colors.

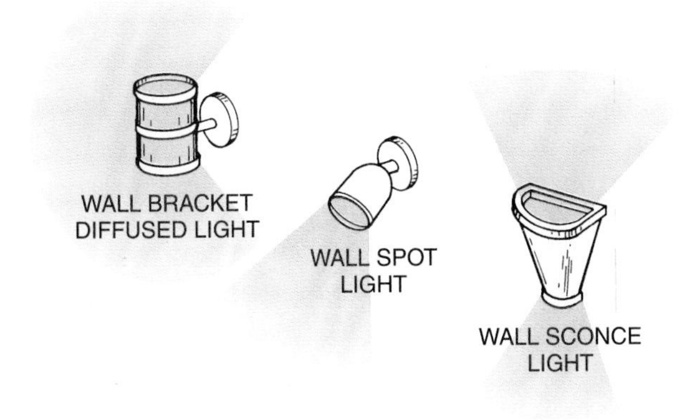

Fig. 31-15 ■ Diffusing spot and sconce wall fixtures.

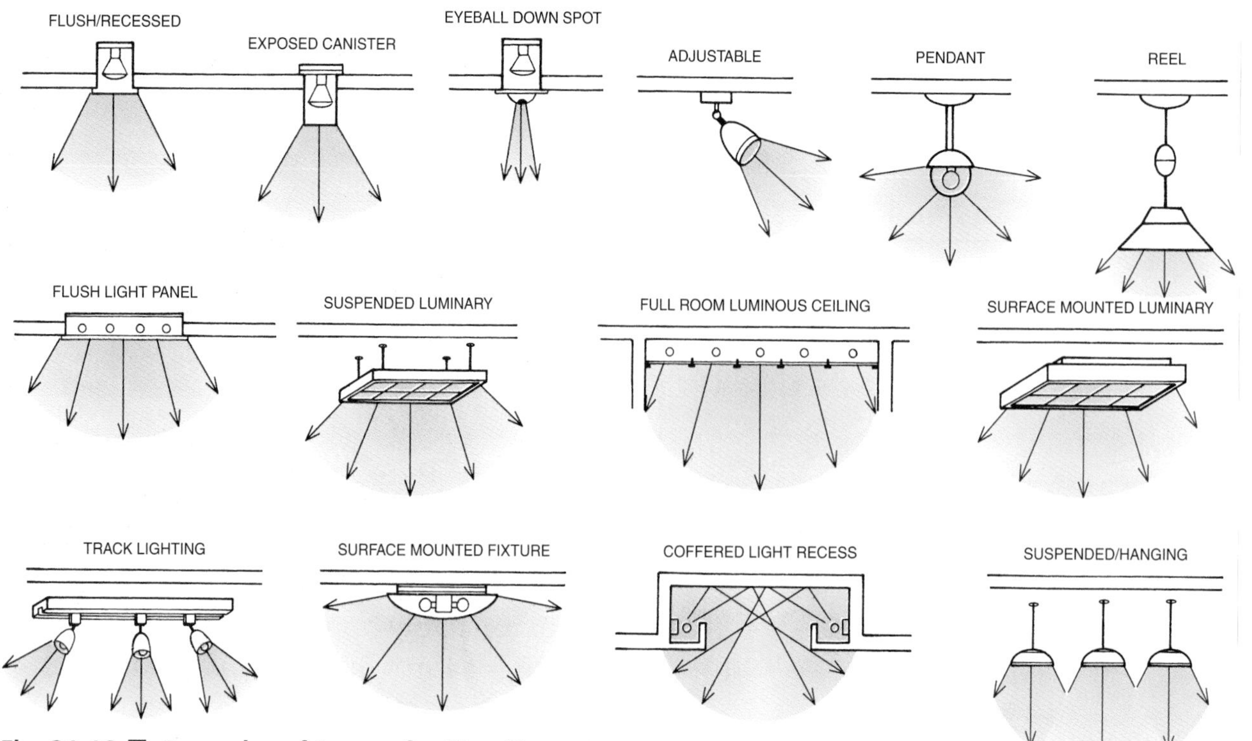

Fig. 31-16 ■ Examples of types of ceiling fixtures.

Because track light units can be moved and rotated, the track should be placed to take full advantage of these features.

When entire ceilings are to be illuminated, fluorescent fixtures are ceiling mounted. Translucent or open mesh panels are suspended below the fixtures. The position of the fixtures should be shown on the electrical plan. The suspended luminous ceiling should be drawn. See Fig. 31-17.

EXTERIOR LIGHTING FIXTURES Waterproof spotlights, flood lights, and wall bracket lights are used on exterior walls for both general lighting and decorative lighting. See Fig. 31-18. Exterior wall lights are often connected to motion detectors for security purposes. Lighting fixtures are used for landscaping, driveways, and walkways. These fixtures are designed to direct light at any angle to illuminate design features. Some fixtures, such as

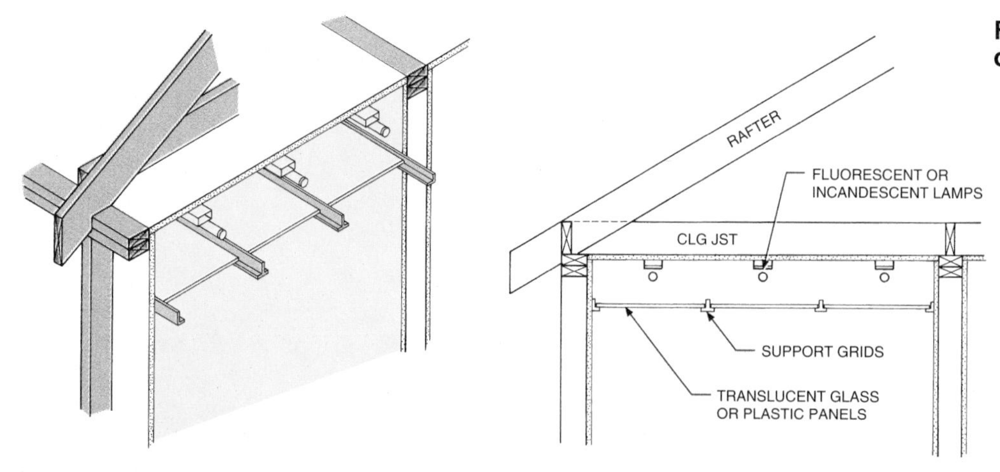

Fig. 31-17 ■ Luminous ceiling design.

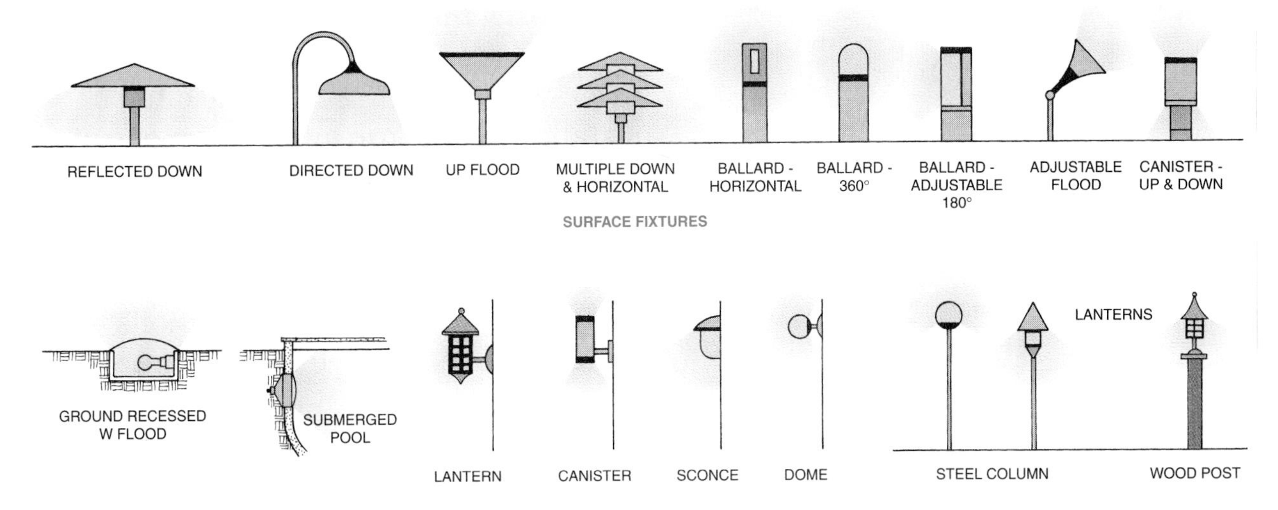

Fig. 31-18 ■ Examples of types of exterior lighting fixtures.

post lamps (lanterns), are designed to emit light in all directions. Other post (ballard) lights are designed with shields that can be adjusted to direct light 360 degrees or to any smaller segment. Swimming pool lights can also be used effectively for landscape lighting since the entire pool becomes a large light source when illuminated.

Developing and Drawing Electrical Plans

Wiring methods are regulated by building codes, and wiring is approved and installed by licensed electricians. However, wiring plans are prepared by designers. For large structures, a consulting electrical contractor may prepare the final detailed electrical plans. Electrical plans include data on the type and location of all fixtures, devices, switches, and outlets.

Fig. 31-19 Electrical fixture and device listing.

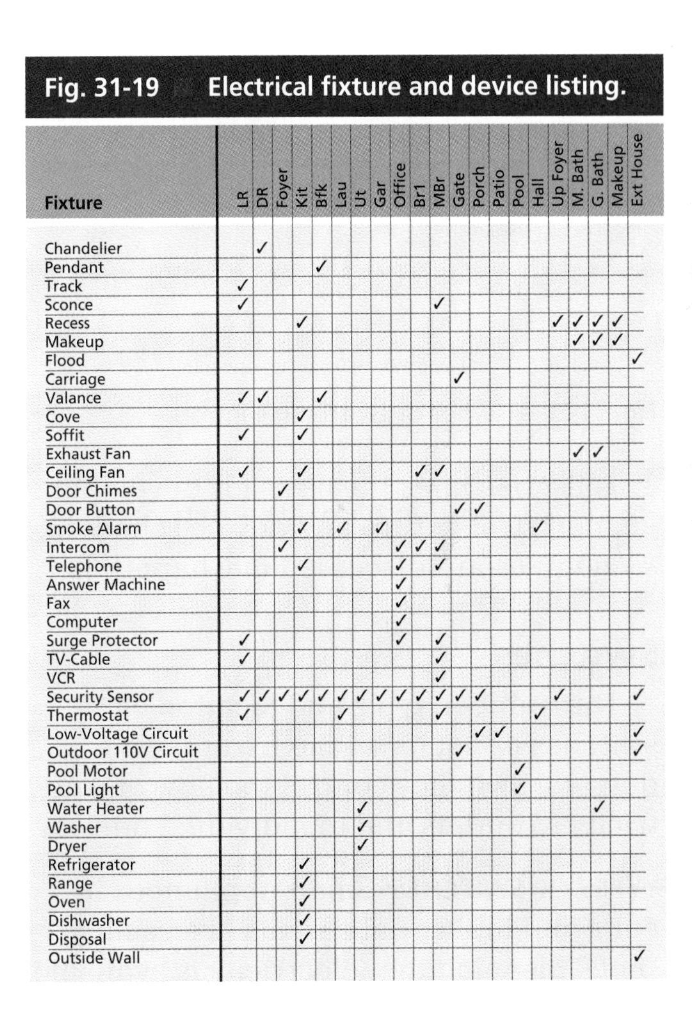

Fixture	LR	DR	Foyer	Kit	Bfk	Lau	Ut	Gar	Office	Br1	MBr	Gate	Porch	Patio	Pool	Hall	Up Foyer	M. Bath	G. Bath	Makeup	Ext House
Chandelier		✓																			
Pendant					✓																
Track	✓																				
Sconce	✓										✓										
Recess				✓													✓	✓	✓	✓	
Makeup																		✓	✓	✓	
Flood																					✓
Carriage												✓									
Valance	✓	✓			✓																
Cove				✓																	
Soffit	✓			✓																	
Exhaust Fan																		✓	✓		
Ceiling Fan	✓			✓						✓	✓										
Door Chimes			✓																		
Door Button												✓	✓								
Smoke Alarm				✓		✓		✓								✓					
Intercom			✓						✓	✓	✓										
Telephone				✓					✓		✓										
Answer Machine									✓												
Fax									✓												
Computer									✓												
Surge Protector	✓								✓		✓										
TV-Cable	✓										✓										
VCR											✓										
Security Sensor	✓	✓	✓	✓	✓	✓	✓	✓	✓	✓	✓	✓	✓				✓				✓
Thermostat	✓					✓					✓					✓					
Low-Voltage Circuit													✓	✓							✓
Outdoor 110V Circuit												✓									✓
Pool Motor															✓						
Pool Light															✓						
Water Heater							✓												✓		
Washer							✓														
Dryer							✓														
Refrigerator				✓																	
Range				✓																	
Oven				✓																	
Dishwasher				✓																	
Disposal				✓																	
Outside Wall																					✓

Fixture and Device Selection

Before placing fixture locations on a floor plan, the number and type of fixtures needed for each room should be determined and listed. See Fig. 31-19. In addition to lighting fixtures, all electrical or electronic devices should also be listed.

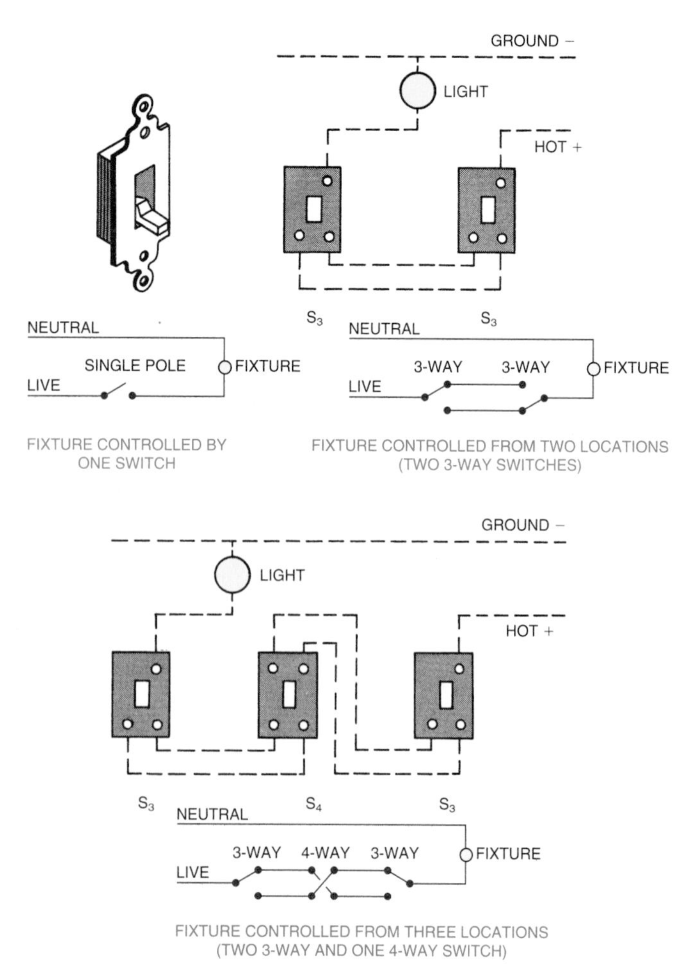

Fig. 31-20 ■ Types of switching controls.

This list becomes the basis for developing an electrical fixture and device schedule. (Schedules are discussed in Chapter 35.)

Switches

The number, type, and location of switches depends on the fixtures and devices. Switches control the flow of electricity to outlets and to individual devices.

TYPES OF SWITCHES Small-appliance circuits and individual circuits are usually "hot," meaning that electricity is available in the outlet at all times. Lighting circuits, however, may be either hot or controlled by switches. See Fig. 31-20. *Single-pole switches* control one fixture, device, or outlet. To control lights from two different switches, a *three-way* switching circuit (three wires and two switches) is used. A three-way switching circuit is often installed for the top and bottom of stairways. Many types of switch mechanisms are used to control circuits. See Fig. 31-21.

SWITCH LOCATIONS Switch symbols are located on floor plans. Connections to the outlet, fixture, or device each controls are shown with a dotted line. See Fig. 31-22. Use the following guidelines in planning switch locations:

1. Include a switch for all structural fixtures and devices that need to be turned on or off.
2. Indicate the height of all switches (usually 4′ above floor level).
3. Locate switches on the latch side of doors, no closer than 2½″ from the casing. See Fig. 31-23.
4. Exceptions to the standard should be dimensioned on the plan or elevation drawing.
5. Select the type of switch, switch mechanism, switch plate cover, and type of finish for each switch.
6. Plan a switch to control at least one light in each room.
7. Use three-way switches to control lights at the ends of stairwells, halls, and garages.
8. Locate garage door-closer switches at the house entry and within reach inside the garage door.
9. Control bedroom lights with a three-way switch at the entry and at the bed.
10. Use timer switches for garage general lighting, bathroom exhaust fans, and heatlights.
11. Use three-way switches for all large rooms that have two exits. Use four-way switches for rooms with more than two exits.
12. Use automatic switches on closet and storage areas.